AXONS AND
BRAIN ARCHITECTURE

AXONS AND BRAIN ARCHITECTURE

Edited by

KATHLEEN S. ROCKLAND

Department of Anatomy & Neurobiology,
Boston University School of Medicine, Boston, MA, USA
and
Cold Spring Harbor Laboratory, NY, USA

AMSTERDAM • BOSTON • HEIDELBERG • LONDON
NEW YORK • OXFORD • PARIS • SAN DIEGO
SAN FRANCISCO • SINGAPORE • SYDNEY • TOKYO
Academic Press is an imprint of Elsevier

Academic Press is an imprint of Elsevier
125, London Wall, EC2Y 5AS.
525 B Street, Suite 1800, San Diego, CA 92101-4495, USA
225 Wyman Street, Waltham, MA 02451, USA
The Boulevard, Langford Lane, Kidlington, Oxford OX5 1GB, UK

Notices
Knowledge and best practice in this field are constantly changing. As new research and experience broaden
our understanding, changes in research methods, professional practices, or medical treatment may become
necessary.

Practitioners and researchers must always rely on their own experience and knowledge in evaluating and
using any information, methods, compounds, or experiments described herein. In using such information
or methods they should be mindful of their own safety and the safety of others, including parties for whom
they have a professional responsibility.

To the fullest extent of the law, neither the Publisher nor the authors, contributors, or editors, assume any
liability for any injury and/or damage to persons or property as a matter of products liability, negligence
or otherwise, or from any use or operation of any methods, products, instructions, or ideas contained in the
material herein.

ISBN: 978-0-12-801393-9

British Library Cataloguing-in-Publication Data
A catalogue record for this book is available from the British Library.

Library of Congress Cataloging-in-Publication Data
A catalog record for this book is available from the Library of Congress.

For Information on all Academic Press publications
visit our website at http://store.elsevier.com/

Typeset by MPS Limited, Chennai, India
www.adi-mps.com

Printed and bound in the United States of America.

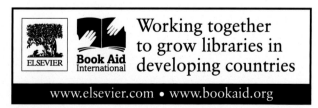

Working together
to grow libraries in
developing countries

www.elsevier.com • www.bookaid.org

Front Cover: The central image, of Ariadne's labyrinth, is reproduced with the gracious permission of
Erwin Reißmann (www.mymaze.de).

Contents

I

MICROCIRCUITRY

1. Axonal Projection of Olfactory Bulb Tufted and Mitral Cells to Olfactory Cortex

KENSAKU MORI

2. The Primate Basal Ganglia Connectome As Revealed By Single-Axon Tracing

MARTIN PARENT AND ANDRÉ PARENT

3. Comparative Analysis of the Axonal Collateralization Patterns of Basal Ganglia Output Nuclei in the Rat

E. MENGUAL, L. PRENSA, A. TRIPATHI, C. CEBRIÁN AND S. MONGIA

4. Anatomy and Development of Multispecific Thalamocortical Axons: Implications for Cortical Dynamics and Evolution

FRANCISCO CLASCÁ, CÉSAR PORRERO, MARIA JOSÉ GALAZO, PABLO RUBIO-GARRIDO AND MARIAN EVANGELIO

5. Geometrical Structure of Single Axons of Visual Corticocortical Connections in the Mouse

IAN O. MASSÉ, PHILIPPE RÉGNIER AND DENIS BOIRE

List of Contributors

Katrin Amunts Institute of Neuroscience and Medicine (INM-1), Research Center Jülich, Jülich, Germany; C. & O. Vogt Institute of Brain Research, University of Düsseldorf, Düsseldorf, Germany

Markus Axer Institute of Neuroscience and Medicine (INM-1), Research Center Jülich, Jülich, Germany

Kevin Beier Departments of Biological Sciences and Psychiatry and Behavioral Sciences, Stanford, CA, USA

Peter Bloomfield MRC Clinical Sciences Centre, Imperial College London, London, UK

Denis Boire Département d'anatomie, Université du Québec à Trois-Rivières, Quebec, Canada

Dirk Bucher Federated Department of Biological Sciences, New Jersey Institute of Technology and Rutgers University, University Heights, Newark, NJ, USA

Alison J. Canty Wicking Dementia Research and Education Centre, School of Medicine, University of Tasmania, Hobart, Australia

Marie Carlén Department of Neuroscience, Karolinska Institutet, Stockholm, Sweden

C. Cebrián Division of Neurosciences, Center for Applied Medical Research (CIMA), Universidad de Navarra, Pamplona, Spain; Departamento de Anatomía, Facultad de Medicina, Universidad de Navarra, Pamplona, Spain

Francisco Clascá Department of Anatomy & Graduate Program in Neuroscience, School of Medicine, Autónoma de Madrid University, Madrid, Spain

Markus Cremer Institute of Neuroscience and Medicine (INM-1), Research Center Jülich, Jülich, Germany

Christiaan P.J. de Kock Center for Neurogenomics and Cognitive Research, Neuroscience Campus Amsterdam, VU University Amsterdam, The Netherlands

Vincenzo De Paola MRC Clinical Sciences Centre, Imperial College London, London, UK

Tim B. Dyrby Danish Research Centre for Magnetic Resonance, Centre for Functional and Diagnostic Imaging and Research, Copenhagen University Hospital Hvidovre, Hvidovre, Denmark

Timothy J. Edwards Queensland Brain Institute, The University of Queensland, Brisbane, Australia

Robert Egger Computational Neuroanatomy Group, Max Planck Institute for Biological Cybernetics, Tuebingen, Germany; Graduate School of Neural Information Processing, University of Tuebingen, Tuebingen, Germany

Marian Evangelio Department of Anatomy & Graduate Program in Neuroscience, School of Medicine, Autónoma de Madrid University, Madrid, Spain

Maria José Galazo Department of Anatomy & Graduate Program in Neuroscience, School of Medicine, Autónoma de Madrid University, Madrid, Spain

Ralf Galuske Systems Neurophysiology, Department of Biology, TU Darmstadt, Darmstadt, Germany

David Gräßel Institute of Neuroscience and Medicine (INM-1), Research Center Jülich, Jülich, Germany

Federico W. Grillo MRC Centre for Developmental Neurobiology, Institute of Psychiatry, Psychology and Neuroscience, Kings College London, London, UK

Giorgio M. Innocenti Department of Neuroscience, Karolinska Institutet, Stockholm, Sweden; Brain and Mind Institute, EPFL, Lausanne, Switzerland

Zoltán Kisvárday MTA-Debreceni Egyetem, Neuroscience Research Group, Debrecen, Hungary

Carol A. Mason Department of Neuroscience, College of Physicians and Surgeons, Columbia University, New York, NY, USA; Departments of Pathology & Cell Biology, and Ophthalmology, College of Physicians and Surgeons, Columbia University, New York, NY, USA

Ian O. Massé Département d'anatomie, Université du Québec à Trois-Rivières, Quebec, Canada

E. Mengual Division of Neurosciences, Center for Applied Medical Research (CIMA), Universidad de Navarra, Pamplona, Spain; Departamento de Anatomía, Facultad de Medicina, Universidad de Navarra, Pamplona, Spain

S. Mongia Departamento de Anatomía, Facultad de Medicina, Universidad de Navarra, Pamplona, Spain

Laura R. Morcom Queensland Brain Institute, The University of Queensland, Brisbane, Australia

Kensaku Mori Department of Physiology, Graduate School of Medicine, The University of Tokyo, Tokyo, Japan; Japan Science and Technology Agency, CREST, Tokyo, Japan

Farzad Mortazavi Department of Anatomy and Neurobiology, Boston University School of Medicine, Boston, MA, USA

Rajeevan T. Narayanan Computational Neuroanatomy Group, Max Planck Institute for Biological Cybernetics, Tuebingen, Germany

Marcel Oberlaender Computational Neuroanatomy Group, Max Planck Institute for Biological Cybernetics, Tuebingen, Germany; Digital Neuroanatomy, Max Planck Florida Institute for Neuroscience, Jupiter, FL, USA; Bernstein Center for Computational Neuroscience, Tuebingen, Germany

Nicola Palomero-Gallagher Institute of Neuroscience and Medicine (INM-1), Research Center Jülich, Jülich, Germany

André Parent   Department of Psychiatry and Neuroscience, Faculty of medicine, Université Laval, Quebec City, Quebec, Canada

Martin Parent Ph.D   Department of Psychiatry and Neuroscience, Faculty of medicine, Université Laval, Quebec City, Quebec, Canada

César Porrero   Department of Anatomy & Graduate Program in Neuroscience, School of Medicine, Autónoma de Madrid University, Madrid, Spain

L. Prensa   Division of Neurosciences, Center for Applied Medical Research (CIMA), Universidad de Navarra, Pamplona, Spain; Departamento de Anatomía, Facultad de Medicina, Universidad de Navarra, Pamplona, Spain

Philippe Régnier   Département d'anatomie, Université du Québec à Trois-Rivières, Quebec, Canada

Linda J. Richards   Queensland Brain Institute, The University of Queensland, Brisbane, Australia; School of Biomedical Sciences, The University of Queensland, Brisbane, Australia

Alard Roebroeck   Faculty of Psychology and Neuroscience, Maastricht University, Maastricht, The Netherlands

Douglas L. Rosene   Department of Anatomy and Neurobiology, Boston University School of Medicine, Boston, MA, USA

Pablo Rubio-Garrido   Department of Anatomy & Graduate Program in Neuroscience, School of Medicine, Autónoma de Madrid University, Madrid, Spain

Philipp Schlömer   Institute of Neuroscience and Medicine (INM-1), Research Center Jülich, Jülich, Germany

Kerstin E. Schmidt   Brain Institute, Federal University of Rio Grande do Norte (UFRN), Natal, Brazil

Arne Seehaus   Faculty of Psychology and Neuroscience, Maastricht University, Maastricht, The Netherlands

Austen A. Sitko   Department of Neuroscience, College of Physicians and Surgeons, Columbia University, New York, NY, USA

A. Tripathi   Division of Neurosciences, Center for Applied Medical Research (CIMA), Universidad de Navarra, Pamplona, Spain

Van J. Wedeen   Martinos Center for Biomedical Imaging, Department of Radiology, Massachusetts General Hospital, Harvard Medical School, Boston, MA, USA

Roger Woods   Ahmanson-Lovelace Brain Mapping Center, UCLA, Los Angeles, CA, USA; David Geffen School of Medicine, UCLA, Los Angeles, CA, USA

Karl Zilles   Institute of Neuroscience and Medicine (INM-1), Research Center Jülich, Jülich, Germany; Department of Psychiatry, Psychotherapy and Psychosomatics, RWTH University Aachen, Aachen, Germany; JARA Jülich-Aachen Research Alliance, Translational Brain Medicine, Germany, Jülich, Germany

Preface

ax·on

'ak͵sän/

noun. mid-19th century (denoting the body axis): from Greek *axōn* "axis."

the long threadlike part of a nerve cell along which impulses are conducted from the cell body to other cells (Google).

Axons, and long-distance axons in particular, are attracting increasing attention as the anatomical reality underlying MR imaging techniques. They are also a vital component of microcircuitry. Their contribution to microcircuitry has been relatively neglected, owing to the incompatability of *in vitro* assays with long axons, but it can be expected to figure prominently in the ongoing efforts of whole brain analysis of connectivity (connectome). A major goal of this volume is to bring together in one source an interrelated treatment of single axons both from the perspective of microcircuitry and as substrates of larger scale organization (tractography and brain architecture). We hope by highlighting the span from microcircuitry, axon guidance, and dynamics, to tractography and brain architecture, that this will serve as a convenient reference and facilitate communication between what have tended to be separate communities.

For the sake of focus, the volume concentrates on mammalian systems, largely overlooking the extensive work on "wiring diagrams" of *Caenorhabditis elegans* and *Drosophila*. Myelination, axon transport, and pathological conditions are also only minimally addressed.

Section I is devoted to anatomical investigations of connections at the single axon level, including implications for cortical architecture in particular. A broad coverage of cortical and subcortical connections from different species serves to highlight recurring spatial patterns of divergence, convergence, and collateralization. A background chapter is included to review neuroanatomical tracing techniques in general, with focus on the newer, still developing virus-based tracers.

Section II addresses mechanisms of axon dynamics in normal and pathological conditions, and the temporal aspect (chronnectome) of axonal morphology and circuitry. Section III covers axon guidance in development: how these long distance axons identify and connect with postsynaptic targets, and sort along intermediate points.

Section IV focuses on imaging and tractography. Three chapters discuss histological parameters in relation to high-resolution MRI/Diffusion imaging; and the final chapter describes in detail the relatively new technique of polarized light imaging. This yields high-resolution visualization of nerve fibers in postmortem brains, by the birefringent property of myelin.

The labyrinth on the cover was made by Erwin Reißman, a retired surveyor. It primarily is meant to suggest the axon as an Ariadne's thread and guide through the labyrinth, but is also evocative of a myelinated axon. It was to some extent inspired by a comment of G.H. Bishop: "As a member of the only species capable of comprehending, or even of speculating on its own origins, and looking from this height down along the trunk of our sprawling family tree, no less a filament than the nerve fiber itself will serve as the thread of Ariadne along which may be retraced the course of our brain's evolution."

Reference

Bishop, G.H., 1965. My life among the axons. Annu. Rev. Physiol. 27, 1–20.

Foreword

This book is both timely and exciting in the context of the massive efforts dedicated worldwide (in Europe, United States, Japan, and China) to deciphering connectivity patterns in the brain; i.e., the wiring diagram (connectome) between various brain regions and within local circuits at the synaptic level. With the increasing use of powerful new methods for staining, tracing, stimulating, recording and imaging neuronal processes (e.g., CLARITY, optogenetics, light-sheet imaging), it is becoming increasing clear that brain connectivity is specific, and this is abundantly in evidence at the level of single axons, both at long distances and in local microcircuits.

This raises a few fundamental questions: How do axons navigate so precisely into the appropriate regions and find their target cells under normal conditions? What goes wrong with the route axons take in pathological conditions? What are the functional implications of such intricate circuitry for the processing of spatiotemporal sequences of neural events? For the most part, the consequences of specific innate connectivity for learning, for coding and decoding sensory information, and for implementing specific computations are far from being understood. This book provides an important foundation that will assist and expedite future theoretical advances toward this end.

This volume emphasizes that axons are dynamic devices, especially in the physiological realm. Axonal "hot regions" such as the axon initial segment (AIS), the nodes of Ranvier, and the axonal boutons (release sites), all are capable of undergoing plastic changes. In these sites, the spatial profile, density, and type of ion channels may go through temporal and spatial modifications—affecting the patterning of spikes at the AIS and modifying the impact (the read out) of these spikes at the axon's synaptic release sites.

A key new insight about axons is that a given train of spikes, generated at the AIS, could be interpreted differently by its many postsynaptic (dendritic) "listeners." For a given axon, some synapses might be powerful (generating large postsynaptic potentials (PSPs)) and other synapses might generate small or even no PSPs. Some synapses could have short-term depressing properties and other synapses, arising from the same axon, are facilitatory. These short-term dynamics are themselves remarkably plastic, due to dynamic machinery at both the presynaptic axonal side and at the postsynaptic (dendritic) side. The energy consumption of the axonal spikes is yet another new and fascinating topic of investigation—demonstrating that out of the 20 W or so that our brain consumes, the axonal activity is energetically expensive.

The chapters in this book clarify that certain structural motifs, at both global and local scales, are innate and that others are learned. "Running" on this structural foundation are electrical signals (digital in axons and mostly analog in dendrites) representing in our brain the external world and our individual interpretations of it. Axons with their astute (although sometimes fragile, especially in certain diseases) "all or none" spikes serve as a major component in this representation/interpretation scheme. We have somewhat neglected this key component of neurons, as the neuroscience field became fascinated with dendrites—the major input regions of neurons. Now, this comprehensive book beautifully demonstrates that axons and dendrites, and their connecting synapses, work in orchestration to generate the "music" of the brain—a music that continuously changes–largely due to the morphological and biophysical flexibility of axons.

One of many personal inspirations that I took from this book has to do with the development of future computers/robots, behaving in a challenging and unexpected environment. To date, we have built computing machines with essentially static prewired connectivity. For these machines to actually compete with and aspire to replicate the amazing capabilities of the brain in solving real-life (e.g., action-perception) tasks, it seems wise to learn from the brain; that is, combine "innate" wirings (e.g., between particular brain/machine regions, such as the visual and the motor systems) and, at the same time, leave a significant degree of freedom and a crucial reservoir of modifiablity (including in the messages that are transmitted via these wires). We can look forward to brain-inspired machines that rewire themselves both locally and globally, as these seem best fit to handle tasks that require efficient learning and resourceful computations.

Idan Segev
The Hebrew University of Jerusalem, Jerusalem, Israel

Acknowledgments

It is a pleasure to thank the Editorial Staff at Elsevier for professional and courteous guidance. Natalie Farra, Acquisitions Editor, piloted the initial launch of this project, and Kathy Padilla, Editorial Project Manager, expertly steered us through all the twists and shoals along the way. The cover design was skillfully montaged by Mark Rogers. The central image, of Ariadne's labyrinth, is reproduced with gracious permission of Erwin Reißmann (www.mymaze.de).

Many thanks also to the students, postdocs, and colleagues over the years who have joined in the work represented by these 18 chapters, and without whose dedication and enthusiasm, little of this would have been possible.

Introduction

Axons figure in axoplasmic transport as substrates of the nerve impulse, as pathfinders in development, as the "wires" in connectivity wiring diagrams. They are at once key players in basic molecular processes, related to axon transport, and presynaptic components of neuronal circuitry and larger scale neural networks. Except for local axons of interneurons and other nonprojecting neurons, axons are long—typically many millimeters in length. This simple fact makes them less accessible to *in vitro* investigations than the more spatially delimited dendrites, and accordingly has led to a comparative neglect despite their central importance.

While research on axons covers as many levels as neuroscience proper, (Waxman et al., 1995; Bagnard, 2007; Feldmeyer and Lubke, 2010; Goldberg and Traktenberg, 2012) this volume is focused on neuroanatomical questions and techniques, largely in relation to a mesoscale of connectivity (i.e., light microscopic). For the sake of general orientation, this short Introduction will briefly address early history of the axon, experimental techniques, and basic questions relating to axons *per se*.

I.1 EARLY HISTORY

Axons have a central place in the history of neuroscience. As macroscopic fiber bundles, axons were already identified by Vesalius in the sixteenth century, predating by several hundred years the identification of neuron cell bodies. The Renaissance investigators in fact sound remarkably modern in their observations of white matter: "To say that the white matter is but a uniform substance like wax in which there is no hidden contrivance, would be too low an opinion of nature's finest masterpiece" (Niels Stenson, quoted in Schmahmann and Pandya, 2006 and see their section "The Era of Gross Dissection").

Closer to the present time, axons were at the heart of at least three major waves of discovery. One is cell biological research on the phenomenon and mechanisms of axonal transport. The early stage of this work coincided with World War II and the effort to improve recovery after peripheral nerve damage. The simple device of nerve constriction demonstrated characteristic swelling, beading, and coiling proximal to the constriction. This was correctly interpreted as an indication of what was initially called "axoplasmic flow" (later modified to "axonal transport," to connote the active nature of the "flow") and inaugurated decades of concentrated research regarding its components, mechanisms, and significance (see Grafstein and Forman, 1980 for an early review).

Second is the pioneering physiological work on the propagation of the nerve impulse and its capstone in the great Hodgkin–Huxley model. Part of this stage is summarized, in the refreshing context of close-up personal experience, in Bishop (1965). Again, the reader will recognize distinctly modern resonances on issues that remain under active discussion, among others: the mystery of unmyelinated axons, the functional significance of the spectrum of small, medium, and large axons; and the increase in the proportion of large fibers in evolution (cf. Perge et al., 2012; Wang et al., 2008). The tone is extraordinarily open-minded and questioning. We hear Bishop, who has spent his research career as a neurophysiologist, pondering, "What is a nerve fiber for, anyway?.... Conduction of an impulse is in fact somewhat incidental to another essential functioning of a neuron, however useful as a sign that the neuron has functioned. Where does one come out if he looks at the neuron as a secretory organ?... To wit: the prime function of a neuron is to produce and apply to other tissues a chemical activator" (Bishop, 1965, p. 12).

Third, in the anatomical domain, axons were of course the focus of heated controversies between "neuronists" and "reticularists" at the end of the nineteenth century. The debate, as is well-known, was resolved in favor of the neuronists, of whom Ramon y Cajal has been generally acknowledged as a key representative. The story has been frequently told, but I would refer the reader to the thoughtful essays of Guillery (2005, 2007). Guillery (Deiters and Guillery, 2013) also has coauthored a biographical account of Otto Deiters who, by fine microdissection of a single large neuron from the ventral horn of the spinal cord, correctly distinguished between the cell's multiple dendrites and its single axon. In effect, this was the first observation of the neuron as a polarized structure, and set the stage for subsequent rigorous work by Cajal and others in support of the neuron doctrine.

The time of Ramon y Cajal and Camillo Golgi, at the beginning of the twentieth century (taking 1906, the year of their shared Nobel Prize award as a reference point), ushered in the highly productive application of the Golgi silver impregnation technique, which lasted well beyond mid-century. Despite the very real limitations of the technique, investigators such as Lorento de No and F. Valverde in Spain, Szentagothai in Hungary, and in the United States, the Scheibels, Ray Guillery, E.G. Jones, and J.S. Lund, among many others, constructed a substantial data set of single axon visualization across the brain of several species. To this day, Golgi preparations are one of the few techniques, with immunohistochemistry, that can be used effectively to visualize cellular and neuropil components in the human brain (e.g., Marin-Padilla, 2015).

I.2 FROM 1970S: THE AGE OF TECHNICAL ADVANCES

I.2.1 Tracer Injections

In the early 1970s, connectivity experiments were transitioning to tracer substances that were transported anterogradely (^{3}H amino acids and autoradiography) or retrogradely (horseradish peroxidase (HRP)) through physiological mechanisms of axonal transport (Morecraft et al., 2015). This was a welcome improvement over the earlier lesion–degeneration methods; but one negative side effect was a decreased attention to the actual projecting axons. Extracellular tracer injections, even if involving only a small volume of cortex (less than 250 µm in diameter), result in what is in effect an average of large numbers of retrogradely filled cells or anterogradely labeled axons and their terminations. The axons of retrogradely labeled neurons are typically not labeled, or labeled only proximally; and many of the anterograde tracers, especially the first generation (e.g., WGA-HRP and autoradiography) result in only dustlike terminal label. Tracers that do produce Golgi-like detail of axon fibers and terminals (PHA-L, biocytin, BDA, and most recently, viral vectors) require laborious collection of histological sections in uninterrupted series and even more laborious reconstruction of labeled profiles across the series—an effort not commonly undertaken. These shortcomings persist to the present, although newer tracers are more successful in demonstrating projections in their entirety (see Section I), and the problem of three-dimensional visualization is being vigorously addressed in Chapter 10 by Beier and Chapter 17 by Mortazavi et al.

I.2.2 Single Axon Visualization

The next step in single axon analysis, after Golgi staining, was the advance to *in vivo* and *in vitro* techniques, where anatomical visualization was achieved by microinjection of HRP in the white matter, by intracellular injection (first, of HRP and subsequently of biocytin or BDA), or by juxtacellular injections (see Section I). Intracellular injections were often subsequent to microelectrode physiological recording, with the obvious advantage of thereby producing a combined data set of physiological and anatomical characteristics.

The approach of intraaxonal recording and HRP fills was instrumental in elucidating basic structural–functional correlations in several systems. Investigations of the retinogeniculocortical projections were among the earliest results. Physiologically identified retinofugal axons were found to have distinct phenotypes and target-specific arborizations that clearly differed in spatial geometry and numbers of boutons (Bowling and Michael, 1980). The projections from lateral geniculate nucleus to visual cortex were mapped by the same method in nonhuman primate (Blasdel and Lund, 1983) and cat (Freund et al., 1985; Humphrey et al., 1985). In both species, distinct structural features were demonstrated that correlated with the known physiological properties of magno- and parvocellular geniculocortical projections; namely, focused parvocellular axon arbors versus the more divergent magnocellular associated, respectively, with color and form sensitivity or transient, contrast-sensitive responses.

Among many other results from this or similar approaches are visualization of: thalamocortical axons to the somatosensory cortex (by microinjections of HRP in the white matter: DeFelipe et al., 1986; Garraghty and Sur, 1990); corticostriatal axons in rat (Cowan and Wilson, 1994); corticothalamic terminations from cat auditory cortex (Ojima et al., 1992). In one groundbreaking study, Katz (1987) injected retrograde tracer *in vivo* to prelabel two populations of projection neurons, and then visualized and intracellularly filled these in a subsequent *in vitro* experiment. The slice preparation did not preserve long distance projection axons, but did allow stunning detail of local intracortical collaterals. In particular, laminar patterns of local collaterals and of dendritic arbors were found to be target-specific, related to extrinsic projections to the claustrum or thalamus.

More recently, combined physiological–anatomical intracellular labeling has become less common. In part, these experiments are not particularly well suited to the investigation of higher order cortical areas, where the neuronal populations are more intermixed, less accessible to intracellular recording and filling, and results are more difficult to interpret than for primary or early sensory cortical areas. For nonhuman primates, another factor in the waning popularity of these experiments is that the awake behaving chronic preparation is now preferred to old-style acute physiological experiments in anesthetized animals. Anatomical studies may be returning, with the renewed interest in the circuitry substrates underlying behaviors and the increasingly routine use of the more controlled optogenetic methods.

The advent of reliably Golgi-like anterograde tracers, such as BDA (~1995) and a repertoire of anterogradely transported viral vectors (see Chapter 10 and Morecraft et al., 2015), gave rise to a small number of anatomical studies where projections were mapped at the level of individual axons. Since axons were typically labeled extracellularly, by relatively large injections, usually only their distal portions, beyond the zone of high density labeling, could be successfully analyzed. Individual axons continue to be investigated, especially in rodents, by the juxtacellular recording-labeling technique (see Section I, and Chapter 2); and injections of low titre Sindbis virus have been particularly successful in demonstrating whole-axon arborizations, again, mainly in rodents (Chapters 1, 3, and 4).

I.3 AXONAL PHENOTYPES IN NEUROANATOMY

Axons are complex in morphology, cell biology, and physiology, in ways that are only partially understood. Axons and their specializations are clearly bound up with the most basic properties of neural organization, from propagation of the nerve impulse to network organization. Elucidating the detailed neuroanatomical phenotypes, as revealed by single axon resolution (see Section I), has functional significance in several domains.

One is the visualization of branching patterns in relation to network organization. In rodents and to some extent primates, extrinsic axons branch (collateralize) to multiple targets. Injections of two or three retrogradely transported, color-differentiated tracers are a popular way of investigating this collateralization. However, injections, even if large, may not involve the relevant termination sites in all the recipient areas. As these are often limited to a territory smaller than 250 μm, the risk of false negatives is high (Rockland, 2013). In addition, while retrograde tracers can identify the cell bodies of collateralizing neurons, they do not show the actual branch points nor the differential density and patterns of termination in the several targets. Visualization of specific branch points is relevant for understanding fasciculation and white matter composition, and thus both for interpreting diffusion images and for better understanding of white matter pathologies.

A second domain is the generation of quantitative data to support neuroanatomically realistic models. Specific data on branching (e.g., branching angle, branch caliber) directly relate to calculations of conduction velocity; and specific data on the distal terminal portions (e.g., size and topology of terminal arbors, and size and number of terminal specializations) provide baseline parameters relevant for microcircuitry and cross-area and cross-species comparisons (Shepherd, 2004 and Chapters 5–7 and 15). For quasiregular structures, such as the cerebellum and hippocampus, a substantial body of morphometric data is available; but even in those cases, major gaps such as subtype specificity and variability remain (Ropireddy et al., 2011). How these parameters are associated with short- or long-term dynamics is a further major area of research (Chapter 12).

A third domain is the categorization into different "types." A few axons are morphologically distinctive, with recognizable phenotypes that are often phylogenetically conserved. Cerebellar parallel fibers are identifiable by their T-shaped branch and narrowly divergent arbors; cerebellar climbing fibers by virtue of their characteristic association with Purkinje cells. Hippocampal "mossy fibers" are distinguishable on the basis of their large boutons, neuronal origin from dentate granular cells, and postsynaptic targets. Corticostriatal afferents are loosely "cruciform" (Cowan and Wilson, 1994, and Chapter 9 in Shepherd, 2004).

Corticothalamic afferents constitute two large groups, in this case based on bouton size (large or small), the shape of terminal arbors (focused or divergent), and the neurons of origin. Classically, large boutons, preferentially targeting proximal sites of postsynaptic neurons, were termed "type 2" (type II or round large (RL)) and small boutons were termed "type 1" (type I or small round (SR)) and targeted mainly distal postsynaptic locations (Ogren and Hendrickson, 1979; Schwartz et al., 1991; Crick and Koch, 1998; Rouiller and Welker, 2000). Work in the 1990s established a further dissociation, according to neurons of origin, such that type 1 (small bouton) originated from neurons in layer 6 and type 2 (large bouton) originated from neurons in layer 5. The two types were considered as having qualitatively different postsynaptic influence (Sherman and Guillery, 2006), as either "driving" (large bouton) or "modulatory" (small bouton). With shifting focus on the physiological roles of the two populations, the favored nomenclature has been revised (Lee and Sherman, 2011) such that "type 1" now often designates "driving" (i.e., classical type 2, RL) and "type 2" designates "modulatory (i.e., classical type 1, SR).

Bouton size, arbor shape, and parent cell identity also dissociate amygdalocortical and amygdalothalamic connections. Namely, the former have small boutons and a divergent arbor, while the latter have large boutons and a small arbor (Miyashita et al., 2007). Corticocortical projections have been broadly classified on the basis of parent cell location, preference of termination layer, and shape of terminal arbor; but there are less obvious differences in bouton size, with the exception of some large neurons, primarily in layer 5. More subtle differences, related to number of beaded versus stalked terminations (respectively, boutons *en passant* and boutons *terminaux*) are still under investigation (Chapters 5 and 6).

Issues of types and diversity of subtypes, based on axonal morphology or correlation with parent cells, are further discussed in several chapters in Section I, as is the significance of different spatial configurations of the distal arbor.

I.4 AXONAL SUBDOMAINS

While the present volume is geared to a relatively meso, light, microscopic scale of organization, important new directions on axon structure and function are emerging from rapidly progressing work on subcellular domains and patterning. The fact of axonal compartments is not new. Myelin and nodes of Ranvier were discovered in the nineteenth century (by, respectively, R. Virchow and L.-A. Ranvier); modern electron microscopic and molecular investigations have provided a wealth of further detail on protein composition (Debanne et al., 2011). Axonal microstructural specializations include differential spacing of nodes of Ranvier; the large heterogeneity of axon initial segments of different neuronal populations, as assayed by subunit composition; density and subcompartmental distribution of voltage-gated sodium and potassium channels (Lorincz and Nusser, 2008); and very recently, the discovery of a periodic actin ring-like organization (D'Este et al., 2015). These aspects can be expected to have implications for microcircuitry, subtypes, and brain architecture; and the cross-disciplinary integration of microstructure data with network organization will be an exciting next phase of brain research.

I.5 OUTLOOK

Looking back, for axon research as for the neuroscience field as a whole, there has been appreciable and impressive progress. The next stages are hard to predict. There will be more data (big data) and better techniques. With the increase in data, the need for synthesis and new theories becomes ever more apparent. Possibly, as expressed in many of the chapters in this book, we are at a crossroads point, where old ideas look less compelling, but newer formulations are not yet in place. Much to do, in moving forward.

Readings

For further in-depth treatment of related topics, the reader is referred to several excellent monographs and selected references, listed below.

Bagnard, D., 2007. *Axon Growth and Guidance*, Advances in Experimental Medicine and Biology. Springer Science, NY.

Bishop, G.H., 1965. My life among the axons. Annu. Rev. Physiol. 27, 1–20.

Blasdel, G.G., Lund, J.S., 1983. Termination of afferent axons in macaque striate cortex. J. Neurosci. 3, 1389–1413.

Bowling, D.B., Michael, C.R., 1980. Projection patterns of single physiologically characterized optic tract fibres in cat. Nature 286, 899–902.

Cowan, R.L., Wilson, C.J., 1994. Spontaneous firing patterns and axonal projections of single corticostriatal neurons in the rat medial agranular cortex. J. Neurophysiol. 71, 17–32.

Crick, F., Koch, C., 1998. Constrains on cortical and thalamic projections: the no-strong loops hypothesis. Nature 391, 245–250.

Debanne, D., Campanac, E., Bialowas, A., Carlier, E., Alcarez, G., 2011. Axon physiology. Physiol. Rev. 91, 555–602.

D'Este, E., Kamin, D., Gottfert, F., El-Hady, A., Hell, S.W., 2015. STED nanoscopy reveals the ubiquity of subcortical cytoskeleton periodicity in living neurons. Cell Rep. 10, 1246–1251.

DeFelipe, J., Conley, M., Jones, E.G., 1986. Long-range focal collateralization of axons arising from corticocortical cells in monkey sensory-motor cortex. J. Neurosci. 6, 3749–3766.

Deiters, V.S., Guillery, R.W., 2013. Otto Friedrich Karl Deiters. J. Comp. Neurol. 521, 1929–1953.

Feldmeyer, D., Lubke, J.H.R., 2010. New Aspects of Axonal Structure and Function. Springer, NY.

Freund, T.F., Martin, K.A., Whitteridge, D., 1985. Innervation of cat visual areas 17 and 18 by physiologically identified X- and Y-type thalamic afferents. I. Arborization patterns and quantitative distribution of postsynaptic elements. J. Comp. Neurol. 242, 263–274.

Garraghty, P.E., Sur, M., 1990. Morphology of single intracellularly stained axons terminating in area 3b of macaque monkeys. J. Comp. Neurol. 294, 583–593.

Goldberg, J.L., Trakhtenberg, E.F. (Eds.), 2012. Axon growth and regeneration. Int. Rev. Neurobiol. Part 1 105, 2–218.

Grafstein, B., Forman, D.S., 1980. Intracellular transport in neurons. Physiol. Rev. 60, 1168–1283.

Guillery, R.W., 2005. Observations of synaptic structures: origins of the neuron doctrine and its current status. Philos. Trans R. Soc. Lond. B. 360, 1281–1307.

Guillery, R.W., 2007. Relating the neuron doctrine to the cell theory. Should contemporary knowledge change our view of the neuron doctrine? Brain Res. Rev. 55, 411–421.

Humphrey, A.L., Sur, M., Uhlrich, D.J., Sherman, S.M., 1985. Projection patterns of individual X- and Y-cell axons from the lateral geniculate nucleus to cortical area 17 in the cat. J. Comp. Neurol. 233, 159–189.

Katz, L.C., 1987. Local circuitry of identified projection neurons in cat visual cortex brain slices. J. Neurosci. 7, 1223–1249.

Lee, C.C., Sherman, S.M., 2011. On the classification of pathways in the auditory midbrain, thalamus and cortex. Hear. Res. 276, 79–87.

Lorincz, L., Nusser, Z., 2008. Cell-type-dependent molecular composition of the axon initial segment. J. Neurosci. 28, 14329–14340.

Marin-Padilla, M., 2015. Human cerebral cortex Cajal-Retzius neuron: development, structure and function. A Golgi study. Front. Neuroanat 9, 21.

Miyashita, T., Ichinohe, N., Rockland, K.S., 2007. Differential modes of termination of amygdalothalamic and amygdalocortical projections in the monkey. J. Comp. Neurol. 502, 309–324.

Morecraft, R.J., Ugolini, G., Lanciego, J.L., Wouterlood, F.G., Pandya, D.N., 2015. Classic and contemporary neural tract-tracing techniques. In: Johansen-Berg, H., Behrens, T.E.J. (Eds.), Diffusion MRI, second ed. Elsevier, Amsterdam, pp. 359–400.

Ogren, M.P., Hendrickson, A.E., 1979. The morphology and distribution of striate terminals in the inferior and lateral subdivisions of the Macaca monkey pulvinar. J. Comp. Neurol. 188, 179–199.

Ojima, H., Honda, C.N., Jones, E.G., 1992. Characteristics of intracellularly injected infragranular pyramidal neurons in cat primary auditory cortex. Cereb. Cortex 2, 197–216.

Perge, J.A., Niven, J.E., Mugnaini, E., Balasubramanian, V., Sterling, P., 2012. Why do axons differ in caliber? J. Neurosci. 32, 626–638.

Rockland, K.S., 2013. Collateral branching of long-distance cortical projections in monkey. J. Comp. Neurol. 521, 4112–4123.

Ropireddy, D., Scorcioni, R., Lasher, B., Buzsáki, G., Ascoli, G.A., 2011. Axonal morphometry of hippocampal pyramidal neurons semi-automatically reconstructed after *in vivo* labeling in different CA3 locations. Brain Struct. Funct. 216, 1–15.

Rouiller, E.M., Welker, E., 2000. A comparative analysis of the morphology of corticothalamic projections in mammals. Brain Res. Bull. 53, 727–741.

Schmahmann, J.D., Pandya, D.N., 2006. Fiber Pathways of the Brain. Oxford University Press, NY.

Schwartz, M.L., Dekker, J.J., Goldman-Rakic, P.S., 1991. Dual mode of corticothalamic synaptic termination in the mediodorsal nucleus of the rhesus monkey. J. Comp. Neurol. 309, 289–304.

Sherman, S.M., Guillery, R.W., 2006. Exploring the Thalamus and Its Role in Cortical Function, second ed. The MIT Press, Cambridge, MA.

Shepherd, G.M., 2004. The Synaptic Organization of the Brain, fifth ed. Oxford University Press, NY.

Wang, S.S.H., Shultz, J.R., Burish, M.J., Harrison, K.H., Hof, P.R., Towns, L.C., et al., 2008. Functional trade-offs in white matter axonal scaling. J. Neurosci. 28, 4047–4056.

Waxman, S.G., Kocsis, J.D., Stys, P.K., 1995. The Axon. Structure, Function, and Pathophysiology. Oxford University Press, NY.

MICROCIRCUITRY

Overview

Individual axons in their distal, terminal portions have been extensively investigated as presynaptic components of circuits within their various target structures. In their full entirety, including local collaterals and potentially multiple branches, they participate in broader networks; that is, the anatomical substrate of brain organization. Ideally, we want to visualize the totality of an axon (its parent neuron, arbors and terminations, and postsynaptic neurons) in space and time, to learn how the distributed information is used and coordinated in various behaviors. At the network level, however, relatively little information is available about the function and even the overall light microscopic configuration of axons in their entirety.

The first four chapters in this section review axon phenotypes in three different systems; namely, olfactory, basal ganglia, and thalamocortical (TC). In Chapter 1, Kensaku Mori discusses the strikingly different phenotypes of the two major olfactory projection neurons; namely, spatially focal projections of tufted cells and spatially dispersed projections of mitral cells. In this relatively well-studied system, the two phenotypes can be tightly and differentially associated with the inhalation–exhalation respiratory cycle, and are shown to have distinct but complementary spatiotemporal roles.

The next two chapters, by Martin and Andre Parent and by Elisa Mengual and colleagues, elaborate on axonal configurations within the several target structures of the basal ganglia. Notably, fine analysis shows morphological variability within each of the major outputs, in terms of whether neurons terminate in one area alone or branch to several. As remarked by both research groups, the diversity can be taken as a sign of exquisitely precise interactions in the context of complex spatiotemporal sequences across the interconnected network.

In Chapter 4, Francisco Clasca and colleagues present evidence for a substantial revision of the traditional distinction between topographically precise, area-specific TC axons versus divergent TC axons that had been thought to diffusely target the superficial layers of many areas. They demonstrate "multispecific" TC axons that branch in defined patterns to separate cortical areas. From this, a novel architecture emerges, with implications for the dynamics, plasticity, and evolution of TC-based networks.

The chapter by Denis Boire and colleagues focuses on corticocortical axons in the mouse. They demonstrate that projection foci, when dissected to the level of the contributing axons, are averages of highly diversified individual axons, presumably each with different integration properties. From this, the authors stress that a single cortical connection is not a single functional channel. Further implications for computational structure, connectional strengths (the distinction of "driving" vs. "modulatory"), and hierarchical organization are considered.

Issues of brain architecture continue as a theme in the chapter by John Anderson and Kevan Martin, where the authors review an extensive body of work on corticocortical axons and interrogate established ideas on cortical organization (Serial Processing and Lateral thinking). They consider what synaptic data can tell us about design principles of cortical connections and the interactions of local and long-distance connections. As in previous chapters, these authors also return to the issue of diversity: "What has been lost in these generalizations, however, is the variance [within] the projections.... This variance is instantly clear from the composite LM drawing shown in Fig. 2...."

The relation between single axons and populational patterns is further treated in Chapters 7 (Zoltan Kisvarday) and 8 (Kerstin Schmidt). Both authors focus on the still mysterious system of patchy intrinsic connections. This system corresponds to local, intraareal collaterals and terminations of pyramidal neurons. These local collaterals are a general cortical feature, except that the patchiness is not apparent in rodents. The patches are approximately the same size and spacing as functional domains in visual cortex; and both Kisvarday and Schmidt review the extensive work to spatially coregister anatomical patches with core visual features such as orientation selectivity. As they both point out, the relationship is biased but not absolute, raising questions as to the complexity of anatomical mechanisms underlying feature selectivity: there is evidence for both iso- and cross-orientation connectivity. Results are further discussed in relation to perceptual processes, physiological response properties in visual circuits, and several different functional models. Kerstin Schmidt also proposes a close functional and anatomical association of the lateral intrinsic system and callosal connections, and discusses the comparable actions on physiological responses.

Issues of functional organization are also developed in Chapter 9 by Oberlaender and colleagues working on mouse barrel cortex. This chapter is representative of hybrid experimental-statistical, probabilistic approaches (reverse engineering) to connectivity that aim toward incorporating quantitative data into a dense cortex model for predicting local and long-range synaptic inputs. Using an interactive software environment called NeuroNet, the authors group cortical neurons into 10 basic excitatory types that share common morphological and physiological properties; in particular, somadendritic morphology, TC innervation, ongoing/sensory-evoked spiking, and inter- and intralaminar intrinsic axonal projections.

Finally, in Chapter 10, Kevin Beier presents a summary of tracing techniques, with a particular emphasis on the newer virus-based methods. Low-titre Sindbis virus, for example, is fast becoming an alternative to juxtacellular labeling, at least in rodents.[1] Special mention is also given to the emerging transsynaptic methods, which offer great promise for mapping out interconnected sets of neurons. Beyond their use in tracing experiments, viral vectors enable activity-related manipulations (i.e., optogenetics); and as fluorescent tags, they enable whole-brain axon tracing, a promising alternative to traditional, stacked reconstructions from two-dimensional histology slices.

In summary, this Section presents several examples of whole axon configurations. Main points are that (i) these differ significantly and to some extent are "fingerprint" recognizable for specific projections; (ii) within a single projection, there are typically a range of different axon morphologies; (iii) the relation to functional architectures (such as orientation selectivity) is not straightforward and may recruit multiple mechanisms. Single axon analysis, as the reader will appreciate, superbly reveals divergence within a target and across targets. In some cases, as for the olfactory projections, axonal configurations can be associated with a relatively detailed functional role. In many others, while some design principles can be inferred, the detailed functional circuitry requires further investigation.

---

[1] The development of Sindbis virus as a tracer has been used to advantage in several publications by Takeshi Kaneko and collaborators. Unfortunately, owing to illness, Prof. Kaneko could not contribute a chapter to this volume.

1

Axonal Projection of Olfactory Bulb Tufted and Mitral Cells to Olfactory Cortex

Kensaku Mori[1,2]

[1]Department of Physiology, Graduate School of Medicine, The University of Tokyo, Tokyo, Japan
[2]Japan Science and Technology Agency, CREST, Tokyo, Japan

1.1 TUFTED CELLS AND MITRAL CELLS ARE PROJECTION NEURONS IN THE OLFACTORY BULB, CONVEYING ODOR INFORMATION TO THE OLFACTORY CORTEX

Olfactory perception depends critically on respiration cycles. The central olfactory system in the brain receives odor information of the external world during the inhalation phase (on-line phase) of the respiration cycle, whereas it is isolated from the external world during the exhalation phase (off-line phase) (Kepecs et al., 2006; Mainland and Sobel, 2006; Mori and Manabe, 2014; Mori et al., 2013). Therefore, a major question in the exploration of the olfactory system in the mammalian brain is *how* and *in which timing during a single inhalation–exhalation cycle* the central olfactory system transforms external odor information into emotional responses and motivated behavior outputs.

Figure 1.1 illustrates the characteristic structural organization of the afferent pathways of the olfactory system. During the inhalation phase, odor molecules (odorants) are inhaled into the nasal cavity and detected by odorant receptors expressed on the cilial surface membrane of olfactory sensory neurons in the nasal epithelium (Buck and Axel, 1991). Mouse olfactory epithelium demonstrates approximately 1000 different odorant receptor species, and each olfactory sensory neuron expresses only one functional odorant receptor gene (one cell–one receptor rule) (Axel, 1995; Buck, 2000). Olfactory sensory neurons send the odor information via their axons (olfactory axons) to the glomeruli that cover the surface of the main olfactory bulb, the first information processing center in the mammalian olfactory system (Mori et al., 1999; Mori and Sakano, 2011; Shepherd et al., 2004).

The olfactory bulb contains two major categories of projection neurons that differ in morphology and functional properties: tufted cells (with smaller cell bodies and shorter dendrites, shown red in Figure 1.1) and mitral cells (with larger cell bodies and longer dendrites, shown blue in Figure 1.1) (Mori, 1987; Shepherd et al., 2004). Individual tufted cells and mitral cells project a primary dendrite to a single glomerulus and receive excitatory synaptic inputs from olfactory axons on their terminal tufts of the primary dendrite within the glomerulus. Both types of projection neurons extend lateral (secondary) dendrites to the external plexiform layer (EPL) and make numerous dendrodendritic reciprocal synapses mainly with granule cells, forming local circuits in the olfactory bulb.

Tufted cells and mitral cells project long axons to a variety of areas in the olfactory cortex, and thus convey odor information from the olfactory bulb to the olfactory cortex. A particularly noteworthy anatomical difference between tufted cells and mitral cells is their pattern of axonal projection to the olfactory cortex, which will be the major theme of this chapter.

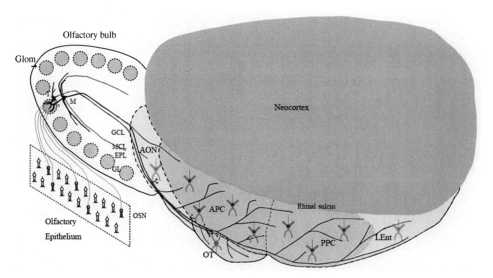

FIGURE 1.1 **A schematic diagram (lateral view of rodent brain) illustrating the afferent sensory pathways of the olfactory system including axonal projection of olfactory bulb tufted cells and mitral cells to the olfactory cortex.** Olfactory sensory neurons (OSNs) expressing a same type of odorant receptor converge their axons to a few fixed glomeruli (glom) in the olfactory bulb. Within each glomerulus, olfactory axons form excitatory synapses on the dendritic tufts of tufted cells (T, shown red) and mitral cells (M, shown blue). The projection neurons in the olfactory bulb extend their axons to many areas of the olfactory cortex including the anterior olfactory nucleus (AON), anterior piriform cortex (APC), posterior piriform cortex (PPC), olfactory tubercle (OT), cortical amygdaloid nuclei (not shown), and lateral entorhinal area (LEnt). GL, glomerular layer; EPL, external plexiform layer; MCL, mitral cell layer; GCL, granule cell layer.

Tufted cells and mitral cells differ not only in their morphology but also in their responses to odor inhalation (Fukunaga et al., 2012; Igarashi et al., 2012; Nagayama et al., 2004). Firstly, odor concentration threshold for inducing spike responses is much lower in tufted cells than in mitral cells. When the concentration of a stimulus odor is gradually increased, a low concentration odor induces spike responses only in tufted cells of the activated glomerulus, whereas a higher concentration odor causes spike discharges in both tufted cells and mitral cells. Second, tufted cells and mitral cells differ in the firing frequencies of their odor responses. Tufted cells show high-frequency burst spike discharges, whereas mitral cells respond with low-frequency burst spike discharges. Finally, but most importantly, tufted cells and mitral cells show characteristic difference in their signal timing in reference to the inhalation–exhalation sniff cycle. Tufted cells start to send signals to the olfactory cortex about 50 ms earlier than mitral cells. This surprising observation suggests that tufted cells and mitral cells convey odor information to the olfactory cortex at partly overlapping but distinct time windows during each respiration cycle.

The olfactory cortex is defined as those areas that receive direct synaptic inputs from the olfactory bulb projection neurons (Neville and Haberly, 2004; Price, 1973). In rodents, the olfactory cortex covers the largest part of the ventral surface of the forebrain (Figure 1.1). The olfactory cortex includes the anterior olfactory nucleus (AON), tenia tecta (TT), dorsal peduncular cortex (DPC), anterior piriform cortex (APC), posterior piriform cortex (PPC), nucleus of the lateral olfactory tract (nLOT), anterior cortical amygdaloid nucleus (ACo), posterolateral cortical amygdaloid nucleus (PLCo), lateral entorhinal area (LEnt), the ventral agranular insula (vAI, not shown in Figure 1.2), and olfactory tubercle (OT) (Figure 1.2). These areas can be classified into the following five distinct groups: (1) olfactory peduncle areas (AON, TT, and DPC) which are key cortical relay stages connecting the olfactory bulb with the other areas of the olfactory cortex (Brunjes et al., 2005; Haberly and Price, 1978), (2) piriform cortex (APC and PPC) which plays a crucial role for the perception and memory of odor objects (Neville and Haberly, 2004; Wilson and Sullivan, 2011), (3) cortical amygdaloid nuclei (nLOT, ACo, and PLCo) which connect with amygdaloid nuclei, (4) non-paleocortical areas including the vAI and LEnt, the latter having reciprocal connections with the hippocampal formation, and (5) OT which is part of the ventral striatum.

A majority of areas in the olfactory cortex have a pyramidal cell-based cortical organization (Figure 1.1) typically with three distinct layers (layers I, II, and III) (Neville and Haberly, 2004). In the APC and PPC, cell bodies of semilunar cells (SLs) and superficial pyramidal cells (SPs) are packed in layer II, while cell bodies of deep pyramidal cells (DPs) are distributed in layer III (see Figure 1.13). The semilunar and pyramidal cells extend apical dendrites superficially to layer I and receive excitatory synaptic inputs from axons of tufted and mitral cells in superficial sublayer of layer I (layer Ia). However, the AON has only two layers: layer I which receives afferent inputs from the olfactory bulb and layer II which contains cell bodies of pyramidal cells (Brunjes et al., 2005).

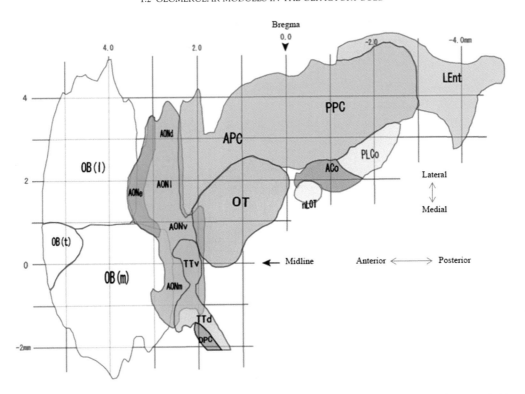

FIGURE 1.2 **Flattened unrolled map of the olfactory bulb and the olfactory cortex areas (ventral view).** The map of olfactory cortex areas is reconstructed based on the coronal sections of the mouse brain shown by Paxinos and Franklin (2001). The olfactory bulb map is reconstructed by OCAM-labeled coronal sections shown by Nagao et al. (2000). ACo, anterior cortical amygdaloid nucleus; AONd, anterior olfactory nucleus pars dorsalis; AONe, anterior olfactory nucleus pars externa; AONl, anterior olfactory nucleus pars lateralis; AONm, anterior olfactory nucleus pars medialis; AONv, anterior olfactory nucleus pars ventralis; APC, anterior piriform cortex; DPC, dorsal peduncular cortex; LEnt, lateral entorhinal area; nLOT, nucleus of the lateral olfactory tract; OB(l), lateral map of the olfactory bulb; OB(m), medial map of the olfactory bulb; OB(t), tongue-like region of the olfactory bulb; OT, olfactory tubercle; PLCo, posterolateral cortical amygdaloid nucleus; PPC, posterior piriform cortex; TTd, dorsal tenia tecta; TTv, ventral tenia tecta. Different areas of the olfactory cortex are shown by different colors. *Source: Modified from Mori and Manabe (2014).*

The OT is unique among the olfactory cortex areas; although the OT has laminated cortical structure, it contains a striatum-like organization with GABAergic medium spiny neurons as principal neurons (Millhouse and Heimer, 1984; Mogenson et al., 1980). Cell bodies of the medium spiny neurons are densely packed in layer II and sparsely distributed in layer III. The medium spiny neurons extend dendrites to layer I and receive afferent inputs from the olfactory bulb in layer Ia.

1.2 GLOMERULAR MODULES IN THE OLFACTORY BULB

1.2.1 Glomerular Convergence of Olfactory Axons

Before discussing the structural and functional differentiation between tufted cells and mitral cells, we review briefly the basic knowledge of axonal connection of olfactory sensory neurons with the projection neurons in the olfactory bulb. The mammalian main olfactory bulb has a cortical organization consisting of six well-defined layers (from the surface to the depth): the olfactory nerve layer (ONL), glomerular layer (GL), EPL, mitral cell layer (MCL), internal plexiform layer (IPL), and granule cell layer (GCL) (Figure 1.3). The ONL covers the surface of the main olfactory bulb and contains numerous intermingled bundles of olfactory axons that originate from olfactory sensory neurons in the olfactory epithelium. An individual olfactory sensory neuron gives rise to a single unbranched axon that passes the ONL and forms terminal branches within a single glomerulus.

An astonishing feature of olfactory axon projection is that axons of olfactory sensory neurons expressing a given type of odorant receptor converge to a fixed projection site, forming a glomerulus in each of two olfactory sensory maps in the mammalian olfactory bulb (Mombaerts, 2006; Mombaerts et al., 1996; Mori and Sakano, 2011). In other words, individual glomeruli receive converging axonal inputs from several thousands of olfactory sensory neurons

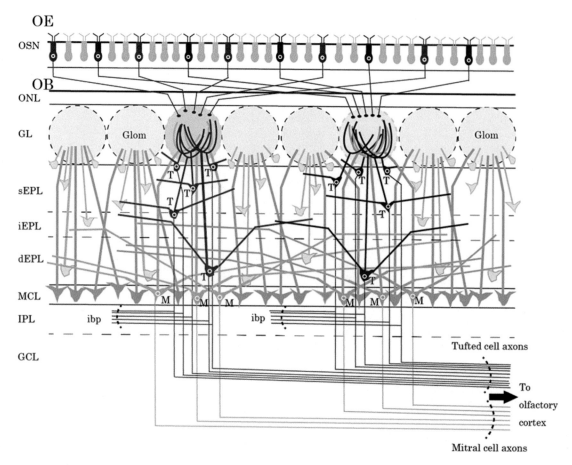

FIGURE 1.3   **Glomerular modules and projection neurons in the mammalian olfactory bulb.** In the mouse olfactory bulb (OB), an individual glomerulus (Glom) is innervated by thousands of axons of olfactory sensory neurons (OSNs) that express a single type of odorant receptor. In addition, an individual glomerulus is innervated by primary dendrites of about 50 sister tufted cells (T) and about 20 sister mitral cells (M). An individual glomerulus together with the associated olfactory axons, sister tufted cells, and sister mitral cells forms a structural and functional module (glomerular module) as shown by red/orange (left module) or dark/light blue (right module). Layers in the olfactory bulb: ONL, olfactory nerve layer; GL, glomerular layer; sEPL, superficial sublayer of the external plexiform layer; iEPL, intermediate sublayer of the external plexiform layer; dEPL, deep sublayer of the external plexiform layer; MCL, mitral cell layer; IPL, internal plexiform layer; GCL, granule cell layer. ibp, intrabulbar projection of axon collaterals of tufted cells; OE, olfactory epithelium.

that express an identical type of odorant receptor (Figures 1.1 and 1.3). Therefore, an individual glomerulus is an input channel representing a single type of odorant receptor. This organization of olfactory axon projection is called as "one glomerulus–one receptor rule." Since each tufted or mitral cell projects a primary dendrite to a single glomerulus, an individual glomerulus together with its associated tufted and mitral cells form a structural and functional module (glomerular module), each module representing a single type of odorant receptor (Figure 1.3).

1.2.2 Odorant Receptor Maps in the Main Olfactory Bulb

In mice, each olfactory bulb contains about 1800 glomeruli, and an individual odorant receptor is typically represented by a stereotypical pair of glomeruli located at two different sites, one at the rostrodorsolateral half (the lateral map) and the other at the caudoventromedial half (the medial map) of the olfactory bulb (Axel, 1995; Mombaerts et al., 1996; Ressler et al., 1994; Vassar et al., 1994). The spatial arrangement of nearly 900 glomerular modules in the lateral and medial maps of the mouse olfactory bulb forms two odorant receptor maps that are arranged in mirror symmetric fashion (Mori et al., 1999; Mori and Sakano, 2011; Nagao et al., 2000).

Since a given odorant typically activates multiple types of odorant receptors, the odor information detected by olfactory sensory neurons is topographically represented as the pattern of activated glomeruli in each of the two odorant receptor maps of the olfactory bulb. In other words, a given odorant is coded by a combination of activated odorant receptors at the molecular level and by a combination of activated glomeruli (or glomerular modules) at the

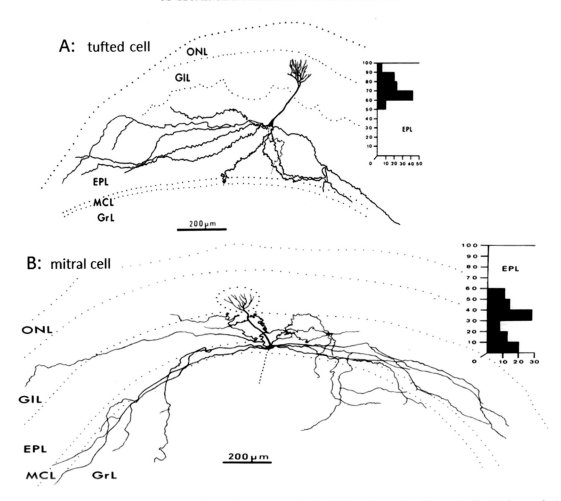

FIGURE 1.4 **Dendritic arborization pattern of a tufted cell (A) and a mitral cell (B) in the rabbit olfactory bulb.** (A) Camera lucida drawing of a middle tufted cell. The lateral dendrites of the middle tufted cell arborize mainly in the superficial sublayer of the external plexiform layer (EPL). The histogram (inset) shows the percentile distribution of lateral dendrites of the middle tufted cell in the sublayers of the EPL. Ordinate indicates the distance (in %) from the deepest part of the EPL. Zero percent indicates the deepest part, while 100% indicates the most superficial part of the EPL. The mitral cell in (B) extends lateral dendrites to the intermediate and deep sublayers of the EPL. *Source: Modified from Mori et al. (1983).*

neuronal circuit level (Axel, 1995; Mori et al., 1999). This is also the case for the coding of odorant mixtures. Thus, regardless whether it is a mono-molecular odor or a multi-molecular odor (odorant mixtures), an inhaled odor is coded by a combination of activated glomerular modules in each of the lateral and medial maps of the olfactory bulb.

Each functional pair of glomerular modules (one in the lateral map and the other in the medial map) is precisely and reciprocally linked via the axon collaterals of tufted cells, which are called as "intrabulbar projections" (Figure 1.3) (Belluscio et al., 2002; Cummings and Belluscio, 2010; Igarashi et al., 2012; Liu and Shipley, 1994; Lodovichi et al., 2003; Schoenfeld et al., 1985; Zhou and Belluscio, 2008). Although mitral cells emit axon collaterals within the granule cell layer of the olfactory bulb, they do not give rise to the intrabulbar projections. These results suggest that tufted cells, but not mitral cells, participate in the coordination of the activity of the functional pairs of glomerular modules.

1.2.3 Each Glomerular Module Contains Several Subtypes of Projection Neurons

As described above, projection neurons in the olfactory bulb are classified into two major categories: tufted cells and mitral cells (Figures 1.3 and 1.4) (Cajal, 1955; Macrides and Schneider, 1982; Macrides et al., 1985; Mori, 1987; Mori et al., 1983; Orona et al., 1984). Tufted cells are further classified into three subtypes. External tufted cells (projection neuron type) have small cell bodies in the deepest part of the GL or at the most superficial part of the EPL and project short secondary dendrites at the superficial sublayer of the EPL (sEPL) (Figures 1.3 and 1.5). Middle tufted cells exhibit their cell bodies in the sEPL or intermediate sublayer of the EPL (iEPL) (Figures 1.3, 1.4A, and 1.5).

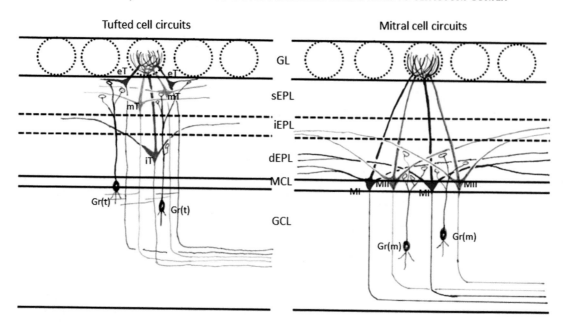

FIGURE 1.5   **A schematic diagram illustrating tufted cell circuits (left) and mitral cell circuits (right) in the olfactory bulb.** External tufted cells (eT), middle tufted cells (mT), and internal tufted cells (iT) extend lateral dendrites to the superficial sublayer of the EPL (sEPL) and the intermediate sublayer of the EPL (iEPL) and make dendrodendritic reciprocal synaptic connections with tufted cell–targeting granule cells (Gr(t)), forming tufted cell circuits. Mitral cells project lateral dendrites to the deep sublayer of the EPL (dEPL) and intermediate sublayer of the EPL (iEPL) and make dendrodendritic reciprocal synaptic connections with mitral cell–targeting granule cells (Gr(m)), forming mitral cell circuits. Type I mitral cells (MI) extend lateral dendrites to the dEPL, while type II mitral cells (MII) project lateral dendrites to iEPL.

They extend secondary dendrites to the sEPL and iEPL. Internal tufted cells have cell bodies in the deep sublayer of the EPL (dEPL) and extend long secondary dendrites to the iEPL.

Large cell bodies of mitral cells are aligned in the thin MCL. Mitral cells are further classified into two subtypes: type-I and type-II mitral cells (Figure 1.5, right) (Orona et al., 1984). A majority of mitral cells are type-I cells that extend long secondary dendrites to the dEPL (Figure 1.4B). The other subtype is type-II mitral cells that project long secondary dendrites mainly to the iEPL.

As illustrated in Figure 1.3, dozens of sister tufted cells and sister mitral cells extend primary dendrites to a same glomerulus. It was estimated that in the rabbit olfactory bulb a single glomerulus receives primary dendrites from an average of 68 sister tufted cells and 24 sister mitral cells (Allison and Warwick, 1949). In the mouse olfactory bulb, it has been recently estimated that a single glomerular module receives primary dendrites from about 50 sister tufted cells and 20 sister mitral cells (Kikuta et al., 2013). When a glomerular module is activated by the inhalation of odor-containing air, *how* and *in which timing of a single sniff cycle* do sister tufted cells and mitral cells convey odor signals to the olfactory cortex?

1.3 DENDRODENDRITIC RECIPROCAL SYNAPTIC INTERACTIONS BETWEEN PROJECTION NEURONS AND GRANULE CELLS IN THE OLFACTORY BULB

1.3.1 Tufted Cell Circuits and Mitral Cell Circuits in the Olfactory Bulb

Tufted cells and mitral cells receive the olfactory sensory inputs in the glomerulus and first process the odor signals in the local circuits within the glomerulus. The projection neurons further process the odor signals in the local circuits of the EPL and finally send the sensory information to the olfactory cortex. For the local processing in the EPL, tufted cells and mitral cells extend lateral (secondary) dendrites to the EPL and form numerous dendrodendritic reciprocal synapses with granule cells, axon-less GABAergic interneurons. The dendrodendritic reciprocal synapse consists of mitral/tufted-to-granule excitatory synapse and granule-to-mitral/tufted inhibitory synapse (Rall et al., 1966; Shepherd et al., 2004). In addition, lateral dendrites of mitral cells form dendrodendritic reciprocal synapses with parvalbumin-positive interneurons in the EPL (Huang et al., 2013; Kato et al., 2013; Miyamichi et al., 2013; Toida et al., 1996). The parvalbumin-positive interneurons may also be involved in the local processing.

Granule cells whose cell bodies are in the superficial part of the GCL (superficial granule cells) tend to form many branches of apical dendrites in the iEPL and sEPL. In contrast, granule cells with their cell bodies in the deep part of the GCL (deep granule cells) tend to restrict their branches of apical dendrites within the dEPL and iEPL (Greer, 1987; Mori et al., 1983; Orona et al., 1983). Since tufted cells project lateral dendrites to the sEPL and iEPL, tufted cells may form dendrodendritic reciprocal synapses preferentially with superficial granule cells. The subtype of granule cells that has dendrodendritic synaptic connections with tufted cell dendrites is thus called as "tufted cell–targeting granule cells" (Gr(t) in Figure 1.5). The local circuits formed by tufted cell dendrites and tufted cell–targeting granule cells are called as "tufted cell circuits" (Figure 1.5, left).

Mitral cells extend several long lateral dendrites to the iEPL and dEPL. Mitral cells thus form dendrodendritic reciprocal synapses preferentially with deep granule cells. The granule cells with dendrodendritic connections with mitral cell dendrites are called as "mitral cell–targeting granule cells" (Gr(m) in Figure 1.5). The local circuits formed by mitral cell dendrites and mitral cell–targeting granule cells are referred to as "mitral cell circuits" (Figure 1.5, right).

These observations raise the possibility that tufted cells and mitral cells process the olfactory sensory information via distinct local circuits in the olfactory bulb; tufted cells via tufted cell circuits and mitral cells via mitral cell circuits.

1.3.2 Odor Inhalation-Induced Gamma Oscillations

What is the functional role of the tufted cell circuits and mitral cell circuits? Rats and mice typically show fast sniff (>5 Hz) during active exploratory behavior while they breathe at slower frequencies (<3 Hz) during quiet wakefulness (Manabe and Mori, 2013; Mori and Manabe, 2014). During both fast and slow sniffs, each odor inhalation induces prominent sine wave–like oscillations of local field potentials in the olfactory bulb. The frequencies of the main components of these oscillations range from 30 Hz to more than 100 Hz. Oscillations within this frequency range are called as "gamma-range oscillations" (Buzsaki, 2006).

Synchronization of olfactory bulb activity with the respiration cycle was noted by Adrian over 70 years ago (Adrian, 1942). He reported that each inhalation of odor-containing air induced prominent gamma oscillations of local field potentials in the hedgehog olfactory bulb. Since then, a number of researchers studied the odor inhalation–induced gamma oscillations in the mammalian olfactory bulb for understanding their functional role in odor information processing (Adrian, 1942; Bressler, 1984; Buonviso et al., 2003; Cenier et al., 2008; Freeman, 1975; Mori and Takagi, 1977; Neville and Haberly, 2003; Rosero and Aylwin, 2011).

Dendrodendritic reciprocal synaptic interactions between tufted/mitral cells and granule cells participate in the generation of the gamma oscillatory local field potentials in the olfactory bulb (Friedman and Strowbridge, 2003; Lagier et al., 2004; Mori and Takagi, 1977; Rall and Shepherd, 1968; Shepherd et al., 2004). Furthermore, tufted cells and mitral cells show odor inhalation–induced spike discharges that are phase-locked with the gamma oscillations (Kashiwadani et al., 1999; Mori and Takagi, 1977). In other words, tufted cell circuits and mitral cell circuits are thought to participate in the generation of odor inhalation–induced gamma oscillations in the olfactory bulb.

1.3.3 Sniff-Paced Fast and Slow Gamma Oscillations

To address the question how tufted cell circuits and mitral cell circuits generate sniff-paced gamma oscillatory potentials in the olfactory bulb, temporal structure of the sniff-induced gamma oscillations was studied in the olfactory bulb of freely behaving rats (Manabe and Mori, 2013). Figure 1.6 shows concurrently recorded respiratory rhythm (monitored by placing a thermocouple into the nasal cavity) and the local field potentials in the GCL of the rat olfactory bulb. For this analysis, it is critical to align the gamma oscillatory responses in reference to the respiration cycle. Each sniff induces early-onset fast gamma oscillations (at the frequencies ranging from 65 to 100 Hz) (shown by orange F) that are nested at the inhalation phase, followed by later-onset slow gamma oscillations (40–65 Hz) (shown by blue S) that are nested at the transition phase from inhalation to exhalation (Lepousez and Lledo, 2013; Manabe and Mori, 2013).

Since tufted cells show early-onset high-frequency burst discharges while mitral cells show later-onset low-frequency burst discharges (Figure 1.7), we compared the time windows of the sniff-paced fast and slow gamma oscillations with those of sniff-paced burst discharges of tufted cells and mitral cells (Manabe and Mori, 2013). The results suggest that although different tufted cells show various time courses of spike responses, the time window of the early-onset fast gamma oscillations corresponds well with that of the early-onset high-frequency burst discharges of a total population of tufted cells. Based on this observation, we suggest that sniff-paced early-onset fast gamma oscillations (F in Figure 1.6) are generated mainly by tufted cell circuits.

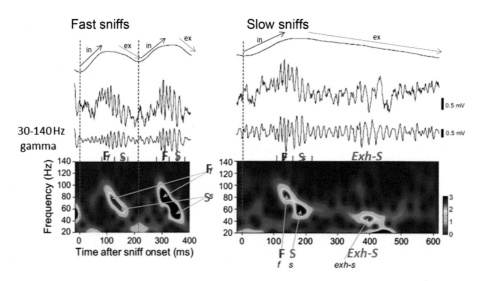

FIGURE 1.6   **Sniff-paced fast and slow gamma oscillatory activities in the olfactory bulb.** Concurrent recordings of breathing (topmost trace; inhalation is upward), local field potential (LFP) in the granule cell layer of the olfactory bulb (middle trace), and gamma oscillations of the LFP (band-pass filtered 30–140 Hz; bottom trace) during two fast sniffs (left) and one slow sniff (right). Wavelet analysis of the LFP is shown at the bottom. Dashed lines indicate sniff onset. Fast gamma oscillation is shown by F, and slow gamma oscillation is shown by S. Slow gamma oscillation during the exhalation periods is shown by Exh-S. *Source: Modified from Manabe and Mori (2013).*

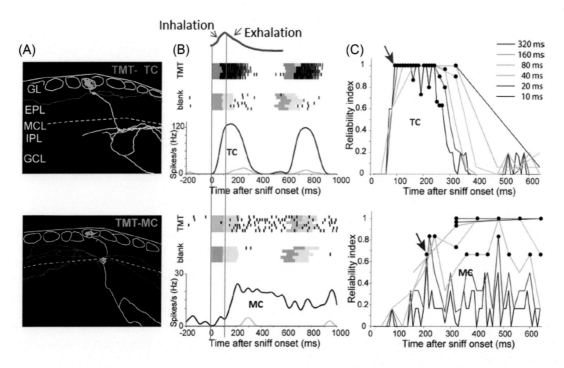

FIGURE 1.7   **Tufted cells and mitral cells show odor-induced spike responses during different phases of the respiration cycle.** (A) Morphological reconstruction of TMT-responsive external tufted cell (upper figure) and TMT-responsive mitral cell (lower figure). Cell bodies, axons, primary dendrites, and lateral dendrites are indicated by blue, yellow, green, and magenta, respectively. (B) Timing of spike responses of the two cells shown in (A). Top and middle, Raster plots of spike trains evoked by the inhalation of TMT (red or blue; top plots) or by blank air (gray; middle plots). Each row corresponds to a single trial. The uppermost trace indicates the respiration (upward swing, inhalation; downward swing, exhalation). The darker and lighter colored shadings behind rasters also indicate respiration pattern. Bottom, peristimulus time histogram of spike responses to TMT stimulation (red or blue) and blank (gray). Time 0 is shown by aligning at the first onset of odor inhalation. (C) Plots of response reliability index of the cells in (A). *Source: Modified from Igarashi et al. (2012).*

The later-onset slow gamma oscillations (S in Figure 1.6) that start at the inhalation–exhalation transition phase correspond in timing and frequency with the later-onset low-frequency burst discharges of a total population of mitral cells. We therefore speculate that the later-onset slow gamma oscillations are largely caused by the activity of mitral cell circuits. In other words, the tufted cell circuits may function as early-onset fast gamma oscillators and the mitral cell circuit as a later-onset slow gamma oscillators. These results suggest that tufted cell circuits and mitral cell circuits are active at overlapped but distinct time windows during each respiratory cycle.

1.3.4 Signal Timing of Tufted Cells and Mitral Cells in Reference to the Respiration Phase

Figure 1.8 summarizes the temporal relations among respiration pattern (monitored by a thermocouple in the nose), the spike activities of tufted cells and mitral cells, and spinal and cranial respiratory motor outputs during the normal respiratory cycle in an awake resting rat (Mori and Manabe, 2014). On top of the opposing phases of nasal airflow (inhalation and exhalation phases), the respiratory motor outputs indicate three behaviorally distinct phases of respiration: inspiration (I), post-inspiration (post-I, E1, or passive expiration), and stage 2 of expiration (E2, late or active expiration) (Smith et al., 2013).

For measuring the time window of spike responses, we recorded tufted cells and mitral cells that respond to a fox odor 2,4,5-trimethylthiazoline (TMT) or to a spoiled food odor 2-methylbutyric acid (2MBA) (see Section 1.4). As exemplified in Figure 1.7B and C, a majority of TMT-responsive tufted cells start to show high-frequency spike discharges several tens of milliseconds after the onset of the inhalation of TMT odor and continue to show burst discharges during the inhalation phase and the post-inspiration (post-I) phase. These tufted cells cease to discharge during the stage-2 expiration phase. In a striking contrast, a majority of TMT-responsive mitral cells start to show odor-induced burst discharges later at the end of the inspiration phase or at the beginning of post-I phase. Furthermore, some mitral cells continue to show low-frequency burst discharges during the stage 2 of expiration phase.

Although such comparison of signal timing between tufted cells and mitral cells was examined only in TMT-responsive cells and 2MBA-responsive cells, the above results suggest that tufted cells and mitral cells convey odor

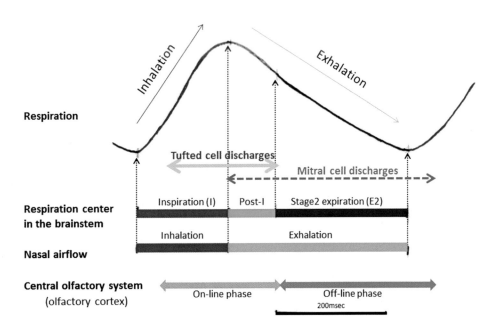

FIGURE 1.8 **The breathing cycle and its phases.** The uppermost trace shows the respiration pattern of a rat during awake resting. The respiration is monitored by a thermocouple placed in the nasal cavity. Upward swing (shown red) indicates inhalation and downward swing (green and blue) shows exhalation. Approximate time window of spike activities of tufted cells is shown by a double-headed orange arrow, while that of mitral cells by a double-headed blue arrow (broken line). While nasal air flow shows two opposing phases (inhalation and exhalation phases), the respiratory motor outputs indicate three distinct phases: inspiration phase (I), post-inspiration (post-I, E1, or passive expiration), and stage 2 of expiration (E2, late or active expiration). The central olfactory system may be at the on-line phase during the inhalation and post-I, while may be at the off-line phase during stage 2 of exhalation phase.

information to the olfactory cortex during overlapping but distinct time windows of the inhalation–exhalation sniff cycle; tufted cells send odor information to the olfactory cortex with early-onset fast gamma synchronization during the inhalation and post-I phases, whereas mitral cells send their signals to the olfactory cortex with later-onset slow gamma oscillations during the post-I phase. There is about 50 ms time lag from the onset of tufted cell discharges to the onset of mitral cell discharges. In addition, some mitral cells convey signals to the olfactory cortex during the latter part of the exhalation phase when tufted cells are silent.

The analysis of spike responses of distinct subtypes of tufted cells suggests that odor inhalation sequentially activate external tufted cells, middle tufted cells, and internal tufted cells in this order. Therefore, we speculate the following sequential shift of recruitment of olfactory bulb projection neurons according to cell size (Nagayama et al., 2014). During the inhalation of odor containing air, direct synaptic inputs from olfactory axons in the glomerulus depolarize the apical dendritic tufts of external and middle tufted cells and start to generate burst spike responses first in external tufted cells and then in middle tufted cells. The activated external and middle tufted cells then activate the tufted cell circuits via the tufted-to-granule dendrodendritic excitatory synapses in the sEPL.

At a slightly later phase of the inhalation, internal tufted cells and type II mitral cells may be recruited and generate burst discharges which activate dendrodendritic synapses in the iEPL. Finally, around the onset of the post-I phase, type I mitral cells start to show spike discharges and activate mitral cells circuits via the mitral-to-granule dendrodendritic synapses in the dEPL.

The characteristic functional differences particularly in the signal timing among different types and subtypes of bulbar projection neurons prompted us to examine the difference in the pattern of axonal projection of these neurons to the olfactory cortex.

1.4 TOTAL VISUALIZATION OF AXONAL ARBORIZATION OF INDIVIDUAL FUNCTIONALLY CHARACTERIZED TUFTED AND MITRAL CELLS

In parallel with the functional difference between tufted cells and mitral cells, do they differ also in the axonal projection pattern to the olfactory cortex? To address the question, we tried more than 30 years ago to visualize the axonal projection of individual tufted cells and mitral cells in the rabbit olfactory bulb using the method of intracellular injection of horseradish peroxidase (Ojima et al., 1984). Although we observed and traced axons of the labeled mitral cells to the caudal part of the APC, we failed to visualize all the axons of individual bulbar neurons presumably because of the large size of the rabbit olfactory cortex. Recently, we tried again to visualize the axons of individual tufted cells or mitral cells of the mouse olfactory bulb, because the mouse olfactory cortex is similar in cortical structure but much smaller than the rabbit olfactory cortex. Using the juxtacellular electroporation method to label individual functionally characterized tufted cells or mitral cells in the mouse olfactory bulb, we succeeded to trace the labeled axons to all parts of the mouse olfactory cortex (Igarashi et al., 2012).

In order to functionally characterize the recorded bulbar projection neurons and visualize their axonal projections, an experimental procedure was developed consisting of the method of optical imaging of intrinsic signals, juxtacellular single-cell recording and labeling methods, the tyramide signal amplification method, and digital three-dimensional (3D) reconstruction technique (Igarashi et al., 2012) (Figure 1.9). Since this is a critical but complex procedure, we describe each of these methods in detail for those who are not familiar with these methods. For those who are not interested in the methodological considerations, Section 1.4 can be skipped.

1.4.1 Methodological Considerations

Different glomerular modules represent diverse types of odorant receptors (Mori et al., 1999; Mori and Sakano, 2011), indicating that projection neurons belonging to different glomerular modules have distinct odor tuning specificity. Therefore, to clarify the difference in the functional properties and axonal projection pattern between tufted cells and mitral cells, it is ideal to compare the properties between tufted cells and mitral cells belonging to a same glomerulus. However with the conventional electrophysiological recording and single-cell labeling methods using glass micropipettes, it is extremely difficult to selectively record from a tufted cell and a mitral cell associated with a same functionally identified glomerulus and to label the tufted cell and the mitral cell with different marker dyes. So we chose second best procedure: comparing tufted cells and mitral cells belonging to a small cluster of glomeruli that respond to a specific odorant.

Another technical difficulty is that, if a same marker dye is used to label more than two projection neurons in a single olfactory bulb, it is very hard to distinguish the axon collaterals of one neuron from those of other neurons

Optical imaging of intrinsic signals

Single-cell recording

Single-cell electroporation

3D-reconstruction of whole dendrites and axons of single bulbar neurons

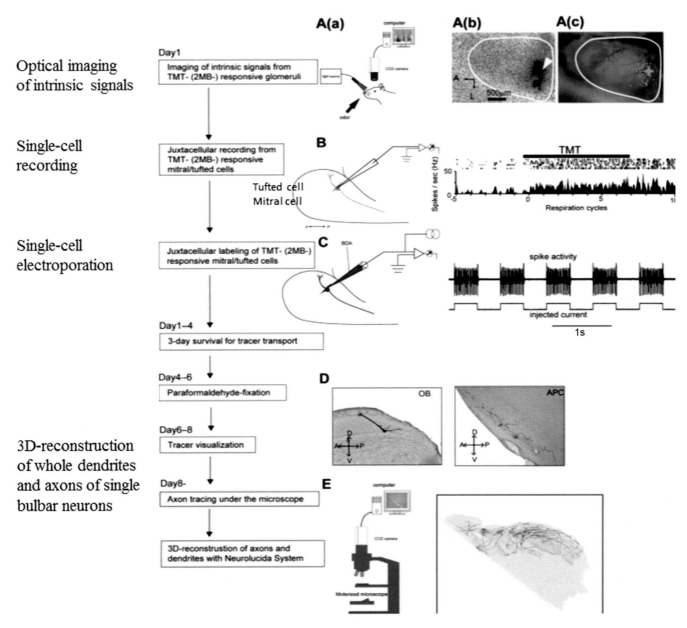

FIGURE 1.9 **Experimental procedure for dye labeling of individual functionally characterized tufted and mitral cells for 3D reconstruction of whole axons.** From top to the bottom: the sequence of optical imaging of intrinsic signals (A), single cell recording (B), single-cell electroporation (C), tracer visualization (D), and 3D reconstruction of axons (E). A(a), the method of intrinsic signal imaging; A(b) and A(c), dark spots indicate TMT-responsive glomeruli in the exposed dorsal surface (white line) of the mouse olfactory bulb. A recording micropipette was lowered near the glomerulus (arrowhead). A, anterior; L, lateral.; B, juxtacellular single-cell recording from a TMT-responsive MC. Spike responses to TMT odor are shown with raster representation (right, top traces) and a peristimulus time histogram (bottom graph). (C) Juxtacellular single-cell electroporation. Current-induced firing of the neuron (top trace) is perfectly locked to the positive current pulses (400 ms on/off; bottom trace) applied to the recorded neuron, indicating successful electroporation. (D) Visualized cell body and dendrites of the labeled mitral cell in the OB and axon collaterals in the layer I of the anterior piriform cortex (APC). (E) 3D reconstruction of whole dendrites and axons of single mitral cell.

in the olfactory cortex. So, to ensure that all labeled axons were derived from a single neuron, only one neuron was labeled per each olfactory bulb.

We focused on the projection neurons associated with the cluster of glomeruli that respond to the fox odor TMT in dorsal class II domain or a spoiled food odor 2MBA in the dorsal class I domain (Kobayakawa et al., 2007; Matsumoto et al., 2010; Mori and Sakano, 2011).

Foxes, cats, and weasels are dangerous predators to mice and rats. To detect the odors of the predators even at a remote position, rats and mice have innate neuronal circuits that can detect predator odors at very low concentrations. A representative predator odor is TMT which is secreted from the anal glands of foxes (Buron et al., 2007; Rosen et al., 2008). Although mice and rats in our laboratory have never experienced to encounter a fox, they can detect the odor TMT and show fear responses. Although TMT-responsive glomeruli are distributed both in the dorsal class II domain and ventral domain of the glomerular map in the rodent olfactory bulb, the fear responses to TMT are induced only by glomeruli in the dorsal class II domain (Kobayakawa et al., 2007).

Fatty acids and alkyl amines are released from spoiled foods and cause aversive responses in rodents. A cluster of glomeruli that respond to both fatty acids and alkyl amines is located in the dorsal class I domain of the glomerular map (Takahashi et al., 2004a,b). Using the method of optical imaging of intrinsic signals (Rubin and Katz, 1999; Uchida et al., 2000), we first identified the cluster of glomeruli that respond to the inhalation of TMT or 2MBA in the dorsal surface of the mouse olfactory bulb.

1.4.2 Intrinsic Signal Imaging and Juxtacellular Recordings

As shown in Figure 1.9A, optical imaging of intrinsic signals revealed the position of a few TMT-responsive glomeruli in the dorsal class II domain of the mouse olfactory bulb (Figure 1.9A(b) and A(c)). Several 2MBA-responsive glomeruli were identified in the dorsal class I domain of the olfactory bulb. We then penetrated glass micropipette (tip diameter, 0.5–1.0 μm; impedance, 40–80 MΩ) at or near the position of TMT-responsive (or 2MBA-responsive) glomeruli and made extracellular single-cell recordings of spike activity from individual tufted cells or mitral cells (Figure 1.9B). Functional properties of the recorded neurons were determined by characterizing the spike responses to the inhalation of TMT (or 2MBA) (Figure 1.9B, right).

1.4.3 Juxtacellualr Labeling

The glass micropipette is filled with 0.1 M potassium acetate (pH 7.5) containing an anterograde tracer (12% biotinylated dextran amine, BDA-3000). After characterizing the cellular response, we carefully advanced the glass micropipette toward the recorded neuron until the spike amplitude (positive potential) became more than 1 mV. We speculate that this step is important for successful electroporation of BDA that requires tight attaching of the micropipette tip to the surface membrane of the recorded neuron.

Single-cell labeling of individual functionally identified neurons was performed using the juxtacellular electroporation technique (Pinault, 1996). After the recording of odor responses, neurons were electroporated to inject BDA by passage of positive current pulses (0.5–1.5 nA; 400 ms duration; 50% duty cycle) through the bridge circuit of the amplifier (Figure 1.9C). The firing of the recorded neuron was continuously monitored during and after the current injection to ensure that the neuron was steadily held throughout the session. In the cases of successful electroporation of BDA, the applied positive currents induced burst discharges of the recorded neuron only during the current injection (Figure 1.9C).

1.4.4 Histochemistry

For tracing axons in the entire olfactory cortex, the current injection was continued for 30 min. After a survival period of 3 days for tracer transport to the tips of long axons, the mice were administered an overdose of anesthetic and perfused transcardially with saline followed by a fixative containing 4% paraformaldehyde and 0.5% glutaraldehyde in phosphate buffer (0.1 M, pH 7.4).

Brains were embedded in egg yolk, postfixed overnight, and cryoprotected in 30% sucrose in PB for 2–3 days. Consecutive parasagittal sections (50 μm thickness) were obtained from individual brain hemispheres using a cryostat. Using the free-floating method, sections were processed for the visualization of BDA according to the avidin–biotin–peroxidase complex histochemical protocol. To see the fine axonal branches, it is essential to amplify the signal using a tyramide signal amplification protocol (Furuta et al., 2009). The tracer was revealed by reaction with 0.05% diaminobenzidine, 0.5% nickel sulfate, and 0.007% hydrogen peroxide in 0.05 M Tris buffer, pH 7.6, for 5 min. Sections were mounted, dried overnight, and counterstained with Nuclear Fast Red (Figure 1.9D).

1.4.5 3D Reconstruction

Cells were defined as mitral cells if their large cell bodies were in the MCL. Tufted cells were defined if their somas were located in the EPL (Macrides and Schneider, 1982; Mori et al., 1983; Shepherd et al., 2004). The labeled tufted cells and mitral cells were three-dimensionally reconstructed using a computer-assisted system (Neurolucida) (Figure 1.9E, left; Figure 1.10). Axons, dendrites, cell bodies, brain surfaces, and cell layers in each region within the olfactory bulb and olfactory cortex were traced. In samples with injections of the tracer for 30 min, almost all axons were continuously traced up to the caudal end of the LEnt, suggesting that nearly complete axons of mitral cells were successfully visualized. To simplify the tracing procedures, only axon collaterals were traced to assess the projection patterns of individual neurons, and bouton structures were not registered. This step of axon tracing requires long hard works; a single mitral cell usually required 2 months for completing 3D reconstruction of axons and dendrites.

The tracing output from Neurolucida was imported into MATLAB for further analyses. To place neurons in identical coordinates and minimize the distortions originating from sectioning and reconstruction, reconstructed neurons were linearly transformed and fitted into reference brain coordinates. A reference brain was constructed by reconstructing the reference mouse brain sections in the Allen Brain Atlas (http://mouse.brain-map.org/). The three landmarks used for transformations were: the end of the LOT, the dorsomedial edge of the AON pars externa, and the ventromedial edge of the AON pars externa. These landmarks were selected because (1) they were distributed at the periphery of the tufted cell axon-targeted areas and (2) they could be uniquely defined and were unaffected

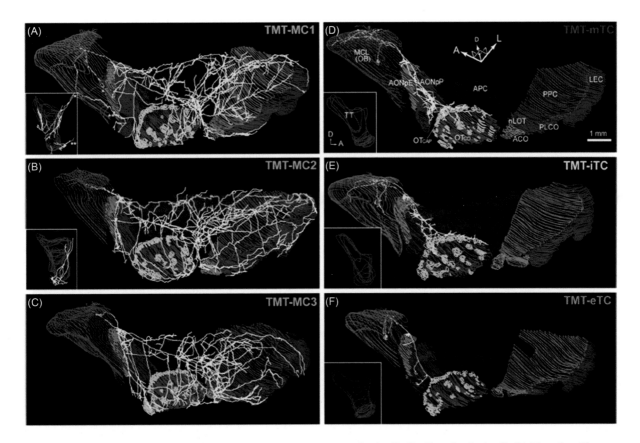

FIGURE 1.10 **Distinct pattern of axonal projection of TMT-responsive mitral cells (A–C) and tufted cells (D–F) to the olfactory cortex.** The 3D reconstruction of the brain containing the labeled cell was rotated to a ventrolateral view. Dendrites (magenta) and axons (yellow) are shown with the mitral cell layer of the OB and layer II of each area of the OC. Inset, The medial view of axons in the tenia tecta, which is hidden in the ventrolateral view. (D–F) Axonal projection of a TMT-responsive middle tufted cell (mTC, D), an internal tufted cell (iTC, E), and an external tufted cell (etc, F) to the olfactory cortex. ACO, anterior cortical amygdaloid nucleus; AONpE, pars externa of the anterior olfactory nucleus; AONpP, pars principalis of anterior olfactory nucleus; APC, anterior piriform cortex; LEC, lateral entorhinal cortex; MCL, mitral cell layer; OT, olfactory tubercle; OTCAP, cap part of olfactory tubercle; OTCO, cortical part of olfactory tubercle; nLOT, nucleus of the lateral olfactory tract; PLCO, posterolateral cortical amygdaloid nucleus; PPC, posterior piriform cortex; TT, tenia tecta; A, anterior; D, dorsal; L, lateral. *Source: Modified from Igarashi et al. (2012).*

by rotation of the brain. The coordinate system was defined as giving the ventromedial edge of the pars externa of the AON coordinates of [0, 0, 0] in the (x (posterior), y (dorsal), and z (lateral)) directions (in micrometers).

Individual sample brains containing a labeled neuron were then fitted to the reference brain with regard to the above landmarks using affine (linear) transformation (translation, rotation, scaling, and shear). For the ventrolateral-view presentation, transformed brains were then rotated 60° around the x-axis, 30° around the y-axis, and 25° around the z-axis (Figure 1.10). This projection has minimum overlap of axonal projections in the olfactory cortex for samples stained in this study, except for the tenia tecta. For the tenia tecta, axons were presented in a medial view.

1.5 AXONAL PROJECTION OF TUFTED CELLS AND MITRAL CELLS TO THE OLFACTORY CORTEX

1.5.1 Comparison Between Tufted Cells and Mitral Cells

Classical studies of gloss anatomy with retrograde tracers or electrophysiology have shown that tufted cells send axons only to the anterior areas of the olfactory cortex (AON, anterolateral part of the OT, and the anteroventral part of the APC), whereas mitral cells project axons to all areas of the olfactory cortex (Haberly and Price, 1977; Luskin and Price, 1982; Scott, 1981). The newly developed procedure of juxtacellular labeling of functionally characterized projection neurons enabled us to visualize and reconstruct almost entire axons of individual tufted cells and mitral cells and thus to compare between tufted cells and mitral cells the detailed pattern of axonal projection to the olfactory cortex. In Figure 1.10, the ventrolateral view of the mouse basal forebrain reconstruction was used to represent 3D reconstructions of axons. Because the tenia tecta is hidden in the medial wall of the basal forebrain, axons in the tenia tecta are shown in a separate medial view (Figure 1.10A, B, and D–F, bottom left insets).

Individual TMT-responsive mitral cells (Figure 1.10A–C) have their cell bodies and dendrites in the dorsal part of the olfactory bulb and extend a single main axon that runs through the LOT at the surface of the AON and APC. The main axon in the LOT emits a large number of collaterals that run dorsolaterally or ventromedially. In the olfactory peduncle areas of the olfactory cortex, the axon collaterals of the mitral cells form a cluster of branches at dorsolateral corner of the AON pars externa. The main axons of the mitral cells also emit collaterals that project to the anterolateral and dorsal parts of the AON but do not reach the posteroventral and medial parts of the AON. Some axon collaterals extend ventrally and terminate in the tenia tecta.

In the ventral part of the APC, the main axons of the TMT-responsive mitral cells emit several axon collaterals, some of which project ventrally and extend to the medial and posterior parts of the OT. From the posterior part of the APC, many axon collaterals extend branches posteriorly for long distances through the PPC, and some of these branches reach the LEnt. The axon collaterals of the mitral cells cover a large portion of the PPC. The axons also project to the nLOT, ACo, and PLCo. Numerous *en passant* boutons occur along the course of axon collaterals (Igarashi et al., 2012).

We made 3D reconstruction of whole axons of four TMT-responsive mitral cells and three 2MBA-responsive mitral cells. All the TMT- and 2MBA-responsive mitral cells show in common the following characteristic pattern of axonal projection (Figures 1.10 and 1.11):

1. Mitral cell axons project to almost all the areas of the olfactory cortex, although axon collateral density varies among target areas.
2. Mitral cells form several clusters of axon collaterals in the AON and anteromedial part of the APC, while they project axon collaterals diffusely over most part of the APC and PPC (Figure 1.10A–C). Thus, axons of individual mitral cells show a biased and at the same time highly distributed projection pattern in the APC and PPC.

Now let us look at the axonal projection pattern of TMT-responsive tufted cells (Figure 1.10D–F). At a glance of the reconstructions, it is noted that axons of TMT-responsive tufted cells show projection pattern strikingly different from those of TMT-responsive mitral cells (Figures 1.10 and 1.11). Similarly, axonal projection pattern of 2MBA-responsive tufted cells differs dramatically from that of 2MBA-responsive mitral cells (Igarashi et al., 2012).

The main axon of TMT-responsive middle tufted cell (Figure 1.10D) and that of internal tufted cell (Figure 1.10E) form axon collateral arbors at the dorsolateral corner of the AON pars externa. The main axons further project caudally and form dense axon collateral plexuses at the posteroventral part of the AON, the medial edge of the anterior part of the APC, and the anterolateral part of the OT. In agreement with prior anatomical and electrophysiological studies (Haberly and Price, 1977; Luskin and Price, 1982; Scott, 1981), the TMT-responsive tufted cells have their

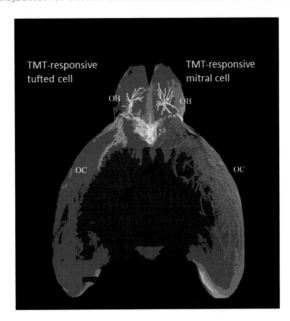

FIGURE 1.11 **Comparison of axonal projection pattern between a TMT-responsive tufted cell (left) and a TMT-responsive mitral cell (right) (ventral view of the mouse brain).** In the tufted cell (left) reconstructed dendrites in the olfactory bulb (OB) are indicated by yellow, while axons by green on a transparent view of the brain. In the mitral cell (right), axons and dendrites are indicated by magenta and yellow, respectively. Note that individual tufted cells project axons to focal targets only in the olfactory peduncle areas, rostroventral parts of the APC, and the lateral edge of the olfactory tubercle, whereas individual mitral cells send axons dispersedly across whole areas of the olfactory cortex (OC). *Source: Modified from Nagayama et al. (2014).*

axonal projection limited to the anterior areas of the olfactory cortex and do not send axons to the more posterior areas of the olfactory cortex (Figure 1.10D–F).

In summary, reconstruction of entire axons of TMT- and 2MBA-responsive tufted cells indicates the following four characteristics of projection (Figures 1.10 and 1.11):

1. Axons of these tufted cells project only to the anterior areas of the olfactory cortex (AON, APC, and OT) and do not project to the more posterior areas.
2. The tufted cells form a patchy and dense cluster of axon terminals at a specific position of the AON pars externa and some clusters in the posteroventral part of the AON.
3. The tufted cells form several dense clusters of axon terminals along the border between the cap compartments of the OT (Meyer and Wahle, 1986) and the medial edge of the APC.
4. The tufted cells do not project axons to the medial and posterior areas of the OT.

Among different subtypes of TMT- or 2MBA-responsive tufted cells, external tufted cells tended to project fewer axon collaterals in the anterior areas (Figures 1.10F), but all the tufted cell subtypes project axons selectively to the focal targets in the AON, APC, and OT.

1.5.2 Distinct Pattern of Axonal Projection of Tufted Cells and Mitral Cells May Relate to Their Functional Differentiation

Given that individual TMT- or 2MBA-responsive tufted cells project axons densely to focal targets only in the AON, the medial edge of APC, and the cap compartments of OT, the early-onset signals of tufted cells may be sent selectively to these focal targets with fast gamma synchronization. In other words, sister tufted cells associated with a given activated glomerulus may fire synchronously at fast gamma frequency during the inhalation phase conveying specific signals selectively to the spatially restricted targets in the anterior parts of the olfactory cortex.

One of the focal targets of axons of TMT-responsive tufted cells is the dorsolateral corner of the AON pars externa (Figure 1.10). At the focal target, axons of TMT-responsive tufted cells form high-density bushes of terminal branches. The terminal branches of the TMT-responsive tufted cells do not occur in any other part of the AON pars externa. Previous studies have shown a topographical relation between the position of tufted cells in the glomerular map of the olfactory bulb and the position of their axon terminals in the AON pars externa (Schoenfeld and Macrides, 1984; Yan et al., 2008). The AON pars externa has a U-shaped structure and position within it is thought to represent specific category of odorants (Kikuta et al., 2010). It has been shown that individual neurons in the AON pars externa show sniff-paced burst discharges and can distinguish the right or left position of an odor source by referencing odor signals from the right and left nostrils. Based on these results, we speculate that at the early phase of a sniff cycle, the tufted cells rapidly send the TMT signal to the topographically fixed position (dorsolateral corner) of the AON pars externa to compute the right or left localization of the TMT odor source (Kikuta et al., 2010).

Another example of a focal target of TMT-responsive tufted cells is a relatively small region in the posteroventral subdivision of the AON. Pyramidal cells in this subdivision of the AON give rise to Ib association fibers that terminate in the layer Ib of ventral part of the APC (Ekstrand et al., 2001; Haberly and Price, 1978; Neville and Haberly, 2004; Price, 1973). Therefore, the early-onset signals of TMT-responsive tufted cells may be relayed by the pyramidal cells in the posteroventral subdivision of AON and then sent via Ib association fibers to the ventral APC. The sensory afferent pathway originating from tufted cells and being relayed by Ib association fibers of the AON is called as "tufted cell axon—AON Ib association fiber pathway" (Mori et al., 2013). TMT-responsive tufted cells also send axons directly to the focal targets in the ventral parts of APC. Therefore, via both the direct route and the tufted cell axon—AON Ib association fiber pathway, early-onset TMT signals of tufted cells are sent selectively to the focal targets of the ventral parts of the APC.

Interestingly, pyramidal cells in the ventral APC project axons to the ventrolateral orbital cortex (VLO) which also receive nociceptive inputs (Ekstrand et al., 2001), the signals that report dangerous tissue damage. These results suggest that tufted cells carrying information about predator odor send the early-onset signal with fast gamma synchronization via the ventral APC to the VLO. Early-onset fast gamma synchronization might be advantageous in rapidly conveying the predator odor information through the tufted cell-(posteroventral AON)-ventral APC-VLO pathway at the early phase of the sniff cycle.

The most distinctive targets of axonal projection of TMT-responsive tufted cells are the cap compartments of the OT. The cap compartments are located in the lateral part of the OT and contain small size medium spiny neurons (dwarf cells) as principal neurons in layer II (Fallon et al., 1978; Hosoya and Hirata, 1974; Millhouse and Heimer, 1984). The TMT-responsive tufted cells emit many axon collaterals that form patchy dense plexuses of terminals at the surface (layer Ia) of the cap compartments (Figure 1.12). Interestingly, the layer Ia is much thicker than layer Ib at the surface of the cap compartment, suggesting that the apical dendrites of the medium spiny neurons in the cap compartments are innervated mainly by direct afferent inputs from the olfactory bulb. 2MBA-responsive tufted cells also project their axons to and form dense plexuses of terminals at the cap compartments of the OT.

The OT appears to be involved in mediating a variety of odor-induced behaviors and to contain modules for specific motivated behaviors such as "appetitive motivation module for obtaining foods" and "aversive motivation module for avoiding dangers." Because the fox odor TMT causes fear responses and the spoiled food odor 2MBA induces aversive responses in mice and rats, we speculate that the cap compartments are part of the "aversive motivation module" receiving specific odor signals that inform dangers. The TMT-responsive tufted cells or 2MBA-responsive tufted cells may convey specific signals of the predator odor or spoiled food odor, respectively, with early-onset fast gamma synchronization to the cap compartments of the OT.

In striking contrast to the selective axonal projection of the tufted cells to the focal targets, individual TMT-responsive or 2MBA-responsive mitral cells project axons in a dispersed manner to nearly all areas of the olfactory cortex, including nearly all parts of the piriform cortex (Igarashi et al., 2012). In addition, sister mitral cells belonging to a given glomerulus project axons to the piriform cortex in a highly dispersed pattern, with their terminals distributed throughout the piriform cortex (Ghosh et al., 2011; Nagayama et al., 2010; Sosulski et al., 2011). Given that mitral cell circuits generate later-onset slow gamma oscillatory activity, we speculate that sister mitral cells belonging to an activated glomerulus provide a mechanism for the dispersion of later-onset slow gamma oscillatory activity across whole parts of the piriform cortex and even across many different areas of the olfactory cortex (Mori et al., 2013). TMT-responsive mitral cells appear to spread their signals to virtually all the pyramidal cells irrespective of their position in the piriform cortex.

1.5.3 Axons of Tufted Cells and Mitral Cells Form Synaptic Terminals in the Most Superficial Layer of the Olfactory Cortex

Tufted cells and mitral cells project afferent axons to the most superficial layer (layer Ia, shadowed yellow in Figure 1.13) of each area of the olfactory cortex. Figure 1.13 illustrates the laminar organization and principal neuron

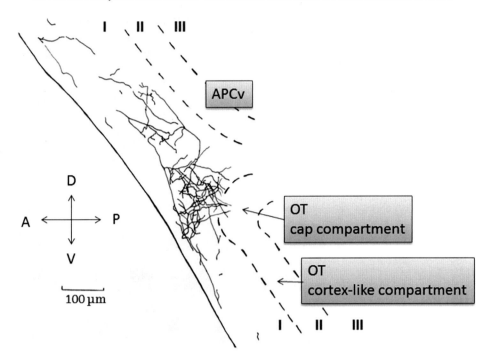

FIGURE 1.12 **Dense axonal plexus of a TMT-responsive tufted cell at the boundary between anterior piriform cortex (APC) and olfactory tubercle (OT).** Axons of the TMT-responsive tufted cell form dense plexus of branches at the most medial part of the ventral APC (APCv) and the cap compartment of the olfactory tubercle. I, layer I of the OT and APC; II, layer II of the OT and APC (indicated by broken lines); III, layer III of the OT and APC. A, anterior; P, posterior; D, dorsal; V, ventral. *Source: Unpublished data obtained by Dr. Nao Ieki.*

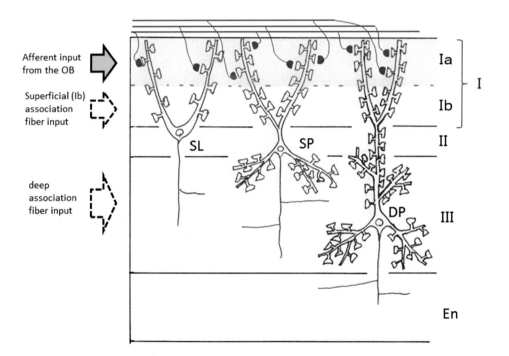

FIGURE 1.13 **Axons of tufted cells and mitral cells terminate on apical dendritic branches of pyramidal cells in layer Ia of the olfactory cortex.** Dendritic distribution of three types of principal neurons, semilunar cell (SL), superficial pyramidal cell (SP), and deep pyramidal cell (DP), is shown in reference to the layers of the piriform cortex. I, layer I; II, layer II; III, layer III; En, endopiriform nucleus. Layer I is subdivided into layer Ia and layer Ib. The three types of principal neurons extend apical dendritic branches to layer I and receive afferent inputs from tufted cells and mitral cells on the dendritic spines in layer Ia (shown by yellow). The apical dendritic branches of the principal neurons receive Ib association fiber inputs in layer Ib. Superficial pyramidal cells extend basal dendrites from their soma to the superficial part of layer III, while deep pyramidal cells extend basal dendritic branches from apical dendritic trunk and somata in layer III. These pyramidal cells receive deep association fiber inputs in layers II and III.

types in the piriform cortex (Neville and Haberly, 2004). SLs are pyramidal-type neurons having only minor or no basal dendritic branches. SLs have cell bodies in the superficial part of layer II and receive excitatory synaptic inputs from axons of tufted and mitral cells on the spines of apical dendritic branches in layer Ia. SPs have pyramid-shaped cell bodies in the deeper part of layer II while cell bodies of DPs are scattered in layer III. These pyramidal cells extend apical dendritic branches superficially to layer I and receive the afferent inputs from the olfactory bulb on the dendritic spines in layer Ia. Therefore a distinctive feature of synaptic organization in the neuronal circuits of the piriform cortex is the restricted laminar termination of olfactory bulb afferents on the most distal parts of the apical dendrites of the pyramidal cells.

Combining the knowledge of axonal projection patterns and functional properties of tufted cells and mitral cells, we speculate the following scenario of olfactory signal transfer from the olfactory bulb to the piriform cortex. During the inhalation and post-I phases, olfactory sensory inputs from tufted cell—AON Ib association fiber pathway first activate the excitatory synapses on the apical dendrites of the selective target pyramidal cells and then inputs from mitral cells later activate the synapses on the apical dendrites of nearly whole pyramidal cells in the piriform cortex (Mori et al., 2013).

1.6 GAMMA OSCILLATION COUPLING BETWEEN OLFACTORY BULB AND OLFACTORY CORTEX

If the tufted cells first convey olfactory sensory signals to the focal targets and then the mitral cells later send the signals dispersedly to all parts of the piriform cortex, it is natural to expect that tufted cell inputs arrive in the piriform cortex earlier than mitral cell inputs. Early-onset fast gamma synchrony of tufted cell activity may be transmitted first to the piriform cortex. After a delay of approximately 50 ms, the later-onset slow gamma synchrony of mitral cell activity would be subsequently transmitted to the piriform cortex.

To address the question whether mitral cell inputs lag behind tufted cell inputs in the piriform cortex, Manabe and others made local field potential recordings simultaneously in the olfactory bulb and the anterior piriform cortex of freely behaving rats (Mori et al., 2013). As shown in Figure 1.14, local field potentials in the anterior piriform cortex show inhalation-induced fast and slow gamma oscillations. Furthermore, the inhalation-induced fast and slow gamma oscillations in the anterior piriform cortex phase-couple those in the olfactory bulb. As expected, the early-onset fast gamma oscillations in the anterior piriform cortex correspond in timing and frequency with the early-onset fast gamma oscillations in the olfactory bulb, which are mainly generated by tufted cell circuits. The later-onset slow gamma oscillations in the anterior piriform cortex correspond in timing and frequency with later-onset slow gamma oscillations in the olfactory bulb, which are mainly generated by mitral cell circuits. These observations are in good agreement with the notion that during the inhalation and post-I phases, early-onset sensory signals are conveyed to the pyramidal cells in the piriform cortex via the fast gamma-synchronized activity of tufted cells and later-onset sensory signals are then sent to the piriform cortex by the slow gamma-synchronized activity of mitral cells.

In addition to the glutamatergic pyramidal neurons, anterior piriform cortex contains GABAergic inhibitory interneurons which include large horizontal cells and layer Ia neurogliaform cells in layer I (Bekkers and Suzuki, 2013; Neville and Haberly, 2004). These inhibitory interneurons receive direct inputs from the olfactory bulb afferents and provide feed-forward inhibition to pyramidal cells. Because pyramidal cells in the piriform cortex receive the mono-synaptic gamma-oscillatory excitatory inputs from tufted and mitral cells and di-synaptic gamma-oscillatory inhibitory inputs via the inhibitory interneurons, the inhalation-induced gamma oscillatory activity in the piriform cortex may be generated by both the monosynaptic excitatory synaptic inputs and di-synaptic inhibitory inputs to the pyramidal cells.

1.7 TUFTED CELLS MAY PROVIDE SPECIFICITY-PROJECTING CIRCUITS WHEREAS MITRAL CELLS GIVE RISE TO DISPERSEDLY PROJECTING "BINDING" CIRCUITS

What is the functional role of the tufted cell pathway and mitral cell pathway in odor information processing in the olfactory cortex? Extrapolating the aforementioned knowledge of TMT- or 2MBA-responsive tufted cells to whole bulbar projection neurons, we speculate that tufted cells provide specificity-projecting circuits which send signals from specific odorant receptors via their early-onset fast gamma-synchronized discharges to selective focal targets in the AON, APC, and OT (Figures 1.10, 1.11, and 1.15). We further speculate that the odor-tuning specificity

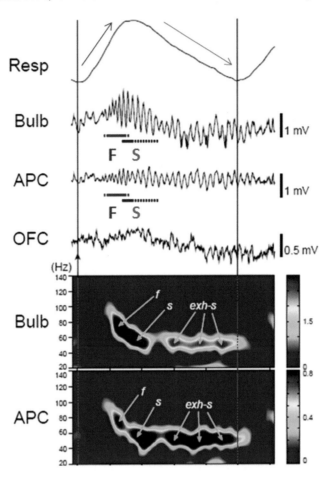

FIGURE 1.14   **Gamma oscillation coupling between the olfactory bulb and olfactory cortex during inhalation.** Concurrent recordings of respiration (Resp, upward swing indicates inhalation), local field potential in the granule cell layer of the olfactory bulb (Bulb), that in layer III of the anterior piriform cortex (APC), and that in the deep layer of orbitofrontal cortex (OFC) during micro-arousal. Wavelet analysis of the local field potentials is shown in the lower two figures. Vertical lines and upward arrows indicate sniff onset. Early-onset fast gamma oscillation is shown by F(f) and red horizontal line, and later-onset slow gamma oscillation by S(s) and blue horizontal line. Slow gamma oscillation during the exhalation phase is shown by exh-s. *Source: Modified from Mori et al. (2013).*

of individual pyramidal cells in the anterior piriform cortex (Litaudon et al., 2003; Poo and Isaacson, 2011; Yoshida and Mori, 2007) is determined primarily by the tufted cell axon—AON Ib association fiber pathways (Mori et al., 2013). Different tufted cell pathways associated with different glomerular modules convey sensory signals of distinct odorant receptors in parallel to the olfactory cortex. However, it is not clear whether and how the signals from different tufted cell pathways are combined and integrated at the stages of the AON, APC, and OT.

In the visual, auditory, and somatosensory systems, sensory receptor–specific afferent pathways to the sensory cortices constitute the essential part of the parallel processing of external sensory inputs for the perception of objects (Kaas, 1993; Van Essen and Gallant, 1994). Similarly odorant receptor–specific tufted cell pathways may be essential for the parallel processing of sensory signals for the perception of odor objects (Neville and Haberly, 2004; Wilson and Sullivan, 2011).

In contrast to the idea of specificity-projection pathways, sister mitral cells associated with a given glomerular module do not project to focal targets but disperse their signals widely and diffusely to nearly all pyramidal cells of the piriform cortex (Ghosh et al., 2011; Igarashi et al., 2012; Sosulski et al., 2011) (Figures 1.10, 1.11, and 1.15). For what functional purpose do sister mitral cells distribute their signals widely throughout the piriform cortex? Since mitral cells extend long lateral dendrites and have extensive dendrodendritic reciprocal synaptic connections, mitral cells belonging to different glomerular modules are able to synchronize their discharges at the slow gamma frequency when co-activated by an odor inhalation (Kashiwadani et al., 1999).

A large population of mitral cells that fires synchronously during odor inhalation may provide later-onset gamma-synchronized synaptic inputs to nearly all pyramidal cells of the piriform cortex (Figure 1.15). Although individual

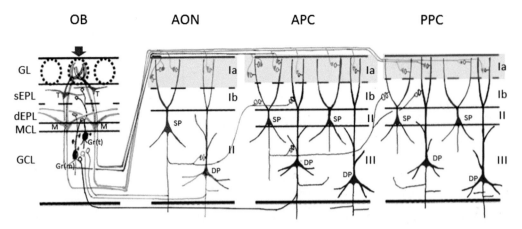

FIGURE 1.15 **Tufted cell pathway and mitral cell pathway in the large-scale olfactory bulbo-cortico-bulbar networks.** In the olfactory bulb, (OB) tufted cells (T, shown by orange) and tufted cell–targeting granule cells (Gr(t)) form tufted cell circuits, while mitral cells (M, shown by green) and mitral cell–targeting granule cells (Gr(m)) form mitral cell circuits. Tufted cells project axons (orange lines) to focal targets in the olfactory peduncle areas including the anterior olfactory nucleus (AON). Mitral cells project axons (green lines) dispersedly to nearly all areas of the olfactory cortex including AON, anterior piriform cortex (APC), and posterior piriform cortex (PPC). Layers in the olfactory bulb: GL, glomerular layer; sEPL, superficial sublayer of EPL; dEPL, deep sublayer of the EPL; MCL, mitral cell layer; GCL, granule cell layer. Layers in the AON: Ia, layer Ia (shaded orange); Ib, layer Ib; II, layer II. Layers in the APC and PPC: Ia, layer Ia (shaded orange); Ib, layer Ib; II, layer II; III, layer III. SP, superficial pyramidal cell; DP, deep pyramidal cell. Pyramidal cells in the AON and APC give rise to massive top-down inputs to the granule cell layer of the olfactory bulb.

pyramidal cells in the piriform cortex receive only weak input from individual mitral cell axons (Davison and Ehlers, 2011), nearly simultaneous arrival of the synchronized inputs from many mitral cells may effectively summate their EPSPs and thereby strongly modulate pyramidal cell activity in synchrony with the slow gamma oscillatory inputs. We therefore hypothesize that the gamma-synchronized coincident inputs from many mitral cell axons coordinate the response timing of pyramidal cells that are spatially distributed across whole parts of the piriform cortex.

The mitral cell–induced synchronized activity of pyramidal cells across whole parts of the piriform cortex may play a key role in "binding" together the spike activities of numerous co-activated pyramidal cells with different odor tuning specificity. The "binding" of the spike activities of pyramidal cells may play an important role in the integration of molecular features of odorants, as proposed for the gamma-synchronized bindings of visual cortex neurons for visual feature integration (Singer, 1999; Singer and Gray, 1995).

During the odor inhalation, spike activities of pyramidal cells in the piriform cortex are presumably driven at two distinct time windows first by the tufted cell–Ib association fiber inputs and then by the combination of the tufted cell inputs and mitral cell inputs. The activity of pyramidal cells during early-onset fast gamma oscillations in the APC may be due to the tufted cell axon—Ib association fiber inputs, while that during later-onset slow gamma oscillation is induced by the combination of preceding tufted cell axon—Ib associational fiber inputs and later-onset synchronized inputs from many mitral cells. If so, the later-onset synchronized inputs from mitral cells may cause profound synchronization of those pyramidal cells that had been depolarized by preceding early-onset inputs from tufted cell axon—Ib association axon inputs, but have little influence on those pyramidal cells that had not been depolarized by the tufted cell axon—Ib association axon inputs.

Further experiments are needed to examine the above scenarios of the possible functional role of tufted cell and mitral cell pathways in the odor information processing in the neuronal circuits of the olfactory cortex.

1.8 OLFACTORY BULBO-CORTICO-BULBAR NETWORKS

In previous sections, we focused on the afferent connections from the olfactory bulb to the olfactory cortex. It should be noted however that the number of afferent axons is much smaller than the number of centrifugal top-down axons from the olfactory cortex to the olfactory bulb (Neville and Haberly, 2004). In contrast to the afferent connections, little is known about the connectivity pattern and functional roles of the top-down inputs. Axons of pyramidal cells in the AON and APC emit numerous axon collaterals that give rise to recurrent association fibers projecting to other areas of the olfactory cortex and massive top-down projections to the olfactory bulb (Figure 1.15). Since one of the major targets of the top-down projections is granule cells that have dendrodendritic reciprocal synaptic connections with tufted or mitral cells, the neuronal circuits in the olfactory bulb and olfactory cortex are reciprocally connected and form large-scale olfactory bulbo-cortico-bulbar networks (Figure 1.15).

In this large-scale networks, signal transfer from the olfactory bulb to the olfactory cortex occur mainly during the inhalation and post-I phases, although some mitral cells send signals during the stage 2 of exhalation phase. When in the respiration cycle do the top-down signals travel from the olfactory cortex to the olfactory bulb? We noted in the freely behaving mice that the AON and APC occasionally show slow gamma oscillations during the long exhalation phase (exh-s in Figure 1.14). These slow gamma oscillations during the exhalation phase in the AON and APC typically phase-couple with those in the olfactory bulb, indicating a rich communication between these cortical areas and olfactory bulb during the off-line exhalation phase (see Figure 1.14). These results suggest that the bottom-up transfer of early-onset tufted cell signals and later-onset mitral cell signals dominate during the on-line inhalation phase, whereas top-down signals from the olfactory cortex to the olfactory bulb occur mainly during the off-line exhalation phase.

The structural organization and functional property of the olfactory bulbar-cortico-bulbar networks (Figure 1.15) resemble those of the thalamo-cortico-thalamic networks in the mammalian brain (Mori et al., 2013). Particularly, the clear differentiation in function and axonal projection pattern between tufted cells and mitral cells closely resembles the differentiation of thalamocortical neurons into core neurons and matrix neurons (Jones, 2001, 2009). Further studies of the olfactory bulbo-cortico-bulbar networks including the top-down projections from the olfactory cortex are necessary to understand the functional roles of the large-scale networks in the processing olfactory information.

1.9 CONCLUSIONS

It is surprising that despite the similarity in dendritic morphology between tufted cells and mitral cells, the two types of projection neurons have distinct pattern of axonal projection to the olfactory cortex and play distinct functional roles in processing of olfactory sensory inputs. Via dendrodendritic reciprocal synaptic connections with distinct types of granule cells, tufted cells and mitral cells form distinct local circuits within the olfactory bulb. Tufted cells emit axon collaterals in the internal plexiform layer and give rise to intrabulbar projections connecting lateral and medial maps of the olfactory bulb, but mitral cells do not. Tufted cells project axons selectively to focal targets in the olfactory peduncular areas, APC, and OT, whereas mitral cells project axons dispersedly to all parts of the piriform cortex and all areas of the olfactory cortex. Odor concentration threshold for activating tufted cells is much lower than that of mitral cells. In response to odor inhalation, tufted cells show high-frequency burst discharges while mitral cells show lower frequency burst discharges. Most importantly, tufted cells and mitral cells convey their signals to the olfactory cortex at different time windows during the inhalation–exhalation respiration cycle.

Based on these observations, we propose a hypothesis that tufted cells and mitral cells function complimentarily both in time and space and play distinct but complimentary functional roles in odor information processing in the large-scale olfactory bulbo-cortico-bulbar networks, particularly in the "bindings" of molecular features of odorants at the neuronal circuits of the olfactory bulb and olfactory cortex. The proper function of the large-scale neuronal networks that include the olfactory bulb and olfactory cortex may require the coordinated function of the tufted cell pathway and mitral cell pathway with distinct axonal connectivity logic.

Acknowledgments

I thank Drs Shin Nagayama, Kei Igarashi, Hiroyuki Manabe, Nao Ieki, Myonho An, Kimiya Narikiyo, and other members of the Department of Physiology, Graduate School of Medicine, the University of Tokyo for performing the key experiments described in this chapter and for valuable discussions. This work was supported by JSPS KAKENHI Grant Number 23240046.

References

Adrian, E.D., 1942. Olfactory reactions in the brain of the hedgehog. J. Physiol. 100 (4), 459–473.
Allison, A.C., Warwick, R.T., 1949. Quantitative observations on the olfactory system of the rabbit. Brain 72 (Pt. 2), 186–197.
Axel, R., 1995. The molecular logic of smell. Sci. Am. 273 (4), 154–159.
Bekkers, J.M., Suzuki, N., 2013. Neurons and circuits for odor processing in the piriform cortex. Trends Neurosci. 36 (7), 429–438.
Belluscio, L., Lodovichi, C., Feinstein, P., Mombaerts, P., Katz, L.C., 2002. Odorant receptors instruct functional circuitry in the mouse olfactory bulb. Nature 419, 296–300.
Bressler, S.L., 1984. Spatial organization of EEGs from olfactory bulb and cortex. Electroencephalogr. Clin. Neurophysiol. 57 (3), 270–276.
Brunjes, P.C., Illig, K.R., Meyer, E.A., 2005. A field guide to the anterior olfactory nucleus (cortex). Brain Res. Brain Res. Rev. 50 (2), 305–335.
Buck, L., 2000. The molecular architecture of odor and pheromone sensing in mammals. Cell 100, 611–618.

Buck, L., Axel, R., 1991. A novel multigene family may encode odorant receptors: a molecular basis for odor recognition. Cell 65 (1), 175–187.

Buonviso, N., Amat, C., Litaudon, P., Roux, S., Royet, J.P., Farget, V., et al., 2003. Rhythm sequence through the olfactory bulb layers during the time window of a respiratory cycle. Eur. J. Neurosci. 17, 1811–1819.

Buron, G., Hacquemand, R., Pourie, G., Lucarz, A., Jacquot, L., Brand, G., 2007. Comparative behavioral effects between synthetic 2,4,5-trimethyl-thiazoline (TMT) and the odor of natural fox (*Vulpes vulpes*) feces in mice. Behav. Neurosci. 121 (5), 1063–1072.

Buzsaki, G., 2006. Rhythms of the Brain. Oxford University Press, New York, NY.

Cajal, S.R., 1955. In: Kraft, L.M. (Ed.), Studies on the Cerebral Cortex Year Book Publishers, Chicago, IL.

Cenier, T., Amat, C., Litaudon, P., Garcia, S., Lafaye de Micheaux, P., Liquet, B., et al., 2008. Odor vapor pressure and quality modulate local field potential oscillatory patterns in the olfactory bulb of the anesthetized rat. Eur. J. Neurosci. 27 (6), 1432–1440.

Cummings, D.M., Belluscio, L., 2010. Continuous neural plasticity in the olfactory intrabulbar circuitry. J. Neurosci. 30 (27), 9172–9180.

Davison, I.G., Ehlers, M.D., 2011. Neural circuit mechanisms for pattern detection and feature combination in olfactory cortex. Neuron 70 (1), 82–94.

Ekstrand, J.J., Domroese, M.E., Johnson, D.M., Feig, S.L., Knodel, S.M., Behan, M., et al., 2001. A new subdivision of anterior piriform cortex and associated deep nucleus with novel features of interest. J. Comp. Neurol. 434 (3), 289–307.

Fallon, J.H., Riley, J.N., Sipe, J.C., Moore, R.Y., 1978. The islands of Calleja: organization and connections. J. Comp. Neurol. 181 (2), 375–395.

Freeman, W.J., 1975. Mass Action in the Nervous System. Academic Press, New York, NY.

Friedman, D., Strowbridge, B.W., 2003. Both electrical and chemical synapses mediate fast network oscillations in the olfactory bulb. J. Neurophysiol. 89 (5), 2601–2610.

Fukunaga, I., Berning, M., Kollo, M., Schmaltz, A., Schaefer, A.T., 2012. Two distinct channels of olfactory bulb output. Neuron 75 (2), 320–329.

Furuta, T., Kaneko, T., Deschenes, M., 2009. Septal neurons in barrel cortex derive their receptive field input from the lemniscal pathway. J. Neurosci. 29 (13), 4089–4095.

Ghosh, S., Larson, S.D., Hefzi, H., Marnoy, Z., Cutforth, T., Dokka, K., et al., 2011. Sensory maps in the olfactory cortex defined by long-range viral tracing of single neurons. Nature 472 (7342), 217–220.

Greer, C.A., 1987. Golgi analyses of dendritic organization among denervated olfactory bulb granule cells. J. Comp. Neurol. 257 (3), 442–452.

Haberly, L.B., Price, J.L., 1977. The axonal projection patterns of the mitral and tufted cells of the olfactory bulb in the rat. Brain Res. 129 (1), 152–157.

Haberly, L.B., Price, J.L., 1978. Association and commissural fiber systems of the olfactory cortex of the rat. II. Systems originating in the olfactory peduncle. J. Comp. Neurol. 181 (4), 781–807.

Hosoya, Y., Hirata, Y., 1974. The fine structure of the "dwarf-cell cap" of the olfactory tubercle in the rat's brain. Arch. Histol. Jpn. 36 (5), 407–423.

Huang, L., Garcia, I., Jen, H.I., Arenkiel, B.R., 2013. Reciprocal connectivity between mitral cells and external plexiform layer interneurons in the mouse olfactory bulb. Front. Neural Circuits 7, 32.

Igarashi, K.M., Ieki, N., An, M., Yamaguchi, Y., Nagayama, S., Kobayakawa, K., et al., 2012. Parallel mitral and tufted cell pathways route distinct odor information to different targets in the olfactory cortex. J. Neurosci. 32 (23), 7970–7985.

Jones, E.G., 2001. The thalamic matrix and thalamocortical synchrony. Trends Neurosci. 24 (10), 595–601.

Jones, E.G., 2009. Synchrony in the interconnected circuitry of the thalamus and cerebral cortex. Ann. N. Y. Acad. Sci. 1157, 10–23.

Kaas, J.H., 1993. The functional organization of somatosensory cortex in primates. [Review]. Ann. Anat. 175 (6), 509–518.

Kashiwadani, H., Sasaki, Y.F., Uchida, N., Mori, K., 1999. Synchronized oscillatory discharges of mitral/tufted cells with different molecular receptive ranges in the rabbit olfactory bulb. J. Neurophysiol. 82 (4), 1786–1792.

Kato, H.K., Gillet, S.N., Peters, A.J., Isaacson, J.S., Komiyama, T., 2013. Parvalbumin-expressing interneurons linearly control olfactory bulb output. Neuron 80 (5), 1218–1231.

Kepecs, A., Uchida, N., Mainen, Z.F., 2006. The sniff as a unit of olfactory processing. Chem. Senses 31 (2), 167–179.

Kikuta, S., Sato, K., Kashiwadani, H., Tsunoda, K., Yamasoba, T., Mori, K., 2010. Neurons in the anterior olfactory nucleus pars externa detect right or left localization of odor sources. Proc. Natl. Acad. Sci. U. S. A. 107 (27), 12363–12368.

Kikuta, S., Fletcher, M.L., Homma, R., Yamasoba, T., Nagayama, S., 2013. Odorant response properties of individual neurons in an olfactory glomerular module. Neuron 77 (6), 1122–1135.

Kobayakawa, K., Kobayakawa, R., Matsumoto, H., Oka, Y., Imai, T., Ikawa, M., et al., 2007. Innate versus learned odour processing in the mouse olfactory bulb. Nature 450, 503–508.

Lagier, S., Carleton, A., Lledo, P.M., 2004. Interplay between local GABAergic interneurons and relay neurons generates gamma oscillations in the rat olfactory bulb. J. Neurosci. 24 (18), 4382–4392.

Lepousez, G., Lledo, P.M., 2013. Odor discrimination requires proper olfactory fast oscillations in awake mice. Neuron 80 (4), 1010–1024.

Litaudon, P., Amat, C., Bertrand, B., Vigouroux, M., Buonviso, N., 2003. Piriform cortex functional heterogeneity revealed by cellular responses to odours. Eur. J. Neurosci. 17 (11), 2457–2461.

Liu, W.L., Shipley, M.T., 1994. Intrabulbar associational system in the rat olfactory bulb comprises cholecystokinin-containing tufted cells that synapse onto the dendrites of GABAergic granule cells. J. Comp. Neurol. 346 (4), 541–558.

Lodovichi, C., Belluscio, L., Katz, L.C., 2003. Functional topography of connections linking mirror-symmetric maps in the mouse olfactory bulb. Neuron 38 (2), 265–276.

Luskin, M.B., Price, J.L., 1982. The distribution of axon collaterals from the olfactory bulb and the nucleus of the horizontal limb of the diagonal band to the olfactory cortex, demonstrated by double retrograde labeling techniques. J. Comp. Neurol. 209 (3), 249–263.

Macrides, F., Schneider, S.P., 1982. Laminar organization of mitral and tufted cells in the main olfactory bulb of the adult hamster. J. Comp. Neurol. 208 (4), 419–430.

Macrides, F., Schoenfeld, T.A., Marchand, J.E., Clancy, A.N., 1985. Evidence for morphologically, neurochemically and functionally heterogeneous classes of mitral and tufted cells in the olfactory bulb. Chem. Senses 10 (2), 175–202.

Mainland, J., Sobel, N., 2006. The sniff is part of the olfactory percept. Chem. Senses 31 (2), 181–196.

Manabe, H., Mori, K., 2013. Sniff rhythm-paced fast and slow gamma oscillations in the olfactory bulb: relation to tufted and mitral cells and behavioral states. J. Neurophysiol. 110, 1593–1599.

Matsumoto, H., Kobayakawa, K., Kobayakawa, R., Tashiro, T., Mori, K., Sakano, H., 2010. Spatial arrangement of glomerular molecular-feature clusters in the odorant-receptor class domains of the mouse olfactory bulb. J. Neurophysiol. 103 (6), 3490–3500.

Meyer, G., Wahle, P., 1986. The olfactory tubercle of the cat. I. Morphological components. Exp. Brain Res. 62 (3), 515–527.

Millhouse, O.E., Heimer, L., 1984. Cell configurations in the olfactory tubercle of the rat. J. Comp. Neurol. 228 (4), 571–597.

Miyamichi, K., Shlomai-Fuchs, Y., Shu, M., Weissbourd, B.C., Luo, L., Mizrahi, A., 2013. Dissecting local circuits: parvalbumin interneurons underlie broad feedback control of olfactory bulb output. Neuron 80 (5), 1232–1245.

Mogenson, G.J., Jones, D.L., Yim, C.Y., 1980. From motivation to action: functional interface between the limbic system and the motor system. Prog. Neurobiol. 14 (2-3), 69–97.

Mombaerts, P., 2006. Axonal wiring in the mouse olfactory system. Annu. Rev. Cell. Dev. Biol. 22, 713–737.

Mombaerts, P., Wang, F., Dulac, C., Chao, S.K., Nemes, A., Mendelsohn, M., et al., 1996. Visualizing an olfactory sensory map. Cell 87 (4), 675–686.

Mori, K., 1987. Membrane and synaptic properties of identified neurons in the olfactory bulb. Prog. Neurobiol. 29 (3), 275–320.

Mori, K., Manabe, H., 2014. Unique characteristics of the olfactory system. In: Mori, K. (Ed.), The Olfactory System Springer, Tokyo, pp. 1–18.

Mori, K., Sakano, H., 2011. How is the olfactory map formed and interpreted in the mammalian brain? Annu. Rev. Neurosci. 34, 467–499.

Mori, K., Takagi, S.F., 1977. Inhibition in the olfactory bulb: dendrodendritic interaction and their relation to the induced waves. In: Katsuki, K., Sato, M., Takagi, S., Oomura, Y. (Eds.), Food Intake and Chemical Senses University of Tokyo Press, Tokyo, pp. 33–43.

Mori, K., Kishi, K., Ojima, H., 1983. Distribution of dendrites of mitral, displaced mitral, tufted, and granule cells in the rabbit olfactory bulb. J. Comp. Neurol. 219 (3), 339–355.

Mori, K., Nagao, H., Yoshihara, Y., 1999. The olfactory bulb: coding and processing of odor molecule information. Science 286, 711–715.

Mori, K., Manabe, H., Narikiyo, K., Onisawa, N., 2013. Olfactory consciousness and gamma oscillation couplings across the olfactory bulb, olfactory cortex, and orbitofrontal cortex. Front. Psychol. 4, 743.

Nagao, H., Yoshihara, Y., Mitsui, S., Fujisawa, H., Mori, K., 2000. Two mirror-image sensory maps with domain organization in the mouse main olfactory bulb. Neuroreport 11, 3023–3027.

Nagayama, S., Takahashi, Y.K., Yoshihara, Y., Mori, K., 2004. Mitral and tufted cells differ in the decoding manner of odor maps in the rat olfactory bulb. J. Neurophysiol. 91 (6), 2532–2540.

Nagayama, S., Enerva, A., Fletcher, M.L., Masurkar, A.V., Igarashi, K.M., Mori, K., et al., 2010. Differential axonal projection of mitral and tufted cells in the mouse main olfactory system. Front. Neural Circuits 4 pii: 120. (http://dx.doi.org/10.3389/fncir.2010.00120. eCollection 2010).

Nagayama, S., Manabe, H., Igarashi, K.M., Mori, K., 2014. Parallel tufted cell and mitral cell pathways from the olfactory bulb to the olfactory cortex. In: Mori, K. (Ed.), The Olfactory System Springer, Tokyo, pp. 133–160.

Neville, K.R., Haberly, L.B., 2003. Beta and gamma oscillations in the olfactory system of the urethane-anesthetized rat. J. Neurophysiol. 90 (6), 3921–3930.

Neville, K.R., Haberly, L.B., 2004. Olfactory cortex. In: Shepherd, G.M. (Ed.), The Synaptic Organization of the Brain Oxford University Press, New York, NY, pp. 415–454.

Ojima, H., Mori, K., Kishi, K., 1984. The trajectory of mitral cell axons in the rabbit olfactory cortex revealed by intracellular HRP injection. J. Comp. Neurol. 230 (1), 77–87.

Orona, E., Scott, J.W., Rainer, E.C., 1983. Different granule cell populations innervate superficial and deep regions of the external plexiform layer in rat olfactory bulb. J. Comp. Neurol. 217 (2), 227–237.

Orona, E., Rainer, E.C., Scott, J.W., 1984. Dendritic and axonal organization of mitral and tufted cells in the rat olfactory bulb. J. Comp. Neurol. 226 (3), 346–356.

Paxinos, G., Franklin, K.B.J., 2001. The Mouse Brain in Stereotaxic Coordinates. Academic Press, London.

Pinault, D., 1996. A novel single-cell staining procedure performed in vivo under electrophysiological control: morpho-functional features of juxtacellularly labeled thalamic cells and other central neurons with biocytin or neurobiotin. J. Neurosci. Methods 65 (2), 113–136.

Poo, C., Isaacson, J.S., 2011. A major role for intracortical circuits in the strength and tuning of odor-evoked excitation in olfactory cortex. Neuron 72 (1), 41–48.

Price, J.L., 1973. An autoradiographic study of complementary laminar patterns of termination of afferent fibers to the olfactory cortex. J. Comp. Neurol. 150 (1), 87–108.

Rall, W., Shepherd, G.M., 1968. Theoretical reconstruction of field potentials and dendrodendritic synaptic interactions in olfactory bulb. J. Neurophysiol. 31 (6), 884–915.

Rall, W., Shepherd, G.M., Reese, T.S., Brightman, M.W., 1966. Dendrodendritic synaptic pathway for inhibition in the olfactory bulb. Exp. Neurol. 14 (1), 44–56.

Ressler, K.J., Sullivan, S.L., Buck, L.B., 1994. Information coding in the olfactory system: evidence for a stereotyped and highly organized epitope map in the olfactory bulb. Cell 79 (7), 1245–1255.

Rosen, J.B., Pagani, J.H., Rolla, K.L., Davis, C., 2008. Analysis of behavioral constraints and the neuroanatomy of fear to the predator odor tri-methylthiazoline: a model for animal phobias. Neurosci. Biobehav. Rev. 32 (7), 1267–1276.

Rosero, M.A., Aylwin, M.L., 2011. Sniffing shapes the dynamics of olfactory bulb gamma oscillations in awake behaving rats. Eur. J. Neurosci. 34 (5), 787–799.

Rubin, B.D., Katz, L.C., 1999. Optical imaging of odorant representations in the mammalian olfactory bulb. Neuron 23 (3), 499–511.

Schoenfeld, T.A., Macrides, F., 1984. Topographic organization of connections between the main olfactory bulb and pars externa of the anterior olfactory nucleus in the hamster. J. Comp. Neurol. 227 (1), 121–135.

Schoenfeld, T.A., Marchand, J.E., Macrides, F., 1985. Topographic organization of tufted cell axonal projections in the hamster main olfactory bulb: an intrabulbar associational system. J. Comp. Neurol. 235 (4), 503–518.

Scott, J.W., 1981. Electrophysiological identification of mitral and tufted cells and distributions of their axons in olfactory system of the rat. J. Neurophysiol. 46 (5), 918–931.

Shepherd, G.M., Chen, W.R., Greer, C.A. (Eds.), 2004. Olfactory Bulb Oxford University Press, New York, NY.

Singer, W., 1999. Neuronal synchrony: a versatile code for the definition of relations? Neuron 24 (1), 49–65. 111–125.

Singer, W., Gray, C.M., 1995. Visual feature integration and the temporal correlation hypothesis. Annu. Rev. Neurosci. 18, 555–586.

Smith, J.C., Abdala, A.P., Borgmann, A., Rybak, I.A., Paton, J.F., 2013. Brainstem respiratory networks: building blocks and microcircuits. Trends Neurosci. 36 (3), 152–162.

Sosulski, D.L., Bloom, M.L., Cutforth, T., Axel, R., Datta, S.R., 2011. Distinct representations of olfactory information in different cortical centres. Nature 472 (7342), 213–216.

Takahashi, Y.K., Kurosaki, M., Hirono, S., Mori, K., 2004a. Topographic representation of odorant molecular features in the rat olfactory bulb. J. Neurophysiol. 92, 2413–2427.

Takahashi, Y.K., Nagayama, S., Mori, K., 2004b. Detection and masking of spoiled food smells by odor maps in the olfactory bulb. J. Neurosci. 24, 8690–8694.

Toida, K., Kosaka, K., Heizmann, C.W., Kosaka, T., 1996. Electron microscopic serial-sectioning/reconstruction study of parvalbumin-containing neurons in the external plexiform layer of the rat olfactory bulb. Neuroscience 72 (2), 449–466.

Uchida, N., Takahashi, Y., Tanifuji, M., Mori, K., 2000. Odor maps in the mammalian olfactory bulb: domain organization and odorant structural features. Nat. Neurosci. 3, 1035–1043.

Van Essen, D.C., Gallant, J.L., 1994. Neural mechanisms of form and motion processing in the primate visual system. Neuron 13 (1), 1–10.

Vassar, R., Chao, S.K., Sitcheran, R., Nunez, J.M., Vosshall, L.B., Axel, R., 1994. Topographic organization of sensory projections to the olfactory bulb. Cell 79 (6), 981–991.

Wilson, D.A., Sullivan, R.M., 2011. Cortical processing of odor objects. Neuron 72 (4), 506–519.

Yan, Z., Tan, J., Qin, C., Lu, Y., Ding, C., Luo, M., 2008. Precise circuitry links bilaterally symmetric olfactory maps. Neuron 58 (4), 613–624.

Yoshida, I., Mori, K., 2007. Odorant category profile selectivity of olfactory cortex neurons. J. Neurosci. 27 (34), 9105–9114.

Zhou, Z., Belluscio, L., 2008. Intrabulbar projecting external tufted cells mediate a timing-based mechanism that dynamically gates olfactory bulb output. J. Neurosci. 28 (40), 9920–9928.

2

The Primate Basal Ganglia Connectome As Revealed By Single-Axon Tracing

Martin Parent and André Parent

Department of Psychiatry and Neuroscience, Faculty of medicine, Université Laval,
Quebec City, Quebec, Canada

2.1 OVERVIEW OF BASAL GANGLIA ORGANIZATION

Neuronographic methods based on the axonal transport of various molecular markers have been of a great help in our attempt to decipher the major organizational features of the basal ganglia. The application of such procedures to this set of subcortical structures, which play a major role in the control of motor behavior as well as in sensorimotor integration, has revealed that the basal ganglia are reciprocally linked to the cerebral cortex via a relay in the thalamus. The cortico-basal ganglia-thalamo-cortical loop comprises the following sequentially arranged elements: (1) the striatum, composed of the caudate nucleus (CD), putamen (PUT) and nucleus accumbens (also termed ventral striatum); (2) the pallidum or globus pallidus (GP), comprising an external (GPe) and an internal (GPi) segment, as well as a ventral region; (3) the substantia nigra (SN) pars reticulata (SNr); and (4) the ventral tier motor thalamic nuclei (comprising ventral anterior and ventral lateral nuclei, abbreviated VA), whose premotor neurons convey the information that has been processed through the basal ganglia back to the cerebral cortex (Parent and Hazrati 1995; Smith et al., 1998; Gerfen and Bolam, 2010; Schmitt et al., 2014). The activity of each component of the basal ganglia is modulated by numerous structures located at the margin of the main axis and which provide neurochemical inputs capable of either blocking or facilitating the flow of neural information along the basal ganglia. Among these ancillary structures are: (1) the subthalamic nucleus (STN); (2) the pars compacta of the SN (SNc); (3) the centromedian-parafascicular thalamic complex (CM/Pf); (4) the dorsal raphe nucleus; and (5) the pedunculopontine tegmental nucleus (PPN).

The striatum is the major input structure and the largest integrative component of the basal ganglia. Its main afferent projections come from the cerebral cortex, the CM/Pf, and the SNc. The glutamatergic excitatory corticostriatal projections are very massive and impose upon the striatum a functional organization that is maintained throughout the main axis of the basal ganglia. The sensorimotor cortex projects principally to the postcommissural sector of the putamen, the associative cortex to the head of the caudate nucleus and the limbic cortex to the nucleus accumbens and adjoining structures. On such basis, the striatum can be subdivided into *sensorimotor*, *associative* and *limbic* territories. Striatal neurons located in each of these distinct functional territories play a complementary role in respect to motor behavior. Neurons in the associative territory are chiefly concerned with the anticipation and planning of movement, those in the sensorimotor territory with its execution, and the ones in the limbic territory deal more specifically with the motivational and emotional aspects of motor behavior (DeLong and Georgopoulos, 1981).

Following its complex processing within the striatum, neural information is conveyed at various levels of the neuraxis through the GPi and the SNr, which are the two major output nuclei of the basal ganglia. These structures exert a tonic GABAergic inhibitory influence upon the excitatory premotor neurons located in the VA thalamic nuclei and in the brainstem PPN. Some striatofugal inputs also target the GPe and, from there, the neural information is conveyed principally to the STN, which is reciprocally linked with the GPe (Parent and Hazrati, 1995; Smith et al., 1998; Gerfen and Bolam, 2010).

It is commonly believed that cortical information is transmitted through the basal ganglia in a parallel manner along a direct and an indirect route. The direct pathway is described as originating from striatal neurons enriched in GABA, substance P/dynorphin and the dopamine receptor D1 and which project monosynaptically to the GPi/SNr. The indirect pathway is believed to arise from striatal neurons that contain GABA/enkephalin and express the dopamine receptor D2; their influence is conveyed to the GPi/SNr polysynaptically via a sequence of connections involving the GPe and the STN. This sequence comprises: (1) a GABAergic inhibitory projection from the striatum to GPe; (2) an inhibitory GABAergic projection from the GPe to the STN; and (3) an excitatory glutamatergic projection from the STN to the GPi/SNr. Imbalance between the activity in the direct and indirect pathways and the resulting alterations in the GPi/SNr are thought to account for the hypo- and hyperkinetic features of basal ganglia disorders (Alexander et al., 1986; Albin et al., 1989; Smith et al., 1998; Gerfen and Bolam, 2010; Freeze et al., 2013; Cazorla, et al., 2014).

In an attempt to unravel the complexities of the basal ganglia organization in nonhuman primates, we initiated two decades ago a series of single-axon labeling investigations in both Old World and New World monkeys. During the course of these studies, we paid attention to the structures that form the main axis of the basal ganglia, that is, the motor cortex (Parent and Parent, 2006), the striatum (Parent et al., 1995; Lévesque and Parent, 2005), the GPe (Sato et al., 2000a), and the GPi (Parent et al., 2001). We also used the same approach to investigate two major control structures of the basal ganglia, namely the STN (Sato et al., 2000b) and the CM-Pf (Parent and Parent, 2005). The image that emerges from these single-axon tracing studies is somewhat different from that of a strict binomial parallel-processing system. Instead, the primate basal ganglia appear to be part of a widely distributed neuronal network, whose neuronal elements are endowed with a highly patterned set of axon collaterals, a morphological feature that allows each basal ganglia component to interact with one another in an exquisitely precise manner.

The present chapter provides an overview of the results of our single-axon tracing studies of this finely tuned network in nonhuman primates in the hope that such knowledge will help understanding the complex spatiotemporal sequence of neural events that ensures the flow of cortical information through the basal ganglia, whose alteration leads to major psychomotor disabilities.

2.2 EXPERIMENTAL PROCEDURES

In our attempt to decipher the complexities of the basal ganglia connectivity we used a powerful microiontophoretic injection procedure that labels small subsets of electrophysiologically identified neurons in a Golgi-like manner (Pinault, 1996). This method reveals the firing pattern of the neurons to be injected, while yielding a detailed view of the somatodendritic domain as well as the entire axonal arborization, which can thus be reconstructed from serial sagittal sections with the help of a camera lucida and a computerized image-analysis system. The results reported herein stem from studies undertaken in adult cynomolgus monkeys (*Macaca fascicularis*) or adult squirrel monkeys (*Saimiri sciureus*) of both sexes. The surgical and animal care procedures adhered to the guidelines for the use and care of experimental animals of the Canadian Council of Animal Care and the experimental protocol is described in detail elsewhere (Parent and Parent, 2005, 2006). In brief, the animals were first anesthetized with a mixture of ketamine/xylazine and their head placed in a specifically designed stereotaxic apparatus. They were then maintained under propofol anesthesia while microiontophoretic injections of either biotin dextran amine (BDA) or biocytin (N-biotinyl-L-lysine) were being made in different portions of each major component of the basal ganglia and associated thalamic and cortical regions. The various targets included the primary motor cortex (M1), the striatum, the GPe, the GPi, the STN and the CM/Pf thalamic complex. Most injections were made bilaterally and the targets reached with the help of the stereotaxic coordinates that were corrected for individual variations by ventriculographic measurements.

The neuronal labeling was carried out with fine glass micropipettes that were filled with the neuronal tracer (BDA or biocytin) and used to monitor the extracellular activity of the neuronal populations encountered during the penetration of the micropipette. The tracer was delivered microiontophoretically and, after a survival period of 2–8 days, the animals were deeply anesthetized with sodium pentobarbital and perfused transcardially with saline solution followed by 4% paraformaldehyde fixative and 10% sucrose solution. The brains were dissected out and cut along the sagittal or frontal plane at 70 µm with a freezing microtome. The sections were collected serially in phosphate buffer saline and processed for the visualization of BDA or biocytin according to the avidin–biotin-peroxidase method with diaminobenzidine as the chromogen. To help identifying nuclei and structures that harbored labeled neurons and axons, most sections were counterstained for cytochrome oxidase (Parent and Parent, 2005, 2006).

All sections were mounted on gelatin-coated slides, dehydrated in graded alcohol, cleared in toluene, and coverslipped with Permount. They were examined under a Nikon light microscope equipped with a camera lucida. Details of single BDA- or biocytin-labeled neurons were drawn at 200X and 400X magnifications. The terminal fields and cell body of the labeled neurons were mapped at lower magnifications to determine their topographic location. Photomicrographs were digitally captured and handled with the Adobe Photoshop software. Most axons were entirely reconstructed along either the sagittal or frontal planes with the help of a camera lucida, and the Neurolucida computerized image-analysis system was used to reconstruct in three dimensions the entire axonal trajectory and terminal arborization of typical basal ganglia neurons. The same computerized system was employed to gather quantitative estimates of the number of terminal boutons and total axonal length of reconstructed neurons (Parent and Parent, 2005, 2006).

2.3 CORTICOSTRIATAL PROJECTIONS

The cortical mantle projects heavily and topographically to the striatum, the input from the sensorimotor cortex being particularly extensive and bilaterally organized. The first information regarding the organization of this massive projection was obtained following the examination of Golgi-stained material in mice by Ramón y Cajal, who concluded that the corticostriatal projection was made of collaterals of fibers descending in the internal capsule (Ramón y Cajal, 1909/1911). This view was supported by more recent electrophysiological recording and intracellular labeling experiments (Donoghue and Kitai, 1981; Cowan and Wilson, 1994) in rats. The corticostriatal projection in rodents is currently viewed as arising from two distinct types of projection neurons: (1) the corticostriatal pyramidal tract (PT) neurons, whose long-range axon emits collaterals to the striatum before reaching brainstem and spinal cord motor centers, and (2) the corticostriatal intratelencephalic (IT) neurons, whose axon also provides collaterals to the striatum and elsewhere in the forebrain, but does not leave the telencephalon (Donoghue and Kitai, 1981; Cowan and Wilson, 1994; Lévesque et al., 1996, Reiner et al., 2003; Reiner, 2010; Rockland, 2013). In rodents, the PT and IT neurons were reported to arborize differently within the striatum (Cowan and Wilson, 1994). Individual corticostriatal axons appears to be distributed in a manner such that they contact a maximum number of striatal neurons but make minimum contacts with each one of them (Zheng and Wilson, 2002). Electron microscopic studies have revealed that, despite the fact that cortical synapses represent approximately 80% of all synapses in the striatum (Pasik et al., 1976), a single striatal neuron is believed to receive relatively few such synapses from restricted cortical areas (Somogyi et al., 1981). Corticostriatal afferents make asymmetrical synaptic contacts principally on the head of distal dendritic spines, but some contacts on the dendritic shaft of the spiny neurons have also been reported (Kemp and Powell, 1971; Bouyer et al, 1984).

In primates, the corticostriatal projections were first visualized following autoradiographic tracing studies in cynomolgus monkeys, and it was described as forming a distinct system solely dedicated to the striatum (Künzle, 1975). This view is supported by retrograde cell labeling (Jones et al., 1977) and electrophysiological investigations in monkeys (Bauswein et al., 1989; Turner and DeLong, 2000). The latter studies concluded that collaterals of long-range corticofugal axons to the putamen were infrequent and suggested that the putamen receives a cortical message that is significantly different from that sent down to the brainstem and spinal cord in primates.

Such a possible difference between rodents and primates in the organizaton of the corticostriatal projection prompted us to investigate, at the single cell level, the trajectory and arborization of corticostriatal axons arising from the primary motor cortex (M1) in primates. Corticostriatal projections arising from the digit area of M1 in cynomolgus monkeys were studied after labeling small pools of neurons with BDA (Parent and Parent, 2006). The unilateral or bilateral microiontophoretic injections, centered upon layer V, were made under electrophysiological guidance. Layer V of M1 was easily recognizable by the charcateristic bursting firing pattern of its neurons under ketamine-xylazine anesthesia (Figure 2.1). A total of 42 anterogradely labeled corticofugal axons were entirely reconstructed from serial frontal sections. Based on their target sites and axonal branching patterns, we identified 5 different types of corticofugal neurons in the primate M1: (1) neurons with an axon that targets principally the ipsilateral striatum (n = 21); (2) neurons with a long-range axon projecting toward the brainstem that emits collaterals to the ispilateral striatum (n = 10); (3) neurons with an axon that arborizes within the contralateral striatum (n = 6); and (4) neurons that project to the subthalamic nucleus and red nucleus, without providing collaterals to the striatum (n = 5).

Neurons of the first type were the most abundant (Table 2.1). Their cell body lies mostly within the upper part of layer V and their axon remains uniformly thin and unbranched throughout its trajectory to the ipsilateral striatum. It follows a tortuous course within the subcortical white matter before entering the putamen through its dorsal or lateral aspects. Once into the putamen, the major axonal branch breaks out into 2–5 smaller collaterals that arborize

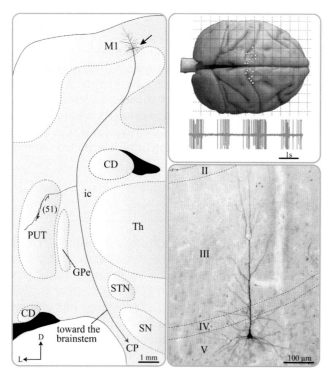

FIGURE 2.1 **Corticostriatal projections.** The left panel provides a frontal plane reconstruction of the arborization of a *type 2* corticostriatal neuron (arrow), whose thick axon provides thin collaterals to the putamen (PUT) before descending into the brainstem. The number of axonal varicosities observed in the PUT is indicated in parentheses. The right panel shows a dorsal view of the brain of a cynomolgus monkey, with the location of the various injection sites (white dots) in the primary motor cortex (M1), as well as the firing pattern and morphological characteristics of the M1-labeled neuron that gave rise to the axon depicted on the left.

scarcely but widely in the dorsolateral sector of the putamen that lies posterior to the anterior commissure, a region commonly referred to as the postcommissural putamen. These small intraputamenal axonal branches display rather equally spaced varicosities of very small size, and their distal ends often display pedunculated varicosities. Some neurons of the first type have an axon that also reaches the claustrum. In these cases, the main axonal branch courses through the lateral aspect of the putamen and emits a thin collateral that pierces the putamen laterally and arborizes sparsely in the dorsolateral sector of the claustrum.

Neurons of the second type (Figure 2.1) have a cell body slightly larger and more deeply located in layer V than those of the first type. The typical axon of these neurons descends through the subcortical white matter and remains unbranched until it reached the corona radiata. It emits a thin collateral that enters the medial aspect of the putamen and remains unbranched until it reaches the dorsolateral sector of the structure. There, it forms several very thin and long collaterals displaying elongated varicosities of the *boutons-en-passant* type. The major axonal branch remains in the internal capsule, becomes faintly labeled and difficult to follow further downward than the caudal pole of the SN. It does not appear to emit collaterals in the SN itself nor in the subthalamic nucleus.

Neurons of the third type have an axon that follows a very long but rather straightforward trajectory through the corpus callosum before reaching the contralateral putamen. Along its way to the putamen, the axon courses around or directly through the caudate nucleus, where it emits axonal varicosities of the *en passant* type, but the bulk of axon terminals occurs within the dorsolateral sector of the contralateral postcommissural putamen. In general, the number of axonal terminals that such contralaterally projecting axons yield in the putamen is higher than that of ipsilaterally projecting corticostriatal axons. Some corticostriatal axons that cross the midline through the corpus callosum emit a collateral that ascends toward the contralateral M1. Because of the very weak staining of their callosal segment, we could not link with certainty the ipsilateral and the contralateral portions of the decussating corticostriatal axons, so that it is impossible to know if these axons arborize bilaterally in the striatum or emit collaterals to other subcortical structures.

Neurons of the fifth type have an axon that emits collaterals in the subthalamic nucleus and/or the red nucleus before invading the cerebral peduncle to descend more caudally in the brainstem. These collaterals arborize rather discretely in the lateral sector of the subthalamic nucleus and the parvicellular division of the red nucleus, but their main descending axon could not be followed further than the midbrain level.

These results argue against the idea that corticostriatal projection in primates arises from a distinct set of neurons whose axons are dedicated solely to the striatum. Instead, our findings indicate that M1 neurons in cynomolgus monkeys have a dual access to the striatum: many corticostriatal neurons possess a relatively short and very thin

TABLE 2.1   Axonal Branching Patterns of Primate Basal Ganglia and Associated Nuclei

Structure, axon number (*n*), types	Principal targets and degrees of axonal arborization[a]	Percentage of traced axons
M1 (n = 37)		
Type 1	PUT: **4+**	**50%**
Type 2	PUT: **2+**; Axons heading toward BS	**24%**
Type 3	Contralateral CD: **2+**; Contralateral PUT: **3+**	**14%**
Type 4	STN: **2+**; RN: **2+**	**12%**
CM (n = 20)		
Type 1	PUT: **4+**; Rt: **1+**	**30%**
Type 2	M1 and PM: **5+**; Rt: **1+**	**35%**
Type 3	PUT: **3+**; M1 and PM: **5+**; Rt: **2+**	**35%**
Pf (n = 9)		
Type 1	CD: **4+**	**55%**
Type 2	M1 and PM: **4+**	**11%**
Type 3	CD: **3+**; M1 and PM: **3+**	**34%**
CD (n = 15)[B]		
Type 1	GPe: **5+**	**13%**
Type 2	GPe: **2+**; GPi: **2+**; SNr: **3+**	**87%**
PUT (n = 21)[B]		
Type 1	GPe: **5+**	**15%**
Type 2	GPe: **2+**; GPi: **4+**; SNr: **2+**	**85%**
GPe (n = 76)		
Type 1	GPi: **3+**; STN: **3+**; SNr: **2+**	**13%**
Type 2	GPi: **4+**; STN: **4+**	**18%**
Type 3	STN: **4+**; SNr: **2+**	**53%**
Type 4	CD: **2+**; PUT: **4+**	**16%**
GPi (n = 60)		
Type 1	VA: **4+**; PPN: **2+**	**25%**
Type 2	VA: **4+**; CM: **3+**; Pf: **2+**; PPN: **3+**	**27%**
Type 3	LH: **5+**	**28%**
Type 4	LH: **4+**; AV and AM: **3+**	**20%**
STN (n = 75)		
Type 1	GPe: **3+**; GPi: **3+**; SNr: **3+**	**21%**
Type 2	GPe: **4+**; SNr: **2+**	**3%**
Type 3	GPe: **4+**; GPi: **4+**	**48%**
Type 4	GPe: **3+**	**11%**
Type 5	PUT: **3+**	**17%**

[a]*Degrees of axonal collateralization (without intranuclear branching): (1+) very weak; (2+) weak; (3+) moderate; (4+) high; (5+) very high.*
[b]*Data derived from both squirrel and cynomolgus monkeys.*

axon that arborizes almost exclusively into the striatum, but others are endowed with a thick, long, brainstem-oriented axon that provides collaterals to the striatum. Still other corticostriatal neurons target the contralateral striatum through axons that cross the midline through the corpus callosum. Hence, M1 neurons in primates appear to be able to influence the striatum through a specifically dedicated channel as well as through collaterals of long-range axons, whereby they can send to the striatum a copy of the motor signal conveyed to the brainstem and/or spinal cord. Furthermore, the presence of corticofugal axons that innervate the subthalamic nucleus and/or the red nucleus, without providing collaterals to the striatum indicate that it is not all descending pyramidal axons that innervate the striatum. These axons are most likely part of the so-called *hyperdirect pathway*, which consist of M1 neurons projecting directly to the STN where they provide a somatotopic body-map representation that complements the one from the supplementary motor area (Nambu et al., 1996; Chu et al., 2015).

We cannot entirely rule out the possibility that some corticostriatal axons projecting solely to the striatum (first type) might, in fact, be long-range axon collaterals (second type) arising very close to the parent cell body. However, the existence of a direct corticostriatal projection system in primates is ascertained by the fact that at least two axons that arborize solely within the striatum were connected to their parent cell body. These two axons remained uniformly thin and unbranched along their entire sinuous course to the striatum, as it was the case for the other corticostriatal axons of the first type that could not be linked to their parent cell body. Axons that project collaterally to the striatum are thicker than those that project solely to the striatum and they remain unbranched until they reach the corona radiata. It is only at this low level that they emit a collateral that runs toward the striatum and this pattern was the same for all long-range axons that provide collaterals to the striatum.

The axons of the three types of corticostriatal neurons terminate chiefly in the dorsolateral sector of the post-commissural putamen, which represents the core of the so-called sensorimotor striatal territory (see Section 2.1). These findings concur with the results of earlier neuroanatomical studies, which have shown that the corticostriatal projection from primary motor and sensorimotor cortices in monkeys projects exclusively to the putamen where a somatotopic representation of the leg, arm and face occurs in the form of obliquely arranged strips. A precise topographic distribution of the cortical motor input in the sensorimotor striatal territory is important because this territory harbors neurons directly involved in the execution of movements.

Altogether, these findings indicate that the organization of the corticostriatal projection in primates shares both similarities and differences with that in rodents. For example, the second type of corticostriatal neurons identified in primates, that is, cortical neurons innervating the striatum through collaterals of long-range axon coursing towards the brainstem, appears to correspond to the so-called pyramidal tract (PT) corticostriatal neurons in rodents (Gerfen and Bolam, 2010, Reiner, 2010; Rockland, 2013). Furthermore, despite their incomplete labeling, corticostriatal neurons of the third type, that is, neurons innervating the contralateral striatum in primates, are likely to correspond to the so-called intratelencephalic (IT) corticostriatal neurons disclosed in rodents (Gerfen and Bolam, 2010, Reiner, 2010). In contrast, corticostriatal neurons of the first type, that is, those that target chiefly the primate striatum by means of a thin and poorly arborized axon, do not appear to have equivalent in rodents.

2.4 THALAMOSTRIATAL PROJECTIONS

The thalamus is the second most important source of striatal afferents, after the cerebral cortex. Although the centromedian-parafascicular complex (CM-Pf) – the major component of the caudal intralaminar nuclei – is the main source of the thalamostriatal projection, many other thalamic nuclei also contribute to the striatal innervation. Indeed, retrograde cell labeling experiments have shown that the striatum receives inputs from virtually all "non-specific" intralaminar and midline nuclear groups, as well as from various "specific" and "association" nuclei, such as the ventral anterior, ventral lateral, mediodorsal, and lateral posterior nuclei (see Smith et al., 2014). These studies also revealed that the thalamostriatal projection is topographically organized, with a rostrocaudal and dorsoventral correspondence between the intralaminar nuclei and their main projection sites in the striatum.

Studies undertaken with the fluorescence retrograde double-labeling technique in primates have shown that the intralaminar projections to the caudate nucleus and the putamen arise from two distinct neuronal populations that are organized quite differently in the rostral and caudal intralaminar nuclei (Parent et al., 1983). In the rostral intralaminar nuclei, small clusters of cells projecting to the caudate nucleus are closely intermingled with clusters of cells projecting to the putamen, whereas in the caudal intralaminar nuclei, the two cell populations are clearly segregated: neurons projecting to the putamen being confined to the CM, those projecting to the caudate nucleus to the Pf (Parent et al., 1983). Anterograde labeling studies have revealed that the projections from the CM terminate principally in the sensorimotor striatal territory, whereas those from the Pf target the associative territory (Sadikot et al., 1992). In both of these territories, CM-Pf fibers arborize in a highly heterogeneous manner, with terminal fields having the form of distinct patches or elongated bands (Sadikot et al., 1992). Electrophysiological, retrograde double labeling and single-axon tracing studies have shown that thalamostriatal axons emit collaterals that arborize within broad cortical areas (Deschênes et al., 1996; Smith et al., 2014). However, although fairly numerous in the rostral intralaminar nuclei, these branching neurons appear to be much less abundant in the CM-Pf (Sadikot et al., 1992). Electron microscopic studies have demonstrated that thalamostriatal axons from the CM-Pf form asymmetric synapses principally on dendritic shafts and, less abundantly, on dendritic spines of medium spiny projection neurons of the striatum (Sadikot et al., 1992, Smith et al., 2014).

Our single-axon tracing studies of the CM-Pf projections in cynomolgus monkeys have shed new light on the organization of this major thalamostriatal projection in primates (Parent and Parent, 2005). Microiontophoretic injections of BDA combined with electrophysiological recording in the primate CM-Pf (Figure 2.2) led to Golgi-like labeling of large perikarya from which emerged several long, poorly branched and sparsely spinous dendrites. Three-dimensional reconstruction revealed that the somatodendritic domain of both CM and Pf neurons had its greatest extent along the sagittal plane. A total of 29 axons were traced entirely from their parent cell body to their terminal fields: 20 of them had their cell body in the CM and 9 in the Pf. Based on their target sites and axonal branching patterns, we identified 3 different neuronal types in the primate CM nucleus: (1) neurons whose axon projects profusely to the putamen (n = 6); (2) neurons that arborize within the motor and premotor cortex (n = 7); and (3) neurons that target both the putamen and the cerebral cortex (n = 7) (Table 2.1).

Neurons of the first type have an axon that exits the CM through its anterior part and courses through the ventral posteromedial and/or the ventral posteroinferior thalamic nuclei and then crosses the reticular thalamic nucleus and bifurcated within the internal capsule. Half of the axons give rise to a collateral that arborizes poorly in the reticular thalamic nucleus. The major axonal segment runs through the globus pallidus to finally arborize within the dorsolateral sector of the postcommissural putamen. At the putamen level, each major axonal branch breaks out into three to five smaller collaterals. These thin branches divide into numerous shorter collaterals that bear typical pedunculated or club-like varicosities organized as dense clusters of axon terminals. These clusters form oblique bands restricted to specific domains of the dorsolateral region of the postcommissural putamen. The striatal arborization is significantly more extended along the sagittal plane than the frontal plane. One CM neuron of this type has a total axonal length of 7.59 cm and yields 3,139 varicosities in the putamen, whereas two others provide a collateral to the claustrum.

Neurons of the second type project sparsely to the motor and premotor cortical areas, but do not arborize in the striatum. The axon of these neurons exits the CM nucleus dorsally en route to the internal capsule and remains unbranched until it reaches the corona radiata through which it ascends to the motor cortex, where it arborizes

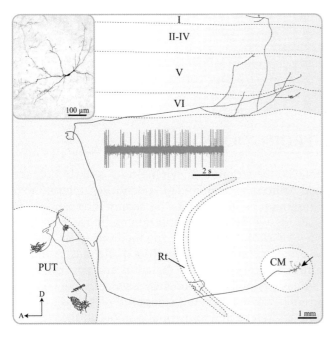

FIGURE 2.2 **Thalamostriatal projections.** A sagittal plane reconstruction of a *type 3* thalamostriatal neuron of the centromedian thalamic nucleus (CM) in a cynomolgus monkey. The cell body of this neuron (arrow) is located within the core of the CM and its axon provides a few collaterals to the reticular thalamic nucleus (Rt), but arborizes principally in the postcommissural sector of the putamen (PUT) and in the deep layers of the primary motor cortex. A photomicrograph depicting the somatodendritic domain of this labeled neuron and a sample of its firing pattern are also provided.

moderately in layers V and VI, and much less profusely in layer I. The axon of one of these neurons has a total axonal length of 6.14 cm and provides 130 varicosities in layer V, 168 varicosities in layer VI, and 58 varicosities of the *en passant* type in the subcortical white matter.

Neurons of the third type (Figure 2.2) have an axon that supplies the reticular thalamic nucleus with small collaterals. The main axonal branch reaches the internal capsule, passes through the GPe and divides into two collaterals within the external medullary lamina. One of these collaterals enters the dorsolateral sector of the postcommissural putamen, where it branches and forms four major clusters of closely packed terminal varicosities. The other collateral climbs up to the cortical white matter, sweeps caudally and, after a rather long and linear course beneath the cortical gray matter, invades the motor cortex. At this level, the terminal axonal arborization is diffuse, but still restricted to layers I, V, and VI.

The Pf harbors 3 types of projection neurons similar to those in the CM nucleus. Out of the 9 entirely reconstructed Pf neurons, 5 projected only to the striatum, 1 only to the cerebral cortex, and 3 to both the cerebral cortex and the striatum. In contrast to CM neurons that target the dorsolateral sector of the postcommissural putamen, Pf neurons project principally to the head of the caudate nucleus. Axons that target solely the striatum exit the Pf dorsally, invade the internal capsule, and course along the dorsal surface of the GPe. They then penetrate the putamen, giving rise to a short collateral that sweeps ventrally and arborizes poorly in the putamen and in the GPe. The main axonal segment continues dorsally and finally reaches the caudate nucleus, where it breaks out into several short terminal collaterals that form several dense clusters of pedunculated axonal varicosities scattered throughout the head of the structure. The main axonal branch of the Pf neuron that projects to both cerebral cortex and striatum reaches the prefrontal cortex after a long but rather straightforward course through the internal capsule, the head of the caudate nucleus, and the white matter underlying the prefrontal cortex. Along its way, the axon sends a short collateral that arborizes poorly in the GPe. As the axon passes through the caudate nucleus it emits some striatal collaterals that are poorly arborized by comparison to those that emerge from Pf neurons that target solely the striatum.

These findings reveal that, in contrast to the situation in rats, where virtually all neurons of the Pf nucleus (the rodent homologue of the primate CM-Pf) project to both striatum and cortex (Deschênes et al., 1996), the primate CM/Pf harbors three types of highly patterned projection neurons: neurons that project to the striatum only, others that target the cerebral cortex only, and still others that arborize in both striatum and cerebral cortex. All axons that targeted the striatum were found to arborize either in the sensorimotor territory (CM neurons) or associative territory (Pf neurons), a finding that confirms the overall topographical organization of the thalamostriatal projections in primates determined earlier by bulk injections of anterograde tracers (Nakano et al., 1990; Sadikot et al., 1992). The thalamostriatal projection originating from CM/Pf neurons is part of a three-synaptic loop that involves: (1) the striatopallidal projection, which arises from neurons located in the sensorimotor striatal territory and project to the core of the GPi (see Section 2.5); (2) the pallidothalamic projection, which originates from neurons in the core of the GPi and project to the CM/Pf thalamic complex (see Section 2.6.2); and (3) the thalamostriatal projection, which stems from CM neurons that project back to the sensorimotor striatal territory (Nauta and Mehler, 1966; Sadikot et al., 1992; Parent et al., 2001). The fact that a significant number of CM neurons innervating the sensorimotor striatal territory also project to the cerebral cortex suggests that this loop is part of an open circuitry rather than a closed one. Hence, CM neurons endowed with an axon that projects to both striatum and cortex can no longer be considered as simple feedback elements; their branched axon allows them to influence neurons in the sensorimotor striatal territory directly, as well as indirectly by means of a relay in the motor cortex, which projects back to the sensorimotor striatal territory.

2.5 STRIATOFUGAL PROJECTIONS

Despite its large size and mediolateral parcellation in primates, the striatum is cytologically a fairly homogeneous structure. It is composed of a multitude of medium-sized neurons with spine-laden dendrites, referred to as medium-spiny neurons, and a smaller proportion of aspiny interneurons. In rodents, the aspiny interneurons are said to represent not more than 3% of the total neuronal populations, whereas this proportion might reach as much as 25% in the striatum of human and nonhuman primates (see Oorschot, 1996, 2013). The medium spiny neurons act as the major output elements of the striatum as well as the principal targets of most striatal afferents. Although they receive inputs from a wide variety of sources, including the cerebral cortex, the thalamus and brainstem, the medium spiny striatal neurons project to a rather limited set of basal ganglia components, essentially the GPe, the GPi and the SNr. In rodents, the striatofugal projection is currently viewed as being composed of a direct and an indirect pathway originating from distinct medium-spiny neurons present in about equal number in the striatum

(Gerfen and Bolam, 2010). The striatal projection neurons that form the direct pathway (the "striatonigral neurons") have an axon that targets chiefly the output structures of the basal ganglia, that is, the entopeduncular nucleus (the rodent homologue of the primate GPi) and the SNr. In contrast, the axon of the striatal neurons at the origin of the indirect pathways (the "striatopallidal neurons") arborizes solely within the globus pallidus (the rodent homologue of the primate GPe). This concept of the organization of the striatofugal projection is supported by the results of several anatomical, single-cell electrophysiological and optogenetic studies in rodents (Kawaguchi et al., 1990, 1995; Smith et al., 1998; Gerfen and Bolam, 2010; Chuhma et al., 2011), but other single-axon tracing investigations in rats have revealed that most striatal projection neurons possess an axon that arborizes in the 3 main striatal targets (Wu et al., 2000; Fujiyama et al., 2011).

The results of our single-axon tracing investigations in cynomolgus monkeys (Parent et al., 1995) and squirrel monkeys (Lévesque and Parent, 2005) have helped in understanding the organization of the striatofugal projection in primates. We used the same methodological approach in both sets of experiments, except that only the putamen was targeted in cynomolgus monkeys, whereas both the caudate nucleus and the putamen were aimed at in squirrel monkeys. In monkeys that received injections of either biocytin or BDA in the striatum, a few small neuronal pools comprising about 5–10 labeled cells per pool were visualized in each striatum. In some animals, the labeling procedure produced Golgi-like images of the somatodendritic domain of the injected neurons, together with a detailed view of their entire axonal arborization, including the recurrent local collaterals.

In cynomolgus monkeys, a total of 9 putamenofugal axons were analyzed in detail; they are very fine and display numerous irregularly spaced varicosities along their course. Short bulb-like appendages are also encountered as striatofugal axons course through the globus pallidus. All striatofugal fibers penetrate the GP by piercing the external medullary lamina, where they give rise to long collaterals arborizing along the inner surface of the lamina. The parent axons continue their course medially to arborize either in the GPe alone (1 axon) or, more commonly, in the GPe and GPi (8 axons). Axons that head toward the GPi emit numerous collaterals along the outer border of the internal medullary lamina before leaving the GPe. The majority (7/9) of the axons that arborize profusely in the GPi have a main branch that traverse the peduncular region of the internal capsule and terminate in the form of poorly arborized processes in the SNr (Figure 2.3). Axons that provide collaterals to the three striatal recipient structures have a preferential target site where they arborize profusely and formed typical "woolly fibers", which are composed of several thin collaterals that closely entwine the unstained core of dendrites belonging to pallidal neurons. The majority of putamenofugal axons arborize in the three major target structures of the striatum, where each axon display at least two distinct arborization patterns: they form elongated bands aligned along the borders of the medullary laminae in the globus pallidus, whereas nigral terminal fields appear as dense plexuses in the SNr. These bands and plexuses are distributed according to a highly precise rostrocaudal sequence and they appear to be

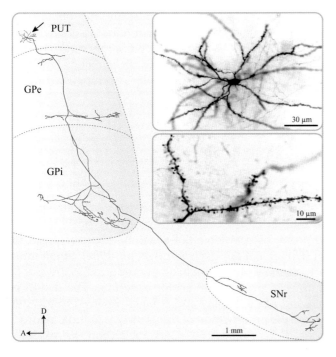

FIGURE 2.3 **Striatofugal projections.** Schematic drawing providing a 2-D rendition of the trajectory and patterns of arborization of a typical *type 2* striatofugal axon plotted on a parasagittal plane. This axon, which emerged from a neuron (arrow) located in the central portion of the postcommissural putamen (PUT) of a cynomolgus monkey, arborized within the three main targets of the striatum. Its somatodendritic domain and spine-laden dendrites are depicted photographically in the upper right portion of the figure.

interrelated with one another. Their presence can be taken to indicate that the striatum has multiple representations at both pallidal and nigral levels (Parent and Hazrati, 1995).

In squirrel monkeys, 27 striatofugal axons – 15 from the caudate nucleus and 12 from the putamen –were totally reconstructed from their parent cell bodies in the striatum to their most distant terminal fields outside the striatum. The terminal boutons of individually labeled axon in each striatal target were counted to assess the relative strength of the various axonal branches. Some striatofugal axons provided as many terminals in GPe as in GPi and SNr together. Most frequently, however, caudatofugal axons yielded a much larger number of terminals at the nigral level than at the pallidal level, whereas the inverse was true for axons of putamen neurons.

Thirteen caudatofugal axons were found to arborize in the GPe, GPi, and SNr, whereas two others innervated the GPe only. At pallidal levels, the caudatofugal axons branch in the dorsomedial part of GPe and GPi, and, along their course through the GP, the main axons exhibit large varicosities. At nigral levels, the axons course throughout the entire anterior–posterior axis of the SNr, having numerous *en passant* isolated varicosities. We estimate that caudate nucleus neurons yield a mean of 22% of their total number of terminal boutons in the GPe, 19% in the GPi, and 59% in the SNr (Table 2.1). In contrast, to the caudatofugal axons, the putamenofugal axons arborize more profusely at pallidal than at nigral levels, but most of them (10/12) branch in the three main striatal targets (GPe, GPi, and SNr), as is the case for the majority of caudatofugal axons. The axonal branching pattern of putamenofugal fibers in the SNr is similar to that of the caudatofugal axons, except that the number of axon terminals is significantly smaller. Two putamenofugal axons arborize only in the GPe, but the number of their terminal varicosities in the GPe is much larger than that of putamenofugal axons that branch into GPe, GPi, and SNr. We estimate that putamen neurons provide a mean of 28% of their total number of terminal boutons in GPe, 55% in GPi, and 17% in SNr (Table 2.1).

The data gathered from our cynomolgus and squirrel monkeys studies reveal the highly collateralized and widely distributed feature of the striatofugal projection, a morphological feature that allows individual striatal neurons to send efferent copies of the same neural information to virtually all striatal targets. These findings are in agreement with single-cell labeling studies in rats, which have reported that the vast majority of striatal neurons project to more than one striatal target site (Kawagushi et al. 1990; Wu et al., 2000; Fujiyama et al., 2011). However, striatofugal axons that branch into more than one striatal recipient structures appear more abundant in primates than in rodents; about 90% of axons traced in the primate studies exhibited this feature, compared with 60% in rodents (Kawagushi et al. 1990; Wu et al., 2000; Fujiyama et al., 2011).

The quantitative estimate of the number of axon terminals yielded by each individual axons reveals that, although the axon of the great majority of striatal projection neurons arborize within the three striatal target nuclei, neurons located in the caudate nucleus project more abundantly to SNr, whereas those in the putamen target chiefly at GPi. This organizational pattern might have important functional implications. The fact that cortical inputs to caudate nucleus and putamen derive respectively from association and sensorimotor regions (see Section 2.1) suggests that the two major output structures of the basal ganglia are involved in the processing of different functional modalities and, as such, cannot be considered equivalent to one another. Based on electrophysiological findings obtained in primates, it appears that GPi neurons are mainly involved in the control of basic movement parameters, whereas SNr neurons are more concerned with events related to memory, attention, or movement preparation (Wichmann and Kliem, 2004). Altogether, these findings suggest that the cortico-caudato-nigral pathway is chiefly involved in the planning of motor acts, whereas the cortico-putameno-pallidal pathway is principally concerned with their execution.

2.6 PALLIDOFUGAL PROJECTIONS

2.6.1 From the External Pallidum

Although they occupy a markedly different position in the basal ganglia circuitry, the two segments of the globus pallidus in primates are virtually undistinguishable from one another on the basis of the morphological characteristics of their neuronal constituents. Both pallidal segments are composed of large (20–60 µm in diameter) fusiform cells with thick, smooth dendrites that are as long as 1000 µm (DiFiglia et al., 1982; Yelnik et al., 1984). Small interneurons (12–21 µm) with relatively short dendrites and a short, sparsely branching axon are reported in the globus pallidus of rodents (Kita, 2010), but the presence of interneurons in primate GP is still uncertain. The dendrites of the large (principal) pallidal neurons are sparsely branched and often display appendages, referred to as "thin processes" or "complex endings" (François et al., 1984). The dendritic arborization of the principal pallidal neurons in macaque monkeys is characteristically discoidal, with a mean dimension of 1500 × 1000 × 250 µm (Yelnik et al.,

1984). These dendritic disks lie parallel to the lateral border of the GP, with their largest surface perpendicular to the incoming striatal axons, so that they are ideally positioned to intercept the entire proximal arborization of the striatopallidal axons (Percheron et al., 1984; Kita, 2010).

Earlier anatomical studies have portrayed the external segment of the globus pallidus (GPe) as being locked in a closed ancillary loop with the subthalamic nucleus (Nauta and Mehler, 1966; Carpenter, 1981). Today, most authors insist on the fact that the GPe conveys information from the striatum to the basal ganglia output nuclei (GPi and SNr) via a relay in the subthalamic nucleus, and thus acts as a major relay nucleus in the indirect pathway (see review by Gerfen and Bolam, 2010). However, besides its massive projection to the STN, the GPe in primates projects to several other structures, including the striatum, the GPi, the SNr and the pedunculopontine tegmental nucleus (see Sato et al., 2000a). Therefore, the GPe appears to exert a much more widespread and varied influence upon the various components of the basal ganglia than originally thought (Gittis et al., 2014).

The role of the GPe as a central integrator in basal ganglia circuitry has received further support from our single-axon tracing study of the efferent connections of this structure in cynomolgus monkeys (Sato et al., 2000a). In this investigation 76 axons were traced from the injection sites in the GPe to their multiple terminal fields in various structures. Thirty of these 76 axons could be directly connected to their parent cell body in the GPe; the remaining 46 axons had branching patterns similar to those that could be connected to their cell body of origin. The axons that emerged from Golgi-like labeled GPe neurons departed from either the cell body or a primary dendrite. Before leaving the GPe some axons gave off 1–3 thin axon collaterals that arborized profusely within the entire somatodendritic domain and sometimes even beyond the limit of the soma and dendrites of their parent neuron. These local collaterals exhibited numerous varicosities of various sizes reminiscent of *boutons-en-passant* or *boutons terminaux*. Four distinct types of GPe neurons were recognized on the basis of their axonal targets: (1) neurons projecting to the GPi, STN, and SNr (n = 10); (2) neurons projecting to the GPi and STN (n = 14); (3) neurons projecting to the STN and SNr (n = 40); and (4) neurons projecting to the striatum (n = 12) (Table 2.1).

The axons of the first two types run medially and caudally for a short distance within the GPe before piercing the internal medullary lamina to enter the GPi (Figure 2.4). They travel principally in a rostrocaudal direction within the GPi and give off 1–4 collaterals along their course. The number and length of collaterals in the GPi vary

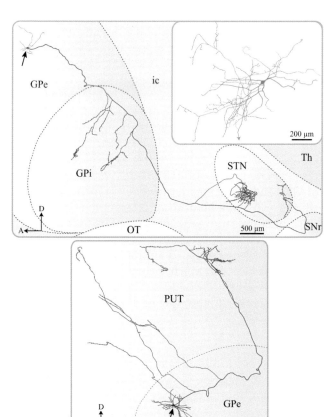

FIGURE 2.4 **Pallidofugal projections from the external pallidum.** Sagittal plane reconstructions of a *type 2* (upper panel) and a *type 4* (lower panel) neuron (arrows) of the external pallidal segment (GPe) in cynomolgus monkeys. The axon of the type 2 neuron provides collaterals to the both internal pallidum (GPi) and subthalamic nucleus (STN), whereas that of the type 4 neuron arborizes essentially within the putamen (PUT). A reconstruction of the somatodendritic domain (green) and the profuse network of local axon collaterals (orange) of type 2 neuron is shown in the right portion of the upper panel.

from one axon to the other. Most of the long collaterals course parallel to the internal medullary lamina, whereas short collaterals often form terminal arborizations near their parent axons. After leaving the GPi, the main axon of most of these neurons traverses the internal capsule and branches into several collaterals just before entering the STN. A few other axons branch directly within the GPi itself. Collaterals penetrate the STN from its anterior and/or ventral aspects and form fine terminal fields within a rather limited sector of the nucleus. Axons of the second type terminate in the STN, whereas axons of the first type had branches that run further caudally and enter the SNr through its rostral pole. Occasionally, axonal branches course between the SNr and STN to enter the SNr through its dorsal part. At the SNr level, the GPe axonal branches become very thin and their pattern of arborization is often difficult to delineate. In cases of intensely labeled axons, however, typical pericellular terminal arborizations can be visualized in the SNr.

Axons of the third type reach their targets either by coursing through the GPi and then piercing the internal capsule, or by entering immediately within the internal capsule, thus completely avoiding the GPi. In both cases the axons terminate in a rather dense and highly focused field in the STN, whereas they arborize more loosely and more widely in the SNr.

The GPe axons of the fourth type terminate in the caudate nucleus or the putamen (Figure 2.4). They branch either within the GPe itself or just after entering the striatum. Typically, pallidostriatal axons course parallel to the lateral medullary lamina and emit 4–5 long and thin collaterals that enter the striatum perpendicularly to the lateral medullary lamina. These collaterals ramify into a few thinner collaterals that run dorsally and laterally in the striatum. These terminal branches cover a rather wide area of the striatum, compared to the limited terminal fields of other GPe neurons in the STN.

Terminal branches of all 4 types of GPe axons exhibit numerous varicosities reminiscent of *boutons en passant* or *boutons terminaux*. These boutons are slightly larger than 1 μm in diameter and often occur in close apposition to non-labeled cell bodies and proximal dendrites in their respective target sites. Such a pericellular type of terminal arborization is particularly obvious at the GPi level, but is also present in the STN and SNr. In contrast, terminal branches of GPe axons arborizing in the striatum display terminal specializations, typical of boutons *en passant*, that are smaller than those seen in the other target structures of the GPe and do not form pericellular arrangements. Instead, the pallidostriatal fibers display a typical *bouton en passant* type of terminal arborization.

Evidence for pallidostriatal projections has been obtained in various species with different methods (Staines et al., 1981; Beckstead, 1983; Bevan et al., 1998; Koshimizu et al., 2013). Whereas in monkeys the pallidostriatal neurons appear to belong to a separate population of neurons that does not project to other major GPe targets, intracellular or juxtacellular labeling experiments in rodents indicate that pallidal neurons projecting to the striatum also provide collaterals locally in the globus pallidus as well as in the major GPe recipient sites, that is, the STN, SNr and entopeduncular nucleus (Kita and Kitai, 1994; Bevan et al., 1998). The pallidostriatal neurons in rodents were found to target both medium spiny projection neurons and aspiny interneurons, principally those that display parvalbumin immunoreactivity (Bevan et al., 1998). Hence, rather than providing a simple feedback mechanism, their complex morphological characteristics suggest that the pallidostriatal neurons play a major integrative role in basal ganglia circuitry, in both rodents and primates (Kita, 2010; Voorn, 2010).

Our single-axon tracing study in the cynomolgus reveal that over 80% of all GPe axons traced provide collaterals to the STN, but none of these axons project only to the STN. We estimate that about half of all GPe axons branch to the STN and SNr, a smaller percentage branch to the STN and GPi, and a still smaller number branch to the STN, SNr, and GPi (Table 2.1). These findings reveal that, along with the STN, GPe neurons also target the SNr and the GPi through a highly collateralized axonal system. This type of anatomical organization implies that a copy of the neural information that is sent by GPe neurons to their reciprocal counterparts in the STN is conveyed to the two major output structures of the basal ganglia (GPi/SNr). Hence, much more than a simple relay in the indirect pathways, the primate GPe stands out as a major integrative structure that can affect virtually all components of the basal ganglia. Through their widespread and highly collateralized axons, individual GPe neurons are able to provide information to the STN, as well as to directly modulate, or even block, the flow of corticostriatal information at basal ganglia output levels.

2.6.2 From the Internal Pallidum

The fact that the GPi – or its rodent homologue the entopeduncular nucleus – is one of the two major output structures of the basal ganglia is supported by a multitude of retrograde and anterograde labeling studies, which have identified its major targets: (1) the ventral tier motor nuclei of the thalamus (VA); (2) the centromedian thalamic nucleus (CM); (3) the lateral habenular nucleus (LH); and (4) the brainstem pedunculopontine tegmental nucleus

(PPN) (see Parent and Hazrati, 1995; Smith et al., 1998; Gerfen and Bolam, 2010). Moreover, the suggestion that pallidothalamic projections might be collateralized (Nauta and Mehler, 1966) was supported by antidromic invasion experiments and retrograde double-cell labeling investigations, and these were further shown to be, at least in part, bilaterally distributed (Parent et al., 1999).

Our single-axon tracing study of the GPi in cynomolgus monkeys, undertaken under electrophysiological guidance, has confirmed some of these findings, while revealing novel aspects of the organization of the pallidofugal projections in primates. Electrophysiologically, the GPi is recognized easily by the characteristic spontaneous firing pattern of its neurons, which includes high frequency tonic discharges (70–120 spikes per second), not interrupted by long periods of silence (Figure 2.5). Based on axonal arborization criteria, the primate GPi is composed of four types of neurons, which target the followings structures: (1) the VA and PPN (n= 15); (2) the VA, PPN, and CM-Pf (n= 16); (3) the LH only (n= 17); and (4) the LH and the anterior (AV and AM) thalamic nuclei (n= 12). Occasionally, the axons of GPi neurons emit one or two thin collaterals that arborize poorly within, and sometimes beyond, the somatodendritic domain of their parent neuron. These local collaterals exhibit varicosities of various sizes reminiscent of *boutons-en-passant* or *boutons terminaux*. The main axon of type 1and 2 neurons branches either within the GPi or just after leaving it. In contrast, axons of types 3 and 4 remain unbranched, at least until they reach the thalamus. All type 3 and 4 axons exit the GPi through the lenticular fasciculus, whereas type 1 and 2 axons emerge either through the ansa lenticularis or the lenticular fasciculus, irrespective of the position of their parent cell body in the GPi.

Neurons of both types 1 and 2 abound preferentially within the large central portion of the GPi, but type 2 neurons prevail ventrally in the caudalmost portion of the GPi. In contrast, type 3 and 4 neurons are largely confined to the periphery of the GPi, as well as along the accessory medullary lamina; they represent approximately 10% of the total number of GPi neurons. The axons of type 1 neurons are long and widely arborized, with one main branch that ascends to the VA and another one that descends without further branching to the PPN. The thalamic portion of the axon subdivides successively into 10–15 long and thin branches that cover most of the pallidal thalamic territory. As they approach the end of their trajectory, these long branches break out into numerous shorter terminal collaterals that form typical glomerular-like clusters composed of a multitude of very thin, highly varicose and closely intermingled axon collaterals. The terminal field in the PPN consists of a small number of varicose terminal branches that emit a few collaterals at right angles from the main axonal branch. These short terminal collaterals display numerous varicosities larger than the thalamic ones. The axons of type 2 neurons arborize more profusely at thalamic levels than those of type 1 neurons and, in addition, innervate the CM/Pf. The axon collaterals do not form

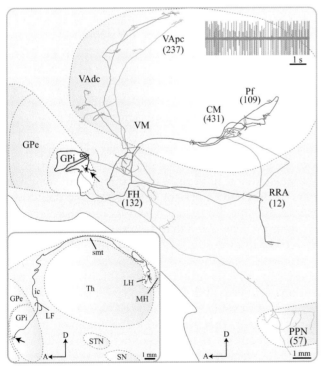

FIGURE 2.5 **Pallidofugal projections from the internal pallidum.** A sagittal plane reconstruction of a *type 2* neuron of the internal pallidal segment (GPi) in cynomolgus monkey, whose cell body is located within the core of the structure (arrow). The axon of this neuron has a total length of 27.22 cm. It arborizes profusely within the ventral tier thalamic nucleus (VA), the centromedian-parafascicular thalamic complex (CM-Pf) and the pedunculopontine tegmental nucleus (PPN). Different colors have been used to help distinguishing the various branches of this single-labeled axon, and the numbers of axonal varicosities yielded by the different axonal branches are indicated in parentheses. A sample of the firing pattern of this labeled neuron is also provided. The drawing in the lower left portion of the figure is a sagittal plane reconstruction of a *type 3* neuron, whose cell body lies at the periphery of the GPi (arrow). The axon of this neuron ascends within the stria medullaris thalami (stm) and arborizes exclusively within the lateral habenular nucleus (LH).

dense plexuses in the CM/Pf as they do in the VA. Instead, they become thin and varicose well before terminating and form typical loops oriented along the longest extent of CM/Pf. They emit several short and varicose collaterals that run at right angle to the main axonal trunk and these thin varicose terminal fibers are often seen in close contact with unlabeled cell bodies in the CM/Pf. Most of these fibers arborize within the medial part of the CM and the lateral part of the Pf. Axons of type 1 and 2 neurons display an astonishing degree of collateralization and a remarkably wide distribution, with a total length that ranged from 6 to 27 cm. Some of these axons also provide a small number of collaterals to the Forel field H1 and the retrorubral regions (Figure 2.5).

Axons of type 3 neurons exit through the lenticular fasciculus, traverse obliquely the lateral hypothalamic area by coursing along the inferior thalamic peduncle, ascend along the rostral pole of the thalamus to enter directly within the stria medullaris (Figure 2.5). From there, the fibers follow a rather straight course to the habenula, where they branch into three major collaterals that arborize within most of the LH, leaving the medial habenular nucleus completely devoid of labeled axonal varicosities. At LH levels, the collaterals become highly varicose and form typical pericellular arrangements around the small, round and closely packed cell bodies in the LH. Many type 3 axons emit one major axonal branch that crossed the midline within the interhabenular commissure and innervated the contralateral LH. The trajectory and branching pattern of type 4 axons are similar to those of type 3 axons, except that they provide 1–3 major collaterals to the anterior thalamic nuclei on their way to the LH. These collaterals arborize principally in the anterior ventral (AV) and anterior median (AM) thalamic nuclei. Overall, the pallidal innervation of the anterior nuclei was much less dense than that of the LH.

This single-axon tracing study reveals that, in contrast to other basal ganglia components, the GPi is composed of only two groups of projection neurons: (1) a large number of centrally located neurons that project to motor thalamic and brainstem centers (types 1 and 2), and a smaller number of peripherally distributed neurons whose axon arborizes densely in the lateral habenular nucleus (types 3 and 4), which is a major limbic system relay nucleus. Virtually all pallidothalamic axons that we traced are collaterals of single-axons that also innervated the PPN and, in about 50% of the cases, the CM/Pf as well. This form of organization implies that single pallidal neurons can send efferent copies of the same neural information to small cell clusters in the ventral tier thalamic nuclei, and to more diffuse cell populations in the PPN and CM/Pf. This finding confirms previous indirect evidence obtained by means of fluorescence double-retrograde labeling investigations in squirrel monkeys (Parent and De Bellefeuille, 1982; Parent et al., 1999).

A typical pallidothalamic axon gives rise to 5–10 cluster-like terminal plexuses that are distributed widely in the ventral tier nuclei, a mode of organization that allows single pallidothalamic axons to transmit the same neural information to small groups of about 20 thalamocortical cells dispersed throughout the pallidal thalamic territory. The neural information can then be processed in parallel by each of these small cell clusters before being conveyed to the cerebral cortex. This pattern is the opposite of the serial type of organization favored by most classical somatosensory systems, which carry a specific type of information that is transmitted to a single thalamic unit, which in turn projects to a specific cortical domain.

In their pioneering study of the efferent connections of the primate lentiform nucleus, Walle J. H. Nauta (1916–1994) and William R. Mehler (1926–1992) were the first to report a projection from the GPi to the CM-Pf in primates and the first also to suggest that this pathway derives from collaterals of fibers that arborize in the ventral tier thalamic nuclei (Nauta and Mehler, 1966). Our findings provide direct evidence that the pallidointralaminar projection does indeed arise from collaterals of single axons that project to VA, but also send descending projections to the PPN. The pallidointralaminar fibers terminate massively in the CM, and less heavily in the lateral part of the Pf, and the two nuclei are link by loops of long and varicose collaterals of pallidointralaminar axonal branches.

The pallidotegmental projection differs in many respects from the pallidothalamic and pallidointralaminar projections. It is principally composed of axons that remain poorly branched and uniformly thick throughout their long trajectory toward the PPN, where they arborize in both the pars diffusa and compacta of the PPN. The pallidotegmental terminal field is much more focused than the widespread thalamic innervation and it comprises terminal collaterals bearing large varicosities that closely surround PPN neurons. The fibers that descend to the PPN remain largely unbranched, whereas the pallidothalamic innervation derives from multiple, long and thin collaterals that branch frequently. Since nerve conduction velocity is directly proportional to the axon diameter and inversely proportional to the degree of axonal branching, it may be presumed that PPN neurons receive GPi information well before thalamic neurons. This spatiotemporal mode of organization of single pallidofugal axons is important because the PPN is known to project back to many basal ganglia structures, principally the SNc, the STN, and the GPi.

A double retrograde fluorescent labeling studies in primates has indicated that the pallidohabenular neurons form a ring around pallidal neurons that project, according to various combinations, to the three other target structures of the GPi (Parent and De Bellefeuille, 1982). A similar study in the rat revealed that the numerous pallidohabenular neurons occupy the rostral two thirds of the entopeduncular nucleus, whereas less abundant neurons that project to the ventrolateral thalamus, Pf and PPN are all confined to the caudal third of the entopeduncular nucleus (Van der Kooy and Carter, 1981). The functional significance of such a difference between primates and rodents regarding the relative number and topographical distribution of the pallidohabenular neurons is at present unknown. Nevertheless, our single-axon tracing study reveals that, although less prominent than in rodents, the pallidohabenular projection is a significant component of the pallidal outflow in primates. The existence of a pallidohabenular projection in both rodents and primates suggests that the basal ganglia may be able to influence neural mechanisms other than ones strictly related to motor behavior.

2.7 SUBTHALAMOFUGAL PROJECTIONS

The STN is classically known as being reciprocally linked with the GPe, with which it forms one of the many closed ancillary loops that are characteristic of the basal ganglia circuitry. More recently, the position of the STN in the basal ganglia connectivity has been reassessed so that it is now viewed as a major relay nucleus in the so-called indirect striatofugal pathway (see Gerfen and Bolam, 2010). Hence, the STN is portrayed as a nucleus that conveys striatal information from the GPe to the two basal ganglia output nuclei (GPi and SNr). However, besides its projection to the GPi and SNr, the STN is now known to project back to the GPe and to send less prominent inputs to the striatum and the pedunculopontine tegmental nucleus (PPN) (Carpenter, 1981; Jackson and Crossman, 1981; Kita and Kitai, 1987; Parent and Smith, 1987; Nakano et al., 1990; Smith et al., 1990; Shink et al., 1996; Koshimizu et al., 2013).

Golgi and intracellular cell-labeling studies in various species have shown that the STN harbors principally Golgi type I projection neurons with a fusiform or ovoid cell body from which emerge 4 to 7 primary dendrites, which divide into several long, thin, sparsely spinous dendrites that form an ellipsoidal dendritic field oriented parallel to the main axis of the nucleus (see Sato et al., 2000a; Gerfen and Bolam, 2010). Most STN axons emerge directly from the cell body and some of them form intrinsic collaterals that exhibit *boutons en passant* or *boutons terminaux*. Such intrinsic collateral systems seem rather frequent in rodents (Kita et al. 1983), but relatively rare in primates (Yelnik and Percheron, 1979). Evidence that axons of STN neurons branch to both the GP and the SN was first obtained following electrophysiological studies with antidromic invasion methods in rats (Deniau et al. 1978). This view was supported in rodents by antidromic invasion studies (Hammond et al., 1983; Granata and Kitai, 1989), as well as by retrograde double-labeling investigations (Van der Kooy and Hattori, 1980) and intracellular labeling experiments (Kita et al., 1983). By comparison, the patterns of axonal collateralization in primate STN neurons are relatively poorly known.

Direct evidence for axonal branching of STN neurons in primates was obtained following a detailed single-axon tracing study undertaken in cynomolgus monkeys (Sato et al., 2000b). A total of 75 axons were traced from multiple injection sites in the STN. Out of this number, 52 axons could be directly connected to their parent cell body, whereas the remaining 23 axons had branching patterns similar to those that could be connected to their soma of origin. Five distinct types of STN projection neurons have been disclosed on the basis of their axonal targets, which were: (1) SNr, GPi, and GPe (n = 12); (2) SNr and GPe (n = 2); (3) GPi and GPe (n = 36); (4) GPe only (n = 8); and (5) striatum (n = 13). The most abundant of these STN neurons are those projecting to the GPi and GPe (type 3), whereas the least frequent are those targeting the SNr and GPe (type 2) (see Table 2.1).

Axons of the first and second types have a similar course; they bifurcate into a rostral and a caudal branch within the STN, or just after leaving the nucleus (Figure 2.6). The caudal branch runs caudomedially to enter the dorsal aspect of the SN and courses caudally through the SNc, giving off several thin collaterals to the SNr, ventrally. The rostral branch travels either dorsally along the ventral aspect of the thalamus or ventrally along the lateral border of the subthalamic nucleus before piercing the internal capsule to reach the GP through its caudal aspect. Type 1 axons give off a few collaterals in the GPi and branch more abundantly in the GPe, whereas type 2 axons have a caudal branch coursing along most of the rostrocaudal extent of the SNr, but whose terminal arborization could not be visualized in detail because of its faint labeling. Their rostral branch travels throughout the GPi and GPe, but arborizes moderately only in the GPe.

Axons of the third and fourth types have only a rostral branch that aims at the GP (Figure 2.6). It courses either ventrally along the lateral border of the STN or dorsally along the ventral aspect of the thalamus. Type 3 axons give off several short or long collaterals in the GPi before terminating in the GPe, whereas type 4 axons traverse the GPi without emitting any collateral and arborize rather poorly within the GPe. Type 5 axons do not penetrate directly the ventral or caudal poles of the GP, but run for a long distance within the internal capsule en route to

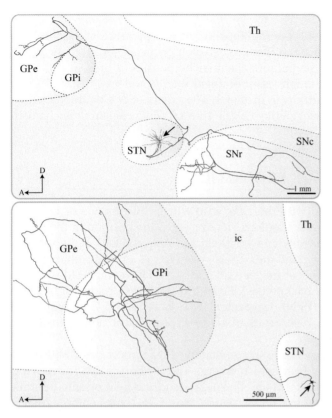

FIGURE 2.6 **Subthalamofugal projections.** Sagittal plane reconstructions of a *type 1* (upper panel) and a type 3 (lower panel) neuron of the subthalamic nucleus (STN) in cynomolgus monkeys. The cell body of these two neurons are located within the core of the STN (arrows). The axon of the type 1 neuron provides collaterals to the internal (GPi) and external (GPe) pallidal segments, as well as to the substantia nigra pars reticulate (SNr), whereas that of the type 3 neuron remains confined to the GPe and GPi.

the putamen. These axons can be followed up to the border of the putamen, where the staining becomes too faint to allow the visualization of their terminal arborization. The STN axons display a terminal arborization that is strikingly similar in all target structures: their terminal branches are very fine and exhibit numerous small (<1 µm) varicosities indicative of *boutons-en-passant* and *boutons terminaux*, which are rarely seen in close contact with a cell body in any of the STN target sites.

This single-axon tracing study shows that, in terms of patterns of axonal arborization, the primate STN harbors a highly heterogeneous neuronal population. This departs from the idea of a homogeneous nucleus composed of morphologically identical neurons with an axon that bifurcates to the GPi and the SNr. Neurons of the STN with an axon that branches to both GPi and SNr have been visualized in the cynomolgus monkey, but all of these neurons were found to send collaterals to the GPe as well. Furthermore, these type I neurons represent only about one-fourth of the total number of STN labeled neurons. Our results indicate that STN neurons endowed with only a rostrally coursing axonal branch (types 3, 4, and 5) are about three times more numerous than STN neurons displaying a rostral and caudal axonal branches (types 1 and 2) (Table 2.1). Furthermore, the group of rostrally projecting STN neurons is itself highly heterogeneous: it comprises neurons projecting to both GPe and GPi (type 3), which represent nearly half of all STN labeled neurons, as well as neurons whose axon arborizes within the GPe only (type 4) or courses toward the striatum (type 5). The fact that the majority of STN neurons project to both GPi and GPe brings us back to earlier anatomical findings, which have emphasized the reciprocal connection of the STN and the GP (Nauta and Mehler, 1966; Carpenter, 1981), although only the GPe appears to project back to the STN (see Section 2.6.1).

These findings do not deny the importance of the STN projection to the substantia nigra. In rodents, the diameter of the subthalamopallidal axonal branch was found to be larger than that of the subthalamonigral branch, a finding that was used to explain the much higher conduction velocity of the subthalamopallidal branch compared to that of the subthalamonigral branch (Kita et al., 1983). However, no significant difference was noted in terms of axonal diameter between the ascending and descending projections of STN neurons in monkeys. Furthermore, in cases of optimal labeling, subthalamonigral axons were found to arborize profusely within the SNr, in accordance with previous indications obtained with anterograde tract-tracing techniques in rodents (Kita and Kitai, 1987) and primates (Smith et al., 1990).

Projections of the STN to the caudal brainstem, principally the PPN, could not be visualized in our single-axon tracing study probably because survival periods were too short to label entirely such long axons in the primate brain. The same technical limitation may have precluded the visualization of the axonal arborization of the STN axons that coursed toward the striatum. Subthalamostriatal projections have been described earlier by means of retrograde cell-labeling and anterograde fiber-tracing methods in rats (Kita and Kitai, 1987), cats (Beckstead, 1983), and monkeys (Smith et al., 1990). In monkeys, the projections to the putamen and caudate nucleus were found to arise from the dorsolateral and ventromedial aspects of the STN, respectively (Parent and Smith 1987). Hence, these two sectors of the primate STN were considered to represent distinct functional territories: *sensorimotor* dorsolaterally and *associative* ventromedially (Parent and Smith, 1987). The subthalamostriatal axons followed a long, linear course and branched infrequently at the striatal level (Smith et al., 1990), but it is not yet known if single STN axons arborize in both the caudate nucleus and the putamen.

Altogether, these results reveal that the STN is a remarkably heterogeneous nucleus that harbors several subtypes of highly branched projection neurons. This anatomical feature allows single STN neurons to exert a multifaceted effect on GPe neurons, with which they are reciprocally connected, as well as on neurons of the two major output structures of the basal ganglia (GPi/ SNr).

2.8 BASAL GANGLIA CONNECTOME AND NEURODEGENERATIVE DISEASES

Our single-axon tracing studies in cynomolgus and squirrel monkeys reveal that axonal collateralization is one of the major organizational features of the basal ganglia in primates. These results, which are based on a detailed analysis of more than 300 neurons from all basal ganglia nuclei whose axon was entirely reconstructed from serially collected sections, demonstrate that each major component of the primate basal ganglia, together with basal ganglia-related cortical, thalamic, subthalamic and brainstem centers, is composed of various types of neurons with distinct axonal arborization patterns. In the light of our stereological three-dimensional investigations, the basal ganglia stand out as a widely distributed neural network that extends over at least three distinct levels of the neuraxis and whose individual elements are endowed with a highly patterned set of axon collaterals. This architecture enables each basal ganglia component to interact with one another, as well as with some major cortical, thalamic, subthalamic and brainstem centers, in an exquisitely precise manner. The partial elucidation of this complex microcircuitry has helped to better understand how neural information flows along the cortico-basal ganglia-thalamo-cortical loop, particularly in respect to the spatiotemporal sequence of neural events responsible for both normal and abnormal motor behavior.

The long, widespread and highly collateralized axon of basal ganglia projection neurons allows exquisitely precise interactions between the various nodal points of a widespread neuronal network. However, the maintenance of this unique morphological trait imposes upon neurons a metabolic burden that may render them highly vulnerable to neurodegenerative, neurotoxic or ischemic insults. This is particularly the case for pallidal neurons, which degenerate massively in cases of toxic encephalopathy. Toxic lesions involving mainly the pallidal complex can produce severe signs of psychic akinesia, without significant motor deficits (see Laplane et al., 1984). This finding indicates that the basal ganglia, and principally the GP, have important non-motor functions that may be severely altered by toxic or hypoxic insults.

Other pathological conditions that affect the basal ganglia can produce severe motor disturbances. Such is the case of Huntington's chorea and Parkinson's disease, characterized clinically by hyperkinetic and hypokinetic symptoms, respectively. The pathological hallmark of Huntington's chorea is a massive atrophy of the striatum due to a degeneration of the medium spiny projection neurons, the majority of which possess a long axon that arborizes in the three major striatal target structures. In contrast, aspiny striatal interneurons, which are endowed with a short, locally arborizing axon, are specifically spared in the striatum of Huntingtonian patients (Cicchetti et al., 2000). These findings indicate that striatal neurons with a long and widely arborized axon that projects outside the striatum are particularly sensitive to neurodegenerative mechanisms at play in Huntington's disease.

In Parkinson's disease, it is the substantia nigra that is affected by neurodegenerative processes, which specifically target the dopaminergic neurons present within the pars compacta of the structure (see Hornykiewicz, 2008). The nigrostriatal dopaminergic pathway has long been considered as a monolithic entity. However, tract-tracing and immunohistochemical studies in monkeys rendered parkinsonian by means of the neurotoxin MPTP suggest that this major striatal afferent is instead composed of different neuronal subsystems, each having its own site of origin, its specific terminal fields within the striatum, and its particular degree of vulnerability to neurodegenerative mechanisms (Parent and Lavoie, 1993). The most sensitive subsystems are those composed of dopaminergic neurons whose axon arborizes widely within the sensorimotor territory of the striatum. Much less vulnerable are subsystems formed by dopaminergic neurons endowed with a relatively short axon that arborizes within various extrastriatal structures, particularly the GPi, and much less so in the striatum (Parent and Lavoie, 1993).

These findings indicate that striatal and nigral neurons that possess a long and widely arborized axon are particularly vulnerable to neurodegenerative mechanisms at play in Huntington's chorea and Parkinson's disease. They also raise the possibility that the neuronal losses seen in Huntington's chorea and Parkinson's disease are the result of a dying back phenomenon, the initial pathological insults occurring not in the structures that degenerate (striatum and substantia nigra pars compacta), but in their main target sites. Detailed single-axon tracing studies of the nigrostriatal dopaminergic projection in primates, under both normal and parkinsonian conditions, are needed to better understand the role of this major striatal afferent system in the functional organization of the basal ganglia, as well as in the pathogenesis of Parkinson's disease.

2.8.1 Abbreviations

AM anterior medial thalamic nucleus;
AV anterior ventral thalamic nucleus;
BS brainstem;
CD caudate nucleus;
CM centromedian thalamic nucleus;
CP cerebral peduncle;
FH Forel's field H;
GP globus pallidus;
GPe external segment of GP;
GPi internal segment of GP;
ic internal capsule;
LF lenticular fasciculus;
LH lateral habenular nucleus;
M1 primary motor cortex;
MH medial habenular nucleus;
OT optic tract;
Pf parafascicular thalamic nucleus;
PM premotor cortex
PPN pedunculopontine tegmental nucleus;
PUT putamen;
RN red nucleus
RRA retrorubral area;
Rt reticular thalamic nucleus;
smt stria medullaris thalami;
SN substantia nigra;
SNc pars compacta of SN;
SNr pars reticulata of SN;
STN subthalamic nucleus;
Th thalamus;
VA ventral tier thalamic nuclei;
VAdc densicellular part of VA;
VApc parvicellular part of VA;
VM ventromedial thalamic nucleus;

References

Albin, R.L., Young, A.B., Penney, J.B., 1989. The functional anatomy of basal ganglia disorders. Trends Neurosci. 12, 366–375.

Alexander, G.E., DeLong, M.R., Strick, P.L., 1986. Parallel organization of functionally segregated circuits linking basal ganglia and cortex. Ann. Rev. Neurosci. 9, 357–381.

Bauswein, E., Fromm, C., Preuss, A., 1989. Corticostriatal cells in comparison with pyramidal tract neurons: contrasting properties in the behaving monkey. Brain Res. 493, 198–203.

Beckstead, R., 1983. A pallidostriatal projection in the cat and monkey. Brain Res. Bull. 11, 629–632.

Bevan, M.D., Booth, P.C.A., Eaton, S.A., Bolam, J.P., 1998. Selective innervation of neostriatal interneurons by a subclass of neuron in the globus pallidus of the rat. J. Neurosci. 18, 9438–9452.

Bouyer, J.J., Park, D.H., Joh, T.H., Pickel, V.M., 1984. Chemical and structural analysis of the relation between cortical inputs and tyrosine hydroxylase-containing terminals in rat neostriaturn. Brain Res. 302, 267–275.

Carpenter, M.B., 1981. Anatomy of the corpus striatum and brain stem integrating systems. In: Brooks, V.B. (Ed.), Handbook of Physiology, Vol. 2, The Nervous System American Physiological Society, Bethesda, pp. 947–995.

Cazorla, M., de Carvalho, F.D., Chohan, M.O., Shegda, M., Chuhma, N., Rayport, S., et al., 2014. Dopamine D2 receptors regulate the anatomical and functional balance of basal ganglia circuitry. Neuron 81, 153–164.

Chu, H.Y., Atherton, J.F., Wokosin, D., Surmeier, D.J., Bevan, M.D., 2015. Heterosynaptic regulation of external globus pallidus inputs to the subthalamic nucleus by the motor cortex. Neuron 85, 364–376.

Chuhma, N., Tanaka, K.F., Hen, R., Rayport, S., 2011. Functional connectome of the striatal medium spiny neurons. J. Neurosci. 31, 1183–1192.

Cicchetti, F., Prensa, L., Wu, Y., Parent, A., 2000. Chemical anatomy of striatal interneurons in normal individuals and in patients with Huntington's disease. Brain Res. Rev. 34, 80–101.

Cowan, R.L., Wilson, C.J., 1994. Spontaneous firing patterns and axonal projections of single corticostriatal neurons in the rat medial agranular cortex. J. Neurophysiol. 71, 17–32.

DeLong, M.R., Georgopoulos, A.P., 1981. Motor functions of the basal ganglia. In: Brookhart, J.M., Mountcastle, V.B., Brooks, V.B., Geiger, S.R. (Eds.), Handbook of Physiology, Sect. 1. The Nert:Ous System, Vol. 2, Motor Control, Part 2 American Physiological Society, Bethesda, pp. 1017–1061.

Deniau, J., Hammond, C., Chevalier, G., Féger, J., 1978. Evidence for branched subthalamic nucleus projections to substantia nigra, entopeduncular nucleus and globus pallidus. Neurosci. Lett. 9, 117–121.

Deschênes, M., Bourassa, J., Doan, V.D., Parent, A., 1996. A single-cell study of the axonal projections arising from the posterior intralaminar thalamic nuclei in the rat. Eur. J. Neurosci. 8, 329–343.

DiFiglia, M., Pasik, P., Pasik, T., 1982. A Golgi ultrastructural study of the monkey globus pallidus. J. Comp. Neurol. 212, 53–75.

Donoghue, J.P., Kitai, S.T., 1981. A collateral pathway to the neostriatum from corticofugal neurons of the rat sensory-motor cortex: an intracellular HRP study. J. Comp. Neurol. 201, 1–13.

François, C., Percheron, G., Yelnik, J., Heyner, S., 1984. A Golgi analysis of the primate globus pallidus. I. Inconstant processes of large neurons, other neuronal types, and afferent axons. J. Comp. Neurol. 227, 182–199.

Freeze, B.J., Kravitz, A.V., Hammack, N., Berke, J.D., Kreitzer, A.C., 2013. Control of basal ganglia output by direct and indirect pathway projection neurons. J. Neurosci. 33, 18531–18539.

Fujiyama, F., Sohn, J., Nakano, T., Furuta, T.Y., Nakamura, K.C., Matsuda, W., et al., 2011. Exclusive and common targets of neostriatofugal projections of rat striosome neurons: a single neuron-tracing study using a viral vector. Eur. J. Neurosci. 33, 668–677.

Gerfen, C.R., Bolam, P.J., 2010. The neuroanatomical organization of the basal ganglia. In: Steiner, H., Tseng, K.Y. (Eds.), Handbook of Basal Ganglia Structure and Function Academic Press/Elsevier, London, pp. 3–28.

Gittis, A.H., Berke, J.D., Bevan, M.D., Chan, C.S., Mallet, N., Morrow, M.M., et al., 2014. New roles for the external globus pallidus in basal ganglia circuits and behavior. J. Neurosci. 34, 15178–15183.

Granata, A.R., Kitai, S.T., 1989. Intracellular analysis of excitatory subthalamic inputs to the pedunculopontine neurons. Brain Res. 488, 57–72.

Hammond, C., Shibazaki, T., Rouzaire-Dubois, B., 1983. Branched output neurons of the rat subthalamic nucleus: electrophysiological study of the synaptic effects on identified cells in the two main target nuclei, the entopeduncular nucleus and the substantia nigra. Neuroscience 9, 511–520.

Hornykiewicz, O., 2008. Basic research on dopamine in Parkinson's disease and the discovery of the nigrostriatal dopamine pathway: the view of an eyewitness. Neurodegenerative Dis. 5, 114–117.

Jackson, A., Crossman, A.R., 1981. Basal ganglia and other afferent projections to the peribrachial region in the rat: a study using retrograde and anterograde transport of horseradish peroxidase. Neuroscience 6, 1537–1549.

Jones, E.G., Coulter, J.D., Burton, H., Porter, R., 1977. Cells of origin and terminal distribution of corticostriatal fibers arising in the sensorymotor cortex of monkeys. J. Comp. Neurol. 173, 53–80.

Kawaguchi, Y., Wilson, C.J., Emson, P.C., 1990. Projection subtypes of rat neostriatal matrix cells revealed by intracellular injection of biocytin. J. Neurosci. 10, 3421–3438.

Kawaguchi, Y., Wilson, C.J., Augood, S.J., Emson, P.C., 1995. Striatal interneurons: chemical, physiological and morphological characterization. Trends Neurosci. 18, 527–535.

Kemp, J.M., Powell, T.P.S., 1971. The site of termination of afferent fibers in the caudate nucleus. Phil. Trans. R. Soc. 262, 413–427.

Kita, H., 2010. Organization of the globus pallidus. In: Steiner, H., Tseng, K.Y. (Eds.), Handbook of Basal Ganglia Structure and Function Academic Press/Elsevier, London, pp. 233–247.

Kita, H., Kitai, S.T., 1987. Efferent projections of the subthalamic nucleus in the rat: light and electron microscopic analysis with the PHA-L method. J. Comp. Neurol. 260, 435–452.

Kita, H., Kitai, S.T., 1994. The morphology of globus pallidus projection neurons in the rat: an intracellular staining study. Brain Res. 636, 308–319.

Kita, H., Chang, H.T., Kitai, S.T., 1983. The morphology of intracellularly labeled rat subthalamic neurons: a light microscopic analysis. J. Comp. Neurol. 215, 245–257.

Koshimizu, Y., Fujiyama, F., Nakamura, K.C., Furuta, T., Kaneko, T., 2013. Quantitative analysis of axon bouton distribution of subthalamic nucleus neurons in the rat by single neuron visualization with a viral vector. J. Comp. Neurol. 521, 2125–2146.

Künzle, H., 1975. Bilateral projections from precentral motor cortex to the putamen and other parts of the basal ganglia. An autoradiographic study in Macaca fascicularis. Brain Res. 88, 195–209.

Laplane, D., Baulac, M., Widlocher, D., Dubois, B., 1984. Pure psychic akinesia with bilateral lesions of basal ganglia. J. Neurol. Neurosurg. Psychiat. 47, 377–385.

Lévesque, M., Parent, A., 2005. The striatofugal fiber system in primates: a reevaluation of its organization based on single-axon tracing studies. Proc. Natl. Acad. Sci. USA 102, 11888–11893.

Lévesque, M., Charara, A., Gagnon, S., Parent, A., Deschênes, M., 1996. Corticostriatal projections from layer V cells in rat are collaterals of long-range corticofugal axons. Brain Res. 709, 311–315.

Nakano, K., Hasegawa, Y., Tokushige, A., Nakagawa, S., Kayahara, T., Mizuno, N., 1990. Topographical projections from the thalamus, subthalamic nucleus and pedunculopontine tegmental nucleus to the striatum in the Japanese monkey. Macaca Fuscata. Brain Res. 537, 54–68.

Nambu, A., Takada, M., Inase, M., Tokuno, H., 1996. Dual somatotopical representations in the primate subthalamic nucleus: evidence for ordered but reversed body-map transformations from the primary motor cortex and the supplementary motor are. J. Neurosci. 16, 2671–2683.

Nauta, W.J.H., Mehler, W.R., 1966. Projections of the lentiform nucleus in the monkey. Brain Res. 1, 3–342.

Oorschot, D.E., 1996. Total number of neurons in the neostriatal, pallidal, subthalamic, and substantia nigral nuclei of the rat basal ganglia: a stereological study using the Cavalieri and optical disector methods. J. Comp. Neurol. 366, 580–599.

I. MICROCIRCUITRY

Oorschot, D.E., 2013. The percentage of interneurons in the dorsal striatum of the rat, cat, monkey and human: a critique of the evidence. Basal Ganglia 3, 9–24.

Parent, A., De Bellefeuille, L., 1982. Organization of efferent projections from the internal segment of globus pallidus in primate as revealed by fluorescence retrograde labeling method. Brain Res. 245, 201–213.

Parent, A., Hazrati, L.N., 1995. Functional anatomy of the basal ganglia. I. The cortico-basal ganglia-thalamo-cortical loop. Brain Res. Rev. 20, 91–127.

Parent, A., Lavoie, B., 1993. The heterogeneity of the mesostriatal dopaminergic system as revealed in normal and parkinsonian monkeys. Adv. Neurol. 60, 25–33.

Parent, A., Smith, Y., 1987. Organization of efferent projections of the subthalamic nucleus in the squirrel monkey as revealed by retrograde labeling methods. Brain Res. 436, 296–310.

Parent, A., Mackey, A., De Bellefeuille, L., 1983. The subcortical afferents to caudate nucleus and putamen in primate: a fluorescence retrograde double labeling study. Neuroscience 10, 1137–1150.

Parent, A., Charara, A., Pinault, D., 1995. Single striatofugal axons arborizing in both pallidal segments and in the substantia nigra in primates. Brain Res. 698, 280–284.

Parent, M., Parent, A., 2005. Single-axon tracing and three-dimensional reconstruction of centre médian-parafascicular thalamic neurons in primates. J. Comp. Neurol. 481, 127–144.

Parent, M., Parent, A., 2006. Single-axon tracing study of the corticostriatal projections arising from primary motor cortex in primates. J. Comp. Neurol. 496, 202–216.

Parent, M., Lévesque, M., Parent, A., 1999. The pallidofugal projection system in primates: evidence for neurons that branch ipsilaterally and contralaterally to the thalamus and brainstem. J. Chem. Neuroanat. 16, 153–165.

Parent, M., Lévesque, M., Parent, A., 2001. Two types of projection neurons in the internal pallidum of primates: single-axon tracing and three-dimensional reconstruction. J. Comp. Neurol. 439, 162–175.

Pasik, P., Pasik, T., DiFiglia, M., 1976. Quantitative aspects of neuronal organization in the neostriatum of the macaque monkey. In: Yahr, M.D. (Ed.), The Basal Ganglia Raven Press, New York, pp. 57–90.

Percheron, G., Yelnik, J., Francois, C., 1984. A Golgi analysis of the primate globus pallidus. III. Spatial organization of the striatopallidal complex. J. Comp. Neurol. 227, 214–227.

Pinault, D., 1996. A novel single-cell staining procedure performed in vivo under electrophysiological control: morpho-functional features of juxtacellularly labeled thalamic cells and other neurons with biocytin or neurobiotin. J. Neurosci. Meth. 65, 113–136.

Ramón y Cajal, S., 1909/1911. Histologie du Système Nerveux de l'Homme et des Vertébrés. Maloine, Paris.

Reiner, A., 2010. Organization of corticostriatal projection neuron types. In: Steiner, H., Tseng, K.Y. (Eds.), Handbook of Basal Ganglia Structure and Function Academic Press/Elsevier, London, pp. 323–339.

Reiner, A., Jiao, Y., Del Mar, N., Laverghetta, A.V., Lei, W.L., 2003. Differential morphology of pyramidal tract-type and intratelencephalically projecting-type corticostriatal neurons and their intrastriatal terminals in rats. J. Comp. Neurol. 457, 420–440.

Rockland, K.S., 2013. Collateral branching of long-distance cortical projections in monkeys. J. Comp. Neurol. 521, 4112–4123.

Sadikot, A.F., Parent, A., François, C., 1992. Efferent connections of the centromedian and parafascicular thalamic nuclei in the squirrel monkey: a PHA-L study of subcortical projections. J. Comp. Neurol. 315, 137–159.

Sato, F., Lavallée, P., Lévesque, M., Parent, A., 2000a. Single-axon tracing study of neurons of the external segment of the globus pallidus in primate. J. Comp. Neurol. 417, 17–31.

Sato, F., Parent, M., Lévesque, M., Parent, A., 2000b. Axonal branching pattern of neurons of the subthalamic nucleus in primates. J. Comp. Neurol. 424, 142–152.

Schmitt, O., Eipert, P., Kettlitz, R., Leßmann, F., Wree, A., 2014. The connectome of the basal ganglia. Brain Struct. Funct..http://dx.doi.org/10.1007/s00429-014-0936-0.

Shink, E., Bevan, M.D., Bolam, J.P., Smith, Y., 1996. The subthalamic nucleus and the external pallidum: two tightly interconnected structures that control the output of the basal ganglia in the monkey. Neuroscience 73, 335–357.

Smith, Y., Hazrati, L.N., Parent, A., 1990. Efferent projections of the subthalamic nucleus in the squirrel monkey as studied by the PHA-L antero-grade tracing method. J. Comp. Neurol. 294, 306–323.

Smith, Y., Bevan, M.D., Shink, E., Bolam, J.P., 1998. Microcircuitry of the direct and indirect pathways of the basal ganglia. Neuroscience 86, 353–387.

Smith, Y., Galvan, A., Ellender, T.J., Doig, N., Villalba, R.M., Huerta-Ocampo, I., et al., 2014. The thalamostriatal system in normal and diseased states. Front. Syst. Neurosci. 8 (5).http://dx.doi.org/10.3389/fnsys.2014.00005.

Somogyi, P., Bolam, J.P., Smith, A.D., 1981. Monosynaptic cortical input and local axon collaterals of identified striatonigral neurons. A light and electron microscopic study using the Golgi-peroxidase transportdegeneration procedure. J. Comp. Neurol. 195, 567–584.

Staines, W.A., Atmadja, S., Fibiger, H.C., 1981. Demonstration of a pallidostriatal pathway by retrograde transport of HRP-labeled lectin. Brain Res. 206, 446–450.

Turner, R.S., DeLong, M.R., 2000. Corticostriatal activity in primary motor cortex of the macaque. J. Neurosci. 20, 7096–7108.

Van der Kooy, D., Carter, D.A., 1981. The organization of the efferent projections and striatal afferents of the entopeduncular nucleus and adjacent areas in the rat. Brain Res. 211, 15–36.

Van der Kooy, D., Hattori, T., 1980. Single subthalamic nucleus neurons project to both the globus pallidus and substantia nigra in rat. J. Comp. Neurol. 192, 751–768.

Voorn, P., 2010. Projections from pallidum to striatum. In: Steiner, H., Tseng, K.Y. (Eds.), Handbook of Basal Ganglia Structure and Function Academic Press/Elsevier, London, pp. 249–257.

Wichmann, T., Kliem, M.A., 2004. Neuronal activity in the primate substantia nigra pars reticulata during the performance of simple and memory-guided elbow movement. J. Neurophysiol. 91, 815–827.

Wu, Y., Richard, S., Parent, A., 2000. The organization of the striatal output system: a single-cell juxtacellular labeling study in the rat. Neurosci. Res. 38, 49–62.

Yelnik, J., Percheron, G., 1979. Subthalamic neurons in primates: a quantitative and comparative analysis. Neuroscience 4, 1717–1743.

Yelnik, J., Percheron, G., François, C., 1984. A Golgi analysis of the primate globus pallidus. II. Quantitative morphology and spatial orientation of dendritic arborizations. J. Comp. Neurol. 227, 200–213.

Zheng, T., Wilson, C.J., 2002. Corticostriatal combinatorics: the implications of corticostriatal axonal arborizations. J. Neurophysiol. 87, 1007–1017.

I. MICROCIRCUITRY

3

Comparative Analysis of the Axonal Collateralization Patterns of Basal Ganglia Output Nuclei in the Rat

E. Mengual[1,2], L. Prensa[1,2], A. Tripathi[1],
C. Cebrián[1,2] and S. Mongia[2]

[1]Division of Neurosciences, Center for Applied Medical Research (CIMA),
Universidad de Navarra, Pamplona, Spain [2]Departamento de Anatomía, Facultad de Medicina,
Universidad de Navarra, Pamplona, Spain

3.1 INTRODUCTION

Axonal collateralization appears to be one of the major organizational features of the basal ganglia (Parent et al., 2000). Each component of the basal ganglia comprises different types of neurons displaying distinct axonal collateralization patterns (Matsuda et al., 2009; Parent et al., 2000; Prensa and Parent, 2001; Tripathi et al., 2010, 2013). The degree of axonal arborization of each type of neurons varies significantly in the different targets, and this notion is essential to properly assess the relative "strength" of each of these collateral projections. Moreover, the pattern of termination of these axons is likely to reflect the intrinsic organization of the innervated structures. This suggests that information on the axonal branching pattern of extrinsic afferents closely correlates with data on the intrinsic organization of the target structures (Bevan et al., 1997). A better knowledge of the finely tuned and widely distributed basal ganglia output networks is essential to understand the complex spatiotemporal sequence of neural events that ensures the flow of pallidal and nigral information at thalamic and brainstem levels.

Dextrans are biologically inert hydrophilic polysaccharides widely used for anatomical studies of the central nervous system connectivity (Reiner et al., 2000). These molecules are transported anterogradely and retrogradely in neurons and their axons, dextrans of smaller molecular weight being transported more quickly than larger ones. Moreover, 3 kDa dextrans seem to transport more effectively in a retrograde direction than do 10 kDa dextrans. The dextran amines conjugated to biotin are called *biotinylated dextran amines* (BDAs). The mechanism of BDA uptake appears to be pinocytotic for intact neurons, but it can also be taken up by damaged neurons and axons.

BDA 10 KDa has been demonstrated to be an effective and highly consistent anterograde tracer. The ease of development of BDA label makes it a superior choice over tracers like PHA-L and CTB which require multistep processing before visualization (Schmued et al., 1990; Wouterlood et al., 1990). It is easy to make small and well-defined injections of BDA, thereby confining the injection to the region whose projections are being investigated (Cebrián et al., 2005; Cebrián and Prensa, 2010; Reiner et al., 2000; Tripathi et al., 2010), which is an important feature for studies aimed at determining the topographic arrangement of projection neurons within a given region. There is a wide window of post-injection survival interval as compared to biocytin or neurobiotin which is catabolized quickly (Izzo, 1991; Kita and Armstrong, 1991; Lapper and Bolam, 1991). The detailed labeling of axon terminals provided by the BDA 10 KDa facilitates the characterization of their morphological features, such as the presence of axonal enlargements or varicosities and/or pedunculated terminals.

BDA 10 KDa has been widely used to produce small deposits of tracer confined to a given structure of the nervous system, resulting in histologically visualized axons that can be fully traced both in rats and primates (Cebrián et al., 2005; Cebrián and Prensa, 2010; Kita and Kita, 2012; Parent and Parent, 2007; Reiner et al., 2000; Tripathi et al., 2010). Axons labeled with this anterograde tracer have shown rich patterns of axon collaterals that can be traced to their distal terminal arbor without any fading, at least when a sufficiently high concentration of BDA is used (Tripathi et al., 2013). However, small volume BDA deposits might preclude the complete filling of all axon collaterals and terminals. To test this possibility, we carried out tracer injections using a replication-defective Sindbis virus, which produces a large amount of green fluorescent protein within the infected neuron without volume constraint (Fujiyama et al., 2011; Matsuda et al., 2009), and compared the axonal collateralization patterns as rendered by the two tracers (Tripathi et al., 2013). The branching pattern of a dorsal striatofugal axon obtained using the virus was totally similar to those of two axons traced with BDA, one from the dorsal striatum reported in the literature (Figure 3.1A and B) (Wu et al., 2000), and another from the ventral striatum using our injection parameters (Figure 3.1C) (Tripathi et al., 2010). Therefore, we have confirmed the sensitivity of BDA and the reliability of the collateralization patterns obtained in our studies using this tracer. In contrast, studies comparing collateralization patterns of axons from a different projection system, like nigrostriatal axons that branch profusely at their terminal target field, using BDA and Sindbis virus have shown a higher sensitivity of the latter compared to BDA in filling the complete terminal arbors and their most distal axon terminals (see discussion in Matsuda et al., 2009). However, the axons described here did not show that profuse terminal arborization even with the virus.

In this chapter, we will compare results from our labs on patterns of axonal collateralization of two basal ganglia output nuclei in the rat, the substantia nigra reticulata and ventral pallidum (VP), obtained using single-axon tracing techniques, in order to establish potential similarities and distinct features between the two. In addition we will compare this analysis with available data on the entopeduncular nucleus (EP) and the globus pallidus (GP), with the aim of finding potential parallelisms that may provide further insights into basal ganglia circuitry at the single axon level. This chapter will focus on the rat basal ganglia, although findings in the cat or primate will be discussed when appropriate.

3.2 BASAL GANGLIA OUTPUT NUCLEI IN THE RAT

The basal ganglia are a set of serially interconnected structures that in close association with the cerebral cortex elicit goal-directed behavior through a final output directed either toward the thalamus and back to the cortex, or toward the brainstem (Haber and Gdowski, 2004; Redgrave, 2007; Takakusaki, 2008). Information arising from different cortical functional domains is conveyed along several parallel circuits to the input station of the basal ganglia, the striatum, from where it is relayed either directly or via intermediate nuclei to several output nuclei which in turn, innervate either the thalamus or brainstem, or both (Alexander et al., 1986; DeLong and Wichmann, 2007). Along these interconnected stations, the information is processed in a largely segregated manner according to the initial functional domains, such that inputs from somatomotor and associative/executive cortical areas primarily target the dorsal striatum, formed by the caudate nucleus and putamen, whereas motivational and emotional (classically regarded as "limbic") information with origin in the prefrontal cortex, hippocampus and amygdala, reaches preferentially the ventral striatum, comprised by the nucleus accumbens (Acb) and olfactory tubercle (Tu). This functional dorsolateral to ventromedial segregation (Haber and Gdowski, 2004) can be further simplified into (1) dorsal (somatomotor and associative) and (2) ventral (emotional/motivational) basal ganglia circuits. However, interconnections between parallel circuits also take place in order to elicit a coordinated response (Joel and Weiner, 1994; Kelly and Strick, 2004). The final basal ganglia output exits from the so-called output nuclei, which are the EP in rat and cat (or internal GP—GPi in primates), and the substantia nigra *pars reticulata* (SNr), within the dorsal basal ganglia circuit, and the VP, in the ventral one (Heimer, 1978). The common targets of the three output nuclei are the thalamus and brainstem (Haber and Gdowski, 2004).

The EP and SNr receive direct inhibitory projections from a subset of dorsal striatal neurons containing substance P (SP), and in turn send inhibitory projections to either thalamic or brainstem tegmental neurons; thus the net result of basal ganglia activation along this "direct" pathway is disinhibition of the targets. A different subset of dorsal striatofugal neurons containing enkephalin (Enk) projects indirectly to the output nuclei, by innervating first an intrinsic component of the basal ganglia, the GP. This then projects to the subthalamic nucleus (STh), another basal ganglia intermediate component. The STh finally sends excitatory projections to the EP and SNr. Thus, this "indirect"

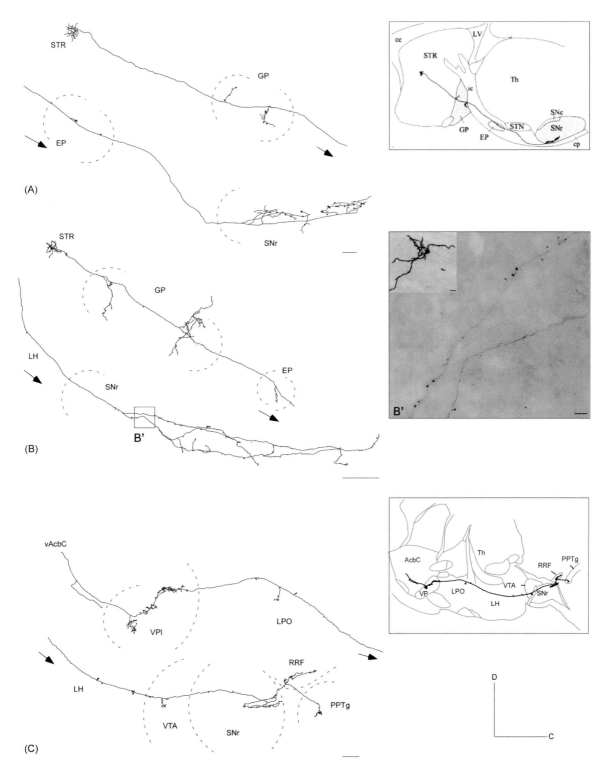

FIGURE 3.1 Comparison between axonal collateralization patterns of striatal neurons obtained using BDA 10k and Sindbis virus. (A) Axonal collateralization pattern of a single striatal neuron after a iontophoretic injection of BDA in the rat. Right inset schematically shows location of the source neuron and the several termination sites. (B) Axonal collateralization pattern of a single striatal axon after injection of Sindbis virus in a C57 BL6 mouse followed by immunoperoxidase conversion of GFP (for the metholodogy, see Aransay et al., 2015). This axon shows a remarkable similarity in branching pattern, target structures, and most importantly, in regard to the density of collaterals, to the one filled with BDA (A), providing evidence of the validity of BDA 10k for this type of analysis. The area within the red box is shown at a higher magnification in (B′). The inset shows the cell body after GFP was converted to a black precipitate for purposes of prolonged viewing and reconstruction. (C) Axonal collateralization pattern of a rat ventral striatal neuron after a iontophoretic injection of BDA. This axonal branching pattern is again remarkably similar to the one obtained with the Sindbis virus in terms of organization, target nuclei, and density of collaterals in the homologue ventral basal ganglia components. Right inset schematically shows location of the tracer deposit, trajectory of the axon, and its termination sites. Arrows in (A)–(C) indicate segments of a single axon (separated for the sake of formatting). AcbC, nucleus accumbens, core; cp, cerebral peduncle; EP, entopeduncular nucleus; GP, globus pallidus; ic, internal capsule; LV, lateral ventricle; PPTg, pedunculopontine tegmental nucleus; RRF, retrorubral field; SNc, substantia nigra *pars compacta*; SNr, substantia nigra *pars reticulate*; STR, striatum; VP, ventral pallidum; VPl, ventral pallidum, dorsolateral compartment. Scale bars: (A) and (C), 200 μm; B′, 20 μm; inset, 40 μm. *Source: Reprinted from (A) Wu et al. (2000), with permission from Elsevier; (C) Tripathi et al. (2010), with permission from Wiley & Sons.*

pathway, upon activation of dorsal striatal projection neurons, leads to thalamic or tegmental inhibition (Smith et al., 1998). The combination of facilitatory and inhibitory inputs to different sets of thalamocortical neurons results in specific goal-directed behaviors (Smith et al., 1998).

Likewise, the ventral striatum sends projections to a ventral pallidal component, the VP, which innervates limbic-related thalamic nuclei and brainstem targets, acting as the output nucleus of ventral basal ganglia circuits. However, both SP- and Enk-containing projection neurons in the ventral striatum project to the VP (Lu et al., 1998), where direct and indirect pathways are harder to identify (Sesack and Grace, 2010). The analysis and comparison of the axonal arborization patterns displayed by the dorsal (EP and GP) and ventral (VP) pallidal neurons may contribute to clarify whether the well-established direct and indirect pathways of dorsal basal ganglia circuits have ventral homologues in the rat.

3.3 AFFERENT AND EFFERENT CONNECTIONS OF THE BASAL GANGLIA OUTPUT NUCLEI

- EP and SNr

The EP and the SNr receive projections predominantly from the dorsal striatum or caudate putamen nucleus (CPu) in the rat, specifically from medium spiny projection neurons (MSNs) belonging to the direct pathway (Smith et al., 1998). Dorsal striatofugal axons projecting to the EP and SNr can be labeled in the same striatal areas (Kawaguchi et al., 1990; Wu et al., 2000). Consistent with this, single striatofugal axons collateralized in both EP and SNr (37% of the total axons, Wu et al., 2000), suggesting that largely the same information is conveyed to these two basal ganglia output nuclei (Kawaguchi et al., 1990; Wu et al., 2000). At the same time, 26% of striatofugal axons branched within the SNr but did not collateralize to the EP (Wu et al., 2000). As the rat EP is a small nucleus compared to the SNr (3.200 vs. 26.300 neurons, respectively; Oorschot, 1996), this suggests that SNr may represent the major output nucleus of the dorsal basal ganglia circuits.

As an output nucleus, the EP predominantly targets the thalamus and brainstem. Consistent with its striatal input originating almost exclusively in the dorsal striatum and scarcely at all from the ventral striatum (Fink-Jensen and Mikkelsen, 1989; Heimer et al., 1991; Tripathi et al., 2010; Usuda et al., 1998; Zahm and Heimer, 1993), EP axons innervate the motor thalamus, mainly the ventromedial nucleus (VM), a ventral region of the ventral anterior nucleus (VA), and the parafascicular nucleus (PF) (Carter and Fibiger, 1978). In addition, the EP sends abundant projections to the lateral habenular complex (LHb) (Herkenham and Nauta, 1977; van der Kooy and Carter, 1981), a structure that belongs to the reward network, and is tightly connected with the VP. Regarding the brainstem, EP axons mostly target the pedunculopontine tegmental nucleus (PPTg), terminating in an area medial to the cholinergic subpopulation (Jackson and Crossman, 1981; Steininger et al., 1992), as well as the associated laterodorsal tegmental nucleus (LDTg) (Mongia et al., 2015).

The SNr also receives predominantly striatal afferents from the dorsal striatum (Gerfen, 1985); however, studies of single ventral striatofugal axons have shown that they also provide a substantial collateralization to SNr, in contrast with the EP (Tripathi et al., 2010). This suggests a broader functional spectrum for SNr, ranging from somatomotor-related functions to associative and motivational roles. The thalamic targets of the SNr are the motor thalamus comprised by the VM, VA, and the ventral lateral nucleus (VL), and also several rostral and caudal intralaminar and midline nuclei, as well as the mediodorsal nucleus (MD) (Beckstead et al., 1979; Bentivoglio et al., 1979; Deniau and Chevalier, 1992; Faull and Mehler, 1978; Gerfen et al., 1982; Sakai et al., 1998). Regarding the brainstem, SNr mainly projects to the superior colliculus (SC), the PPTg, and the periaqueductal gray (PAG) (Beckstead et al., 1981; Faull and Mehler, 1978), three sensorimotor and motivational structures connected with lower brainstem systems (McHaffie et al., 2005; Winn, 2006). The PPTg represents a prominent brainstem target of the basal ganglia outflow, which in turn projects back to the basal ganglia (Lee et al., 2000). Most of the nigropedunculopontine fibers in the rat terminate in the medial two-thirds of the *pars dissipata*; and the projection is mostly ipsilateral with a small contralateral component (Spann and Grofova, 1991). The SC also receives a prominent projection from the SNr, which supports the oculomotor function assigned to the basal ganglia (Hikosaka et al., 2000) and is unique in relation to the other basal ganglia output nuclei. This projection is also mostly ipsilateral (Bickford and Hall, 1992; Gerfen et al., 1982; Redgrave et al., 1992; Williams and Faull, 1985). The elegant study of Redgrave et al. (1992) revealed that the nigrocollicular pathway is composed of segregated channels that link regionally segregated population of cells within SNr with spatially separated terminations within the SC and might control different aspects of head movement in rats (Redgrave et al., 1992). Actually nigrotectal terminals make synaptic contact on tectospinal neurons in the intermediate grey layers of the SC (Bickford and Hall, 1992), exerting a GABAergic inhibitory influence on them (Hikosaka and Wurtz, 1985).

• The VP

The VP processes emotional and motivational information and also forms part of a specific network implicated in reward and addiction (Haber and Knutson, 2010; McGinty et al., 2011; Root et al., 2015). This pallidal component receives a massive input from the ventral striatum, representing its main relay within the ventral basal ganglia circuit (Haber et al., 1985; Heimer et al., 1987, 1991). Two different subterritories have been defined within the Acb according to histochemical and hodological criteria, a central Acb subterritory or Acb core (AcbC) and a medial peripheral Acb or Acb shell (AcbSh) (Heimer et al., 1991; Zaborszky et al., 1985). The cortical and amygdalar information is then conveyed downstream along three different subcircuits that originate in the AcbC, AcbSh, and Tu, and preferentially project to the dorsolateral VP or VPl, ventromedial VP or VPm, and the rostral extension of the VP or VPr, respectively (Groenewegen et al., 1993; Heimer et al., 1987, 1991; Tripathi et al., 2010; Zahm and Heimer, 1993). This connectional segregation is largely maintained further downstream as VPl projects mainly to STh and SNr, whereas VPm and VPr preferentially innervate the LHb, MD, and the ventral tegmental area (VTA) (Zahm et al., 1996). Thus, the ventral striatopallidal system can be additionally segregated into a "motor" subcircuit arising from the AcbC (AcbC-VPl-STh-SNr subcircuit), and a "limbic" subcircuit arising from the AcbSh (AcbSh-VPm-MD/VTA) (Kalivas et al., 1999).

3.4 AFFERENT AND EFFERENT CONNECTIONS OF THE GP

The GP is an intrinsic or core nucleus within the basal ganglia circuits that receives projections preferentially from the Enk-containing MSNs of the dorsal striatum, forming part of the indirect pathway (Smith et al., 1998). Actually, 36% of striatofugal axons arborize only in the GP (Wu et al., 2000). The GP projection neurons target mainly the STh which in turn projects to the output nuclei, EP and SNr. However, single axon studies of GP neurons have shown that only a subset of GP neurons projecting to the STh exclusively arborize in this nucleus, whereas a large proportion of them additionally targets either SNr or both SNr and EP (Bevan et al., 1998; Kita and Kitai, 1994). Studies differentiating rostral versus intermediate and caudal regions in GP have reported that the rostral GP also targets the reticular thalamic nucleus (Rt), the caudate nucleus, putamen, and SNr (Nauta, 1979; Shammah-Lagnado et al., 1996), whereas the caudal GP shows quite different targets projecting first, more sparingly to STh and SNr, and second, specifically innervating the caudal intralaminar thalamic nuclei, the caudal Rt, the auditory cortex, PPTg, and the substantia nigra *pars lateralis* (SNl) (Moriizumi and Hattori, 1992; Schmued et al., 1989; Shammah-Lagnado et al., 1996). Finally, another pallidal subpopulation has been recently identified and characterized comprising Enk-positive neurons that arborize only and profusely in the striatum (Abdi et al., 2015; Mallet et al., 2012).

3.5 COLLATERALIZATION PATTERNS OF SINGLE AXONS FROM THE BASAL GANGLIA OUTPUT NUCLEI

What have single-axon tract-tracing studies in the VP, SNr, and EP taught us about basal ganglia circuits in the rat? These studies have confirmed in most cases previous results of dual or triple retrograde labeling studies that had investigated the potential simultaneous innervation of two or three target structures from an output nucleus. Additionally, they have revealed the complete axonal wiring architecture of single axons including all their target structures, as well as the density of the arbor in each of them, its areal distribution, etc. Finally, the assessment of the relative percentages of each type of axon within a given nucleus provides a gross estimation of its cell composition, in relation to its different targets. Some of that information from each of the basal ganglia output nuclei is further commented on below.

3.6 AXONAL BRANCHING PATTERNS OF SNR NEURONS

Similarly to other basal ganglia nuclei, the output neurons of the SNr are endowed with highly collateralized axons, which in addition, display a great diversity of branching patterns (Cebrián et al., 2005). In our study analyzing 30 single axons in SNr, three main branching patterns were distinguished according to their target areas: (1) axons projecting to the thalamus only (~30%), (2) axons innervating the brainstem only (~33%), and (3) axons collateralizing in both the thalamus and brainstem (~37%) (Cebrián et al., 2005).

3.6.1 Thalamus-Only Projecting Axons

The SNr neurons giving rise to thalamus-only projecting axons were distributed widely across the nucleus and represented approximately 30% of the total traced axons (Cebrián et al., 2005; Kha et al., 2001) (Figures 3.2A and B and 3.3B). The axons entered the thalamus either ventrally through the VM or caudally, generally through the PF. There seems no consistent correlation between the axonal trajectories and the localization of their parent cell bodies within SNr (Cebrián et al., 2005). Once in the thalamus, the axons either innervated the VM-VL complex—and often the intralaminar paracentral nucleus (PC) or PF as well (Cebrián et al., 2005; Kha et al., 2001), or targeted the MD and the PC/central medial (CM), the VM or the PF (Cebrián et al., 2005). Interestingly, VM-VL complex only received ipsilateral innervation. In contrast, axon collaterals in MD (mostly targeting its central part, with few inconstant fibers entering also the lateral part) and the intralaminar nuclei originated either from the ipsilateral or contralateral SNr. When the innervation was bilateral, it was generally denser ipsilaterally (Cebrián et al., 2005; Gulcebi et al., 2012). The set of thalamic nuclei innervated by single SNr axons emphasizes the idea that SNr can act simultaneously upon thalamic nuclei involved in motor control and, through the intralaminar nuclei, in sensory integration.

3.6.2 Brainstem-Only Projecting Axons

SNr neurons that only innervated the brainstem targeted mainly the deep layers of the SC and the deep mesencephalic nucleus (DpMe), and represented approximately 33% of the total traced axons (Figure 3.3A and B) (Cebrián et al., 2005). These findings confirm previous reports on nigrotectal projections (Redgrave et al., 1992) which are responsible for eye movements, both ipsilaterally and contralaterally (Basso and Wurtz, 2002; Liu and Basso, 2008). The increase in neuronal activity observed in the SC following the inhibition of SNr neurons (Chevalier et al., 1981, 1985; Karabelas and Moschovakis, 1985) is believed to be caused by a monosynaptic GABAergic-mediated disinhibition of the striatonigral pathway (Chevalier et al., 1985; Williams and Faull, 1985). The present findings extend the previous results by showing that these axons generally collateralized in the DpMe, a midbrain reticular region which it is thought to facilitate the interhemispheric regulation of the output of the basal ganglia through conspicuous contralateral projections to the thalamus and SC (Rodriguez et al., 2001). Some of these axons provided a bilateral innervation, traversing the midline through the commissure of the SC, confirming previous studies (Gerfen et al., 1982). The analysis of single axons, however, has unequivocally shown bilateral innervations of SC and DpMe, and that these were denser ipsilaterally (Cebrián et al., 2005). In addition, occasional collaterals or varicosities were also found in other nuclei of the mesopontine tegmentum like the lateral parabrachial nucleus (LPB), subpeduncular tegmental nucleus (SPTg), and the cuneiform nucleus (CnF) (Cebrián et al., 2005).

3.6.3 Thalamus- and Brainstem-Projecting Axons

In addition to thalamus-only and brainstem-only projecting axons, 37% of axons were found to bifurcate to innervate both the thalamus and the brainstem (Figures 3.2C and 3.3B). Two distinct collateralization patterns were observed within these SNr axons: a first subgroup that collateralized predominantly in SC and PPTg, and another that collateralized in PAG (Cebrián et al., 2005). Interestingly, VM/VL and PF were more frequently innervated simultaneously with SC and PPTg, suggesting that these thalamic nuclei might collaborate in oculomotor functions (Sakai et al., 1998), whereas those collateralizing in MD, CM, and PC also collateralized in PAG (Cebrián et al., 2005).

Double retrograde tracer studies had already reported the existence of SNr axons projecting to the thalamus and SC (Bentivoglio et al., 1979) or to SC and the pontine reticular formation (Yasui et al., 1995). Also a dual projection to SC and PPTg had been previously suggested (Jackson and Crossman, 1981). The present studies reveal a connectional arrangement at the single axon level where the basal ganglia via SNr control both separately and simultaneously thalamic and brainstem targets like the SC, which are implicated in the control of eye movements at those different levels: Recall that the SC directly projects to the same intralaminar nuclei (Krout et al., 2001). Furthermore, single SNr axons also collateralize in a third structure implicated in eye movement, the PPTg. SNr projections to PPTg had been described previously (Grofova and Zhou, 1998; Spann and Grofova, 1991), and electrophysiological studies have demonstrated the monosynaptic inhibitory effect of SNr on PPTg neurons (Granata and Kitai, 1991; Kang and Kitai, 1990; Noda and Oka, 1984; Takakusaki et al., 2011, 1997). Thus, a single SNr axon may also simultaneously control a third motor center at a lower level than SC, which in turn projects upstream to the SC (Hall et al., 1989; Harting and Van Lieshout, 1991; Krauthamer et al., 1995). Thus, the present studies reveal a wiring redundancy that starting from a lower hierarchical level, progressively reaches higher level centers. Furthermore, the output innervation of the SNr to the PPTg overlaps with the ascending projections from PPTg to

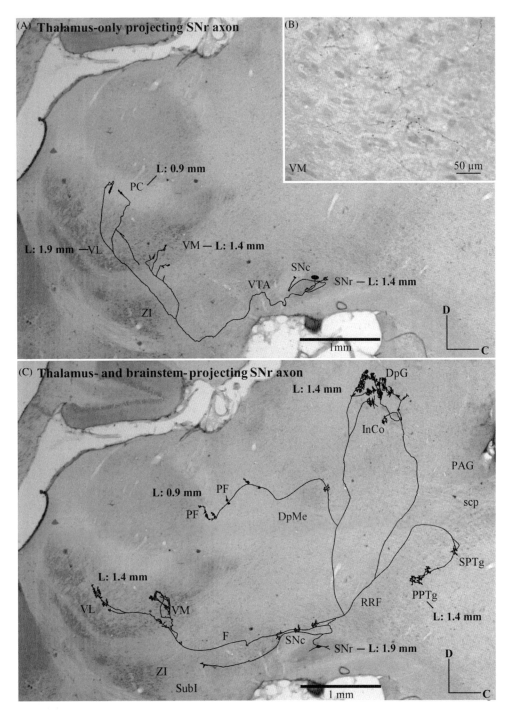

FIGURE 3.2 (A) Sagittal section of the rat brain immunoreacted for the calcium-binding protein Calbindin D-28k (CB). Superimposed is the trajectory and pattern of axonal branching of a thalamus-only projecting SNr axon. This fiber traverses the ZI and penetrates the thalamus through the VM. (B) Photomicrograph showing a portion of the terminal arborization among the CB-ir neurons of the VM, of the axon illustrated in (A). (C) Photomicrograph of a sagittal section of the rat brain stained for CB to show the trajectory and pattern of axonal branching of a thalamus- and brainstem-projecting SNr axons, whose cell body is located in the dorsal half of the SNr. This fiber targets simultaneously the PPTg, the deep layers of SC, and the thalamus through its collaterals. Note that the branch that innervates the PPTg arises from the collateral that branches in the SC and that this axon innervates both the ventral tier thalamic nuclei and the PF. (L) indicates the mediolateral coordinates at which axons in (A) and (C) arise from the SNr to innervate their target structures, according to the Paxinos and Watson Atlas (1998). DpMe, deep mesencephalic nucleus; DpG, deep gray layer of the superior colliculus; F, nucleus of the fields of Forel; InCo, intercollicular nucleus; PAG, periaqueductal gray matter; PC, paracentral thalamic nucleus; PF, parafascicular thalamic nucleus; PPTg, pedunculopontine tegmental nucleus; RRF, retrorubral field; scp, superior cerebellar peduncle *(brachium conjunctivum)*; SNc, substantia nigra *pars compacta*; SNr, substantia nigra *pars reticulate*; SPTg, subpeduncular tegmental nucleus; VM, ventral medial thalamic nucleus; VTA, ventral tegmental area; ZI, zona incerta. *Source: Adapted from Cebrián et al. (2005; License Number 3557700464016).*

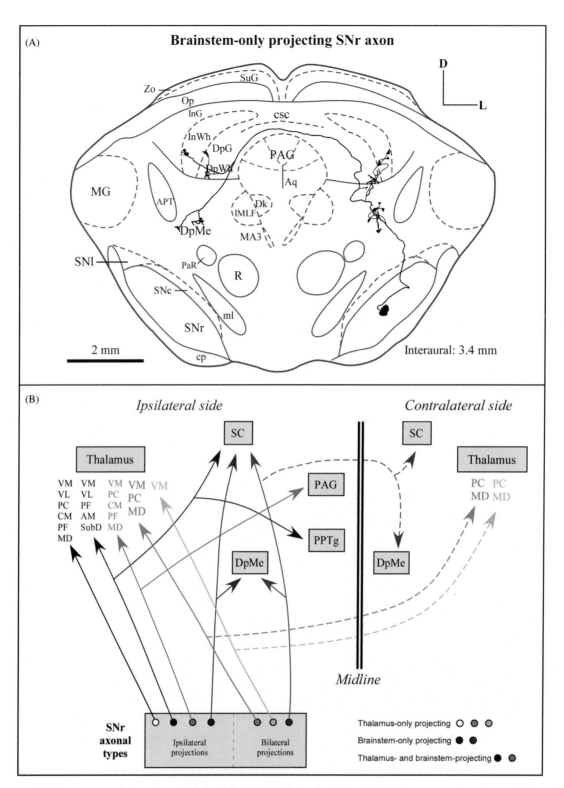

FIGURE 3.3 (A) Reconstruction in coronal section of a brainstem-only projecting axon arising from a tracer injection in the SNr. This fiber innervates bilaterally the DpMe and the deep layers of the SC. The interaural plane (I) at which the axon exited the SNr and targeted the various thalamic nuclei is indicated, according to the atlas of Paxinos and Watson (1998). (B) Schematic representation of the axonal branching patterns of SNr neurons and their main target structures. Ipsilateral projections are indicated with continuous lines whereas contralateral projections are depicted with dashed lines. AM, anteromedial thalamic nucleus; APT, anterior pretectal nucleus; Aq, aqueduct (Sylvius); CM, central medial thalamic nucleus; cp, cerebral peduncle; csc, commissure of the superior colliculus; Dk, Darkschewitch nucleus; DpMe, deep mesencephalic nucleus; DpG, deep gray layer of the superior colliculus; DpWh, deep white layer of the superior colliculus; IMLF, interstitial nucleus of the medial longitudinal fasciculus; InG, intermediate gray layer of the superior colliculus; InWh, intermediate white layer of the superior colliculus; MD, mediodorsal thalamic nucleus; MG, medial geniculate nucleus; ml, medial lemniscus; Op, optic nerve layer of the superior colliculus; PAG, periaqueductal gray matter; PaR, pararubral nucleus; PF, parafascicular thalamic nucleus; PPTg, pedunculopontine tegmental nucleus; R, red nucleus; SC, superior colliculus; SNc, substantia nigra *pars compacta*; SNl, substantia nigra *pars lateralis*; SNr, substantia nigra *pars reticulate*; SubD, submedius thalamic nucleus, dorsal part; VL, ventral lateral thalamic nucleus; VM, ventral medial thalamic nucleus.

the basal ganglia, thus establishing reciprocal loops that exert excitatory influences on the pallidal, subthalamic, and nigral cells (Winn, 2006).

SNr neurons can influence other structures involved in the control of eye movement besides the SC and the PPTg. For example, SNr also provides axon collaterals to the interstitial nucleus of the medial longitudinal fasciculus and the nucleus of Darkschewitsch (Cebrián et al., 2005), which play an important role in eye movement coordination (Buttner and Buttner-Ennever, 1988). Interestingly, the fact that axonal branches that target the deep layers of the SC also emit collaterals to the lateral aspect of the PF thalamic nucleus suggests that at least some PF thalamostriatal and thalamocortical neurons might participate in visuomotor orienting responses in conjunction with premotor neurons of the SC (Cebrián et al., 2005).

From the subset targeting the PAG, the thalamic branch collateralized in MD, CM, PC, PF, and VM forming terminal fields similar to those provided by thalamus-only projecting axons. The innervation of the PAG was not very profuse, but covered a large sector of this structure. The direct projection from the SNr to the PAG had already been described (Beitz, 1982; Kirouac et al., 2004; Marchand and Hagino, 1983). Although we did not find any brainstem-only projecting axons innervating PAG, the results obtained in relation to the SC/DpMe allow us to hypothesize that a similar control from the SNr upon the thalamus and PAG both singly and simultaneously may also take place, given that, similarly to SC, PAG-thalamic projections have been reported to innervate the same thalamic nuclei (Krout and Loewy, 2000).

The PAG is a tegmental structure involved in defensive behavior and anxiety (Coimbra et al., 2006), consciousness (Watt, 2000), reproductive and maternal behavior (Murphy et al., 1999), and nociception (Guethe et al., 2013; Lei et al., 2014). Interestingly, the PF thalamic nucleus is also involved in nociception (Gao et al., 2012; Wang and Guo, 2009) and the dorsolateral PAG-ascending fibers projecting to the thalamus provide a clear projection to the intralaminar nuclei, something that has been recently related to fear learning (Kincheski et al., 2012). All these data suggest that the PAG and different thalamic structures that are innervated by the same SNr axon might be working together on several behavioral and sensory-motor functions.

3.7 THE VENTRAL PALLIDUM

A total of 105 axons were fully reconstructed after BDA deposits in the three VP compartments (Tripathi et al., 2013). The analysis of the reconstructions revealed three distinct axonal trajectories encompassing three distinct patterns of collateralization: (1) a caudal or caudodorsal trajectory, followed by axons that arborized either in the thalamus or brainstem, or both, consistent with the known preferential targets of basal ganglia output nuclei. This trajectory and arborization pattern was the most frequent one, being displayed by greater than 80% of VP axons (Tripathi et al., 2013). (2) A lateral trajectory, followed by axons that targeted the amygdaloid complex and/or extended amygdala system (Mongia et al., 2015). These were regarded as amygdalopetal axons and represented 15% of the total. (3) A rostrodorsal trajectory that arborized essentially in the cerebral cortex (Mongia et al., 2015). This corticopetal pattern was displayed by just one axon (<1% of VP axons).

This analysis provides two pieces of relevant information: first, a relative estimation of VP neuronal composition in relation to the different neural circuits that arise from the VP. According to this, 84% of VP neurons participate in "classical" basal ganglia circuits as part of the ventral striatopallidal system, 15% are engaged in amygdalopetal circuits, and less than 1% give rise to corticopetal axons. Second, the full arborization patterns of single axons are revealed, demonstrating frequent axon collaterals from one neural circuit to another. This provides the anatomical substrate for functional interactions among the different basal forebrain macrosystems.

3.7.1 VP Neurons Participating in Basal Ganglia Circuits

The vast majority of neurons within each compartment of the VP displayed a typically "pallidal" projection pattern. This consisted of axons that coursed either caudally or caudodorsally, which innervated the brainstem or thalamus, respectively. A third subpopulation bifurcated into two branches, each of which arborized extensively in each of the two targets. An interesting finding regarding these typical "pallidal" axons was that they often collateralized in the rostral sublenticular extended amygdala (SLEAr) (Tripathi et al., 2013) (Figure 3.4A). This nucleus forms part of the extended amygdala system (Tripathi and Mengual, 2012); however, it is frequently targeted by collaterals of VP axons, and also of ventral striatofugal axons from both the AcbC and AcbSh (Tripathi et al., 2010). Therefore, the information processed within the ventral basal ganglia circuit at both accumbal and ventral pallidal levels may access the extended amygdala system via these axon collaterals, enabling functional interactions between the two basal forebrain macrosystems.

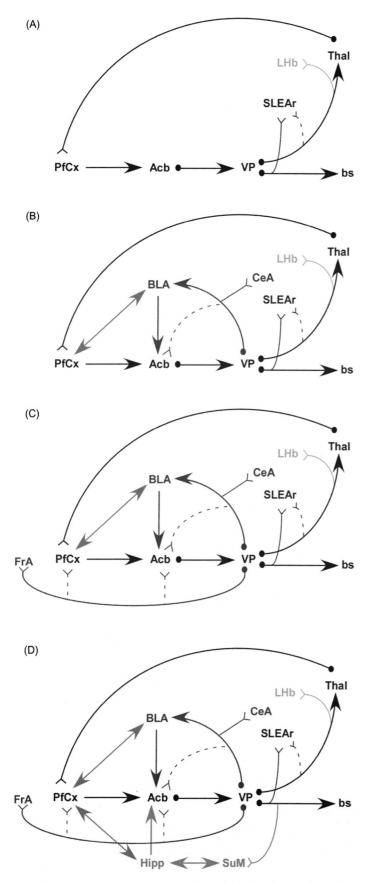

FIGURE 3.4 **Schematic representation of the distinct neural subcircuits originating in the VP based on the axonal branching patterns of VP neurons.** (A) Single ventral pallidothalamic and pallidotegmental axons forming part of the classical ventral basal ganglia loop (depicted in blue). These frequently collateralize in the rostral sublenticular extended amygdala (SLEAr), a component of the extended amygdala system. Via these collaterals, the ventral striatopallidal system may feed into the extended amygdala system. (B) Amygdalopetal axons from the VP preferentially collateralize in the basolateral amygdaloid nucleus (BLA), which heavily innervates the Acb, and frequently also in the CeA. Thus, the ventral striatopallidal system may provide feedback to the BLA via the VP (depicted in red). In addition, the ventral striatopallidal system may modulate the extended amygdala system at its output nucleus, the CeA. (C) Corticopetal axons from the VP innervate the FrA while also providing axon collaterals to the PfCx and Acb (depicted in purple). Thus, the ventral striatopallidal system may reciprocate the cortical input to the ventral striatum via corticopetal VP axons. (D) A third non-reciprocated input to the Acb arises from the hippocampal formation. A potential pathway through which the VP might provide feedback to this structure is via frequent axon collaterals of VP brainstem-projecting axons which innervate the SuM (Tripathi et al., 2013). This nucleus in turn is reciprocally connected with the hippocampal formation (depicted in green). Acb, nucleus accumbens; bs, brainstem; CeA, central amygdaloid nucleus; FrA, frontal association cortex; Hipp, hippocampal formation; LHb, lateral habenula; PfCx, prefrontal cortex; Thal, thalamus; VP, ventral pallidum.

3.7.2 Amygdalopetal Projections from the VP

A subset of VP axons coursing laterally instead of caudally was additionally traced (Mongia et al., 2015). These axons represented 15% of the total VP axons. Their patterns of arborization were totally different to those of the pallidal-like axons: they did not collateralize in any basal ganglia component, but rather innervated either components of the extended amygdala, or the basolateral amygdaloid nucleus (BLA), or both (Mongia et al., 2015) (Figure 3.4B); occasionally some of these axons provided a collateral to the Acb. The BLA provides a massive input to the Acb (Kelley et al., 1982; Russchen and Price, 1984), and these projections are not reciprocated from the ventral striatum. However, since the Acb projects heavily to the VP, and since actually every ventral striatofugal axon collateralized in at least one VP component (Tripathi et al., 2010), these VP-amygdalopetal axons may act as feedback projections to the BLA via the VP (Figure 3.4B). In addition, most of these axons also collateralized in the central amygdaloid nucleus (CeA) (Mongia et al., 2015) (Figure 3.4B). The BLA together with other amygdaloid nuclei sends converging projections to the CeA and to the amygdalohippocampal area, the major functions of which are to execute output commands (Pitkanen et al., 1997). Therefore, the ventral striatopallidal system—again via the VP—also has the connectional substrate to modulate the output of the amygdaloid complex at its final stage, the CeA (Figure 3.4B).

3.7.3 Corticopetal Projections from the VP

One axon out of the 105 traced from VP followed a rostrodorsal trajectory to innervate the cortex. This axon, which represented less than 1% of the total axons in our sample, terminated in the frontal association cortex (FrA) (Mongia et al., 2015), consistent with previous data on basal forebrain corticopetal axons in the rat (Luiten et al., 1987). Interestingly, on its way to FrA this VP axon also provided a few varicosities to the infralimbic and prelimbic cortices (Mongia et al., 2015), areas that in turn are projecting to the Acb (Gabbott et al., 2005; Vertes, 2004). This corticopetal axon also provides feedback to the Acb via moderately dense axon collaterals in both AcbC and AcbSh (Mongia et al., 2015). Thus this axon, similarly to the VP-mediated feedback from Acb to the amygdala (Figure 3.4B), could represent a feedback to the cortical input via the corticopetal neurons in VP (Figure 3.4C). However, the termination of this axon in the motor association cortex—while providing collaterals to prefrontal cortical areas that in turn innervate the Acb—suggests a corticopetal projection composed of reciprocal and non-reciprocal components, similar to that reported for cortico-thalamic projections (Haber and McFarland, 2001; McFarland and Haber, 2002), rather than a direct feedback projection. This is however, highly speculative, as only one corticopetal axon was traced. Finally, if these axon collaterals somehow do subserve a feedback to both the basolateral amygdaloid nuclear complex and the prefrontal cortex, a question arises as to whether any feedback might be provided to the hippocampal formation as well, the other unreciprocated key input to the Acb. Interestingly, 15% of VP brainstem-projecting axons collateralized in the supramammillary nucleus (SuM), a structure reciprocally connected with the hippocampus (Figure 3.4D) (Woolf et al., 1984; Cui et al., 2013). To what extent all these collaterals actually provide a functional feedback needs further investigation.

3.8 DIFFERENCES IN "PALLIDAL-LIKE" PROJECTIONS AMONG THE VP COMPARTMENTS

Brainstem-projecting axons coursed along the lateral hypothalamic area (LH) and most often extended into the midbrain or further caudally into the pons (Tripathi et al., 2013). Thalamus-projecting axons reached the rostral LH where they turned dorsally and entered the thalamus either *via* the *stria medullaris* (sm) or by directly traversing the Rt to collateralize in distinct limbic-related nuclei. Finally, bifurcating axons generally branched at the rostral LH, and each branch followed the same trajectory as non-bifurcating axons did to their final targets (Tripathi et al., 2013). Although projections to the LH from the VP and especially from VPm had been reported previously (Groenewegen et al., 1993; Haber et al., 1985; Zahm, 1989), the analysis of the branching patterns of VP axons specifically reveals that every single VP axon provides at least a few varicosities in the LH if not a moderately dense collateralization as they course along it (Figures 3.5, 3.7, and 3.8).

All three VP compartments gave rise to "pallidal-like" axons of the three subtypes; that is, brainstem-only, thalamus-only, and bifurcating axons. However, an interesting finding was that the relative percentage of each subtype varied across compartments (Table 3.1). Furthermore, a gradation was observed such that most of VPl axons innervated the brainstem only (≈80%), while a minority projected either to the thalamus or both brainstem and thalamus (14% and 7%, respectively). The VPm also preferentially targeted the brainstem, but in this case, it was only 64% of axons,

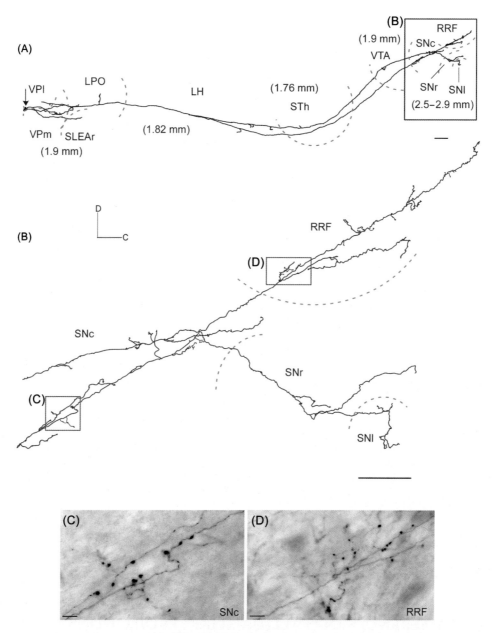

FIGURE 3.5 Pattern of collateralization of a brainstem-projecting VPl axon. (A) Camera lucida reconstruction of the full trajectory and collaterals of an axon exiting from a tracer deposit. Dashed arcs represent estimated boundaries as observed on sagittal sections, while numbers within parentheses indicate approximate mediolateral coordinates. This axon collateralized in the VPm and SLEAr, bifurcated at the LH, and extended caudally to also innervate the VTA, SNc, SNr, SNl, and RRF. The boxed area is shown at higher magnification in (B) (in blue) to illustrate details of the terminal arborization of this axon. Microphotographs of two other boxed areas are shown in (C) and (D), where thin axon branches and varicosities are visible in black, in sections dually stained for CB. Scale bars: (A), 200 μm; (B), 100 μm; (C) and (D), 10 μm. *Source: Reproduced from Tripathi et al. (2013) (Figure 3.6), with permission from Springer.*

TABLE 3.1   Distribution of Arborization Pattern Subtypes Within "Pallidal-Like" Axons from Each VP Compartment

	Pattern subtypes		
Origin of axons	Brainstem only	Thalamus only	Bifurcating axons
VPl (*n* = 14)	11/14 (80%)	2/14 (14%)	1/14 (7%)
VPm (*n* = 50)	32/50 (64%)	13/50 (26%)	5/50 (10%)
VPr (*n* = 20)	6/20 (30%)	7/20 (35%)	7/20 (35%)

while up to 26% of axons targeted the thalamus, and 10% innervated both. Finally in VPr almost equal to one-third of axons belonged to each subtype, with brainstem projecting axons being the smallest percentage of the three (30%).

3.9 VPL AS THE VENTRAL REPRESENTATIVE OF THE INDIRECT PATHWAY

As previously mentioned, 80% of VPl axons course caudally to innervate the brainstem (Table 3.1) and mainly collateralize in the LH, STh, SNr, and VTA: The VTA is the second most frequently targeted structure by these axons (73% of VPl axons) (Figure 3.5). This branching pattern strongly resembles that of GP axons that target the STh, SNr, and EP in different proportions (Bevan et al., 1998; Mallet et al., 2012). Thus, in parallel fashion to those GP axons, the striatal activation of Acb would disinhibit the STh via VPl axons, resulting in an increased inhibition of thalamic and brainstem targets. This represents a ventral homologue of the dorsal indirect pathway (Figure 3.6). At the same time the projections to STh remain largely segregated, the VP afferents being the most medial ones, adjacent to—and partially overlapping—the area of termination of GP afferents (Bevan et al., 1997). For convergence of VP and GP inputs at the cellular level in the STh, see Bevan et al. (1997).

A small proportion of VPl axons instead targeted the thalamus (Table 3.1), collateralizing in the rostral portion of the Rt and in 66% of cases also in the zona incerta (ZI) (Figure 3.6) (Mallet et al., 2012; Tripathi et al., 2013). Despite the small number of VPl axons traced, these findings are consistent with those obtained with transynaptic transport of pseudorabies virus. Those elegantly showed how transynaptic striatal projections to the MD originate largely from the AcbSh via VPm, whereas the sparse AcbC projections to the thalamus via the VPl terminate mainly on the Rt (O'Donnell et al., 1997). Rt neurons are GABAergic, and ZI neurons projecting specifically to the thalamus are also in part GABAergic (Bartho et al., 2002; Mitrofanis, 2005). As a result, the overall balance of ventral striatal activation via the VPl would be disinhibition of the rostral Rt and the subsequent inhibition of the MD, its main thalamic target (O'Donnell et al., 1997). These projections might also explain the excitation observed in 60% of MD cells after electrical stimulation of the VP (Lavin and Grace, 1994). Thus, thalamus-projecting VPl axons would also act as ventral representatives of the indirect pathway (Figure 3.6).

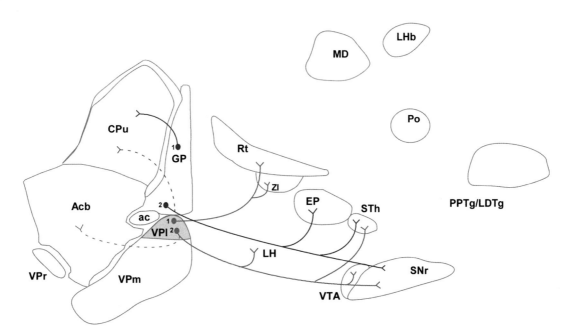

FIGURE 3.6 **Comparison between the basic arborization patterns of GP and VPl axons after single-axon tracing analysis.** Two different branching patterns have been found in the GP: projections to the striatum only (purple axon #1), or caudally directed projections to generally innervate the EP, STh, and SNr (purple axon #2) (Mallet et al., 2012). VPl axons in turn collateralize either in the thalamus, largely in Rt and ZI (orange axon #1), or predominantly in the brainstem, targeting the LH, STh, VTA, and SNr (orange axon #2) (Tripathi et al., 2013). Thus, the GP and VPl comprise a subset of neurons (axons #2) that show a parallel organization and may subserve a parallel function within parallel basal ganglia circuits. Dots represent the cell bodies of projection neurons. CPu, caudate putamen; EP, entopeduncular nucleus; GP, globus pallidus; LDTg, laterodorsal tegmental nucleus; Po, posterior thalamic nucleus; STh, subthalamic nucleus*brachium conjunctivum*; VPl, ventral pallidum, dorsolateral compartment; VPm, ventral pallidum, ventromedial compartment; VPr, ventral pallidum, rostral portion; ZI, zona incerta.

3.10 VPM AND VPR AS VENTRAL REPRESENTATIVES OF THE DIRECT PATHWAY

In contrast to VPl, a larger proportion of VPm and VPr axons target the thalamus (Table 3.1) terminating preferentially in the MD, which has direct projections to the prefrontal cortex (Groenewegen, 1988; Ray and Price, 1992). Thus, VPm and VPr thalamus-projecting axons, in marked difference with the VPl ones, directly act upon thalamocortical neurons, in parallel fashion to the EP within the dorsal basal ganglia circuits (Heimer, 1978). However, the VP projection is directed toward a higher order associative nucleus, instead of to the motor thalamus (VM and VA/VL complex), as in the case of the EP. In addition, all VPm axons traversing through the Rt provided a light innervation to this nucleus (Figure 3.7) (Mallet et al., 2012; Tripathi et al., 2013). This innervation to Rt provides the neural substrate for a potential sequential inhibition of MD neurons—via Rt-MD inhibitory projections—after the disinhibition rendered by ventral striatal activation. Similarly to EP axons, VPm and VPr axons frequently collateralized in the LHb (Figure 3.8) (Tripathi et al., 2013).

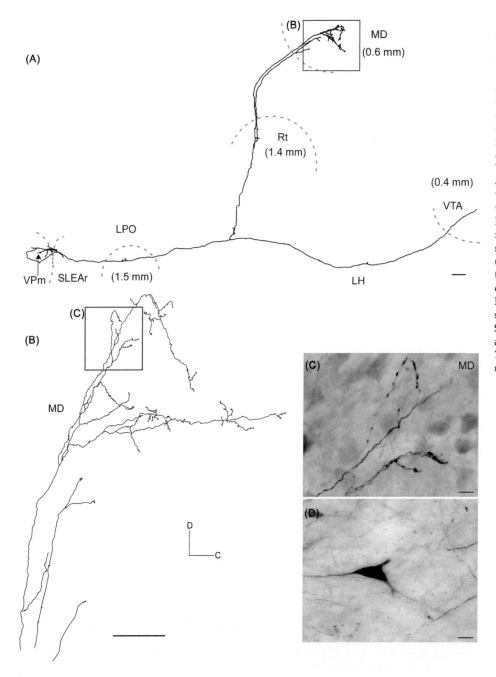

FIGURE 3.7 **Pattern of collateralization of a thalamus- and brainstem-projecting axon from VPm.** (A) Camera lucida reconstruction of the full trajectory and collaterals of an axon linked to a cell body in the VPm (arrow), shown at higher magnification in (D). Dashed arcs represent estimated boundaries as observed on sagittal sections, while numbers within parentheses indicate approximate mediolateral coordinates. This axon bifurcated in the rostral LH and gave off a dorsal branch that traversed the Rt, providing a light innervation, and then arborized profusely in the MD. The branch extending caudally terminated in the VTA. The boxed area is shown at higher magnification (blue) in (B). (B) Detail of the terminal arborization in the MD, which has been rotated counter-clockwise for convenience. Microphotograph of the boxed area is shown in (C) at higher magnification. Scale bars: (A), 200 μm; (B), 100 μm; (C) and (D), 10 μm. *Source: Reproduced from Tripathi et al. (2013) (Figure 3.8), with permission from Springer.*

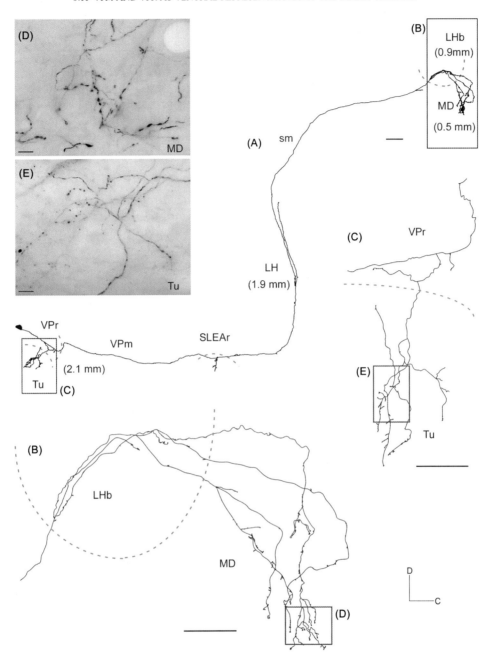

FIGURE 3.8 **Pattern of collateralization of a thalamus-projecting axon from VPr.** (A) Camera lucida reconstruction of the full trajectory and collaterals of an axon directly exiting from a tracer deposit in the VPr. Dashed arcs represent estimated boundaries as observed on sagittal sections, while numbers within parentheses indicate approximate mediolateral coordinates. This axon provided axon collaterals to the Tu and then extended caudally to innervate the SLEAr. At the rostral LH it turned dorsally to enter the sm and thus access the LHb, from where axon collaterals extended ventrally and innervated the MD: The boxed areas are shown at higher magnification in (B) (blue) and (C) (purple). (B) Detail of the terminal arborization in the LHb and MD. (C) Detail of the arborization in the Tu, which has been rotated counter-clockwise for convenience. (D) and (E) Microphotographs of the boxed areas in (B) and (C), respectively, at higher magnification. Tu, olfactory tubercle. Scale bars: (A), 200 mm; (B), 100 mm; (C) and (D), 10 mm. *Source: Reproduced from Tripathi et al. (2013) (Figure 3.9), with permission from Springer.*

A substantial proportion of VPm and VPr axons targeted the brainstem (Table 3.1), where several effector centers are directly innervated by these axons; namely, the VTA, PPTg, the retrorubral field (RRF), and PAG, among others (Tripathi et al., 2013). Thus, both thalamus-projecting and brainstem-projecting axons—including bifurcating axons—are in a position to control thalamocortical neurons and brainstem targets, acting as ventral representatives of the direct pathway.

3.11 POTENTIAL BRANCHING PATTERNS OF THE ENTOPEDUNCULAR PROJECTIONS

To our knowledge, there is only one report on single axons traced from the EP in the rat (*n* = 8 axons, Kha et al., 2000). These axons either targeted only the VM or collateralized also in the VL and PF, but seemed to be confined to the thalamus; in this sense, they could be regarded as "thalamus-projecting only" axons. They show a parallel pattern to that of VPm/VPr axons that reach the thalamus directly via the Rt (Figures 3.7 and 3.9, blue and magenta axons #2). Early retrograde labeling studies investigating collateral projections from the EP found both single and dually labeled neurons in the EP after tracer deposits in the ventral thalamic nuclei and the PPTg (van der Kooy and Carter, 1981).

These findings suggest that similarly to the patterns displayed by VPm and VPr axons, thalamus-projecting only, brainstem-projecting only, and bifurcating axons probably arise from the EP (Figure 3.9, magenta axons #2 and 3, respectively—brainstem-only axons have been omitted for clarity). In contrast, neurons targeting LHb were not dually labeled after injections in any of the other targets, apparently because these belong to a separate cell population located in the rostral two-thirds of the EP (van der Kooy and Carter, 1981). Thus, EP axons projecting to the LHb may display a branching pattern similar to that observed in VPm and VPr neurons, projecting predominantly to the LHb via the sm (Figures 3.8 and 3.9, blue and magenta axons #1). In the case of VPm and VPr,

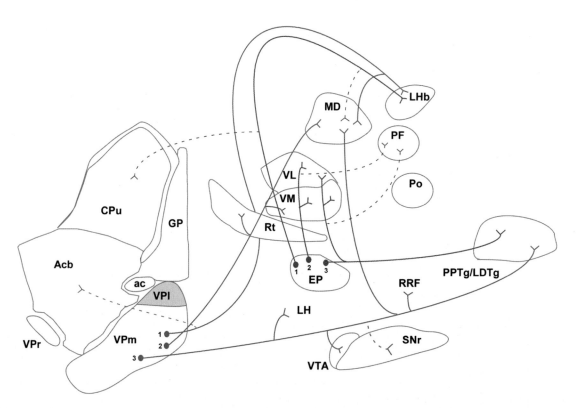

FIGURE 3.9 **Schematic diagram comparing the basic arborization patterns of EP and VPm axons after single-axon tracing studies.** VPm axons display three different branching patterns: one (light blue axon #1) targets the LHb via the sm with occasional collaterals to the Acb; a second one (light blue axon #2) targets the MD and/or the LHb, traversing Rt; and a third one (light blue axon #3) targets the brainstem (Tripathi et al., 2013). In relation to the EP, retrograde labeling studies have reported axons that via the *sm* target the LHb and the striatum. Thus, we propose a potential trajectory and branching pattern (magenta axon #1) for this axons, in parallel fashion to VPm axon #1. Kha et al. (2000) have traced axons from the EP targeting the thalamus only (i.e., magenta axon #2): these axons display a parallel trajectory and targets to, that is, VPm axon #2. Finally, retrograde tracing studies have reported that axons from the EP target simultaneously the PPTg and the thalamus (van der Kooy and Carter, 1981). Thus, we postulate a possible trajectory and branching pattern (magenta axon #3) for EP axons parallel to, that is, VPm axon #3. Axon collaterals depicted in discontinuous lines indicate a lower frequency of occurrence.

however, LHb-projecting axons frequently collateralize in MD as well, another limbic structure implicated in reward (Figure 3.9, blue axon #1). LHb-projecting axons from the primate GPi do not collateralize in the motor thalamus either, but occasionally do in the limbic-related anterior thalamic nuclei (Parent et al., 2001).

Neurons in the LHb discharge when an animal anticipates an aversive outcome or when an expected reward is not forthcoming (Ootsuka and Mohammed, 2015). Both the rostral—limbic—pole of the GPi/EP and the VP play a role in reward-related behavior, via their projections to the LHb (Geisler and Trimble, 2008; Hikosaka et al., 2008). However, a functional gradation is also observed here, as EP-LHb projections terminate in the lateral half of the LHb, or "pallidal" portion, with certain overlap into the medial—limbic—half (Herkenham and Nauta, 1979). VPm and VPl axon varicosities were present throughout LHb but were more concentrated in the medial half, and the terminal varicosities were all in the medial LHb.

3.12 POTENTIAL BRANCHING PATTERNS OF THE GP

Few studies have analyzed the axonal arborization patterns of the rat GP, and these have traced only a limited number of axons (Bevan et al., 1998; Kita and Kitai, 1994; Mallet et al., 2012). Two distinct cell subpopulations have been described according to their distinct branching patterns, neurochemical phenotypes, and electrophysiological properties: one, a subset of neurons regarded as "prototypical" that contain parvalbumin and collateralize in the STh, EP, and/or SNr (Abdi et al., 2015; Mallet et al., 2012), consistent with previous studies (Shammah-Lagnado et al., 1996). These axons are both trajectory- and target-wise, homologous to VPl brainstem-projecting axons (Figure 3.6, compare red and purple axons #2). Furthermore, these GP axons projecting downstream occasionally provided rostral axon collaterals to the striatum, in parallel fashion to VPl axons (Mallet et al., 2012; Tripathi et al., 2013).

The other subpopulation of GP neurons exclusively projects to the striatum in a profuse manner, they are parvalbumin-negative and have a different physiology (Abdi et al., 2015; Mallet et al., 2012). These two types of collateralization patterns have been also reported in the primate GP—or external GP—(GPe) (Sato et al., 2000). However, no ventral striatum-only projecting axons were observed in our sample in the rat (Tripathi et al., 2013).

Previous anterograde tracer studies reported that the rostral GP innervated the rostral Rt, MD, and occasionally the LHb (Gandia et al., 1993; Shammah-Lagnado et al., 1996). Retrograde tracer studies have also reported the presence of retrogradely labeled neurons in the GP after injections in the MD (Churchill et al., 1996; Young et al., 1984; Zahm et al., 1996). However, thus far there are no available data regarding the trajectory or collateralization patterns of these individual axons. The present analysis on VPm axons has revealed that a subset of them traversed the Rt to finally innervate the MD, while providing varicosities or collaterals to Rt on the way (Figure 3.7). Furthermore, the retrogradely labeled neurons described by Young and colleagues after MD injections were abundant in the VP and extended dorsally, being continuous with a scarcer population located in the GP (Young et al., 1984; Zahm et al., 1996). Therefore, we hypothesize that a few cells with collateralization patterns similar to those of VPm and VPr neurons extend dorsally into the GP with a similar function (Figure 3.10, green axon #1 and blue axon # 2); that is, being part of the direct pathway subpopulation within the GP.

Brain targets quite distinct to those classically innervated by the GP have been described for its caudal third; namely, the SNl, peripeduncular area, and terminal nucleus of the accessory optic system as well as PPTg (Jackson and Crossman, 1981; Moriizumi and Hattori, 1992). The caudal intralaminar nuclei and the posterior thalamic nuclei also seem to be innervated by the caudal GP, as well as the caudal Rt (Shammah-Lagnado et al., 1996). As the posterior thalamic nuclei are projecting to the auditory cortex, while the caudal Rt is also associated to this cortex, these authors have suggested that the caudal GP may contribute to the cortico-striatopallido-thalamo-cortical loop (Shammah-Lagnado et al., 1996), in like manner to the VPm. Therefore, based on the branching patterns observed in single axons from the VPm/VPr, we hypothesize that similar direct-pathway neurons may originate from the caudal GP and display a branching pattern similar to those of VPm axons; that is, either thalamus-projecting only or bifurcating axons that simultaneously target the caudal Rt, the posterior thalamic nucleus (Po), and the PPTg (Figure 3.10, green and blue axons #3).

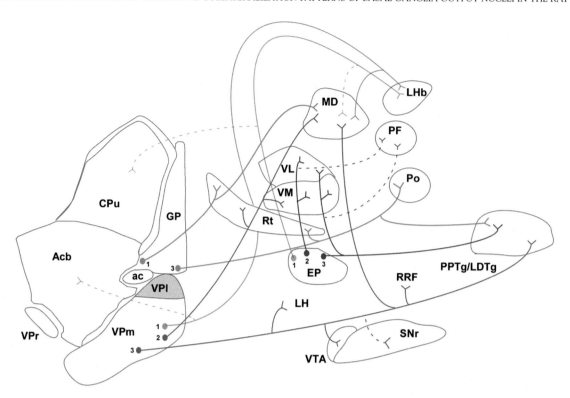

FIGURE 3.10   **Schematic diagram showing potential homologies between GP axons and both VPm and EP axons.** Anterograde and retrograde tracing studies have reported that GP axons innervate Rt and MD, as VPm does. Thus, we propose that a subset of GP axons (green axon #1) displays a similar branching pattern to VPm axons (light blue axon #2), in parallel to that of some EP axons (i.e., magenta axon #2). Likewise, neurons in the caudal GP innervate the posterior thalamic nucleus, PPTg, and the Rt. In view of the reported EP and VPm branching patterns (light blue and magenta axons #3), we propose a potential trajectory and branching pattern for caudal GP axons (green axon #3) resembling that of VPm and EP axons, that may subserve a similar function, that is, modulating the thalamus and cortex. This would contrast with the "classical indirect pathway" GP axons. Whether a common embryological origin gives rise to these three similar cell subpopulations is an open question that would be interesting to address.

3.13 CONCLUSIONS

Studies of axonal collateralization patterns provide a wealth of information about brain architecture at the single axon level. How important each and all of these branches are in terms of function needs to be addressed with other techniques. Furthermore, how these individual and highly diverse mini-networks become "coordinated" as cell populations that may fire rhythmically is a fascinating issue that requires further study.

References

Abdi, A., Mallet, N., Mohamed, F.Y., Sharott, A., Dodson, P.D., Nakamura, K.C., et al., 2015. Prototypic and arkypallidal neurons in the dopamine-intact external globus pallidus. J. Neurosci. 35, 6667–6688.

Alexander, G.E., DeLong, M.R., Strick, P.L., 1986. Parallel organization of functionally segregated circuits linking basal ganglia and cortex. Annu. Rev. Neurosci. 9, 357–381.

Aransay, A., Rodriguez-Lopez, C., Garcia-Amado, M., Clasca, F., Prensa, L., 2015. Long-range projection neurons of the mouse ventral tegmental area: a single-cell axon tracing analysis. Front. Neuroanat. 9, 59. http://www.ncbi.nlm.nih.gov/pubmed/26042000.

Bartho, P., Freund, T.F., Acsady, L., 2002. Selective GABAergic innervation of thalamic nuclei from zona incerta. Eur. J. Neurosci. 16, 999–1014.

Basso, M.A., Wurtz, R.H., 2002. Neuronal activity in substantia nigra pars reticulata during target selection. J. Neurosci. 22, 1883–1894.

Beckstead, R.M., Domesick, V.B., Nauta, W.J., 1979. Efferent connections of the substantia nigra and ventral tegmental area in the rat. Brain Res. 175, 191–217.

Beckstead, R.M., Edwards, S.B., Frankfurter, A., 1981. A comparison of the intranigral distribution of nigrotectal neurons labeled with horseradish peroxidase in the monkey, cat, and rat. J. Neurosci. 1, 121–125.

Beitz, A.J., 1982. The organization of afferent projections to the midbrain periaqueductal gray of the rat. Neuroscience 7, 133–159.

Bentivoglio, M., van der Kooy, D., Kuypers, H.G., 1979. The organization of the efferent projections of the substantia nigra in the rat. A retrograde fluorescent double labeling study. Brain Res. 174, 1–17.

Bevan, M.D., Clarke, N.P., Bolam, J.P., 1997. Synaptic integration of functionally diverse pallidal information in the entopeduncular nucleus and subthalamic nucleus in the rat. J. Neurosci. 17, 308–324.

Bevan, M.D., Booth, P.A., Eaton, S.A., Bolam, J.P., 1998. Selective innervation of neostriatal interneurons by a subclass of neuron in the globus pallidus of the rat. J. Neurosci. 18, 9438–9452.

Bickford, M.E., Hall, W.C., 1992. The nigral projection to predorsal bundle cells in the superior colliculus of the rat. J. Comp. Neurol. 319, 11–33.

Buttner, U., Buttner-Ennever, J.A., 1988. Present concepts of oculomotor organization. Rev. Oculomot. Res. 2, 3–32.

Carter, D.A., Fibiger, H.C., 1978. The projections of the entopeduncular nucleus and globus pallidus in rat as demonstrated by autoradiography and horseradish peroxidase histochemistry. J. Comp. Neurol. 177, 113–123.

Cebrián, C., Prensa, L., 2010. Basal ganglia and thalamic input from neurons located within the ventral tier cell cluster region of the substantia nigra pars compacta in the rat. J. Comp. Neurol. 518, 1283–1300.

Cebrián, C., Parent, A., Prensa, L., 2005. Patterns of axonal branching of neurons of the substantia nigra pars reticulata and pars lateralis in the rat. J. Comp. Neurol. 492, 349–369.

Chevalier, G., Thierry, A.M., Shibazaki, T., Feger, J., 1981. Evidence for a GABAergic inhibitory nigrotectal pathway in the rat. Neurosci. Lett. 21, 67–70.

Chevalier, G., Vacher, S., Deniau, J.M., Desban, M., 1985. Disinhibition as a basic process in the expression of striatal functions. I. The striato-nigral influence on tecto-spinal/tecto-diencephalic neurons. Brain Res. 334, 215–226.

Churchill, L., Zahm, D.S., Kalivas, P.W., 1996. The mediodorsal nucleus of the thalamus in rats—I. Forebrain gabaergic innervation. Neuroscience 70, 93–102.

Coimbra, N.C., De Oliveira, R., Freitas, R.L., Ribeiro, S.J., Borelli, K.G., Pacagnella, R.C., et al., 2006. Neuroanatomical approaches of the tectum-reticular pathways and immunohistochemical evidence for serotonin-positive perikarya on neuronal substrates of the superior colliculus and periaqueductal gray matter involved in the elaboration of the defensive behavior and fear-induced analgesia. Exp. Neurol. 197, 93–112.

Cui, Z., Gerfen, C.R., Young III, W.S., 2013. Hypothalamic and other connections with dorsal CA2 area of the mouse hippocampus. J. Comp. Neurol. 521, 1844–1866.

DeLong, M.R., Wichmann, T., 2007. Circuits and circuit disorders of the basal ganglia. Arch. Neurol. 64, 20–24.

Deniau, J.M., Chevalier, G., 1992. The lamellar organization of the rat substantia nigra pars reticulata: distribution of projection neurons. Neuroscience 46, 361–377.

Faull, R.L., Mehler, W.R., 1978. The cells of origin of nigrotectal, nigrothalamic and nigrostriatal projections in the rat. Neuroscience 3, 989–1002.

Fink-Jensen, A., Mikkelsen, J.D., 1989. The striato-entopeduncular pathway in the rat. A retrograde transport study with wheatgerm-agglutinin-horseradish peroxidase. Brain Res. 476, 194–198.

Fujiyama, F., Sohn, J., Nakano, T., Furuta, T., Nakamura, K.C., Matsuda, W., et al., 2011. Exclusive and common targets of neostriatofugal projections of rat striosome neurons: a single neuron-tracing study using a viral vector. Eur. J. Neurosci. 33, 668–677.

Gabbott, P.L., Warner, T.A., Jays, P.R., Salway, P., Busby, S.J., 2005. Prefrontal cortex in the rat: projections to subcortical autonomic, motor, and limbic centers. J. Comp. Neurol. 492, 145–177.

Gandia, J.A., De Las Heras, S., Garcia, M., Gimenez-Amaya, J.M., 1993. Afferent projections to the reticular thalamic nucleus from the globus pallidus and the substantia nigra in the rat. Brain Res. Bull. 32, 351–358.

Gao, H.R., Shi, T.F., Yang, C.X., Zhang, G.W., Zhang, D., Jiao, R.S., et al., 2012. Dopamine involved in the nociceptive modulation in the parafascicular nucleus of morphine-dependent rat. Neurochem. Res. 37, 428–435.

Geisler, S., Trimble, M., 2008. The lateral habenula: no longer neglected. CNS Spectr. 13, 484–489.

Gerfen, C.R., 1985. The neostriatal mosaic. I. Compartmental organization of projections from the striatum to the substantia nigra in the rat. J. Comp. Neurol. 236, 454–476.

Gerfen, C.R., Staines, W.A., Arbuthnott, G.W., Fibiger, H.C., 1982. Crossed connections of the substantia nigra in the rat. J. Comp. Neurol. 207, 283–303.

Granata, A.R., Kitai, S.T., 1991. Inhibitory substantia nigra inputs to the pedunculopontine neurons. Exp. Brain Res. 86, 459–466.

Groenewegen, H.J., 1988. Organization of the afferent connections of the mediodorsal thalamic nucleus in the rat, related to the mediodorsal-prefrontal topography. Neuroscience 24, 379–431.

Groenewegen, H.J., Berendse, H.W., Haber, S.N., 1993. Organization of the output of the ventral striatopallidal system in the rat: ventral pallidal efferents. Neuroscience 57, 113–142.

Grofova, I., Zhou, M., 1998. Nigral innervation of cholinergic and glutamatergic cells in the rat mesopontine tegmentum: light and electron microscopic anterograde tracing and immunohistochemical studies. J. Comp. Neurol. 395, 359–379.

Guethe, L.M., Pelegrini-da-Silva, A., Borelli, K.G., Juliano, M.A., Pelosi, G.G., Pesquero, J.B., et al., 2013. Angiotensin (5-8) modulates nociception at the rat periaqueductal gray via the NO-sGC pathway and an endogenous opioid. Neuroscience 231, 315–327.

Gulcebi, M.I., Ketenci, S., Linke, R., Hacioglu, H., Yanali, H., Veliskova, J., et al., 2012. Topographical connections of the substantia nigra pars reticulata to higher-order thalamic nuclei in the rat. Brain Res. Bull. 87, 312–318.

Haber, S., Gdowski, M.J., 2004. The basal ganglia. In: Paxinos, G., Mai, J.K. (Eds.), The Human Nervous System, second ed. Elsevier Academic Press, San Diego, CA, pp. 676–738.

Haber, S., McFarland, N.R., 2001. The place of the thalamus in frontal cortical-basal ganglia circuits. Neuroscientist 7, 315–324.

Haber, S.N., Knutson, B., 2010. The reward circuit: linking primate anatomy and human imaging. Neuropsychopharmacology 35, 4–26.

Haber, S.N., Groenewegen, H.J., Grove, E.A., Nauta, W.J., 1985. Efferent connections of the ventral pallidum: evidence of a dual striato pallidofugal pathway. J. Comp. Neurol. 235, 322–335.

Hall, W.C., Fitzpatrick, D., Klatt, L.L., Raczkowski, D., 1989. Cholinergic innervation of the superior colliculus in the cat. J. Comp. Neurol. 287, 495–514.

Harting, J.K., Van Lieshout, D.P., 1991. Spatial relationships of axons arising from the substantia nigra, spinal trigeminal nucleus, and pedunculopontine tegmental nucleus within the intermediate gray of the cat superior colliculus. J. Comp. Neurol. 305, 543–558.

Heimer, L. (Ed.), 1978. The Olfactory Cortex and the Ventral Striatum Plenum Press, Toronto, Canada. 95–188 p.

Heimer, L., Zaborszky, L., Zahm, D.S., Alheid, G.F., 1987. The ventral striatopallidothalamic projection: I. The striatopallidal link originating in the striatal parts of the olfactory tubercle. J. Comp. Neurol. 255, 571–591.

Heimer, L., Zahm, D.S., Churchill, L., Kalivas, P.W., Wohltmann, C., 1991. Specificity in the projection patterns of accumbal core and shell in the rat. Neuroscience 41, 89–125.

Herkenham, M., Nauta, W.J., 1977. Afferent connections of the habenular nuclei in the rat. A horseradish peroxidase study, with a note on the fiber-of-passage problem. J. Comp. Neurol. 173, 123–146.

Herkenham, M., Nauta, W.J., 1979. Efferent connections of the habenular nuclei in the rat. J. Comp. Neurol. 187, 19–47.

Hikosaka, O., Wurtz, R.H., 1985. Modification of saccadic eye movements by GABA-related substances. II. Effects of muscimol in monkey substantia nigra pars reticulata. J Neurophysiol. 53, 292–308.

Hikosaka, O., Takikawa, Y., Kawagoe, R., 2000. Role of the basal ganglia in the control of purposive saccadic eye movements. Physiol. Rev. 80, 953–978.

Hikosaka, O., Sesack, S.R., Lecourtier, L., Shepard, P.D., 2008. Habenula: crossroad between the basal ganglia and the limbic system. J. Neurosci. 28, 11825–11829.

Izzo, P.N., 1991. A note on the use of biocytin in anterograde tracing studies in the central nervous system: application at both light and electron microscopic level. J. Neurosci. Methods 36, 155–166.

Jackson, A., Crossman, A.R., 1981. Basal ganglia and other afferent projections to the peribrachial region in the rat: a study using retrograde and anterograde transport of horseradish peroxidase. Neuroscience 6, 1537–1549.

Joel, D., Weiner, I., 1994. The organization of the basal ganglia-thalamocortical circuits: open interconnected rather than closed segregated. Neuroscience 63, 363–379.

Kalivas, P.W., Churchill, L., Romanides, A., 1999. Involvement of the pallidal-thalamocortical circuit in adaptive behavior. Ann. N. Y. Acad. Sci. 877, 64–70.

Kang, Y., Kitai, S.T., 1990. Electrophysiological properties of pedunculopontine neurons and their postsynaptic responses following stimulation of substantia nigra reticulata. Brain Res. 535, 79–95.

Karabelas, A.B., Moschovakis, A.K., 1985. Nigral inhibitory termination on efferent neurons of the superior colliculus: an intracellular horseradish peroxidase study in the cat. J. Comp. Neurol. 239, 309–329.

Kawaguchi, Y., Wilson, C.J., Emson, P.C., 1990. Projection subtypes of rat neostriatal matrix cells revealed by intracellular injection of biocytin. J. Neurosci. 10, 3421–3438.

Kelley, A.E., Domesick, V.B., Nauta, W.J., 1982. The amygdalostriatal projection in the rat—an anatomical study by anterograde and retrograde tracing methods. Neuroscience 7, 615–630.

Kelly, R.M., Strick, P.L., 2004. Macro-architecture of basal ganglia loops with the cerebral cortex: use of rabies virus to reveal multisynaptic circuits. Prog. Brain Res. 143, 449–459.

Kha, H.T., Finkelstein, D.I., Pow, D.V., Lawrence, A.J., Horne, M.K., 2000. Study of projections from the entopeduncular nucleus to the thalamus of the rat. J. Comp. Neurol. 426, 366–377.

Kha, H.T., Finkelstein, D.I., Tomas, D., Drago, J., Pow, D.V., Horne, M.K., 2001. Projections from the substantia nigra pars reticulata to the motor thalamus of the rat: single axon reconstructions and immunohistochemical study. J. Comp. Neurol. 440, 20–30.

Kincheski, G.C., Mota-Ortiz, S.R., Pavesi, E., Canteras, N.S., Carobrez, A.P., 2012. The dorsolateral periaqueductal gray and its role in mediating fear learning to life threatening events. PLoS One. 7, e50361.

Kirouac, G.J., Li, S., Mabrouk, G., 2004. GABAergic projection from the ventral tegmental area and substantia nigra to the periaqueductal gray region and the dorsal raphe nucleus. J. Comp. Neurol. 469, 170–184.

Kita, H., Armstrong, W., 1991. A biotin-containing compound N-(2-aminoethyl)biotinamide for intracellular labeling and neuronal tracing studies: comparison with biocytin. J. Neurosci. Methods 37, 141–150.

Kita, H., Kitai, S.T., 1994. The morphology of globus pallidus projection neurons in the rat: an intracellular staining study. Brain Res. 636, 308–319.

Kita, T., Kita, H., 2012. The subthalamic nucleus is one of multiple innervation sites for long-range corticofugal axons: a single-axon tracing study in the rat. J. Neurosci. 32, 5990–5999.

Krauthamer, G.M., Grunwerg, B.S., Krein, H., 1995. Putative cholinergic neurons of the pedunculopontine tegmental nucleus projecting to the superior colliculus consist of sensory responsive and unresponsive populations which are functionally distinct from other mesopontine neurons. Neuroscience 69, 507–517.

Krout, K.E., Loewy, A.D., 2000. Periaqueductal gray matter projections to midline and intralaminar thalamic nuclei of the rat. J. Comp. Neurol. 424, 111–141.

Krout, K.E., Loewy, A.D., Westby, G.W., Redgrave, P., 2001. Superior colliculus projections to midline and intralaminar thalamic nuclei of the rat. J. Comp. Neurol. 431, 198–216.

Lapper, S.R., Bolam, J.P., 1991. The anterograde and retrograde transport of neurobiotin in the central nervous system of the rat: comparison with biocytin. J. Neurosci. Methods 39, 163–174.

Lavin, A., Grace, A.A., 1994. Modulation of dorsal thalamic cell activity by the ventral pallidum: its role in the regulation of thalamocortical activity by the basal ganglia. Synapse 18, 104–127.

Lee, M.S., Rinne, J.O., Marsden, C.D., 2000. The pedunculopontine nucleus: its role in the genesis of movement disorders. Yonsei Med. J. 41, 167–184.

Lei, J., Sun, T., Lumb, B.M., You, H.J., 2014. Roles of the periaqueductal gray in descending facilitatory and inhibitory controls of intramuscular hypertonic saline induced muscle nociception. Exp. Neurol. 257, 88–94.

Liu, P., Basso, M.A., 2008. Substantia nigra stimulation influences monkey superior colliculus neuronal activity bilaterally. J. Neurophysiol. 100, 1098–1112.

Lu, X.Y., Ghasemzadeh, M.B., Kalivas, P.W., 1998. Expression of D1 receptor, D2 receptor, substance P and enkephalin messenger RNAs in the neurons projecting from the nucleus accumbens. Neuroscience 82, 767–780.

Luiten, P.G., Gaykema, R.P., Traber, J., Spencer Jr., D.G., 1987. Cortical projection patterns of magnocellular basal nucleus subdivisions as revealed by anterogradely transported *Phaseolus vulgaris* leucoagglutinin. Brain Res. 413, 229–250.

Mallet, N., Micklem, B.R., Henny, P., Brown, M.T., Williams, C., Bolam, J.P., et al., 2012. Dichotomous organization of the external globus pallidus. Neuron 74, 1075–1086.

Marchand, J.E., Hagino, N., 1983. Afferents to the periaqueductal gray in the rat. A horseradish peroxidase study. Neuroscience 9, 95–106.

Matsuda, W., Furuta, T., Nakamura, K.C., Hioki, H., Fujiyama, F., Arai, R., et al., 2009. Single nigrostriatal dopaminergic neurons form widely spread and highly dense axonal arborizations in the neostriatum. J. Neurosci. 29, 444–453.

McFarland, N.R., Haber, S.N., 2002. Thalamic relay nuclei of the basal ganglia form both reciprocal and nonreciprocal cortical connections, linking multiple frontal cortical areas. J. Neurosci. 22, 8117–8132.

McGinty, V.B., Hayden, B.Y., Heilbronner, S.R., Dumont, E.C., Graves, S.M., Mirrione, M.M., et al., 2011. Emerging, reemerging, and forgotten brain areas of the reward circuit: notes from the 2010 Motivational Neural Networks conference. Behav. Brain Res. 225, 348–357.

McHaffie, J.G., Stanford, T.R., Stein, B.E., Coizet, V., Redgrave, P., 2005. Subcortical loops through the basal ganglia. Trends Neurosci. 28, 401–407.

Mitrofanis, J., 2005. Some certainty for the "zone of uncertainty"? Exploring the function of the zona incerta. Neuroscience 130, 1–15.

Mongia, S., Luquin, E., Mengual, E. 2015. Quantitative study of pallidotegmental projections contacting cholinergic, calbindin-, and calretinin-immunoreactive neurons in the rat pedunculopontine and laterodorsal tegmental nuclei. 2015 Neuroscience Meeting Planner. Chicago, IL: Society for Neuroscience, Online.

Moriizumi, T., Hattori, T., 1992. Separate neuronal populations of the rat globus pallidus projecting to the subthalamic nucleus, auditory cortex and pedunculopontine tegmental area. Neuroscience 46, 701–710.

Murphy, A.Z., Shupnik, M.A., Hoffman, G.E., 1999. Androgen and estrogen (alpha) receptor distribution in the periaqueductal gray of the male rat. Horm. Behav. 36, 98–108.

Nauta, H.J., 1979. Projections of the pallidal complex: an autoradiographic study in the cat. Neuroscience 4, 1853–1873.

Noda, T., Oka, H., 1984. Nigral inputs to the pedunculopontine region: intracellular analysis. Brain Res. 322, 332–336.

O'Donnell, P., Lavin, A., Enquist, L.W., Grace, A.A., Card, J.P., 1997. Interconnected parallel circuits between rat nucleus accumbens and thalamus revealed by retrograde transsynaptic transport of pseudorabies virus. J. Neurosci. 17, 2143–2167.

Oorschot, D.E., 1996. Total number of neurons in the neostriatal, pallidal, subthalamic, and substantia nigral nuclei of the rat basal ganglia: a stereological study using the cavalieri and optical disector methods. J. Comp. Neurol. 366, 580–599.

Ootsuka, Y., Mohammed, M., 2015. Activation of the habenula complex evokes autonomic physiological responses similar to those associated with emotional stress. Physiol. Rep., 3.

Parent, A., Sato, F., Wu, Y., Gauthier, J., Levesque, M., Parent, M., 2000. Organization of the basal ganglia: the importance of axonal collateralization. Trends Neurosci. 23, S20–S27.

Parent, M., Parent, A., 2007. The microcircuitry of primate subthalamic nucleus. Parkinsonism Relat. Disord. 13 (Suppl 3), S292–S295.

Parent, M., Levesque, M., Parent, A., 2001. Two types of projection neurons in the internal pallidum of primates: single-axon tracing and three-dimensional reconstruction. J. Comp. Neurol. 439, 162–175.

Paxinos, G., Watson, C. 1998. The Rat Brain in Stereotaxic Coordinates, fourth ed. San Diego, CA.

Pitkanen, A., Savander, V., LeDoux, J.E., 1997. Organization of intra-amygdaloid circuitries in the rat: an emerging framework for understanding functions of the amygdala. Trends Neurosci. 20, 517–523.

Prensa, L., Parent, A., 2001. The nigrostriatal pathway in the rat: a single-axon study of the relationship between dorsal and ventral tier nigral neurons and the striosome/matrix striatal compartments. J. Neurosci. 21, 7247–7260.

Ray, J.P., Price, J.L., 1992. The organization of the thalamocortical connections of the mediodorsal thalamic nucleus in the rat, related to the ventral forebrain-prefrontal cortex topography. J. Comp. Neurol. 323, 167–197.

Redgrave, P., 2007. Basal ganglia. Scholarpedia 2, 1825. http://dx.doi.org/10.4249/scholarpedia.1825.

Redgrave, P., Marrow, L., Dean, P., 1992. Topographical organization of the nigrotectal projection in rat: evidence for segregated channels. Neuroscience 50, 571–595.

Reiner, A., Veenman, C.L., Medina, L., Jiao, Y., Del Mar, N., Honig, M.G., 2000. Pathway tracing using biotinylated dextran amines. J. Neurosci. Methods 103, 23–37.

Rodriguez, M., Abdala, P., Barroso-Chinea, P., Gonzalez-Hernandez, T., 2001. The deep mesencephalic nucleus as an output center of basal ganglia: morphological and electrophysiological similarities with the substantia nigra. J. Comp. Neurol. 438, 12–31.

Root, D.H., Melendez, R.I., Zaborszky, L., Napier, T.C., 2015. The ventral pallidum: subregion-specific functional anatomy and roles in motivated behaviors. Prog. Neurobiol. 2015 Apr 6. pii: S0301-0082(15)00027-1. http://dx.doi.ogr/10.1016/j.pneurobio.2015.03.005.

Russchen, F.T., Price, J.L., 1984. Amygdalostriatal projections in the rat. Topographical organization and fiber morphology shown using the lectin PHA-L as an anterograde tracer. Neurosci. Lett. 47, 15–22.

Sakai, S.T., Grofova, I., Bruce, K., 1998. Nigrothalamic projections and nigrothalamocortical pathway to the medial agranular cortex in the rat: single- and double-labeling light and electron microscopic studies. J. Comp. Neurol. 391, 506–525.

Sato, F., Lavallee, P., Levesque, M., Parent, A., 2000. Single-axon tracing study of neurons of the external segment of the globus pallidus in primate. J. Comp. Neurol. 417, 17–31.

Schmued, L., Phermsangngam, P., Lee, H., Thio, S., Chen, E., et al., 1989. Collateralization and GAD immunoreactivity of descending pallidal efferents. Brain Res. 487, 131–142.

Schmued, L., Kyriakidis, K., Heimer, L., 1990. *In vivo* anterograde and retrograde axonal transport of the fluorescent rhodamine-dextran-amine, Fluoro-Ruby, within the CNS. Brain Res. 526, 127–134.

Sesack, S.R., Grace, A.A., 2010. Cortico-basal ganglia reward network: microcircuitry. Neuropsychopharmacology 35, 27–47.

Shammah-Lagnado, S.J., Alheid, G.F., Heimer, L., 1996. Efferent connections of the caudal part of the globus pallidus in the rat. J. Comp. Neurol. 376, 489–507.

Smith, Y., Bevan, M.D., Shink, E., Bolam, J.P., 1998. Microcircuitry of the direct and indirect pathways of the basal ganglia. Neuroscience 86, 353–387.

Spann, B.M., Grofova, I., 1991. Nigropedunculopontine projection in the rat: an anterograde tracing study with phaseolus vulgaris-leucoagglutinin (PHA-L). J. Comp. Neurol. 311, 375–388.

Steininger, T.L., Rye, D.B., Wainer, B.H., 1992. Afferent projections to the cholinergic pedunculopontine tegmental nucleus and adjacent midbrain extrapyramidal area in the albino rat. I. Retrograde tracing studies. J. Comp. Neurol. 321, 515–543.

Takakusaki, K., 2008. Forebrain control of locomotor behaviors. Brain Res. Rev. 57, 192–198.

Takakusaki, K., Shiroyama, T., Kitai, S.T., 1997. Two types of cholinergic neurons in the rat tegmental pedunculopontine nucleus: electrophysiological and morphological characterization. Neuroscience 79, 1089–1109.

Takakusaki, K., Obara, K., Nozu, T., Okumura, T., 2011. Modulatory effects of the GABAergic basal ganglia neurons on the PPN and the muscle tone inhibitory system in cats. Arch. Ital. Biol. 149, 385–405.

Tripathi A., Mengual E., 2012. Axonal collateralization patterns of the rostral sublenticular extended amygdala in the rat. 8th FENS Forum of Neuroscience. Barcelona, Spain.

Tripathi, A., Prensa, L., Cebrián, C., Mengual, E., 2010. Axonal branching patterns of nucleus accumbens neurons in the rat. J. Comp. Neurol. 518, 4649–4673.

Tripathi, A., Prensa, L., Mengual, E., 2013. Axonal branching patterns of ventral pallidal neurons in the rat. Brain Struct. Funct. 218, 1133–1157.

Usuda, I., Tanaka, K., Chiba, T., 1998. Efferent projections of the nucleus accumbens in the rat with special reference to subdivision of the nucleus: biotinylated dextran amine study. Brain Res. 797, 73–93.

Van der Werf, Y.D., Witter, M.P., Groenewegen, H.J., 2002. The intralaminar and midline nuclei of the thalamus. Anatomical and functional evidence for participation in processes of arousal and awareness. Brain Res. Brain Res. Rev. 39, 107–140.

Vertes, R.P., 2004. Differential projections of the infralimbic and prelimbic cortex in the rat. Synapse 51, 32–58.

Wang, J.P., Guo, Z., 2009. Propofol suppresses activation of the nociception specific neuron in the parafascicular nucleus of the thalamus evoked by coronary artery occlusion in rats. Eur. J. Anaesthesiol. 26, 60–65.

Watt, D.F., 2000. The centrecephalon and thalamocortical integration: neglected contributions of periaqueductal gray. Conscious. Emot. 1, 91–114.

Williams, M.N., Faull, R.L., 1985. The striatonigral projection and nigrotectal neurons in the rat. A correlated light and electron microscopic study demonstrating a monosynaptic striatal input to identified nigrotectal neurons using a combined degeneration and horseradish peroxidase procedure. Neuroscience 14, 991–1010.

Winn, P., 2006. How best to consider the structure and function of the pedunculopontine tegmental nucleus: evidence from animal studies. J. Neurol. Sci. 248, 234–250.

Woolf, N.J., Eckenstein, F., Butcher, L.L., 1984. Cholinergic systems in the rat brain: I. Projections to the limbic telencephalon. Brain Res. Bull. 13, 751–784.

Wouterlood, F.G., Goede, P.H., Groenewegen, H.J., 1990. The *in situ* detectability of the neuroanatomical tracer *Phaseolus vulgaris*-leucoagglutinin (PHA-L). J. Chem. Neuroanat. 3, 11–18.

Wu, Y., Richard, S., Parent, A., 2000. The organization of the striatal output system: a single-cell juxtacellular labeling study in the rat. Neurosci. Res. 38, 49–62.

Yasui, Y., Tsumori, T., Ando, A., Domoto, T., 1995. Demonstration of axon collateral projections from the substantia nigra pars reticulata to the superior colliculus and the parvicellular reticular formation in the rat. Brain Res. 674, 122–126.

Young III, W.S., Alheid, G.F., Heimer, L., 1984. The ventral pallidal projection to the mediodorsal thalamus: a study with fluorescent retrograde tracers and immunohistofluorescence. J. Neurosci. 4, 1626–1638.

Zaborszky, L., Alheid, G.F., Beinfeld, M.C., Eiden, L.E., Heimer, L., Palkovits, M., 1985. Cholecystokinin innervation of the ventral striatum: a morphological and radioimmunological study. Neuroscience 14, 427–453.

Zahm, D.S., 1989. The ventral striatopallidal parts of the basal ganglia in the rat—II. Compartmentation of ventral pallidal efferents. Neuroscience 30, 33–50.

Zahm, D.S., Heimer, L., 1993. Specificity in the efferent projections of the nucleus-accumbens in the rat—comparison of the rostral pole projection patterns with those of the core and shell. J. Comp. Neurol. 327, 220–232.

Zahm, D.S., Williams, E., Wohltmann, C., 1996. Ventral striatopallidothalamic projection: IV. Relative involvements of neurochemically distinct subterritories in the ventral pallidum and adjacent parts of the rostroventral forebrain. J. Comp. Neurol. 364, 340–362.

van der Kooy, D., Carter, D.A., 1981. The organization of the efferent projections and striatal afferents of the entopeduncular nucleus and adjacent areas in the rat. Brain Res. 211, 15–36.

4

Anatomy and Development of Multispecific Thalamocortical Axons: Implications for Cortical Dynamics and Evolution

Francisco Clascá, César Porrero, Maria José Galazo,
Pablo Rubio-Garrido and Marian Evangelio

Department of Anatomy & Graduate Program in Neuroscience, School of Medicine,
Autónoma de Madrid University, Madrid, Spain

In every cortical area, the input conveyed by thalamocortical (TC) axons determines, to a large extent, both the content and dynamics of information flow. Some TC axons act as entry routes for sensory, including viscerosensory, information into the cerebral cortex. Other TC axons feed back to cortex the output signals from multisynaptic loops in the cerebellum and the basal ganglia that are required for volitional movement goal setting, coordination, and adjustment. Still other TC axons control cortical arousal, or may regulate communication between particular sets of cortical areas. Precise knowledge of TC circuits, therefore, is crucial for modeling normal and diseased brain function.

Despite decades of research, the intricacy of TC circuits made it difficult to draw a clear and comprehensive picture of their organizational principles. In part, this owes to the fact that standard connection-tracing methods that work well for simple and topographically arranged pathways yield results that are ambiguous, or even confusing when applied to branched, intermingled, and/or overlapping axonal architectures. A factor further complicating their analysis is that at least some TC cells simultaneously target, in addition to the cerebral cortex, subcortical structures such as the striatum, amygdala, or claustrum.

New techniques that allow the selective labeling of complete single axonal arbors are finally opening the possibility of directly observing with cellular resolution the actual structure of long-range projection pathways, no matter how long and complex. Studies carried out with these techniques are revealing an unsuspected diversity and specificity of TC axon patterns that challenge earlier views and suggest some intriguing functional possibilities.

Below, we first sum up current models of the TC systems and the experimental evidence on which they were based. Then, we review observations from our group and others carried out with single-axon tracing methods, with special focus on what we call here "multispecific" axons; that is, axons that branch to innervate restricted domains, often with different laminar distributions, in separate cortical areas and often in subcortical structures as well. Subsequently, we summarize our observations on the developmental mechanisms that give rise to the multibranched TC architectures. Finally, we consider some testable hypotheses about how these complex axons may impact the dynamics, plasticity, and evolution of cortical networks.

69

4.1 THALAMOFUGAL AXON ARCHITECTURES AS REVEALED BY BULK-TRACING METHODS

Gene expression and fate mapping studies have shown that mammalian projection neurons in all dorsal thalamus nuclei arise in development from the same domain of the alar neuropithelium of a single prosencephalic segment (P2, Puelles and Rubenstein, 2003). Depending on the species, projection cells constitute the only neuronal type or the vast majority of neurons in the thalamus. Reflecting this close clonal origin, adult projection cells across the thalamus are fundamentally similar. They have almost identical gene expression profiles (Murray et al., 2007; Nakagawa and Shimogori, 2012) phenotypic features such as glutamatergic neurotransmission, ion channel expression profile, and general somatodendritic morphology. Even their axon circuit architectures can be seen as basically similar: they extend only ipsilaterally and innervate both prethalamic reticular nucleus (RPtN) and one or more of several alar telencephalic structures; including the cerebral cortex, striatum, lateral amygdala, and/or claustrum.

Upon this basic similarity, several interlinked factors contribute to create phenotypic diversity among adult projection neurons. Neurogenetic gradients and diffusible signals from patterning centers such as the adjacent zona limitans intrathalamica may lay down a first crude histogenetic pattern within the P2 alar neuropithelium (Altman and Bayer, 1988; Jones, 2007; Nakagawa and O'Leary, 2001; Nakagawa and Shimogori, 2012; Suzuki-Hirano et al., 2011). Subsequently, afferent axonal systems that invade at an early stage the developing thalamus, first from subcortical structures and later from the cerebral cortex (Clascá et al., 1995) distributing selectively within it, may provide additional patterning via secreted or membrane-bound signals. In particular, the subcortical afferents that invade the thalamus at a very early embryonic age have been shown to control the survival and specification of particular populations of projection cells (Angelucci et al., 1998; Karlen et al., 2006; Rakic et al., 1991).

In the adult, the most obvious and functionally relevant differences between thalamic projection neurons of different nuclei or nuclear subdivisions are in their axon distribution patterns to subcortical and cortical target structures. Additional differences have been observed in dendritic morphology or in the expression of particular calcium-binding or cell adhesion proteins between projection cells (Clascá et al., 2012; Jones, 2001).

4.1.1 Subcortical Targets of the Thalamic Projection Neurons

Axons from thalamic projection neurons radiate ipsilaterally out of the thalamus, making essentially no contacts with cells in their own or in other nuclei of the dorsal thalamus. Axons first traverse the RPtN and then fan out through the so-called thalamic radiations along the internal capsule (IC), finally reaching the white matter of the cerebral cortex. Throughout this path, thalamic neuron axons follow progressively divergent trajectories that may gently turn or twist, but nevertheless keep a strict topological order (Adams et al., 1997; López-Bendito and Molnár, 2003; Molnár et al., 2012; Vanderhaeghen and Polleux, 2004).

The RPtN is interposed between the thalamus and its telencephalic targets as a cellular lamina that wraps around the anterior and lateral aspects of the thalamus. Axons of thalamic projection neurons pierce the RPtN lamina orthogonally, and as they do, they give off small collateral branches that arborize locally within the nucleus. Anterograde tracer injections in different thalamic nuclei have shown that such thalamo-reticular collaterals distribute in a highly topographic fashion (Jones, 1975). Furthermore small, well-localized tracer injections have revealed subtle sublamination and overlap patterns among functionally related types of cells from the same or neighboring thalamic nuclei (Guillery et al., 1998; Crabtree, 1999).

The striatum is the major subcortical target for thalamic projection neurons. The thalamo-striatal connections arise from a limited number of thalamic nuclei. Interestingly, anterograde tracer injections in thalamus have revealed two different patterns of thalamo-striatal projection. Axons from the posterior intralaminar nuclei (center median–parafascicular complex in primates or parafascicular nucleus in rodents) innervate the entire striatum in a heavy focal and highly topographic manner (Galvan and Smith, 2011). Other, less dense and more overlapping thalamo-striatal projections arise from the anterior intralaminar (central medial, paracentral, and central lateral), midline, and association nuclei such as the dorsomedial and lateral posterior–pulvinar complex. Posterior intralaminar nuclei, particularly the center median nucleus, are sizable in primates. Bulk injections of anterograde tracers in the central portion of the primate center median yield dense anterograde labeling in the striatum but almost none in the cerebral cortex, indicating that projection neurons in this nucleus are exclusively thalamo-striatal. In contrast, anterograde tracer injections in the remaining intralaminar nuclei of primates and rodents consistently produce anterograde labeling both in the striatum and in the cerebral cortex (Berendse and Groenewegen, 1990; Groenewegen and Berendse, 1994; Herkenham, 1986; McFarland and Haber, 2002).

Retrograde tracer injections in the amygdala as well as anterograde tracer injections in the thalamus (Day-Brown et al., 2010; Doron and Ledoux, 1999; Turner and Herkenham, 1991; Ledoux et al., 1990; Linke et al., 2000) revealed the existence of direct thalamo-amygdaloid projections in rodents, carnivores, and primates. The most numerous of these projections arise in some subdivisions of the medial geniculate nucleus. Others arise in the posterior–pulvinar complex and the parvocellular regions of the ventral posterior (VP) complex. Anterograde tracer injections in these thalamic nuclei have shown that thalamo-amygdaloid axons target almost exclusively the lateral amygdaloid nucleus (Jones and Burton, 1976b; Yasui et al., 1984). Interestingly, a double-retrograde labeling experiment in rat demonstrated that at least some of the medial geniculate nucleus cells innervating the amygdala simultaneously innervate some temporal areas of the cerebral cortex (Doron and Ledoux, 2000; Namura et al., 1997).

A projection to the dorsal claustrum arising from neurons in the central lateral thalamic nucleus was demonstrated in carnivores and scandentia (Carey and Neal, 1986; Kaufman and Rosenquist, 1985). Whether such thalamo-claustral axons extend to cortex is yet unknown.

4.1.2 TC Pathways: Tangential Distribution Patterns

All neocortical areas, as well as the parahippocampal cortices, subiculum, and field CA1 of the hippocampus are innervated by thalamic axons. Far from following an unequivocal point-to-point correspondence between the three-dimensional nuclear mass of thalamic cells and the two-dimensional cortex mantle, however, most TC pathways are remarkably divergent/convergent.

Connection-tracing studies have consistently shown that only a few thalamic nuclei distribute their axons with point-to-point topographic order within a single cortical area. TC pathways from the dorsal lateral geniculate (DLG), ventral medial geniculate (VMG), VP, and ventral lateral (VL) nuclei are among the ones that come closest to this definition (Agmon et al., 1995; Brandner and Redies, 1990). Remarkably, these nuclei have historically been the favorite models for studies of TC function because they receive powerful and topographic inputs from the retina or brainstem sensory nuclei that can be readily evoked, even under experimental anesthesia. As a result, their rather atypical axonal wiring to cortex is still widely regarded as representative of TC circuits in general.

The majority of thalamic nuclei, however, innervate the cortex in much more divergent/convergent patterns. Bulk injections of anterograde tracers into rodent and carnivore thalamic nuclei such as the ventral medial–ventral anterior, lateral posterior–pulvinar, posterior, suprageniculate (Sg), dorsal medial geniculate (DMG), as well as the midline interanteromedial (IAM) and reuniens (Re) nuclei (Herkenham, 1978, 1979, 1980; Vertes et al., 2006), or their primate equivalents (Burton and Jones, 1976) have shown that their TC axons reach a variable number of cortical areas, adjacent or not, and often cover very large portions of the cerebral hemisphere. This widespread divergence of axons is confirmed by the repeated labeling of neuronal bodies within the above nuclei following retrograde tracer injections in distant points of the cerebral cortex (Rubio-Garrido et al., 2009, 2014). Interestingly, the retrograde tracer injections also show that these divergent connections overlap extensively with each other in any given point of the cerebral cortex, as well as with the projections from the topographic nuclei mentioned above. As a result, every neocortical area receives multiple convergent TC inputs, in characteristic proportions of abundance (Clascá et al., 1997; Rubio-Garrido et al., 2009).

The focal versus diffuse TC distribution patterns in cortex were first made by Lorente de No (1938) based on Golgi stain observations of postnatal mice cortex. Lorente noted that, among the axons presumed to be of thalamic origin running tangentially in the subcortical white matter, some (which he called "specific," labeled "a" and "b" in Figure 4.1) turned upward to end within a single cortical spot. Others (which he called "nonspecific," labeled "c" and "d" in Figure 4.1) innervated that cortical spot by means of an ascending collateral branch while the main trunk continued toward other cortical regions. In subsequent decades, microelectrode studies that were able to record from the distal segments of identified thalamic axons in the subcortical white matter and to label them with horseradish peroxidase confirmed the presence of both kinds of general axonal TC morphology (see below).

Retrograde double-labeling methods based on the accumulation in the same neuronal body of two distinguishable tracers injected in separate regions of the nervous system were proposed in the 1980s as a tool to demonstrate the existence of branched axonal connections (Rosina et al., 1980). Applied to the study of TC axonal architectures, such techniques revealed that at least some axons from particular thalamic nuclei reach more than one cortical point (Minciacchi et al., 1986; Spreafico et al., 1987). The double-retrograde labeling strategy was also used to estimate the maximal tangential spread of individual TC axons (Rausell and Jones, 1995). In most cases, however, such studies detected very few (<5%) double-labeled TC cells, leading some researchers to conclude that branched axons TC innervating separate areas were exceptional (Kishan et al., 2008). The significance of these observations, however, is

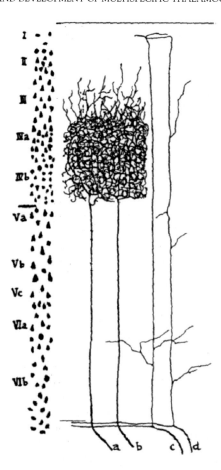

FIGURE 4.1   **Rafael Lorente de No's (1938) depiction of "specific thalamic afferents" (a and b) and "nonspecific or pluriareal thalamic afferents" (c and d) as observed in a single section (of unspecified thickness) from a 11-day-old mouse auditory cortex stained with the Golgi method.** Modified from the original figure. For reference, a diagrammatic depiction of cortical layering as visualized with Nissl stain is displayed in the left column. Lorente noted that the nonspecific axons "are of thalamic origin and innervate at least two adjacent cortical fields," and that "during their course in the white matter they give off collaterals ascending into the cortex and reaching as far as layer 1."

unclear, because double-retrograde labeling is open to many possible false-negative concerns. Significant differences in the effectiveness of the two retrograde tracers applied in a given experiment can lead to a gross underestimation of the prevalence of axon branching; for example, due to asymmetries in the size of axonal arborizations in the injected areas, ambiguity in the co-localization analysis due to overlap of the excitation–emission spectra of the fluorescent tracers used, or, if the two axonal arborizations are spatially restricted, simple lack of accuracy in the tracer injection placement. Moreover, even in the best possible situation, the actual usefulness of these methods is limited to qualitative analysis of simple (binary) branching patterns.

4.1.3 TC Pathways: Radial Distribution Patterns

Another prominent element of diversity among TC axonal architectures is the selective distribution of their terminal arborizations to different cellular layers along the radial dimension of the cerebral cortex. Given the strictly layered arrangement of cortical cell somas and dendrites, TC axons arborizing in different layers of a given cortical zone are bound to synapse onto different cell populations and/or different dendritic domains of the same cells. Selective laminar distribution of afferents is not limited to TC axons, but also applies to other afferents to cortex and is today recognized as a fundamental wiring principle of cortical circuits that supports the parallel flow and precise integration of top-down and bottom-up processing streams of information among cortical areas (Bastos et al., 2012; Crick and Koch, 1998; Felleman and Van Essen, 1991).

Lorente de No (1938) had also noted that, upon entering the cortex, the "specific" axons concentrate their terminal arborization in layer 4, leaving additional, simpler branches in layer 6 (Figure 4.1). In contrast, the branched "nonspecific" TC arborizations were more evenly spread across cortical layers and reached up to layer 1 (L1).

Subsequent studies with anterograde tracer injections in different thalamic nuclei in several mammal species confirmed that connections arising from different nuclei terminate in the cortex according to different laminar and/ or sublaminar patterns (Frost and Caviness, 1980; Herkenham, 1980; Killackey and Ebner, 1972, 1973; Jones and Burton, 1976a). Reviewing his extensive collection of precise tritiated amino acids injections in different nuclei of rat thalamus, Herkenham (1986) proposed the distinction of three main laminar TC patterns: a) Projections terminating heavily in layer 4, with additional branches in L6 ("Layer 4+6 type"); b) projections directed to layer 5 ad 6, usually in a weak and diffuse manner ("Layer 5+6 type") and; and c) projections targeting L1 with additional branches in deeper cortical layers, usually L3 or L5, but never L4 ("Layer 1+ type). This author also tentatively mapped the origin of each type of projection in the thalamus. He noted that injections in a great many thalamic nuclei yielded a "layer 1+" TC labeling pattern. Interestingly, injections in Re yielded similar "L1+"-type labeling in the stratum lacunosum-moleculare of hippocampal sector CA1 and subiculum as well (Herkenham, 1980; Vertes et al., 2006; Wouterlood et al., 1990). Injections in the "topographic" sensory relay nuclei (see above) produced a "layer 4+6" labeling pattern. Anterograde tracers injected in the intralaminar nuclei typically produced an L5+6 pattern (Berendse and Groenewegen, 1991; Herkenham, 1986).

An intriguing observation from these anterograde labeling studies was that, following injections in a nucleus innervating several areas, the laminar pattern of the terminal labeling often varied markedly among the target areas (Herkenham, 1986). However since even the smallest of these tracer injections involved large numbers of TC neurons, it remained unclear, whether the different patterns reflected the presence in the nucleus of different TC neuron populations or, alternatively, of axons simultaneously arborizing into different layers in each area.

Other studies examined TC laminar diversity by comparing the distributions of cells retrogradely labeled in the thalamus by tracer deposits limited to cortical L1 versus deposits extended into deeper layers. Such studies found that many nuclei throughout the thalamus contain neurons that innervate L1, but that the topographic nuclei (DLG, VMG, VP, and VL) virtually lack such cells. The mean size of neurons backlabeled in some nuclei by L1 deposits was different from that of cells labeled by deposits in deeper layers, further indicating the presence of two TC cell populations with different patterns of laminar arborization (Avendaño et al., 1990; Carey et al., 1979a,b; Rausell and Avendaño, 1985). Subsequent studies in macaques that added immunolabeling for calcium-binding proteins to the above retrograde labeling strategy found that the cells projecting to L1 express calbindin, whereas cells that avoid L1 and innervate L4 and L6 usually express parvalbumin (Hashikawa et al., 1991; Rausell et al., 1992; Hendry and Yoshioka, 1994). Summing up these observations, Jones (1998, 2001) proposed as a fundamental principle of TC organization (Jones, 2007) the existence of two major types of TC projection cells: (1) a large population spread across most nuclei of the thalamus, which he called "matrix" cells, characterized by expressing calbindin and having widely divergent axons that target the superficial cortical layers, and (2) cells limited to some thalamic nuclei that express parvalbumin and innervate topographically cortical L4–L3. Subsequent studies in rodents and New World primates have consistently found calbindin expression in most TC cells projecting to the superficial cortical layers; however, parvalbumin expression in L4-projecting TC cells has been found only in macaques (Rubio-Garrido et al., 2007, 2014). Moreover, recent single-axon labeling studies (see below) have shown that the same axon may actually have "Matrix"-like or "Core"-like laminar distributions in different areas, indicating that such concepts need revision.

4.2 THALAMOFUGAL AXON ARCHITECTURES: SINGLE-AXON LABELING STUDIES

Given the limitations of bulk retrograde and anterograde labeling methods, the complex architectures of TC projections began to be clarified only with the advent of techniques specifically developed to label individual axons.

The first images of individual TC axons were obtained by labeling them with anterograde tracers microinjected in the subcortical white matter or in the thalamus (Ferster and LeVay, 1978; Freund et al., 1985; Garraghty et al., 1987; Humphrey et al., 1985; Jensen and Killackey, 1987a; Rockland et al., 1999; Shinoda and Kakei, 1989). These techniques, however, allowed reconstructing only distal axon fragments.

The introduction of juxtacellular methods for recording from and loading single cells with biocytin or dextrans (Pinault, 1996) achieved for the first time labeling of both the cell soma, dendrites and axon of individual long-range projection neurons. Application of this technique to the study of TC neurons, however, has been relatively limited, both because of its technical difficulty and because of ever-present concerns about incomplete labeling (Deschênes et al., 1996, 1998; Gauriau and Bernard, 2004; Gheorghita et al., 2006; Monconduit and Villanueva, 2005; Noseda et al., 2011; Oberlaender et al., 2012; Parent and Parent, 2005). More recently, the introduction of virus vectors able to drive

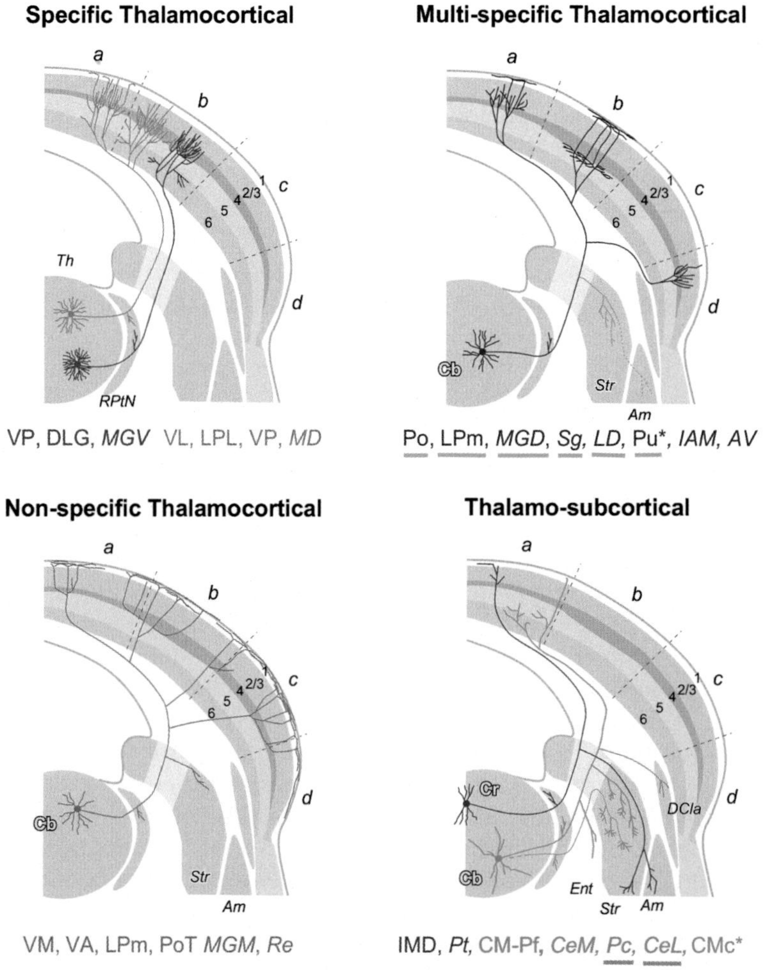

FIGURE 4.2 **Schematic summary of the various general thalamic projection neuron morphotypes, represented on an idealized mammalian forebrain.** The thalamic nuclei containing each cell type are indicated in abbreviations. Note that some cytoarchitectonic nuclei contain more than one type of cell. As in Table 4.1, we italicize the nuclei for which there is yet no direct evidence from single-cell labeling studies, only bulk anterograde tracing data. Within each group of nuclei, identified by a particular color, some nuclei are underlined using a second color to indicate that their neurons consistently extend an additional collateral to a particular subcortical target structure (indicated using second color in the diagram). Numbers indicate cortical layers. Four idealized cerebral cortex areas, labeled "a"–"d" and separated by dashed lines are shown. Yellow case letters indicate selective expression of calbinin (cb) or calretinin (Cr) in some neuron types. See Table 4.1 for nuclei abbreviations. Other abbreviations: Am, amygdala; DCla, dorsal claustrum; Ent, entopeduncular nucleus/lateral globus pallidus; RPtN, reticular prethalamic nucleus; Str, striatum; Th, thalamus.

in adult neurons high levels of expression of labeled proteins that are directed to the axon (Chamberlin et al., 1998; Furuta et al., 2001) has allowed, for the first time, the consistent visualization of complete individual TC axons, no matter how long and complex their branching (Kuramoto et al., 2009, 2015; Nakamura et al., 2015; Ohno et al., 2012).

Although still limited to few nuclei, the single-cell labeling studies are revealing an unsuspected degree of diversity and complexity in TC cells (Figure 4.2; Table 4.1). To date, all but one such study, including our own observations, have been carried out in rodents; for this reason we refer below mainly to nuclear nomenclature of the rodent thalamus.

4.2.1 Basic TC Neuron Axon Morphotypes

As a rule, all thalamic projection neurons extend their axons out of the thalamus without having intra-thalamic branches. The only reported exception seems to be neurons in the posterior intralaminar (CM–Pf) nuclei of rat (Deschênes et al., 1996), but not of the equivalent primate nuclei (Parent and Parent, 2005). Axons then traverse the reticular nucleus and give off one or a few thin collateral branches that exit perpendicularly from the main axon

TABLE 4.1   Morphological Traits used to Define Thalamic Projection Axon Types

	Nuclei where present	Somal expression of Ca²⁺-binding proteins	Major axonal domain	Subcortical collaterals			Cortical arborization	
				RPtN	Str	Am/DCla	No. of terminal domains	Layers
Specific thalamocortical (STC)	VPM, *MGV*	–	Cortical	+	–	–	1	4–3, 6
	VL, LPL, VPMc, *VPL, DLG (x,y), MD*	–	Cortical	+	–	–	1–3 (adjacent)	4–3, 6
Multispecific thalamocortical (MTC)	Po, LPm, LD, *MGD, Sg, Spf, Pu (Primate)*	Calb (70–90%)	Cortical	+	++	Am	2–4	1,3,4,5a
	IAM, AV, DLG(w)	Calb (70–90%)	Cortical	+	–	–	2–4	1,3,4,5a
Nonspecific thalamocortical (NTC)	VM, VA, LPm, PoT, *MGM, Re*	Calb (>90%)	Cortical	+	+	Am	>5	1
Thalamosubcortical (TS)	PVA, *Pt*	Calretinin	Subcortical	+	+++	Am	1–2	1,2/3
	CM-Pf, CeM	Calb (50%)	Subcortical	+	+++	–	1–2	1,5,6
	Pc, *CeL*	Calb (50%)	Subcortical	+	+++	DCla	1–2	5,6
	CMc (primate)	Calb (<70%)	Subcortical	+	+++	–	–	–

Morphological traits used to define thalamic projection neuron types. The single-cell labeling studies from which the data are derived are listed along with the spelling of each nucleus abbreviation. We italicize those nuclei for which the axon branching patterns suggested by small bulk anterograde tracer injections still await direct confirmation by single-cell tracing methods (see text). For clarity, differences in dendrite morphology (summarized in Figure 4.2) are not listed here. Abbreviations: AM, amygdala; *AV, anteroventral nucleus;* Calb, calbindin; *CeL,* central lateral nucleus (Deschênes et al., 1996); CeM, central medial nucleus (Deschênes et al., 1998); CM–Pf, center median–parafascicular nucleus (Parent and Parent, 2005); CMc (primate), core region of center median nucleus (Parent and Parent, 2005); DCla, dorsal claustrum; *DLG (w), dorsal lateral geniculate, W or equivalent cells; DLG (x,y), dorsal lateral geniculate; X and Y cells, or equivalent cells in other species; IAM, interanteromedial nucleus;* LD, laterodorsal nucleus (Noseda et al., 2011); LPL, lateral posterolateral nucleus (Nakamura et al., 2015); LPm, lateral posteromedial nucleus (Nakamura et al., 2015); *MD, mediodorsal nucleus; MGD, medial geniculate nucleus, dorsal division; MGM, medial geniculate nucleus, medial division; MGV, medial geniculate nucleus, ventral division; Pc, paracentral nucleus;* Po, posterior nucleus, rostral part, (Deschênes et al., 1998; Noseda et al., 2011; Ohno et al., 2012.); PoT, posterior nucleus, triangular part (Gauriau and Bernard, 2004; Ohno et al., 2012.); *Pt, paratenial nucleus; Pu, pulvinar complex (primates);* PVA, anterior paraventricular nucleus (Komlósi et al., 2015); *Re, reuniens nucleus;* RPtN, reticular prethalamic nucleus; *Sg, suprageniculate nucleus; Spf, subparafascicular nucleus;* Str, striatum; VA, ventral anterior nucleus (Kuramoto et al., 2009); VL, ventral lateral nucleus (Kuramoto et al., 2009); VM, ventromedial nucleus (Kuramoto et al., 2015; Monconduit and Villanueva, 2005); VPM, ventroposterior medial nucleus (Gheorghita et al., 2006; Noseda et al., 2011; Oberlaender et al., 2012); *VPL, ventroposterior lateral nucleus.*

trunk and form relatively localized and simple arbors that expand along a plane parallel to the surface of RPtN (Figure 4.4).

A large majority of thalamic projection neurons innervate the cerebral cortex; thus "thalamic projection" and "thalamocortical" may seem equivalent terms. However, there is at least one clear exception that, in our view, justifies keeping both concepts separate: cells in the core of the center median nucleus (CMc) of primates do not target the cortex and innervate only the striatum (Parent and Parent, 2005). Such cells have not yet been observed in rodents (Deschênes et al., 1996).

After exiting the IC, the main trunks of thalamic projection neuron axons turn dorsally and then extend to the cortical white matter in trajectories that run roughly parallel to cortex. In axons targeting a single cortical spot, the axon trunk takes an upward turn near its target and ascends to form the terminal arborization. In axons with multiple cortical targets, collaterals always emerge along the tangential trajectory as one or more perpendicular branches that leave the trunk at roughly right angles to form intracortical terminal arborizations. There is a wide and continuous spectrum of variation in the number and tangential separation of such first-order axonal branches, as well as in the density and laminar distribution in the cortex of their terminal arborizations.

Neurons in the VP, VL, DLG, VMG, lateral posterior lateral (LPL), and probably also the mediodorsal nucleus (MD) arborize densely in L4 and lower L3 and, less densely, also in L5b–L6a. Interestingly, virtually identical laminar patterns have been also observed in feedforward cortico-cortical connections (Coogan and Burkhalter, 1993; Crick and Koch, 1998; Felleman and Van Essen, 1991; Markov et al., 2013). LPL and MD axons also arborize in L1. In most cases, axons often divide several times as they cross layers 6 and 5. Axons from the rostral part of VP and MGV concentrate all their branches in a single L4–L3 terminal spot, with a less dense focus of L6 terminal branching underneath. White matter and L6 branches of axons from caudal VP, DGL, and VL remain more dispersed in

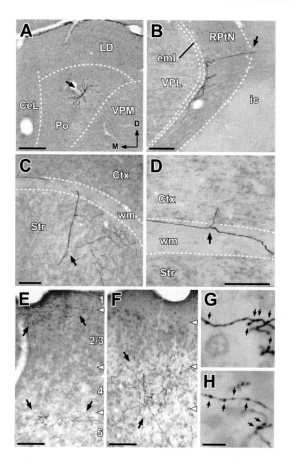

FIGURE 4.3   **A single thalamic projection neuron of the posterior nucleus (Po) of an adult mouse labeled by Sindbis-Pal-GFP RNA electroporation.** Brightfield images of 50-μm thick sections show different portions of the same cell made visible by nickel-enhanced immunostaining against GFP. Thionin counterstain. (A) The cell body (arrow) and part of the dendritic bush can be seen in the posterior thalamic nucleus (Po). (B) Collateral arborization in the reticular prethalamic nucleus (RPtN). The arrow indicates the axon shaft extended across the nucleus. (C) Collateral arborization in the striatum (Str). The arrow indicates the point at which a striatal branch is seen emerging from the thick axonal shaft, and that continues in the white matter (wm) toward more medially situated areas. (D) Axon running tangentially in the superficial cortical white matter/layer 6b and giving off a collateral branch (arrow) that extends up into the overlying cortex (Ctx). Compare with Figure 4.1. (E) Terminal arborizations in the primary somatosensory cortex. Multiple branch segments are visible in layer 1 (upper arrows) and layer 5a (lower arrows). (F) Terminal arborizations in the ectorhinal cortex are indicated by arrows. (G) and (H) High-magnification details of the terminal arborizations in the same two cortical regions. Note the numerous varicosities (arrowheads), suggestive of en-passant synaptic specializations. Other Abbreviations: Cel, central lateral nucleus; eml, external medullary lamina; ic, internal capsule; LD, laterodorsal nucleus; VPL, ventral posterolateral nucleus; VPM, ventral posteromedial nucleus. Bars: A, 250 μm; B–F, 100 μm; G and H, 10 μm.

the tangential dimension of the cortex and usually give rise to several separate terminal spots within the same or even in adjacent areas.

Observations of distal axon fragments from equivalent nuclei in primates and carnivores reveal similar patterns, although primate DLG axons originating from different projection neuron subtypes display an even greater sublamination of their terminal arbors within layers 4 and 3b (Fitzpatrick et al., 1983; Lund, 1988). Another characteristic of these neurons is that they consistently have not been observed to give off branches to the striatum or to other subcortical telencephalic structures. Neurons with this general axonal morphotype seem to correspond to the classic "specific thalamic afferents" of Lorente de No. For this reason, we will refer to them, from here on, as Specific Thalamocotical (STC) projection neurons.

In rodents, axons from the posterior medial nucleus (Po) and medial lateral posterior (LPM) nuclei give off several (usually two to four) branches at separate points of their subcortical tangential segment (Figure 4.3). These ascending branches then enter the cortex and cross L6 leaving virtually no collaterals there. In some areas the terminal arborizations are mostly concentrated in L4–L3, whereas in other areas they may be located in L5a and outer L1 (sublayer 1a). These multiple first-order branches may be quite far apart in the cortical hemisphere. The lateral extent of the terminal arbors from these branches is similar to that of the STC axons. Arborizations of the same axon in one area may be markedly more profuse than in other areas. Although not yet studied with single-cell labelling methods, the available anterograde bulk tracing data suggests that axons form from the MGD, Sg, lateral dorsal (LD), subparafascicular (Spf), anteroventral (AV), and IAM nuclei may share this general morphology.

In addition to innervating the cerebral cortex, Po and LPM usually emit collaterals in the striatum. These striatal branches often form loose and extensive terminal arbors there. Comparable single-axon data are not yet available for the other nuclei that we have included in this group; however, evidence from bulk anterograde tracer injections (Allen Mouse Brain Connectivity Atlas (Internet); Van der Werf et al., 2002) strongly suggest that these nuclei also give off collateral arborizations to the striatum. In addition, PoM axons often extend into the lateral amygdaloid nucleus. It is probable that Spf and MGD axons also send similar collaterals down to the amygdala (Doron and Ledoux, 2000). Given the multiple yet selective tangential and laminar targeting of several cortical domains and often of subcortical structures as well, we propose for neurons with this basic axonal morphotype the denomination of "multispecific thalamocortical" (MTC) projection cells.

The axon trunks from neurons in the VM, VA, and some neurons of the lateral posterior medial (LPM) or the caudal part of the posterior nucleus (PoT) have along their white matter path numerous (5–12), separate collateral branches that innervate the cerebral cortex. As they ascend through the cortical layers, such branches follow roughly radial trajectories and leave few or virtually no terminal arborizations in L6–L2, although they may divide in several branches that continue their radial ascent up to L1. As they reach sublayer 1a, all branches take a 90° turn to extend tangentially within the sublayer. There, they form "flat" extensive subpial arbors. Despite being made of relatively sparse and mostly straight branches, these L1 arbors contain in fact a large amount of axonal membrane and boutons. The lateral spread of such subpial arborizations is far more extensive than the localized domains innervated by STC and MTC axons. In fact, at least in rodents, the combined L1 surface innervated by the various tangential arbors from just one of these axons can encompass a large portion of the cerebral hemisphere (Kuramoto et al., 2015; Nakamura et al. 2015). In addition to innervating the cortex, some of these axons also give off a collateral branch to the striatum. Because of the shared traits of having a widespread tangential dispersion and predominant innervation in sublayer 1a in the cerebral cortex, we suggest referring to the axons from VM, VA, LPM, and PoT neurons as "nonspecific thalamocortical" (NTC) projection neurons. Although not yet explored with single-cell tracing methods, the available data suggest that the medial division of medial geniculate (MGM) and the Re nuclei can be included in this axonal morphotype (Herkenham, 1986; Vertes et al., 2006).

In rodents, most neurons in the intralaminar nuclei (central medial, CeM; paracentral, Pc; central lateral, CL; and center median–parafascicular, CM–Pf) innervate the cerebral cortex (Figure 4.2). Their cortical arborizations, however, are poorly branched and remain largely confined to L6 and L5 with an occasional simple branch shooting to L1 (Deschênes et al., 1996; Parent and Parent, 2005). In contrast, these same axons innervate profusely and topographically the striatum by means of large collateral branches. Some intralaminar nuclei axons occasionally have a small, poorly branched collaterals to the entopeduncular nucleus. The arborizations coming from the "anterior intralaminar" group of nuclei (CeM, Pc, and CL) tend to be more widespread within the striatum than those coming from "posterior intralaminar" group (Cm–Pf) axons, which instead form several focal clusters of terminal branches. Interestingly, the only single-cell labeling study in primates published to date indicates that axons arising from the neurons in their large CMc innervate the striatum but do not extend to the cortex (Parent and Parent, 2005).

We have observed that some cells in CL and Pc also send an axonal branch to the dorsal claustrum (see also Carey and Neal, 1986; Kaufman and Rosenquist, 1985; Real et al., 2006; Van der Werf et al., 2002). Remarkably, such thalamo-claustral branches are different from those directed to the striatum and arise perpendicularly from the axon trunk immediately beyond the lateral edge of IC (i.e., in the pallial domain of the telencephalon).

As suggested by anterograde bulk-tracing studies, recent observations (Komlósi et al., 2015) indicate that neurons in midline nuclei such as the anterior paraventricular (PVA) and paratenial (Pt) innervate both the cortex and striatum in a pattern similar to that of the intralaminar nuclei. Interestingly, their striatal branches target only the ventral striatum and adjacent olfactory tubercle, and often continue down into the central amygdaloid nuclei. Some PVA neurons, in fact, only innervate these subcortical structures and not the cortex.

Since the intralaminar, PVA, and Pt projection neurons have most of their axon arborizations directed to subcortical structures rather than to the cerebral cortex, we refer to these cells as "thalamosubcortical" (TS). Moreover, as pointed out above, at least some of these TS neurons actually do not extend their axons to the cortex (Figure 4.2).

4.2.2 Correlations Between Axonal and Somatodendritic Morphologies

Single-cell labeling techniques enable now to correlate, for the same neuron, the axonal morphotype with the somatodendritic domain morphology and/or expression of particular chemical markers.

Thalamic projection cells have 3–15 principal dendrites arranged in multipolar or roughly bipolar patterns. These principal dendrites split proximally into a variable number of higher order branches that keep the same roughly radial arrangement. Dendrites carry a variable number of short spines (see Jones, 2007 for a review). Although the picture is still fragmentary, there seems to be a general correlation between the number of higher order dendrite branches in a given cell and whether its axon innervates the cortex in a more dense and focal pattern. Thus, cells with STC axons have significantly more numerous higher order radial dendrites than neurons with MTC axons, and the latter in turn have more higher order radial dendrites than do the cells with NTC axons. The cells with TS axons have the fewest dendrites. An inverse trend seems to be present regarding the maximum dendritic length, that is, cells with STC axons having the shortest dendrites and the cells with TS axons the longest ones (Kuramoto et al., 2009, 2015; Ohno et al., 2012; Nakamura et al., 2015).

Single-cell studies (Kuramoto et al., 2009, 2015; Nakamura et al., 2015; Ohno et al., 2012) and bulk retrograde tracer studies (Hashikawa et al., 1991; Molinari et al., 1995; Rausell et al., 1992; Rubio-Garrido et al., 2007) combined with inmunolabeling for specific calcium-binding proteins also show that, as a rule, cells with NTC and MTC axons express calbindin and do not express parvalbumin. In contrast, cells with STC axons consistently fail to express calbindin; however, in particular nuclei of macaques they were reported to express parvalbumin (Hashikawa et al., 1991; Molinari et al., 1995; Rausell et al., 1992) or the alpha subunit of the type II calmodulin-dependent protein kinase (Hendry and Yoshioka, 1994).

Calcium binding protein expression is diverse among TS neurons in different nuclei. For example, calbindin is expressed in the lateral part of the rodent Pf as well as in some cells of the anterior intralaminar nuclei, but not in medial Pf or IMD neurons (Table 4.1).

4.3 DEVELOPMENTAL DIFFERENTIATION OF TC AXON ARCHITECTURES

Over the past two decades, TC axons have been a favorite model system for developmental studies of the cellular and genetic mechanisms that regulate growth, navigation, and synapse formation of long-range axonal connections (for reviews see Molnár et al., 2012; Sur and Rubenstein, 2005). However, most such studies did not considered differences in wiring architectures, essentially regarding all axons as it they were of the STC type.

Overall, these investigations have revealed that thalamic axon growth cones begin their extension shortly after thalamic neurons are born (around embryonic day "E" 13–14, in rats). They exit en masse the dorsal thalamus (prosomere 2, Pr2) in a ventral and rostral direction toward the telencephalon. They are repelled from the midline by Slit proteins expressed at high levels by the Pr2 subventricular zone, acting on Robo receptors expressed by the thalamic axons. Growth cones cross first the zona limitans intrathalamica (future external medullary lamina, Figure 4.4A) and then the alar region of prosomere 3 (future RPtN, Figure 4.4A and B). All TC growth cones then make a sharp laterodorsal turn toward the medial ganglionic eminence (MGE) of the ventral telencephalon. This turn is again mediated by chemorepulsion by Slits expressed in the hypothalamus (Braisted et al., 2009). Subsequently, growth cones are able to extend across the nonpermissive MGE mantle layer by growing along a stream of medially migrating lateral ganglionic eminence (LGE) cells that opens a permissive "corridor" across MGE (Bielle et al., 2011; López-Bendito et al., 2006; asterisk in Figure 4.4A).

Growth cones continue extending dorsolaterally across the LGE up until reaching its lateral edge, the so-called pallio-subpallial boundary (PSPB, double asterisk in Figure 4.4A), and opening up the future IC. In rodents, growth cones cover the path from the thalamus to the PSPB quite rapidly, and by E15 many are already exiting the outer edge of IC. Within IC, axons remain bundled in tightly parallel trajectories that directly reflect the general topology of the regions of the dorsal thalamus from which each axon originates (Molnar et al., 1998a). Such highly ordered trajectories have been shown to result from receptor–ligand interactions of the growth cones with guidance molecules (Semaphorins, Ephrins, and Netrins) that are expressed as a succession of spatial gradients in the "corridor" and in the LGE. Such interactions sort out the cones along a general frontal-to-occipital axis within the developing IC (review Molnar et al., 2012). This stereotyped ordering is still recognizable in the adult thalamic axons as they traverse the IC.

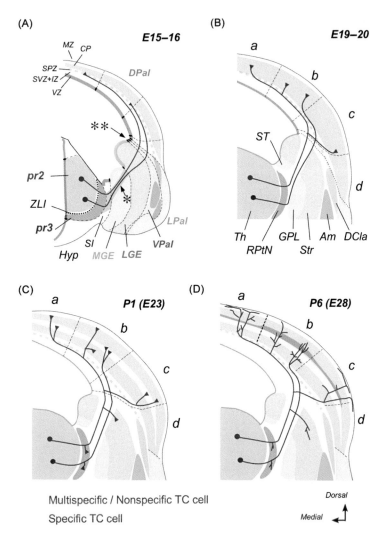

FIGURE 4.4 **Diagrammatic sequence of specific thalamocortical and multispecific/nonspecific thalamocortical axon development.** Idealized transversal sections of the prenatal (A and B) and postnatal (C and D) rat forebrain are represented. The filled circles represent the neuronal cell body in the thalamus (Th). Triangles at the tip of the lines indicate patent axonal growth cones. The structures derived from the same histogenetic compartments are labeled with shades of the same color. (*) Corridor compartment; (**) Pallio-subpallial boundary. In (A)–(C) the general trajectory of the pial processes of radial precursors at the borders of prospective cortical areas is indicated by dashed lines. In (A) the position of radial precursor cell bodies near the ventricular lumen is represented by black points. In (D) the dashed lines indicate definitive area boundaries. Numbers preceded by "E" indicate rat embryonic age in days, and numbers preceded by "P" the age in postnatal days. Note that the growth of collateral branches from the multispecific/nonspecific axon to *both* cortical and subcortical structures follows the arrangement of radial precursor processes. Note also the delayed sprouting of collateral branches to lateral cortical areas (B), and the even more delayed sprouting of branches to the striatum (Str) and reticular prethalamic nucleus (RPtN; (C)). Other abbreviations: Am, amygdala; CP, critical plate; DCla, dorsal claustrum; DPal, dorsal pallium; GPL, lateral globus pallidus; Hyp, hypothalamus; LGE, lateral ganglionic eminence; LPal, lateral pallium; MGE, medial ganglionic eminence; MZ, marginal zone; pr2, prosomere 2; Pr3, prosomere 3 (prethalamic eminence); SI, substantia innominata/perirreticular nucleus; SPZ, subplate zone; ST, stria terminalis; Str, striatum; SVZ+ IZ, subventricular zone + intermediate zone fiber layer; VPAL, ventral pallium; VZ, ventricular zone; ZLI, zona limitans intrathalamica.

The development of thalamic projections to the RPtN and to the striatum has only been investigated to date in axons from the Po nucleus (whose neurons are of MTC or NTC type) in rats (Galazo et al., 2008). The Po axons remain extended for almost a week (E14–22) across the developing RPtN and striatum without giving off branches (Figures 4.4A and B and 4.5A, B, D, E). Branches finally sprout at E23 to innervate these two structures (Figures 4.4C and 4.5C, G). If one considers that the reciprocal reticulothalamic connection invades the thalamus at E15 (Molnár et al., 1998a,b), thalamoreticular collateral sprouting appears to be remarkably delayed. Likewise, the Po axons that eventually innervate the striatum remain in close vicinity to their target cells for many days before they sprout the thalamostriatal branches. It is not yet known if TS axons, that have the striatum as their major target, behave similarly.

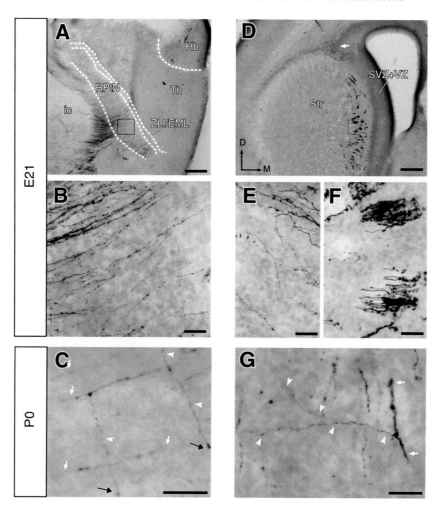

FIGURE 4.5 **Development of thalamoreticular and thalamostriatal branches from Po nucleus (MTC) axons.** (A) Axons labeled after an *ex vivo* biocytin injection in the Po nucleus at embryonic day (E21) can be seen traversing the reticular prethalamic nucleus (RPtN) and entering the internal capsule (ic). (B) High-magnification view of labeled thalamocortical axons in the reticular nucleus at E21 (from inset in (A)). Note that axons show no branches. (C) High-magnification detail of labeled thalamocortical axons in the reticular nucleus at postnatal day 0 (E22, "P0"). The image corresponds to a field equivalent to that in (B) and has the same orientation. Interstitial branches (indicated by white arrowheads) can be seen extending perpendicularly from the main trunks of MTC axons (also indicated by white arrows). Note that the branches are capped by growth cones (black arrows), which grow parallel to the main axis of the RPtN. (D) Labeled axons traversing the mantle layer of the developing striatum (Str) in the same experiment. Note that the axons are widely dispersed in numerous small fiber bundles. At the level of the striato-pallial border (white arrow), some axons can be seen extending into the pallial white matter. (E) and (F) High-magnification views of labeled axons in the striatal mantle layer. Note that at this age, axons do not yet show collateral branches in the striatum. (G) High-magnification view of labeled axons in the striatum at P0. An interstitial branch can be seen extending from a thicker parent axon trunk (indicated by arrowheads and arrows, respectively). Note that the branch is itself divided into two subbranches that extend toward lateral zones of the striatum. Other abbreviations: Hb, habenula; ST, stria terminalis; ZLI/EML, zonal limitans intrathalamica/external medullary lamina; Th, thalamus. Bars: (A) and (D), 250 μm; B, E, F, and G, 25 μm; C, 20 μm.

4.3.1 Thalamic Projection Axon Growth and Branching in the Developing Cortex

The PSPB is a gene expression boundary that separates LGE from the ventral and lateral pallia (VPal, LPal, Figure 4.4A). There is evidence that TC axons may interact at this level with a convergent system of descending axons from early-differentiated neurons of the dorsal pallium (DP, Meyer et al., 1998; Molnár et al., 1998a,b) and even fasciculate on this substrate to reach their correct targets within DP (Chen et al., 2012; Molnar et al., 1998a). Such fasciculation may also help to explain the remarkable swift and orderly "radiation" of TC axons along the intermediate zone (IZ) of DP, despite their trajectories running orthogonal to the general layout of the DP radial precursor processes.

In rats, TC growth cones begin exiting the IC around E15 (Galazo et al., 2008; Molnár et al., 1998a,b). In ferrets, the same event occurs around E28–31 (Clascá and Sur, 1996). As they fan out under the developing cortex, however, TC axons remain tightly confined within IZ forming a "flat" radiating bundle (Figure 4.6). The precise position of

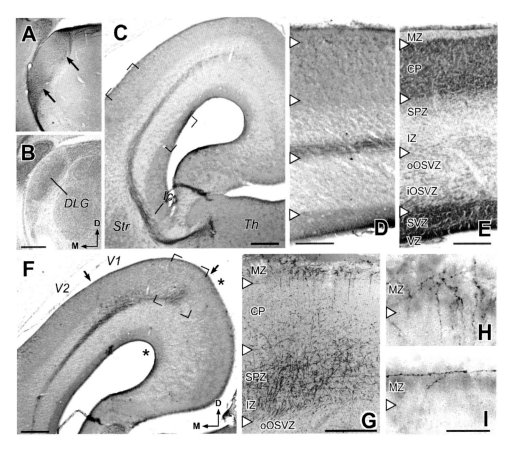

FIGURE 4.6 **Anterogradely labeled geniculocortical axons in an early postnatal P4 or "E46" in postcoital days ferret.** (A) and (B) Two iontophoretic deposits of choleratoxin subunit B (arrows) stereotaxically placed within the dorsal lateral geniculate nucleus (DLG) and an adjacent zone of the Pulvinar nucleus. An inmunolabeled section is shown in (A), and the adjacent Nissl-stained section is shown in (B) for reference. (C) Coronal section showing a compact bundle of labeled geniculocortical axons extending through the internal capsule (ic) into the deep layers (future white matter) of the developing cortex. (D) Detail of the cortex region framed in (C). (E) Comparison with an adjacent Nissl-stained section shows that the axon bundle occupies the deepest part of the intermediate zone (IZ). (F) Coronal section though the location of the prospective visual cortex (V1; delineated by arrows). It should be kept in mind that, at this age, the ferret V1 cortical plate (CP) contains only the neurons that will be in the adult infragranular layers (Reillo et al., 2011). Note that geniculocortical axons defasciculate, turn superficially and form two separate plexuses in the subplate zone, at locations retinotopically correspondent to the two injection sites in DLG. Note also that the most medially extending geniculocortical axons do not extend beyond the medial edge of V1 (asterisk). (G) Detail of the region framed in (F): A profuse plexus made of relatively simple branches is present in the subplate zone (SPZ). Note that many straight axons or axon branches probably from the Pulvinar (which in the adult targets L1), extend radially through the cortical plate without giving off any branch in it, and reach the marginal zone (MZ, future cortical sublayer 1a) where they turn 90° to form a tangential subpial plexus. (H) and (I) High-magnification detail of the tangential subpial axons and their branches. Note that the morphology of these terminal axons is already very similar to that of the adult, while the cortical plate remains without innervation from the geniculocortical axons. Other abbreviations: iOSVZ, inner compartment of the outer subventricular zone; oOSVZ, outer compartment of the outer subventricular zone; SVZ, subventricular zone; Th, thalamus. Bars: (A)–(C) and (F), 500 μm; (D) and (E), 250 μm; (H) and (I), 100 μm.

the TC axonal fascicle is best visualized in ferrets (Figure 4.6C–F), where the layers are thicker and cytoarchitecturally distinct (Reillo et al., 2011). Interestingly, all these TC axons fan out of the IC in gently arched, non-crossing trajectories toward their corresponding target regions in the frontal, dorsal parietal, occipital, and temporal cortices (Figures 4.4 and 4.5A). The growth cones that reach the dorsal parietal or occipital areas do not extend further medially to innervate the cingulate cortex; instead TC growth cones directed to cingulate areas in the medial aspect of the hemisphere radiate first rostrally and then circumvent the rostral end of the lateral ventricle to finally navigate backward along the cingulum bundle of the white matter to reach their targets (Galazo et al., 2008).

Within the IZ bundle, the axons directed to more lateral portions of the cortex systematically occupy a more superficial position than those directed to dorsal or occipital cortical regions (Figure 4.4; Molnár et al., 1998a,b). As the bundle extends dorsally, the most superficial growth cones defasciculate in succession and turn to enter the subplate zone (SPZ). In this way, growth cones superficial in IZ enter the subplate of prospective lateral neocortical areas, whereas axons running deepest in IZ enter the subplate at the dorsomedial edge of parietal and occipital areas, near the border with the cingulate cortex (asterisk in Figure 4.6F).

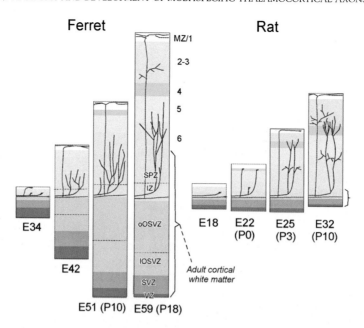

FIGURE 4.7   **Species-specific differences in the developmental sequence of TC axon invasion and arborization into the cerebral cortex.**
Highly simplified MTC (blue) and STC (red) thalamocortical axons are depicted on an idealized "column" of developing prospective visual cortex. This MTC axon can arise from DLG W cells, or from adjacent lateral-posterior pulvinar complex cells. The STC axons that could arise from X or Y ferret cell or equivalent cells in rat cortical layers are indicated by numbers. Developing white matter/cortical precursor compartments are labeled as in Figure 4.6. Ages are indicated in embryonic days ("E") throughout for consistency. The corresponding postnatal age in days is indicated within parentheses. Note that in ferrets, but not in rats, axons directed to layer 1 (MZ/L1) reach this layer at a very early age by crossing the cortical plate layers without branching. These axons soon form a dense, adult-like layer 1 plexus; in contrast, layer-4-directed specific axons delay in the subplate zone (SPZ) and deep CP for about a month (after E61/P30) before even beginning their slow process of arborization in this layer. Interestingly, the L3 branches of layer-1-directed axons develop within the same timeframe of layer 4 invasion by STC axons.

Studies in rat axons from the VP nucleus directed to the posteromedial barrel subfield of the somatosensory cortex (Catalano et al., 1991; Erzurumlu and Jhaveri, 1990, 1992; López-Bendito and Molnár, 2003; Molnár et al., 1998a,b) have shown that, at about E17, their growth cones begin defasciculating from the IZ bundle, and then take an ascending radial direction and begin a much slower advance through SPZ and the deep tiers of the cortical plate (future L6). During this period (E17–23), axons sprout several short branches, probably in part temporary, forming a transient plexus in SPZ–L6. The tangential precision of growth cone defasciculation and SPZ–L6 radial invasion of these and other STC axons are remarkable (Agmon et al., 1993; Molnar et al., 1998a).

After a period of minimal radial progression that in rats lasts about 6 days (E19–P2), branches from the SPZ–L6 plexus finally extend up to L4, their main adult target layer (Figure 4.7). In the case of the axons coming from the anterior part of the mouse VP, the precision of this invasion is reportedly extreme, with axons entering directly only into their appropriate "barrel" L4 domain (Agmon et al., 1995; Figure 4.4D). Over the following weeks, numerous additional local branchlets sprout form these branches within the L4 terminal domain, giving rise to the definitive clustered morphology typical of adult VP axons by about P30 (Jensen and Killackey, 1987a; Figure 4.2). The morphology of MGV in mice primary auditory cortex suggests that they might follow a similar developmental pattern (Lorente de No, 1922). The formation of two or more adjacent focal terminal arbors within the same area, as observed in the case of DLG, caudal VP, and VL axons (Ferster and LeVay, 1978; Garraghty and Sur, 1990; Kuramoto et al., 2009) may result from a greater local dispersion of the branches extending from SPZ–L6 into L4.

The slow progression across SPZ and infragranular layers is believed to reflect an active process of sorting and local remodeling. This would contribute to the observed rearrangements within the subcortical white matter and deep cortical layers that refine the precise intra-areal topography of STC axons in supragranular layers (Adams et al., 1997; Agmon et al., 1993; Catalano and Shatz, 1998; López-Bendito and Molnár, 2003; Shimogori and Grove, 2005). In this process, extension, retraction, and/or branching may depend on interactions with secreted or membrane-bound signals present in the deep cortical layers (López-Bendito and Molnár, 2003), as well as on activity-dependent mechanisms that may be mediated by early, transient synaptic contacts (Friauf and Shatz, 1991; Hanganu et al., 2002; Higashi et al., 2005).

Studies in rat Po and VM nucleus axons have shown that the MTC and NTC axonal architectures are the product of a partly different developmental program (Galazo et al., 2008). In these axons, the principal growth cone first navigates all the way up to dorsally situated areas. Later (about 24 h in rats) collateral branches begin sprouting at right angles from the main axon trunk, always growing toward SPZ (Figure 4.4B). It is known that this secondary ("interstitial") type of branching is regulated by specific mechanisms, different from those controlling the splitting of the leading growth cone (Portera-Cailliau et al., 2005; Szebenyi et al., 2001). Our observations show that in MTC and NTC axons such mechanisms must be activated at one or several spots in the axon's IZ segment. Remarkably, there seem to be little randomness or initial exuberance in this process, as branches are seen from the start only under their appropriate prospective cortical target zones (compare Figures 4.2 and 4.4; Galazo et al., 2008). Branch extension from IZ to the deep cortical layers is strictly parallel to the known arrangement of radial glia/subventricular zone precursor processes (Figure 4.4B and C), strongly suggesting that the growth cones that extend the branches may be actually fasciculating on such processes.

Cortical areas situated superficial to the putamen and claustrum (ventral parietal and insular) are targeted by branches that follow closely the "s"-shaped trajectory of the radial glial processes linking the ventricular lumen with the pial surface of these lateral areas (Figure 4.4B–D). This observation, along with the generalized radial growth of the interstitial branches, strongly suggests an interaction between these branches and radial glia processes. In contrast, in the distal cortical region primarily targeted by the leading growth cone, the corresponding cortical arborization is directly built by the cone turning toward the cortical plate and growing into it. These distal arbors seem to grow slightly faster than those in lateral regions (Galazo et al., 2008).

Given the "static" nature of the observations in anatomical tracing studies, it is possible that branch sprouting and terminal arborization is a highly dynamic process rather than a simple linear progression. Several studies noted occasional collateral branches in TC axons under distant or "inappropriate" areas (Bicknese et al., 1994; Catalano et al., 1996; Ghosh and Shatz, 1992; Molnár et al., 1998a,b), but regarded the branching to be transient. Since the above studies mainly examined nuclei where "unbranched" STC neurons predominate, it is indeed possible that most of the collateral branches that they observed were actually retracted (Bruce and Stein, 1988; Catalano et al., 1996; Krug et al., 1998; Naegele et al., 1988). In any case, taken together with the adult single-axon tracing studies discussed above (Figure 4.2), it is now clear that many of the interstitial branches that sprout from Po and VM axons are not transient, but become full-grown intracortical arbors in the adult (Kuramoto et al., 2009, 2015; Ohno et al., 2012). Interstitial axonal branching as a mechanism to reach multiple targets is common in some forebrain axon systems (O'Leary and Terashima, 1988; Szebenyi et al., 2001), but until recently had not been considered in TC axons (Galazo et al., 2008). The location at which interstitial sprouting occurs may depend on local signals acting on the primary growth cone at the time of its transit through IZ, or on signals acting later on the axon shaft (Szebenyi et al., 2001; Tang and Kalil, 2005). Candidate signaling molecules include secreted or membrane-bound proteins such as LAMP (Mann et al., 1998), PSA-NCAM (Yamamoto et al., 2000), Netrin 1 (Dent et al., 2004), Anosmin 1 (Soussi-Yanicostas et al., 2002), or some growth factors and their receptors (Szebenyi et al., 2001; Yamamoto and Hanamura, 2005).

While the terminal arborizations of STC axons remain largely confined to L4-L3, occasional branch tips may extend up to L1. However, such branches never form complex arborizations within L1. In contrast, most MTC arbors and all the arbors from NTC axons form profuse, bidimensional and bouton-rich terminal arborizations within L1. Interestingly, their branches remain confined to the outer half of L1 ("sublayer 1a"), which is the direct adult derivative of the embryonic MZ, situated immediately below the basal lamina and the meninges. Moreover it is remarkable that during development, axons directed to L1 extend radially across the cortical plate (future layers 6-2) with little or no branching in it. In rodents, the radially ascending growth cones of the MTC/NTC layer 1 directed axons have been observed to pause for about 48 h upon arrival at upper border of the cortical plate then take a 90° turn (probably for changing from a radial glia-guided to a meningeal/basal lamina mode of guidance/adhesion; Borrell and Marin, 2006) and then begin a fast process of subpial extension and branching within MZ/Sublayer 1a (Galazo et al., 2008; Portera-Cailliau et al., 2005). In any case, the layer-specific targeting of the MTC and NTC axon arbors directed to L1 (Galazo et al., 2008; Kichula and Huntley, 2008), strongly suggests that they respond to different guidance signals, or react to the same signals in a different manner, than L4-directed STC axons (Maruyama et al., 2008; Yamamoto et al., 2000).

Additional evidence for fundamental differences in the mechanisms that govern the development of STC axons vs. those of MTC/NTC axons is suggested by the different timing of their "waiting" in the SPZ and of their layer-specific terminal branch growth. Such diffferences are not evident in the fast-developing rodent brain (Galazo et al., 2008). However, in the slower developing cortex of carnivores, such as ferrets and cats, they are striking (Figure 4.7). In carnivores both types of axons arrive at their target areas together, about E28–31. In the prospective visual cortex, STC axons from the Lateral Geniculate Nucleus delay for several weeks within the SPZ/L6. In contrast, MTC axons from the Pulvinar nucleus extend directly across the cortical plate, arborize within L1, and even make functional synapses there several weeks before the STC axons from the Dorsal Lateral Geniculate even begin entering L4 (Figures 4.6 and 4.7; Clascá and Sur, 1996; Kato et al., 1983; see also bulk-labeling data in cats by Kato, 1987; Shatz and Luskin, 1986). Likewise, available studies in other mammals with protracted cortical development such as primates (Rakic, 1977) or marsupials (Marotte, Leamey and Waite, 1997; Sheng et al., 1991) also noted broad differences between the timing in the first appearance of anterograde TC labeling in L1 or L4. Such markedly different axonal interactions with the same cortical tissue hint at fundamental phenotypic differences among thalamic projection neurons. Moreover, as SPZ neurons are thought to play a critical role as transient scaffolding of TC circuits (Kanold and Luhmann, 2010), it is tempting to speculate that differences in the nature of interactions with SPZ neurons could contribute to the distinct connectivity of L4-directed STC axon arbors in the adult cortex.

4.3.2 Developmental Plasticity of TC Axon Growth and Branching

In recent years, experimental studies have begun exploring the developmental mechanisms that control the tangential (area) and radial (layer) targeting of TC terminal arborizations by means of specific genetic manipulations *in vivo*. These studies have yielded tantalizing clues about the specificity and plasticity of the processes that control the terminal wiring of TC axons.

Fibroblast growth factor 8 (Fgf8) is a diffusible morphogen protein that is normally expressed early in cortical development (E10–12) in the frontomedial pole of the cerebral hemisphere and imparts "frontal" identity to DP neuronal precursors of the neuroepithelium in a dose-dependent manner (Figure 4.8A). *In utero* electroporation of plasmids driving abnormally high expression levels of Fgf8 in selected spots of the developing cerebral hemisphere has been thus used to modify the normal location and relative size of cortical areas (Fukuchi-Shimogori and Grove, 2001). FgF8 overexpression in its normal (frontal) location at E10.5 causes an expansion of frontal areas and a caudalward shift of an otherwise normal primary somatosensory cortex (S1, including the somatotopic whiskwerpad cortical representation "Wp"; Figure 4.8B). Bulk tracing with carbocyanines shows that VPM axons simply extend in the IZ for a longer distance until reaching the caudally displaced Wp and, once there, defasciculate in SPZ and arborize in L6 and L4 exactly as control axons (Shimogori and Grove, 2005).

If the electroporation is carried out 1 day later (E11.5) when SPZ and deep L6 neurons are already born and committed, frontal overexpression of Fgf8 affects only the neurons destined to L5–L2. As a consequence, SPZ and L6 neurons of the Wp occupy a normal location, but the L5–L2 neurons with Wp identity are situated more caudally (Figure 4.8C). Thalamic axons defasciculate from the IC bundle under the normally located SPZ and then form the usual plexus in the overlying L6 cells. However, instead of then ascending radially to the overlying L4 as do normal axons, they extend along L6 up to the caudally located Wp superficial layers. This pattern is reminiscent of some horizontally extended intracortical arborizations observed in MTC and NTC axons (Figure 4.2).

Finally, if Fgf8 expression is induced ectopically in the occipital pole of the cerebral hemisphere, the result is mirror-duplication of frontal and parietal areas, including the the Wp. Interestingly, the two whiskerpads thus formed are often situated at some distance from each other, separated by an interposed patch of parietal/auditory cortex (Figure 4.8D). In the subplate of the first Wp, VP axons split. One branch ascends into L6 and innervates L4 topographically, as normal axons, while the other branch continues caudally in the IZ/SPZ up to the second Wp, where they also topographically innervate L6 and L4, albeit according to the "inverted" pattern of this whiskerpad.

Although the bulk carbocyanine tracing did not allow Shimogori and Grove (2005) a complete reconstruction of single axons, we find the last experiment discussed above particularly remarkable for four reasons. First, it shows that the distinction between STC axons projecting to one or several sites in L4 is shallow (not cell-autonomous), as it may emerge at any time, from the simple local duplication of adequate positioning signals in the cortex. Moreover, it suggests that a similar mechanism may be iterated more times to produce the multibranched architectures of MTC and NTC axons. Second, it indicates that, despite having reached their normal target subplate (that of the rostral Wp), TC axons are still able to sense guidance signals coming from the caudally situated WP, which are powerful enough to induce their sprouting of a caudally directed branch. The observation that these branches elongate as a bundle along the IZ (Figure 4.8D) may indicate that such a signal is associated with area-specific descending pioneer axons from the caudal whiskerpad zone (Meyer et al., 1998; Molnár et al., 1998a,b), and that the TC branches fasciculate on them. Third, together with the observations in adult axons, the experiment resolves parsimoniously a long-standing conundrum: how TC axons following in strictly divergent, noncrossing trajectories can innervate topographically two or more areas with mirror topography, as is common in the cortex (Adams et al., 1997). And, finally, the experiment suggests a ready cellular mechanism for providing

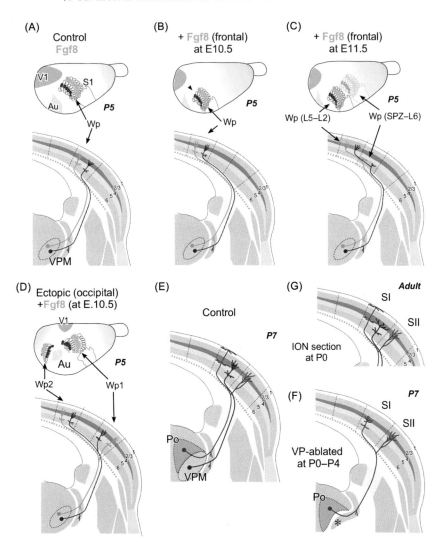

FIGURE 4.8 **Effects of manipulating in development the target area position and number, axon convergence, or global input levels on the structure of the thalamic terminal arborizations in the cerebral cortex.** (A) Top: a dorsal view of the cortical areas of a 5-day-old (P5) mouse cerebral hemisphere showing the position of the primary somatosensory cortex (S1), primary visual cortex (V1), and primary auditory cortex (A1). The vibrissal follicle representation in S1 (whiskerpad, Wp) is indicated by an arrow. Two separate vibrissal rows are colored in red and green, for reference. A gradient of yellow shade decreasing caudally from the frontal pole represents the normal diffusion of fibroblast growth factor 8 (Fgf8), a morphogen protein that induces the commitment of cortical neural precursors to the formation of rostral cerebral cortex areas. Bottom: the diagram showing the path from thalamus to cortex of two simplified specific TC axons from the ventroposterior medial nucleus (VPM) ending topographically in Wp. (B) Overexpression of Fgf8 at its normal frontal location at embryonic day 10.5 (E10.5) before subplate (SPZ) and layer 6 (L6) neurons are generated. This manipulation expands the portion of the cerebral hemisphere occupied by frontal areas, while the portion occupied by occipital/temporal cortices (V1, A1) shrinks. As a result, the Wp develops in a position that is more caudal in the cerebral hemisphere than in control animals (compare to A). The diagram below shows that VPM axons react by extending for longer distances within SPZ to the new location and innervate it normally. (C) Overexpression of Fgf8 at its normal frontal location at E11.5, after SPZ and L6 neurons have been generated. Wp SPZ and L6 cells are situated at their normal position, but the Wp L5–L2 neurons differentiate from a more caudal portion of the neuropithelium, once Fgf8 overexpression is induced. VPM axons react to the mismatch by forming a normal L6 plexus and then extending horizontally within layer 6 to reach the caudally displaced L4 barrels. (D) Ectopic (occipital) expression of Fgf8 induces a mirror-image duplication of Wp. VPM axons that normally do not branch in SPZ now branch systematically and provide topographical innervation to the normal and the supernumerary WP, forming in both, normal L4 and L6 terminal arbors. (E) Idealized diagram depicting the convergence and complementary laminar patterns in Wp of VPM (specific TC type) and posterior nucleus (Po, multispecific type) axons. (F) Genetic ablation of VPM neurons at early postnatal ages P0–P4 induces the arbors of Po in S1 to take a VPM-like laminar termination pattern. (G) Abolition of peripheral sensory input (postnatal infraorbital nerve section, ION) to both VPM and Po neurons causes minor changes in Po axon architecture, but markedly modifies VPM L4 terminal arborizations. *Source: Based on the data (A)–(D) Shimogori and Grove (2005); (E) and (F) Pouchelon et al. (2014); (G) Frangeul et al. (2014).*

adequate thalamic input to expanding or duplicating cortical areas during evolution. In this regard, since the Wp cortex is almost doubled, it would be very interesting to investigate whether the two whiskerpad representations are innervated by the same amount of VP axonal processes, or whether a fixed, homeostatic total amount of VP terminal axon length is divided between the two.

Recent genetic manipulation studies have explored the possible influences between TC axons of different types innervating the same area on each other's arborization patterns. A transgenic mouse line engineered to express from P0 onwards a toxic protein that selectively kills all VP neurons while sparing the adjacent Po neurons provides a clean experiment to examine possible competitive/cooperative postnatal interactions between STC and MTC axon arborizations developing within the same cortical target area (S1 in this case). (Pouchelon et al., 2014; Figure 4.8E and F). The surprising result is that Po axons take on a VP-like appearance, arborizing profusely in layer 4, a layer that they normally avoid, forming functional synapses on L4 neurons, and no longer innervate L1. In a related study (Figure 4.8G), we compared the effects of completely eliminating from birth the sensory input to Wp (by infraorbital nerve section) on the final laminar patterning of multispecific Po and specific VP axons in the adult (Frangeul et al., 2014). Remarkably, despite the early and complete sensory deprivation, Po axons were not significantly modified. In contrast, VP L4 arborizations became loosely spread in L4 and did not accumulate in barrels, as previously described (Van der Loos and Woolsey 1973; Jensen and Killackey, 1987b); in addition, quantification of axonal length revealed a modest decrease in L4. These experiments may be interpreted to indicate that, at least in the rodent Wp, TC laminar patterning is resilient to global changes in synaptic activity, but responds rapidly to the imbalance and/or synapse vacating by one of the various TC axon types that innervate the area.

Overall, developmental plasticity studies indicate that the architecture of distal TC arborizations depends, to a significant extent, on non-cell autonomous mechanisms of local signaling from the target cortical cells, and that, consequently, it is partly modifiable both in tangential distribution and laminar pattern even at advanced stages of development. This conclusion is consistent with the observation that single MTC axons may innervate several areas, each with different laminar patterns (Figure 4.2).

Keeping in mind the close clonal origin in the diencephalic neuroepithelium of all thalamic projection neurons, comparison of the developmental sequence (Figure 4.4) and plasticity (Figure 4.8) of their axons with the observed diversity in adult axon architectures (Figure 4.2) make it tempting to speculate that, to a large extent, adult diversity may result from small variations in the sensitivity of thalamic cells to the local cues their axons encounter along their path. Such cues would be able to induce, upon a single early common pattern, the sprouting or retraction of particular axon segments or branches at later developmental stages. Depending on the timing and location, the same mechanisms might produce relatively local changes (e.g., layer arborization in a given cortical area) or more global changes (e.g., TS vs. STC axon morphotypes). Such a scenario is reminiscent of the situation observed in cortical layer 5 pyramidal cells that achieve highly diverse axonal projection patterns (Jones, 1984; Larsen et al., 2007) through the regulation of secondary branch sprouting and segment retraction in different targets (O'Leary and Terashima, 1988).

On the other hand, the consistency of some correlations between particular axonal architectures and somato-dendritic morphologies and/or calcium-binding protein expression patterns (Figure 4.2, Table 4.1) points to the existence of some degree of deeper cell diversity among thalamic projection neurons, perhaps dependent on their precise origin within the alar neuroepithelium of prosomere 2 (Puelles and Rubenstein, 2003). For example, correlation of calbindin expression and cortical arborization in cortical L1 has been consistently found in equivalent nuclei of rodents and primates (Jones, 2001; Rausell et al., 1992; Rubio-Garrido et al., 2007, 2014). Likewise, in rodents and primates, the TS axonal architecture is associated with longer and simpler dendrites, whereas the STC architecture is associated with the most numerous and profusely branched, but overall shorter, dendrites.

4.4 CONCLUDING REMARKS: FUNCTIONAL IMPLICATIONS OF THE DIVERSE AXONAL ARCHITECTURES

The neuroanatomical studies reviewed above show that the thalamus is wired to cortical and subcortical forebrain circuits through multiple, complex, and often superimposed axonal architectures. Evidence from direct experimental comparisons of the effects of axons with different architectures on the same brain region is still very limited (see, e.g., Petreanu et al., 2009; Frangeul et al., 2014). However, some experimentally testable functional predictions may already be derived from the structural data now available.

The TS axons are mainly or exclusively focused on the striatum. Important differences in terminal arborization pattern, cell-type targeting, and synaptic organization have been described between the TS axons coming from the CM–Pf and those coming from the anterior intralaminar nuclei (Galvan and Smith, 2011). Interestingly, terminal arborizations from the anterior intralaminar nuclei axons turn out to be structurally identical to those of the thalamostriatal branches of MTC axons. Since the CM–Pf neurons are highly responsive to attention-related sensory stimuli and have strong connections to striatal cholinergic interneurons, it is believed that they may play a role in basal ganglia–mediated learning, behavioral switching, and reinforcement. Input from anterior intralaminar and MTC axons may have a more global and modulatory role centered on striatal medium spiny neurons.

STC axon architectures seem ideally suited to convey spatially precise, high-resolution representations of sensory receptor surfaces or multisynaptic cerebellar loops. Convergence of their terminal branches on specialized postsynaptic cells in layer 4 makes these axons powerful feed-forward drivers of cortical activity (Bruno and Sakmann, 2006; Petreanu et al., 2009; Sherman, 2012).

By having most of their terminals within cortical sublayer 1a, NTC axons preferentially access the spinous apical dendrites of L2–3 and some L5 pyramidal cells. The L2 and L3 pyramidal cells are the main source of feed-forward cortico-cortical connections, both short-range and long-range, including callosal connections (Markov et al., 2013). Among layer 5 pyramidal cells, only those projecting to subcortical targets appear to have apical dendritic tufts in sublayer 1a (Larsen et al., 2007). By modulating membrane potential through these highly plastic and dynamic apical dendritic L1 tufts (Larkum et al., 2004; Sjöström and Häusser, 2006), NTC axons might modify cortico-cortical communication and subcortical output over vast cortical districts. In this regard, it is interesting to consider that, because of the nuclei from which they originate, NTC axons preferentially convey the output "go" signals from the multisynsptic basal ganglia loops, or high-saliency sensory signals such as pain, loud noise, etc.

The evidence from the single-cell reconstruction studies and bulk-tracing data reviewed here indicates that neurons with MTC axons may be the most abundant type of cell across the thalamus in rodents, and probably even more so in primates. This observation is relevant not only because it is at odds with currently held notions about the basic layout of TC circuits, but also because it raises some tantalizing functional implications.

Since axonal arbors from the same MTC axon innervate several distant cortical domains and often the striatum, and since action potentials in myelinated axons propagate in all-or-none fashion, with virtually no filtering (Segev and Schneidman, 1999) the same message may reach various spatially separate target areas. However, since the axonal arborizations then may target different layers in the various areas, thus contacting different elements of the local circuits, the same input may be computed in parallel, but in different/complementary ways in the various areas.

In addition, because MTC axons are relatively thick and myelinated, action potential arrival in the various target areas must happen almost synchronously. In this way, they might bias the coherence of membrane potential oscillations (subthreshold and/or suprathreshold) in the various target areas. Dynamically emerging synchronous membrane potential oscillations have been proposed as a key permissive mechanism for the selective routing of cortico-cortical communication between distant cortical regions involved in a given perceptual, cognitive, or motor task (review Fries, 2009; Singer, 2011). Activation of particular MTC neuronal groups may thus be a pivotal structural substrate for the fast emergence and reconfiguration of large and spatially distributed assemblies of synchronized cortical cells upon a stable skeleton of anatomical connections. Recent studies on the primate lateral pulvinar nucleus, which probably consists almost entirely of MTC-type projection neurons, have in fact shown that pulvinar activation is necessary for attention-dependent exchange of information between extrastriate cortical areas (Saalmann et al., 2012).

During development, TC axons are a crucial contributor to the specification of cerebral cortical areas (Rakic et al., 1991; Sur and Rubenstein, 2005; Pouchelon et al., 2014). TC axon interactions with the developing cortex occur first by means of simple cell–cell interactions, and later through synaptic activity. Since MTC axons establish synapses in two or more separate areas, and often in the striatum, well before long-distance cortico-cortical or corticostriatal connections mature, it is conceivable that the synchronized activity produced by MTC axons in their target areas contributes to the consolidation and/or refinement of the late-developing pathways. In this way, it is also possible that small genetic variants that modify the pattern of MTC collateral sprouting during development may cascade into large-scale, heritable changes in the organization of multiarea cortical networks.

The repeated observation that a collateral MTC arbor in one given area is markedly more profuse than other arbors of the same axon in other areas suggests that action potentials running through the parent axon may have stronger effects in some areas than in others. Alternatively, the simpler branches may have little functional impact in the adult, but represent an extensive lifelong reservoir of long-range connections, already in place, ready for terminal branch growth in response to new learning demands, and even for functional compensation after adult lesions.

Acknowledgments

The authors wish to thank Dr. Mriganka Sur for stimulating discussions and support for initial experiments that were instrumental in launching this line or research, and Dr. Carlos Avendaño for extended discussions of some of the ideas presented here.

Financial Support: MINECO (Spain) (BFU2010-19695) and the European Union Seventh Framework Programme (FP7/2007–2013) under grant agreement no. 604102 (Human Brain Project).

References

Adams, N.C., Lozsádi, D.A., Guillery, R.W., 1997. Complexities in the thalamocortical and corticothalamic pathways. Eur. J. Neurosci. 9, 204–209.

Agmon, A., Yang, L.T., O'Dowd, D.K., Jones, E.G., 1993. Organized growth of thalamocortical axons from the deep tier of terminations into layer IV of developing mouse barrel cortex. J. Neurosci. 13, 5365–5382.

Agmon, A., Yang, L.T., Jones, E.G., O'Dowd, D.K., 1995. Topological precision in the thalamic projection to neonatal mouse barrel cortex. J. Neurosci. 15, 549–561.

Allen Mouse Brain Connectivity Atlas (Internet). Available from: <http://connectivity.brain-map.org/>.

Altman, J., Bayer, S.A., 1988. Development of the rat thalamus: I. Mosaic organization of the thalamic neuroepithelium. J. Comp. Neurol. 275, 346–377.

Angelucci, A., Clascá, F., Sur, M., 1998. Brainstem inputs to the ferret medial geniculate nucleus and the effect of early deafferentation on novel retinal projections to the auditory thalamus. J. Comp. Neurol. 400, 417–439.

Avendaño, C., Stepniewska, I., Rausell, E., Reinoso-Suárez, F., 1990. Segregation and heterogeneity of thalamic cell populations projecting to superficial layers of posterior parietal cortex: a retrograde tracer study in cat and monkey. Neuroscience 39, 547–559.

Bastos, A.M., Usrey, W.M., Adams, R.A., Mangun, G.R., Fries, P., Friston, K.J., 2012. Canonical microcircuits for predictive coding. Neuron 76, 695–711.

Berendse, H.W., Groenewegen, H.J., 1990. Organization of the thalamostriatal projections in the rat, with special emphasis on the ventral striatum. J. Comp. Neurol. 299, 187–228.

Berendse, H.W., Groenewegen, H.J., 1991. Restricted cortical termination fields of the midline and intralaminar thalamic nuclei in the rat. Neuroscience 42, 73–102.

Bicknese, A.R., Sheppard, A.M., O'Leary, D.D., Pearlman, A.L., 1994. Thalamocortical axons extend along a chondroitin sulfate proteoglycanen-riched pathway coincident with the neocortical subplate and distinct from the efferent path. J. Neurosci. 14, 3500–3510.

Bielle, F., Marcos-Mondejar, P., Keita, M., Mailhes, C., Verney, C., Nguyen Ba-Charvet, K., et al., 2011. Slit2 activity in the migration of guidepost neurons shapes thalamic projections during development and evolution. Neuron 69, 1085–1098.

Borrell, V., Marin, O., 2006. Meninges control tangential migration of hem-derived Cajal-Retzius cells via CXCL12/CXCR4 signaling. Nat. Neurosci. 9, 1284–1293.

Braisted, J.E., Ringstedt, T., O'Leary, D.D., 2009. Slits are chemorepellents endogenous to hypothalamus and steer thalamocortical axons into ventral telencephalon. Cereb. Cortex 19 (Suppl. 1), i144–i151.

Brandner, S., Redies, H., 1990. The projection from medial geniculate to field AI in cat: organization in the isofrequency dimension. J. Neurosci. 10, 50–61.

Bruce, L.L., Stein, B.E., 1988. Transient projections from the lateral geniculate to the posteromedial lateral suprasylvian visual cortex in kittens. J. Comp. Neurol. 278, 287–302.

Bruno, R.M., Sakmann, B., 2006. Cortex is driven by weak but synchronously active thalamocortical synapses. Science 312, 1622–1627.

Burton, H., Jones, E.G., 1976. The posterior thalamic region and its cortical projection in New World and Old World monkeys. J. Comp. Neurol. 168, 249–301.

Catalano, S.M., Robertson, R.T., Killackey, H.P., 1991. Early ingrowth of thalamocortical afferents to the neocortex of the prenatal rat. Proc. Natl. Acad. Sci. U.S.A. 88, 2999–3003.

Catalano, S.M., Robertson, R.T., Killackey, H.P., 1996. Individual axon morphology and thalamocortical topography in developing rat somato-sensory cortex. J. Comp. Neurol. 367, 36–53.

Catalano, S.M., Shatz, C.J., 1998. Activity-dependent cortical target selection by thalamic axons. Science 281, 559–562.

Carey, R.G., Fitzpatrick, D., Diamond, I.T., 1979a. Thalamic projections to layer I of striate cortex shown by retrograde transport of horseradish peroxidase. Science 203, 556–559.

Carey, R.G., Fitzpatrick, D., Diamond, I.T., 1979b. Layer I of striate cortex of *Tupaia glis* and *Galago senegalensis*: projections from thalamus and claustrum revealed by retrograde transport of horseradish peroxidase. J. Comp. Neurol. 186, 393–437.

Carey, R.G., Neal, T.L., 1986. Reciprocal connections between the claustrum and visual thalamus in the tree shrew (*Tupaia glis*). Brain Res. 386, 155–168.

Chamberlin, N.L., Du, B., de Lacalle, S., Saper, C.B., 1998. Recombinant adeno-associated virus vector: use for transgene expression and antero-grade tract tracing in the CNS. Brain Res. 793, 169–175.

Chen, Y., Magnani, D., Theil, T., Pratt, T., Price, D.J., 2012. Evidence that descending cortical axons are essential for thalamocortical axons to cross the pallial-subpallial boundary in the embryonic forebrain. PLoS One 7 (3), e33105. http://dx.doi.org/10.1371/journal.pone.0033105

Clascá, F., Angelucci, A., Sur, M., 1995. Layer-specific programs of development in neocortical projection neurons. Proc. Natl. Acad. Sci. U.S.A. 92, 11145–11149.

Clascá, F., Llamas, A., Reinoso-Suárez, F., 1997. Insular cortex and neighboring fields in the cat: a redefinition based on cortical microarchitecture and connections with the thalamus. J. Comp. Neurol. 384, 456–482.

Clascá, F., Rubio-Garrido, P., Jabaudon, D., 2012. Unveiling the diversity of thalamocortical neuron subtypes. Eur. J. Neurosci. 35, 1524–1532.

Clascá, F., Sur, M., 1996. Cortical specification: thalamic projections to layer I develop weeks ahead of projections to layer IV in ferrets. Soc. Neurosci. Abstr. 22, 1014.

Coogan, T.A., Burkhalter, A., 1993. Hierarchical organization of areas in rat visual cortex. J. Neurosci. 13, 3749–3772.

Crabtree, J.W., 1999. Intrathalamic sensory connections mediated by the thalamic reticular nucleus. Cell. Mol. Life Sci. 56, 683–700.

Crick, F., Koch, C., 1998. Constraints on cortical and thalamic projections: the no-strong-loops hypothesis. Nature 391, 245–250.

Day-Brown, J.D., Wei, H., Chomsung, R.D., Petry, H.M., Bickford, M.E., 2010. Pulvinar projections to the striatum and amygdala in the tree shrew. Front. Neuroanat. 2010 Nov 15 (4), 143. http://dx.doi.org/10.3389/fnana.2010.00143 eCollection 2010.

Dent, E.W., Barnes, A.M., Tang, F., Kalil, K., 2004. Netrin-1 and semaphorin 3a promote or inhibit cortical axon branching, respectively, by reorganization of the cytoskeleton. J. Neurosci. 24, 3002–3012.

Deschênes, M., Bourassa, J., Parent, A., 1996. Striatal and cortical projections of single neurons from the central lateral thalamic nucleus in the rat. Neuroscience 72, 679–687.

Deschênes, M., Veinante, P., Zhang, Z.W., 1998. The organization of corticothalamic projections: reciprocity versus parity. Brain Res. Rev. 28, 286–308.

Doron, N.N., Ledoux, J.E., 1999. Organization of projections to the lateral amygdala from auditory and visual areas of the thalamus in the rat. J. Comp. Neurol. 412, 383–409.

Doron, N.N., Ledoux, J.E., 2000. Cells in the posterior thalamus project to both amygdala and temporal cortex: a quantitative retrograde double-labeling study in the rat. J. Comp. Neurol. 425, 257–274.

Erzurumlu, R.S., Jhaveri, S., 1990. Thalamic axons confer a blueprint of the sensory periphery onto the developing rat somatosensory cortex. Brain Res. Dev. Brain Res. 56, 229–234.

Erzurumlu, R.S., Jhaveri, S., 1992. Emergence of connectivity in the embryonic rat parietal cortex. Cereb. Cortex 2, 336–352.

Felleman, D.J., Van Essen, D.C., 1991. Distributed hierarchical processing in the primate cerebral cortex. Cereb. Cortex 1 (Suppl. 1), 1–47.

Ferster, D., LeVay, S., 1978. The axonal arborizations of lateral geniculate neurons in the striate cortex of the cat. J. Comp. Neurol. 182, 923–944.

Fitzpatrick, D., Itoh, K., Diamond, I.T., 1983. The laminar organization of the lateral geniculate body and the striate cortex in the squirrel monkey (*Saimiri sciureus*). J. Neurosci. 3, 673–702.

Frangeul, L., Porrero, C., Garcia-Amado, M., Maimone, B., Maniglier, M., Clascá, F., et al., 2014. Specific activation of the paralemniscal pathway during nociception. Eur. J. Neurosci. 39, 1455–1464.

Freund, T.F., Martin, K.A., Whitteridge, D., 1985. Innervation of cat visual areas 17 and 18 by physiologically identified X- and Y-type thalamic afferents. I. Arborization patterns and quantitative distribution of postsynaptic elements. J. Comp. Neurol. 242, 263–274.

Friauf, E., Shatz, C.J., 1991. Changing patterns of synaptic input to subplate and cortical plate. J. Neurophysiol. 66, 2059–2071.

Fries, P., 2009. Neuronal gamma-band synchronization as a fundamental process in cortical computation. Annu. Rev. Neurosci. 32, 209–224.

Frost, D.O., Caviness Jr., V.S., 1980. Radial organization of thalamic projections to the neocortex in the mouse. J. Comp. Neurol. 194, 369–393.

Fukuchi-Shimogori, T., Grove, E.A., 2001. Neocortex patterning by the secreted signaling molecule FGF8. Science 294, 1071–1074.

Furuta, T., Tomioka, R., Taki, K., Nakamura, K., Tamamaki, N., Kaneko, T., 2001. *In vivo* transduction of central neurons using recombinant sindbis virus: Golgi-like labeling of dendrites and axons with membrane-targeted fluorescent proteins. J. Histochem. Cytochem. 49, 1497–1507.

Galazo, M.J., Martinez-Cerdeño, V., Porrero, C., Clascá, F., 2008. Embryonic and postnatal development of the layer I-directed ("matrix") thalamocortical system in the rat. Cereb. Cortex 18, 344–363.

Galvan, A., Smith, Y., 2011. The primate thalamostriatal systems: anatomical organization, functional roles and possible involvement in Parkinson's disease. Basal Ganglia 1, 179–189.

Garraghty, P.E., Sur, M., 1990. Morphology of single intracellularly stained axons terminating in area 3b of macaque monkeys. J. Comp. Neurol. 294, 583–593.

Garraghty, P.E., Frost, D.O., Sur, M., 1987. The morphology of retinogeniculate X- and Y-cell axonal arbors in dark-reared cats. Exp. Brain Res. 66, 115–127.

Gauriau, C., Bernard, J.F., 2004. Posterior triangular thalamic neurons convey nociceptive messages to the secondary somatosensory and insular cortices in the rat. J. Neurosci. 24, 752–761.

Gheorghita, F., Kraftsik, R., Dubois, R., Welker, E., 2006. Structural basis for map formation in the thalamocortical pathway of the barrelless mouse. J. Neurosci. 26, 10057–10067.

Ghosh, A., Shatz, C.J., 1992. Pathfinding and target selection by developing geniculocortical axons. J. Neurosci. 12, 39–55.

Groenewegen, H.J., Berendse, H.W., 1994. The specificity of the "nonspecific" midline and intralaminar thalamic nuclei. Trends Neurosci. 17, 52–57.

Guillery, R.W., Feig, S.L., Lozsádi, D.A., 1998. Paying attention to the thalamic reticular nucleus. Trends Neurosci. 21, 28–32.

Hanganu, I.L., Kilb, W., Luhmann, H.J., 2002. Functional synaptic projections onto subplate neurons in neonatal rat somatosensory cortex. J. Neurosci. 22, 7165–7176.

Hashikawa, T., Rausell, E., Molinari, M., Jones, E.G., 1991. Parvalbumin- and calbindin-containing neurons in the monkey medial geniculate complex: differential distribution and cortical layer specific projections. Brain Res. 544, 335–341.

Hendry, S.H., Yoshioka, T., 1994. A neurochemically distinct third channel in the macaque dorsal lateral geniculate nucleus. Science 264, 575–577.

Herkenham, M., 1978. The connections of the nucleus reuniens thalami: evidence for a direct thalamo-hippocampal pathway in the rat. J. Comp. Neurol. 177, 589–610.

Herkenham, M., 1979. The afferent and efferent connections of the ventromedial thalamic nucleus in the rat. J. Comp. Neurol. 183, 487–517.

Herkenham, M., 1980. Laminar organization of thalamic projections to the rat neocortex. Science 207, 532–535.

Herkenham, M., 1986. New perspectives on the organization and evolution of nonspecific thalamocortical projections, 1986. In: Jones, E.G. (Ed.), Cerebral Cortex, vol. 5 Plenum Press, New York, NY, pp. 1985.

Higashi, S., Hioki, K., Kurotani, T., Kasim, N., Molnár, Z., 2005. Functional thalamocortical synapse reorganization from subplate to layer IV during postnatal development in the reeler-like mutant rat (shaking rat Kawasaki). J. Neurosci. 25, 1395–1406.

Humphrey, A.L., Sur, M., Uhlrich, D.J., Sherman, S.M., 1985. Projection patterns of individual X- and Y-cell axons from the lateral geniculate nucleus to cortical area 17 in the cat. J. Comp. Neurol. 233, 159–189.

Jensen, K.F., Killackey, H.P., 1987a. Terminal arbors of axons projecting to the somatosensory cortex of the adult rat. I. The normal morphology of specific thalamocortical afferents. J. Neurosci. 11, 3529–3543.

Jensen, K.F., Killackey, H.P., 1987b. Terminal arbors of axons projecting to the somatosensory cortex of the adult rat. II. The altered morphology of thalamocortical afferents following neonatal infraorbital nerve cut. J. Neurosci. 7, 3544–3553.

Jones, E.G., 1975. Some aspects of the organization of the thalamic reticular complex. J. Comp. Neurol. 162, 285–308.

I. MICROCIRCUITRY

Jones, E.G., 1984. Laminar distribution of cortical efferent cells. In: Jones, E.G., Peters, A. (Eds.), Cerebral Cortex, Vol. 1: Cellular Components of the Cerebral Cortex Plenum Press, New York, NY, pp. 521–555.

Jones, E.G., 1998. Viewpoint: the core and matrix of thalamic organization. Neuroscience 85, 331–345.

Jones, E.G., 2001. The thalamic matrix and thalamocortical synchrony. Trends Neurosci. 24, 595–601.

Jones, E.G., 2007. The Thalamus, second ed. Cambridge University Press, New York.

Jones, E.G., Burton, H., 1976a. Areal differences in the laminar distribution of thalamic afferents in cortical fields of the insular, parietal and temporal regions of primates. J. Comp. Neurol. 168, 197–247.

Jones, E.G., Burton, H., 1976b. A projection from the medial pulvinar to the amigdala in primates. Brain Res. 104, 142–147.

Kanold, P.O., Luhmann, H.J., 2010. The subplate and early cortical circuits. Annu. Rev. Neurosci. 33, 23–48.

Karlen, S.J., Kahn, D.M., Krubitzert, L., 2006. Early blindness results in abnormal corticocortical and thalamocortical connections. Neuroscience 142, 843–858.

Kato, N., Kawaguchi, S., Yamamoto, T., Samejima, A., Miyata, H., 1983. Postnatal development of the geniculocortical projection in the cat: electrophysiological and morphological studies. Exp. Brain Res. 51, 65–72.

Kato, N., 1987. The postnatal development of thalamocortical projections from the pulvinar in the cat. Exp. Brain Res. 68, 533–540.

Kaufman, E.F., Rosenquist, A.C., 1985. Efferent projections of the thalamic intralaminar nuclei in the cat. Brain Res. 335, 257–279.

Kichula, E.A., Huntley, G.W., 2008. Developmental and comparative aspects of posterior medial thalamocortical innervation of barrel cortex in mice and rats. J. Comp. Neurol. 509, 239–258.

Killackey, H., Ebner, F., 1972. Two different types of thalamocortical projections to a single cortical area in mammals. Brain Behav. Evol. 6, 141–169.

Killackey, H., Ebner, F., 1973. Convergent projection of three separate thalamic nuclei on to a single cortical area. Science 179, 283–285.

Kishan, A.U., Lee, C.C., Winer, J.A., 2008. Branched projections in the auditory thalamocortical and corticocortical systems. Neuroscience 154, 283–293.

Komlósi, G., Porrero, C., Barthó, P., Babiczky, Á., Dávid, Cs., Barsy, B., et al., 2015. Quantitative control of arousal via the midline thalamic nuclei. Soc. Neurosci. Abstr. 167.13/V2.

Krug, K., Smith, A.L., Thompson, I.D., 1998. The development of topography in the hamster geniculo-cortical projection. J. Neurosci. 18, 5766–5776.

Kuramoto, E., Furuta, T., Nakamura, K.C., Unzai, T., Hioki, H., Kaneko, T., 2009. Two types of thalamocortical projections from the motor thalamic nuclei of the rat: a single neuron-tracing study using viral vectors. Cereb. Cortex 19, 2065–2077.

Kuramoto, E., Ohno, S., Furuta, T., Unzai, T., Tanaka, Y.R., Hioki, H., et al., 2015. Ventral medial nucleus neurons send thalamocortical afferents more widely and more preferentially to layer 1 than neurons of the ventral anterior–ventral lateral nuclear complex in the rat. Cereb. Cortex 25, 221–235.

Larkum, M.E., Senn, W., Lüscher, H.R., 2004. Top-down dendritic input increases the gain of layer 5 pyramidal neurons. Cereb. Cortex 14, 1059–1070.

Larsen, D.D., Wickersham, I.R., Callaway, E.M., 2007. Retrograde tracing with recombinant rabies virus reveals correlations between projection targets and dendritic architecture in layer 5 of mouse barrel cortex. Front. Neural Circuits, 2008 Mar 28 (1), 5. http://dx.doi.org/10.3389/neuro.04.005.2007.eCollection 2007.

LeDoux, J.E., Farb, C., Ruggiero, D.A., 1990. Topographic organization of neurons in the acoustic thalamus that project to the amygdala. J. Neurosci. 10, 1043–1054.

Linke, R., Braune, G., Schwegler, H., 2000. Differential projection of the posterior paralaminar thalamic nuclei to the amygdaloid complex in the rat. Exp. Brain Res. 134, 520–532.

López-Bendito, G., Cautinat, A., Sánchez, J.A., Bielle, F., Flames, N., Garratt, A.N., et al., 2006. Tangential neuronal migration controls axon guidance: a role for neuregulin-1 in thalamocortical axon navigation. Cell 125, 127–142.

López-Bendito, G., Molnár, Z., 2003. Thalamocortical development: how are we going to get there? Nat. Rev. Neurosci. 4, 276–289.

Lorente de No, R., 1922. La corteza cerebral del ratón (La corteza acústica). Trabajos del Laboratorio de Investigaciones Biológicas de la Universidad de Madrid, 20, 41–78. English translation in: Somatosensory and Motor Research, 9, 3–36.

Lorente de No, R., 1938. Cerebral cortex: architecture, intracortical connections, motor projections. In: Fulton, J. (Ed.), Physiology of the Nervous System Oxford University Press, London, pp. 291–340.

Lund, J.S., 1988. Anatomical organization of macaque monkey striate visual cortex. Ann. Rev. Neurosci. 11, 253–288.

Mann, F., Zhukareva, V., Pimenta, A., Levitt, P., Bolz, J., 1998. Membrane-associated molecules guide limbic and nonlimbic thalamocortical projections. J. Neurosci. 18, 9409–9419.

Markov, N.T., Ercsey-Ravasz, M., Van Essen, D.C., Knoblauch, K., Toroczkai, Z., Kennedy, H., 2013. Cortical high-density counterstream architectures. Science 342, 1238406.

Marotte, L.R., Leamey, C.A., Waite, P.M., 1997. Timecourse of development of the wallaby trigeminal pathway: III. Thalamocortical and cortico-thalamic projections. J. Comp. Neurol. 387, 194–214.

Maruyama, T., Matsuura, M., Suzuki, K., Yamamoto, N., 2008. Cooperative activity of multiple upper layer proteins for thalamocortical axon growth. Dev. Neurobiol. 68, 317–331.

McFarland, N.R., Haber, S.N., 2002. Thalamic relay nuclei of the basal ganglia form both reciprocal and nonreciprocal cortical connections, linking multiple frontal cortical áreas. J. Neurosci. 22, 8117–8132.

Meyer, G., Soria, J.M., Martínez-Galán, J.R., Martín-Clemente, B., Fairén, A., 1998. Different origins and developmental histories of transient neurons in the marginal zone of the fetal and neonatal rat cortex. J. Comp. Neurol. 397, 493–518.

Minciacchi, D., Bentivoglio, M., Molinari, M., Kultas-Ilinsky, K., Ilinsky, I.A., Macchi, G., 1986. Multiple cortical targets of one thalamic nucleus: the projections of the ventral medial nucleus in the cat studied with retrograde tracers. J. Comp. Neurol. 252, 106–129.

Molinari, M., Dell'Anna, M.E., Rausell, E., Leggio, M.G., Hashikawa, T., Jones, E.G., 1995. Auditory thalamocortical pathways defined in monkeys by calcium-binding protein immunoreactivity. J. Comp. Neurol. 362, 171–194.

Molnár, Z., Adams, R., Blakemore, C., 1998a. Mechanisms underlying the early establishment of thalamocortical connections in the rat. J. Neurosci. 18, 5723–5745.

I. MICROCIRCUITRY

Molnár, Z., Adams, R., Goffinet, A.M., Blakemore, C., 1998b. The role of the first postmitotic cortical cells in the development of thalamocortical innervation in the reeler mouse. J. Neurosci. 18, 5746–5765.

Molnár, Z., Garel, S., López-Bendito, G., Maness, P., Price, D.J., 2012. Mechanisms controlling the guidance of thalamocortical axons through the embryonic forebrain. Eur. J. Neurosci. 35, 1573–1585.

Monconduit, L., Villanueva, L., 2005. The lateral ventromedial thalamic nucleus spreads nociceptive signals from the whole body surface to layer I of the frontal cortex. Eur. J. Neurosci. 21, 3395–3402.

Murray, K.D., Choudary, P.V., Jones, E.G., 2007. Nucleus- and cell-specific gene expression in monkey thalamus. Proc. Natl. Acad. Sci. U.S.A. 104, 1989–1994.

Naegele, J.R., Jhaveri, S., Schneider, G.E., 1988. Sharpening of topographical projections and maturation of geniculocortical axon arbors in the hamster. J. Comp. Neurol. 277, 593–607.

Nakagawa, Y., O'Leary, D.D., 2001. Combinatorial expression patterns of LIM-homeodomain and other regulatory genes parcellate developing thalamus. J. Neurosci. 21, 2711–2725.

Nakagawa, Y., Shimogori, T., 2012. Diversity of thalamic progenitor cells and postmitotic neurons. Eur. J. Neurosci. 35, 1554–1562.

Nakamura, H., Hioki, H., Furuta, T., Kaneko, T., 2015. Different cortical projections from three subdivisions of the rat lateral posterior thalamic nucleus: a single-neuron tracing study with viral vectors. Eur. J. Neurosci. 41, 1294–1310.

Namura, S., Takada, M., Kikuchi, H., Mizuno, N., 1997. Collateral projections of single neurons in the posterior thalamic region to both the temporal cortex and the amygdala: a fluorescent retrograde double-labeling study in the rat. J. Comp. Neurol. 384, 59–70.

Noseda, R., Jakubowski, M., Kainz, V., Borsook, D., Burstein, R., 2011. Cortical projections of functionally identified thalamic trigeminovascular neurons: implications for migraine headache and its associated symptoms. J. Neurosci. 31, 14204–14217.

Oberlaender, M., Ramirez, A., Bruno, R.M., 2012. Sensory experience restructures thalamocortical axons during adulthood. Neuron 74, 648–655.

Ohno, S., Kuramoto, E., Furuta, T., Hioki, H., Tanaka, Y.R., Fujiyama, F., et al., 2012. A morphological analysis of thalamocortical axon fibers of rat posterior thalamic nuclei: a single neuron tracing study with viral vectors. Cereb. Cortex 22, 2840–2857.

O'Leary, D.D., Terashima, T., 1988. Cortical axons branch to multiple subcortical targets by interstitial axon budding: implications for target recognition and "waiting periods". Neuron 1, 901–910.

Parent, M., Parent, A., 2005. Single-axon tracing and three-dimensional reconstruction of centre median-parafascicular thalamic neurons in primates. J. Comp. Neurol. 481, 127–144.

Petreanu, L., Mao, T., Sternson, S.M., Svoboda, K., 2009. The subcellular organization of neocortical excitatory connections. Nature 457, 1142–1145.

Pinault, D., 1996. A novel single-cell staining procedure performed *in vivo* under electrophysiological control: morpho-functional features of juxtacellularly labeled thalamic cells and other central neurons with biocytin or neurobiotin. J. Neurosci. Methods 65, 113–136.

Portera-Cailliau, C., Weimer, R.M., De Paola, V., Caroni, P., Svoboda, K., 2005. Diverse modes of axon elaboration in the developing neocortex. PLoS Biol. 3, e272.

Pouchelon, G., Gambino, F., Bellone, C., Telley, L., Vitali, I., Lüscher, C., et al., 2014. Modality-specific thalamocortical inputs instruct the identity of postsynaptic L4 neurons. Nature 511, 471–474.

Puelles, L., Rubenstein, J.L., 2003. Forebrain gene expression domains and the evolving prosomeric model. Trends Neurosci. 26, 469–476.

Rausell, E., Avendaño, C., 1985. Thalamocortical neurons projecting to superficial and to deep layers in parietal, frontal and prefrontal regions in the cat. Brain Res. 347, 159–165.

Rausell, E., Jones, E.G., 1995. Extent of intracortical arborization of thalamocortical axons as a determinant of representational plasticity in monkey somatic sensory cortex. J. Neurosci. 15, 4270–4288.

Rausell, E., Bae, C.S., Viñuela, A., Huntley, G.W., Jones, E.G., 1992. Calbindin and parvalbumin cells in monkey VPL thalamic nucleus: distribution, laminar cortical projections, and relations to spinothalamic terminations. J. Neurosci. 12, 4088–4111.

Rakic, P., 1977. Prenatal development of the visual system in rhesus monkey. Philos. Trans. R. Soc. B: Biol. Sci. 278, 245–260.

Rakic, P., Suñer, I., Williams, R.W., 1991. A novel cytoarchitectonic area induced experimentally within the primate visual cortex. Proc. Natl. Acad. Sci. U.S.A. 88, 2083–2087.

Real, M.A., Dávila, J.C., Guirado, S., 2006. Immunohistochemical localization of the vesicular glutamate transporter VGLUT2 in the developing and adult mouse claustrum. J. Chem. Neuroanat. 31, 169–177.

Reillo, I., de Juan Romero, C., García-Cabezas, M.Á., Borrell, V., 2011. A role for intermediate radial glia in the tangential expansion of the mammalian cerebral cortex. Cereb. Cortex 21, 1674–1694.

Rockland, K.S., Andresen, J., Cowie, R.J., Robinson, D.L., 1999. Single axon analysis of pulvinocortical connections to several visual areas in the macaque. J. Comp. Neurol. 406, 221–250.

Rosina, A., Provini, L., Bentivoglio, M., Kuypers, H.G., 1980. Ponto-neocerebellar axonal branching as revealed by double fluorescent retrograde labeling technique. Brain Res. 195, 461–466.

Rubio-Garrido, P., Pérez de Manzo, F., Clascá, F., 2007. Calcium-binding proteins as markers of layer-I projecting vs. deep layer-projecting thalamocortical neurons: a double-labeling analysis in the rat. Neuroscience 149, 242–250.

Rubio-Garrido, P., Pérez de Manzo, F., Porrero, C., Galazo, M.J., Clasca, F., 2009. Thalamic input to apical dendrites in neocortical layer 1 is massive and highly convergent. Cereb. Cortex 19, 2380–2395.

Rubio-Garrido, P., Nascimento, E.S., de Lima R.R.M., Córdoba-Claros, M.A., Porrero, C., Cavalcante, J.S., et al., 2014. Quantitative mapping of thalamic inputs to layer 1 of frontal, parietal and occipital cortices in the marmoset monkey. FENS Forum Abst. http://fens2014.meetingxpert.net/FENS_427/poster_100603/program.aspx

Saalmann, Y.B., Pinsk, M.A., Wang, L., Li, X., Kastner, S., 2012. The pulvinar regulates information transmission between cortical areas based on attention demands. Science 337, 753–756.

Segev, I., Schneidman, E., 1999. Axons as computing devices: basic insights gained from models. J. Physiol. Paris 93, 263–270.

Shatz, C.J., Luskin, M.B., 1986. The relationship between the geniculocortical afferents and their cortical target cells during development of the cat's primary visual cortex. J. Neurosci. 6, 3655–3668.

Sheng, X.M., Marotte, L.R., Mark, R.F., 1991. Development of the laminar distribution of thalamocortical axons and corticothalamic cell bodies in the visual cortex of the wallaby. J. Comp. Neurol. 307, 17–38.

Sherman, S.M., 2012. Thalamocortical interactions. Curr. Opin. Neurobiol. 22, 575–579.

I. MICROCIRCUITRY

Shimogori, T., Grove, E.A., 2005. Fibroblast growth factor 8 regulates neocortical guidance of area-specific thalamic innervation. J. Neurosci. 25, 6550–6560.

Shinoda, Y., Kakei, S., 1989. Distribution of terminals of thalamocortical fibers originating from the ventrolateral nucleus of the cat thalamus. Neurosci. Lett. 96, 163–167.

Singer, W., 2011. Dynamic formation of functional networks by synchronization. Neuron 69, 191–193.

Sjöström, P.J., Häusser, M., 2006. A cooperative switch determines the sign of synaptic plasticity in distal dendrites of neocortical pyramidal neurons. Neuron 51, 227–238.

Soussi-Yanicostas, N., de Castro, F., Julliard, A.K., Perfettini, I., Chedotal, A., Petit, C., 2002. Anosmin-1, defective in the X-linked form of Kallmann syndrome, promotes axonal branch formation from olfactory bulb output neurons. Cell 109, 217–228.

Spreafico, R., Barbaresi, P., Weinberg, R.J., Rustioni, A., 1987. SII-projecting neurons in the rat thalamus: a single- and double-retrograde-tracing study. Somatosens. Res. 4, 359–375.

Sur, M., Rubenstein, J.L., 2005. Patterning and plasticity of the cerebral cortex. Science 310, 805–810.

Suzuki-Hirano, A., Ogawa, M., Kataoka, A., Yoshida, A.C., Itoh, D., Ueno, M., et al., 2011. Dynamic spatiotemporal gene expression in embryonic mouse thalamus. J. Comp. Neurol. 519, 528–543.

Szebenyi, G., Dent, E.W., Callaway, J.L., Seys, C., Lueth, H., Kalil, K., 2001. Fibroblast growth factor-2 promotes axon branching of cortical neurons by influencing morphology and behavior of the primary growth cone. J. Neurosci. 21, 3932–3941.

Tang, F., Kalil, K., 2005. Netrin-1 induces axon branching in developing cortical neurons by frequency-dependent calcium signaling pathways. J. Neurosci. 25, 6702–6715.

Turner, B.H., Herkenham, M., 1991. Thalamoamygdaloid projections in the rat: a test of the amygdala's role in sensory processing. J. Comp. Neurol. 313, 295–325.

Van der Loos, H., Woolsey, T.A., 1973. Somatosensory cortex: structural alterations following early injury to sense organs. Science 179, 395–398.

Van der Werf, Y.D., Witter, M.P., Groenewegen, H.J., 2002. The intralaminar and midline nuclei of the thalamus. Anatomical and functional evidence for participation in processes of arousal and awareness. Brain Res. Brain Res. Rev. 39, 107–140.

Vanderhaeghen, P., Polleux, F., 2004. Developmental mechanisms patterning thalamocortical projections: intrinsic, extrinsic and in between. Trends Neurosci. 27, 384–391.

Vertes, R.P., Hoover, W.B., Cristina do Valle, A., Sherman, A., Rodriguez, J.J., 2006. Efferent projections of reuniens and rhomboid nuclei of the thalamus in the rat. J. Comp. Neurol. 499, 768–796.

Wouterlood, F.G., Saldana, E., Witter, M.P., 1990. Projection from the nucleus reuniens thalami to the hippocampal region: light and electron microscopic tracing study in the rat with the anterograde tracer Phaseolus vulgaris-leucoagglutinin. J. Comp. Neurol. 296, 179–203.

Yamamoto, N., Hanamura, K., 2005. Formation of the thalamocortical projection regulated differentially by BDNF- and NT-3-mediated signaling. Rev. Neurosci. 16, 223–231.

Yamamoto, N., Inui, K., Matsuyama, Y., Harada, A., Hanamura, K., Murakami, F., et al., 2000. Inhibitory mechanism by polysialic acid for lamina-specific branch formation of thalamocortical axons. J. Neurosci. 20, 9145–9151.

Yasui, Y., Itoh, K., Mizuno, N., 1984. Projections from the parvocellular part of the posteromedial ventral nucleus of the thalamus to the lateral amygdaloid nucleus in the cat. Brain Res. 292, 151–155.

5

Geometrical Structure of Single Axons of Visual Corticocortical Connections in the Mouse

Ian O. Massé, Philippe Régnier and Denis Boire

Département d'anatomie, Université du Québec à Trois-Rivières, Quebec, Canada

5.1 INTRODUCTION

5.1.1 Connectomics of the Mouse Cortex

Since the recognition of neurons as basic units in the nervous system, the wiring of the brain has been a major field of investigation. The advent of experimental tract tracing with compounds such as HRP, and now a host of molecules and viral vectors, has provided tools by which connections of the brain can be visualized in great detail. "Connectomics" emerged as a specific field of investigation in search of the wiring diagram of the brain. Connectomes can be described at different levels of organization: macro-, meso-, and microscale (Oh et al., 2014; Sporns et al., 2005). The macroscale provides information on the composition and paths of the larger fiber tracts of the brain. The mesoscale level of analysis is a description of cellular connectivity. There have been several attempts for the development of the mouse connectome at this level of analysis such as the development of the Allen Brain Connectivity Atlas,[1] the Mouse Connectome Project,[2] and the Brain Architecture Project.[3] The microscale analysis describes connectivity of the brain at the synapse level (see Lichtman et al., 2014). Clearly a wiring diagram or connectome for each of these levels would be necessary to understand brain structure (DeFelipe, 2010).

Network analysis of the brain connectivity is highly useful in revealing global connectivity patterns (see Sporns, 2011). The more simple units of these networks have been described as motifs (Sporns, 2011). These have been described as "networks-within-networks" or subgraphs that form the basic structural alphabet of basic brain circuitry (Sporns, 2011). These motifs can be considered at several levels of organization. At a very fine local scale, there is a highly organized, non-random connectivity between specific cells types within an area in the cerebral cortex. For example, there are such structured local subnetworks in the visual cortex of rodents (Song et al., 2005; Yoshimura and Callaway, 2005; Yoshimura et al., 2005) and some subnetworks seem to be related to particular functions such as orientation selectivity (Hofer et al., 2011; Ko et al., 2011), to cell lineage (Ohtsuki et al., 2012; Yu et al., 2012), or to cortical output streams to the striatum, superior colliculus, and thalamic nuclei (Brown and Hestrin, 2009). This level of connectomics is within the scope of a mesoscale analysis but is still widely neglected compared to the vast efforts to establish the cortical interareal connectome (Oh et al., 2014; Zingg et al., 2014). Structural and functional diversity of both local and long distance axons within a given path would greatly increase the diversity of the edges

---

[1] http://connectivity.brain-map.org/.
[2] http://www.mouseconnectome.org/.
[3] http://www.brainarchitecture.org/.

Axons and Brain Architecture.
DOI: http://dx.doi.org/10.1016/B978-0-12-801393-9.00005-0

in cortical networks and would reflect more accurately the diversity of parallel functional streams within a single connection (Budd and Kisvarday, 2012).

The mesoscale interareal connectome of the mouse cerebral cortex is now becoming quite well documented, and network analysis has shown the main hubs and subnetworks (Oh et al., 2014; Zingg et al., 2014). However, the axonal level of study in large-scale connectomics and network analysis of the brain is generally neglected, in part because of the need for uninterrupted serial sections as well as the difficulty in achieving single axonal labeling (see Rockland, 2002 for more discussion). The single axon level of analysis has been greatly developed in the study of interareal connectivity in monkeys (Rockland, 1989, 1992, 1995, 2002, 2013; Rockland et al., 1994, 1999; Rockland and Virga, 1989, 1990). These studies of the single axon level of connections between visual areas in the monkey have contributed greatly to defining geometric and laminar patterns of feedback and feedforward projections and in establishing the basic concepts of cortical hierarchical organizations. Although there is now a great effort for understanding the mouse cortical connectome (Oh et al., 2014; Zingg et al., 2014), gradually resulting in increased knowledge of visual cortical organization in the mouse (Wang and Burkhalter, 2007; Wang et al., 2011, 2012), there is yet no study of single axon structure of the interareal corticocortical connections.

In this chapter, we report and summarize structural features of corticocortical axons in the mouse. More specifically, we studied afferent and efferent connections of the visual cortex to assess the diversity of axonal structure within a single connection and also to see whether there are large differences between functionally distinct pathways. These data show our initial steps in studying the visual pathways of the mouse cortex. Since this is an overview of the material that we presently have in hand, it is obviously incomplete, and some pathways have been given more emphasis than others. Our goal has been to show the main similarities and differences with visual cortical axons in the monkey, to explore the morphological diversity of corticocortical axons, and to enunciate some functional and computational properties that can by derived from their relevant morphological features.

5.1.2 Connectivity of the Mouse Visual Cortex

In the early literature, V1 in mouse was shown to be surrounded by at least five distinct visual extrastriate areas (Rose, 1929). Tracing and electrophysiological recordings have demonstrated that V1 projects to several distinct sites in the cortices lateral and medial to it (Olavarria et al., 1982; Olavarria and Montero, 1989; Rose, 1929; Wagor et al., 1980; Wang and Burkhalter, 2007) and rat (Montero et al., 1973). In addition, neurofilament staining delineated monocular and binocular V1, and two lateral and five medial extrastriate areas (Van Der Gucht et al., 2007). Continuing studies in mice now provide convincing evidence for at least nine different extrastriate areas surrounding V1 (Andermann et al., 2011; Glickfeld et al., 2013, 2014; Marshel et al., 2011; Roth et al., 2012; Wang and Burkhalter, 2007; Wang et al., 2012). As in primates, the output of the visual cortex of the mouse is organized in dorsal and ventral streams of information processing, with AL and LM being the two gateways to these pathways, respectively (Wang et al., 2011). Neurons in AL have a greater orientation or direction selectivity and are tuned to lower spatial frequencies than those in AM (Marshel et al., 2011). Other studies confirm that extrastriate visual areas receive inputs from functionally distinct neurons of V1 (Glickfeld et al., 2013; Matsui and Ohki, 2013). Thus, evidence for parallel processing can be ascertained at least in V1 for these functional properties, even though the neurons in the primary visual cortex are not grouped together in functionally homogeneous modules as in monkeys.

The full extent of the cortical and subcortical connections of the mouse visual cortex will not be described here. There are several accounts that provide detailed descriptions of the connections (Charbonneau et al., 2012; Wang et al., 2011, 2012). The object here is to highlight just a few of these, at the single axon level of study.

The visual cortex projects to an estimated 25 cortical targets, but the main projections are certainly to the surrounding extrastriate areas (Wang et al., 2012). The strongest projections are to the lateral and medial visual extrastriate areas and parietal association cortices (Wang et al., 2012). These projections all appear to be highly topographic. One important feature of the mouse visual cortex projections is that they are much more widespread than in monkeys. V1 projects to all visual cortical areas in the mouse (Olavarria and Montero, 1989; Wang et al., 2012); also, extrastriate areas of the ventral and dorsal streams are much more interconnected than in monkeys (Wang et al., 2012). This could indicate less specialized functional properties for each area. For example, the two more important output projections of V1 are to LM and AL. Because LM maintains strong projections to LI with subsequent projections leading to temporal cortical areas, LM is considered as a portal to ventral stream in mice (Wang et al., 2011). However, the situation is complex in that LM contains neurons with low sensitivity for spatial frequency and also many neurons tuned to high temporal frequencies (Marshel et al., 2011). This could suggest that LM is a "divided gateway" to dorsal and ventral streams in mice (Wang et al., 2012). The high cortical interconnectivity between these two streams (Wang et al., 2012) might be an indication that both the primary and extrastriate visual cortices of the mouse maintain less specialized functional connections with each of the extrastriate areas (Andermann et al., 2011; Marshel et al., 2011). From this, the question arises of whether, at the axonal level, we might expect less stereotypical axonal morphologies within each projection and more diverse functional connections within a single projection?

5.1.3 Intermodal Connections of the Primary Visual Cortex

In addition to connections with extrastriate visual cortices, the primary visual cortex in rodents maintains many projections with non-visual sensory areas as well as with other non-sensory cortices (Charbonneau et al., 2012; Miller and Vogt, 1984; Wang and Burkhalter, 2007). It is striking that in several rodent species many direct connections are found between primary sensory cortices (Budinger et al., 2006, 2008; Budinger and Scheich, 2009; Campi et al., 2010; Charbonneau et al., 2012; Henschke et al., 2014; Stehberg et al., 2014; Wang et al., 2012). Such intermodal projections have also been described in the opossum (Dooley et al., 2013; Kahn et al., 2000, 2006; Wang et al., 2012). In primates, very few projections have been shown directly between primary sensory cortices (Clavagnier et al., 2004; Falchier et al., 2002; Rockland and Ojima, 2003). This connectivity pattern raises questions about the strictly visual quality of the primary cortex in the mouse (Budinger et al., 2006, 2008; Campi et al., 2010; Charbonneau et al., 2012; Hirokawa et al., 2008; Laramée et al., 2011; Miller and Vogt, 1984; Paperna and Malach, 1991).

These intermodal connections of the visual cortex challenge the hierarchical model of cortical connectivity. Intermodal projections from auditory and somatosensory cortices exhibit features of feedback projections in the mouse (Charbonneau et al., 2012; Henschke et al., 2014). In the gerbil, the projection from the primary visual cortex to the auditory cortex has features of a feedforward projection, and its projection to the somatosensory cortex has features of a lateral projection (Henschke et al., 2014). We also have evidence for a similar lateral type of connection for the projection of the visual cortex to the somatosensory cortex in mice (Massé et al., in preparation). Thus, the definition of types of connections, based on the laminar distribution of retrogradely labeled neurons and terminals, might not clearly reflect the interactions between primary sensory cortices. Also to consider is the possibility that primary sensory cortices might not always be at the same hierarchical level.

5.1.4 Hierarchical Organization of Cortical Connections

Cortical connections have been classified as feedforward and feedback projections (Felleman and Van Essen, 1991; Rockland and Pandya, 1979). These types of projections are the neuroanatomical basis for the classification of cortical areas in a hierarchical order. Feedforward projections originate mainly in supragranular layers 1 to 3 and end mainly in layer 4, whereas feedback projections originate in infragranular layers and terminate in the most superficial layers of the cortex (Felleman and Van Essen, 1991; Rockland and Pandya, 1979). The two types of cortical connections have also been described in rodents (Coogan and Burkhalter, 1990, 1993; Yamashita et al., 2003). However, feedforward projections in rodents, while originating mostly from supragranular layers, have terminal fields that are not as focused on granular layers as in primates; namely, the axon terminals are observed throughout all cortical layers as a column of terminals and preterminal axonal arbors.

The laminar distribution of labeling following cortical injections of anterograde tracers represents the bulk labeling of all the axons of a connection; but single axons can enter their cortical target and arborize in one specific layer or in a combination of layers that does not necessarily correspond to the bulk projection pattern. Analysis at the single axon level is thus highly desirable to assess the diversity of the individual building blocks of corticocortical connections.

In carrying out single axonal reconstructions of corticocortical connections of the mouse, we here demonstrate (1) that there is a wide range of axonal morphologies with respect to arbor structure, density of swellings, and bifurcations and (2) that there are both common and different features as compared with single axons in corticocortical projections in primates.

5.1.5 Computational Properties of Axons

Single axon morphology can be taken as a map of the spatial distribution of information to one or several targets in the brain and an integral part of neural processing, as summarized below (see Debanne, 2004 for review; Innocenti et al., 1994; Tettoni et al., 1998).

Several morphological features are relevant to axonal computational capabilities. Irregularities such as abrupt changes in diameter, as these occur at branch points and varicosities, are crucial in determining propagation failure, reflection, delays, and changes in spike amplitude in axons (Debanne, 2004). This means that signals carried by axons might not always arrive at all the terminal branches, and also that these signals might not be synchronized. Therefore, in addition to the branching structure of axons, variations in diameter are important features of spatio-temporal filtering (Manor et al., 1991). Detailed quantitative analysis of single axons including the number of branch points in different types of mammalian axons is essential for the construction of realistic models of connectivity (Debanne, 2004).

At axonal bifurcations points, when the impedance of the parent and daughter branches are matched, diameters of axons were proposed to follow a 3/2 power rule when membrane capacitance is assumed equal (Goldstein and Rall, 1974; Rall, 1959):

$$d^{\frac{3}{2}}_{mother} = \left(d^{\frac{3}{2}}_{daughter\,1} + d^{\frac{3}{2}}_{daughter\,2}\right)$$

In particular, axon potentials fail to propagate where axon diameters change abruptly over quite short distances. Such sharp changes in diameter occur at large, *en passant* boutons and at bifurcations where daughter branches have very different diameters. From the studies of Rall, a decision rule for axon potential propagation from a mother root axon to the two daughter branches has been proposed. A determined geometric ratio (GR):

$$GR = \left(d^{\frac{3}{2}}_{daughter\,1} + d^{\frac{3}{2}}_{daughter\,2}\right)/d^{\frac{3}{2}}_{mother}$$

is the ratio between the sum of the diameter of the two daughter branches (or many other branches if more than two occur) and the mother branch, to which a 3/2 power exponent is applied. When the GR is unity, the diameter of each daughter branch is 63% of the parent branch, there is a perfect impedance match between the input and the output, and propagation occurs to both daughter branches. Large values of GR introduce delays but also distort the shape of individual spikes at the branch point. The spike recovers its shape in the daughter branches but is delayed (Manor et al., 1991). When GR>1, the mother branch might not be able to provide sufficient current for the axon potential to propagate to both branches (Goldstein and Rall, 1974). When the 1<GR<10, a delay would be introduced beyond the branch point. When this GR is less than unity, axon potential propagation is accelerated.

In complex axonal arbors, action potentials propagate through numerous inhomogeneities and branch points, and propagation delays can thus be introduced throughout the arbor. The velocity of a spike is reduced in locations of increasing axonal diameter and is increased when the diameter of the axon decreases. However, in the two situations, a greater delay is introduced by a diameter increase than by a decrease in axonal diameter (Manor et al., 1991). Therefore greater number of varicosities will always increase propagation delay. The longer the axon and greater the density of varicosities, the greater the delay (see Bucher and Goaillard, 2011).

In order to assess the potential for spatiotemporal filtering of single axons in our dataset of the mouse cortico-cortical connections, the GRs for all the bifurcations in all the axons were computed. The ratios are compared to previous estimates of such ratios in other systems. They are also analyzed with respect to the mapping of these ratios along the length of the axons.

Overall, the total length, density of varicosities, and GRs in mouse corticocortical connections will illustrate the diversity of axonal structures both between and within these connections.

5.2 SINGLE AXON STRUCTURE IN MOUSE CORTEX

5.2.1 Axon Reconstruction Methods

Cortical axons were reconstructed following anterograde tracer injections in several cortical sites, mainly the primary visual cortex but also auditory cortex and primary somatosensory cortex, both within and outside of the barrel field representation of the whiskers (S1BF), in C57Bl6/J mice (Charles River, Montreal, QC). Animals were treated in accordance with the regulations of the Canadian Council for the Protection of Animals, and the study was approved by the Animal Care Committee of the Université du Québec à Trois-Rivières. All animals were adults when sacrificed.

Iontophoretic injections of 10% solution of high-molecular-weight BDA (10 kDa) in phosphate-buffered saline (PBS) (Molecular Probes; Cedarlane Laboratories, Ontario, Canada) were performed through glass micropipettes (20 µm tip diameter) with a 1.5 µA positive current under a 7-s duty cycle applied for 10 min. After a 7-day survival period, the brains were harvested and frozen. Serial 50-µm-thick coronal sections were prepared using a freezing microtome and processed with an avidin–biotin complex solution (ABC Vectastain-Elite) and nickel intensification. Sections were counterstained with bisbenzimide to identify the cortical areas and layers.

Axons were reconstructed with the Neurolucida software (MBF Biosciences, Williston VT). All reconstructions were performed at high power with a 100× oil-immersion objective (N.A. 1.4) under an Olympus microscope. Three-dimensional reconstructions were undertaken with the aim of obtaining the most complete axonal trees. Axons were selected form the target areas. First, a random branch was selected, and we further proceeded with reconstructions bidirectionally, toward the terminals and then back to the origins, adding branches as they came along. Much care was taken in recording the axonal diameters. Axonal segments that were cut off at the top or bottom of sections were located and continued on adjacent sections. This required spatial matching of these adjacent sections. This was done until either individual branches ended in the middle of a section, or until the continuation could not be matched on an adjacent section. Many axons were reconstructed from the injection site throughout the whole of their arborization, but axonal diameter and all other measurements were analyzed only from the point where they entered the cortical targets. The axons in the reciprocal projection between the primary visual cortex and the somatosensory barrel field were reconstructed only from their entry in their gray matter target from the white matter.

All the dendrograms (see Figures 5.2 to 5.6) shown here represent the axonal arbors in their distal portion, from their entry into their target area, and not from the injection sites. When axons travelled into an area from the gray matter, they are shown from their entry in the target area as they cross interareal borders of the cortex. The density of labeling at the injection sites was too intense to allow for tracing individual axons back to neuronal cell bodies. We show here only the long-range axons without their local axonal collaterals.

In the Neurolucida software, a Z shrinkage correction was applied to correct for the significant section thickness reduction due to dehydration. The nominal section thickness of 50 µm results typically in 20-µm thick sections when mounted, dried, and coverslipped. This correction applies to Z (thickness of sections) values and total length of an axon and its segments, not to the X and Y dimensions of the sections, and does not change the diameters of reconstructed axons segments.

The presence of axonal swellings was recorded along the full extent of the reconstructed axons. This allowed for a quantification of the total number of swellings and for the estimations of their size. For the whole length of the reconstructed axons, varicosities were identified, and their largest diameter was measured. Varicosities were identified as localized *en passant* swellings that appeared thicker than the adjacent axonal segments or terminal axonal swellings. They appeared more darkly stained, as the density of reaction product chromogen seems greater than in the thinner flanking axonal segments.

The reconstructed three-dimensional axonal structures are represented here as dendrograms (see Figures 5.2–5.6). These are two-dimensional, flattened "tree-diagram" representations, generated by the Neurolucida software (Microbrightfeld Bioscience, Williston, VT). The origin (less distal portion of the axon) for all the trees is at the left side of the dendrograms. The branches of the axons are horizontal lines, and the lengths are to scale as showing the linear path distance from the origin. On all the figures, the scale is at the bottom of the figures with the dendrograms. On the dendrograms, the branch points indicate where axons bifurcate. The vertical distances between branches are not to scale; they are arbitrary displacements, sufficiently offset so that all the branches can be drawn without overlap. All the *en passant* and terminal swellings on the axons are labeled by a red dot. All the branches of the axons were included in this analysis, even if they were very short and gave rise to no further branching. Short branches were quite numerous and can be seen in the dendrograms as very small vertical displacement of the main trunk.

The purpose of these dendrogram graphs is to show the basic features of the three-dimensional structure of the axons in a simplified two-dimensional graph. This allows to compare the complexity of the axons in terms of the richness of bifurcations and also to quickly visualize the location along the axons length of these bifurcations.

Several quantitative parameters were obtained in order to compare the axonal morphologies. (1) The total length of the axons was obtained from the Neurolucida software (Figure 5.7). (2) Each axonal swelling being either *en passant* boutons or terminal endings was charted and labeled with a marker. The total number of these swellings was then given by the software. (3) The linear density of swellings per 100 µm for each axon was then calculated (dens) (see also Figure 5.8A). (4) The initial diameter of the axons was also calculated. Axonal diameter changes are frequent and occur over short distances. In order to have an estimate of axonal diameter that is not just a point estimate, we calculated a weighted average diameter for what we defined in our analysis as the initial 50 µm of each axon. These diameters are not the caliber of the axons as they emerge from the neuronal cell bodies but rather the caliber of the axons distally, as they enter their targets, as described above.

The Neurolucida software encodes axons as a succession of small segments. Each of these is attributed coordinates in three-dimensional space, a length and a diameter. These segments correspond to the interval between two mouse clicks as the observer traces the axons. These clicks will be done to record changes in direction and/or diameter of the axons. The data were used to calculate the average diameter over the first 50 µm of the tracing, weighted by the length of the short segments recorded over this distance. This gives a better estimate of the axon thickness than a measurement at a single location (see Figure 5.7C).

(5) Particular attention was given here to axonal bifurcations. These constitute local diameter inhomogeneities that have functional consequences on spike propagation. In order to assess the GR at these branch points, the parent (or mother) branch diameter was taken for the axonal segment, just before the bifurcation, and the diameters of the two daughter segments, just after the bifurcations. These are much more local diameter estimates than the estimates for the axonal caliber at the origin of the traced reconstruction. This is necessary because the GR reflects a very local size ratio between the thickness of the daughter and the parent branch. There ratios were calculated for all the bifurcations in all these axons, even for the short side branches that did not give further branches.

In order to see how these parameters were distributed along the length of the axonal trees, the GR values were plotted as a function of their distance from the origin (see Figures 5.2–5.6). To see how the diameter of axons changes along the extent of their arborization, the diameter of the mother branch (D_M) was also plotted against the distance on the axon from its origin (d) (see Figures 5.2–5.6).

5.2.2 Individual Axons Projecting from Area V1 to Extrastriate Visual Areas AL and LM

Many of these axons come from cases included in another study of extrastriate projections of the mouse (Laramée et al., 2014). The pattern of labeling was as previously described (Coogan and Burkhalter, 1990, 1993; Dong et al., 2004; Olavarria and Montero, 1989; Yamashita et al., 2003). Projections appear as a columnar area of terminal labeling in distinct patches of cortical areas surrounding the primary visual cortex, typical of feedforward visual projections in the mouse and terminating throughout layers 2–3 to 6, including layer 4. Single axon analysis, however, clarifies that the overall laminar projection patterns to AL (Figure 5.1A) and to LM (Figure 5.1B) result from the superposition of axons of different morphologies.

Nineteen axons were reconstructed in LM and six were reconstructed in AL. All the axons in these projections travelled from V1 exclusively through the gray matter, in layers 5 and 6. No axons in these reconstructions travelled in the white matter (see below). One axon projecting to AL continued further to the auditory cortex (this branch is shown in Figure 5.1A). Two axons projecting from V1 to LM continued on to the perirhinal cortex. No other branches were found. We did not have, in this sample, axons that bifurcated to send projections to other extrastriate areas. The projection to LM however could be confounding. There are no clear cytoarchitectonic criteria to distinguish the extrastriate areas from one another in the mouse.

In the axons going to LM in the lateral portion of V2, one axon had a particularly strange configuration (see orange axon in Figure 5.1B). On its path in infragranular layers, it bifurcated to emit one unbranched linear collateral, and another that emitted two vertical poorly branched collaterals that are directed toward the pial surface. We cannot rule out the possibility that the more medial portion of this patch of labeling is in LM (left branch on the figure) and that the more lateral collateral might be in the more lateral area LI. In both these projections, the axons were poorly branched in the infragranular layers 5 and 6, but rather travelled in infragranular layers for some distance, and then directed a branch toward the pial surface. Again, there was little branching within infragranular layers and layer 4. More significant branching was observed in supragranular layers. The complexity of branching was quite variable between axons in AL (Figure 5.2) and LM (Figure 5.3). This is more apparent in the illustration of LM because of the larger sample of axons. Figure 5.1 gives some idea of the morphological diversity within a single patch of axons in the same animal.

These axonal morphologies in a feedforward visual projection are very different from those reported in primates (Rockland and Virga, 1990). For example, the axons from V1 to AL and LM do not show significant branching in layer 4. In the projection to V2, axons of V1 in the monkey branch in the top half of layer 4 and bottom of layer 3 and some form two or three spatially separate terminal clusters either in layer 4 or layers 3 and 4 (Rockland and Virga, 1990). In our sample, there are numerous *en passant* boutons on these axons as they go through layer 4 but no branching therein. Few axons in this sample formed several branching clusters. For example, in LM, two axons show two branches (the orange and yellow axons in Figure 5.1B).

Dendrogram representations of the axons in the projection to AL and LM are shown in Figures 5.2 and 5.3, respectively. The dendrograms enable a simple comparison of their structure and complexity. We did not perform here a quantitative analysis of their topological properties as in the study of Binzegger et al. (2005); but the dendrogram representations follow similar rules. The analysis done by Binzegger et al., although similar to ours, focuses on local axon collaterals of the cat visual cortex; and the branching structure of this local collateral portion of the axons is very different to the structures shown here.

Axons are displayed in the order of the number of bifurcations; and it is readily apparent that the branching complexity increases from the top to bottom graphs in these figures. The number of bifurcations range from 28 to

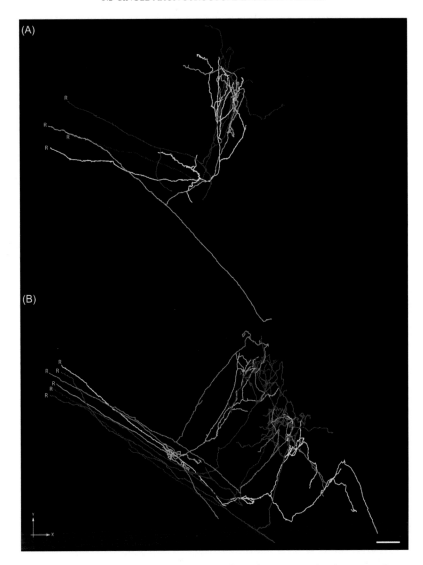

FIGURE 5.1   Three-dimensional axonal reconstruction of the projection from the primary visual cortex to the exstrastriate visual areas AL (A) and LM (B). Four different individual axons in the same projection column in AL of the same animal are shown here. Similarly six different individual axons in a large projections patch in LM in a different mouse than above. The root (R) of these axons represents the place where the axons enter these areas and not the location of the injection sites. These axons traveled in layers 5 and 6 and not in the white matter. These illustrations show that the overall laminar projection pattern is the result of the sum of axons of different morphologies. Scale bar = 100 μm.

131 in axons in AL (Figure 5.2) and from 8 to 96 for those targeting LM (Figure 5.3). The distribution of bifurcations along the length also varies greatly between axons; and a common pattern within and between these two projections is not self-evident. For example, the very simple axon in AL (Figure 5.2A) has few bifurcations (28) that are distributed more or less evenly along the length of the axons, whereas the more simple axons in LM (Figure 5.3A and B) also have few bifurcations (8 and 28, respectively) that are more distal in the axon. In these simple axons, the more proximal bifurcations were only very short side branches that did not produce further branching. These short side branches were quite abundant in all the axons in our sample and can be seen on the dendrograms as dead-end bifurcations with a vertical displacement of the main branch. Other studies omitted in their analysis these very short branches (Tettoni et al., 1998). Including these short branches made it difficult for us to calculate branching order, so this aspect will not be discussed here. It was necessary, however, for further study of other properties of bifurcations, and thus we opted to include the short branches in these diagrams.

Very few axons divided early to produce two or three second-order branches; these appeared further along the axons. One axon in LM has such a structure (Figure 5.3E). The main point here is that there is a diversity of branching patterns in both the feedforward projections of area V1 to LM and AL.

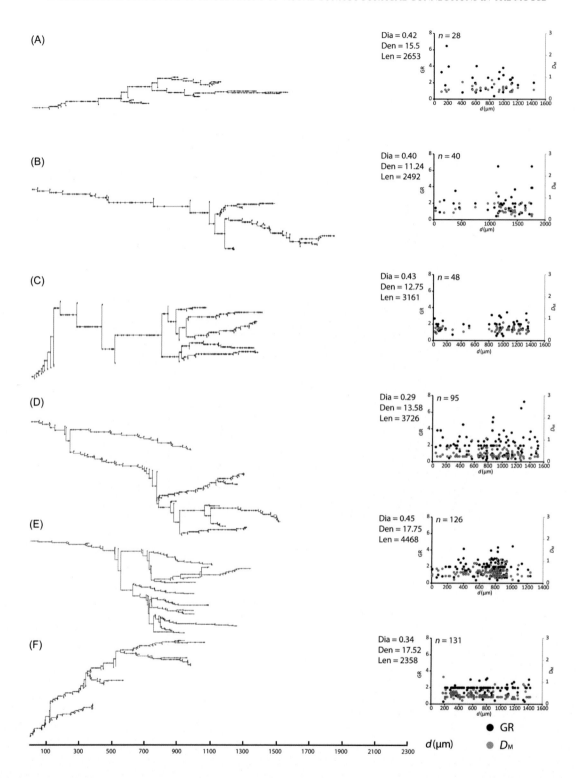

FIGURE 5.2 Dendrogram representation of axonal geometry in the projection from V1 to AL extrastriate area. All the dendrograms of the reconstructed axons are shown here. On each dendrogram, the red dots are local swellings. Each axon is reconstructed for its point of entry in the target region of interest. The distance (*d*) is measured from this entry point (see scale at the bottom of each display). Axons are displayed in the order of the number of bifurcations in this portion of the dendrogram (*n*). On the right hand side of each graph are given the weighted average diameter of the initial 500 μm of the axons at the entry point (Dia), the density of axonal swelling (number of swelling/100 μm) (Den), and the total length of the axon (Len). On the far right, diagrams of the relationship between geometric ratios (GRs; black dots, scale on the left hand side) and diameter of the axons just before each bifurcation along their path (D_M, gray dots, scale on the right hand side of the graph) are shown as a function of distance along the axon from its origin (*d*).

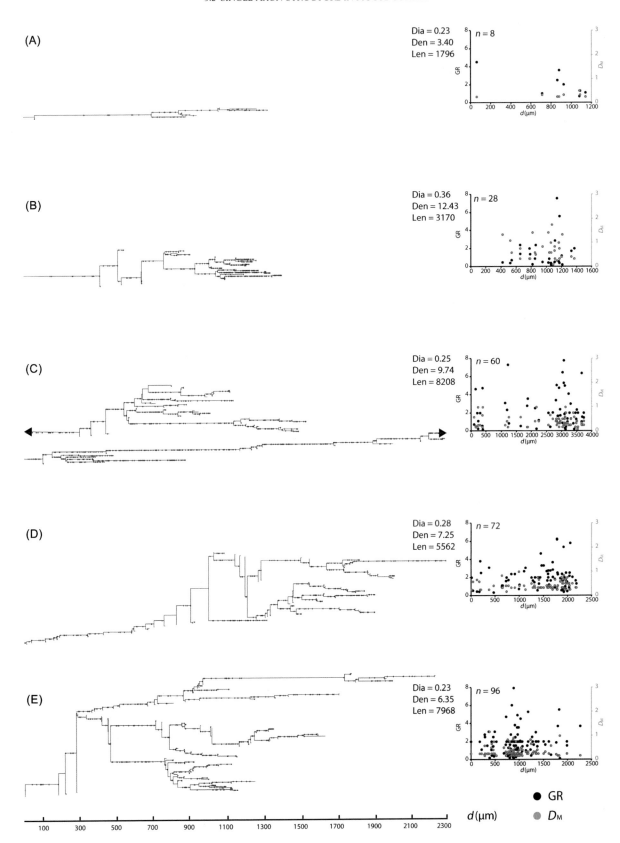

FIGURE 5.3 Dendrogram representation of axonal geometry in the projection from V1 to LM extrastriate area. Only a sample of the reconstructed axons is shown here. In the whole sample of these axons, the number of bifurcations ranged from 8 to 96. The five axons shown here were chosen to represent the lowest and greatest number of bifurcations and intermediate values in branching and length (legends as in Figure 5.2).

5.2.3 Intermodal Projections

5.2.3.1 Cortical Hierarchy of Visuo-Tactile Cortical Connections

BDA injections in the primary visual cortex and in the barrel field of the somatosensory cortex produced substantial labeling of reciprocal projections between these primary sensory cortices. These projections are noteworthy in several aspects. In particular, in a hierarchical model of cortical connectivity, one would expect connections between the two low-level cortices to have neither clear features of feedforward or feedback projections (see Crick and Koch, 1998). We have shown previously that the projection from S1 to V1 originates from a greater proportion of infragranular layer neurons and therefore has a negative layer index typical of a feedback projection (Charbonneau et al., 2012).

The laminar distribution of anterogradely labeled axonal terminals in V1 following BDA injections in S1BF is in agreement with this feedback pattern in showing labeling in infragranular and supragranular layers and avoiding layer 4. There was also more significant labeling in layer 1 than in feedforward projections (Massé et al., in preparation). This is commensurate with previous reports of feedback projections in rodent cortex (Coogan and Burkhalter, 1990, 1993; Dong et al., 2004; Olavarria and Montero, 1989; Yamashita et al., 2003).

In the reciprocal projection from V1 to the S1BF, a different pattern was observed. Retrogradely labeled neuronal cell bodies following injections of the B fragment of cholera toxin in the S1BF were found in the supragranular and infragranular layers of the somatosensory cortex, in similar proportions, yielding layer indices close to 0 (Massé et al., in preparation), such as would suggest a lateral projection between two cortices of the same hierarchical level. The anterograde labeling in V1 following BDA injections in the barrel field was also typical of a feedback projection in avoiding layer 4 (references above). This questions the hierarchical positions of primary sensory cortices in their reciprocal projections. A similar pattern has been reported in the gerbil (Henschke et al., 2014).

5.2.3.2 Single Axon Structure in Visuo-Tactile Cortical Connections

In the axons projecting from V1 to S1BF (Figure 5.4) and from S1BF to V1 (Figure 5.5), there was, as in the feedforward projections of V1 to LM and AL, a range of the number of bifurcations and arbor complexity. We can more clearly see, however, that there are quite short axons (Figures 5.4A–D and 5.5A, D, and E). These arborizations are mainly in infragranular layers. There were also axons of greater length with more elaborate branching in supragranular layers (Figures 5.4E and F and 5.5F). These more simple motifs define an overall typical feedback labeling in mouse cortical projections. Some axons had very long and poorly or moderately branched structures extending for a total length greater than 4000 μm (Figures 5.4E and 5.5F). The long branch in axons of Figure 5.4E travelled for at least 2000 μm in layer 1 of the barrel field. This branching structure has a similarly divergent feature in common with cortical visual feedback axons in the monkey (Rockland, 1994; Rockland and Knutson, 2000).

5.2.3.3 Single Axons Structure in Audio Visual Connections

For the sake of comparison, we show here a very limited sample of axons in the intermodal projection from the auditory cortex to V1 (Figure 5.6A–E) and to lateral V2 (V2L). The exact terminal location of these axons is difficult to determine but might likely be in LM and/or LI.

The range of the number of bifurcations in these axons is more restricted compared with what we observed in the projections of V1 to extrastriate and somatosensory cortices, although the small sample precludes any strong conclusion about the complexity of these axons. Nevertheless, it is worth remarking that they all seem to be quite simple branching structures and are less divergent than those observed in the axons between V1 and S1BF (Figures 5.4 and 5.5). The anterograde labelling in the projections to V2L and V1 (and also V2M) is also quite typical of feedback projections (see Figure 5.8; Charbonneau et al., 2012). Layer indices in these projections in gerbil are in agreement with this pattern (Henschke et al., 2014). We did not observe, in this small sample, any axons that travelled long distances in layer 1, as we did in the projection from V1 to S1BF. The substantial amount of anterograde labelling in layer 1 of V2L and V1, however, strongly suggests that such axons should be part of these projections.

5.2.3.4 Are There Two Modes of Cortical Connectivity?

Long-range connections between cortical areas of different sensory modalities raise further questions at the axonal level of connectivity in the cortex. A recent proposal suggests that corticocortical projections spread radially in two modes. Specifically, efferent projections from primary sensory cortices would spread for considerable distances, crossing into other primary sensory cortices, and evoke subthreshold sensory activations (Frostig et al., 2008; Stehberg et al., 2014). These long-range intermodal projections would coexist with what have been considered the more traditional long-range projections that travel through white matter to areas of the same sensory modality and evoke suprathreshold neuronal activity (Stehberg et al., 2014). These authors propose a

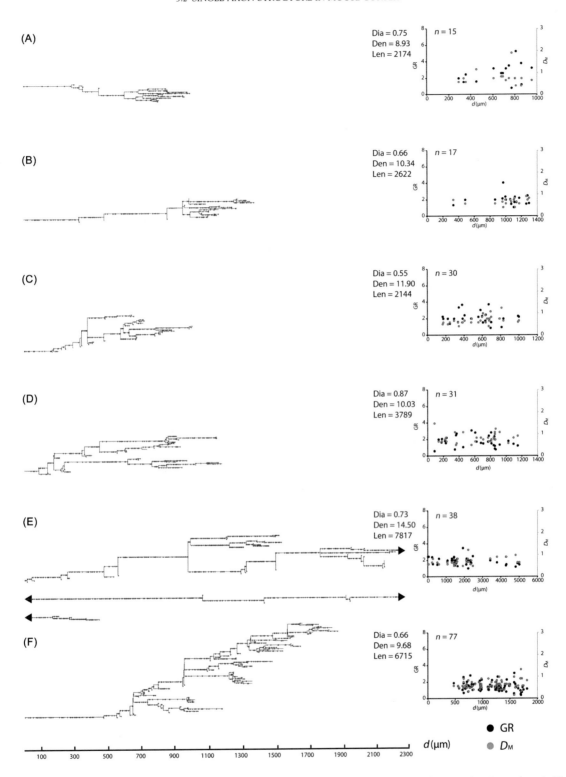

FIGURE 5.4   Dendrogram representation of axonal geometry in the projection from V1 to the barrel field of S1 (legends as in Figure 5.2).

hybrid model of cortical connections in which the traditional feedforward and feedback projections through white matter, that are the basis for the hierarchical organization of cortical system, might coexist with a more diffuse long-range projections that spread into other sensory domains (Stehberg et al., 2014). This proposal implies that axons connecting primary sensory cortices with their surrounding secondary areas, or with higher level areas in the cortical hierarchy but within a same sensory modality, could differ from the connections that directly link primary sensory cortices.

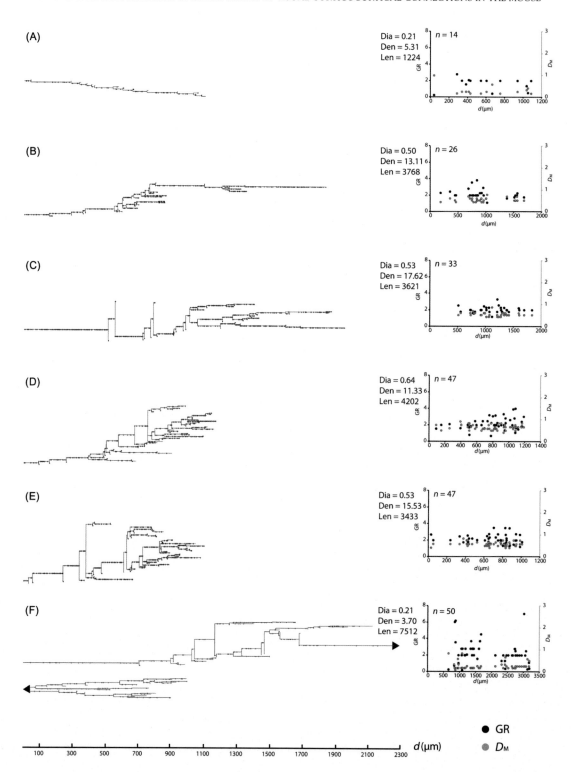

FIGURE 5.5   Dendrogram representation of axonal geometry in the projection from the barrel field of S1 to V1 (legends as in Figure 5.2).

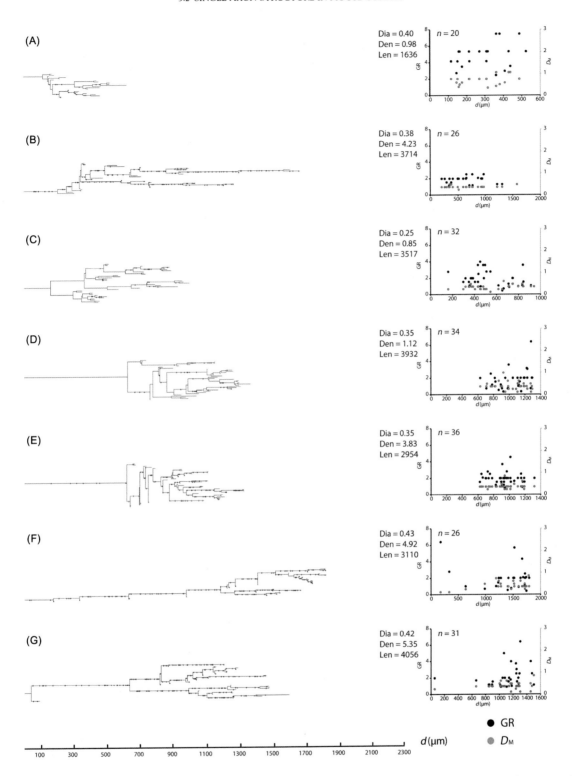

FIGURE 5.6 Dendrogram representation of axonal geometry in the projection from A1 to V1 (axons A–E) and in the projection from A1 to V2L (axons F and G) (legends as in Figure 5.2).

From our data, we see no evidence for such a duality. Firstly, all the axons linking the primary visual cortex with the extrastriate areas LM and AL travelled in our material in the grey matter, whereas all the axons that linked the primary visual cortex and the somatosensory barrel field travelled in the white matter. The axons between A1 and V2L and V1 also travelled in the white matter. The location of the axons therefore does not seem to be related to either a feedback–feedforward hierarchical mode of connectivity or to a diffuse intermodal network of connections. Second, there are no structural differences here that would suggest that the projections from the primary visual cortex to extrastriate cortices are strictly suprathreshold, while those between different sensory modalities are subthreshold.

The number of boutons could be taken as an indication of the strength of a projection and could be relevant to this aspect. The density of boutons was greater on the axons projecting from V1 to the extrastriate area AL (14.6 boutons/100 μm) than on those projecting to LM (6.2 boutons/100 μm) (Figure 5.8A). Moreover, the density of boutons was also greater in the axons linking the primary visual cortex with the barrel field (10.9 boutons/100 μm) (see Figure 5.8A); and the density of boutons was quite low in the efferent axons from the auditory cortex (2.2 boutons/100 μm in A1 to V1 and 5.1 boutons/100 μm in A1 to V2L). If projections between primary sensory cortices evoke subthreshold activation, one could expect lower bouton densities on these axons. This is not the case in the present data (see Figure 5.8). Evidently, other aspects need to be considered since the density of boutons of one individual axon does not account for the density of the projection, and the density of boutons is not uniform throughout the whole arbor of each individual axon.

Finally, axonal diameter can also be an indication of functional differences. The diameter of the axons of the projections to the extrastriate areas AL and LM was significantly smaller than those of the reciprocal projections between V1 and S1 (Figure 5.7C). Therefore, single axon morphology does not support a dual hybrid model of cortical connectivity.

Our diameter measurements of varicosities demonstrate clear differences. The combined effect of size and number can be assumed to contribute to the projectional efficacy. This rather "democratic" view of cortical connectivity has recently been challenged by the functional differences between the two "classes" of cortical glutamatergic synaptic contacts (Sherman and Guillery, 2013). These two categories of terminals appear to be on two distinct types of corticothalamic axons that originate from distinct populations of cortical neurons (Bourassa et al., 1995; Rouiller and Welker, 2000). This does not appear to be the case in the cortex. Individual axons are decorated with a range of bouton sizes (not shown) that are commensurate with those observed by Sherman.

There are at least two major classes of postsynaptic glutamatergic corticothalamic projections (see Sherman and Guillery, 2013 for extended discussion; Viaene et al., 2011a). The class 1 inputs are considered as drivers and class 2 is considered to have a modulatory function. There are significant physiological and morphological features that distinguish these two glutamatergic inputs especially well investigated for corticothalamic projections (see Sherman and Guillery, 1996 for extended discussion; Sherman and Guillery, 1998, 2006), but also reported for thalamocortical (Viaene et al., 2011a,b) and corticocortical (Covic and Sherman, 2011; De Pasquale and Sherman, 2011, 2013) projections. Mainly, class 1 inputs activate ionotropic receptors and morphologically are associated with large terminals. Class 2 inputs activate metabotropic receptors and are associated with small terminals (see Sherman and Guillery, 2013 for review). In the corticothalamic projections, these two classes of terminals are found on two very distinct axon types. The class 1 terminals are on thick axons that originate from layer 5 pyramidal neurons. These were known in the older literature as type II axon terminals. Class 2 terminals are on thin axons issued by neurons of layer 6 and were termed type I axon terminals in the older literature (Bourassa et al., 1995). There is evidence that the types of corticothalamic axons are phylogenetically conserved (Rouiller and Welker, 2000) in cats (Bajo et al., 1995; Baldauf et al., 2005; Crick and Koch, 1998; Huppé-Gourgues et al., 2006; Li et al., 2003; Rouiller et al., 1998; Van Horn and Sherman, 2004; Vidnyanszky et al., 1996; Wang et al., 2010, 2002) and monkeys (Feig and Harting, 1998; Rockland, 1996; Rouiller et al., 1998).

In the mouse cortex, there is evidence that class 1 terminals are found in the larger and class 2 on the smaller boutons (Covic and Sherman, 2011; Viaene et al., 2011a). However, there is no evidence that there are two distinct classes of axons. Although the data is not shown here, all the axons in our sample have mainly quite small *en passant* or terminal boutons, and only very few large boutons, and, importantly, there are individual axons that have a wide range of boutons sizes. We found very few large terminals on the axons projecting from V1 to LM and AL, and surprisingly more numerous larger terminals in the axons from V1 to the barrel field (data not shown). In addition, the reciprocal projection from the barrel field to V1 contained a very limited population of larger terminals, demonstrating a particular asymmetry in this reciprocal projection (Massé et al., in preparation).

A recent proposal put forth the hypothesis that class 1 driving and class 2 modulatory projections in the visual system of primates might express selectively VGLUT2 and VGLUT1, respectively (Balaram et al., 2013). Whether this proposal applies to feedforward and feedback projections in the cortex, and how this would apply to the expression

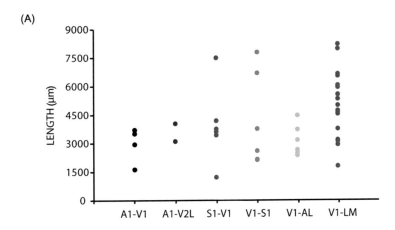

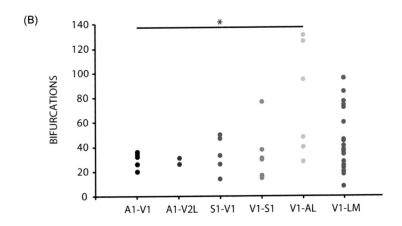

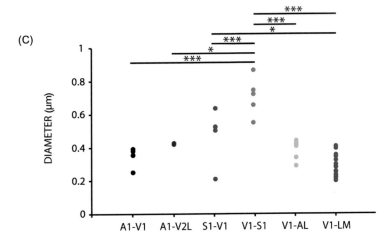

FIGURE 5.7 Morphological features of single axons in some afferent and efferent projections of the primary visual cortex and extrastriate cortices of the mouse. (A) Total length of the axons; (B) total number of bifurcation in each axon; and (C) diameter of the initial portions of the axon. All the measurements are taken from these axons as they enter their target.

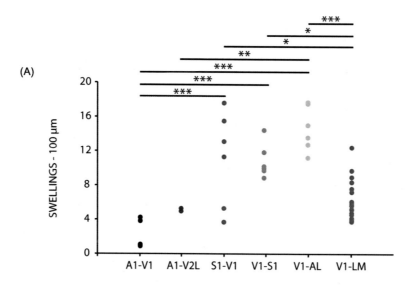

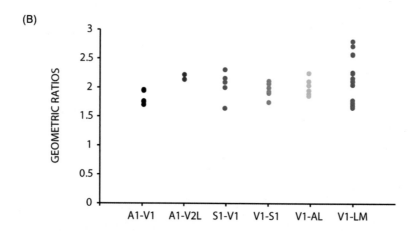

FIGURE 5.8   Morphological features of single axons in some afferent and efferent projections of the primary visual cortex and extrastriate cortices of the mouse. (A) Average density of axonal swelling over the entire length of each individual axons. The total number of swellings charted on each axon was divided by the total length of the axons to obtain these density estimates. (B) The average GRs for each individual axons within each projections.

of the vesicular transporters in a single axon is not yet clear. Firstly it is not at all convincing that all feedforward cortical projections have a driving class 1 type effect on their targets and that feedback projections are essentially modulatory and class 2. Indeed both types of responses were observed in the reciprocal projections between the primary and secondary auditory cortex (Covic and Sherman, 2011) and in the reciprocal projections between the primary visual and lateral extrastriate cortex (De Pasquale and Sherman, 2011). Secondly, large terminals were found in both directions in the reciprocal connections between primary and secondary auditory cortices (Covic and Sherman, 2011). Since we find a range of terminal sizes on single axons and very few large terminals on any given axon, it would be interesting to see how these vesicular glutamate transporters are expressed within a single neuron and how these proteins might be selectively localized within the arbor of single axons. If the correlation between terminal classes and expression of glutamate transporters holds for corticocortical connections, we could expect that there are at least some neurons that might express both these proteins and that they would be selectively located in particular terminals. There is evidence for colocalization of these glutamate transporters in cortical neurons and single axonal terminals in mice (Nakamura et al., 2007). The predominance of small terminals suggests a large contingent of axons expressing only VGLUT1 and a more modest contingent of neurons expressing both transporters. This hypothesis has yet to be tested.

5.2.4 Are There Distinct Types of Corticocortical Axons?

Since there are very few studies of interareal corticocortical axons in rodents at the level of single axon structure, it is difficult to propose with strong conviction that there are distinct types of axonal morphologies. One study proposed two categories of ipsilateral (Santiago et al., 2010) and callosal axons (Rocha et al., 2007) in the projections from S1 to S2 of a South American rodent. A cluster analysis of a large sample of axonal segments in the cortex showed two groups. The type I axonal segments had fewer branch points per millimeter than what was designated as type II segments (2.77 vs. 4.54 branch points/mm) and, as a corollary, evidently had less numerous and longer segments within the axonal fragments. In the callosal (Rocha et al., 2007) and ipsilateral projection of S1 to S2 (Santiago et al., 2010), type I fragments were more frequent in the hindlimb than in the forelimb representations. The authors suggested that these axonal "types" might represent two distinct functional processing streams. The main drawback for these studies is that this definition of axonal types is not derived from complete reconstructions of axonal arbors. The authors reconstructed terminal arbors in 100-μm thick sections, without attempting to follow these through adjacent sections. Therefore, their analysis bears only on terminal portions of axons that might not extend tangentially for more that the distance imposed by the 100 μm section thickness. Notably, these terminal arbors could be parts of the same axon.

To reinforce this cautionary note: there are several examples in our own sample of axons where bouton density and density of branch points seem different for different arbors of the same axon. Cluster analysis is an interesting tool for classifying objects in multidimensional parameter space but it does have certain limitations. "Types," or what seem distinct species, will be shown as clusters of objects in this analysis. It does not allow distinguishing between distinct discontinuous categories and a gradual continuum of structures. In the previous study, we performed a cluster analysis in an attempt to classify cortical layer 5 pyramidal neurons that project to the primary visual cortex of the mouse (Laramée et al., 2013). A cluster analysis indeed suggested several categories or types of neurons. However, these were not *distinct* types, but a continuum of morphologies. A representation of the objects under study in reduced space in the first few dimensions of a principal component analysis is required to show whether the "types" shown by a cluster analysis are distributed along a continuum or form distinct populations (Laramée et al., 2013). The above-mentioned studies on axonal segments do not show the distribution of these objects in the few most important principal components and do not have the power to be affirmative about the distinctness of these axon types.

We described here a range of axonal morphologies, but we cannot clearly classify them as distinct "types." The typology-from-topology of axons remains an outstanding question of the deep structure–function relationship of axonal structures.

5.2.5 Computational Properties of Axons

5.2.5.1 Axon Diameter

Axon diameter is an important variable in determining conduction speed. Along with axon length, this determines transmission delays and synchronization of impulses to the different targets of a single axon, as well as how delays compare between individual axons within a single connection between two cortical areas (Tomasi et al., 2012). Previous studies show that corticothalamic projections originating from different areas have different diameters (Caminiti et al., 2009, 2013; Tomasi et al., 2012). Similarly, corticocortical projections comprise axons of different diameters in relation to different source and target cortical areas. For example, parietal cortical projections comprise thinner axons when directed to cortical area 24 than to areas 4 or 6 in monkeys (Innocenti et al., 2014). These authors further suggested that axons with different diameters originate from different neuronal populations and proceed in parallel to different targets (Innocenti et al., 2014). Since each projection appears to comprise axons with different diameters (see e.g., Anderson and Martin, 2002), an auxiliary conclusion is that a spectrum of conductions delays will ensue to each target. This illustrates the underlying complexity of a single projection as described at the mesopic (vs. microscopic single axon) scale.

The present results have also shown that the axons of different cortical projections can have different diameters. The initial diameter of axons in the projection from V1 to S1BF was greater than that of the axons in all the other projections shown here (Figure 5.7C). Whether an axon's diameter is related to whether it travels through the gray or white matter of the cortex, or to the length of a connection can be discussed by comparing these several projections. That is, the initial diameter of the axons projecting from the visual cortex to areas LM and Al does not appear to be significantly different. All the axons of the projection from V1 to AL and LM travelled within the grey matter and all were quite thin. This contrasts with the axons in the reciprocal projections between V1 and the S1 barrel field, and in the projections between visual and auditory cortices. These all travelled within the white matter. This shows that thin axons can be found in both the white and grey matter. Only thin axons may travel within the grey matter, as in the case of the connections between V1 and LM and AL. We have no examples of thick axons in the grey matter.

Distance could also be a factor. The shorter projections shown here travel through the gray matter whereas the longer projections travel in the white matter. Whether there is a relationship between axon diameter and length of a connection and its location in white or grey matter is not obvious here. Little is known about gray matter axons and why they are there. A recent study has shown that motor cortex (M1) callosal axons travel to contralateral targets through the white matter, but concurrently they also emit collaterals to the ipsilateral barrel field and secondary somatosensory cortices, and these travel through layer 6 in the grey matter (Watakabe et al., 2014). Corticocortical projections in the monkey generally travel through the white matter (Schmahmann and Pandya, 2009). This might support that distance, and possibly space optimization, is related to where axons travel in the cortex.

The interconnections between V1 and the barrel field might serve to illustrate the independence of axon diameter and distance. Being reciprocal projections, we might assume that the distances travelled by these axons are the same (but see Rockland and Virga, 1990), but the axons of the projections from V1 to S1BF were significantly thicker that those from S1BF to V1 (Figure 5.7C). Other features of these projections suggest a functional asymmetry of this reciprocal connection. Indeed, there were more numerous larger terminal swellings in the projection from V1 to S1BF (Massé et al., in preparation). In addition, as pointed out earlier, the laminar distribution of neurons contributing to the projections shows a lateral profile in the projection from V1 to S1BF and a clear feedback profile in the projection from S1BF to V1 as was also demonstrated in the gerbil (Henschke et al., 2014).

There is a wide range of axons diameters in the projections from V1 to S1 and LM and also in the projection from S1BF to V1. This diversity would be in agreement with the proposal that the projection from V1 to 18a in the rat is composed of multiple parallel pathways with different conduction velocities ranging between 0.1 and 0.5 m/s (Domenici et al., 1995). Another study estimates this range from 0.3 to 0.8 m/s (Nowak et al., 1997). The feedforward projection from V1 to V2 in monkey also comprises axons with different conduction velocities ranging from 3 to 6 m/s, showing that this cortical connection is also composed of axonal populations that will produce a spectrum of activation delays to their respective targets (Girard et al., 2001). This supports a multiplex model of cortical connectivity in which a single projection can comprise a range of axons issued by functionally differentiated populations of neurons.

5.2.5.2 Varicosities

Local axonal swellings more commonly known as varicosities being terminal or *en passant* boutons are common features of axons. They are found along the axonal path from their origin to their site of termination as well as along the terminal branches in terminal sites. The size and density of these varicosities varies widely. Very small boutons, less than 1 μm, have been described on unmyelinated axons and larger boutons up to 5 μm in diameter were shown on hippocampal mossy fibers (Blackstad and Kjaerheim, 1961; Raastad and Shepherd, 2003; Shepherd et al., 2002). The size of axonal swellings vary greatly and have been shown to vary two- to fivefold compared to axons thickness on which they are found (Florence and Casagrande, 1987; Gerfen et al., 1987; Jensen and Killackey, 1987).

In the sample of axons we reconstructed, there is a wide range of bouton density in axons both within a single interareal projection and between projections (Figure 5.8A). There is also a wide range in the total number of varicosities on individual axon. Some axons are quite short and have low varicosity density. For example, the smallest axon in the projection from A1 to V1 (Figure 5.6A) has a length of 1636 μm and a density of 0.98 varicosities per 100 μm and a total of 16 varicosities. At the other end of the spectrum, a very long axon (7817 μm) in the projection from V1 to S1BF (Figure 5.4E) had a quite high density of varicosities (14.50/100 μm) bearing a total number of 1133 varicosities. Total number of varicosities was also quantified on corticocortical visual axons in monkey. Between 200 and 1000 boutons were estimated on individual feedforward axons from V1 to V2 (Rockland and Virga, 1990) or from V2 to V4 in monkey (Rockland, 1992) and between 164 and 644 varicosities on feedback axons from V2 to V1 in monkey (Rockland, 1994). An average of 300 varicosities were found on callosal axons in the cat (Tettoni et al., 1998). These values are lower than the 1339 and 1547 varicosities found on somatosensory thalamocortical axons in the normal and barrelless mice, respectively (Tettoni et al., 1998), and the 2000–3000 boutons found on geniculocortical axons in the cat (Ferster and LeVay, 1978; Humphrey et al., 1985) and monkey (Freund et al., 1989). Thalamocortical axons have more numerous boutons, longer end branches than callosal axons. The difference in the density and number of varicosities between callosal and thalamocortical projections was interpreted as an indication that thalamocortical axons have to convey a strong synaptic input onto a restricted territory, in contrast to callosal axons that distribute a rather weak input over a broad territory (Tettoni et al., 1998).

Corticocortical axons in our current dataset just might be somewhere in between these two functional constraints. For most of them, the total number of varicosities is commensurate with the number shown for callosal axons by the Innocenti group and could suggest a common role in conveying a weaker input than thalamocortical projections.

5.2.5.3 GRs of Axonal Bifurcations

There is very little information for the diameter of cortical axons at branching points within their terminal arbor. We measured here the diameter of the parent (mother) branch and the diameter of the two daughter branches at all the bifurcations of all the axons we reconstructed. The average GRs for the axons (Figure 5.8B) show values close to 2 indicating that, on average, the diameter of the daughter branches is similar to the parent branch and that a small delay in propagation could be introduced here.

GRs of axonal bifurcations were studied in very few axonal types and animals: lobster motor axons (GR = 0.97) (Grossman et al., 1979), crayfish (GRs between 0.78 and 1.968) (Smith, 1980), and peripheral axons of the leech (GR~20) (Baccus, 1998). This aspect of axonal morphology and physiology has been given little attention in the mammalian central nervous system. Diameters at bifurcations of thalamocortical and corticocortical axons in cats showed GR values that varied between 0.78 and 1.98 (Deschenes and Landry, 1980).

In Figures 5.2–5.6, we show plots of the GRs at each bifurcation as a function of distance along the branching structure, and also the diameter of the mother branch along this distance. First, there is no significant relationship between the parent branch diameter (D_M) and distance along the axons (d). There is no indication either that axons become thinner as they branch toward their terminal segments. In fact, the diameters of all the axons we show here appear quite stable along their arbor.

Within a single axon, a range of GR values was found. In many axons we reconstructed, the GR values within a single axon are quite homogeneous. For example, in the reciprocal projection between V1 and S1BF, the GR values are close to 2 and quite stable along the branching structure in 5 of the 6 reconstructed axons in the projection from S1BF to V1 (Figure 5.5A–E) and also quite stable and close to 2 in most axons in the projection from V1 to S1BF (Figure 5.4B–F). Conversely, within these same projections, we find axons with bifurcations with quite high GR values of 6 and greater (Figures 5.4A and 5.5F).

There is no clear relationship between distance along the axon and GR values in all the projections we show here. In addition, we see no relationship between axon branching complexity (or number of bifurcations) and GR values. For example, in several projections, some of the most simple axons have quite homogeneous GRs with values close to 2 (as in axon A of the projections from S1BF to V1 Figure 5.5) and some of these have bifurcations with more heterogeneous GR values that reach higher values between 6 and 8 (see axon A in the projection from V1 to S1BF Figure 5.4, and A in the projection from A1 to V1 Figure 5.6, and axon A in the projection from V1 to AL Figure 5.2). Moreover, there are, within the more complex axons in our sample, cases with quite homogeneous GR values close to 2 as in axon F of the projection from V1 to S1BF (Figure 5.4) and axon F in the projection from V1 to AL (Figure 5.2); and some other complex axons with very wide range of GR values. The most striking of such cases are seen in the most complex axons in the projection from V1 to LM (Figure 5.3).

Overall, the diversity of the axons we have here within and between projections with respect to the spatial distribution of the branch points and the GRs is striking, further supporting the need to take into account the structure of single axons in the design of a cortical connectome and network structure. In these models of connectivity, projections are simple edges that are not differentiated. This level of analysis will be crucial in the development of theoretical models of cortical connectivity.

If axons behave as a perfect cable, the sum of the impedance of daughter branches is equivalent to that of the parent branch, and the GR is equal to unity. In a hypothetical axon that would behave as a perfect cable throughout is whole length and in which all the bifurcations would have a GR value of 1, the diameter of the daughter branches would be progressively smaller. This is clearly not what is observed in our material. On the contrary, average GRs are often close to 2 indicating that the diameter of daughter branches is the same as the diameter of the mother branch. In such a situation, the cortical axons are not perfect cables but could be represented as a single cable that would have a larger diameter at branch points. This would be equivalent to a short thickening of the axons at the bifurcations. As shown above, axon diameters are quite stable along the length of their arbor and branch points and local swellings would have similar effects on impulse propagation. We examined here whether there might be a pattern in the impedance of daughter branches and the parent branch and see how divergent cortical axons might be from perfect cables. This relationship is shown in Figure 5.9. There is no relationship between the sum

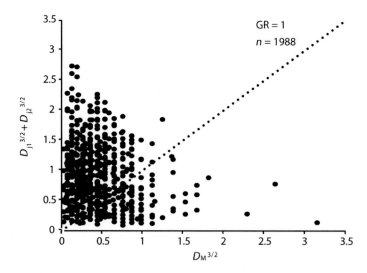

FIGURE 5.9   Impedance matching between the mother branch (D_M) and the sum of the daughter branches (D_{j1} and D_{j2}) according to the power law of Ralls (Rall, 1959). Each dot is an axonal bifurcation. The bifurcations of all the reconstructed axons are plotted here ($n=1988$). Axons that are perfect cables would have bifurcations that have a GR of 1, as shown by the dotted line. Note that GR values are above and below unity in thinner axons ($D_M<1$), whereas they are only less than one in thicker axons.

of the impedance of the daughter branches as a function of that of the parent branch. The line of perfect cables as shown when GR = 1 might foster, however, interesting questions on axonal structure. The GR values of the smaller axons when the parent branch diameter is approximately 1 μm are below 1 and many bifurcations have GR values greater than 1. Bifurcations in which the parent branches are larger than 1.5 μm however never have GR greater than unity. This represents situations where thin axons bifurcate in either smaller (GR<1) or larger (GR>1) daughter branches whereas thicker axons always bifurcate in two smaller branches and never in larger branches. The plot of the values for all the branches of the reconstructed axons suggests a limit condition rather than a correlation between these values.

5.3 CONCLUSIONS

From the data and analysis presented here, we can confidently comment on two important aspects of corticocortical connectivity in the visual cortices of the mouse. Firstly, there is diversity in axonal structure, within a single projection between two cortical areas. Given the current understanding that the projections of the mouse visual cortex to extrastriate areas are functionally distinct channels, the diversity of axons could suggest that these functional channels are not homogeneous. The functional differences between visual extrastriate areas could arise though several mechanisms. One possibility is a greater axonal arborization of the primary visual cortex neurons that are functionally matched to the target (Glickfeld et al., 2013). We show here that in the projection to and from V1, there are axons that branch more profusely than others and the range of branching node numbers is great. In addition there is a wide range of boutons density on axons. This stresses the notion that a given cortical connection is a complex assembly of functional channels.

Second, the data presented here also supports the underlying diversity of structure, and possibly of function, between interareal connections. As a corollary of the diversity within a connection, different connections might comprise axons issued by different subsets of neurons. Further study of axonal structure and laminar origin of specific axons will be needed to understand the organization of functional differences between areas and between feedback and feedforward lines of projection.

The mouse cortical connectome might be confounded with this greater functional diversity of each edge. The direction of information flow is not sufficient to describe these edges. A multiplex approach to edges will be necessary in developing models of cortical connectivity. The cortical connectome as it is presently foreseen as a complete set of interareal connections at the mesopic scale is certainly but the tip of the iceberg of a more comprehensive functional connectome of the cortex.

Acknowledgments

This work was supported by discovery grants of the Natural Sciences and Engineering Research Council of Canada (NSERC) and the Canadian Foundation for Innovation to DB. Financial support from the Réseau de la Recherche en Santé de la Vision (RRSV-FRQS) and institutional support from UQTR are also acknowledged. Thanks to Karen Tong and Stéphanie Ross for their excellent work in axon reconstructions and their contribution was invaluable. KT, SR, and PR were supported by internship grants from NSERC and RRSV-FRQS. Very special thanks to Kathleen Rockland for critical comments and enlightening discussions.

References

Andermann, M.L., Kerlin, A.M., Roumis, D.K., Glickfeld, L.L., Reid, R.C., 2011. Functional specialization of mouse higher visual cortical areas. Neuron 72, 1025–1039.

Anderson, J.C., Martin, K.A., 2002. Connection from cortical area V2 to MT in macaque monkey. J. Comp. Neurol. 443, 56–70.

Baccus, S.A., 1998. Synaptic facilitation by reflected action potentials: enhancement of transmission when nerve impulses reverse direction at axon branch points. Proc. Natl. Acad. Sci. U. S. A. 95, 8345–8350.

Bajo, V.M., Rouiller, E.M., Welker, E., Clarke, S., Villa, A.E., de, R.Y., et al., 1995. Morphology and spatial distribution of corticothalamic terminals originating from the cat auditory cortex. Hear Res. 83, 161–174.

Balaram, P., Hackett, T.A., Kaas, J.H., 2013. Differential expression of vesicular glutamate transporters 1 and 2 may identify distinct modes of glutamatergic transmission in the macaque visual system. J. Chem. Neuroanat. 50-51, 21–38.

Baldauf, Z.B., Chomsung, R.D., Carden, W.B., May, P.J., Bickford, M.E., 2005. Ultrastructural analysis of projections to the pulvinar nucleus of the cat. I: Middle suprasylvian gyrus (areas 5 and 7). J. Comp. Neurol. 485, 87–107.

Binzegger, T., Douglas, R.J., Martin, K.A., 2005. Axons in cat visual cortex are topologically self-similar. Cereb. Cortex 15, 152–165.

Blackstad, T.W., Kjaerheim, A., 1961. Special axo-dendritic synapses in the hippocampal cortex: electron and light microscopic studies on the layer of mossy fibers. J Comp. Neurol. 117, 133–159.

Bourassa, J., Pinault, D., Deschenes, M., 1995. Corticothalamic projections from the cortical barrel field to the somatosensory thalamus in rats: a single-fibre study using biocytin as an anterograde tracer. Eur. J. Neurosci. 7, 19–30.

Brown, S.P., Hestrin, S., 2009. Intracortical circuits of pyramidal neurons reflect their long-range axonal targets. Nature 457, 1133–1136.

Bucher, D., Goaillard, J.M., 2011. Beyond faithful conduction: short-term dynamics, neuromodulation, and long-term regulation of spike propagation in the axon. Prog. Neurobiol. 94, 307–346.

Budd, J.M., Kisvarday, Z.F., 2012. Communication and wiring in the cortical connectome. Front. Neuroanat 6, 42.

Budinger, E., Scheich, H., 2009. Anatomical connections suitable for the direct processing of neuronal information of different modalities via the rodent primary auditory cortex. Hear Res. 258, 16–27.

Budinger, E., Heil, P., Hess, A., Scheich, H., 2006. Multisensory processing via early cortical stages: connections of the primary auditory cortical field with other sensory systems. Neuroscience 143, 1065–1083.

Budinger, E., Laszcz, A., Lison, H., Scheich, H., Ohl, F.W., 2008. Non-sensory cortical and subcortical connections of the primary auditory cortex in Mongolian gerbils: bottom-up and top-down processing of neuronal information via field AI. Brain Res. 1220, 2–32.

Caminiti, R., Ghaziri, H., Galuske, R., Hof, P.R., Innocenti, G.M., 2009. Evolution amplified processing with temporally dispersed slow neuronal connectivity in primates. Proc. Natl. Acad. Sci. U. S. A. 106, 19551–19556.

Caminiti, R., Carducci, F., Piervincenzi, C., Battaglia-Mayer, A., Confalone, G., Visco-Comandini, F., et al., 2013. Diameter, length, speed, and conduction delay of callosal axons in macaque monkeys and humans: comparing data from histology and magnetic resonance imaging diffusion tractography. J. Neurosci. 33, 14501–14511.

Campi, K.L., Bales, K.L., Grunewald, R., Krubitzer, L., 2010. Connections of auditory and visual cortex in the prairie vole (Microtus ochrogaster): evidence for multisensory processing in primary sensory areas. Cereb. Cortex 20, 89–108.

Charbonneau, V., Laramée, M.E., Boucher, V., Bronchti, G., Boire, D., 2012. Cortical and subcortical projections to primary visual cortex in anophthalmic, enucleated and sighted mice. Eur. J. Neurosci. 36, 2949–2963.

Clavagnier, S., Falchier, A., Kennedy, H., 2004. Long-distance feedback projections to area V1: implications for multisensory integration, spatial awareness, and visual consciousness. Cogn. Affect. Behav. Neurosci. 4, 117–126.

Coogan, T.A., Burkhalter, A., 1990. Conserved patterns of cortico-cortical connections define areal hierarchy in rat visual cortex. Exp. Brain Res. 80, 49–53.

Coogan, T.A., Burkhalter, A., 1993. Hierarchical organization of areas in rat visual cortex. J. Neurosci. 13, 3749–3772.

Covic, E.N., Sherman, S.M., 2011. Synaptic properties of connections between the primary and secondary auditory cortices in mice. Cereb. Cortex 21, 2425–2441.

Crick, F., Koch, C., 1998. Constraints on cortical and thalamic projections: the no-strong-loops hypothesis. Nature 391, 245–250.

De Pasquale, R., Sherman, S.M., 2011. Synaptic properties of corticocortical connections between the primary and secondary visual cortical areas in the mouse. J. Neurosci. 31, 16494–16506.

De Pasquale, R., Sherman, S.M., 2013. A modulatory effect of the feedback from higher visual areas to V1 in the mouse. J. Neurophysiol. 109, 2618–2631.

Debanne, D., 2004. Information processing in the axon. Nat. Rev. Neurosci. 5, 304–316.

DeFelipe, J., 2010. From the connectome to the synaptome: an epic love story. Science 330, 1198–1201.

Deschenes, M., Landry, P., 1980. Axonal branch diameter and spacing of nodes in the terminal arborization of identified thalamic and cortical neurons. Brain Res. 191, 538–544.

Domenici, L., Harding, G.W., Burkhalter, A., 1995. Patterns of synaptic activity in forward and feedback pathways within rat visual cortex. J. Neurophysiol. 74, 2649–2664.

Dong, H., Wang, Q., Valkova, K., Gonchar, Y., Burkhalter, A., 2004. Experience-dependent development of feedforward and feedback circuits between lower and higher areas of mouse visual cortex. Vision Res. 44, 3389–3400.

Dooley, J.C., Franca, J.G., Seelke, A.M., Cooke, D.F., Krubitzer, L.A., 2013. A connection to the past: *Monodelphis domestica* provides insight into the organization and connectivity of the brains of early mammals. J. Comp. Neurol. 521, 3877–3897.

Falchier, A., Clavagnier, S., Barone, P., Kennedy, H., 2002. Anatomical evidence of multimodal integration in primate striate cortex. J. Neurosci. 22, 5749–5759.

Feig, S., Harting, J.K., 1998. Corticocortical communication via the thalamus: ultrastructural studies of corticothalamic projections from area 17 to the lateral posterior nucleus of the cat and inferior pulvinar nucleus of the owl monkey. J. Comp. Neurol. 395, 281–295.

Felleman, D.J., Van Essen, D.C., 1991. Distributed hierarchical processing in the primate cerebral cortex. Cereb. Cortex 1, 1–47.

Ferster, D., LeVay, S., 1978. The axonal arborizations of lateral geniculate neurons in the striate cortex of the cat. J. Comp. Neurol. 182, 923–944.

Florence, S.L., Casagrande, V.A., 1987. Organization of individual afferent axons in layer IV of striate cortex in a primate. J. Neurosci. 7, 3850–3868.

Freund, T.F., Martin, K.A., Soltesz, I., Somogyi, P., Whitteridge, D., 1989. Arborisation pattern and postsynaptic targets of physiologically identified thalamocortical afferents in striate cortex of the macaque monkey. J. Comp. Neurol. 289, 315–336.

Frostig, R.D., Xiong, Y., Chen-Bee, C.H., Kvasnak, E., Stehberg, J., 2008. Large-scale organization of rat sensorimotor cortex based on a motif of large activation spreads. J. Neurosci. 28, 13274–13284.

Gerfen, C.R., Herkenham, M., Thibault, J., 1987. The neostriatal mosaic: II. Patch- and matrix-directed mesostriatal dopaminergic and non-dopaminergic systems. J. Neurosci. 7, 3915–3934.

Girard, P., Hupe, J.M., Bullier, J., 2001. Feedforward and feedback connections between areas V1 and V2 of the monkey have similar rapid conduction velocities. J. Neurophysiol. 85, 1328–1331.

Glickfeld, L.L., Andermann, M.L., Bonin, V., Reid, R.C., 2013. Cortico-cortical projections in mouse visual cortex are functionally target specific. Nat. Neurosci. 16, 219–226.

Glickfeld, L.L., Reid, R.C., Andermann, M.L., 2014. A mouse model of higher visual cortical function. Curr. Opin. Neurobiol. 24, 28–33.

Goldstein, S.S., Rall, W., 1974. Changes of action potential shape and velocity for changing core conductor geometry. Biophys. J. 14, 731–757.

Grossman, Y., Parnas, I., Spira, M.E., 1979. Differential conduction block in branches of a bifurcating axon. J. Physiol. 295, 283–305.

Henschke, J.U., Noesselt, T., Scheich, H., Budinger, E., 2014. Possible anatomical pathways for short-latency multisensory integration processes in primary sensory cortices. Brain Struct. Funct 220, 955–977.

Hirokawa, J., Bosch, M., Sakata, S., Sakurai, Y., Yamamori, T., 2008. Functional role of the secondary visual cortex in multisensory facilitation in rats. Neuroscience 153, 1402–1417.

Hofer, S.B., Ko, H., Pichler, B., Vogelstein, J., Ros, H., Zeng, H., et al., 2011. Differential connectivity and response dynamics of excitatory and inhibitory neurons in visual cortex. Nat. Neurosci. 14, 1045–1052.

Humphrey, A.L., Sur, M., Uhlrich, D.J., Sherman, S.M., 1985. Projection patterns of individual X- and Y-cell axons from the lateral geniculate nucleus to cortical area 17 in the cat. J. Comp. Neurol. 233, 159–189.

Huppé-Gourgues, F., Bickford, M.E., Boire, D., Ptito, M., Casanova, C., 2006. Distribution, morphology, and synaptic targets of corticothalamic terminals in the cat lateral posterior-pulvinar complex that originate from the posteromedial lateral suprasylvian cortex. J. Comp. Neurol. 497, 847–863.

Innocenti, G.M., Lehmann, P., Houzel, J.C., 1994. Computational structure of visual callosal axons. Eur. J. Neurosci. 6, 918–935.

Innocenti, G.M., Vercelli, A., Caminiti, R., 2014. The diameter of cortical axons depends both on the area of origin and target. Cereb. Cortex 24, 2178–2188.

Jensen, K.F., Killackey, H.P., 1987. Terminal arbors of axons projecting to the somatosensory cortex of the adult rat. I. The normal morphology of specific thalamocortical afferents. J. Neurosci. 7, 3529–3543.

Kahn, D.M., Huffman, K.J., Krubitzer, L., 2000. Organization and connections of V1 in *Monodelphis domestica*. J. Comp. Neurol. 428, 337–354.

Karlen, S.J., Kahn, D.M., Krubitzer, L., 2006. Early blindness results in abnormal corticocortical and thalamocortical connections. Neuroscience 142, 843–858.

Ko, H., Hofer, S.B., Pichler, B., Buchanan, K.A., Sjostrom, P.J., Mrsic-Flogel, T.D., 2011. Functional specificity of local synaptic connections in neocortical networks. Nature 473, 87–91.

Laramée, M.E., Kurotani, T., Rockland, K.S., Bronchti, G., Boire, D., 2011. Indirect pathway between the primary auditory and visual cortices through layer V pyramidal neurons in V2L in mouse and the effects of bilateral enucleation. Eur. J. Neurosci. 34, 65–78.

Laramée, M.E., Rockland, K.S., Prince, S., Bronchti, G., Boire, D., 2013. Principal component and cluster analysis of layer V pyramidal cells in visual and non-visual cortical areas projecting to the primary visual cortex of the mouse. Cereb. Cortex 23, 714–728.

Laramée, M.E., Bronchti, G., Boire, D., 2014. Primary visual cortex projections to extrastriate cortices in enucleated and anophthalmic mice. Brain Struct. Funct. 219, 2051–2070.

Li, J., Wang, S., Bickford, M.E., 2003. Comparison of the ultrastructure of cortical and retinal terminals in the rat dorsal lateral geniculate and lateral posterior nuclei. J. Comp. Neurol. 460, 394–409.

Lichtman, J.W., Pfister, H., Shavit, N., 2014. The big data challenges of connectomics. Nat. Neurosci. 17, 1448–1454.

Manor, Y., Koch, C., Segev, I., 1991. Effect of geometrical irregularities on propagation delay in axonal trees. Biophys. J. 60, 1424–1437.

Marshel, J.H., Garrett, M.E., Nauhaus, I., Callaway, E.M., 2011. Functional specialization of seven mouse visual cortical areas. Neuron 72, 1040–1054.

Matsui, T., Ohki, K., 2013. Target dependence of orientation and direction selectivity of corticocortical projection neurons in the mouse V1. Front. Neural Circuits 7, 143.

Miller, M.W., Vogt, B.A., 1984. Direct connections of rat visual cortex with sensory, motor and association cortices. J. Comp. Neurol. 226, 184–202.

Montero, V.M., Rojas, A., Torrealba, F., 1973. Retinotopic organization of striate and peristriate visual cortex in the albino rat. Brain Res. 53, 197–201.

Nakamura, K., Watakabe, A., Hioki, H., Fujiyama, F., Tanaka, Y., Yamamori, T., et al., 2007. Transiently increased colocalization of vesicular glutamate transporters 1 and 2 at single axon terminals during postnatal development of mouse neocortex: a quantitative analysis with correlation coefficient. Eur. J. Neurosci. 26, 3054–3067.

Nowak, L.G., James, A.C., Bullier, J., 1997. Corticocortical connections between visual areas 17 and 18a of the rat studied *in vitro*: spatial and temporal organisation of functional synaptic responses. Exp. Brain Res. 117, 219–241.

Oh, S.W., Harris, J.A., Ng, L., Winslow, B., Cain, N., Mihalas, S., et al., 2014. A mesoscale connectome of the mouse brain. Nature 508, 207–214.

Ohtsuki, G., Nishiyama, M., Yoshida, T., Murakami, T., Histed, M., Lois, C., et al., 2012. Similarity of visual selectivity among clonally related neurons in visual cortex. Neuron 75, 65–72.

Olavarria, J., Montero, V.M., 1989. Organization of visual cortex in the mouse revealed by correlating callosal and striate-extrastriate connections. Vis. Neurosci. 3, 59–69.

Olavarria, J., Mignano, L.R., Van Sluyters, R.C., 1982. Pattern of extrastriate visual areas connecting reciprocally with striate cortex in the mouse. Exp. Neurol. 78, 775–779.

Paperna, T., Malach, R., 1991. Patterns of sensory intermodality relationships in the cerebral cortex of the rat. J. Comp. Neurol. 308, 432–456.

Raastad, M., Shepherd, G.M., 2003. Single-axon action potentials in the rat hippocampal cortex. J. Physiol. 548, 745–752.

Rall, W., 1959. Branching dendritic trees and motoneuron membrane resistivity. Exp. Neurol. 1, 491–527.

Rocha, E.G., Santiago, L.F., Freire, M.A., Gomes-Leal, W., Dias, I.A., Lent, R., et al., 2007. Callosal axon arbors in the limb representations of the somatosensory cortex (SI) in the agouti (*Dasyprocta primnolopha*). J. Comp. Neurol. 500, 255–266.

Rockland, K.S., 1989. Bistratified distribution of terminal arbors of individual axons projecting from area V1 to middle temporal area (MT) in the macaque monkey. Vis. Neurosci. 3, 155–170.

Rockland, K.S., 1992. Configuration, in serial reconstruction, of individual axons projecting from area V2 to V4 in the macaque monkey. Cereb. Cortex 2, 353–374.

Rockland, K.S., 1994. The organisation of feed-back connections from area V2 (18) to V1 (17). In: Peters, A., Rockland, K.S. (Eds.), Cerebral Cortex Plenum Press, New York, NY and London, pp. 261–299.

Rockland, K.S., 1995. Morphology of individual axons projecting from area V2 to MT in the macaque. J. Comp. Neurol. 355, 15–26.

Rockland, K.S., 1996. Two types of corticopulvinar terminations: round (type 2) and elongate (type 1). J. Comp. Neurol. 368, 57–87.

Rockland, K.S., 2002. Visual cortical organization at the single axon level: a beginning. Neurosci. Res. 42, 155–166.

Rockland, K.S., 2013. Collateral branching of long-distance cortical projections in monkey. J. Comp. Neurol. 521, 4112–4123.

Rockland, K.S., Pandya, D.N., 1979. Laminar origins and terminations of cortical connections of the occipital lobe in the rhesus monkey. Brain Res. 179, 3–20.

Rockland, K.S., Virga, A., 1989. Terminal arbors of individual "feedback" axons projecting from area V2 to V1 in the macaque monkey: a study using immunohistochemistry of anterogradely transported *Phaseolus vulgaris*-leucoagglutinin. J. Comp. Neurol. 285, 54–72.

Rockland, K.S., Virga, A., 1990. Organization of individual cortical axons projecting from area V1 (area 17) to V2 (area 18) in the macaque monkey. Vis. Neurosci. 4, 11–28.

Rockland, K.S., Knutson, T., 2000. Feedback connections from area MT of the squirrel monkey to areas V1 and V2. J. Comp. Neurol. 425, 345–368.

Rockland, K.S., Ojima, H., 2003. Multisensory convergence in calcarine visual areas in macaque monkey. Int. J. Psychophysiol. 50, 19–26.

Rockland, K.S., Saleem, K.S., Tanaka, K., 1994. Divergent feedback connections from areas V4 and TEO in the macaque. Vis. Neurosci. 11, 579–600.

Rockland, K.S., Andresen, J., Cowie, R.J., Robinson, D.L., 1999. Single axon analysis of pulvinocortical connections to several visual areas in the macaque. J. Comp. Neurol. 406, 221–250.

Rose, M., 1929. Cytoarchitecktonischer atlas der Grosshirnrinde der maus. J. Psychol. Neurol., 1–51.

Roth, M.M., Helmchen, F., Kampa, B.M., 2012. Distinct functional properties of primary and posteromedial visual area of mouse neocortex. J. Neurosci. 32, 9716–9726.

Rouiller, E.M., Welker, E., 2000. A comparative analysis of the morphology of corticothalamic projections in mammals. Brain Res. Bull. 53, 727–741.

Rouiller, E.M., Tanne, J., Moret, V., Kermadi, I., Boussaoud, D., Welker, E., 1998. Dual morphology and topography of the corticothalamic terminals originating from the primary, supplementary motor, and dorsal premotor cortical areas in macaque monkeys. J. Comp. Neurol. 396, 169–185.

Santiago, L.F., Rocha, E.G., Santos, C.L., Pereira Jr., A., Franca, J.G., Picanco-Diniz, C.W., 2010. S1 to S2 hind- and forelimb projections in the agouti somatosensory cortex: axon fragments morphological analysis. J. Chem. Neuroanat. 40, 339–345.

Schmahmann, J., Pandya, D., 2009. Fiber Pathways of the Brain. Oxford University Press, Oxford.

Shepherd, G.M., Raastad, M., Andersen, P., 2002. General and variable features of varicosity spacing along unmyelinated axons in the hippocampus and cerebellum. Proc. Natl. Acad. Sci. U. S. A. 99, 6340–6345.

Sherman, S.M., Guillery, R.W., 1996. Functional organization of thalamocortical relays. J. Neurophysiol. 76, 1367–1395.

Sherman, S.M., Guillery, R.W., 1998. On the actions that one nerve cell can have on another: distinguishing "drivers" from "modulators". Proc. Natl. Acad. Sci. U. S. A. 95, 7121–7126.

Sherman, S.M., Guillery, R.W., 2006. Exploring the Thalamus and its Role in Cortical Function. MIT Press, Cambridge, MA.

Sherman, S.M., Guillery, R.W., 2013. Functional Connections of Cortical Areas. A New View from the Thalamus. MIT Press, Cambridge, MA and London, England.

Smith, D.O., 1980. Morphological aspects of the safety factor for action potential propagation at axon branch points in the crayfish. J Physiol. 301, 261–269.

Song, S., Sjostrom, P.J., Reigl, M., Nelson, S., Chklovskii, D.B., 2005. Highly nonrandom features of synaptic connectivity in local cortical circuits. PLoS Biol. 3, e68.

Sporns, O., 2011. Networks of the Brain. MIT Press, Cambridge, MA.

Sporns, O., Tononi, G., Kotter, R., 2005. The human connectome: a structural description of the human brain. PLoS Comput. Biol. 1, e42.

Stehberg, J., Dang, P.T., Frostig, R.D., 2014. Unimodal primary sensory cortices are directly connected by long-range horizontal projections in the rat sensory cortex. Front. Neuroanat. 8, 93.

Tettoni, L., Gheorghita-Baechler, F., Bressoud, R., Welker, E., Innocenti, G.M., 1998. Constant and variable aspects of axonal phenotype in cerebral cortex. Cereb. Cortex 8, 543–552.

Tomasi, S., Caminiti, R., Innocenti, G.M., 2012. Areal differences in diameter and length of corticofugal projections. Cereb. Cortex 22, 1463–1472.

Van Der Gucht, E., Hof, P.R., Van, B.L., Burnat, K., Arckens, L., 2007. Neurofilament protein and neuronal activity markers define regional architectonic parcellation in the mouse visual cortex. Cereb. Cortex 17, 2805–2819.

Van Horn, S.C., Sherman, S.M., 2004. Differences in projection patterns between large and small corticothalamic terminals. J. Comp. Neurol. 475, 406–415.

Viaene, A.N., Petrof, I., Sherman, S.M., 2011a. Synaptic properties of thalamic input to layers 2/3 and 4 of primary somatosensory and auditory cortices. J. Neurophysiol. 105, 279–292.

Viaene, A.N., Petrof, I., Sherman, S.M., 2011b. Synaptic properties of thalamic input to the subgranular layers of primary somatosensory and auditory cortices in the mouse. J. Neurosci. 31, 12738–12747.

Vidnyanszky, Z., Borostyankoi, Z., Gorcs, T.J., Hamori, J., 1996. Light and electron microscopic analysis of synaptic input from cortical area 17 to the lateral posterior nucleus in cats. Exp. Brain Res. 109, 63–70.

Wagor, E., Mangini, N.J., Pearlman, A.L., 1980. Retinotopic organization of striate and extrastriate visual cortex in the mouse. J. Comp. Neurol. 193, 187–202.

Wang, C., Huang, J.Y., Bardy, C., FitzGibbon, T., Dreher, B., 2010. Influence of 'feedback' signals on spatial integration in receptive fields of cat area 17 neurons. Brain Res. 1328, 34–48.

Wang, Q., Burkhalter, A., 2007. Area map of mouse visual cortex. J. Comp. Neurol. 502, 339–357.

Wang, Q., Gao, E., Burkhalter, A., 2011. Gateways of ventral and dorsal streams in mouse visual cortex. J. Neurosci. 31, 1905–1918.

Wang, Q., Sporns, O., Burkhalter, A., 2012. Network analysis of corticocortical connections reveals ventral and dorsal processing streams in mouse visual cortex. J. Neurosci. 32, 4386–4399.

Wang, S., Eisenback, M.A., Bickford, M.E., 2002. Relative distribution of synapses in the pulvinar nucleus of the cat: implications regarding the "driver/modulator" theory of thalamic function. J. Comp. Neurol. 454, 482–494.

Watakabe, A., Takaji, M., Kato, S., Kobayashi, K., Mizukami, H., Ozawa, K., et al., 2014. Simultaneous visualization of extrinsic and intrinsic axon collaterals in Golgi-like detail for mouse corticothalamic and corticocortical cells: a double viral infection method. Front. Neural Circuits 8, 110.

Yamashita, A., Valkova, K., Gonchar, Y., Burkhalter, A., 2003. Rearrangement of synaptic connections with inhibitory neurons in developing mouse visual cortex. J. Comp. Neurol. 464, 426–437.

Yoshimura, Y., Callaway, E.M., 2005. Fine-scale specificity of cortical networks depends on inhibitory cell type and connectivity. Nat. Neurosci. 8, 1552–1559.

Yoshimura, Y., Dantzker, J.L., Callaway, E.M., 2005. Excitatory cortical neurons form fine-scale functional networks. Nature 433, 868–873.

Yu, Y.C., He, S., Chen, S., Fu, Y., Brown, K.N., Yao, X.H., et al., 2012. Preferential electrical coupling regulates neocortical lineage-dependent microcircuit assembly. Nature 486, 113–117.

Zingg, B., Hintiryan, H., Gou, L., Song, M.Y., Bay, M., Bienkowski, M.S., et al., 2014. Neural networks of the mouse neocortex. Cell 156, 1096–1111.

6

Interareal Connections of the Macaque Cortex: How Neocortex Talks to Itself

J.C. Anderson and K.A.C. Martin

Institute of Neuroinformatics, University of Zurich and ETH Zurich, Zurich, Switzerland

6.1 INTRODUCTION

The past century has seen numerous attempts to analyze the structure of neocortex in the hope of divining some universal principles of organization that will provide key insights into the complexity of neocortical function. The thread running through these investigations is the conviction that the local circuits of the neocortex are built using the same basic set of neuronal elements connected by the same rules to a common basic circuit. This implies that the basic mechanisms that control corticogenesis are common to the whole neocortical sheet in all species (Douglas and Martin, 2012). Indeed, the gradient of neurogenesis and migration seen developmentally (Dehay et al., 1993; Rakic, 2002; Smart, 1983) is consistent with the recent observation (Charvet et al., 2015) that across species, rostral areas tend to have fewer cells per unit surface area than caudal areas, and that the main increases are due to an increase in the proportion of cells in layers 2–4 compared to layers 5 and 6. It should be noted, however, that there are significant differences between rodents and primates in the development and organization of the germinal zones for corticogenesis (Kennedy and Dehay, 1993; Dehay and Kennedy, 2007), which is perhaps not surprising given the profound differences in duration of corticogenesis in primates compared to mice.

While the notion that there is a basic uniformity of neocortex is venerable, the question of what actually is "uniform" has been much more difficult to define. Attempts to reveal the underlying structure of local circuits from anatomical studies have been exceedingly challenging (e.g., Ramon y Cajal, 1888; Lorento de No, 1949; Lund et al. 1979,1981; Szentágothai, 1978). In contrast, the view down the microelectrode has had some advantages over the microscope in suggesting simplifying principles. Hubel and Wiesel (1974) speculated that their physiological investigations of monkey visual cortex revealed that "the machinery may be roughly uniform over the whole striate cortex, the difference being in the inputs. A given region of cortex simply digests what is brought to it, and the process is the same everywhere.... It may be that there is a great developmental advantage in designing such machinery once only, and repeating it over and over monotonously, like a crystal." Hubel and Wiesel here offered the important insight that it is the afferents of a particular area, rather than the local circuits, that are the key determinants of the functional specialization of different cortical areas. In their view, the operation of the monotonously repeated local circuit was to make a serial-wise addition and subtraction of receptive field properties through the layer-specific connections within a cortical column.

That the serial model of the local circuit introduced by Hubel and Wiesel (1962) has endured is a tribute to their deep insights, but it offered rather few specifics on the underlying neuronal basis of the cortical activity they recorded. The translation of their receptive field hierarchy into an actual neuronal circuit, however, was achieved through the structure–function studies of Gilbert and Wiesel (1979; Wiesel and Gilbert, 1983), who interpreted the neuronal circuit for V1 of the cat as the structural basis of the serial textbook functional model of Hubel and Wiesel (1962). Importantly for the present discussion, it appears that the circuit of interlaminar connections summarized by Wiesel and Gilbert (1983) can be found in all cortical areas of all species so far examined (Douglas and Martin, 2004). This circuit of course is static and accounts only for the feedforward excitatory component of the local circuit.

Inspired by the engineering methods of systems identification, Douglas and colleagues (Douglas et al., 1989; Douglas and Martin, 1991) set out on a different tack to characterize the dynamical behavior and organization of local circuits in the cat's visual cortex. Instead of mapping receptive fields, they recorded the intracellular "impulse response" of neurons in cat visual cortex *in vivo* and identified them by injecting them with HRP (Douglas et al., 1989; Douglas and Martin, 1991). An approximation to an impulse was obtained by electrical simulation of the thalamic afferents. They noted that all neurons in cat area 17 responded to a pulse stimulation of the thalamic afferents with a short excitatory postsynaptic potential (EPSP) followed by a long inhibitory postsynaptic potential (IPSP). This pattern was first seen by Phillips (1959) following antidromic activation of Betz cells in the cat's motor cortex. Pharmacological manipulations of the neurons with agonists and antagonists of the inhibitory neurotransmitter GABA revealed the presence of strong recurrent excitation, especially in the superficial layers, that was balanced with a proportional recurrent inhibition. Through computer simulations, Douglas and colleagues explored the circuit configurations that were consistent with their physiological results and proposed the result as a "canonical circuit" for neocortex (Douglas et al., 1989; Douglas and Martin, 1991). Their canonical circuit incorporated the major neuronal types and their intra- and interlaminar connections within the major laminar divisions of the neocortex, together with input from the thalamic core nuclei. The canonical circuit differed fundamentally from Gilbert and Wiesel's in its emphasis—not so much on serial processing, but on the recurrent excitatory and inhibitory processing of a tiny excitatory input from the thalamus. These principles derived from physiological and modeling studies were supported by our detailed structural studies that produced a connection matrix for cat V1 (Binzegger et al., 2004). A successful model of macaque and human FEF based on the cat connection matrix provided more support for the hypothesis of a canonical circuit for neocortex (Heinzle et al., 2007). Experimentally much more is required to establish whether the form of the circuit is universal, especially for the "agranular" cortex, but important work in this direction has now begun (Godlove et al., 2014; Barbas and Garcia-Cabezas, 2015; Beul and Hilgetag, 2015).

6.2 FEEDFORWARD VERSUS FEEDBACK

We come back to the question of how cortical areas come to be functionally specialized if they share the same basic local circuit? Can the long-range connections from subcortical regions like the thalamus and other cortical areas account for the functional differences between areas? We now know a great deal about the subtypes of connections and the laminar location of the pyramidal cells that form these interareal projections. The vast majority of excitatory neurons are pyramidal cells, most of which possess an efferent axon (e.g., Martin and Whitteridge, 1984). Determining how and where these efferents terminate, what terminal patterns they form, and what targets they form synapses with is an enormous endeavor. The early work in this field used degeneration techniques to provide answers (Nauta and Gygax, 1951). Using reduced silver techniques, degenerating axon terminals could be identified with precision enough to reveal that the principal thalamic projection was to cortical layer 4 (Kuypers et al., 1965), but this technique proved problematic for distinguishing between retrograde and anterograde degeneration (Powell and Cowan, 1964), because the rates of degeneration overlapped considerably. A significant advance was to use axonally transported neuronal tracers, of which there are many: radioactive amino acids (e.g., tritiated proline), enzymes like horseradish peroxidase (HRP), amines and lectins, fluorescent dyes like diamidino yellow and fast blue, biotinylated dextran amines (BDAs), toxins (e.g., cholera toxin B), and, more recently, various viruses like adenovirus or rabies.

In 1979, Rockland and Pandya made a very significant observation when using tritiated amino acids to label the terminal arbors of interareal projections in monkey visual cortex (Figure 6.1). Working in the occipital lobe of macaques, they observed if an area projected more anteriorly, its axons arborized in layer 4 of their target areas, whereas areas that projected to more posterior areas, formed terminal arborizations in layer 1. Such lamina specificity for termination sites had been seen earlier (Kuypers et al., 1965), but now could be seen as a fundamental feature of cortical organization. Rockland and Pandya (1979) went on to interpret their observations by referring to the layer 4 termination site as part of a "feedforward" pathway, analogous to the feedforward projections from the core thalamic nuclei, and the layer 1 termination site as part of a "feedback" pathway. The principle analogy for this was the bidirectional connection between the dorsal lateral geniculate nucleus (dLGN) of the thalamus and the primary visual cortex where the "feedforward" and "feedback" components of the connection are unambiguous. The feedforward projection from the dLGN terminates principally in cortical layer 4. The feedback projection from cortex to the thalamus originates in the deep layers of cortex (layer 6), where the "feedback" corticocortical projections also originate.

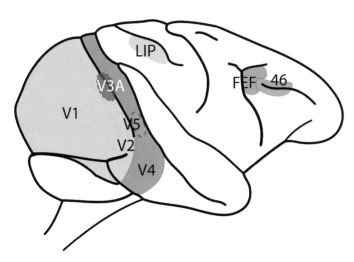

FIGURE 6.1 Schematic drawing of a macaque brain, lateral view, showing regions highlighted in this chapter. Each area examined is shown in a different color: V1, yellow; V2, bright green; V3A, transparent dark green; V4, purple; V5 or MT, light green; LIP, light blue; FEF, deeper blue; area 46, red. Areas V3A and V5 or MT are not visible from the surface of the brain and are shown with boundaries of dotted lines.

There is an important third pattern of connectivity between areas that is referred to as "lateral" or "columnar" (Maunsell and Van Essen, 1983). In this pattern all laminae are innervated with no apparent lamina preference. This kind of projection usually occurs between neighboring, higher order areas. Originally tentatively referred to as an "intermediate" projection, it proved highly variable from experiment to experiment in that it was not always identified (Maunsell and Van Essen, 1983). It seems likely that the different patterns depend on what subgroups of projection cells have been labeled. These anatomical observations offer a logic of relationships between cortical areas that has been extensively used for ordering the areas in a hierarchy.

6.3 HIERARCHY

Maunsell and Van Essen (1983) seized the opportunity provided by Rockland and Pandya (1979) to sketch a hierarchy of cortical processing according to the laminar patterns of origin and termination of the interareal connections. Their proposal has been consistently exploited ever since (Friedman, 1983; Maunsell and Van Essen, 1983; Felleman and Van Essen, 1991; Young, 1992; Jouvé et al., 1998; Sporns et al., 2000; Barone et al., 2000; Ts'o et al., 2001). Felleman and Van Essen (1991) extended this by examining the reports of all available publications with a description of an interareal connection. They selected some 32 visual areas and 7 visual association areas amounting to 305 pathways. From this pool of reported data they gleaned the appropriate information to build a hierarchical model of visual and visual association cortex accounting for some 55% of the cortex and produced a summary figure that showed the relationships between cortical areas over 10 levels of "processing." It encapsulated all the known cortical connections in all the elegance of an engineering wiring diagram. It has become one of the most iconic brain wiring diagrams since Ramon y Cajal's.

The Felleman and Van Essen diagram is incomplete (Hilgetag et al., 1996; Markov and Kennedy, 2013; Beul et al., 2014). The information they mined came from published papers, but also abstracts and short reports, some of which were in direct conflict with one another. The majority of connections showed bidirectionality, an observation noted earlier in the monkey (Tigges et al., 1973; Rockland and Pandya, 1979), but it should be noted that some reports that did not report a reciprocal innervation simply did not look for it (Felleman and Van Essen, 1991). Although defined cortical areas are regarded as homogeneous with regard to their inputs and outputs, this is a convenient assumption; for the location of the injection within a single area may itself be a significant factor in determining the pattern of labeling seen (Zeki, 1978; Falchier et al., 2002). Another generalization is that feedforward projections are more restricted in the extent of the target they innervate, whereas feedback projections show much more divergence in the spread of their terminal branches (Krubitzer and Kaas, 1989; Shipp and Zeki, 1989). From their single axonal reconstructions, Rockland (1997) inferred that the terminal arbors of feedback projections must cross functionally defined domains, such as ocular dominance or cytochrome oxidase (CO) columns.

Many projections terminate in more than one lamina, for example, the feedforward projection from V1 to MT, which terminates primarily in layer 4, but also layer 6 (Rockland, 1989; Figure 6.2). Such feedforward projections tend to have a secondary lamina of innervation that is consistently smaller than that of the primary lamina. Another example is the feedback pathway from V4 to V2 that shows a primary termination zone in layers 1 and 2 and a secondary termination zone in layer 6, where there are fewer collaterals than the primary termination layers. The same trend is apparent for the magno and parvocellular dLGN projections to V1 in the macaque (Blasdel and Lund, 1983; Freund et al., 1989), and the X and Y dLGN projections in cat V1(Freund et al., 1985).

Much of the detailed light microscopy of single interareal axons of macaque cortex has been studied by Rockland and colleagues using neuronal tracers (Rockland and Pandya,1979; Rockland, 1989, 1992, 1994, 1995; Rockland and Virga, 1989, 1990; Rockland et al., 1994; Suzuki et al., 2000). Their extensive body of work amplifies what we know about the location of interareal axon terminals by providing greater detail necessary for a list of observations that include: (i) Feedforward axons terminate in layer 4, but sometimes focus on upper layer 4 and lower layer 3. (ii) One branch of a single axon can selectively innervate layer 3 and another branch of the same axon selectively innervates layer 4. (iii) Collaterals may mostly innervate layer 4, but can also appear in layer 5 (V1 to V2, Rockland and Virga, 1990). (iv) Collaterals can be seen to target layers 1 and 2 (V2 to V4, Rockland, 1992, 1997). (v) V1 projections to neighboring V2 are mainly not bistratified (Rockland and Virga, 1990) nor is V2 to V5 (Rockland, 1995). (vi) The feedback pathways terminate principally in layer 1 but also in layer 2 and layer 5 or 6. In some cases layer 3 may also be targeted (V4 to V1 and V2, Rockland, 1994).

6.4 DISTANCE RULE

The retrograde labeling method has provided another key structural feature for defining the relationship of cortical areas. The feedforward pathway tends to originate from the pyramidal cells of the superficial or supragranular layers, with a minority contribution from the deep or infragranular layers. Conversely, feedback pathways originate from cells of the deep or infragranular layers with a minority contribution of supragranular cells (Rockland and Pandya, 1979; Barone et al., 2000). Exploiting this fact, and using the relatively easy and reliable technique of retrograde labeling of cell bodies with fluorescence, it became clear that physically or functionally close areas were strongly interconnected, as reflected in the number of labeled cells bodies and their laminar location. This strength of connections tends to lessen with distance from the injection site, and the ratio of the contribution of superficial and deep layers also changes systematically (Kennedy and Bullier, 1985; Salin and Bullier, 1995; Rockland, 1997). These features gave rise to a "distance rule" (Markov and Kennedy, 2013), which in its quantitative form made it possible to construct a hierarchy or relationships between the visual areas. The hierarchy for Barone et al. (2000) was based upon counting cell bodies in the different layers in the areas of origin after making large injections of fluorescent retrograde label in areas V1 and V4, thus generating the fraction of labeled neurons contributed by the superficial layer neurons (SLN%).

Of course, neither the hierarchy nor the feedforward/feedback principle along with detail of anatomical connection tells us how all this interareal connectivity functions. Indeed, an alternate view is that the major routing of interareal communication in cortex is via subcortical structures, not the direct corticocortical pathways (Sherman and Guillery, 1998, 2011; Guillery and Sherman, 2002).

6.5 DRIVERS AND MODULATORS

Guillery and Sherman's hypothesis is that connections can be parsed into two basic types—those that "drive" their targets and those that do not and only "modulate" them—and that this difference is visible in the morphology of the axons (Guillery and Sherman, 2002; Sherman and Guillery, 1998, 2011). Their model example is the feedforward input pathway from thalamus to cortex, which they characterize as a "driver" of cells in layer 4, while the feedback from cortex to thalamus is characterized as a "modulator." The notion is that the "driver" transmits receptive field properties and activates the postsynaptic cortical cell under suitable stimulus conditions, whereas the "modulator" changes the gain of the response. In the classical receptive field, the modulator path might serve to amplify the response while simultaneously suppressing unsuited stimuli in the non-classical receptive field (Hupé et al., 1998).

The same driver–modulator interpretation can be applied to corticocortical connections as they ascend and descend the cortical hierarchy. Crick and Koch (1998) used the self-same organization as seen by Rockland and

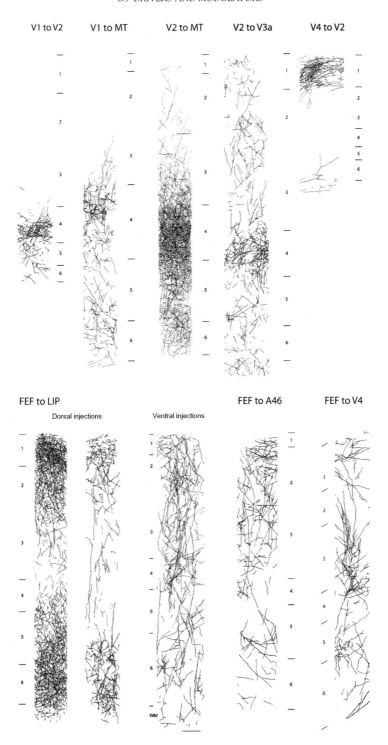

FIGURE 6.2 Light microscopic reconstructions of the axon projections from (top row, left to right) areas V1 to V2, V1 to MT (or V5), V2 to MT, V2 to V3a, V4 to V2; and (bottom row, left to right) areas FEF to LIP (dorsal injection), FEF to LIP (ventral injection), FEF to area 46, and FEF to V4. The axons were labeled with biotinylated dextran amine (BDA) and reconstructed from 80 μm thick sections of cortex. In the case of FEF to V4, two adjacent sections were used. Lamina boundaries as indicated. The FEF to LIP reconstruction is composed of three strips of cortex from LIP. The two lower left strips are from the termination zone of a BDA injection in dorsal FEF, the leftmost is from a more central region of the zone, and the next right is taken from a more peripheral site of the same zone. The next right (middle) strip of LIP cortex is from a BDA injection in ventral FEF. Scale bar, 100 μm.

Pandya (1979) in their light microscopic studies, with layer 4 receiving the driver input and layer 1 receiving the modulatory input. They proposed that there should be no "strong" loops between areas, so that no two areas would be able to drive each other (Crick and Koch, 1998).

6.6 ROUTING RULES

The observations of Rockland and colleagues have become part of the bedrock of understanding that guides us through the maze of interareal connections of macaque cortex. Subsequent experiments aimed at examining the synaptic targets of the projecting axons provided light microscopic data (Anderson et al., 1998, 2011; Anderson and Martin, 2002, 2005, 2006, 2009) that support the observations of Rockland and colleagues (Rockland and Pandya, 1979; Rockland, 1989, 1992, 1994, 1995, 1997; Rockland and Virga, 1989, 1990; Rockland and Douglas, 1993; Rockland et al., 1994). Like Rockland and Pandya (1979), we show that axons originating from caudal areas of occipital cortex and projecting to more anterior areas, terminated in the middle layers, especially layer 4, but in the reverse direction terminated mostly in layer 1. This is one basis for the designation of projections as feedforward or feedback (Maunsell and Van Essen, 1983; Felleman and Van Essen, 1991) and hence as "bottom-up" and "top-down" processing within cortex. These concepts have been vital in developing computational models at many levels, from, for example, models of object recognition (Grossberg and Mingolla, 1985; Li, 1998; Mumford, 1992) to whole brain function (Bastos et al., 2012).

What has been lost in these generalizations, however, is the variance between the projections, whether feedforward or feedback. This variance is instantly clear from the composite LM drawing shown in Figure 6.2, which illustrates the laminar pattern obtained from axon labeling after bulk injection into a number of key projections, namely: V1 to V2, V1 to MT, V2 to MT, V2 to V3a, V4 to V2, FEF to LIP (both dorsal and ventral projections), FEF to area 46, and FEF to V4.

Rockland and colleagues reported on the spread of axons beyond the principle target zone (e.g., Rockland and Virga, 1990; Rockland, 1992). This spread was most noticeable in our studies when the label was dense or when axon processes were reconstructed from several serial sections and then superimposed together (Figure 6.2). The appearance of labeled axons in all layers of V4 after injecting BDA into the frontal eye fields (FEFs), for example, does not agree with the Felleman and Van Essen classification of FEF to V4 being a feedback pathway. Instead, from our axon labeling it resembled a columnar or "lateral" connection, which is consistent with the classification of Markov and Kennedy (2013) using the percentage of labeled supragranular layer neurons (SLN%) method of hierarchical determination.

Two patterns of labeling were found in the projection from FEF to the lateral intraparietal area (LIP). In one pattern the labeled profiles innervate all laminae and especially layer 1, but clearly not upper layer 4 and lower layer 3. The second pattern was one in which all laminae were equally innervated (Figure 6.2). One possible source of difference could be related to whether the injection of tracer was made in dorsal or ventral FEF, which is the case in the injections made in one of the animals in Figure 6.2. Dorsal FEF has been shown to be involved in driving small saccadic eye movements, while the ventral FEF drives larger saccades, as well as smooth pursuit eye movements (MacAvoy et al., 1991). The labeling of profiles in the above (Figure 6.2) sections was so dense that individual axons could not be reconstructed. Rockland (1997) noted, however, that at the level of individual axons innervation could pass from lower layer 4 to upper layer 3, "almost indiscriminately" (Rockland, 1997; Budd, 1998).

6.7 SYNAPSES OF INTERAREAL PATHWAYS

In the electron microscope we were able to confirm that, with the exception of one pathway from V2 to MT (described below), all the boutons we identified at light microscopic level formed synapses. The large boutons (~1 μm diameter), such as those described for V1 to MT (Anderson et al., 1998), or the layer 1 feedback axons connecting V2 to V1 (Anderson and Martin, 2009), formed multiple synapses (three or four synapses), but the majority formed only one synapse per bouton (Figure 6.3). In some cases (e.g., V1 to MT, V2 to MT, V2 to V1), the boutons are often large enough to engulf entirely the synaptic spine head within its volume (Figures 6.4 and 6.5; Anderson et al., 1998). The more complete the engulfing, the more removed from "view" is the synapse. Up to 50% of the V1 to MT boutons had a spine embedded in them (Figures 6.6 and 6.7). Large boutons of the V1 to MT projection could enfold up to three spines in this way. Some of the larger boutons even formed two synapses with the same target—as when a bouton was lodged between a dendritic shaft and its spine (Anderson et al., 1998). A similar

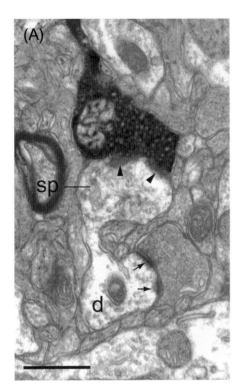
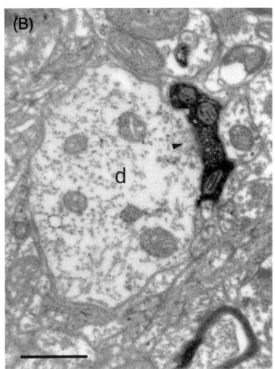

FIGURE 6.3   Electron micrograph of boutons in layers 1 and 2/3 of area LIP labeled after an injection of BDA in FEF. (A) A labeled bouton in layer 1 forms an asymmetric synapse (solid arrowheads) with a large spine (so) that can be followed back to the parent dendrite (d). (B) A labeled bouton in layer 2/3 forms an asymmetric synapse (solid arrowhead) with a large diameter dendritic shaft (d). The dendrite formed asymmetric synapses with unidentified boutons and contained numerous mitochondria. These features are characteristic of the smooth dendrites of GABAergic neurons. Scale bar, 0.5 μm.

pattern was seen for the projection from V2 to MT (Anderson and Martin, 2002). In this case, some of the labeled axons projected to layer 1 and completely surrounded the spine head as if with a thin coating of labeled bouton (Figures 6.6 and 6.7 Anderson and Martin, 2002). These boutons all appear to be filled with vesicles and contained mitochondria though, perhaps surprisingly, less than half appeared to form a synaptic specialization.

The great majority of labeled boutons formed by interareal projections were small (~0.5 μm diameter), densely filled with vesicles and mitochondria, and formed their synapses with a dendritic spine. Most of the boutons formed by the corticocortical projections are of the *en passant* type (~80%) with the exception being a slender axonal type seen in the feedback projection from V2 to V1, which arborized in layer 1 and formed mainly *terminaux* type boutons (70%). Interestingly, a preponderance of bouton *terminaux* is a characteristic of the axons of corticothalamic pyramidal cells in the cortex of cat and monkey (Martin and Whitteridge, 1984; Anderson et al., 1993) (Figure 6.4).

As might be expected from their origin from pyramidal cells, all the synapses formed by interareal axons are asymmetric (excitatory) synapses. The advantage of serial section reconstruction is that the morphology and size of the postsynaptic density (PSD) can be precisely measured. The largest and most complex synapses were formed with spines, whereas the synapses formed with dendritic shafts were smaller and, viewed *en face*, their postsynaptic densities tended to be disks rather than the horseshoe or annular shapes seen for many spine synapses. The smallest and simplest synapses were formed with somata. Target type, however, was not the sole arbiter of synapse size: axon type was also a significant factor. For example, in the feedback connection from V2 to V1, we labeled two axons with very different morphologies that innervated layer 1 (Figure 6.4) and whose synapses bracketed the range of sizes. The smallest synapse areas (mean 0.032 μm²) belonged to a fine axon that formed many termineaux boutons, while the largest synapses (mean 0.19 μm²) were formed by the grape-like clusters of boutons of an axon originating in V2 and forming synapses in layer 1 of V1. The PSD of small synapses appears as a disk when viewed *en face*, whereas the larger synapses, particularly those formed with spines, include complex perforations when viewed *en face* (see Figures 6.8-6.10 of Anderson and Martin, 2009).

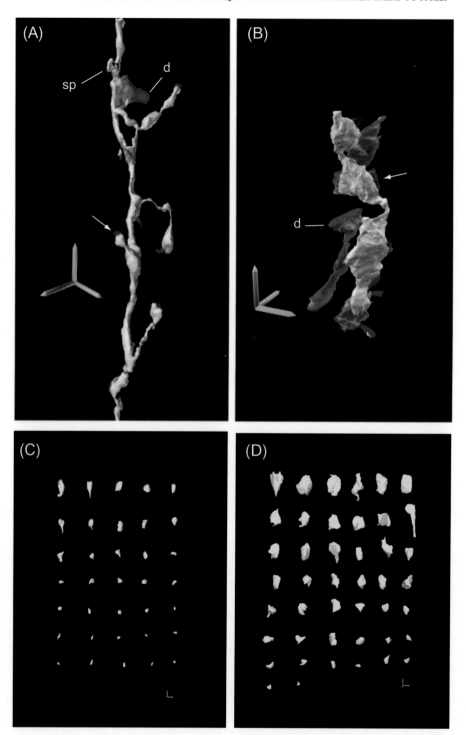

FIGURE 6.4 Three-dimensional reconstruction from serial ultrathin sections of boutons in layer 1 of V1 after a PHA-L injection into V2. (A) Thin axon (blue) showing numerous projecting *boutons terminaux* each forming a single synapse in most cases with a small spine (transparent brown, e.g., arrow). A *bouton terminaux* branches to form two boutons, one of which forms a synapse with a dendritic shaft (d, transparent mauve). A spine (so) projects from the uppermost surface of the dendrite and forms another synapse with the labeled axon. This same spine also forms a symmetric synapse with an unidentified bouton (not shown). (B) Thick axon (blue) showing a string of three *en passant* boutons and their synaptic targets; spines (e.g., arrow, transparent brown) and dendritic shafts (e.g., d, transparent mauve). Postsynaptic densities, often perforated, are shown in yellow. The two uppermost boutons each form two synapses, one each with a spine and the second with a dendritic shaft that passes between the two boutons. The dendrite also receives an asymmetric synapse from an unidentified bouton. The lower bouton forms four synapses, two with spines and two with a dendritic shaft. One of the spines can be traced back to the parent dendrite that bores more spines and forms an asymmetric synapse with the shaft. (C) and (D) Reconstructed spines found postsynaptic to labeled boutons of thin axon (C) and thick axon (D). The spines have been ordered by size. Each spine has been oriented to present its broadest face. Scale bar, 2 μm.

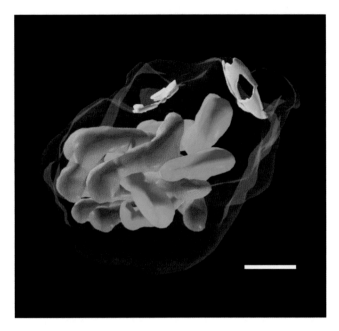

FIGURE 6.5 A bouton from layer 6 of V5 labeled by BDA/PHA-L injection into V1, reconstruction from serial EM sections rendered with transparent skins to show solid mitochondria and postsynaptic specializations (yellow). The colored structures are individual mitochondria (pink, blue, green, and orange). The bouton formed two synapses, both with spines, and with perforated synaptic specializations (yellow). Scale bar, 0.5 μm.

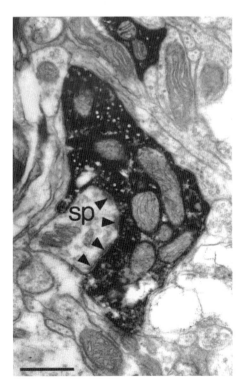

FIGURE 6.6 Electron micrograph of a bouton located in layer 6 of area V5 (MT) labeled after an injection of BDA/PHA-L in area V1. A spine (so) embedded within a large labeled bouton, containing spine apparatus, forms an asymmetric synapse with the bouton. A perforated postsynaptic density (solid arrowheads) is formed with the spine head. Scale bar, 0.5 μm.

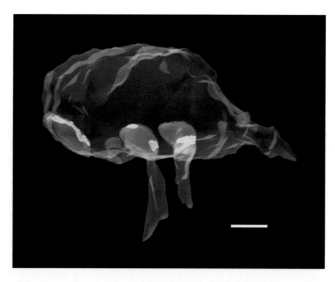

FIGURE 6.7   Three-dimensional reconstruction from serial ultrathin sections of a synaptic bouton in V5, labeled by injection of BDA/PHA-L in V1. Two spines (brown) are enveloped by a large bouton (transparent gray) and the face of a third spine is pressed into the surface of the bouton. The asymmetric synapses have complex and fragmented or discontinuous postsynaptic densities (yellow). Scale bar, 0.5 μm.

There are also very consistent laminar differences in the size of synapses formed by the same projections. Feedforward projections, like that from V1 to V2 or V2 to V5, have layer 4 as their principal target, where they make larger synapses on average than they do for their secondary projections to layer 3 or 6, which involve collaterals of the same axons. Similarly in the V4 to V2 feedback projection, the labeled synapses of the primary target lamina were substantially larger (mean= 0.117 μm^2) than those made in the secondary lamina (mean= 0.073 μm^2). A similar trend was evident in the synapse size of axons originating from FEF and terminating in areas V4, 46, and LIP. They formed their largest synapses in layer 1, while those in layer 4 were the smallest, which might indicate that layer 1 is the primary target of certain FEF axons in these areas.

6.8 SPINY (EXCITATORY) NEURONS AS TARGETS

By partial reconstruction from series of ultrathin sections, one can reconstruct the targets of the different pathways and identify whether the neurons are smooth (GABAergic) or spiny, using established criteria (Ahmed et al., 1997; Somogyi et al., 1983; Peters and Saint Marie, 1984). The pathways and their excitatory targets are summarized in Figure 6.8. Clearly spines are the dominant target, but a fraction of the targets are dendritic shafts and, very occasionally, cell bodies. Some of the dendritic shaft targets were characteristic of pyramidal cells in that they bore spines, contained few mitochondria, showed few variations in diameter, and had few synapses on the shaft. Synapses were also formed with dendritic shafts that contained numerous mitochondria, were studded with asymmetric synapses, formed swellings or beads, and bore no spines (Figure 6.11). Dendrites with these characteristics are typical of smooth or GABAergic neurons and therefore are putative inhibitory cells (Somogyi et al., 1983; Peters and Saint Marie, 1984; Kisvárday et al., 1985; Ahmed et al., 1997). The proportion of synapses formed with the dendrites of smooth cells ranges from 5% to 26%. This range compares with the proportion of cortical neurons that are GABAergic in the neocortex, that is, 15–25% (Beaulieu and Colonnier, 1985; Hendry et al., 1987; Beaulieu et al., 1992; Gabbott and Bacon, 1996).

6.9 SMOOTH (INHIBITORY) CELLS AS TARGETS

The projection of V1 to layer 4 of MT showed the largest proportion of smooth dendrites as targets (26%). The large boutons associated with the V1 to MT axons formed synapses with a number of somata of smooth neurons, which typically contained large numbers of mitochondria and other cellular organelles, rough endoplasmic

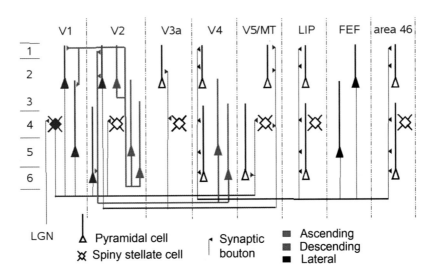

FIGURE 6.8 Schematic representation summarizing the excitatory targets of labeled boutons examined in this study. The majority of synapses are formed with spines belonging to pyramidal and spiny stellate cells. The different paths are color coded; feedforward or ascending in blue, feedback or descending in red, and lateral or columnar in black. Laminae are indicated to the left. Vertical dashed lines represent cortical area boundaries.

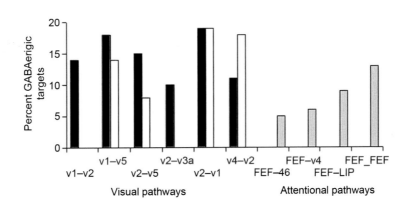

FIGURE 6.9 Histogram of the fraction of postsynaptic targets showing features that are characteristic of cells containing GABA, from visual (left) and "attentional" pathways originating in area FEF (right). Targets of the visual cortex (left) are divided into primary lamina of termination (black) and secondary laminae (white). The labeled termination pattern after FEF injections appeared lateral or columnar, innervating all lamina equally.

reticulum, deep nuclear invaginations and formed numerous unidentified asymmetric synapses (Peters and Saint Marie, 1984). The size of the synapses formed between these boutons and their target somata was small, though there could be up to five separate PSDs formed between a single bouton and a single soma (Anderson et al., 1998). We found more GABAergic targets associated with the primary lamina of termination (V1 to MT: 18% in layer 4 vs. 14% in layer 6; V4 to V2: 11% in layer 1 vs. 6% in layer 6). The secondary lamina in both these cases is layer 6 and so it resembles the thalamic innervation of V1 by X and Y cell axons, which also have a small, secondary innervation of layer 6 in both cat and monkey (Hendrickson et al., 1978; Blasdel and Lund, 1983; Freund et al., 1985) (Figure 6.9).

Reconstruction of some of the mitochondria from the V1 to MT boutons showed that each bouton contained a few mitochondria, which looped and branched and could occupy up to 22% of the volume of the bouton (Figure 6.5). There appears to be a significant role for mitochondria in synaptic transmission (Nicholls and Åkerman, 1982; Herrington et al., 1996; Tang and Zucker, 1997; Xu et al., 1997; Peng, 1998) and so it is noteworthy that invariably mitochondria were found in even the smallest boutons (Figure 6.3).

In the rat, the projection of primary visual cortex to the secondary visual cortex or the lateromedial area indicated that 87% of the targets were spines in the primary lamina of innervation, while in the secondary lamina 100% of the targets were spines (Gonchar and Burkhalter, 2003). Furthermore, pyramidal cells were the most frequent target of the feedback projection in rat (98%) and less so in the feedforward projection (90%; Johnson and Burkhalter, 1996). Since the remaining targets were assumed to be GABAergic neurons, the implication was that inhibition has a more important role in both the feedforward pathway and in the primary lamina of innervation of the feedback pathways. This is not entirely consistent with the monkey data, however, as feedback pathways in the visual cortex have consistently high numbers of GABAergic targets with 19% for V2 to V1 and 11% and 18% for V4 to V2 in layers 1 (primary) and 2 (secondary), respectively (Figure 6.9).

6.10 INFLUENCE OF SYNAPSE NUMBER AND LOCATION

One puzzle with the interareal projections is that that they contribute very few synapses to the excitatory neurons in their target areas (e.g., ~0.16% of asymmetric synapses in LIP layer 1, Anderson et al., 2011). The fraction of GABAergic targets among these projections can also be vanishingly small. Descriptions of the axons of the feedforward pathway indicated that they terminate in relatively tight clusters (200–250 μm diameter) with an increase in patch size as one progresses further into extrastriate cortex from V1 (Yoshioka et al., 1992; Lund et al., 1993; Rockland, 1997). Feedback terminal patches are described as spreading over millimeters, although, as Rockland (1997) pointed out, these descriptions are not derived from single axon fillings, which might show a more restricted picture. We did observe, however, that individual axons in the V4 to V2 feedback pathway traversed several millimeters through layer 1 (Anderson and Martin, 2006). Layer 1 of the V1 to V5 pathway is also populated by thick and thin axons traversing horizontally (Anderson et al., 1998). The functional role of the different inputs converging on a single target area has been difficult to tease out. Lesion or inactivation studies have revealed varying degrees of dependence of higher areas upon lower cortical areas (Rodman et al., 1989; Girard et al., 1991, 1992). This variance could be a reflection of their degree of connectivity to subcortical structures (e.g., Rodman et al., 1989).

It is evident from our data that the proportion of dendritic shafts forming synapses with an afferent projection is greatest when layer 4 is the principal lamina of innervation. This may reflect the composition of the neuropil or simply that there is a bias for dendritic shafts (of both smooth and spiny neurons), at least in layer 4. In the cat primary visual cortex, and approximately 50% of all excitatory (asymmetric) synapses they receive are formed on the shafts, not spines, of their dendrites (Anderson et al., 1994). Saint-Marie and Peters (1985) however found only 5–25% of asymmetric synapses were formed with the dendritic shafts of spiny stellate cells in macaque V1. Interestingly, layer 4 of monkey V1 contains 45% of the total number of neurons in V1, but only 35% of synapses (O'Kusky and Colonnier, 1982), suggesting that the dendritic trees of layer 4 spiny neurons are far more restricted in number of branches and extent than are the pyramidal neurons that populate other layers. By contrast in macaque V2, pyramidal cells are the major spiny cell type in layer 4 (Lund et al., 1981).

The reconstruction of extensive segments of the axon can be very informative. One prime instance occurred by careful observation in the light microscope of an axon originating in FEF and traversing layer 3 in V4. We noted that it uncharacteristically produced a very localized spray of fine axon terminals and when this curiosity was examined in the electron microscope, we discovered that the single axon formed 11 asymmetric synapses with spines of a very thick (2 μm diameter) dendrite (Anderson and Martin, 2009). This is the most numerous excitatory-to-excitatory synaptic connection thus far described for neocortex, but it should be emphasized that this is a unique exemplar in our *oeuvre*. The size and orientation of the dendrite suggested that it was the apical dendrite of a large layer 3 or 5 pyramidal cell where the synchronous activation of 11 synapses is likely to produce a very large unitary EPSP (Figure 6.10).

6.11 SERIAL PROCESSING AND LATERAL THINKING

The attractions of a hierarchical view of brain processing have been evident since the advent of the neuron doctrine (Waldeyer-Hartz, 1891) and Cajal's law of dynamic polarization (1897). The advantage of hierarchies of processing is perhaps best spelled out in Barlow's "Neuron Doctrine" (Barlow, 1972), where the "feature detectors" become increasingly selective as one advances from primary visual cortex through a tier of extrastriate areas. These differences between cortical areas in terms of receptive field properties make them attractive abstractions for models of cortex (Riesenhuber and Poggio, 1999). For many theorists, however, the feedforward or bottom-up approach of a rigid serial hierarchy has too many limitations for functions other than object recognition. Thus more recently

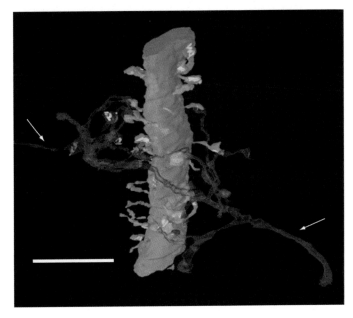

FIGURE 6.10   Reconstruction of axon (red) and target dendrite located in layer 2/3 of area V4. The axon was *labeled by a BDA injection in area FEF. The main trajectory of the axon is right to left (arrows) and* is roughly orthogonal to the radially aligned large caliber dendrite (grey). All reconstructed *dendritic* spines *formed* a synapse (green). At the point of intersection of the dendrite and the axon the latter produces a cluster of boutons. Eleven of these *labeled boutons formed synapses with spines coming from the reconstructed dendrite. Ten more of the labeled boutons formed synapses with spines (not shown) that could not be traced back to the dendrite. One labeled bouton formed a synapse with a small dendrite and one formed no synapse. The large dendrite shows many features characteristic of the apical dendrite of a large pyramidal neuron. Scale bar, 5 µm.*

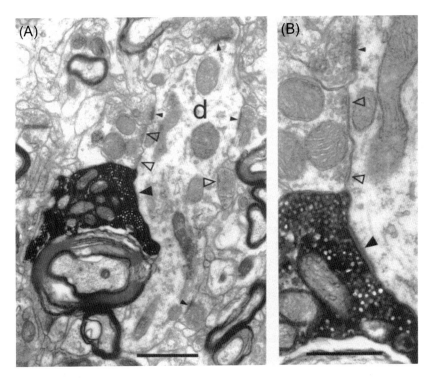

FIGURE 6.11   Electron micrograph of labeled synaptic bouton in contact with dendrites of smooth cell. (A) Large bouton filled with vesicles, located in layer 6 of V5, labeled after BDA injection in V1. It forms an asymmetric synapse (large solid arrowhead) with a dendritic shaft. The beaded dendrite contains numerous mitochondria and forms many synapses of the symmetric (open arrowhead) and asymmetric (small solid arrowhead) types with unidentified boutons. These features are consistent with those of neurons with smooth GABAergic dendrites. (B) A magnified view of (A) and the two types of synapse—symmetric and asymmetric—for comparison. Scale bars, 1 and 0.5 µm.

Bayesian models have been in the ascendant, because they offer a general framework for cortical computation across the wide swathe of perception, cognition, and action (Lee and Mumford, 2003; Friston, 2005; Bastos et al., 2012).

Most of the patterns of arborizations of axons originating from area FEF appear to have the lateral or columnar appearance, indicating they connect areas at the same level of the hierarchy (Maunsell and Van Essen, 1983). This is the case even when, paradoxically, one or more levels of the hierarchy have been traversed, such as FEF to LIP, or V4 (Felleman and Van Essen, 1991). Feedforward and feedback pathways by definition terminate in distinct laminae and therefore only have access to specific populations of cells or different segments of the dendritic trees of these cells, for example, the apical tuft of pyramidal cells in layer 1. The laminar specificity of the projections, along with the branching pattern of the dendritic tree, is the major constraint on synaptic position. Thus, we can say that the fact of massive innervation of layer 1 by cells providing feedback gives special functional prominence to the tufts of the apical dendrites, which are the principal source of the spines in layer 1. In these cases, input arriving from cells in higher or equally placed positions within the hierarchy may act as a gain control, rather than generating specific receptive field properties (Angelucci and Bressloff, 2006).

One perennial puzzle is the very small number of synapses provided by any single projection. Take for example, the FEF projection to layer 1 of LIP, area 46, and V4 where FEF afferents form a few tenths of a percent of all the excitatory synapses (Anderson et al., 2011) in layer 1 of LIP, area 46, and V4. This is an extremely low proportion, even in relation to the numerically small dLGN input to primary visual cortex in cat and monkey (Ahmed et al., 1994; Latawiec et al., 2000), which typically comes in at about 5–12% of the excitatory synapses in the principal layer of termination. Since the dLGN synapses are depressing the dLGN alone could not sustain a spike discharge beyond an initial transient (Banitt et al., 2007), from which we can infer that intracortical excitatory pathways are necessarily involved to provide the additional gain needed.

6.12 PATHWAYS OF ATTENTION

Attention is generally thought of as the means of enhancing the salience of a feature, hence the notion of a "searchlight of attention" (Crick, 1984; Schall and Hanes, 1993; Schall, 2002). Accumulating evidence, however, has directed thinking away from this "searchlight" concept toward one of competition (Desimone, 1998). For example if two objects evoke similar responses in the RF of a V2 cell, attending to one object biases the responses to the attended object (Reynolds et al., 1999). Electrical stimulation of macaque FEF has indicated the importance of FEF in driving shifts in visual attention, as both receptive field properties in V4 and the monkey's perceptual behavior show attention-like changes during electrical stimulation (Moore and Fallah, 2001, 2004). During directed attention, neurons of areas 46 and V4 that generated narrow spikes were found to increase their firing relative to the neurons that generated wide spikes (Mitchell et al., 2007). The narrow spikes were thought to originate from parvalbumin positive inhibitory neurons, and so taken together, these observations suggested that FEF has selective control of inhibitory neurons in target areas 46, V4, and LIP. Normalization models of attention rely on divisive inhibition (Reynolds and Heeger, 2009) that, in the case of visual attention, would be under control of the attentional signal provided by the projections from FEF. An alternative mechanism suggested by Fries et al. (2001) is that attention involves synchronization in the firing of neurons, which would be produced by fast inhibition. Again these inhibitory neurons would need to be independently controlled by an attentional signal.

The structural basis for these hypothesized mechanisms, however, is not evident. Instead of a specific excitatory connection from FEF to the parvalbumin positive inhibitory neurons in V4, the data indicate that smooth (inhibitory) neurons as a whole form only a small fraction (5–16%) of the targets of the FEF projections to area 46, V4, and LIP: the dominant targets are the pyramidal neurons (Anderson et al., 2011).

Yet another hypothesized mechanism of attention is that the signal from FEF reduces the correlations in the firing of neurons in their target area and so allows the uncorrelated noise to be averaged out (Cohen and Maunsell, 2009; Mitchell et al., 2009). Since inhibition increases synchrony (e.g., Fries et al., 2001), a reduction in noise correlations would require a reduction in inhibition (Mitchell et al., 2007). As indicated above, however, the structural data show that pyramidal cells are the major targets of the FEF projection, so the means of exerting a selective control of a circuit to reduce the excitatory drive to inhibitory neurons, or selectively inhibit inhibitory neurons, is missing. Consequently, we are unable yet to reconcile the structural data with the various hypotheses as to the mechanisms of attention. This may in part be due to the fact that the structural studies have had to average together all the output neurons of a pathway. Progress to a better understanding of the underlying mechanisms may come from a finer differentiation of the component neurons that link two areas.

6.13 CONCLUSIONS

We now have a deeper understanding of some of the synaptic relationships of the interareal pathways found in cortex:

1. All the labeled synapses formed by the long-distance axons examined in this study are excitatory, and the vast majority of them are formed with the spines of pyramidal cells.
2. One surprising finding, however, is the tiny fraction of synapses that these key pathways contribute to their target layers. Thus more than 80% of the excitatory synapses and 100% of the inhibitory synapses within a given cortical areas are formed locally by the neurons of that area.
3. Feedforward and feedback projections typically innervate one principal lamina and one secondary lamina. In the principal lamina, they form a higher fraction of their synapses with GABAergic neurons than in the secondary lamina. The area of the PSD is larger in the principal lamina than in the secondary lamina.
4. The feedforward, feedback, and lateral long-distance connections that innervate the recurrent local circuits clearly have a significant influence despite their small numbers. Even in the exceptional case where a single axon formed 11 synapses with a single dendrite (Anderson et al., 2011), this is only a fraction of a percent of the excitatory synapses a large pyramidal cell receives.
5. Typically, one long-distance axon will form only a few synapses with any single target cell.

While it is perhaps logical that a constraint of "wire minimization" might lead to the sparse interareal connections we observe, it leaves us with the conundrum of how so few synapses exert such significant effects. One solution we have proposed is that the local recurrent circuit amplifies spatiotemporal patterns of excitation generated by long-distance connections—both thalamic and interareal (Douglas and Martin, 2004, 2012). The specific computations within an area thus result from the specific patterns of input from the long-distance projections that are shaped by the local patterns of excitatory and inhibitory activity.

References

Ahmed, B., Anderson, J.C., Douglas, R.J., Martin, K.A.C., Nelson, J.C., 1994. Polyneuronal innervation of spiny stellate neurons in cat visual cortex. J. Comp. Neurol. 341, 39–49.

Ahmed, B., Anderson, J.C., Martin, K.A.C., Nelson, J.C., 1997. Map of the synapses onto layer 4 basket cells of the primary visual cortex of the cat. J. Comp. Neurol. 380, 230–242.

Anderson, J.C., Martin, K.A.C., 2002. Connection from cortical area V2 to MT in macaque monkey. J. Comp. Neurol. 443, 56–70.

Anderson, J.C., Martin, K.A.C., 2005. Connection from cortical area V2 to V3A in macaque monkey. J. Comp. Neurol. 488, 320–330.

Anderson, J.C., Martin, K.A.C., 2006. Synaptic connection from cortical area V4 to V2 in macaque monkey. J. Comp. Neurol. 495, 709–721.

Anderson, J.C., Martin, K.A.C., 2009. The synaptic connections between cortical areas V1 and V2 in macaque monkey. J. Neurosci. 29, 11283–11293.

Anderson, J.C., Martin, K.A.C., Whitteridge, D., 1993. Form, function, and intracortical projections of neurons in the striate cortex of the monkey Macacus nemestrinus. Cereb. Cortex 3, 412–420.

Anderson, J.C., Douglas, R.J., Martin, K.A.C., Nelson, J.C., 1994. Map of the synapses formed with the dendrites of spiny stellate neurons of cat visual cortex. J. Comp. Neurol. 341, 25–38.

Anderson, J.C., Binzegger, T., Martin, K.A.C., Rockland, K.S., 1998. The connection from cortical area V1 to V5: a light and electron microscopic study. J. Neurosci. 18, 10525–10540.

Anderson, J.C., Martin, K.A.C., Kennedy, H., 2011. Pathways of attention: synaptic relationships of frontal eye field to V4, lateral intraparietal cortex, and area 46 in macaque monkey. J. Neurosci. 31, 10872–10881.

Angelucci, A., Bressloff, P.C., 2006. Contribution of feedforward, lateral and feedback connections to the classical receptive field center and extra-classical receptive field surround of primate V1 neurons. Prog. Brain Res. 154, 93–120.

Banitt, Y., Martin, K.A.C., Segev, I., 2007. A biologically realistic model of contrast invariant orientation tuning by thalamocortical synaptic depression. J. Neurosci. 27, 10230–10239.

Barbas, H., Garcia-Cabezas, M.A., 2015. Motor cortex layer 4: less is more. Trends Neurosci. 38, 259–261.

Barlow, H.B., 1972. Single units and sensation: a neuron doctrine for perceptual psychology? Perception, 1371–1394.

Barone, P., Batardiere, A., Knoblauch, K., Kennedy, H., 2000. Laminar distribution of neurons in extrastriate areas projecting to visual areas V1 and V4 correlates with the hierarchical rank and indicates the operation of a distance rule. J. Neurosci. 20, 3263–3281.

Bastos, A., Usrey, W.M., Adams, R.A., Mangun, G.R., Fries, P., Friston, K., 2012. Canonical microcircuits for predictive coding. Neuron 76, 695–711.

Beaulieu, C., Colonnier, M., 1985. A laminar analysis of the number of round-asymmetrical and flat-symmetrical synapses on spines, dendritic trunks, and cell bodies in area 17 of the cat. J. Comp. Neurol. 231, 180–189.

Beaulieu, C., Kisvárday, Z., Somogyi, P., Cynader, M., Cowey, A., 1992. Quantitative distribution of GABA-immunopositive and -immunonegative neurons and synapses in the monkey striate cortex (area 17). Cereb. Cortex 2, 295–309.

Beul, S.F., Hilgetag, C.C., 2015. Towards a "canonical" agranular cortical microcircuit. Front. Neuroanat. 8, 165.

Beul, S.F., Grant, S., Hilgetag, C.C., 2014. A predictive model of the cat cortical connectome based on cytoarchitecture and distance. Brain Struct. Funct. July 26. [Epub ahead of print].

Binzegger, T., Douglas, R.J., Martin, K.A.C., 2004. A quantitative map of the circuit of cat primary visual cortex. J. Neurosci. 24, 8441–8453.

Blasdel, G.G., Lund, J.S., 1983. Termination of afferent axons in macaque striate cortex. J. Neurosci. 7, 1389–1413.

Budd, J.M.L., 1998. Extrastriate feedback to primary visual cortex in primates: a quantitative analysis of connectivity. Proc. Roy. Soc. B 265, 1037–1044.

Charvet, C.J., Cahalane, D.J., Finlay, B.L., 2015. Systematic, cross-cortex variation in neuron numbers in rodents and primates. Cereb. Cortex 25, 147–160.

Cohen, H.R., Maunsell, J.H.R., 2009. Attention improves performance primarily by reducing interneuronal correlations. Nat. Neurosci. 12, 1594–1601.

Crick, F., 1984. Function of the thalamic reticular complex: the searchlight hypothesis. Proc. Natl. Acad. Sci. U.S.A. 81, 4586–4590.

Crick, F., Koch, C., 1998. Constraints on cortical and thalamic projections: no-strong-loops hypothesis. Nature 391, 245–250.

Dehay, C., Kennedy, H., 2007. Cell-cycle control and cortical development. Nat. Rev. Neurosci. 8, 438–450.

Dehay, C., Giroud, P., Berland, M., Smart, I., Kennedy, H., 1993. Modulation of the cell cycle contributes to the parcellation of the primate visual cortex. Nature 366, 464–466.

Desimone, 1998. Visual attention mediated by biased competition in extrastriate visual cortex. Philos. Trans. R. Soc. Lond. B Biol. Sci. 353, 1245–1255.

Douglas, R.J., Martin, K.A.C., 1991. A functional circuit for cat visual cortex. J. Physiol. 440, 735–769.

Douglas, R.J., Martin, K.A.C., 2004. Neuronal circuits of the neocortex. Annu. Rev. Neurosci. 27, 419–451.

Douglas, R.J., Martin, K.A.C., 2012. Behavioral architecture of the cortical sheet. Curr. Biol. 22, R1033–1038.

Douglas, R.J., Martin, K.A.C., Whitteridge, D., 1989a. A canonical circuit for neocortex. Neural Comput. 1, 480–488.

Douglas, R.J., Martin, K.A.C., Whitteridge, D., 1989b. A functional microcircuit for cat visual cortex. J. Physiol. 440, 735–769.

Falchier, A., Clavagnier, S., Barone, P., Kennedy, H., 2002. Anatomical evidence of multimodal integration in primate striate cortex. J. Neurosci. 22, 5749–5759.

Felleman, D.J., Van Essen, D.C., 1991. Distributed hierarchical processing in the primate cerebral cortex. Cereb. Cortex 1, 1–47.

Freund, T.F., Martin, K.A.C., Whitteridge, D., 1985. Innervation of cat visual areas 17 and 18 by physiologically identified X- and Y-type thalamic afferents. I. Arborization patterns and quantitative and quantitative distribution of postsynaptic elements. J. Comp. Neurol. 242, 263–274.

Freund, T., Martin, K.A.C., Soltesz, I., Somogyi, P., Whitteridge, D., 1989. Arborisation pattern and postsynaptic targets of physiologically identified thalamocortical afferents in striate cortex of the macaque monkey. J. Comp. Neurol. 289, 315–336.

Friedman, D.P., 1983. Laminar patterns of termination of corticocortical afferents in the somatosensory system. Brain Res. 273, 147–151.

Fries, P., Reynolds, J.H., Rorie, A.E., Desimone, R., 2001. Modulation of oscillatory neuronal synchronization by selective visual attention. Science 291, 1560–1563.

Friston, K., 2005. A theory of cortical response. Philos. Trans. R. Soc. Lond. B Biol. Sci. 360, 815–836.

Gabbott, P.L.A., Bacon, S.J., 1996. Local circuit neurons in the medial prefrontal cortex (areas 24a,b,c, 25 and 32) in the monkey: II. Quantitative areal and laminar distributions. J. Comp. Neurol. 364, 609–636.

Gilbert, C.D., Wiesel, T.N., 1979. Morphology and intracortical projections of functionally characterized neurons in cat visual cortex. Nature 5718, 120–125.

Girard, P., Salin, P.A., Bullier, J., 1991. Visual activity in areas V3a and V3 during reversible inactivation of area V1 in the monkey. J. Neurophysiol. 66, 1493–1503.

Girard, P., Salin, P.A., Bullier, J., 1992. Response selectivity of neurons in area MT of the macaque monkey during reversible inactivation of area V1. J. Neurophysiol. 67, 1437–1446.

Godlove, D.C., Maier, A., Woodman, G.F., Schall, J.D., 2014. Microcircuitry of agranular frontal cortex: testing the generality of the canonical circuit. J. Neurosci. 34, 5355–5369.

Gonchar, Y., Burkhalter, A., 2003. Distinct GABAergic targets of feedforward and feedback connections between lower and higher areas of rat visual cortex. J. Neurosci. 23, 10904–10912.

Grossberg, S., Mingolla, E., 1985. Neural dynamics of perceptual grouping: textures boundaries and emergent segmentations. Percept. Psychophys. 38, 141–171.

Guillery, R.W., Sherman, S.M., 2002. Thalamic relay functions and their role in corticocortical communication: generalizations from the visual system. Neuron 33, 163–175.

Heinzle, J., Hepp, K., Martin, K.A.C., 2007. A microcircuit model of the frontal eye fields. J. Neurosci. 27, 9341–9353.

Hendrickson, A.E., Wilson, J.R., Ogren, M.P., 1978. The neuroanatomical organization of pathways between the dorsal lateral geniculate nucleus and visual cortex in old world and new world primates. J. Comp. Neurol. 182, 123–136.

Hendry, S.H., Schwark, H.D., Jones, E.G., Yan, J., 1987. Numbers and proportions immunoreactive neurons in different areas of monkey cerebral cortex. J. Neurosci. 7, 1503–1519.

Herrington, J., Park, Y.B., Babcock, D.F., Hille, B., 1996. Dominant role of mitochondria in clearance of large Ca^{2+} loads from rat adrenal chromaffin cells. Neuron 16, 219–228.

Hilgetag, C.C., O'Neil, M.A., Young, M.P., 1996. Indeterminate organization of the visual system. Science 271, 776–777.

Hubel, D.H., Wiesel, T.N., 1962. Receptive fields, binocular interaction and functional architecture in the cats visual cortex. J. Physiol. 160, 105–154.

Hubel, D.H., Wiesel, T.N., 1974. Uniformity of monkey striate cortex: a parallel relationship between field size, scatter and magnification factor. J. Comp. Neurol. 158, 295–305.

Hupé, J.M., James, A.C., Payne, B.R., Lomber, S.G., Girard, P., Bullier, J., 1998. Cortical feedback improves discrimination between figure and background by V1, V2 and V3 neurons. Nature 394, 784–787.

Johnson, R.R., Burkhalter, A., 1996. Microcircuitry of forward and feedback connections within rat visual cortex. J. Comp. Neurol. 368, 383–398.

Jouvé, B., Rosenthiehl, P., Imbert, M., 1998. A mathematical approach to the connectivity between the cortical visual areas of the macaque monkey. Cereb. Cortex 8, 28–39.

Kennedy, H., Bullier, J., 1985. Double-labeling investigation of the afferent connectivity to cortical areas V1 and V2 of the macaque monkey. J. Neurosci. 10, 2815–2830.

Kennedy, H., Dehay, C., 1993. Cortical specification of mice and men. Cereb. Cortex 3, 171–186.

I. MICROCIRCUITRY

Kisvárday, Z.F., Martin, K.A.C., Whitteridge, D., Somogyi, P., 1985. Synaptic connections of intracellularly filled clutch cells, a type of small basket neuron in the visual cortex of the cat. J. Comp. Neurol. 241, 111–137.

Krubitzer, L.A., Kaas, J.H., 1989. Cortical integration of parallel pathways in the visual system of primates. Brain Res. 478, 161–165.

Kuypers, H.G.J.M., Szwarcbart, M.K., Mishkin, M., Rosvold, H.E., 1965. Occipitotemporal corticocortical connections in the rhesus monkey. Exp. Neurol. 11, 245–262.

Latawiec, D., Martin, K.A.C., Meskenaite, V., 2000. Termination of the geniculocortical projection in the striate cortex of macaque monkey: a quantitative immunoelectron microscopic study. J. Comp. Neurol. 419, 306–319.

Lee, T.S., Mumford, D., 2003. Hierarchical Bayesian inference in the visual cortex. J. Opt. Soc. Am. A. Opt. Image Sci. Vis. 20, 1434–1448.

Li, Z., 1998. A neural model of contour integration in the primary visual cortex. Neural Comput. 10, 903–940.

Lorento de No R., 1949. Cerebral cortex: architecture, intracortical connections, motor projections. In: Physiology of the Nervous System, by Fulton. Oxford Medical Publications. Oxford University Press, New York.

Lund, J.S., Hendrickson, A.G., Ogren, M.P., Tobin, E.A., 1981. Anatomical organization of primate visual cortex area VII. J. Comp. Neurol. 202, 19–45.

Lund, J.S., Henry, G.H., MacQueen, C.L., Harvey, A.R., 1979. Anatomical organization of the primary visual cortex (area 17) of the cat. A comparison with area 17 of the macaque monkey. J. Comp. Neurol. 184, 599–618.

Lund, J.S., Yoshioka, T., Levitt, J.B., 1993. Comparison of intrinsic connectivity in different areas of macaque monkey cerebral cortex. Cereb. Cortex 3, 148–162.

MacAvoy, M.G., Gottlieb, J.P., Bruce, C.J., 1991. Smooth-pursuit eye movement representation in the primate frontal eye field. Cereb. Cortex 1, 95–102.

Markov, N.T., Kennedy, H., 2013. The importance of being hierarchical. Curr. Opin. Neurobiol. 23, 1–8.

Martin, K.A.C., Whitteridge, D., 1984. Form, function and intracortical projections of spiny neurons in the striate cortex of the cat. J. Physiol. Lond. 353, 463–504.

Maunsell, J.H.R., Van Essen, D.C., 1983. The connections of the middle temporal visual area (MT) and their relationship to a cortical hierarchy in the macaque monkey. J. Neurosci. 3, 2563–2586.

Mitchell, J.F., Sundberg, K.A., Reynolds, J.H., 2007. Differential attention-dependent response modulation across cell classes in macaque visual area V4. Neuron 55, 131–141.

Mitchell, J.F., Sundberg, K.A., Reynolds, J.H., 2009. Spatial attention decorrelates intrinsic activity fluctuations in macaque area V4. Neuron 63, 879–888.

Moore, T., Fallah, M., 2001. Control of eye movements and spatial attention. Proc. Natl. Acad. Sci. U.S.A. 98, 1273–1276.

Moore, T., Fallah, M., 2004. Microstimulation of the frontal eye field and its effects on covert spatial attention. J. Neurophysiol. 91, 152–162.

Mumford, D., 1992. On the computational architecture of the neocortex. II. The role of cortico-cortical loops. Biol. Cybern. 66, 241–251.

Nauta W., Gygax P.A., 1951. Silver impregnation of degenerating axon terminals in the central nervous system: (1) Technic (2) Chemical notes. 26, 5–11.

Nicholls, D.G., Åkerman, K.E.O., 1982. Mitochondrial calcium transport. Biochim. Biophys. Acta 683, 57–88.

O'Kusky, J., Colonnier, M., 1982. A laminar analysis of the number of neurons, glia, and synapses in the adult cortex (area 17) of adult macaque monkeys. J. Comp. Neurol. 210, 278–290.

Peng, Y.-Y., 1998. Effects of mitochondrion on calcium transients at intact presynaptic terminals depend on frequency of nerve firing. J. Neurophysiol. 80, 186–195.

Peters, A., Saint Marie, R.L., 1984. Smooth and sparsely spinous non-pyramidal cells forming local axonal plexuses. In: Jones, E.G., Peters, A. (Eds.), Cerebral Cortex, vol. 1. Cellular Components of the Cerebral Cortex Plenum Press, New York, NY, pp. 419–445.

Phillips, C.G., 1959. Actions of antidromic pyramidal volleys on single Betz cells in the cat. Q. J. Exp. Physiol. Cogn. Med. Sci 44, 1–25.

Powell, T.P.S., Cowan, W.M., 1964. A note on retrograde fibre degeneration. J. Anat. Lond. 98, 579–585.

Rakic, P., 2002. Pre- and post-developmental neurogenesis in primates. Clin. Neurosci. Res. 2, 29–39.

Ramon y Cajal, 1888. In Garcia-Lopez, P., Garcia-Marin, V., Freire, M., 2007. The discovery of dendritic spines by Cajal in 1888 and its relevance in the present neuroscience. Prog. Neurobiol. 83, 110–130.

Reynolds, J.H., Heeger, D.J., 2009. The normalization model of attention. Neuron 61, 168–185.

Reynolds, J.H., Chelazzi, L., Desimone, R., 1999. Competitive mechanisms subserve attention in macaque areas V2 and V4. J. Neurosci. 19, 1736–1753.

Riesenhuber, M., Poggio, T., 1999. Hierarchical models of object recognition in cortex. Nat. Neurosci. 2, 1019–1025.

Rockland, K.S., 1989. Bistratified distribution of terminal arbors of individual axons projecting from area V1 to middle temporal area (MT) in the macaque monkey. Vis. Neurosci. 3, 155–170.

Rockland, K.S., 1992. Configuration in serial reconstruction of individual axons projecting from area V2 to V4 in the macaque monkey. Cereb. Cortex 5, 353–374.

Rockland, K.S., 1994. The organization of feedback connections from area V2 (18) to V1 (17). In: Peters, A., Rockland, K.S. (Eds.), Cerebral Cortex, vol. 10. Primary Visual Cortex in Primates Plenum Press, New York, NY, pp. 261–299.

Rockland, K.S., 1995. Morphology of individual axons projecting from area V2 to MT in the macaque. J. Comp. Neurol. 355, 15–26.

Rockland, K.S., 1997. Elements of cortical hierarchy revisited. In: Rockland, K.S., Kaas, J.H., Peters, A. (Eds.), Cerebral Cortex, vol. 12. Exstrastriate Cortex in Primates Plenum Press, New York, NY, pp. 243–293.

Rockland, K.S., Douglas, K.L., 1993. Excitatory contacts of feedback connections in layer 1 of area V1: an EM biocytin study in the macaque. Neurosci. Abst. 19, 424.

Rockland, K.S., Pandya, D.N., 1979. Lamina origins and terminations of cortical connections of the occipital lobe of the rhesus monkey. Brain Res. 179, 3–20.

Rockland, K.S., Virga, A., 1989. Terminal arbors of individual "feedback" axons projecting from area V2 to V1 in the macaque monkey: a study using immunohistochemistry of anterogradely transported *Phaseolus vulgaris*-leucoagglutinin. J. Comp. Neurol. 285, 54–72.

Rockland, K.S., Virga, A., 1990. Organization of individual cortical axons projecting from area V1 (area 17) to V2 (area 18) in the macaque monkey. Vis. Neurosci. 4, 11–28.

I. MICROCIRCUITRY

Rockland, K.S., Saleem, K.S., Tanaka, K., 1994. Divergent feedback connections from areas V4 and TEO in the macaque. Vis. Neurosci. 11, 579–600.

Rodman, H.R., Gross, C.G., Albright, T.D., 1989. Afferent basis of visual response properties in area MT of the macaque. II. Effects of striate cortex removal. J. Neurosci. 10, 2033–2050.

Saint-Marie, R.L., Peters, A., 1985. The morphology and synaptic connections of spiny stellate neurons in monkey visual cortex (area 17): a Golgi-electron microscopic study. J. Comp. Neurol. 233, 213–235.

Salin, P.A., Bullier, J., 1995. Corticocortical connections in the visual system: structure and function. Physiol. Rev. 75, 107–154.

Schall, J.D., 2002. The neural selection and control of saccades by the frontal eye field. Philos. Trans. R. Soc. Lond. B Biol. Sci. 357, 1073–1082.

Schall, J.D., Hanes, D.P., 1993. Neural basis of saccade target selection in frontal eye field during visual search. Nature 366, 467–469.

Sherman, S.M., Guillery, R.W., 1998. On the actions that one nerve cell can have on another: distinguishing 'drivers' from 'modulators'. Proc. Natl. Acad. Sci. U.S.A. 95, 7121–7126.

Sherman, S.M.1., Guillery, R.W., 2011. Distinct functions for direct and transthalamic corticocortical connections. J. Neurophysiol. 106, 1068–1077.

Shipp, S., Zeki, S., 1989. The organization of connections between area V5 and V1 in macaque monkey visual cortex. Eur. J. Neurosci. 1, 309–332.

Smart, I.H., 1983. Three dimensional growth of the mouse isocortex. J. Anat. 137, 683–694.

Somogyi, P., Kisvárday, Z.F., Martin, K.A.C., Whitteridge, D., 1983. Synaptic connections of morphologically identified and physiologically characterized large basket cells in the striate cortex of cat. Neuroscience 10, 261–294.

Sporns, O., Tononi, G., Edelman, G.M., 2000. Theoretical neuroanatomy: relating anatomical and functional connectivity in graph and cortical connection matrices. Cereb. Cortex 10, 127–141.

Suzuki, W., Saleem, K.S., Tanaka, K., 2000. Divergent backward projections from the anterior part of the inferotemporal cortex (area TE) in the macaque. J. Comp. Neurol. 422, 206–228.

Szentágothai, J., 1978. The neuron network of the cerebral cortex: a functional interpretation. Proc. R. Soc. Lond. B Biol. Sci. 201, 219–248.

Tang, Y.-G., Zucker, R.S., 1997. Mitochondrial involvement in post-tetanic potentiation of synaptic transmission. Neuron 18, 483–491.

Tigges, J., Spatz, W.B., Tigges, M., 1973. Reciprocal point-to-point connections between parastriate and striate cortex in the squirrel monkey (*Saimiri*). J. Comp. Neurol. 148, 481–489.

Ts'o, D.Y., Wang Roe, A., Gilbert, C.D., 2001. A hierarchy of the functional organisation for color, form and disparity in primate visual area V2. Vis. Res. 41, 1333–1349.

Waldeyer-Hartz, HWG von, 1891. Über einige neuere Forschungen im Gebiete der Anatomie des Centralnervensystems. Deutsch. Med. Wschr., 17.

Wiesel, T.N., Gilbert, C.D., 1983. The Sharpey-Schafer lecture. Morphological basis of visual cortical function. Q. J. Exp. Physiol. 68, 525–543.

Xu, T., Naraghi, M., Kang, H., Neher, E., 1997. Kinetic studies of Ca^{2+} binding and Ca^{2+} clearance in the cytosol of adrenal chromaffin cells. Biophys. J. 19, 532–545.

Yoshioka, T., Levitt, J.B., Lund, J.S., 1992. Intrinsic lattice connections in macaque monkey visual cortical area V4. J. Neurosci. 12, 2785–2802.

Young, M.P., 1992. Objective analysis of the topological organization of the primate cortical visual system. Nature 358, 152–155.

Zeki, S., 1978. Uniformity and diversity of structure and function in rhesus monkey prestriate visual cortex. J. Physiol. 277, 273–290.

Topography of Excitatory Cortico-cortical Connections in Three Main Tiers of the Visual Cortex: Functional Implications of the Patchy Horizontal Network

Zoltán Kisvárday

MTA-Debreceni Egyetem, Neuroscience Research Group, Debrecen, Hungary

7.1 INTRODUCTION

The discovery of a patchy, discontinuous network of intrinsic connections in the mammalian cortex (Rockland and Lund, 1982) has attracted the considerable interest of neuroanatomists, neurophysiologists, and brain modelers. Even at present, it is a point of as great attention as it was 40 years ago. Although the topology and some of the spatial parameters of patchy cortico-cortical connections vary from area to area in the brain, the clustered nature of the neural network can be found in many mammalian species (for review, see Levitt and Lund, 2002; Douglas and Martin, 2004; Voges et al., 2010a,b). Therefore, they can be supposed to represent a basic architectural feature of the brain organization. Our aim is to provide an overview of this system considering its topography in relation to the functional compartmentalization of the cortex, ultimately delineating its putative role through examples excerpted from the literature and also to introduce some theoretical considerations. For consistency, we focus on the primary visual cortex since most functional anatomical data are available only for this brain area. We will deal only with the patchy lateral network because, despite the wealth of information available, it is still not clear what general principles of cortical processing this network subserves. A specific key issue that the patchy connectivity system has been less investigated in cortical tiers below layers 2 and 3, mostly due to technical difficulties and the resulting ambiguity in interpreting the results. Indeed, only sporadic information is available for the functional connectivity of the middle and lower tiers of the cortex. Therefore, the present account aims at discussing available data on the functional connectivity of the patchy network including deep layers.

7.2 METHODICAL CONSIDERATIONS

Long-range lateral connections have been investigated using anterograde and retrograde neuronal tracers. These tracers can reveal populations of connections from and to the seed area, that is, the injection site, at single cell resolution. Several of such tracers are available based on a wide array of approaches including from the enzyme horseradish peroxidase to viral tracers. These are capable to reveal even the finest axons from their origin to most

135

distal end. In addition, biotinylated molecules (e.g., biotinylated dextrane amines, biocytin, and neurobiotin) and fluorescent dextrans and carbocyanides (e.g., DiI) have also been widely used for the purpose of neuron labeling. It is worth to note that only a few neuronal tracers have a unique property of being exclusively either anterograde or retrograde; most tracers are transported to some extent in both directions along axons and dendrites. The most intensively investigated brain areas belong to visual cortices. In reference to this, there are several studies dealing with visual cortices in different species such as treeshrew (Rockland and Lund, 1982; Bosking et al., 1997), ferret (Rockland, 1985; Wallace and Bajwa, 1991), cat (Gilbert and Wiesel, 1983; Martin and Whitteridge, 1984; Matsubara et al., 1985; Luhmann et al., 1990; Kisvárday and Eysel, 1992; Löwel and Singer, 1992; Kisvárday et al., 1997; Buzás et al., 2006), monkey (Rockland, 1985; Lund et al., 1993; Yoshioka et al., 1996; Malach et al., 1993; Sincich and Blasdel, 2001; Angelucci et al., 2002; Tanigawa et al., 2005), and human (Burkhalter and Bernardo, 1989). There are two important constraints that determine the method of choice while investigating long-range lateral connections. One of them is concerned with dimensions in the plane parallel to the cortical surface while the other is about the columnar structure of the cortex. Regarding the former, due to the relatively large brain area wherein cortico-cortical connections arborize, the geometrical anisotropies deriving primarily from the curved nature of the cortical mantle need to be taken into account. Therefore highly convoluted brains are more difficult to investigate than those with smooth surface. The mammalian species with relatively smooth cortical surface (e.g., ferret, treeshrew, galago, and owl monkey) are more favorable as experimental subjects than those with highly folded brain (e.g., cat, old-world monkeys, and human).

Mapping the distribution of functional attributes, such as orientation preference, direction preference, ocularity, or spatial frequency preference in large cortical regions requires imaging techniques with high spatial resolution (in a millimeter range) across large cortical areas. In this regard, the most adequate imaging tools are intrinsic signal optical imaging (Grinvald et al., 1986) and voltage sensitive dye imaging (Blasdel and Salama, 1986). Both of these methods meet the spatial requirement necessary to image a huge cortical area (several tens of mm^2). Since the focal plane of the camera optics is usually very narrow, in an optimum condition the data acquisition should be made from flat cortical regions, otherwise the processing of the results would be quite cumbersome.

Inherent to the optical imaging methods, the signal derived and averaged across several cortical layers, typically, from layer 1 to layer 4 (Tian et al., 2011), is prone to statistical error associated with photon scatter (Polimeni et al., 2005). There are two important points to mention regarding the data interpretation. First, due to the limited depth of optical signals, an application dealing with deep layer structures must rely on the assumption that the spatial arrangement of the functional attributes follows a columnar organization through the entire cortical thickness. Although this has been validated for some functional properties, such as orientation selectivity (Hubel and Wiesel, 1962; Murphy and Sillito, 1986) in lower visual cortical areas, only limited information is available in higher visual cortical areas for the columnar arrangement of other functional features. Therefore, any optical imaging performed on deep layer structures, that is, below 600 μm (or layer 4), must rely on extrapolation of signals obtained from more superficial layers. Second, the columnar structure of the cortex accommodates its curvature, so that the trajectories of white matter input fibers and cell body columns show a curved organization pattern (Figure 7.1A). This raises serious questions about the accuracy of the mapping of a functional property using optical imaging and mapping the same functional property by using electrophysiological recording of single- or multiunits. Clearly, careful validation of the two approaches and their relationship is desired (Figure 7.1B), especially when imaging is carried out in a non-flat cortical region.

Just as critical issue is matching the functional maps with anatomical reconstruction of axons and dendrites within the same specimen. Several factors require extreme care. Tissue shrinkage and potential nonlinear distortions—due to certain factors such as fixation, sectioning, and handling the specimen—can severely affect the overall tissue quality. For a proper aligning of the sections (anatomical reconstruction) with functional maps, use of a common reference coordinate system is obligatory. This can only be done by natural and artificial landmarks detectable *in vivo* as well as in the histological sections. For example, blood vessels running in the pia mater can readily be imaged and the same pattern will be recognizable in horizontal sections taken subsequently from the perfused brain, at the level of the cortical surface. In addition, blood vessels of the pia mater which dive into the gray matter can be identified because in general these vessels have a perpendicular course to the brain surface, meaning that their spatial layout changes little from section to section. Alternatively, sham penetrations can be made at several locations within the mapped cortical region. For example, empty glass micropipettes driven into the cortex at known anatomical coordinates typically result in a small diameter (10–50 um in diameter) tissue damage that is readily visible in the histological specimen. By using either or both of the above aligning methods, a precise alignment between the functional map and the histological data can be obtained, in the range of 10–40 μm (see Figure 7.3 and Yousef et al., 1999).

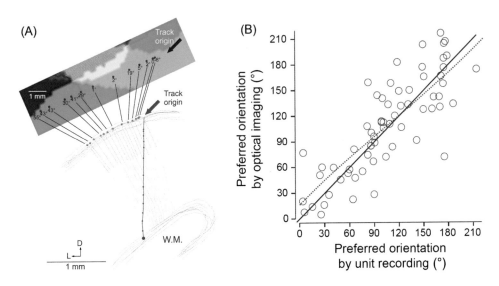

FIGURE 7.1 Preferred orientation revealed by optical imaging versus unit recordings. (A) Histological reconstruction of an electrode track (red line) along which unit recordings (red spot) were obtained at different depths of the visual cortex. Large red spot represents the deepest electrode position where a small electrolytic lesion was made for determining electrode tip position. Asterisks indicate surface projection of individual recording sites. The trajectory of apical dendrites and cell body columns representing the columnar organization was obtained in coronal sections stained with cresyl violet. The same surface projections are marked on the enlarged orientation map (red spots). Numbers indicate difference between electrophysiological and corresponding optical imaging orientation values in degrees. Note that the deeper the penetration is the larger the difference. (B) Scatter plot of orientation preference obtained by unit recordings and optical imaging. Data derived from layer 6. Red dotted line represents linear fit of the observed data, black line represents unity line. L, lateral; D, dorsal; W.M., white matter. (B) Courtesy of Dr. Fuyuki Karube.

7.3 LAMINAR DISTRIBUTION OF LONG-RANGE LATERAL CONNECTIONS

7.3.1 Upper Tier—Layers 2–3

Early concepts on the functional correlate of the long-range patchy system emphasized orientation preference as the most plausible functional feature to follow the anatomical distribution of patches in layer 3 (Mitchison and Crick, 1982). Indeed, electrophysiological data using cross-correlation analysis of recorded neurons spaced up to several millimeters laterally showed a strong coupling when their orientation preferences were similar (Michalski et al., 1983; Ts'o et al., 1986; Schwarz and Bolz, 1991). Contemporary anatomical data obtained for the cat visual cortex concluded on the basis of spatial parameters of cortico-cortical connections that these implied a relationship with the functional architecture of the cortex and in particular linked columns of similar orientations. Single axon reconstructions along with population labeling showed that the average patch size has a diameter of approximately 200–400 μm (Gilbert and Wiesel, 1983; Luhmann et al., 1986; Kisvárday and Eysel, 1992), a dimension that matches with the width of an iso-orientation column. Furthermore, the average inter-patch distance is 600–1100 μm (Luhmann et al., 1986; Löwel et al., 1987; Kisvárday and Eysel, 1992) corresponding to the average hypercolumn distance wherein the orientation preference completes a full cycle (Hubel and Wiesel, 1974).

Importantly, the above functional and anatomical data provide strong support to the early recognition that the spatial organization of lateral connections is ideally suited to promote spatially extended excitatory convergence between cells of similar orientations for constructing elongated receptive fields (RFs) (Bolz and Gilbert, 1989; Gilbert, 1977; Gray et al., 1989). Inevitably, these data provide some experimental support to the concept of "like connects to like" and through this idea, further explains perceptual phenomena and a number of psychophysical observations where long-range connections are expected to play a key role (for review see Gilbert, 1992; Kapadia et al., 1995). However, more recent anatomical data (Kisvárday et al., 1997; Schmidt et al., 1997; Buzás et al., 2006; Martin et al., 2014), and functional data (Chavane et al., 2011; Huang et al., 2014), casts doubt on the strong belief that the main role of long-range patchy axons is to link similar functional columns. Close inspection of patchy axons of pyramidal cells revealed only a modest bias toward iso-orientation instead of a hard core specificity. Our pool of layers 2–3 pyramidal cells possessed diverse distribution patterns in relation to orientation maps. These distributions varied widely around the distribution characterizing the entire set of cells (Buzás et al., 2006) (Figure 7.2).

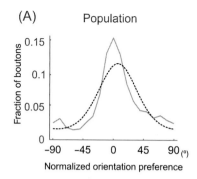
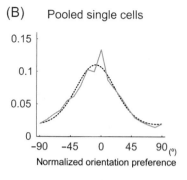

FIGURE 7.2   (A) Orientation distribution (grey line) of superficial layer boutons labeled from a focal extracellular injection into layers 2–3. (B) Orientation distribution of boutons of reconstructed single pyramidal cells. Mean relative orientation and orientation selectivity was somewhat narrower for the population data than for the pooled cell data. Dashed lines show the best fit of von Mises distributions. *Source: With permission from Buzás et al. (2006).*

Thus, the above single cell data corroborated the population results. Most of the connections (62%) were directed toward iso-orientation sites, whereas only 38% toward non-iso-orientation, including 15% cross-orientation (Kisvárday et al., 1997). This raises an inevitable question about the possible role of these non-iso-orientation connections. While there is as yet no definitive answer to this question, a simple conjecture they may underlie RF interactions to form object boundaries (Marr and Hildreth, 1980). For example to detect orientation discontinuity (Sillito et al., 1995), curved contours (Samonds et al., 2006) and illusory contours (von der Heydt and Peterhans, 1989), long-range non-iso-orientation connections can be useful.

7.3.1.1 Lateral Connections Are Embedded in Anisotropic Orientation Map

The known anisotropic structure of the orientation map is composed of orientation domains, where a smooth change in orientation preference is detectable along the cortical plane, and orientation centers, where orientation preference shows a rapid change often forming a pinwheel-like arrangement along the same plane (Bonhoeffer and Grinvald, 1991; Blasdel, 1992). Long-range lateral connections must conform to this intrinsic structure of the orientation map. Indeed, combined optical imaging and extracellular tracer injections revealed that the connectivity pattern of orientation domains versus orientation centers differs from each other in many aspects (Figure 7.3).

While injections into orientation domains produce a strongly patchy pattern of retrogradely labeled cells, injections into orientation centers result in a circularly symmetric distribution of labeled somata with a smaller lateral extent (typically about 1.6 mm in radius; Yousef et al., 2001) from the injection center as opposed to domain injections (typically about 3–4 mm in radius). This observation raises several questions about the function and topography of connections originating from orientation centers. Electrophysiology of orientation center neurons showed a similar orientation tuning to that of domain cells, despite the fact that center neurons obviously face several orientations (Maldonado et al., 1997). Some of the studies elaborating this issue reveal a sharp orientation tuning of individual center neurons, and also that their spatial organization is highly precise (Ohki et al., 2006). In fact, the only difference that has been detected so far regards the stronger adaptation capacity of orientation center neurons to drifting oriented grating stimuli than that of orientation domain neurons. A parsimonious explanation is that center neurons have a higher exposure to inputs representing a broad range of orientation and can thereby constitute foci of plasticity (Dragoi et al., 2001). The lesson that has been learned so far for the superficial layer patchy network can be summarized as follows: the diversity that is seen anatomically for the functional topography of connections compares well with the multitude of effects observed electrophysiologically in RF interactions.

7.3.2 Middle Tier—Layer 4

Layer 4 represents the middle tier of the six-layered cortex. It can be also considered as some kind of "origin" of the cortical microcircuitry since it receives the bulk of the thalamic input drive and is sandwiched between superficial and deep cortical layers. Despite the fact that layer 4 plays a pivotal role in hosting primary visual inputs from the thalamus and the fact that the neurons here generate novel RF properties such as orientation selectivity (Gilbert, 1977), little is known about their relationships with the neighboring cortical columns via long-range lateral connections. While local layer 4 connections have been studied using a host of *in vivo* and *in vitro* approaches (Hirsch, 1995;

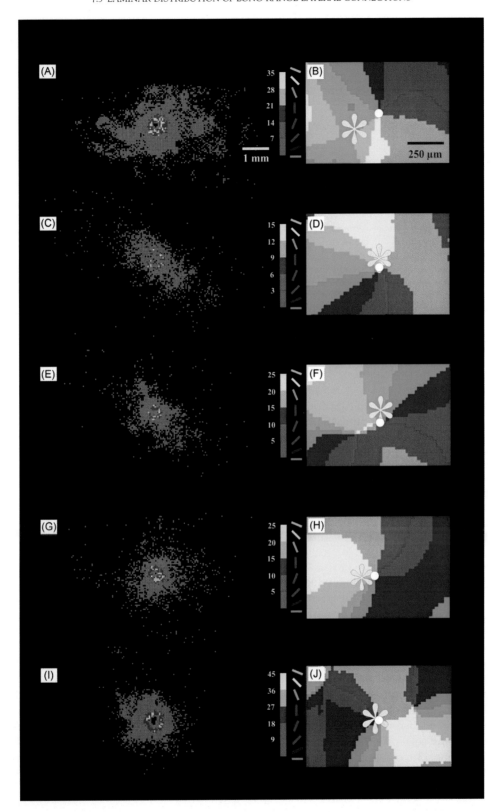

FIGURE 7.3 Retrograde labeling by fluorescent tracers (latex-bead) in layer 3 of the cat primary visual cortex (area 18). Left panels (A, C, E, and G) show cell density maps, right panels (B, D, F, and H) display enlarged view of orientation maps taken from the injection sites (asterisks). Note the loss of patchy character of the labeling as the injection sites approach orientation center zones. Injection into an orientation domain gives rise to an elongated area of labeled patches (A) whereas near-center injections result in quasi-circular non-patchy labeling (G). Color scheme represents cell density (number of cells per pixel), color bars code for orientation preference.

I. MICROCIRCUITRY

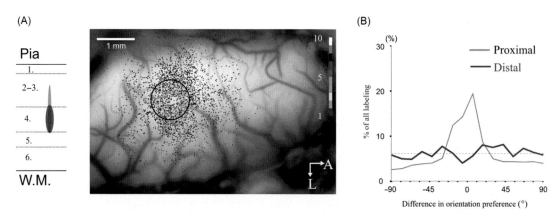

FIGURE 7.4   Orientation topography of retrogradely labeled neuronal cell bodies in layer 4. (A) Inset shows the injection core in red. Distribution of labeled cells superimposed on the surface image of area 18 in the cat. Labeling pattern differs radically from that of layer 3 injections. (B) Normalized orientation distribution of layer 4 projections representing cumulative data of anterograde and retrograde labeling. Proximal projections (grey) show iso-orientation preference whereas distal projections (red) fluctuate around hypothetical even distribution indicated by broken line. Circle in (A) marks border between proximal and distal distance from the core of the injection site. W.M., white matter; A, anterior; L, lateral. *Source: With permission from Yousef et al. (1999).*

Tarczy-Hornoch et al., 1998; Thomson et al., 2002; Bannister and Thomson, 2007), only few attempts that focused on lateral connectivity. In this regard, previous experiments using Golgi impregnation (for review, see Lund, 1988) and small focal injection of horseradish peroxidase (Fitzpatrick et al., 1985) were successful in revealing how far layer 4 axon can terminate but a direct link to functional columns could not be determined.

With the technical advent of *in vivo* functional imaging and subsequent bulk injection of neuronal tracers, quite a different and unexpected organization scheme of layer 4 lateral connections was unraveled in contrast to that of layer 3 (Yousef et al., 1999). Firstly, small focal injections of neuronal tracers resulted in a markedly reduced patchy labeling compared with that of the layer 3 (Figure 7.4). Another intriguing finding was almost complete lack of iso-orientation preference of distal connections (but refer data for other species by Roerig et al., 2003; Mooser et al., 2004). This finding alone addresses a number of questions about the "iso-orientation commitment" of layer 4 as the "home layer" of orientation selectivity. It is difficult to estimate the functional consequence of the relatively strong non-iso-orientation character of excitatory lateral connections, and there is no easy answer to what might be their functional role. Notwithstanding, electrophysiological recordings using the remote GABA inactivation paradigm indicated that lateral cross-orientation inhibitory interactions, including those of layer 4, can influence the sharpening of orientation selectivity via cross-oriented excitatory connections terminating on inhibitory cells (Crook et al., 1998), most likely via activation of remotely located inhibitory interneurons.

The study by Karube and Kisvárday (2011) investigated the functional topography and spatial parameters of layer 4 connections at the single cell level. They carried out extracellular injections of BDA or biocytin into the primary visual cortex that had been mapped for orientation preference using intrinsic signal optical imaging. Two findings are worth mentioning in relation to this work. First, the single cell data comprising pyramidal and spiny stellate cells confirmed the population results as to the presence of non-iso-orientation connections associated with layer 4 axons. Some spiny cells showed an overall iso-orientation preferred axon distribution, while other spiny cells showed a clear preference for non-iso-orientations including cross-orientation. The examples for the different types and cases representing oblique orientation preference are shown in Figure 7.5. For the entire layer 4 data, 52% of all boutons were found to be at iso-orientations, 22% at oblique orientation, and 26% at cross-orientation. On average, the axons of layer 4 spiny cells provided 2.4-times more cross-orientation connections than layer 3 pyramidal cells; and for remote locations 43% of boutons represented cross-orientation (four times more than pyramidal neurons in layer 3). For many layer 4 spiny cells, each bouton patch had a sharp orientation peak although the orientation distribution of all boutons had a broad range. When layer 4 axons are compared quantitatively with data from layers 2–3 and 5 (Buzás et al., 2006), the results clearly show that layer 4 connections provide more cross-orientation preferring patchy connections than do the iso-orientation dominated layers 2–3 and 5 pyramidal cells, confirming the results for layer 4 and layers 2–3 populations (see above). Taken together, 7 of 23 layer 4 cells established connections primarily to cross-orientation sites, typically, via long-range axon branches.

The second important aspect of elongated axons is whether their axis parallels the same orientation preference of the cell body. Such a morphological signature can be interpreted as an association of RFs according to Gestalt

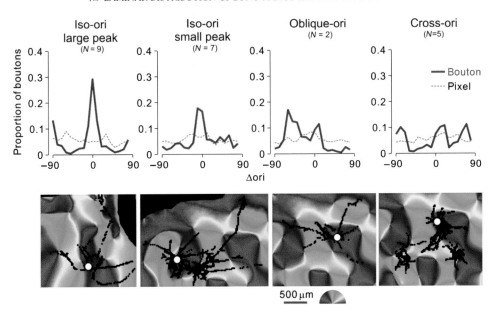

FIGURE 7.5 Normalized orientation distribution of boutons emitted by layer 4 spiny neurons. Graphs show average orientation difference (Δ ori) between boutons and parent somata (red line) and frequency distribution of map pixels (broken line). The four categories from left to right represent, respectively, the transition from iso- to cross-orientation preferred cells. Lower panels show examples of bouton distributions superimposed on orientation map for each category. White dots mark cell body location, black dots indicate boutons. Black shaded areas were excluded from the analysis. *Source: With permission from Karube and Kisvárday (2011).*

rules. In the previous study carried out in treeshrew, the axon arbor of layer 4 neurons was found to align with the preferred orientation (Mooser et al., 2004); in contrast to this no such spatial organization of the axons could be detected in the cat visual cortex (Martin et al., 2014; Karube and Kisvárday, 2011). It is not clear whether the above difference is due to interspecies variability or other factors. Obviously, future investigations should address this issue on a larger sample of layer 4 axon arbors, at the population as well as the single cell level.

Karube and Kisvárday (2011) also investigated the patchy phenomenon of layer 4 axons. They used the mean shift algorithm (Binzegger et al., 2007) to identify bouton patches following objective criteria. For the pool of 23 layer 4 cells, a total of 131 bouton patches were delineated with an average of 6 patches per cell. Interestingly, most bouton patches had a sharp orientation preference as shown for two spiny stellate cells in Figure 7.6. For better visibility, individual axon patches were colored according to the orientation preference, so as to better allow for detecting marked differences between the two cells. All boutons of cell (A) represented a broad orientation range, while those of cell (B) represented chiefly iso-orientation. Regarding the former cell (A), three bouton patches were devoted to iso-orientation and the other four were to cross-orientation. Clearly, these findings corroborate the broad orientation spectrum of layer 4 connections, as revealed by the population labeling, but at the same time reveal new aspects regarding the intricate orientation specificity of the single bouton patches. Axonal patches may differ considerably in size and relative weight. The general trend that emerges from studying such patchy axons is for the greatest bouton density and, typically, largest patch size, to be in the column of the parent soma (Karube and Kisvárday, 2011; Binzegger et al., 2007) (Figure 7.6).

The morphological framework presented here for layer 4 begs the question of why do almost one-third of the lateral connections (7 of 23 cells) prefer non-iso-orientation? Certainly, this result for layer 4 excitatory connections suggests more complex interactions than a simple enhancement of RF features, such as orientation preference (see below).

7.3.3 Lower Tier—Layer 6

An exploration of layer 6 lateral connections is important since it represents one of the chief output layers of the cortex to other cortical areas and subcortical structures (e.g., dLGN, for review, see Sherman and Guillery, 2001). In addition, layer 6 is known to receive inputs from intrinsic cortical connections (for review, see Briggs, 2010), subcortical inputs from the dorsal lateral geniculate nucleus (LGN, LeVay and Gilbert, 1976), and the visual claustrum (LeVay and Sherk, 1981; but see also Da Costa and Martin, 2010). Some layer 6 neurons have been reported

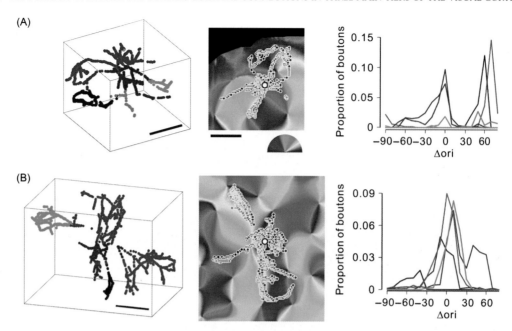

FIGURE 7.6   Cluster analysis of two layer 4 spiny stellate cells and orientation distribution of the individual axon patches. For clarity, bouton patches (7 for cell in (A) and 5 for cell in (B)) are shown by different colors. Middle panels show boutons colored to their respective cluster (small circles) and parent cell body (large circles) overlaid on orientation maps. Scale bar, 1 mm. *Source: With permission from Karube and Kisvárday (2011).*

not to project extrinsically (Katz, 1987; Briggs and Callaway, 2001), but most project subcortically and are known to establish cortico-subcortical loops (Gilbert and Kelly, 1975; LeVay and Sherk, 1981; McCourt et al., 1986; Katz, 1987). Therefore, the connections of layer 6 spiny cells must have an impact on the entire visual signaling pathway, including the home area and other cortical areas, as well as subcortical visual centers (Cudeiro and Sillito, 2006; Sillito et al., 2006; Thomson, 2010).

Lower tier connections have been analyzed using techniques similar to upper- and middle tier connections. Due to the lack of a representative number of reconstructed layer 5 cells, we will deal with the population of layer 6 cortico-cortical connections and their topography in relation to the orientation map. Here we analyzed labeled layer 6 neurons by determining axonal bouton distribution patterns in relation to local orientation map obtained with intrinsic signal optical imaging. To investigate the morphological difference among the cells, we applied a classification scheme based on a cluster analysis (Ward method) using three parameters. The first parameter represented the total length of apical dendrites. Here we calculated the proportion of, "path lengths" from the origin of apical dendrites to each dendritic ending measured along the course of the dendrites. Only the proportion of all the paths longer than 600 μm was calculated. The second parameter represented the median of radial distance (along the line perpendicular to the cortical surface) from the origin of an apical dendrite to each node. The third parameter represented the median of radial distances from the origin of the apical dendrite to each dendritic ending. In this way, we divided the sample cells into two large clusters, referred to as Type A and Type B. The Type A group is composed of cells with short apical dendrites lacking terminal tufts and Type B contains cells with long apical dendrites and with terminal dendritic tufts often reaching upper layers 2–3 (Figure 7.7). Type A was further divided into two subgroups: Type Aa and Type Ab. Type Aa had an apical dendrite which rarely crossed the border between layers 3 and 4. Type Ab also had a short apical dendrite, typically, giving off oblique branches close to the parent soma. Comparing the dendritic morphology of the aforementioned layer 6 cells with published data reported by others, it is still a bit uncertain that Type Aa corresponds to the LGN projecting population, Type Ab projects to other cortical regions, and Type B is linked to the claustrum (among others, Katz, 1987; for reviews, see Thomson, 2010; Briggs, 2010) (Figure 7.8).

Subsequently, the axon distribution of each type in relation to the orientation map was investigated. Since layer 6 is far from the superficial zone where optical imaging signals are acquired, extreme care was taken to include only those layer 6 neurons in the analysis which met our stringent criteria as discussed in Section 7.2. Notably, the central 5° lower visual field representation is located at the top of the lateral gyrus, which is a reasonably flat region of the cat brain. We limited our analyses to this flat zone in order to be reasonably sure that apparent orientation preference of our layer 6 cells is not from a distortion but is faithfully represented by the orientation map. The relationship

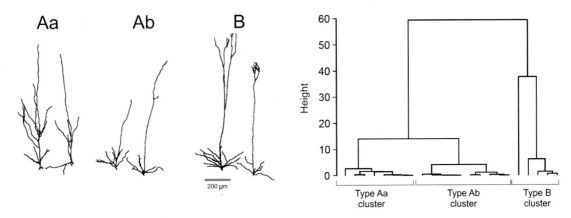

FIGURE 7.7 Three main types of layer 6 spiny cells: Type Aa, Type Ab, and Type B. Cell body and basal dendrites are not shown in order to more clearly emphasize the apical dendrite morphology. Dendrogram by independent cluster analysis (Ward method) revealed three main groups of layer 6 spiny neurons using basic morphological parameters of apical dendrites alone. *Source: Courtesy of Dr. Fuyuki Karube.*

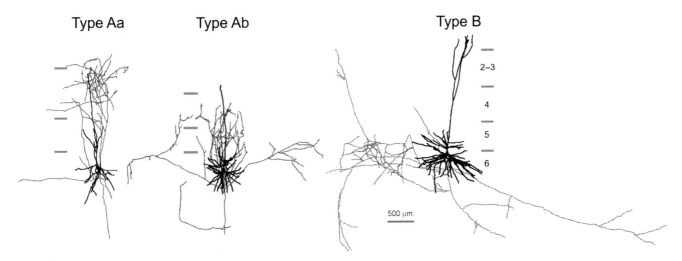

FIGURE 7.8 Examples of reconstructed layer 6 spiny neurons. Somata and dendrites are shown in black, axons in red. For Types Aa, Ab, and B not only the apical dendrites differed but also axonal fields showed distinct morphology (see text for details). Layers are indicated by numbers. *Source: Courtesy of Dr. Fuyuki Karube.*

between single cell bouton distribution and functional maps was analyzed for more than 21 reconstructed cells. In Figure 7.9, an example for each type is shown. It should be noted that although the pool of the reconstructed cells studied herein were labeled extracellularly (see Methodical considerations), we could trace the axons to their distal ends with the exception of a few collaterals, obscured at the injection site. The loss of some axon collaterals at the extracellular injection site can explain why the number of boutons per cell (mean 1399) in our layer 6 sample was smaller than that of intracellularly labeled cells (Binzegger et al., 2004) where no such masking of the labeling is present. Note that the mean dendritic length of both apical and basal dendrites was almost the same in the two studies, indicating that our cell classification based on dendrite morphology is comparable.

The output selectivity in terms of orientation was determined for each labeled cell. To do this, the orientation difference between the parent soma and each bouton was computed on a pixel-by-pixel basis (Δ ori, Δ range: 0–90°). For the sake of simplicity and comparison of the findings with previous data on single cell connections, Δ ori values were divided into iso- (<30°), oblique- (30–60°), and cross- (>60°) orientation categories, and the proportions of iso-, oblique-, and cross-orientation preferred boutons were plotted for each cell (Figure 7.10).

For the entire population of 21 cells, the proportions of iso-, oblique-, and cross-orientation preferred boutons were 52%, 28%, and 20%, respectively. This suggested an iso-orientation bias of a similar magnitude as that of middle (Karube and Kisvárday, 2011; Yousef et al., 1999) and upper tiers (Bosking et al., 1997; Kisvárday et al., 1997; Schmidt et al., 1997; Sincich and Blasdel, 2001; Chisum et al., 2003). However, comparison of the bouton distribution according to the non-iso-orientation category showed that cross-orientation preference was statistically more

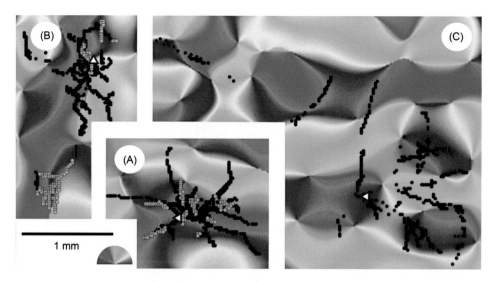

FIGURE 7.9   Distribution of boutons of Type Aa (A), Type Ab (B), and Type B (C) layer 6 spiny neurons to orientation map. Upper tier boutons are indicated by circles filled in grey, and lower tier boutons by circles filled in black. Preferred orientation is indicated by color code. Triangle shows soma position. *Source: Courtesy of Dr. Fuyuki Karube.*

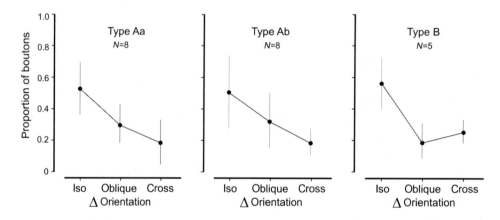

FIGURE 7.10   Orientation preference of bouton distribution obtained for 21 layer 6 spiny cells. Difference of preferred orientation (Δ ori) between soma pixel and bouton pixel was calculated and binned into iso- (<30°), oblique- (30–60°), and cross-(>60°) orientations. Each line represents an average. *Source: Courtesy of Dr. Fuyuki Karube.*

frequent for lower tier projections (and middle tier) than for upper tier projections. The most distal terminations of axons within layers 5–6, however, typically enregistered to a broad range of orientations. The exception was the group of putative claustrum projecting layer 6 cells (Type B), which terminated chiefly in iso-orientation domains (see Figures 7.9C and 7.10 Type B).

The functional role of layer 6 spiny cells must be viewed in the context of the postsynaptic targets. There are at least two functional streams in which layer 6 cells can be strongly implicated. First, they provide main intracortical input to layer 4. Indeed, electron microscopic studies estimated that the axons of layer 6 cells account for 20–40% of all synapses on layer 4 spiny stellate cells (in cat: Ahmed et al., 1994, 1997; Binzegger et al., 2004) as well as on GABAergic cells (cat: McGuire et al., 1984; Somogyi, 1989; Ahmed et al., 1994, 1997; rat: West et al., 2006). This strong input must have a significant impact on layer 4 RFs. For example, Type A cells can improve signal to noise ratio and control synaptic gain of thalamic input as reported in rodents (Olsen et al., 2012; Bortone et al., 2014). Another plausible scenario is that the input serves to sharpen orientation preference via di-synaptic lateral inhibition. Cross-orientation connections terminating in layer 4 offer a structural substrate for this (Figure 7.11). Finally, layer 6 may act on layer 4 by generating novel RF properties, such as end inhibition and length sensitivity (Bolz and Gilbert, 1986; Grieve and Sillito, 1991). In summary, layer 6 well deserves its anatomical name, "multiform layer," composed of several types of neuron. Of these, the three spiny cell types analyzed herein are unlikely to represent the full

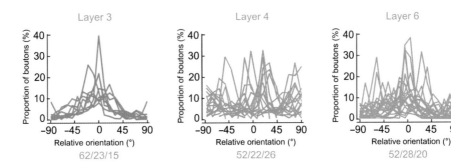

FIGURE 7.11 Summary chart showing normalized orientation preference of lateral connections in the three main tiers of primary visual cortex. Numbers indicate percentile values of iso-, oblique-, and cross-orientations. Each gray line represents data of a single cell.

morphological scale (c.f. van Aerde and Feldmeyer, 2015). The morphological diversity of dendrites and axons in relation to orientation maps indicates that the layer 6 spiny cell population represents a far more complex stage in processing visual information than that envisaged so far.

7.3.4 Summary on Lateral Connections of the Three Main Cortical Tiers

Taken together, the data presented here for lateral connections of axons in the primary visual cortex reveal a morphological diversity and topographical inhomogeneity as matched to the pattern of the most robust functional attribute, orientation selectivity. Through this, we conceive a highly integrative nature for each of the three main tiers of the visual cortex. The general rule that each neuron receives a broad range of inputs from neighboring cortical columns via intra- and interlaminar lateral connections places each tier at a highly versatile stage for processing visual information. At present, there is no unifying model which can take account of the strong structural and functional diversity of axons. From this perspective, a formidable task moving forward will be to develop brain model strategies capable of incorporating or at least addressing the richness of brain architectures that is evident at all system levels.

7.4 ORGANIZATION PRINCIPLES OF PATCHY LATERAL CONNECTIONS

The neuronal organization of the patchy system has incited a great interest not only among the experimentalists but also the modelers. The clustered or patchy phenomenon is one of the most appealing features of cortical network organization in addition to the two basic attributes, cortical layers and columnar architecture. Recently, there have been a number of attempts to trace out the rules that underlie the spatial constraints of patchy connections, in development and interaction with other members of the patch system, as well as the functional implications. Basic questions are those related to patch topology and parameters such as number of cells and network size. These parameters would seem to be directly linked to neuronal types, the intrinsic neuronal morphology, and functional properties of the contributing cells, all of which eventually establish a synaptically coupled network. Model constraints are equally important, since the patchy network must have an operational range and employ a limited number of cells. There are several models aiming to resolve several questions about the patchy cortico-cortical system. In this section, we consider those network models which directly address the organizational signatures and principles of the patchy system and its functional embedding in the cortical machinery.

7.4.1 Spatial Constraints of Superficial Layer Patches

The modular organization of the cerebral cortex is particularly evident in the primary visual cortex, both in terms of anatomical connections and functional architecture. Patchy cortico-cortical connections are generally considered as tightly correlating with the distribution of functional attributes. The key question then becomes, how good is the above correlation. Can this indicate the significance of the structural–functional relationship and, if so, imply the influence of the patchy system in shaping the neuronal response?

A recent study by Muir et al. (2011) investigated the general spatial constraints of the superficial layer patch system using a technique that measures the statistical structure inherent in the arrangement of the anatomically labeled patches and the functional maps, as typically revealed with optical imaging of intrinsic signals. At this point the general assumption is that the anatomical substrate encodes a "statistical expectation" of the cortical response and the latter is manifested, for example, in the orientation angle map. Briefly, the authors employed the so-called Gabriel graphing (a subset of the commonly used Delaunay triangulation) to construct neighbor graphs for the orientation maps in order to answer to their simple but fundamental question, whether the patches and the active regions of the orientation map are randomly distributed. The answer was a clear "no." What they found was an overrepresentation of 60° for the inter-neighboring patch angle and from this a hexagonal arrangement—which is a common biological preference (Levin et al., 2006)—could be conjectured. In fact the arrangement of the lattice was a noisy hexagonal, but importantly, definitely deviated from a random distribution (i.e., Poisson type). The careful conclusion that was drawn emphasized that the patchy connectivity system acts as a physical implementation of statistically congruent cortical response domains. In conjunction with these specific findings, an important interpretation was additionally made: the commonly and often rigidly used "like-connects-to-like" concept is incompatible with the superficial layer patchy system. To back up this conclusion, one should also mention the converging results of a large number of experimental studies (see later) for assessing the orientation preference of patchy lateral connections. They almost universally showed an iso-orientation bias (50–60% of connections), rather than specificity as suggested by the early model of Mitchison and Crick (1982). Before discussing what functional role the iso-orientation biased patchy system might have, we shall look at further structural details of the patchy network down to the single cell level.

7.4.2 Inherent Structural Features of the Patchy Network

Brain modeling is a formidable task because it takes into account a wealth of the biological data that ranges from molecular neurobiology to the diversity of neuronal types, each of them having particular synaptology, connectivity relations, and diversity of functional properties. Regarding the layer 3 patchy network, most of the models, however, have used a highly simplified structural and functional parameter set since the complexity of computation exponentially increases with each model parameter. Nevertheless, there are increasing number of studies employing biologically realistic parameter sets adapted from neuroanatomical and neurophysiological data, with layer-specific connectivity and functional dynamics.

Traditionally, the patchy system is linked to orientation selectivity (Mitchison and Crick, 1982; Gilbert and Wiesel, 1983) and the visual cortex has been referenced as the key assay to study interrelations of the patchy network and map-like functional representations such as orientation selectivity. In fact, mammals (e.g., rodents) without orientation maps are assumed to lack patchy connections in the primary visual cortex (van Hooser et al., 2006), although these creatures do contain orientation selective cells (Girman et al., 1999) and long horizontal collaterals (Burkhalter & Charles, 1990). This conundrum calls for alternative explanations as how orientation selectivity of individual neurons can be accounted for without any evident structural patterning, in other words, how synaptic strength can vary in a patch-like manner without patchy connectivity. The complexity in the relationship between orientation selectivity and underlying structural architectures is represented by primates and the treeshrew, in which the orientation column system is horizontally split by layer 4 containing neurons lacking orientation selectivity (Humphrey et al., 1980).

In previous models, anatomical patches are considered purely randomly distributed along the cortical plane that is surely an oversimplification of their spatial organization. More recent attempts employed stochastic graph theory approaches for simplifying the probabilistic feature of patch topology. Along this line Voges et al. (2010a) analyzed four possibilities for the spatial arrangement of patchy connections against homogeneously distributed synapses (Figure 7.12).

The first type of patchy structure lacks a stringent connectivity rule between the randomly positioned patches (RPs). Patch-to-patch connections are made by chance, and consequently the probability for neuronal convergence into a common distal patch is low. Anatomical data indicate that such a network rule may apply in paleocortical regions such as the pyriform cortex (Johnson et al., 2000). The second type has overlapping patches (OPs) resulting in slabs or elongated termination fields of synapses. The most likely anatomical equivalent of such an organization can be found in primate prefrontal areas 9 and 46 (Lund et al., 1993). The third type corresponds to the shared patches (SPs) model in which any single neuron participating in axonal patches contributes to a subset of a common patchy pattern. This model is compatible with anatomical results revealed in cat sensory cortices (Kisvárday and Eysel, 1992; Ojima et al., 1991). The fourth type is the variant of the SP model, called PP model, it allows partial overlap of patches (PPs) originating from somewhat distally located cells. Considering biological plausibility, most of the elements of the PP model can be validated by data obtained for cat sensory cortices (Kisvárday and Eysel, 1992; Ojima et al., 1991). There are important predictions made by Voges et al. (2010a) from the above models. Notably, patchy

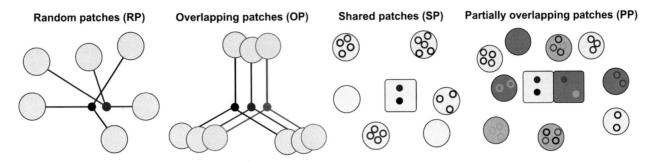

FIGURE 7.12   Possible spatial arrangement of long-range patchy connections. From left to right an increased complexity of the patchy network is present. Large filled circles, patches; small filled circles, nodes; small open circles, single axon patches; and box, node clusters. *Source: Modified from Voges et al. (2010a). See details in the text.*

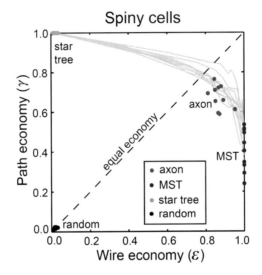

FIGURE 7.13   Axon arbor design including neurons with long-range patchy axon represents a trade-off between spatial and temporal communication cost. Axon arbors (red dots) of ten spiny cells lay on or near to the trade-off curves (gray lines) with a slight bias towards wire minimization relative to equal economy line (broken black line). Which are suboptimal for wire length economy (ε) compared to wire—minimized minimum spanning tree (MST, blue dots) and suboptimal for length economy (Υ) compared with path-minimized star-trees (green dots). Randomized trees (black dots) are close to the origin of the "economy plane" that was generated by the so-called *light approximate spanning tree* algorithm. *Source: With permission from Budd et al. (2010).*

connections exhibit a "small-world effect" (see Watts and Strogatz, 1998; Bassett and Bullmore, 2006) wherein upon formation of more patches the path length becomes smaller. Path length is defined as a temporal cost that is equivalent to the total distance travelled by a notional axon potential from the axon origin to each bouton (Manor et al., 1991). Optimization of the cortical grey matter connections is an important aspect for fast and reliable signal conduction in the brain, which has limited energy resources and space constraints (Attwell and Laughlin, 2001; Chklovskii et al., 2002). However, it has been shown recently for individual pyramidal cells that due to the branching properties of their intracortical axonal trees they actually do not match those of wire-minimized Steiner minimal trees (Budd et al., 2010). Instead it was shown by the same study that path length has a trade-off with wire length, that is, conduction time delay versus total axon volume (Figure 7.13). Thus, the patchy network serving communication between remote cortical columns has a trade-off with the two opposing constraints, temporal and spatial costs.

The study of Voges et al. (2010a) clearly indicates an increase in the functional versatility of the clustered network. The grandeur of cortical dynamics appears as an outcome of patchiness in terms of enhanced fluctuations in spike rates and shapes of power spectrum of population activity. This latter can be readily propagated for large distances particularly in SP and PP models. Importantly, the latter patchy network types perform far better than any other network types with reduced complexity, which are randomly distributed patches or networks without patches (see Figure 7.2; Voges et al., 2010a).

Interestingly, while the realization of this network topology with orientation preference (Mitchison and Crick, 1982) can be made in certain species (monkey: Malach et al., 1994; cat: Kisvárday et al., 1997; treeshrew: Bosking et al., 1997; ferret: Roerig and Chen, 2002), other animal species like rodents do not have orientation columns and the underlying patchy network in primary visual cortices (but see Burkhalter and Charles, 1990; Lohmann and Rörig, 1994), though neurons can be orientation selective. As already mentioned above there is no general explanation for the dichotomy of animal species having long-range connections with or without the above patchy network character. Nonetheless, recent theoretical studies suggest that a relatively small number of network parameters can determine whether orientation preference maps follow a salt-and-pepper organization or the type hallmarked by orientation domains and orientation centers. For example, according to the wire length minimization approach the competing spatial components (Gaussian and constants) of connections can result in radically different map topologies (Koulakov and Chklovskii, 2001). By the same token, network geometry symmetry and the amount of heterogeneity in short- and long-range neuronal interactions can determine which of the organization types prevail (salt-and-pepper: strong, suppressive short-range interactions and strongly broken geometry symmetries; orientation map containing orientation centers: sufficiently homogeneous interactions and strong geometric symmetries) (for review, see Kaschube, 2014). However, as is always the case, the situation is more complex because other modes of patchiness can be found in many cortical areas, typically, in those which are phylogenetically highly conserved across primates, carnivores, and rodents (Ichinohe et al., 2003).

7.4.3 Observing Patches Made of a Single Cell Type

The model by Perin et al. (2013) offers another simple description of the constraints pertaining to the type of patchy network that looks at the phenomenon primarily from a point of view of the dendritic field instead of axonal patches. Their approach is based on the "common neighbor" rule which determines connection probability between any two neurons and the number of other neurons connected to this pair of neurons (Perin et al., 2011). The common neighbor rule based clustering model was exploited, for deriving the relationships between cluster size and cluster number as a function of experimentally determined parameters. One of these parameters is "morphological reach" that is not only related to neuronal density and network size but also to details of the extent of axonal and dendritic arborization. The model was tested for the population of layer 5 pyramidal cells (TTL-5, thick tufted apical dendrites) which are known to possess the largest spread of basal dendrites and are found clustered in the rodents. The results revealed a linear relationship between network size and number of cell clusters; that is, the larger the network the more cell clusters of similar size can fit in. However, an increase in network density (cells/mm^3) is accompanied by larger average cluster size, which in turn follows a lognormal profile (first increasing then decreasing) with network size (Figure 7.14; Perin

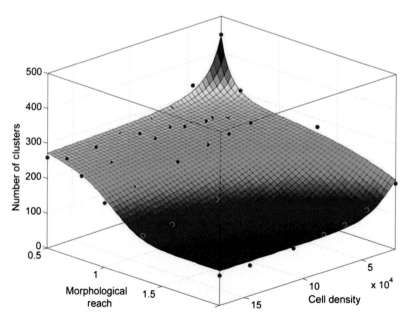

FIGURE 7.14   Network model shows that the number of cell clusters increases unevenly with cell density. Very low cell density causes numerous clusters with few neurons due to strong isolation of cells. Higher cell density produces a plateau in cluster number indicating less dependence on cell density. Then at extreme high cell density, cluster number drops due to a parallel increase in cluster size (not shown). "Morphological reach" follows a similar trend to cell density. *Source: With permission from Perin et al. (2013).*

et al., 2013). Neuron density and morphological reach have a minute effect on the number of clusters, but they show a modest sigmoid relationship with the cluster size. The TTL-5 pyramidal cell network introduced in the model emphasized two important attributes with regard to the patch formation. The neuronal components of the patch show strong dendritic and axonal overlap, and these assemblies are also interlaced in space at the same time. It is suggested that this spatial organization could be an equivalent of synchronous firing by subsets of neurons participating in the same circuitry. An obvious emergent property of this network is that the very same neuron can be shared by many network frequencies. It should be noted that this idea is also shared by other researchers and is compatible with large-scale functional considerations such that how successive groups of neurons may process synchronous activity to form, for example, synfire chain (Abeles, 1991; Kriener et al., 2008).

So far the most exhaustive modeling to explore the precise rules governing how the patchy system is constructed from the axonal arbors of pyramidal cells is that of Muir and Douglas (2011). In Section 7.4.1, we discussed how the same authors could find a modeling solution to embed the patchy network in global functional architectures such as the orientation map (Muir et al., 2011). To that they utilized structural details of five model variants which differ from each other in satisfying basic observations obtained by *in vivo* experiments. By using a set of model constraints, they were able to configure a biologically quasi-realistic model, which is consistent with the known and often robust properties of the patchy pyramidal cell network. The simulations carried out by Muir and Douglas (2011) were based on shifting a hexagonal arrangement of individual axons according to the geometric and connectivity rules which were able to reproduce many essential features of the patchy system observed after bulk injection of neuronal tracers. The numerous achievable criteria for the model such as the limited size of individual patches irrespective of how large is the injection site, the qualitative change in the patch- and the interleaved non-patch locations as a function of radial distance from the injection site, and the reciprocal nature between a set of patches within the network could be achieved. One of the main findings of this study is the clear dependence of the patchy network on multiple connectivity rules. Although the model performs well and explains many features of the superficial layer patchy system but it fails to address some other obvious features. For example, strong morphological diversity and spatial anisotropy of individual axons are common features of real preparations; these important issues could not be addressed by the model. In fact, few studies that visualized and reconstructed patchy axons of single pyramidal neurons in utmost detail proved that axon clusters of individual cells are established in a highly nonrandom manner across the cortical plane (Buzás et al., 2006; Binzegger et al., 2007; Martin et al., 2014). We hypothesize that correlation of patchy axons with activity representation might serve as a tool to better understand the functional role and significance of the patchy system. In Section 7.5.1, we provide an analysis tool which enables unraveling of possible correlations from that the involvement of long-range lateral connections in perceptually relevant Gestalt-like processing can be inferred.

7.5 POSSIBLE FUNCTIONAL ROLE OF THE PATCHY SYSTEM

As discussed above, the functional role of the long-range patchy system has been thought to link regions sharing similar functional properties (Mitchison and Crick, 1982; Gilbert and Wiesel, 1989). Indeed, ever since the patchy intrinsic network could be visualized in V1 using tangential sections, the similarity in spacing of the patches with the iso-orientation columns have led to the commonly accepted view that these are two superimposable systems, observing a rule of "like connects to like." This recognition incited a series of studies in various species that addressed two principal questions as to what determines (i) the scale of the patchy system and (ii) the pattern that is often seen changing from area to area. In this regard, probably the most surprising data so far derive from studies in which the spatio-teamporal properties of single neurons could be related to large-scale activity modification including the patchy system (Chavane et al., 2011; Huang et al., 2014). Recently, Huang et al. (2014) used optogenetic activation of excitatory cells in combination with multiunit recording. Surprisingly, when iso-orientation patches neighboring to the recording site were activated together with visual stimulation, the recorded neurons showed a linear activation independent of orientation preference. Instead of the expected iso-orientation activation, the response magnitude depended only on the distance from the optogenetically activated patch. Another interesting observation was that any combination of the activated patches, whether it be spatial matching of the orientation preference of the recorded cell, or, in contrast, activated cross-orientation zones, had an impact on response intensity rather than changing the tuning properties. These results not only question the long standing view of iso-orientation selectivity of the patchy network, but also raise again the fundamental question: what is their role?

The most parsimonious explanation of the above experiment in failing to show nonlinear effects for optogenetic activation of patches is that the activation paradigm used by Huang did not generate the required balance of excitation and inhibition. The most dramatic perturbation of the optogenetic experimentation compared with the control case lies in the amount of the excitation. Under the visually stimulated conditions (control), probably only a subset of excitatory cells become activated, which are involved in processing the particular visual stimulus. By turning on the optogenetically manipulated excitatory cells without the corresponding inhibitory counterparts, the entire excitatory population is activated. It is conceivable that the net effect of the optogenetic stimulation cannot be different from nonspecific excitation that in fact drives all neurons of the network involved. While the authors of the above study concluded that feedforward projections are necessary for the establishment of orientation tuning properties, it remained unclear whether the orientation tuning observed in layers 2–3 cells derives from layer 4, or whether layers 2–3 lateral connections can additionally shape response characteristics in a context-dependent manner. It is pertinent to mention that the cat data indicate otherwise. That is, pharmacological silencing of a patch in layer 3 does change orientation tuning of the cells recorded at 300 μm laterally in the same layer (Girardin and Martin, 2009). It is thus reasonable to assume that a radical shifting of the balance of excitation and inhibition either by allowing excitation to prevail alone (optogenetic case) or by silencing both (GABA inactivation case) induces prominent changes in the spatiotemporal characteristics of response properties.

7.5.1 Contribution of Lateral Patchy Connections to Contour Integration

The basic hypothesis that we pursue here relies on the simple assumption that the strong preference of connections, as expressed in the density of connections, bears functional importance. Here we discuss evidence pertaining to connectivity rules of single pyramidal cells—possibly also applicable to populations—which comply with the association field concept proposed by Field et al. (1993). This provides a conceptual framework of contour integration that is supported by psychophysical results (Gilbert et al., 1996; Kulikowski et al., 1973) and modeling of spatial integration (Usher et al., 1999). According to the association field concept, common characteristics of neighboring RFs can conform to and facilitate interactions whereby the aggregate RF is endowed with features consistent with the visual percept. So far, the most likely neuronal cell type that meets these criteria is the superficial layers 2–3 pyramidal cell with a long-horizontal patchy axon. A major implication in the above logic is that the elongated axonal field parallels the spatial elongation of RFs of the targeted cells. This type of organization has been shown primarily for populations of cells in superficial layers (Bosking et al., 1997; Blasdel, 1992). However, there are so far no data available for their existence in the deeper layers.

For a better understanding of establishment of long-range connections with Gestalt-like inferences, we need to view them in the context of RF properties and visual field position, that is, retinotopy (c.f. Martin et al., 2014). Therefore, the visual field location of both pre- and postsynaptic components of the network has to be determined and taken into account. The location of each axon terminal (bouton) corresponds to a synaptic contact between pre- and postsynaptic structures and thus they point out the visual field coordinates or the retinotopic position represented by the postsynaptic target cell. Importantly, from this type of correspondence spatial interactions among the network components can be inferred. Pertinently, for a complete analysis of connections all boutons must be considered irrespective of whether they belong to the patches or are situated in the interpatch regions. Using these conjectures in combination with orientation preference of the parent cells and the terminal boutons, pyramidal neurons arranged in patches may in fact facilitate three types of interactions, each representing a particular aspect of contour integration (Figure 7.15). *Iso-orientation* patchy connections represent parallel orientation preference of RF locations at parent cell body and target zones. This representation type, however, does not require strictly a common axis along which the RFs of the presynaptic (parent) cell and the target patches are aligned. The second type is called, *radial* connections which show an orientation preference of target patches arranged in radial fashion to the RF position of the presynaptic (parent) cell. In this case, several orientations are permitted at target patches provided that the orientation axis runs across the retinotopic location of the parent cell body. The third type is *co-axial* connections for which RF locations of both parent cell and target patches are located along the same line in visual space. The orientation preference of target patches may deviate from that of the parent cell. Finally, the "classical" *co-linear* connectivity type is a special type for which the iso-oriented RFs are aligned along the same axis in visual space conforming a radial arrangement, too. Indeed in treeshrew visual cortex (Bosking et al., 1997), population labeling of long-range horizontal axons originating, presumably, from excitatory pyramidal neurons show a good deal of spatial alignment in visual space with same RF orientation preference to facilitate co-linearity. This type of spatial alignment can be taken as evidence for the main axis being in line with the association field concept (Field et al., 1993). Interestingly, in the cat visual cortex, population and single cell data do not manifest such a clear-cut arrangement (Kisvárday et al., 1997; Martin et al., 2014). Instead lateral connections of layer 3 pyramidal neurons

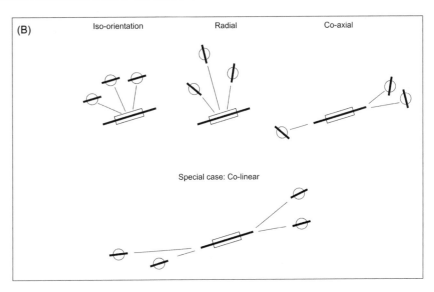

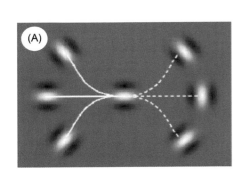

FIGURE 7.15 (A) The association field concept proposed by Field et al. (1993) provides a simplified theoretical mechanism of contour integration using Gestalt rules of alignment/grouping. The prediction is that there should be strong connections between adjacent orientation detectors. These detectors can provide a smooth, continuous curve (thick white line) but weak or no connection if they create a discontinuous curve (broken white line). (B) Connectivity types observed experimentally show a wider variety of connections. They combine information at local receptive field to perceive the contours of an object according to three main categories: iso-orientation, radial, and co-axial with respect to RF location and orientation preference of the presynaptic cell. "Classical" co-linear connectivity pattern is the special case that represents a combination of iso-orientation, radial, and co-axial. Rectangle, RF; black long bar, preferred orientation of the presynaptic cell; black short bar, preferred orientation of postsynaptic target cells; red circles, long-range axonal patches of the presynaptic cell. *Source: (A) With permission from Field et al. (1993).*

in the cat suggest a more diverse role of the patchy long-range system in line with combined optical imaging data and intracellular recordings (Chavane et al., 2011).

7.5.1.1 Spatial Statistical Approach

Most cortico-cortical connections terminate on dendrites and dendritic spines whose spatial distance from the cell body of the target cell follows a Gaussian distribution. The protocol described here is based on estimates of bouton density maps of exemplary layer 3 pyramidal cells. In order to take into account the probability of target distributions, the very first step is to generate a raw bouton map convolved with an annular Gaussian mask, whose width is determined by the normalized dendritic density of a sample of pyramidal cells (Buzás et al., 2006). Only those bouton density peaks are displayed which are above 10% with respect to the maximum peak value (100%, soma location) of the axon (Figure 7.16).

7.5.1.2 Spatial Statistics of Bouton Density to Orientation and Direction Maps

A spatial statistical approach was employed for a detailed analysis of how bouton density distributions are convolved with functional maps (preliminary data by Budd et al., 2010). Figure 7.17A shows a simplified representation of the relative relationship of the pyramidal cell axon (blue) with an orientation map corresponding to iso- (<30°), oblique- (30–60°), and cross- (>60°) orientations where the corresponding regions are shown in white, gray, and black, respectively. The orientation map is then superimposed with the convolved bouton probability density map (red lines) and peaks (black crosses). The orientation preference of the pyramidal cell axon for iso-, oblique-, and cross-orientation categories was calculated by the weighted sum of the convolved maps. It should be added here that although there is an overall preference for iso-orientation regions (86% of peaks, 62% of boutons), this axon sends collaterals to target cross-orientation regions (14% of peaks, 22% of boutons) arguing against the exclusivity of the like-connects-to-like concept. In this regard, the current data corroborate our earlier results obtained for populations of excitatory connections in area 18 of the cat brain displaying only iso-orientation bias of patchy excitatory connections (62% of boutons; Kisvárday et al., 1997; Buzás et al., 2006) rather than strong iso-orientation selectivity.

In a recent study, similar observations were made for the iso-orientation bias of axonal patches of layer 3 pyramidal cells (Martin et al., 2014). For a pool of 33 neurons, the results match with ours in that 60% (but only 60%) of the

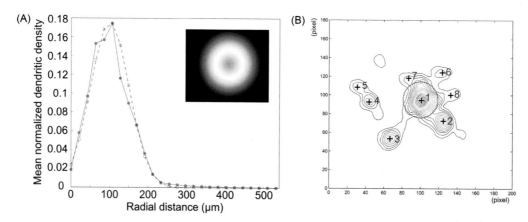

FIGURE 7.16   (A) Since excitatory boutons can terminate on any part of the dendrites, the position of the bouton is always slightly displaced from the true location of the target neuron. To account for this uncertainty we approximated the 2D probability density of the target soma around a single bouton by a Gaussian function rotated in 2D resulting in an annular mask. The mask was then applied to each pixel of the raw bouton map (not shown). In this way, a parsimonious estimate of the distribution of postsynaptic target cells could be constructed. (B) Convolved bouton probability density map of a layer 3 pyramidal cell. Each line represents iso-density levels, and crosses indicate density peaks (numbered). Note that the highest density peak is at the soma-dendritic region (circle, No. 1) but other remote regions also contain density peaks (No. 2–8). *Source: (A) With permission from Buzás et al. (2006).*

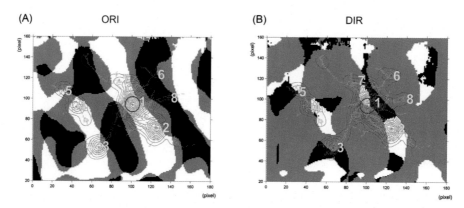

FIGURE 7.17   (A) Local peak (in red circle, peak No. 1) is located in an iso-orientation region surrounded by oblique orientation regions. The distal peaks typically show a clear preference for either iso-orientation (peaks No. 2, 3, 4, 5, 7) or cross-orientation (peaks No. 6 and 8) regions. (B) Local peak (in red circle, peak No. 1) is located in same (SAME) direction region but it also covers a somewhat greater region of opposite (OPP) direction and in between same and opposite (MID). Most distal peaks are located mainly in regions covered by MID direction properties (peaks No. 6, 7, 8) or a mixture including MID (peaks No. 3, 4). The only clear cases are peak No. 2 centered within SAME region and peak No. 5 near to the boundary and covering neighboring SAME and OPP directions. Peak No. 3 shares an OPP and MID regions while peak No. 4 contacts a SAME and OPP regions. Importantly, these plots representing the connectivity data in the cortical space offer fewer inferences for peak relations than those in the visual space (c.f. Figure 7.18).

bouton patches lay in or near orientation domains similar to that of the parent soma. A common sense explanation of the orientation diversity seen in the above and other studies (Malach et al., 1993; Schmidt et al., 1997) might be that such a connectivity scheme representing a large proportion of non-iso-orientation contributes to the detection of focal orientation discontinuities, for example, junctions and corners (i.e., cross-detectors; Sillito et al., 1995); but there may be further explanations for the observed diversity.

Similar to the above orientation preference map, in the relative direction preference map (DIR) the same-direction regions are represented in white and opposite-direction regions in black, while all other directions are shown in gray (Figure 7.17B). In this figure, raw and convolved bouton maps are also overlaid. The direction preference of the pyramidal cell axon for same/opposite/mid-categories was calculated by the weighted sum of the convolved map. Interestingly, most distal peaks are located mainly in the "mid" direction category (peaks 6, 7, 8) or include a mixture with "mid" (peaks 3, 4). When these peaks are visualized in retinotopic plane, a circle type of flow pattern around the parent soma is seen (see yellow circle in left panel of Figure 7.18). This pattern of a clockwise circular movement hints that this presynaptic pyramidal cell may contribute to its detection.

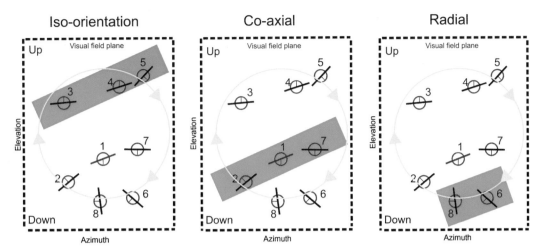

FIGURE 7.18 Projections of axon patches from cortical to visual space. Black long bars show the absolute orientation preference and short black bars show the absolute direction preference measured using optical imaging at the density peak of each axon patch (numbered red circles). Yellow circle with arrowheads represent clockwise circular motion direction according to direction preference of remote bouton density peaks with the exception of peak No. 4 (circular sequence of peaks: 5, 7, 6, 8, 2, 3). Areas highlighted in blue indicate special constellations between seed (patch No. 1) and target patches. Note that Patch No. 5 abide by both quasi-iso-orientation and radial constellation, and patch No. 2 and 7 both co-axial and radial constellations. See details in the text.

7.5.1.3 Axial Rules and Cortical to Visual Space Projection

There is a commonly held belief that patchy connections contribute to the processing of contour integration in the primary visual cortex by implying Gestalt-like interactions. Previous results showing the patchy arrangement of pyramidal cells suggested that each patch connects regions of similar functional attributes, for example, orientation (for review, see Gilbert, 1992). The functionally diverse representation of patches observed here for single axons suggests more complex duties of the patchy system. Instead of connecting solely the functionally similar regions they can terminate at non-iso-orientations; this finding is supported by the data from other investigations (Martin et al., 2014).

We think one of the key issues for understanding how the integration of contour elements can be brought about by patchy connections is the determination of the visual space representation of these connections. Hence, layer 3 patchy axons must be viewed in the context of the visual field location in which their RFs as well as axons are embedded. Below, we will show for an exemplary layer 3 pyramidal cell that the spatial layout of the axon patches/connections can underlie complex RF interactions. We have selected this cell because it was intracellularly labeled and, therefore, we are confident that the entire axonal and dendritic arbor could be visualized. Figure 7.18 attempts to summarize the main findings for the orientation and direction domains. Transformation of the connectivity peaks in the visual plane is a key step for carrying out an in-depth analysis of single-cell patchy connections. Here we followed a simple mathematical approximation for the coordinate transformation from visual space upon the cat area 18 as described by Tusa et al. (1979). Having done this, the RF preferences and the global locations of distal peaks show a thought provoking pattern which challenges Gestalt-like inferences (Figure 7.18). Apart from one peak (No. 4), the RF orientation and direction preference of the peaks form, for the most part, a near-radial pattern indicating that there are preferential connections between cells that lie on the same retinotopic axis. On these grounds, in Figure 7.18 one can recognize a circle-type flow pattern (yellow circle with arrowheads) around the local region (or parent soma, No. 1) with directional preference of the distal peaks pointing clockwise in a virtual rotating motion field. *This pattern of axon connectivity may facilitate the context of a spinning object appearing in one or both eyes.* This functional bias is best revealed when the rotating motion flow travels in phase with intracortical horizontal propagation of cortical activity (see Peyré and Frégnac, 2012). At this point it should be noted that the connections interpreted above do not have to be reciprocal and this notion is in line with experimental observations (Kisvárday and Eysel, 1992) as well as theoretical considerations (Voges et al., 2010b).

There are several other inferred interrelations between the local region (or parent soma, No. 1) and the target regions, which can be summarized as follows: peaks 3, 4, and 5 are aligned in an axis parallel to the preferred axis between peaks 1, 2, and 7, and in this regard they represent *iso-orientation (Rule 1)* relationship. It should be added here that iso-orientation connections can terminate at any retinotopic location. However, peaks 3, 4, and 5 revealed an interesting retinotopic constellation because they could contribute to an excitatory flank onto the RF of the presumed target cells (Figure 7.18). It should be also noted that peaks 1, 2, and 7 display a topographic arrangement

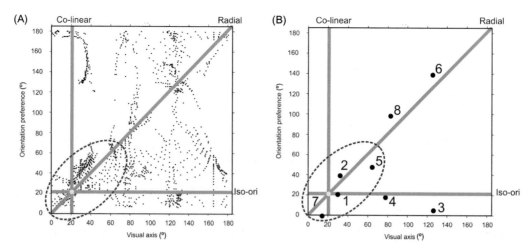

FIGURE 7.19   Scatter plot shows the distribution of individual boutons (A) and bouton peaks (B) of the layer 3 pyramidal cell, also shown in Figure 7.17A. Abscissa shows the orientation of axes running through the point in the visual field represented by the cell body. Ordinate shows the orientation preference measured with optical imaging. The location of the cell body had an orientation preference of 21 degrees (yellow spot at the intersection of the three grey lines). Solid lines correspond to the three connectivity rules along which the majority of connections are organized. Their intersection is the equivalent of the "classical" co-linear connectivity where data points accumulate at high density (red oval). Note the close correspondence between the bouton plot and the density peak plot. Numbering of bouton peaks in (B) refers to that of Figures 7.16–7.18.

where the target RFs are aligned along the retinotopic axis defined by the orientation preference of the seed. This type of arrangement is herein called, *co-axial (Rule 2)*. This arrangement is known from the literature as part of the co-linearity rule. The "classical" spatial co-linear alignment (Bosking et al., 1997) applies when the retinotopic axis is shared with the same orientation preference as is the case for peaks 1, 2, and 7. It has been hypothesized that such a constellation of RFs between the seed and the target populations may permit an efficient construction of complex RFs which are presumed to be essential in contour integration (Kapadia et al., 1995; Stettler et al., 2002). Finally, peaks 6 and 8 reveal a *radial (Rule 3)* relationship with peak 1 since their orientation preference in the visual space extends through the RF position of peak 1 (Figure 7.18). A possible implication of radial connections is to forward non-iso-orientation input to the targets including cross-orientation (i.e., peaks 6 and 8). In this regard, a similar constellation of lateral connections has been assumed to serve the detection of discontinuity features between RFs such as junctions and corners (Sillito et al., 1995; Das and Gilbert, 1997; Schmid, 2008).

In the previous section, we described an analysis based on convoluted peaks of target cell density, which provides a simplified view of the entire bouton distribution in relation to the orientation map and the representation of visual field plane (Figure 7.19). A more detailed analysis of the same data takes into account all boutons individually and assumes equal importance for each bouton. The scatter plots in Figure 7.19 show the result for our exemplary layer 3 pyramidal cell. Absolute *orientation preference* is represented along the *y*-axis and *visual field angle* of the line passing through the presynaptic (pyramidal cell body), and the presumed postsynaptic target cells and their RFs (marked by the boutons) are represented along the *x*-axis. The plots show striking but not entirely unexpected results. Most of the boutons tend to accumulate along three major lines which represent the three connectivity rules as described above. Accordingly, many boutons are seen along the iso-orientation invariant line (at 21°) which equals the electrophysiologically measured orientation preference of the pyramidal cell. Target orientation–independent co-linear connections (visual axis invariant) lie along the vertical line at the preferred orientation of the pyramidal cell. It should be added here that the special case, "classical" co-linearity type of connections are found at the intersection of iso-orientation and co-linear lines. The third line which runs diagonally represents those radial connections which do not depend on the orientation of the source, the pyramidal cell. Obviously, for radial connections, the orientation value of the target RF co-varies with the visual field axis. Importantly, the scatter plots show strong tendencies which are in line with the dominance of the three rules described above. Notably, the majority of boutons peak along the three lines representing the three connectivity rules. It is also noted that the highest density of boutons is found at the intersection of the three lines where boutons (connections) abide by all three rules, which is called classical co-linear. Nonetheless, a substantial proportion of boutons follows only one or the other of the rules indicating that the same pyramidal cell may be involved in several types of RF interactions.

Ideally, the above analytical approach is extended to compile the diversity of data on RF structure, spatiotemporal neuron behavior, functional imaging, and anatomical connectivity. We expect that this kind of result will be exploited in neural network simulations in order to gain better understanding of how cell-to-cell interactions may contribute to higher order signatures of visual information processing.

7.6 OUTLOOK

The present account aimed at emphasizing the importance of long-range cortico-cortical connections that are thought to be involved in context-dependent processing of visual contours. There are numerous questions which remain still unanswered, but we believe that a better understanding of the cortex must rely on analyses exploiting the multiplexing nature of neuron morphology. The axial rule analysis we have introduced above may be criticized as having little relevance with visual feature processing; but we think it is significant that other examples in our experimental material show a similar degree of projection diversity as the pyramidal cell demonstrated here, which projected to a wide variety of orientation. We anticipate that the same topographic diversity exists in relation to other functional modalities such as direction of motion, ocularity, spatial frequency, and spatial phase, each of which possesses its own map-like representation on the cortical sheet.

Acknowledgments

The author thanks Dr. Kathleen Rockland for helpful comments on the manuscript; Drs. Julian ML Budd, Péter Buzás, and Fuyuki Karube for helpful discussions; and Ms. Lívia Kicska for secretarial help. This work was supported by FP7-ICT-2011-7 (No. 269921, BrainScaleS-enlargEU) and KTIA_13_NAP-A-I/7.

References

Abeles, M., 1991. Corticonics: Neural Circuits of the Cerebral Cortex. Cambridge University Press, Cambridge, UK.

Ahmed, B., Anderson, J.C., Douglas, R.J., Martin, K.A., Nelson, J.C., 1994. Polyneuronal innervation of spiny stellate neurons in cat visual cortex. J. Comp. Neurol. 341, 39–49.

Ahmed, B., Anderson, J.C., Martin, K.A., Nelson, J.C., 1997. Map of the synapses onto layer 4 basket cells of the primary visual cortex of the cat. J. Comp. Neurol. 380, 230–242.

Angelucci, A., Levitt, J.B., Walton, E.J., Hupe, J.M., Bullier, J., Lund, J.S., 2002. Circuits for local and global signal integration in primary visual cortex. J. Neurosci. 22, 8633–8646.

Attwell, D., Laughlin, S.B., 2001. An energy budget for signaling in the grey matter of the brain. J. Cereb. Blood Flow Metab. 21, 1133–1145.

Bannister, A.P., Thomson, A.M., 2007. Dynamic properties of excitatory synaptic connections involving layer 4 pyramidal cells in adult rat and cat neocortex. Cereb. Cortex 17, 2190–2203.

Bassett, D.S., Bullmore, E., 2006. Small-world brain networks. Neuroscientist 12, 512–523.

Binzegger, T., Douglas, R.J., Martin, K.A., 2004. A quantitative map of the circuit of cat primary visual cortex. J. Neurosci. 24, 8441–8453.

Binzegger, T., Douglas, R.J., Martin, K.A., 2007. Stereotypical bouton clustering of individual neurons in cat primary visual cortex. J. Neurosci. 27, 12242–12254.

Blasdel, G.G., 1992. Orientation selectivity, preference, and continuity in monkey striate cortex. J. Neurosci. 12, 3139–3161.

Blasdel, G.G., Salama, G., 1986. Voltage-sensitive dyes reveal a modular organization in monkey striate cortex. Nature 321, 579–585.

Bolz, J., Gilbert, C.D., 1986. Generation of end-inhibition in the visual cortex via interlaminar connections. Nature 320, 362–365.

Bolz, J., Gilbert, C.D., 1989. The role of horizontal connections in generating long receptive fields in the cat visual cortex. Eur. J. Neurosci. 1, 263–268.

Bonhoeffer, T., Grinvald, A., 1991. Iso-orientation domains in cat visual cortex are arranged in pinwheel-like patterns. Nature 353, 429–431.

Bortone, D.S., Olsen, S.R., Scanziani, M., 2014. Translaminar inhibitory cells recruited by layer 6 corticothalamic neurons suppress visual cortex. Neuron 82, 474–485.

Bosking, W.H., Zhang, Y., Schofield, B., Fitzpatrick, D., 1997. Orientation selectivity and the arrangement of horizontal connections in tree shrew striate cortex. J. Neurosci. 17, 2112–2127.

Briggs, F., 2010. Organizing principles of cortical layer 6. Front. Neural Circuits 4, 3.

Briggs, F., Callaway, E.M., 2001. Layer-specific input to distinct cell types in layer 6 of monkey primary visual cortex. J. Neurosci. 15, 3600–3608.

Budd, J.M., Kovács, K., Ferecskó, A.S., Buzás, P., Eysel, U.T., Kisvárday, Z.F., 2010. Neocortical axon arbors trade-off material and conduction delay conservation. PLoS Comput. Biol. 6, e1000711.

Burkhalter, A., Bernardo, K.L., 1989. Organization of corticocortical connections in human visual cortex. Proc. Natl. Acad. Sci. U.S.A. 86, 1071–1075.

Burkhalter, A., Charles, V., 1990. Organization of local axon collaterals of efferent projection neurons in rat visual cortex. J. Comp. Neurol. 302, 920–934.

Buzás, P., Kovács, K., Ferecskó, A.S., Budd, J.M., Eysel, U.T., Kisvárday, Z.F., 2006. Model-based analysis of excitatory lateral connections in the visual cortex. J. Comp. Neurol. 499, 861–881.

Chavane, F., Sharon, D., Jancke, D., Marre, O., Frégnac, Y., Grinvald, A., 2011. Lateral spread of orientation selectivity in v1 is controlled by intracortical cooperativity. Front. Syst. Neurosci 5, 4.

Chisum, H.J., Mooser, F., Fitzpatrick, D., 2003. Emergent properties of layer 2/3 neurons reflect the collinear arrangement of horizontal connections in tree shrew visual cortex. J. Neurosci. 23, 2947–2960.

Chklovskii, D.B., Schikorski, T., Stevens, C.F., 2002. Wiring optimization in cortical circuits. Neuron 34, 341–347.

Crook, J.M., Kisvárday, Z.F., Eysel, U.T., 1998. Evidence for a contribution of lateral inhibition to orientation tuning and direction selectivity in cat visual cortex: reversible inactivation of functionally characterized sites combined with neuroanatomical tracing techniques. Eur. J. Neurosci. 10, 2056–2075.

Cudeiro, J., Sillito, A.M., 2006. Looking back: corticothalamic feedback and early visual processing. Trends Neurosci. 29, 298–306.

Da Costa, N.M., Martin, K.A., 2010. Whose cortical column would that be? Front. Neuroanat. 4, 16.

Das, A., Gilbert, C.D., 1997. Distortions of visuotopic map match orientation singularities in primary visual cortex. Nature 387, 594–598.

Douglas, R.J., Martin, K.A., 2004. Neuronal circuits of the neocortex. Neuroscience 27, 419–451.

Dragoi, V., Rivadulla, C., Sur, M., 2001. Foci of orientation plasticity in visual cortex. Nature 411, 80–86.

Field, D.J., Hayes, A., Hess, R.F., 1993. Contour integration by the human visual system: evidence for a local "association field". Vision Res. 33, 173–193.

Fitzpatrick, D., Lund, J.S., Blasdel, G., 1985. Intrinsic connections of macaque striate cortex: afferent and efferent connections of lamina 4C. J. Neurosci. 5, 3329–3349.

Gilbert, C.D., 1977. Laminar differences in receptive field properties of cells in cat primary visual cortex. J. Physiol. 268, 391–421.

Gilbert, C.D., 1992. Horizontal integration and cortical dynamics. Neuron 9, 1–13.

Gilbert, C.D., Kelly, J.P., 1975. The projections of cells in different layers of the cat's visual cortex. J. Comp. Neurol. 163, 81–105.

Gilbert, C.D., Wiesel, T.N., 1983. Clustered intrinsic connections in cat visual cortex. J. Neurosci. 3, 1116–1133.

Gilbert, C.D., Wiesel, T.N., 1989. Columnar specificity of intrinsic horizontal and corticocortical connections in cat visual cortex. J. Neurosci. 9, 2432–2442.

Gilbert, C.D., Das, A., Ito, M., Kapadia, M., Westheimer, G., 1996. Spatial integration and cortical dynamics. Proc. Natl. Acad. Sci. U.S.A. 93, 615–622.

Girardin, C.C., Martin, K.A., 2009. Inactivation of lateral connections in cat area 17. Eur. J. Neurosci. 29, 2092–2102.

Girman, S.V., Sauvé, Y., Lund, R.D., 1999. Receptive field properties of single neurons in rat primary visual cortex. J. Neurophysiol. 82, 301–311.

Gray, C.M., König, P., Engel, A.K., Singer, W., 1989. Oscillatory responses in cat visual cortex exhibit inter-columnar synchronization which reflects global stimulus properties. Nature 338, 334–337.

Grieve, K.L., Sillito, A.M., 1991. A re-appraisal of the role of layer VI of the visual cortex in the generation of cortical end inhibition. Exp. Brain Res. 87, 521–529.

Grinvald, A., Lieke, E., Frostig, R.D., Gilbert, C.D., Wiesel, T.N., 1986. Functional architecture of cortex revealed by optical imaging of intrinsic signals. Nature 324, 361–364.

Hirsch, J.A., 1995. Synaptic integration in layer IV of the ferret striate cortex. J. Physiol. 483, 183–199.

Huang, X., Elyada, Y.M., Bosking, W.H., Walker, T., Fitzpatrick, D., 2014. Optogenetic assessment of horizontal interactions in primary visual cortex. J. Neurosci. 34, 4976–4990.

Hubel, D.H., Wiesel, T.N., 1962. Receptive fields, binocular interaction and functional architecture in the cat's visual cortex. J. Physiol. 160, 106–154.

Hubel, D.H., Wiesel, T.N., 1974. Sequence regularity and geometry of orientation columns in the monkey striate cortex. J. Comp. Neurol. 158, 267–293.

Humphrey, A.L., Norton, T.T., Albano, J.E., 1980. Topographic organization of the orientation column system in the striate cortex of the tree shrew (*Tupaia glis*). I. Microelectrode recording. J. Comp. Neurol. 192, 531–547.

Ichinohe, N., Fujiyama, F., Kaneko, T., Rockland, K.S., 2003. Honeycomb-like mosaic at the border of layers 1 and 2 in the cerebral cortex. J. Neurosci. 23, 1372–1382.

Johnson, D.M., Illig, K.R., Behan, M., Haberly, L.B., 2000. New features of connectivity in piriform cortex visualized by intracellular injection of pyramidal cells suggest that "primary" olfactory cortex functions like "association" cortex in other sensory systems. J. Neurosci. 20, 6974–6982.

Kapadia, M.K., Ito, M., Gilbert, C.D., Westheimer, G., 1995. Improvement in visual sensitivity by changes in local context: parallel studies in human observers and in V1 of alert monkeys. Neuron 15, 843–856.

Kaschube, M., 2014. Neural maps versus salt-and-pepper organization in visual cortex. Curr. Opin. Neurobiol. 24, 95–102.

Katz, L.C., 1987. Local circuitry of identified projection neurons in cat visual cortex brain slices. J. Neurosci. 7, 1223–1249.

Karube, F., Kisvárday, Z.F., 2011. Axon topography of layer IV spiny cells to orientation map in the cat primary visual cortex (area 18). Cereb. Cortex 21, 1443–1458.

Kisvárday, Z.F., Eysel, U.T., 1992. Cellular organization of reciprocal patchy networks in layer III of cat visual cortex (area 17). Neuroscience 46, 275–286.

Kisvárday, Z.F., Tóth, E., Rausch, M., Eysel, U.T., 1997. Orientation-specific relationship between populations of excitatory and inhibitory lateral connections in the visual cortex of the cat. Cereb. Cortex 7, 605–618.

Koulakov, A.A., Chklovskii, D.B., 2001. Orientation preference patterns in mammalian visual cortex: a wire length minimization approach. Neuron 29, 519–527.

Kriener, B., Tetzlaff, T., Aertsen, A., Diesmann, M., Rotter, S., 2008. Correlations and population dynamics in cortical networks. Neural Comput. 20, 2185–2226.

Kulikowski, J.J., Abadi, R., King-Smith, P.E., 1973. Orientational selectivity of grating and line detectors in human vision. Vision Res. 13, 1479–1486.

LeVay, S., Gilbert, C.D., 1976. Laminar patterns of geniculocortical projection in the cat. Brain Res. 113, 1–19.

LeVay, S., Sherk, H., 1981. The visual claustrum of the cat. I. Structure and connections. J. Neurosci. 1, 956–980.

Levin, S., Hutson, M., Ellis, R., 2006. "Tensegrity: The New Biomechanics". Textbook of Musculoskeletal Medicine. Oxford University Press, Oxford, UK.

Levitt, J., Lund, J., 2002. Intrinsic connections in mammalian cerebral cortex. In: Schüz, A., Miller, R. (Eds.), Cortical Areas: Unity and Diversity Taylor & Francis, New York, USA, pp. 133–154.

Lohmann, H., Rörig, B., 1994. Long-range horizontal connections between supragranular pyramidal cells in the extrastriate visual cortex of the rat. J. Comp. Eurol 344, 543–558.

Löwel, S., Singer, W., 1992. Selection of intrinsic horizontal connections in the visual cortex by correlated neuronal activity. Science 255, 209–212.

Löwel, S., Freeman, B., Singer, W., 1987. Topographic organization of the orientation column system in large flat-mounts of the cat visual cortex: a 2-deoxyglucose study. J. Comp. Neurol. 255, 401–415.

Luhmann, H.J., Martínez-Millán, L., Singer, W., 1986. Development of horizontal intrinsic connections in cat striate cortex. Exp. Brain Res. 63, 443–448.

Luhmann, H.J., Singer, W., Martínez-Millán, L., 1990. Horizontal interactions in cat striate cortex: I. Anatomical substrate and postnatal development. Eur. J. Neurosci. 2, 344–357.

Lund, J.S., 1988. Anatomical organization of macaque monkey striate visual cortex. Annu. Rev. Neurosci. 11, 253–288.

Lund, J.S., Yoshioka, T., Levitt, J.B., 1993. Comparison of intrinsic connectivity in different areas of macaque monkey cerebral cortex. Cereb. Cortex 3, 148–162.

Malach, R., Amir, Y., Harel, M., Grinvald, A., 1993. Relationship between intrinsic connections and functional architecture revealed by optical imaging and *in vivo* targeted biocytin injections in primate striate cortex. Proc. Natl. Acad. Sci. U.S.A. 90, 10469–10473.

Malach, R., Tootell, R.B., Malonek, D., 1994. Relationship between orientation domains, cytochrome oxidase stripes, and intrinsic horizontal connections in squirrel monkey area V2. Cereb. Cortex 4, 151–165.

Maldonado, P.E., Gödecke, I., Gray, C.M., Bonhoeffer, T., 1997. Orientation selectivity in pinwheel centers in cat striate cortex. Science 276, 1551–1555.

Manor, Y., Gonczarowski, J., Segev, I., 1991. Propagation of action potentials along complex axonal trees. Model and implementation. Biophys. J. 60, 1411–1423.

Marr, D., Hildreth, E., 1980. Theory of edge detection. Proc. R. Soc. Lond. B. Biol. Sci 207, 187–217.

Martin, K.A.C., Whitteridge, D., 1984. Form, function, and intracortical projections of spiny neurones in the striate visual cortex of the cat. J. Physiol. Lond. 353, 463–504.

Martin, K.A., Roth, S., Rusch, E.S., 2014. Superficial layer pyramidal cells communicate heterogeneously between multiple functional domains of cat primary visual cortex. Nat. Commun. 5, 5252.

Matsubara, J., Cynader, M., Swindale, N.V., Stryker, M.P., 1985. Intrinsic projections within visual cortex: evidence for orientation-specific local connections. Proc. Natl. Acad. Sci. U.S.A. 82, 935–939.

McCourt, M.E., Boyapati, J., Henry, G.H., 1986. Layering in lamina 6 of cat striate cortex. Brain Res. 364, 181–185.

McGuire, B.A., Hornung, J.P., Gilbert, C.D., Wiesel, T.N., 1984. Patterns of synaptic input to layer 4 of cat striate cortex. J. Neurosci. 4, 3021–3033.

Michalski, A., Gerstein, G.L., Czarkowska, J., Tarnecki, R., 1983. Interactions between cat striate cortex neurons. Exp. Brain Res. 51, 97–107.

Mitchison, G., Crick, F., 1982. Long axons within the striate cortex: their distribution, orientation, and patterns of connection. Proc. Natl. Acad. Sci. U.S.A. 79, 3661–3665.

Mooser, F., Bosking, W.H., Fitzpatrick, D., 2004. A morphological basis for orientation tuning in primary visual cortex. Nat. Neurosci. 7, 872–879.

Muir, D.R., Douglas, R.J., 2011. From neural arbors to daisies. Cereb. Cortex 21, 1118–1133.

Muir, D.R., Da Costa, N.M., Girardin, C.C., Naaman, S., Omer, D.B., Ruesch, E., et al., 2011. Embedding of cortical representations by the superficial patch system. Cereb. Cortex 21, 2244–2260.

Murphy, P.C., Sillito, A.M., 1986. Continuity of orientation columns between superficial and deep laminae of the cat primary visual cortex. J. Physiol. 381, 95–110.

Ohki, K., Chung, S., Kara, P., Hübener, M., Bonhoeffer, T., Reid, R.C., 2006. Highly ordered arrangement of single neurons in orientation pinwheels. Nature 442, 925–928.

Ojima, H., Honda, C.N., Jones, E.G., 1991. Patterns of axon collateralization of identified supragranular pyramidal neurons in the cat auditory cortex. Cereb. Cortex 1, 80–94.

Olsen, S.R., Bortone, D.S., Adesnik, H., Scanziani, M., 2012. Gain control by layer six in cortical circuits of vision. Nature 483, 47–52.

Perin, R., Berger, T.K., Markram, H., 2011. A synaptic organizing principle for cortical neuronal groups. Proc. Natl. Acad. Sci. U.S.A. 108, 5419–5424.

Perin, R., Telefont, M., Markram, H., 2013. Computing the size and number of neuronal clusters in local circuits. Front. Neuroanat 7, 1.

Peyré, G., Frégnac, Y., 2012. Editorial: New trends in neurogeometrical approaches to the brain and mind problem. J. Physiol. Paris 106, 171–172.

Polimeni, J.R., Granquist-Fraser, D., Wood, R.J., Schwartz, E.L., 2005. Physical limits to spatial resolution of optical recording: clarifying the spatial structure of cortical hypercolumns. Proc. Natl. Acad. Sci. U.S.A. 102, 4158–4163.

Rockland, K.S., 1985. Anatomical organization of primary visual cortex (area 17) in the ferret. J. Comp. Neurol. 241, 225–236.

Rockland, K.S., Lund, J.S., 1982. Widespread periodic intrinsic connections in the tree-shrew visual cortex. Science 215, 1532–1534.

Roerig, B., Chen, B., 2002. Relationships of local inhibitory and excitatory circuits to orientation preference maps in ferret visual cortex. Cereb. Cortex 12, 187–198.

Roerig, B., Chen, B., Kao, J.P., 2003. Different inhibitory synaptic input patterns in excitatory and inhibitory layer 4 neurons of ferret visual cortex. Cereb. Cortex 13, 350–363.

Samonds, J.M., Zhou, Z., Bernard, M.R., Bonds, A.B., 2006. Synchronous activity in cat visual cortex encodes collinear and cocircular contours. J. Neurophysiol. 95, 2602–2616.

Schmid, A.M., 2008. The processing of feature discontinuities for different cue types in primary visual cortex. Brain Res. 1238, 59–74.

Schmidt, K.E., Goebel, R., Löwel, S., Singer, W., 1997. The perceptual grouping criterion of colinearity is reflected by anisotropies of connections in the primary visual cortex. Eur. J. Neurosci. 9, 1083–1089.

Schwarz, C., Bolz, J., 1991. Functional specificity of a long-range horizontal connection in cat visual cortex: a cross-correlation study. J. Neurosci. 11, 2995–3007.

Sherman, S.M., Guillery, R.W., 2001. Exploring the Thalamus. Academic Press, New York, USA.

Sillito, A.M., Grieve, K.L., Jones, H.E., Cudeiro, J., Davis, J., 1995. Visual cortical mechanisms detecting focal orientation discontinuities. Nature 378, 492–496.

Sillito, A.M., Cudeiro, J., Jones, H.E., 2006. Always returning: feedback and sensory processing in visual cortex and thalamus. Trends Neurosci. 29, 307–316.

Sincich, L.C., Blasdel, G.G., 2001. Oriented axon projections in primary visual cortex of the monkey. J. Neurosci. 21, 4416–4426.

Somogyi, P., 1989. Synaptic organization of GABAergic neurons and GABAA receptors in the lateral geniculate nucleus and visual cortex. In: Lam, D.K.-T., Gilbert, C.D. (Eds.), Neural Mechanisms of Visual Perception Portfolio, Texas, pp. 35–62.

Stettler, D.D., Das, A., Bennett, J., Gilbert, C.D., 2002. Lateral connectivity and contextual interactions in macaque primary visual cortex. Neuron. 36, 739–750.

Tanigawa, H., Wang, Q., Fujita, I., 2005. Organization of horizontal axons in the inferior temporal cortex and primary visual cortex of the macaque monkey. Cereb. Cortex 15, 1887–1899.

Tarczy-Hornoch, K., Martin, K.A., Jack, J.J., Stratford, K.J., 1998. Synaptic interactions between smooth and spiny neurones in layer 4 of cat visual cortex in vitro. J. Physiol. 508, 351–363.

Thomson, A.M., 2010. Neocortical layer 6: a review. Front. Neuroanat. 4, 13.

Thomson, A.M., West, D.C., Wang, Y., Bannister, A.P., 2002. Synaptic connections and small circuits involving excitatory and inhibitory neurons in layers 2-5 of adult rat and cat neocortex: triple intracellular recordings and biocytin labelling in vitro. Cereb. Cortex 12, 936–953.

Tian, P., Devor, A., Sakadžić, S., Dal, A.M., Boas, D.A., 2011. Monte Carlo simulation of the spatial resolution and depth sensitivity of two-dimensional optical imaging of the brain. J. Biomed. Opt. 16, 016006.

Tusa, R.J., Rosenquist, A.C., Palmer, L.A., 1979. Retinotopic organization of areas 18 and 19 in the cat. J. Comp. Neurol. 185, 657–678.

Ts'o, D.Y., Gilbert, C.D., Wiesel, T.N., 1986. Relationships between horizontal interactions and functional architecture in cat striate cortex as revealed by cross-correlation analysis. J. Neurosci. 6, 1160–1170.

Usher, M., Bonneh, Y., Sagi, D., Herrmann, M., 1999. Mechanisms for spatial integration in visual detection: a model based on lateral interactions. Spat. Vis. 12, 187–209.

van Aerde, K.I., Feldmeyer, D., 2015. Morphological and physiological characterization of pyramidal neuron subtypes in rat medial prefrontal cortex. Cereb. Cortex 25, 788–805.

Van Hooser, S.D., Heimel, J.A., Chung, S., Nelson, S.B., 2006. Lack of patchy horizontal connectivity in primary visual cortex of a mammal without orientation maps. J. Neurosci. 26, 7680–7692.

Voges, N., Guijarro, C., Aertsen, A., Rotter, S., 2010a. Models of cortical networks with long-range patchy projections. J. Comput. Neurosci. 28, 137–154.

Voges, N., Schüz, A., Aertsen, A., Rotter, S., 2010b. A modeler's view on the spatial structure of intrinsic horizontal connectivity in the neocortex. Prog. Neurobiol. 92, 277–292.

Von der Heydt, R., Peterhans, E., 1989. Mechanisms of contour perception in monkey visual cortex. I. Lines of pattern discontinuity. J. Neurosci. 9, 1731–1748.

Wallace, M.N., Bajwa, S., 1991. Patchy intrinsic connections of the ferret primary auditory cortex. Neuroreport 2, 417–420.

Watts, D.J., Strogatz, S.H., 1998. Collective dynamics of 'small-world' networks. Nature 393, 440–442.

West, D.C., Mercer, A., Kirchhecker, S., Morris, O.T., Thomson, A.M., 2006. Layer 6 cortico-thalamic pyramidal cells preferentially innervate interneurons and generate facilitating EPSPs. Cereb. Cortex 16, 200–211.

Yoshioka, T., Blasdel, G.G., Levitt, J.B., Lund, J.S., 1996. Relation between patterns of intrinsic lateral connectivity, ocular dominance, and cytochrome oxidase-reactive regions in macaque monkey striate cortex. Cereb. Cortex 6, 297–310.

Yousef, T., Bonhoeffer, T., Kim, D.S., Eysel, U.T., Tóth, E., Kisvárday, Z.F., 1999. Orientation topography of layer 4 lateral networks revealed by optical imaging in cat visual cortex (area 18). Eur. J. Neurosci. 11, 4291–4308.

Yousef, T., Tóth, E., Rausch, M., Eysel, U.T., Kisvárday, Z.F., 2001. Topography of orientation centre connections in the primary visual cortex of the cat. Neuroreport 12, 1693–1699.

8

Do Lateral Intrinsic and Callosal Axons Have Comparable Actions in Early Visual Areas?

Kerstin E. Schmidt

Brain Institute, Federal University of Rio Grande do Norte (UFRN), Natal, Brazil

8.1 INTRODUCTION

In the mammalian brain, the majority of axons between homologous cortical areas in the two hemispheres run through the corpus callosum. Among callosally projecting axons, a great variety in degree of myelination (e.g., Swadlow et al., 1980) and fiber diameter (e.g., Houzel et al., 1994; Houzel and Milleret, 1999) has been described. Evidently, the corpus callosum is a collection of different axonal pathways with different conduction delays and processing speeds (Tomasi et al., 2012) compatible with a broad range of slower or faster, direct or indirect, excitatory as well as inhibitory functions (Bloom and Hynd, 2005). The exact function of those pathways depends on the type of cortical system that is linked, its position in the cortical hierarchy and also on the animal species (Olivares et al., 2001; Caminiti et al., 2009; Innocenti et al., 2014). For early cortical areas, but in some distinction with higher order areas, callosally projecting axons interconnect feature spaces separated at the body's midline (Berlucchi and Rizzolatti, 1968; Manzoni et al., 1989).

Accordingly, several functional types have been investigated. First, callosal actions in higher mammals can—based on an asymmetry in lateral specialization—contribute to a functional asymmetry producing a cognitive feature like, for example, language ability or handedness.

Second, bimanual coordination and stereopsis are functions, which might benefit from integration via callosal connections because of the specific position of those axons within the brain's topography relative to the representation of the sensory periphery.

Finally, callosal connections in early sensory areas might simply and primarily extend the intrinsic network of intracortical short- and/or long-range connections onto the contralateral hemisphere to ensure functional integrity and perpetuate intrahemispheric function over the midline.

For visual areas, this corresponds to the representation of the vertical midline of the visual field. Callosal connections are known to link parts of the two visual areas receiving input from corresponding retinal loci, i.e., in a visuotopic manner, and expressing similar receptive field locations at and close to that midline. In areas 17 and 18, callosal connections have many anatomical and functional features in common with the long, horizontally projecting axon collaterals of pyramidal neurons interconnecting neuronal domains within the same cortical area, i.e., intrinsically, over distances of more than 1 mm. These connections will be referred to as long-range or horizontal intrinsic connections throughout the chapter.

As such, intrinsic and callosal connections, at the same level of cortical hierarchy, could be parts of the same circuit. In this chapter, the contemporary knowledge about anatomy and function of horizontal intrinsic and callosal axons in the early visual cortex of carnivores, mainly cats, will be reviewed. An attempt of an understanding of these two axon types in a more general definition of lateral networks operating in the central visual field will be undertaken. Arguments for a specific callosal function versus continuation of intrahemispheric functions will be discussed.

8.2 ANATOMICAL AND TOPOGRAPHICAL PARTICULARITIES OF LONG-RANGE INTRINSIC AND CALLOSAL AXONS

8.2.1 Long-Range Intrinsic Axons

A recent estimate on neuronal circuits in cat visual cortex based on three-dimensional reconstructions of axons and laminar patterns of bouton distributions came to the conclusion that horizontal intralaminar connections numerically provide the strongest synaptic input to excitatory neurons (Binzegger et al., 2004). In addition, new and sophisticated techniques genetically targeting only GABAergic neurons for photostimulation show that also inhibitory inputs to excitatory circuits arise mainly from the same cortical layer (Kätzel et al., 2011; Fino and Yuste, 2011), though within a smaller diameter of influence. Even more astonishing, a recent report claims that synapses formed by long-range projections from outside the range of a classical functional column make up to 82% of the synaptic inputs and thus by far outnumber local intracolumnar synapses (Stepanyants et al., 2009).

Axons extending horizontally to the cortical surface have long been known for great diversity in their branching patterns but one feature repeatedly attracted attention; namely, their horizontal axon collaterals tend to arborize and terminate in clusters with label-free intervals between them. The first attempts to relate anatomy and function in this system used intracellular injections of horseradish peroxidase (HRP) into electrophysiologically characterized pyramidal neurons in layers II/III and V (Gilbert and Wiesel, 1979). The reconstructed axons branched in regular intervals parallel to the cortical layers for distances up to several millimeters, much longer than would have been guessed from Golgi stains (Gilbert and Wiesel, 1983; Martin and Whitteridge, 1984). Extracellular HRP injections in the tree shrew revealed a regular pattern of retrogradely labeled neurons together with anterogradely labeled axon terminals, named clusters or patches (Rockland and Lund, 1982; Rockland et al., 1982). Patches, prominent in supragranular layers II and IIIA, could measure 200–300 mm in diameter, spread up to 2–3 mm from the injection site and were separated by label-free spaces of about 500 mm. These patterns have been repeatedly observed in various species and areas with different retrograde and anterograde tracers, i.e., tree shrew (Rockland and Lund, 1982; Fitzpatrick, 1996; Bosking et al., 1997), ferret (Rockland, 1985; Durack and Katz, 1996; Ruthazer and Stryker, 1996), cat (Gilbert and Wiesel, 1983, 1989; Martin and Whitteridge, 1984; Luhmann et al., 1986; Luhmann and Prince, 1991; Kisvárday and Eysel, 1992, 1993; Kisvárday et al., 1986, 1994, 1997; Callaway and Katz, 1990, 1992; Albus and Lübke, 1992; Löwel and Singer, 1992; Lübke and Albus, 1992; Boyd and Matsubara, 1994; Galuske and Singer, 1996; Schmidt et al., 1997a,b; Buzás et al., 1998, 2006; Yousef et al., 1999; Binzegger et al., 2007; Martin et al., 2014), macaque and squirrel monkey (Rockland and Lund, 1983; Livingstone and Hubel, 1984; Amir et al., 1993; Malach et al., 1993; Yoshioka et al., 1996; Sincich and Blasdel, 2001), as well as human (Burkhalter and Bernado, 1989) but not, as based on intrinsic horizontal collaterals, in rodents (Burkhalter and Charles, 1990; Van Hooser et al., 2006). Clustering has been further observed among intracortical connections within extrastriate visual areas (Price, 1986; Matsubara et al., 1987; Yoshioka et al., 1992; Amir et al., 1993; Levitt et al., 1993, 1994; Malach et al., 1997; Kisvárday et al., 1997) and among feedforward and feedback connections between visual areas (Wong-Riley, 1979; Montero, 1980; Tigges et al., 1981; Rockland and Lund, 1982; Gilbert and Wiesel, 1983, Livingstone and Hubel, 1987; Sincich and Horton, 2002; Rockland and Pandya, 1981; Zeki and Shipp, 1989; Nakamura et al., 1993; Price et al., 1994; Rockland, 1985; Felleman et al., 1997; Angelucci et al., 2002; Anderson and Martin, 2002, 2005; Banno et al., 2011). It is noteworthy that patchiness is not as pronounced in feedback as in intrinsic connections (e.g., Rockland and Drash, 1996; Rockland and Knutson, 2000, Anderson and Martin, 2006, 2009) but see also (Angelucci et al., 2002). In higher visual areas, mixed organizations containing both modular and continuous connectivity occur (Borra et al., 2010). Also, the span of the connections seems to increase within the cortical hierarchy (Amir et al., 1993, Lund et al., 1993; Angelucci et al., 2002). Although clustered connections have been predominantly studied in the visual cortex, they likely constitute a hallmark feature of cortical organization in carnivores and primates.

8.2.2 Clustering and Feature Selectivity of Long-Range Intrinsic Axons

Early anatomical reports of individual axonal arbors of pyramidal neurons in identified cortical layers suggested that the axon's morphology would predict a specific function (Gilbert and Wiesel, 1979). Theoretical considerations indicated that the regular clustering of long-range connections would be related to the modular functional architecture of primary visual cortex (Mitchison and Crick, 1982). The first systematic relationship with a modular cortical compartment was demonstrated for regions of high cytochrome oxidase (CO) activity in macaque monkeys (Livingstone and Hubel, 1984). Those are specialized neuronal domains exhibiting monocular, color- and low spatial frequency specific responses (Horton and Hubel, 1981; Livingstone and Hubel, 1984; Tootell et al., 1988; Shoham et al., 1997). Biocytin tracing in both squirrel and macaque monkeys confirmed preferential intrinsic connections between CO blobs (Lachica et al., 1993; Levitt et al., 1994; Malach et al., 1994; Yoshioka et al., 1996). Interestingly, in the cat, a link between CO blobs has been observed for callosal connections but not for intrinsic connections (Boyd and Matsubara, 1994).

The first attempts to find functional reasons for clustering delivered contradictory results (Rockland et al., 1982; Matsubara et al., 1987). Cross-correlations of electrophysiological spiking activity revealed coherent activation of distant neurons with similar orientation preference (Ts'o et al., 1986; Hata et al., 1991; Singer and Gray, 1995), suggesting monosynaptic connections between them. Newer tracers like fluorescent latex microspheres and biocytin overcame some of the limitations of HRP (i.e., transneuronal spilling, large injections). Combining targeted tracer injections with multisite electrophysiological recordings, 2-deoxyglucose labeling and/or optical imaging revealed that pyramidal neurons preferring similar orientation preference in supra- or infragranular layers exhibit a higher probability to be interconnected by a long-range intrinsic axon than those preferring othogonal orientations (cat: Gilbert and Wiesel, 1989; Kisvárday et al., 1997; Schmidt et al., 1997a,b; macaque: Malach et al., 1993; squirrel and owl monkey: Sincich and Blasdel, 2001; tree shrew: Bosking et al., 1997).

Superposition of complete axonal arbors with optically imaged maps of orientation preference allowed for detailed analyses on the functional bias of long-range axon terminals (Buzás et al., 1998; Yousef et al., 2001; Kisvárday et al., 2002; Chisum et al., 2003; Rochefort et al., 2009; Karube and Kisvárday, 2011; Martin et al., 2014). All studies of interconnected neurons and bouton distributions underline both selectivity and variability with respect to iso-oriented sites, and confirm an overall moderate bias for connections between neurons of similar orientation preference in supra- and infragranular layers (Malach et al., 1993: 66%; Bosking et al., 1997: 57% on iso-orientation domains within an interval of ±35° from the preference at injection core; Schmidt et al., 1997a,b: 56% on iso ±27°; Kisvárday et al., 1997: area 17, 53% on iso ±30°; Karube and Kisvárday, 2011: area 18, 62% on iso; Martin et al., 2014: area 17, 62% on iso). For layer IV spiny stellate cells, whose lateral extent is more limited, boutons distribute over a wider range of target orientations (Yousef et al., 1999). Bulk tracer injections in pinwheel centers indicated that connections originating in singularities are less selective than those elsewhere in the maps (Yousef et al., 2001). This interpretation is not confirmed by axon reconstructions in three dimensions (Martin et al., 2014), seemingly because individual axons are less selective than the averaged population bias after mass bulk tracer injections has suggested.

8.2.3 Callosal Axons

Each visual cortex processes only information from the contralateral visual hemifield. In cats, all retinotopically organized areas are densely linked across the two hemispheres by callosal axons (Segraves and Rosenquist, 1982a,b). In primary visual cortex, they are densest along the border between area 17 and 18, which also corresponds to the representation of the vertical meridian of the visual field (Innocenti and Fiore, 1976; Houzel et al., 1994; Innocenti et al., 2002). Early on, callosal axons were attributed the role of binding the two visual hemifields separated anatomically but remaining functionally united (Hubel and Wiesel, 1967).

First experiments on their function—cutting the chiasm and additionally removing the callosal input—abolished ipsilateral responses close to the vertical midline (Choudhury et al., 1965). Direct recordings from callosal fibers confirmed that these connections deal mainly with the midline of the visual field (Hubel and Wiesel, 1967). Later, it turned out that neurons near the vertical meridian indeed have bilateral receptive fields that are thus activated by both retinogeniculate and callosal inputs (Berlucchi and Rizzolatti, 1968; Rochefort et al., 2007). Both pathways contain the same information about orientation and direction preference (Houzel and Milleret, 1999; Milleret et al., 2005). Removing callosal input decreases the size of the midline centered receptive fields (Payne, 1994) and removes visual information from the ipsilateral visual field (Berlucchi and Rizzolatti, 1968; Lepore and Guillemot, 1982; Blakemore et al., 1983).

In carnivores' early visual areas, visual callosal connections clearly are preferentially located at areal borders (Innocenti, 1986). This tendency seems to gradually decline with progression along the interconnected visual areas in the cortical hierarchy (cat: Segraves and Rosenquist, 1982b; Payne and Siwek, 1991a; ferret: Innocenti et al., 2002). This might not be unexpected given that receptive field sizes increase and comprehensiveness of retinotopic representation decreases in higher visual areas.

More recent studies claim that connections link border and nonborder areas already in primary visual cortex. That is, callosal neurons in areas 17 and 18 are reported to project to the contralateral transition zone (TZ), whereas

P ←——→ A

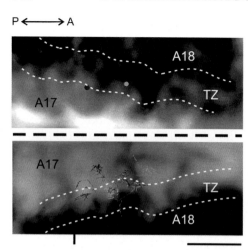

FIGURE 8.1   Terminal arbors of callosal axons. Two ipsilateral dextran amine injection sites (top: blue and red dots) giving rise to clustered callosal axon terminals in cat primary visual cortex. Injections and axons are superimposed with differential maps of ipsi- (top, with injections) and contralateral hemispheres (below) where area 18 (A18) shows dark activation and area 17 (A17) light activation. The transition zone (TZ) is located within the white dashed outline. Maps are obtained by dividing the sum of all single condition maps optically imaged while presenting gratings optimized for A18 activation (0.15 cpd, 4.5 Hz), by the sum of maps imaged while presenting gratings optimized for A17 activation (0.6 cpd, 1.5 Hz). The dashed line indicates the midline between the two hemispheres. Vertical bar at the bottom indicates stereotaxic coordinate AP0. A, anterior, P, posterior. Scale bar, 2 mm. Note that axons are shifted in the same direction and with the same distance as ipsilateral injection sites. *Courtesty of Rochefort et al. (2009), by permission of Oxford University Press.*

the "back" projection from that zone terminates mainly outside that zone (Olavarria, 2001; Alekseenko et al., 2005). Thereby, connections remain visuotopic as they interconnect regions receiving input from the same parts of the retina.

Rochefort et al. (2009) confirm the latter reports of visuotopic interconnectivity. Injections into the TZ give rise to axon arbors in the contralateral area 17, and injections into area 18 or area 17 give rise to label in the contralateral TZ. The distance between the projection fields in the receiving hemisphere matches the distance between ipsilateral injection sites in the TZ indicating a match of the cortical magnification factor (Figure 8.1).

Callosal connections have been frequently reported to be heterotopic meaning that contralateral projections mirror the extended ipsilateral projections—also in other areas than only the homotopically corresponding one—on the same level of hierarchy, but at lower density (e.g., in monkeys: Hedreen and Yin, 1981; Barbas et al., 2005).

8.2.4 Bilateral Representation of the Visual Field's Vertical Meridian

In several mammals, a part of the ipsilateral visual field is represented bilaterally in areas 17 and 18 (e.g., Stone, 1978; Hubel and Wiesel, 1967; Tusa et al., 1978). In the cat's contralateral hemisphere, that ipsilateral representation increases from 4° in central parts to 23° toward upper and lower elevations (Payne, 1990; Payne and Siwek, 1991b) accompanying the visual field magnification and also the extent of the callosally projecting zone. In more lateral-eyed animals like ferrets (Law et al., 1988; White et al., 1999), tree shrews (Bosking et al., 2000), or sheep (Clarke and Whitteridge, 1976), the overlap is even greater but confined to area 17.

It is likely that transcallosal axons cover in many cases even a larger zone than that bilaterally represented stripe (area 18, Innocenti et al., 2002). In summary, there is a composite organization of double feedforward and widespread lateral callosal input in both supra- and infragranular layers. A reasonable conclusion is that callosal connections in early visual areas perpetuate the intrahemispheric lateral network and—under normal circumstances—serve rather modulatory than feedforward driving functions.

Only under special circumstances, like removal of direct geniculocortical input through chiasm transection, callosal connections can activate directly—though less strongly—neurons in the opposite transcallosal zone (Choudhury et al., 1965; Berlucchi and Rizzolatti, 1968; Berardi et al., 1987; Rochefort et al., 2007). For comparison, also long-range intrinsic connections can adopt driving rather than only modulating functions during adult plasticity (Chavane et al., 2011; for review Gilbert et al., 2009).

8.2.5 Clustering and Feature Selectivity of Callosal Axons

In cats, callosal axons originate from and terminate on similar pre- and postsynaptic cell classes in supragranular, less in infragranular layers (for review see Innocenti, 1986). Most fibers stem from pyramidal cells in layers III and VI as well as some spiny stellate cells in upper layer IV (Segraves and Rosenquist, 1982a; Fisken et al., 1975; Innocenti, 1980; Voigt et al., 1988). Single axon reconstructions in cat areas 17 and 18 confirm synaptic boutons mainly in layer III, up to one-third in upper layer IV and fewer boutons in layer V/VI (Figure 8.1, Rochefort et al., 2009). Different from cats, mass tracer injections in ferrets (Innocenti et al., 2002) and also in rodents (Olavarria and Van Sluyters, 1985) indicate that visual callosally projecting neurons in these animals predominate at least numerically in infragranular layers, especially in extrastriate areas.

Similar to what we are proposing to be their "relatives" within the ipsilateral hemisphere, callosal axon arbors terminate in the contralateral hemisphere at more or less regular intervals (Houzel et al., 1994; Aggoun-Zouaoui et al., 1996; Rochefort et al., 2009) and interconnected callosally projecting neuron aggregate in clusters (Voigt et al., 1988; Berman and Payne, 1983; Innocenti, 1986; Boyd and Matsubara, 1994; Schmidt et al., 1997a,b).

In a series of experiments, we used a combination of targeted tracer injection and superposition of the retrogradely labeled cell bodies with 2-deoxyglucose autoradiographs of the map of iso-orientation domains. As these injections were very close to the border between areas 17 and 18, they labeled not only ipsilateral clusters but also neurons projecting transcallosally to the ipsilateral injection site. Those neurons coincided with the pattern of iso-orientation domains with the same precision, as did the intracortical clusters. On average, 56.6 ± 8.8% of the ipsilaterally and 60.4 ± 11.8% of the contralaterally labelled neurons were found in columns of the same orientation preference as the injection site, within an interval of ±27° from the preference at injection core (Schmidt et al., 1997a,b). The distributions of labeled cells with respect to the injection site compartment (defined by orientation preference) in the two hemispheres are not significantly different from each other. This is compatible with the interpretation that callosal connections in that area perpetuate the intracortical network in a continuous fashion (Figure 8.2).

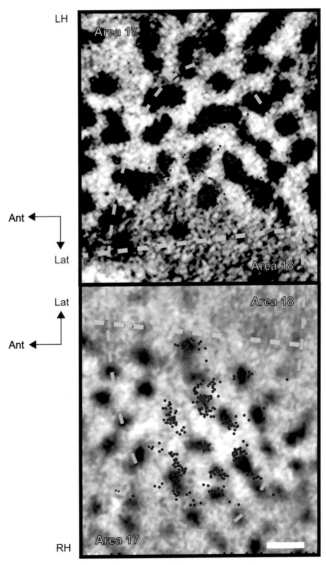

FIGURE 8.2   Feature selectivity of both lateral intrinsic and callosal axons. Projection neurons in right (RH) and left (LH) hemispheres labeled from an injection of red fluorescent microspheres (red star, lower panel) in the right area 17 of a cat. The injection site was targeted to domains of horizontal orientation preference, as identified by intravenous injection of 2-deoxyglucose during visual stimulation with horizontal gratings (0.5 cpd, 2 Hz) as in Schmidt et al. (1997a,b). Area 18 did not take up deoxyglucose because of the high spatial frequency stimulus. Areal borders in the upper and lower panel are aligned to match the visual field representation in the two cortices. Cortical coordinates: ant, anterior; post, posterior. Note that both intrinsically and transcallosally labeled projection neurons tend to aggregate in iso-orientation domains. Scale bar: 1 mm.

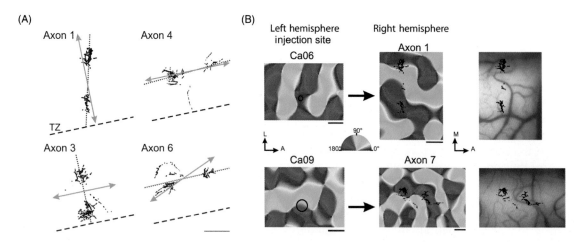

FIGURE 8.3 Axial specificity of callosal axons. (A) Spatial distribution of callosal axon terminals (black dots) relative to the axis of the 17/18 transition zone (black dashed lines). Distributions are rotated to align the TZs with each other. Red dashed lines indicate the principal axis of each axon's terminal field. Double-headed arrows (gray) represent a retinotopic axis in A17 that corresponds to the preferred orientation at the injection site. The examples show axons for which the principal axis of bouton distribution (red dashed lines) was approximately parallel to (axons 1, 4, 6) and in one case perpendicular (axon 3) to the axis derived from orientation preference (double arrows) at injection sites. Scale bar, 500 μm. (B) Contour and axial specificity of callosal connections related to orientation preference. (A) Orientation preference maps obtained by optical imaging, where the injection (black circle = core) is in the left hemisphere, and the boutons (in black) of a corresponding callosal axon are distributed in the right hemisphere. Note that callosal boutons tend to cluster in columns of the same color as the core of the ipsilateral injection site. Note that the axis along which the axons elongate also changes. *Adapted from Rochefort et al. (2009), by permission of Oxford University Press.*

Rochefort et al. (2009) reconstructed callosal axons anterogradely labeled with dextran amines from targeted injections and superimposed the boutons with the optically imaged map in the contralateral TZ. An astonishingly high number of 72 ± 17% boutons clustered within iso-orientation columns within an interval of ±30° from the orientation preference at the injection site's core (Figure 8.3).

However, like Martin et al. (2014) for intrinsic axons, Rochefort also shows that axons cover a wide range of selectivity, where between 50% and 91% of the synaptic boutons of individual callosal axons accumulated in columns of same preference as the parent somata.

Overall, there is abundant evidence that both intrinsic and interhemispheric long-range axons in areas 17 and 18 exhibit some bias to link neurons preferring similar response properties (contour and/or direction preference), compatible with a crude tendency of "like connects to like"; but, importantly, there is also cross talk between neurons of different properties. The exact bouton distribution of a pyramidal neuron may depend on the neuron's location within the map (Yousef et al., 2001; Buzás et al., 2006) or on different functional properties of which only a few have been investigated so far (Martin et al., 2014; for review Schmidt and Löwel, 2002).

Interestingly, in tree shrews, which have lateral eyes, the congruence between intrinsic and callosal circuits seems to be broken and callosal fibers are less specific but still visuotopic (Bosking et al., 2000).

8.2.6 Axial Specificity of Long-Range Lateral and Callosal Axons

More recently, a yet nondescribed similarity between intrinsic and callosal circuits has been systematically explored. It is known for quite some time that long-range horizontal axons exhibit elliptic axonal arbor fields interconnecting not only neurons of the same orientation preference but more specifically those with receptive fields aligning along their prolonged axis of colinearity within the visual field (cat: Schmidt et al., 1997a,b; tree shrew: Bosking et al., 1997; squirrel monkey: Sincich and Blasdel, 2001). This axial selectivity was hypothesized to be the anatomical substrate for the physiological finding (matching psychophysical observations) that responses to optimally oriented but low salient stimuli in the classical receptive field are often facilitated when collinearly aligned contours are presented outside it (for review Polat, 1999).

Surprisingly, axial selectivity was recently discovered in arbors of callosal projection neurons (Rochefort et al., 2009). Boutons covered elongated fields reflecting the fact that axon arbors terminated in two to three clusters aligned along one axis. Like for intrinsic arbors, the axes of those projection fields—when analyzed with respect to their relative position to the 17/18 border,—matched the cortical "image" of the orientation preferred at the ipsilateral injection center. Assuming reciprocity of connections, such a pattern would be expected also for neurons labeled after injection of retrograde tracer.

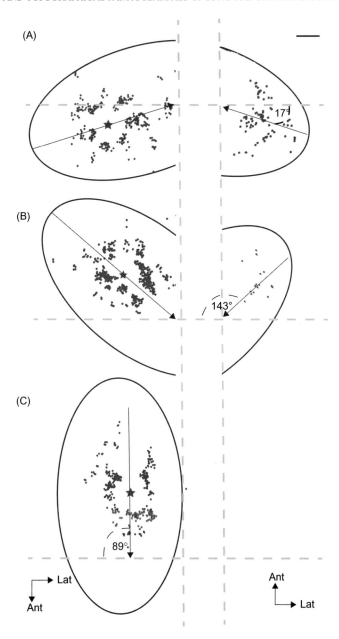

FIGURE 8.4 Axially similar distribution of ipsi- and contralaterally projecting neurons. Spatial distribution of projection neurons retrogradely labeled in both hemispheres after an injection of "beads" in the right area 17 close to the areal border (vertical blue dashed lines). Injections were targeted to domains preferring different contours (A: horizontal, B: oblique, C: vertical). Arrows indicate the main axis of the projection pattern computed as a function of neuron number, and distance to the center of gravity projected on the axis perpendicular to the areal border (horizontal blue dashed lines, as in Schmidt et al., 1997a,b). These axes roughly match the cortical projection of the contour preferred at the injection site. Note that the axes of the callosal patterns mirror the axis of the intrinsic patterns. The injection in (C) did not give rise to transcallosal label (Schmidt et al., unpublished observation). Scale bar: 1 mm.

Indeed, revisiting patterns of retrograde labeling through the corpus callosum produced by targeted injections of beads in cats confirm that idea. Neurons whose receptive fields share the same orientation preference and which are aligned along the same collinear line have a high likelihood to be interconnected and over longer distances in both hemispheres (Schmidt and Löwel, unpublished). Figure 8.4 reveals that neurons projecting through the corpus callosum to that injection site are arranged in a half ellipse mirroring the ellipse in the ipsilateral hemisphere, when aligning the axes of the two 17/18 borders. Such an arrangement is expected from the fact that the visual field representations are mirrored in the two hemispheres (Tusa et al., 1978).

From Figure 8.4, it becomes also apparent that different consequences arise from collinear arrangements for long-range circuits of neurons preferring vertical (A), oblique (B), or horizontal contours (C) and situated close to the 17/18 border. The ellipse for the latter neurons is not perfect but rather interrupted by the 17/18 border (Figure 8.4A). This does not seem to be the case for an injection into a domain preferring vertical contours, which by chance also did not produce transcallosal label in this case (Figure 8.4C). The projection pattern is elongated along a line parallel to the border and thus to the representation of the vertical meridian. There are two reasons for that ellipse to be more pronounced, namely, (i) an areal border does not interrupt the pattern and, (ii) the cortical magnification factor in the cat's central visual field representation is anisotropic for the two cardinal visual field orientations and favors the vertical axis. The latter tendency is more pronounced for area 18, which is much more narrow than area 17, but starts already at the border of area 17 (Tusa et al., 1978; Cynader et al., 1987).

In contrast, ellipses formed by neurons connecting to an injection site of horizontal or oblique orientation preference are continued in the contralateral hemisphere. This would be compatible with the interpretation that the intrinsic network naturally continues into the hemifield representation on the other side and thus through the corpus callosum.

Contour and axial selectivities of lateral connections are in line with Gestalt criteria like common shape (contour selectivity) and colinearity (axial selectivity). In this frame, the above observations foster the hypothesis that both types of lateral connections can equally subserve perceptual grouping (Schmidt et al., 1997a,b), also across the vertical midline of the visual field. As such, they would form part of the same circuit as suggested earlier (Voigt et al., 1988).

In the following, experiments about the functional role of callosal connections for the representation of visual stimuli close to the 17/18 border will be reviewed and further discussed in the framework of that hypothesis.

8.3 FUNCTIONAL IMPACT OF CALLOSAL AXONS ON REPRESENTATIONS AND PROCESSING IN CAT AREAS 17 AND 18

Experiments on the impact of callosal connections between areas 17 and 18 of both hemispheres bear some advantages over experiments on lateral intrinsic circuits. When trying to manipulate lateral intrinsic circuits by pharmacology, lesion, or intracortical stimulation, there is a risk of directly affecting the investigated neuronal responses. This is not the case when probing callosal impact because the site of manipulation is remote from the recording site. However, the outcome of a manipulation experiment depends also critically on the nature of the manipulation.

In earlier studies, the corpus callosum has often been cut through to study the influence of the lack of callosal connections; and animals would have their optic chiasm split for dissociation of signals delivered by either the callosal or the geniculocortical input. This approach supposedly abolishes all axons from the nasal retina and thus the major input to central primary visual cortex. Moreover, after lesions, animals need recover time before any analysis, leaving time for plasticity of connections and synaptic strength to stabilize.

Already Choudhury et al. (1965) applied local cooling of nerve cell tissue within the projecting hemisphere to overcome some of the shortcomings of callosal section. Payne et al. (1991) further developed the technique of thermal inactivation to study visual transcallosal interactions. Cooling inactivation effects are reversible and they are less invasive than sectioning or lesioning. Although inactivating the 17/18-border region does not directly interrupt callosal fibers, it most likely discloses the impact of any interhemispheric projection from that region onto the other hemisphere. Given the anatomy, the most direct route for interhemispheric interactions is through the corpus callosum, and comparative studies deactivating different corticocortical projections indicate that indirect circuits exhibit only marginal effects in cooling studies (Payne and Lomber, 2003; Galuske et al., 2002).

8.3.1 Stimulus Gain Regulation by Callosal Action

In a series of deactivation experiments in ferrets and cats, we recorded optical images, single units, and local field potentials during visual stimulation covering both hemifields, only local ipsilateral receptive fields, or without any stimulation.

In a typical experiment, we first localize the 17/18 border in the contralateral hemisphere by means of optical imaging (Wunderle et al., 2013, 2015; Figure 8.5D). Then, we position a metal cooling loop over the exposed part of the border to inactivate and reactivate callosal input from the TZ and the adjacent central parts of areas 17 and 18.

When deactivating the corresponding contralateral region giving rise to callosally projecting axons in ferrets, optical maps of orientation preference of both areas 17 and 18 obtained by whole-field gratings get weaker and less specific. In Figure 8.5, differential maps between horizontal and vertical domains demonstrate a slightly lower contrast during cooling deactivation and a decay of overall vector strengths (Schmidt et al., 2010). This decay happens

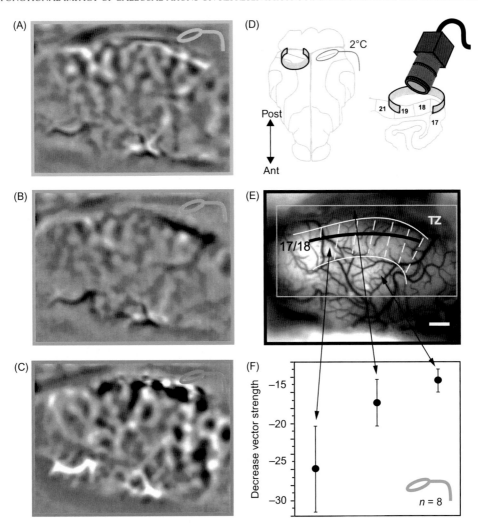

FIGURE 8.5   Unspecific response decrease in the absence of callosal input. Differential activity maps obtained during stimulation with cardinally oriented gratings in ferret areas 17 and 18 while deactivating the topographically corresponding area in the contralateral hemisphere by surface cooling to 2°C. (A) Baseline, light green; (B) cooling, blue; (C) rewarmed, dark green. (D) Typical experimental setup. A cooling loop flooded by chilled methanol is positioned on the left hemisphere while the right early visual areas are covered by a recording chamber for intrinsic signal imaging (or electrophysiology). (E) Blood vessel image of the recorded area with the outlines (white) of the transcallosally connected area (transition zone, TZ); dark horizontal line represents the 17/18 border. (F) Relative change in map vector strength within subregions of the rectangular region (white) of interest outline in (E). Note that responsivity decreases predominantly in the TZ. *Adapted from Schmidt et al. (2010), by permission of Oxford University Press.*

in particular in the TZ, which receives the strongest callosal input (Innocenti et al., 2002; Grigonis et al., 1992). As in cats (Rochefort et al., 2007), the lateral influence spreads into areas 17 and 18. This is in good agreement with the anatomy as, in particular, area 18 in the ferret is extensively transcallosally linked (Innocenti et al., 2002). It is noteworthy that the map layout does not change at all during cooling, in accordance with recent observations in cats (Altavini et al., unpublished).

Electrophysiological unit activity reveals increases and decreases when stimulating both hemifields with a moving bar (Payne et al., 1991; Payne, 1994) or a moving continuous whole-field grating (Makarov et al., 2008; Schmidt et al., 2010). A total of 58% of a population of 491 single units, sampled maximally up to 10° away from the vertical meridian representation, changed their firing rate significantly during cooling deactivation (Wunderle et al., 2013). When presenting moving dot textures (MDTs)—a visual stimulus that activates more neurons but in a less selective manner than gratings (Skottun et al., 1988; Wörgötter and Eysel, 1989)—cooling affects stimulus-evoked responses of substantially more units. Now, 73% of the units significantly drop their firing rate. This is much more than with gratings, and only one unit (0.3%) increases it (Figures 8.6 and 8.7). The difference between gratings and MDTs remains when correcting for differences in contrast and baseline spike rate (Wunderle et al., 2013).

Chance et al. (2002) hypothesized that within active cortical circuits, the overall level of synaptic (background) input to a neuron acts as a gain control signal that modulates responsiveness to an excitatory drive. Thus, naturally, amplification can act better when cortical neurons are (i) stimulated in an unselective manner like with dot textures and (ii) when overall background levels of synaptic activity are not yet saturated, like it might be the case with gratings. This is also in agreement with earlier observations. The spatial range over which neurons integrate increases when the feedforward drive introduced by the stimulus is low, as is the case with low contrast (e.g., Nauhaus et al., 2009), i.e., background activity saturated by a high contrast full-field grating may mask or even impede influences of the lateral network.

It is noteworthy that the levels of rate change caused by cooling deactivation for a particular neuron tested under different visual stimulation (grating, dot, or bar textures) were always highly correlated. This confirms that within the TZ, some neurons are more influenced by callosal input than others. This seems independent of the stimulation protocol, and dependent on their amount of direct or indirect callosal input.

However, our results on that matter strongly indicate that the gain of the input, which is effectually delivered by callosal axons onto a particular neuron, is dynamically adapted to the feedforward drive introduced by visual stimulation. This feedforward drive, in turn, is contributed by the geniculocortical loop, both to that neuron and to the population within which it is embedded through the lateral intrahemispheric network. The contribution of feedforward drive through long-range intrinsic circuits follows from the observation that the global nature of full-field MDTs can be only revealed when we consider the larger context outside the classical receptive field. A stimulus-dependent gain control jointly executed by lateral intrinsic and callosal circuits could amplify small signals such as weakly tuned feedforward input delivered by geniculocortical afferents (Ben-Yishai et al., 1995; Sompolinsky and Shapley, 1997).

8.3.2 Excitation and Inhibition in Callosal Action

8.3.2.1 Long-Range Lateral Axons (Inhibitory)

We need to remember here that most of the long-range intrinsic axons are excitatory and terminate on excitatory neurons (LeVay, 1988; McGuire et al., 1991; Binzegger et al., 2004). As for long-distance extrinsic connections, inhibitory cells are only targeted by 5–20% of the synapses made by long-range intrinsic axons (Fisken et al., 1975; Kisvárday et al., 1986; LeVay, 1988; McGuire et al., 1991).

Although inhibitory neurons constitute about 20% of all cortical neurons, most of them make only local axon collaterals restricted in their lateral extent to the immediately vicinity of the parent neuron (LeVay, 1988; Matsubara, 1988). The characteristic distant clustering of long-range axons is mainly observed in excitatory pyramidal or spiny stellate neurons in layers II/III and V. Among the GABA immunopositive neurons, only a few types such as large basket cells, large multipolar cells and dendrite-targeting cells exhibit long-range axons extending more than 1.5 mm (Somogyi et al., 1983; Kisvárday et al., 1994).

Compared to the excitatory network, the long-range inhibitory network is thought to be less selective, but single basket cell axons can exhibit a very specific long-range topography (Kisvárday et al., 2002). In cat areas 17 and 18, 48% of the inhibitory boutons were located in iso-, 28% in oblique, and 24% in cross-oriented domains (Kisvárday et al., 1997). Thus, the inhibitory network is only weakly selective for orientation preference and extends more locally than the excitatory network—about one-third to one-half the extent of the excitatory one—but probably involves more postsynaptic neurons (Kisvárday et al., 1997; Crook et al., 1998; Stepanyants et al., 2009). Its local impact is considered to be substantial because bouton density in focal clusters is four times more concentrated for smooth than for spiny neurons (Binzegger et al., 2007).

8.3.2.2 Callosal Axons

Because of the above-discussed localized nature of inhibitory networks, we would expect a rather low number of direct inhibitory connections through the corpus callosum. Indeed, the majority of the neurons projecting through the corpus callosum are excitatory. In the literature for both, carnivores and rodents, only very few inhibitory neurons— basket cells—with direct transcallosally projecting axons have been described (Buhl and Singer, 1989; Conti and Manzoni, 1994). Also, the targets of those axons in the receiving hemisphere are mainly of excitatory nature (Voigt et al., 1988; Houzel et al., 1994). In fact, transcallosally projecting axons make preferentially excitatory synapses on pyramidal and spiny stellate cells (e.g., Rochefort et al., 2009). Only some synapses made on inhibitory neurons have been reported (Martin et al., 1983).

We would therefore expect both inhibitory and excitatory impact contributed by visual callosal connections but with a strong bias toward excitation. As both excitatory and inhibitory long-range axons exhibit some selectivity to link neurons of iso-orientation preference, a certain amount of stimulus-dependency should prevail not only in the excitatory but also in the inhibitory actions. Our data from cat visual cortex give evidence for both of these predictions.

In spiking activity of both ferret and cat primary visual cortex, we observe more facilitating than suppressive callosal actions (Schmidt et al., 2010; Wunderle et al., 2013). This is in agreement with previous experiments studying callosal interactions in cats either by cooling deactivation (Payne et al., 1991; Payne, 1994) or by infusing GABA/bicuculline into the contralateral hemisphere (Sun et al., 1994). As discussed in the previous paragraph, this ratio might depend not only on the numeric distribution of excitatory or inhibitory callosally projecting neurons and their synapses but also on other intrinsic or feedback inputs and the actual stimulus-derived feedforward drive through geniculocortical input interacting with the callosal action.

In accordance with that hypothesis, in Wunderle et al. (2013), we observe that the balance of excitation and inhibition contributed via the interhemispheric connections varies greatly with global attributes of the stimulus. With moving high contrast full-field gratings, about 7% of all response changes during cooling, hereafter called callosal actions, exhibit significant inhibitory character, whereas 48 % are significantly excitatory.

With the less salient stimulus of MDTs, not only more cells are affected by cooling but also these cells almost exclusively lose spikes. This points toward predominantly excitatory action through the corpus callosum. It is in contrast to the mixed picture of callosal actions observed with gratings (Figure 8.7). To examine whether the larger excitatory action was disclosed because of an unselective recruitment of many more neurons by dot stimuli than by gratings, we increased the orientation component of the visual stimulus by elongating the dots to form scattered short bars of a certain orientation moving on the screen. The result is much more similar to the one obtained with moving dots than to the one with gratings (Figure 8.6). This allows for the conclusion that the strength and nature of callosal actions exerted on their postsynaptic targets are not easily related to the presence or absence of the orientation component, i.e., an optimal or nonoptimized stimulation. Rather, the excitatory–inhibitory balance seems to change with the local and global composition of the driving stimulus.

Along this line, stimuli completely void of context, like small Gabor gratings tailored on the confines of the receptive field under study, attract more facilitating influence over the corpus callosum than a grating covering 40° × 30° of visual field angle (−31% overall change, Peiker et al., 2013). In contrast to local stimuli, or textures with less structure, whole-field gratings recruit interconnected populations of similar orientation preference over all. They thus do not only evoke more recurrent iso-orientation excitation, but this excitation is probably counter-balanced by recurrent inhibition (Stimberg et al., 2009). When the whole field is uniformly stimulated, part of that inhibition occurs in the contralateral hemisphere. Accordingly, it can be transferred by the few but presumably effective direct inhibitory connections, or indirectly relayed by contacting the local intracortical inhibitory circuits in the contralateral hemisphere, in effect lowering the excitatory impact of callosal fibers. Cooling of that hemisphere will remove that component and reveal release from inhibition.

Although individual suppressive actions on spiking activity in area 17/18 can be strong (Figure 8.7), excitatory impact is always larger than inhibitory callosal impact, in line with the anatomical ratios of excitatory and inhibitory long-range circuits.

In local field potential recordings, many more inhibitory effects are observed with whole-field gratings (Makarov et al., 2008). This supports the hypothesis that transcallosal inhibition becomes only occasionally suprathreshold, with appropriate stimulation.

8.3.3 Feature Selectivity in Callosal Action

We obtained first evidence of a functional proof of "like connects to like" via the corpus callosum in local field potential recordings in area 17 close to the ferret's 17/18 border. Synchronization of ipsilateral field potentials was better supported by uniform as opposed to contrasting stimulation in the two hemifields (Carmeli et al., 2007). This is consistent with excitatory callosal connections between neurons preferring similar features. Also, certain inhibitory actions were more frequently reversed, indicated as an increase in the LFP amplitude during cooling deactivation, when gratings continued collinearly over the visual field midline as opposed to changing orientation or direction of motion (S2 vs. S3 stimulation in Makarov et al., 2008). This is in line with the conclusion from the previous paragraph that a whole-field grating generates inhibition in both hemifield representations, which can be transmitted over the corpus callosum directly or indirectly via selective callosal axons.

Later, we made more detailed observations concerning the subpopulations of orientation-selective neurons in both ferrets and cats. In ferrets, the strength of responses in domains evoked by stimulation of either vertical or horizontal contours is preferentially affected by deactivating of callosal input. The same tendency can be observed

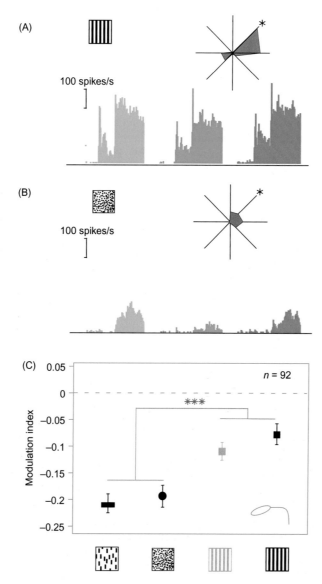

FIGURE 8.6 Stimulus-dependency of callosal action. Example PSTHs (peri-stimulus time histogram) of a single unit recorded in the cat's TZ during cooling of contralateral inputs, during grating (A) or MDT (B) stimulation. Polar plots indicate the neuron's tuning with the different stimuli. (preferred direction indicated by a star) Color code as Figure 8.5. (C) Normalized spike rate changes of 92 units during different visual stimulations (sketches on *x*-axis: high and low contrast grating, moving dot and bar textures), during cooling deactivation of callosal input. *Y*-axis: Modulation index = (rate$_{cool}$ − rate$_{base}$/rate$_{base}$ + rate$_{cool}$). Note that spike rates decrease (negative values) on average less with grating stimulation. *** p < 0.0001, analysis of variance. *Adapted from Wunderle et al. (2013), by permission of Oxford University Press.*

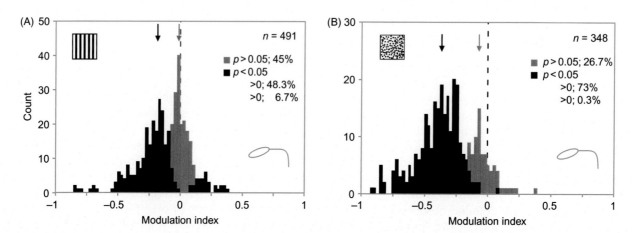

FIGURE 8.7 Stimulus-dependency of excitation and inhibition in callosal action. Distribution of modulation indices for single units recorded in the cat's TZ during grating (A) or MDT (B) stimulation. Black counts, significant rate changes; grey, insignificant; arrows point to the median of the distributions. Negative values indicate excitatory callosal input; positive values indicate release from inhibitory input. *From Wunderle et al. (2013), by permission of Oxford University Press.*

for spiking responses in neuronal populations in cat area 18. Absence of callosal input decreased normalized spike rates more for neurons preferring cardinal, especially horizontal contours, than for those preferring oblique contours (Schmidt et al., 2010; Wunderle et al., 2013; Altavini et al., unpublished). Likewise, stimulus-evoked synchronization in the ipsilateral hemisphere decreased more within populations of cardinally preferring neurons (Carmeli et al., 2007).

It has long been known that, within the central visual field representation, neurons preferring vertical and horizontal contours respond more vigorously and are greater in number (e.g., Li et al., 2003). In agreement, stimulation with cardinal contours evokes larger cortical representations in imaging studies (Chapman and Bonhoeffer, 1998; Coppola et al., 1998; Dragoi et al., 2001; Wang et al., 2003; Yacoub et al., 2008). Also, imaging responses are more vigorous and apparently faster (Wang et al., 2011), and ongoing fluctuations appear to be more often correlated with maps stimulus-derived from cardinal as opposed to oblique contours (Kenet et al., 2003). Interestingly, this population of neurons preferring cardinal contours seems to be more susceptible to the lack of callosal input. One might thus imagine that lateral intrinsic or callosal connections between neurons preferring cardinal contours are more numerous or of stronger impact.

Among the neurons preferring cardinal contours, there are differences between those optimized for vertical contours and those responding preferentially to horizontal contours (Peiker et al., 2013). On average, neurons preferring horizontal contours loose more spikes with the absence of callosal input (Schmidt et al., 2010; Altavini et al., unpublished). This seems to be independent of the stimulation applied as it is observed with both whole-field gratings, isolated Gabor patches and with prestimulus activity (Figure 8.8).

As laid out above, the anatomy of feature-biased connections among the different orientation-selective neuronal subpopulations, i.e., the axial selectivity (Figure 8.4) makes us suspect differences in the functional impact of callosal connections. In the TZ, the vertical axis of the visual field is represented parallel to the areal border. It benefits also from larger magnification (Cynader et al., 1987). The horizontal axis is interrupted by the split of the two hemifields at the areal border and is less magnified. Therefore, the network linking neurons preferring horizontal contours depends much more on its continuation in the contralateral hemisphere than the network linking neurons preferring vertical contours (Figure 8.4). In the visual environment, it is more likely that a horizontal contour or shape within the central visual field crosses the vertical midline than would a vertical shape. For vertical contours, in turn, crossing the vertical meridian is more probable in the case of movement. Therefore, it might be legitimate to argue that neurons processing horizontal image movements—and vertical contours—would greatly benefit from callosal interaction, but depending on the direction of movement.

Although there is no anatomical evidence that neurons sharing similar direction preference are preferentially linked by long-range horizontal or callosal connections, such an arrangement is likely. An influence of callosal connectivity on the processing of motion would be expected in particular for neurons preferring directions of motion crossing the vertical midline (Houzel et al., 2002).

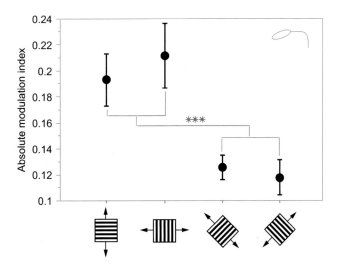

FIGURE 8.8 Feature selectivity of callosal action. Firing rate changes of multiunits in ferret visual cortex during deactivation of callosal input while stimulating with whole-field gratings of different orientations. Only preferred responses are shown. Y-axis: Modulation indices as in Figures 8.6 and 8.7 but averaging excitatory and inhibitory action. Note that absolute changes are bigger for neurons preferring cardinally oriented contours. *p<0.001, Mann-Whitney. *From Schmidt et al. (2010), by permission of Oxford University Press.*

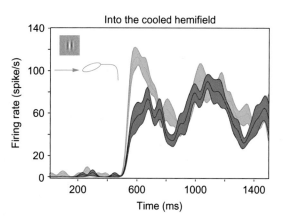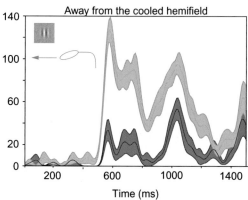

FIGURE 8.9   Direction selectivity of callosal action. PSTHs of a single unit preferring vertical contours, recorded in the cat's TZ during cooling deactivation of callosal input. The neuron was stimulated with a vertically oriented Gabor grating either moving toward the cooled hemifield (left) or away from it (right). Green: baseline; blue: cooling. *Courtesy of C. Peiker.*

In the above-described studies, the issue of direction selectivity in callosal actions has not been directly addressed. However, it has long been observed that callosal influences on the two directions of motion might be asymmetric when applying cooling (Payne et al., 1991). In the ferret studies (Carmeli et al., 2007; Makarov et al., 2008), local field potentials were obtained during stimulation with opposite or orthogonal directions of motion in the two hemifields. In summary, cooling affects the processing of these directionally biased stimuli differently than that of uniform stimuli. Further, spiking of neurons preferring horizontal motion (and thus vertical contours) is differently affected by callosal input than the spiking of neurons preferring other directions of motion (Schmidt et al., 2010). These results are so far compatible with the interpretation that among neurons with the same orientation preference those preferring similar directions of motion are preferentially linked via the corpus callosum, but do not proof it.

In a recent study, we obtained direct evidence for that hypothesis. We positioned small Gabor gratings of vertical or horizontal orientation on previously characterized receptive fields of neurons in the TZ between areas 17 and 18 of cats. The grating within the Gabor patches moved either to the left or right, or up or down but within the borders of the classical receptive field. When the direction of movement of the vertically oriented grating patch matched the movement out of the cooled hemifield, the spiking responses were more strongly impaired than the spiking responses to the opposite direction, into the cooled hemifield (Peiker et al., 2013; Figure 8.9). This was most pronounced when the direction of movement matched the movement preference of the neuron but became also obvious when considering all responses recorded using that stimulus. The asymmetry reflects most likely that neurons preferring the same horizontal direction of motion are preferentially linked by callosal connections and that also other neurons with different direction preferences are embedded within a network of crudely direction-selective connections.

We detect this asymmetry only because we cool one hemisphere while recording the receiving hemisphere and because the direction of movement crosses the vertical midline. The finding also indicates that lateral intracortical connections exhibit some direction selectivity. These connections could explain why cooling impairs movement processing toward the cooled hemifield less. For this direction of movement, direction-selective lateral connections within the receiving hemisphere—which are not interrupted by cooling—can provide input generated by the processing of movement in the intact hemifield (Figure 8.10). Supposedly, we would obtain the opposite finding when cooling the opposite hemisphere. Moreover, if we could reversibly deactivate the upper or the lower hemifield representations without interfering directly with the recorded responses, we would probably detect a clear asymmetry between neurons preferring the up versus down movement.

Interestingly, there is a difference between the unselective responses to up and down movements (see figure 6A in Peiker et al., 2013). This might be linked to the fact that all recordings were made from receptive field locations below the horizontal meridian. As we probably do not make this clear distinction between upper and lower hemifield representations when cooling the contralateral TZ, direction-selective responses in the ipsilateral lower hemifield might be asymmetrically affected by the deactivation procedure.

In conclusion, when investigating lateral interactions in the TZ we might consider two probabilities. First, there exists a local anisotropy among connections between different populations of orientation-selective neurons within this particular part of the visual field representation, caused by the split of the visual field at its midline. This

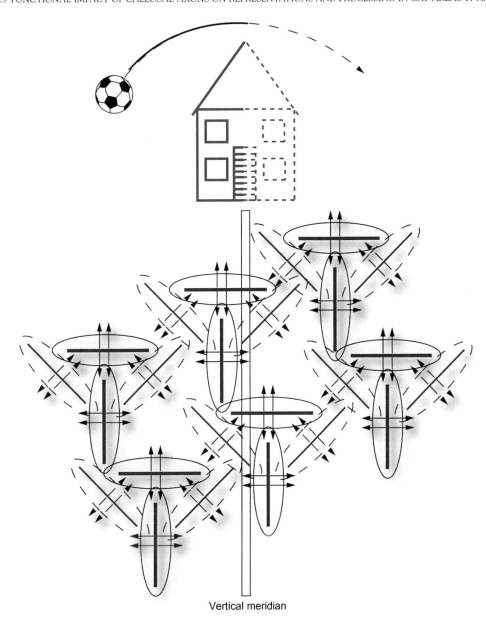

Vertical meridian

FIGURE 8.10   Anticipatory lateral action close to the 17/18 border. Sketch summarizing the putative layout of intrinsic (shadowed) and callosal connections close to the 17/18 border. Note that the anatomical layout predicts a potential to facilitate anticipation along the cortical "image" of frequently occurring contours (pink house) or movement trajectories (black football). Depicted is the special case of features crossing the vertical midline of the visual field (represented at the 17/18 border, vertical bar). Continuous pink lines indicate contour-selective connections; black arrow-headed lines indicate direction-selective connections. Ellipses indicate the axial selectivity of the connection fields. Connections between neurons preferring cardinal contours are drawn thicker to indicate stronger or more numerous connections.

anisotropy shows up because we disrupt the network at the midline's cortical representation in a unilateral manner and present a (potentially) midline-crossing stimulus. Second, there exists a general bias toward connections between neurons preferring cardinal as opposed to more oblique contours, which is not necessarily restricted to the 17/18 border (Figure 8.10).

In any case, spike rate changes observed in callosal deactivation experiments do not drastically affect the orientation and direction selectivity of the studied neurons (Wunderle et al., 2013). The few significant effects are usually observed for direction selectivity indices of cardinally preferring neurons (Schmidt et al., 2010).

When manipulating the intrahemispheric lateral network with local GABA application, Girardin and Martin (2009a) report effective changes in preferred orientation of area 17 single cells. However, this result could not be repeated with local cooling close to the recording sites (Girardin and Martin, 2009b). Overall, changes were relatively tiny and might thus have escaped other investigations of lateral connections, including callosal connections, because, usually, orientations are not sampled with sufficient resolution to answer this question. Usually no differences smaller than 22.5° are tested. It is also conceivable that the more local short-range network was implicated in the former study (Girardin and Martin, 2009a). Distances between GABA infusion site and recording sites were not larger than 500 μm. Within this radius, the inhibitory network might be stronger, so that both excitatory and inhibitory lateral circuits are rather unselective. In agreement with this possibility, inhibiting short-range connections between adjacent columns of orthogonal orientation preference in area 18 by infusion of GABA, resulted in orientation tuning curves that broadened in 65% of the neurons (Crook et al., 1996). However, when deactivating more remote sites of similar preference (Crook et al., 1997)—compatible with the range of long-range connections—broadening occurs only in 5% of the cases.

8.3.4 Ongoing Callosal Action

It is noteworthy that in the experiment described earlier (Peiker et al., 2013), there was no stimulus presented in the deactivated hemifield. Therefore, there was no stimulus-related activation to be delivered by callosal axons, whose impact we would have removed by cooling. Matching our finding, humans with hemianopia show impaired visual integration in their intact visual hemifield, although the intrahemispheric network should be entirely intact (Paramei and Sabel, 2008; Bola et al., 2013). Astonishingly, the selective changes during cooling observed within the different orientation-selective populations of neurons became already apparent in the prestimulus activity. In a follow-up study, we observed that ongoing activity of neurons preferring cardinal orientations suffers more spike loss from the absence of callosal input than ongoing activity of neurons preferring oblique contours (Altavini et al., unpublished).

It seems intuitive that neurons in the ipsilateral hemisphere, which receive visual information of a movement or a form that potentially crosses the vertical midline in the future, forewarn relevant units in the contralateral hemisphere of the soon-to-be-incoming sensory input. This preparative role might proceed even if the stimulus finally will not move or cross into the other visual hemifield. Induced activity by preparative motion was described by Harvey et al. (2009) who stated that there is a predepolarizing wave in cortical layers I–III, which occurs 30 ms prior to the onset of spiking activity and is strongest along the motion trajectory. Jancke et al. (2004) used extracellular recordings in cat area 17 and found that neuronal activity is induced much earlier by a moving stimulus in comparison to a stationary stimulus. The authors suggested that a moving object causes subthreshold activation in the motion trajectory, which might be conveyed by long-range lateral connections. Our findings support this notion when assuming that callosal connections between the area 17/18 borders continue the function of long-range intrinsic connections across the midline (Peiker et al., 2013).

Lateral connectivity might be involved in maintaining a spontaneously active network which preactivates neurons along frequently occurring motion or shape trajectories, via feature-selective long-range connections (Figure 8.10). Accordingly, cortical maps, which have similarities with orientation—predominantly cardinal—preference maps, occur spontaneously (Kenet et al., 2003). A recent modeling study proposes that the biased long-range lateral network would physically encode statistical properties of certain modalities represented in a cortical area (Muir et al., 2011). Indeed, statistical properties of natural scenes as viewed by cats exhibit a bias for cardinal contours, and among these, an additional prevalence for horizontal contours (Betsch et al., 2004). A bias for cardinal orientations was also found in humans after presentation of natural scenes (Nasr and Tootell, 2012).

8.3.5 Multiplicative and Additive Scaling of Callosal Action

Surprisingly, despite significant stimulus-driven and ongoing response changes, neurons largely keep their tuning profiles. To solve this puzzle, Wunderle et al. (2013) approximated the modifications of tuning curves during cooling deactivation of callosal input by a linear model. Most of the changes could be described by multiplication of the tuning curve with a constant factor (Figure 8.11). The bigger the rate decrease, the more does the modification applied to the tuning curve during cooling resemble a multiplication of input rates. Only a minority of units exhibits a pronounced additive shift. These types of shifts were mainly observed with gratings, and then often associated with a release from inhibition, i.e., increase of spike rates during cooling. This points toward a positive relation between additive scaling and inhibition on one hand, and multiplicative scaling and excitation on the other (Wunderle et al., 2013). The transition between the two types of action is probably continuous and—as for the excitatory–inhibitory balance—input dependent. Multiplicative scaling mainly preserves the neuron's response selectivity and explains why tuning indices hardly change during deactivation experiments. It is supposed to be a dominant action mechanism exerted by corticocortical connections. This can be also deduced from other experiments on

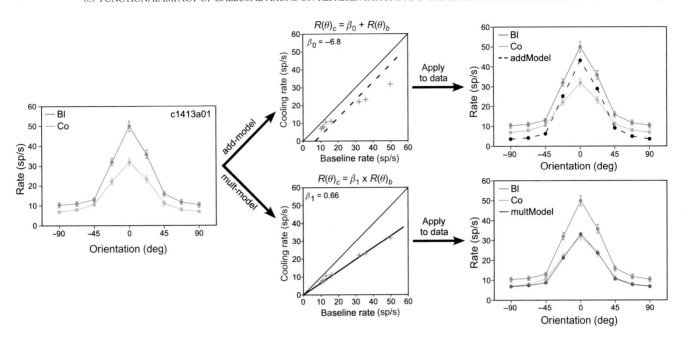

FIGURE 8.11 Callosal action scales responses in a multiplicative manner. The callosal action on a neuron's orientation tuning curve is approximated by a linear model. Upper panel: Model fit by subtraction/addition of spikes to baseline rate during cooling. Lower panel: Model fit by multiplication of the rate with a constant factor. Left: Fit to rates during cooling (y-axis) versus baseline (x-axis). Right: Model fit projected on tuning curves during baseline (Bl, green) and cooling (Co, blue). Dotted lines: additive model; continuous lines: multiplicative model. Note that the multiplicative performs better on the real data than the additive model. *From Wunderle et al. (2013), by permission of Oxford University Press.*

feedback (Wang et al., 2007), contextual modulation (Li et al., 2000), and spatial attention (McAdams and Maunsell, 1999; Treue and Martínez Trujillo, 1999).

8.3.6 Direct and Indirect Callosal Actions

Combining ipsilateral and contralateral input scales the tuning curves of orientation-selective ipsilateral neurons. This could reflect a specific callosal contribution to orientation selectivity or a more general signature of cortical processing or both. To shed light onto this question, Wunderle et al. (2015) studied the impact of callosal connections on contrast response functions (CRF). In summary, callosal inputs can change the contrast sensitivity and/or the responsivity of a certain neuron, i.e., modulate the gain at the input or output, or both in combination. In many cases, the neuron's responsivity changes, expressed as a shift of the response along the y-axis of the CRF—the response axis—because the maximal response increases (response gain). This corresponds to a linear output increase, i.e., the above-described multiplicative scaling of the neuron's evoked response, and can be exerted by an indirect action of callosal connections on intracortical circuits. In other cases, the neuron's sensitivity changes and thereby shifts the response along the x-axis of the CRF—the contrast axis—without affecting the maximal response (contrast gain). This corresponds to a multiplicative scaling of the input, i.e., the stimulus contrast. The contrast gain mechanism presumably requires a direct monosynaptic callosal action on the target neurons. Indeed, we found this mechanism to be particularly strong in supragranular layers (figure 8.5 in Wunderle et al., 2015) where most visual callosal connections in the cat terminate (Houzel et al., 1994; Rochefort et al., 2009). The response gain mechanism is observed in all layers with similar probability. Very frequently, the two mechanisms also mix up. These findings are compatible with the interpretation that the scaling mechanism for a particular neuron results from its position within the intrahemispheric circuit—and the relative contribution of its direct callosal and indirect intrinsic inputs, which were manipulated in our experiments.

8.3.7 Binocularity and Callosal Actions

Neither lateral intrinsic nor callosal connections in normally raised animals link domains of same ocular dominance (Schmidt et al., 1997a,b). Rather than the ocular dominance profile, the matching of visuotopic position by the interconnected neurons seems to have an important role for connectivity. For example, neurons in the two hemispheres receiving input from the same retinal location are selectively linked (Olavarria, 1996, 2001). It follows that there have to be connections between columns driven by the same eye as well as between columns driven by different eyes because the midline region receives input from the two retinae in both hemispheres.

Early cat experiments sectioning the corpus callosum or lesioning the contralateral cortex left us with the impression that callosal afferents contribute extensively to the binocularity of callosal neurons in this species (Dreher and Cottee, 1975; Payne et al., 1980; Blakemore et al., 1983; Yinon et al., 1988). This result was contrasted by work of Payne et al. (1984), Minciacchi and Antonini (1984), and Cynader and Gardner (1987). Finally, a developmental study indicated that there is a postnatal critical period for the influence of corpus callosum section on binocularity (Elberger and Smith, 1985). This was apparently circumvented in the studies reporting only minor impact of callosal section on binocularity. In support, our own preliminary findings in adult cats do not support any significant reduction of binocularity at the 17/18 border in the absence of callosal input (Peiker et al., unpublished).

A significant role for binocularity has been attributed to visual callosal connections in small rodents. In rodents, as opposed to carnivores and primates, the majority of the retinal fibers cross at the chiasm. It is thus not remarkable that there is an enormous impact from interrupting callosal transmission on adult binocular function and its development (for review Pietrasanta et al., 2012). Future studies might thus keep in mind and exploit that visual callosal connections cover a different spectrum in rodents than in carnivores and primates. For frontal-eyed animals, binocularity as such might not depend on the integrity of the callosal network for integration over the vertical midline, but rather on the geniculocortical input as anywhere else in the visual field.

However, it was suggested from studies in humans (Blakemore, 1969; Mitchell and Blakemore, 1970) and animals with disparity-selective neurons (Lepore et al., 1992; Kalberlah et al., 2009) that the corpus callosum takes part in stereoscopic function, because of its unique position within the visual cortex' topography. Stereoscopic functions presumably are more dependent on neurons preferring horizontal image movements, and coding for horizontal disparity. It is to be expected that those units when situated close to the vertical midline representation would be under the influence of callosal interactions. In fact, vertically preferring units disclose more complex influences of callosal input, i.e., more inhibitory impact (Schmidt et al., 2010), direction selectivity across the hemispheres (Peiker et al., 2013), than other units.

Importantly, this does not render callosal connections in these animals special over any other lateral intrinsic connection. Intrinsic connections could also subserve stereopsis in binocular representations of the visual field that are acallosal or more peripheral. Along these lines, split-brained humans exhibit normal stereovision in those parts (Mitchell and Blakemore, 1970).

8.3.8 Comparison Between Feedback and Interhemispheric Circuits

To situate visual callosal connections within the visual circuits, it might be necessary to compare them not only to horizontal intrinsic but also to other extrinsic connections, i.e., the ipsilateral and contralateral feedback connections. Morphologically, visual feedback connections have been reported to exhibit anisotropically arranged clusters (Angelucci et al., 2002). However, on the single cell level, axon terminals are much less clustered (e.g., Rockland and Knutson, 2000; Borra and Rockland, 2011) and less feature-selective than intrinsic connections (Stettler et al., 2002). They can integrate from a more comprehensive part of the visual field than long-range intrinsic circuits and thus might be apt to contribute information from a larger modulatory surround to which intrinsic circuits do not have access to (for review Angelucci and Bullier, 2003; Gilbert and Sigman, 2007; Rockland, 2013). One of the key findings from deactivation studies in behaving animals is that feedback connections are important for differentiating a figure from the background (Bullier et al., 2001).

It was thus surprising that deactivating extrastriate areas in anesthetized cats (posterior middle suprasylvian cortex (pMS), in Galuske et al., 2002; Schmidt et al., 2010, and area 21 in Wang et al., 2000, 2007) already impairs basic responses of neurons in early visual areas affecting orientation or direction selectivity.

The comparison of deactivation studies using grating stimulation reveals that both ipsilateral feedback and interhemispheric connections do not instruct orientation preference or the layout of orientation preference maps. However, loss of spikes and of map vector strengths is much stronger without feedback than without callosal input (vector strength: up to 50% in Galuske et al., 2002 vs. 20–25% in Schmidt et al., 2010). Moreover, changes in direction selectivity of single units and maps related to pMS input are relevant (Galuske et al., 2002; Schmidt et al., 2011), whereas direction selectivity changes related to callosal input are tiny (Wunderle et al., 2013).

It might be mentioned here that visual interhemispheric projections as probed by cutting, lesioning, or cooling include projections from both contralateral area 18 and 17 and heterotopic connections from even more remote cortices to the recipient transcallosal zone. These heterotopic callosal connections could be interpreted as a contralateral feedback (and even feedforward) connection. More likely, however, is that the strongest impact of cooling the contralateral TZ is exerted by homotopic callosal connections from the cooled area.

A comparison of various deactivation studies and the strength of manipulated connections indicates that the impact of deactivating a projection area is roughly proportional to the density of neurons which project to the target area, i.e., area 18 (Payne and Lomber, 2003). Ipsilateral pMS feedback connections to central area 18 are indeed much stronger than any input from the contralateral hemisphere (26.5% vs. 4.7% of all neurons retrogradely labeled with HRP from an area 18 injection), thus explaining partially their stronger functional impact in deactivation studies. However, numerical differences alone might not explain the various quantitative and qualitative functional differences between extrinsic lateral and feedback inputs.

8.4 CONCLUSION

In cat primary visual cortex, both lateral intrinsic as well as callosal connections are of patchy nature and predominantly link groups of neurons with receptive field properties following motifs known from perceptual grouping like similar contour orientation or similar direction of motion or alignment along a collinear axis of the visual field. For neurons situated at the border between areas 17 and 18 consigned to the central visual field representation—the circuits have different anatomical consequences, which depend on their orientation selectivity.

The influence of callosal input on ongoing and visually driven activity in primary visual cortex can be directly investigated by reversible inactivation of the contralateral hemisphere. Effects range from a direct sensitivity change for a specific feature to an indirect and unspecific gain setting, most likely depending on the synaptic distance of the manipulated inputs to the neurons under investigation.

Thereby, excitatory and inhibitory actions always reflect the functional and anatomical selectivity of lateral networks. Influences are stronger between neurons processing similar and collinear as opposed to dissimilar and non-collinear features, and between neurons processing cardinal as opposed to oblique features reflecting a general cardinal bias. Also, neurons preferring horizontal contours or movements are prominent in the cortical processing across the visual field's midline, in agreement with the supposed lateral connectivity pattern among these neurons.

Moreover, feature selectivity can be observed in callosal interactions in the absence of visual stimulation. That is, biased lateral connections help to anticipate meaningful grouping operations within future visual input by biasing spontaneous activity waves to preference patterns occurring frequently in natural scenes.

Summarizing anatomical and functional evidence from intra- and interhemispheric lateral and feedback circuits leads to the conclusion that visual callosal connections extend lateral intrahemispheric connections across the midline. Therefore it seems reasonable to interpret experimental results about the callosal circuit in early visual areas with respect to lateral connectivity in general.

References

Aggoun-Zouaoui, D., Kiper, D.C., Innocenti, G.M., 1996. Growth of callosal terminal arbors in primary visual areas of the cat. Eur. J. Neurosci. 8 (6), 1132–1148.

Albus, K., Lübke, J., 1992. Widespread lateral processes of glial cells in the immature striate cortex of the cat. Dev. Brain Res. 68 (2), 278–281.

Alekseenko, S.V., Toporova, S.N., Makarov, F.N., 2005. Neuronal connection of the cortex and reconstruction of the visual space. Neurosci. Behav. Physiol. 35 (4), 435–442.

Amir, Y., Harel, M., Malach, R., 1993. Cortical hierarchy reflected in the organization of intrinsic connections in macaque monkey visual cortex. J. Comp. Neurol. 334 (1), 19–46.

Anderson, J.C., Martin, K.A.C., 2002. Connection from cortical area V2 to MT in macaque monkey. J. Comp. Neurol. 443 (1), 56–70.

Anderson, J.C., Martin, K.A.C., 2005. Connection from cortical area V2 to V3 A in macaque monkey. J. Comp. Neurol. 488 (3), 320–330.

Anderson, J.C., Martin, K.A.C., 2006. Synaptic connection from cortical area V4 to V2 in macaque monkey. J. Comp. Neurol. 495 (6), 709–721.

Anderson, J.C., Martin, K.A.C., 2009. The synaptic connections between cortical areas V1 and V2 in macaque monkey. J. Neurosci. 29 (36), 11283–11293.

Angelucci, A., Bullier, J., 2003. Reaching beyond the classical receptive field of V1 neurons: horizontal or feedback axons? J. Physiol. Paris 97, 141–154.

Angelucci, A., Levitt, J.B., Walton, E.J.S., Hupe, J.-M., Bullier, J., Lund, J.S., 2002. Circuits for local and global signal integration in primary visual cortex. J. Neurosci. 22 (19), 8633–8646.

Banno, T., Ichinohe, N., Rockland, K.S., Komatsu, H., 2011. Reciprocal connectivity of identified color-processing modules in the monkey inferior temporal cortex. Cereb. Cortex 21 (6), 1295–1310.

Barbas, H., Hilgetag, C.C., Saha, S., Dermon, C.R., Suski, J.L., 2005. Parallel organization of contralateral and ipsilateral prefrontal cortical projections in the rhesus monkey. BMC. Neurosci. 6, 32.

Ben-Yishai, R., Bar-Or, R.L., Sompolinsky, H., 1995. Theory of orientation tuning in visual cortex. Proc. Natl. Acad. Sci. U.S.A. 92 (9), 3844–3848.

Berardi, N., Bisti, S., Maffei, L., 1987. The transfer of visual information across the corpus callosum: spatial and temporal properties in the cat. J. Physiol. 384, 619–632.

Berlucchi, G., Rizzolatti, G., 1968. Binocularly driven neurons in visual cortex of split-chiasm cats. Science 159 (812), 308–310.

Berman, N.E., Payne, B.R., 1983. Alterations in connections of the corpus callosum following convergent and divergent strabismus. Brain. Res. 274 (2), 201–212.

Betsch, B.Y., Einhäuser, W., Körding, K.P., König, P., 2004. The world from a cat's perspective—statistics of natural videos. Biol. Cybern. 90 (1), 41–50.

Binzegger, T., Douglas, R.J., Martin, K.A.C., 2004. A quantitative map of the circuit of cat primary visual cortex. J. Neurosci. 24 (39), 8441–8453.

Binzegger, T., Douglas, R.J., Martin, K.A.C., 2007. Stereotypical bouton clustering of individual neurons in cat primary visual cortex. J. Neurosci. 27 (45), 12242–12254.

Blakemore, C., 1969. Binocular depth discrimination and the nasotemporal division. J. Physiol. 205, 471–497.

Blakemore, C., Diao, Y.C., Pu, M.L., Wang, Y.K., Xiao, Y.M., 1983. Possible functions of the interhemispheric connexions between visual cortical areas in the cat. J. Physiol. 337, 331–349.

Bloom, J.S., Hynd, G.W., 2005. The role of the corpus callosum in interhemispheric transfer of information: Excitation or inhibition? Neuropsychol. Rev. 15, 59–71.

Bola, M., Gall, C., Sabel, B.A., 2013. "Sightblind": perceptual deficits in the "intact" visual field. Front. Neurol. 4, 80.

Borra, E., Rockland, K.S., 2011. Projections to early visual areas v1 and v2 in the calcarine fissure from parietal association areas in the macaque. Front. Neuroanat. 22 (5), 35.

Borra, E., Ichinohe, N., Sato, T., Tanifuji, M., Rockland, K.S., 2010. Cortical connections to area te in monkey: hybrid modular and distributed organization. Cereb. Cortex 20 (2), 257–270.

Bosking, W.H., Zhang, Y., Schofield, B., Fitzpatrick, D., 1997. Orientation selectivity and the arrangement of horizontal connections in tree shrew striate cortex. J. Neurosci. 17 (6), 2112–2127.

Bosking, W.H., Kretz, R., Pucak, M.L., Fitzpatrick, D., 2000. Functional specificity of callosal connections in tree shrew striate cortex. J. Neurosci. 20 (6), 2346–2359.

Boyd, J., Matsubara, J., 1994. Tangential organization of callosal connectivity in the cat's visual cortex. J. Comp. Neurol. 347 (2), 197–210.

Buhl, E.H., Singer, W., 1989. The callosal projection in cat visual cortex as revealed by a combination of retrograde tracing and intracellular injection. Exp. Brain Res. 75 (3), 470–476.

Bullier, J., Hupé, J.M., James, A.C., Girard, P., 2001. The role of feedback connections in shaping the responses of visual cortical neurons. Prog. Brain. Res. 134, 193–204.

Burkhalter, A., Bernado, K.L., 1989. Organization of corticocortical connections in human visual cortex. Proc. Natl. Acad. Sci. U.S.A. 86, 1071–1075.

Burkhalter, A., Charles, V., 1990. Organization of local axon collaterals of efferent projection neurons in rat visual cortex. J. Comp. Neurol. 302 (4), 920–934.

Buzás, P., Eysel, U.T., Kisvárday, Z.F., 1998. Functional topography of single cortical cells: an intracellular approach combined with optical imaging. Brain Res. Protoc. 3 (2), 199–208.

Buzás, P., Kovács, K., Ferecskó, A.S., Budd, J.M.L., Eysel, U.T., Kisvárday, Z.F., 2006. Model-based analysis of excitatory lateral connections in the visual cortex. J. Comp. Neurol. 499 (6), 861–881.

Callaway, E.M., Katz, L.C., 1990. Emergence and refinement of clustered horizontal connections in cat striate cortex. J. Neurosci. 10 (4), 1134–1153.

Callaway, E.M., Katz, L.C., 1992. Development of axonal arbors of layer 4 spiny neurons in cat striate cortex. J. Neurosci. 12 (2), 570–582.

Caminiti, R., Ghaziri, H., Galuske, R., Hof, P.R., Innocenti, G.M., 2009. Evolution amplified processing with temporally dispersed slow neuronal connectivity in primates. Proc. Natl. Acad. Sci. U.S.A. 106 (46), 19551–19556.

Carmeli, C., Lopez-Aguadao, L., Schmidt, K.E., De Feo, O., Innocenti, G.M., 2007. A novel interhemispheric interaction: modulation of neuronal cooperativity in the visual areas. PLoS ONE 2 (12) <http://doi.org/10.1371/journal.pone.0001287>.

Chance, F.S., Abbott, L.F., Reyes, A.D., 2002. Gain modulation from background synaptic input. Neuron 35 (4), 773–782.

Chapman, B., Bonhoeffer, T., 1998. Overrepresentation of horizontal and vertical orientation preferences in developing ferret area 17. Proc. Natl. Acad. Sci. U.S.A. 95 (5), 2609–2614.

Chavane, F., Sharon, D., Jancke, D., Marre, O., Frégnac, Y., Grinvald, A., 2011. Lateral spread of orientation selectivity in V1 is controlled by intracortical cooperativity. Front. Syst. Neurosci. 5, 4.

Chisum, H.J., Mooser, F., Fitzpatrick, D., 2003. Emergent properties of layer 2/3 neurons reflect the collinear arrangement of horizontal connections in tree shrew visual cortex. J. Neurosci. 23 (7), 2947–2960.

Choudhury, B.P., Whitteridge, D., Wilson, M.E., 1965. The function of the callosal connections of the visual cortex. Q. J. Exp. Physiol. Cogn. Med. Sci. 50, 214–219.

Clarke, P.G., Whitteridge, D., 1976. The cortical visual areas of the sheep. J. Physiol. 256 (3), 497–508.

Conti, F., Manzoni, T., 1994. The neurotransmitters and postsynaptic actions of callosally projecting neurons. Behav. Brain. Res. 64 (1-2), 37–53.

Coppola, D.M., White, L.E., Fitzpatrick, D., Purves, D., 1998. Unequal representation of cardinal and oblique contours in ferret visual cortex. Proc. Natl. Acad. Sci. U. S. A. 95 (5), 2621–2623.

Crook, J.M., Kisvárday, Z.F., Eysel, U.T., 1996. GABA-induced inactivation of functionally characterized sites in cat visual cortex (area 18): effects on direction selectivity. J. Neurophysiol. 75 (5), 2071–2088.

Crook, J.M., Kisvárday, Z.F., Eysel, U.T., 1997. GABA-induced inactivation of functionally characterized sites in cat striate cortex: effects on orientation tuning and direction selectivity. Vis. Neurosci. 14 (1), 141–158.

Crook, J.M., Kisvárday, Z.F., Eysel, U.T., 1998. Evidence for a contribution of lateral inhibition to orientation tuning and direction selectivity in cat visual cortex: reversible inactivation of functionally characterized sites combined with neuroanatomical tracing techniques. Eur. J. Neurosci. 10 (6), 2056–2075.

Cynader, M.S., Gardner, J.C., 1987. Mechanisms for binocular depth sensitivity along the vertical meridian of the visual field, 413, 60–74.

Cynader, M.S., Swindale, N.V., Matsubara, J.A., 1987. Functional topography in cat area 18. J. Neurosci. 7 (5), 1401–1413.

Dragoi, V., Turcu, C.M., Sur, M., 2001. Stability of cortical responses and the statistics of natural scenes. Neuron 32 (6), 1181–1192.

Dreher, B., Cottee, L.J., 1975. Visual receptive-field properties of cells in area 18 of cat's cerebral cortex before and after acute lesions in area 17. J. Neurophysiol. 38 (4), 735–750.

Durack, J.C., Katz, L.C., 1996. Development of horizontal projections in layer 2/3 of ferret visual cortex. Cereb. Cortex. 6 (2), 178–183.

Elberger, A.J., Smith, E.L., 1985. The critical period for corpus callosum section to affect cortical binocularity. Exp. Brain. Res. 57 (2), 213–223.

Felleman, D.J., Xiao, Y., McClendon, E., 1997. Modular organization of occipito-temporal pathways: cortical connections between visual area 4 and visual area 2 and posterior inferotemporal ventral area in macaque monkeys. J. Neurosci. 17 (9), 3185–3200.

Fino, E., Yuste, R., 2011. Dense inhibitory connectivity in neocortex. Neuron 69 (6), 1188–1203.

Fisken, R.A., Garey, L.J., Powell, T.P., 1975. The intrinsic, association and commissural connections of area 17 on the visual cortex. Philos. Trans. R. Soc. Lond. B. Biol. Sci. 272 (919), 487–536.

Fitzpatrick, D., 1996. The functional organization of local circuits in visual cortex: insights from the study of tree shrew striate cortex. Cereb. Cortex 6, 329–341.

Galuske, R.A.W., Singer, W., 1996. The origin and topography of long-range intrinsic projections in cat visual cortex: a developmental study. Cereb. Cortex 6 (3), 417–430.

Galuske, R.A.W., Schmidt, K.E., Goebel, R., Lomber, S.G., Payne, B.R., 2002. The role of feedback in shaping neural representations in cat visual cortex. Proc Natl Acad Sci USA 99, 17083–17088.

Gilbert, C.D., Wiesel, T.N., 1979. Morphology and intracortical projections of functionally characterised neurones in the cat visual cortex. Nature 280 (5718), 120–125.

Gilbert, C.D., Wiesel, T.N., 1983. Clustered intrinsic connections in cat visual cortex. J. Neurosci. 3 (5), 1116–1133.

Gilbert, C.D., Wiesel, T.N., 1989. Columnar specificity of intrinsic horizontal and corticocortical connections in cat visual cortex. J. Neurosci. 9 (7), 2432–2442.

Gilbert, C.D., Sigman, M., 2007. Review brain states : top-down influences in sensory processing. Neuron 54, 677–696.

Gilbert, C.D., Li, W., Piech, V., 2009. Perceptual learning and adult cortical plasticity. J. Physiol. 587 (12), 2743–2751.

Girardin, C.C., Martin, K.A.C., 2009a. Cooling in cat visual cortex: stability of orientation selectivity despite changes in responsiveness and spike width. Neuroscience 164 (2), 777–787.

Girardin, C.C., Martin, K.A.C., 2009b. Inactivation of lateral connections in cat area 17. Eur. J. Neurosci. 29 (10), 2092–2102.

Grigonis, A.M., Rayos del Sol-Padua, R.B., Murphy, E.H., 1992. Visual callosal projections in the adult ferret. Vis. Neurosci. 9 (1), 99–103.

Harvey, M.A., Valentiniene, S., Roland, P.E., 2009. Cortical membrane potential dynamics and laminar firing during object motion. Front. Syst. Neurosci. 3, 7.

Hata, Y., Tsumoto, T., Sato, H., Tamura, H., 1991. Horizontal interactions between visual cortical neurones studied by cross-correlation analysis in the cat. J. Physiol. 441, 593–614.

Hedreen, J.C., Yin, T.C., 1981. Homotopic and heterotopic callosal afferents of caudal inferior parietal lobule in *Macaca mulatta*. J. Comp. Neurol. 197 (4), 605–621.

Horton, J.C., Hubel, D.H., 1981. Regular patchy distribution of cytochrome oxidase staining in primary visual cortex of macaque monkey. Nature 292 (5825), 762–764.

Houzel, J.C., Milleret, C., 1999. Visual inter-hemispheric processing: constraints and potentialities set by axonal morphology. J. Physiol. Paris 93, 271–284.

Houzel, J.C., Milleret, C., Innocenti, G., 1994. Morphology of callosal axons interconnecting areas 17 and 18 of the cat. Eur. J. Neurosci. 6 (6), 898–917.

Houzel, J.C., Carvalho, M.L., Lent, R., 2002. Interhemispheric connections between primary visual areas: beyond the midline rule. Braz. J. Med. Biol. Res. 35 (12), 1441–1453.

Hubel, D.H., Wiesel, T.N., 1967. Cortical and callosal connections concerned with the vertical meridian of visual fields in the cat. J. Neurophysiol. 30 (6), 1561–1573.

Innocenti, G.M., 1986. Postnatal development of corticocortical connections. Ital. J. Neurol. Sci. Suppl. 5, 25–28.

Innocenti, G.M., Fiore, L., 1976. Morphological correlates of visual field transformation in the corpus callosum. Neurosci. Lett. 2 (5), 245–252.

Innocenti, G.M., Manger, P.R., Masiello, I., Colin, I., Tettoni, L., 2002. Architecture and callosal connections of visual areas 17, 18, 19 and 21 in the ferret (*Mustela putorius*). Cereb. Cortex 12 (4), 411–422.

Innocenti, G.M., Vercelli, A., Caminiti, R., 2014. The diameter of cortical axons depends both on the area of origin and target. Cereb. Cortex 24 (8), 2178–2188.

Jancke, D., Erlhagen, W., Schöner, G., Dinse, H.R., 2004. Shorter latencies for motion trajectories than for flashes in population responses of cat primary visual cortex. J. Physiol. 556 (Pt. 3), 971–982. <http://doi.org/10.1113/jphysiol.2003.058941>.

Kalberlah, C., Distler, C., Hoffmann, K.P., 2009. Sensitivity to relative disparity in early visual cortex of pigmented and albino ferrets. Exp. Brain. Res. 192 (3), 379–389.

Karube, F., Kisvárday, Z.F., 2011. Axon topography of layer IV spiny cells to orientation map in the cat primary visual cortex (area 18). Cereb. Cortex 21 (6), 1443–1458.

Kätzel, D., Zemelman, B.V., Buetfering, C., Wölfel, M., Miesenböck, G., 2011. The columnar and laminar organization of inhibitory connections to neocortical excitatory cells. Nat. Neurosci. 14 (1), 100–107.

Kenet, T., Bibitchkov, D., Tsodyks, M., Grinvald, A., Arieli, A., 2003. Spontaneously emerging cortical representations of visual attributes. Nature 425 (6961), 954–956.

Kisvárday, Z.F., Eysel, U.T., 1992. Cellular organization of reciprocal patchy networks in layer III of cat visual cortex (area 17). Neuroscience 46 (2), 275–286.

Kisvárday, Z.F., Eysel, U.T., 1993. Functional and structural topography of horizontal inhibitory connections in cat visual cortex. Eur. J. Neurosci. 5 (12), 1558–1572.

Kisvárday, Z.F., Cowey, A., Somogyi, P., 1986. Synaptic relationships of a type of GABA-immunoreactive neuron (clutch cell), spiny stellate cells and lateral geniculate nucleus afferents in layer IVC of the monkey striate cortex. Neuroscience 19 (3), 741–761.

I. MICROCIRCUITRY

Kisvárday, Z.F., Kim, D.S., Eysel, U.T., Bonhoeffer, T., 1994. Relationship between lateral inhibitory connections and the topography of the orientation map in cat visual cortex. Eur. J. Neurosci. 6 (10), 1619–1632.

Kisvárday, Z.F., Tóth, É., Rausch, M., Eysel, U.T., 1997. Orientation-specific relationship between populations of excitatory and inhibitory lateral connections in the visual cortex of the cat. Cereb. Cortex 7 (7), 605–618.

Kisvárday, Z.F., Ferecskó, A.S., Kovács, K., Buzás, P., Budd, J.M.L., Eysel, U.T., 2002. One axon-multiple functions: specificity of lateral inhibitory connections by large basket cells. J. Neurocytol. 31 (3–5 (special issue)), 255–264.

Lachica, E.A., Beck, P.D., Casagrande, V.A., 1993. Intrinsic connections of layer III of striate cortex in squirrel monkey and bush baby: correlations with patterns of cytochrome oxidase. J. Comp. Neurol. 329 (2), 163–187.

Law, M.I., Zahs, K.R., Stryker, M.P., 1988. Organization of primary visual cortex (area 17) in the ferret. J. Comp. Neurol. 278 (2), 157–180.

Lepore, F., Guillemot, J.P., 1982. Visual receptive field properties of cells innervated through the corpus callosum in the cat. Exp. Brain. Res. 46 (3), 413–424.

Lepore, F., Samson, A., Paradis, M.C., Ptito, M., Guillemot, J.P., 1992. Binocular interaction and disparity coding at the 17-18 border: contribution of the corpus callosum. Exp. Brain. Res. 90 (1), 129–140.

LeVay, S., 1988. Patchy intrinsic projections in visual cortex, area 18, of the cat: morphological and immunocytochemical evidence for an excitatory function. J. Comp. Neurol. 269 (2), 265–274.

Levitt, J.B., Lewis, D.A., Yoshioka, T., Lund, J.S., 1993. Topography of pyramidal neuron intrinsic connections in macaque monkey prefrontal cortex (areas 9 and 46). J. Comp. Neurol. 338 (3), 360–376.

Levitt, J.B., Yoshioka, T., Lund, J.S., 1994. Intrinsic cortical connections in macaque visual area V2: evidence for interaction between different functional streams. J. Comp. Neurol. 342 (4), 551–570.

Li, W., Thier, P., Wehrhahn, C., 2000. Contextual influence on orientation discrimination of humans and responses of neurons in V1 of alert monkeys. J. Neurophysiol. 83 (2), 941–954.

Li, B., Peterson, M.R., Freeman, R.D., 2003. Oblique effect: a neural basis in the visual cortex. J. Neurophysiol. 90, 204–217.

Livingstone, M.S., Hubel, D.H., 1984. Specificity of intrinsic connections in primate primary visual cortex. J. Neurosci. 4 (11), 2830–2835.

Livingstone, M.S., Hubel, D.H., 1987. Connections between layer 4B of area 17 and the thick cytochrome oxidase stripes of area 18 in the squirrel monkey. J. Neurosci. 7 (11), 3371–3377.

Löwel, S., Singer, W., 1992. Selection of intrinsic horizontal connections in the visual cortex by correlated neuronal activity. Science 255 (5041), 209–212.

Lübke, J., Albus, K., 1992. Rapid rearrangement of intrinsic tangential connections in the striate cortex of normal and dark-reared kittens: lack of exuberance beyond the second postnatal week. J. Comp. Neurol. 323 (1), 42–58.

Luhmann, H.J., Prince, D.A., 1991. Postnatal maturation of the GABAergic system in rat neocortex. J. Neurophysiol. 65 (2), 247–263.

Luhmann, H.J., Martínez Millán, L., Singer, W., 1986. Development of horizontal intrinsic connections in cat striate cortex. Exp. Brain. Res. 63 (2), 443–448.

Lund, J.S., Yoshioka, T., Levitt, J.B., 1993. Comparison of intrinsic connectivity in different areas of macaque monkey cerebral cortex. Cereb. Cortex. 3 (2), 148–162.

Makarov, V. a, Schmidt, K.E., Castellanos, N.P., Lopez-Aguado, L., Innocenti, G.M., 2008. Stimulus-dependent interaction between the visual areas 17 and 18 of the 2 hemispheres of the ferret (Mustela putorius). Cereb. Cortex 18 (8), 1951–1960.

Malach, R., Amir, Y., Harel, M., Grinvald, A., 1993. Relationship between intrinsic connections and functional architecture revealed by optical imaging and in vivo targeted biocytin injections in primate striate cortex. Proc. Natl. Acad. Sci. U.S.A. 90 (22), 10469–10473.

Malach, R., Tootell, R.B., Malonek, D., 1994. Relationship between orientation domains, cytochrome oxidase stripes, and intrinsic horizontal connections in squirrel monkey area V2. Cereb. Cortex 4 (2), 151–165.

Malach, R., Schirman, T.D., Harel, M., Tootell, R.B.H., Malonek, D., 1997. Organization of intrinsic connections in owl monkey area MT. Cereb. Cortex 7 (4), 386–393.

Manzoni, T., Barbaresi, P., Conti, F., Fabri, M., 1989. The callosal connections of the primary somatosensory cortex and the neural bases of midline fusion. Exp. Brain. Res. 76 (2), 251–266.

Martin, K.A., Whitteridge, D., 1984. Form, function and intracortical projections of spiny neurones in the striate visual cortex of the cat. J. Physiol. 353, 463–504.

Martin, K.A.C., Somogyi, P., Whitteridge, D., 1983. Physiological and morphological properties of identified basket cells in the cat's visual cortex. Exp. Brain. Res. 50 (2–3), 193–200.

Martin, K.A.C., Roth, S., Rusch, E.S., 2014. Superficial layer pyramidal cells communicate heterogeneously between multiple functional domains of cat primary visual cortex. Nat. Commun. 5, 5252.

Matsubara, J.A., Cynader, M.S., Swindale, N.V., 1987. Anatomical properties and physiological correlates of the intrinsic connections in cat area 18. J. Neurosci. 7 (5), 1428–1446.

Matsubara, J. A., 1988. Local, horizontal connections within area 18 of the cat. Prog. Brain Res. 75, 163–72.

McAdams, C.J., Maunsell, J.H., 1999. Effects of attention on orientation-tuning functions of single neurons in macaque cortical area V4. J. Neurosci. 19 (1), 431–441.

McGuire, B.A., Gilbert, C.D., Rivlin, P.K., Wiesel, T.N., 1991. Targets of horizontal connections in macaque primary visual cortex. J. Comp. Neurol. 305 (3), 370–392.

Milleret, C., Buser, P., Watroba, L., 2005. Unilateral paralytic strabismus in the adult cat induces plastic changes in interocular disparity along the visual midline: contribution of the corpus callosum. Vis. Neurosci. 22 (3), 325–343.

Minciacchi, D., Antonini, A., 1984. Binocularity in the visual cortex of the adult cat does not depend on the integrity of the corpus callosum. Behav. Brain. Res. 13 (2), 183–192.

Mitchell, D.E., Blakemore, C., 1970. Binocular depth perception and the corpus callosum. Vision. Res. 10 (1), 49–54.

Mitchison, G., Crick, F., 1982. Long axons within the striate cortex: their distribution, orientation, and patterns of connection. Proc. Natl. Acad. Sci. U.S.A. 79 (11), 3661–3665.

Montero, V.M., 1980. Patterns of connections from the striate cortex to cortical visual areas in superior temporal sulcus of macaque and middle temporal gyrus of owl monkey. J. Comp. Neurol. 189 (1), 45–59.

Muir, D.R., Da Costa, N. M. a, Girardin, C.C., Naaman, S., Omer, D.B., Ruesch, E., et al., 2011. Embedding of cortical representations by the superficial patch system. Cereb. Cortex 21 (10), 2244–2260. <http://doi.org/10.1093/cercor/bhq290>.

Nakamura, H., Gattass, R., Desimone, R., Ungerleider, L.G., 1993. The modular organization of projections from areas V1 and V2 to areas V4 and TEO in macaques. J. Neurosci. 13 (9), 3681–3691.

Nasr, S., Tootell, R.B.H., 2012. A Cardinal orientation bias in scene-selective visual cortex. J. Neurosci. 32 (43), 14921–14926.

Nauhaus, I., Busse, L., Carandini, M., Ringach, D.L., 2009. Stimulus contrast modulates functional connectivity in visual cortex. Nat. Neurosci. 12 (1), 70–76.

Olavarria, J.F., 2001. Callosal connections correlate preferentially with ipsilateral cortical domains in cat areas 17 and 18, and with contralateral domains in the 17/18 transition zone. J. Comp. Neurol. 433 (4), 441–457.

Olavarria, J., Van Sluyters, R.C., 1985. Organization and postnatal development of callosal connections in the visual cortex of the rat. J. Comp. Neurol. 239 (1), 1–26.

Olivares, R., Montiel, J., Aboitiz, F., 2001. Species differences and similarities in the fine structure of the mammalian corpus callosum. Brain. Behav. Evol. 57 (2), 98–105.

Paramei, G.V., Sabel, B.A., 2008. Contour-integration deficits on the intact side of the visual field in hemianopia patients. Behav. Brain. Res. 188 (1), 109–124.

Payne, B.R., 1990. Function of the corpus callosum in the representation of the visual field in cat visual cortex. Vis. Neurosci. 5 (2), 205–211.

Payne, B.R., 1994. Neuronal interactions in cat visual cortex mediated by the corpus callosum. Behav. Brain. Res. 64, 55–64.

Payne, B.R., Siwek, D.F., 1991a. The visual map in the corpus callosum of the cat. Cereb. Cortex. 1 (2), 173–188.

Payne, B.R., Siwek, D.F., 1991b. Visual-field map in the callosal recipient zone at the border between areas 17 and 18 in the cat. Vis. Neurosci. 7 (3), 221–236.

Payne, B.R., Lomber, S.G., 2003. Quantitative analyses of principal and secondary compound parieto-occipital feedback pathways in cat. Exp. Brain. Res. 152 (4), 420–433.

Payne, B.R., Elberger, A.J., Berman, N., Murphy, E.H., 1980. Binocularity in the cat visual cortex is reduced by sectioning the corpus callosum. Science 207 (4435), 1097–1099.

Payne, B.R., Pearson, H.E., Berman, N., 1984. Deafferentation and axotomy of neurons in cat striate cortex: time course of changes in binocularity following corpus callosum transection. Brain. Res. 307 (1–2), 201–215.

Payne, B.R., Siwek, D.F., Lomber, S.G., 1991. Complex transcallosal interactions in visual cortex. Vis. Neurosci. 6, 283–289.

Peiker, C., Wunderle, T., Eriksson, D., Schmidt, A., Schmidt, K.E., 2013. An updated midline rule: visual callosal connections anticipate shape and motion in ongoing activity across the hemispheres. J. Neurosci. 33 (46), 18036–18046.

Pietrasanta, M., Restani, L., Caleo, M., 2012. The corpus callosum and the visual cortex: plasticity is a game for two. Neural Plast. 10p. <http://dx.doi.org/10.1155/2012/838672>.

Polat, U., 1999. Functional architecture of long-range perceptual interactions. Spat. Vis. 12 (2), 143–162.

Price, D.J., 1986. The postnatal development of clustered intrinsic connections in area 18 of the visual cortex in kittens. Brain. Res. 389 (1–2), 31–38.

Price, D.J., Ferrer, J.M., Blakemore, C., Kato, N., 1994. Functional organization of corticocortical projections from area 17 to area 18 in the cat's visual cortex. J. Neurosci. 14 (5 Pt 1), 2732–2746.

Rochefort, N.L., Buzás, P., Kisvárday, Z.F., Eysel, U.T., Milleret, C., 2007. Layout of transcallosal activity in cat visual cortex revealed by optical imaging. NeuroImage 36 (3), 804–821.

Rochefort, N.L., Buzás, P., Quenech'Du, N., Koza, A., Eysel, U.T., Milleret, C., et al., 2009. Functional selectivity of interhemispheric connections in cat visual cortex. Cereb. Cortex 19 (10), 2451–2465.

Rockland, K.S., 1985. Anatomical organization of primary visual cortex (area 17) in the ferret. J. Comp. Neurol. 241, 225–236.

Rockland, K.S., 2013. Collateral branching of long-distance cortical projections in monkey. J. Comp. Neurol. 521 (18), 4112–4123.

Rockland, K.S., Drash, G.W., 1996. Collateralized divergent feedback connections that target multiple cortical areas. J. Comp. Neurol. 373 (4), 529–548.

Rockland, K.S., Knutson, T., 2000. Feedback connections from area MT of the squirrel monkey to areas V1 and V2. J. Comp. Neurol. 425 (3), 345–368.

Rockland, K.S., Lund, J.S., 1982. Widespread periodic intrinsic connections in the tree shrew visual cortex. Science 215 (4539), 1532–1534.

Rockland, K.S., Lund, J.S., 1983. Intrinsic laminar lattice connections in primate visual cortex. J. Comp. Neurol. 216 (3), 303–318.

Rockland, K.S., Pandya, D.N., 1981. Cortical connections of the occipital lobe in the rhesus monkey: interconnections between areas 17, 18, 19 and the superior temporal sulcus. Brain. Res. 212 (2), 249–270.

Rockland, K.S., Lund, J.S., Humphrey, A.L., 1982. Anatomical binding of intrinsic connections in striate cortex of tree shrews (*Tupaia glis*). J. Comp. Neurol. 209 (1), 41–58.

Ruthazer, E.S., Stryker, M.P., 1996. The role of activity in the development of long-range horizontal connections in area 17 of the ferret. J. Neurosci. 16 (22), 7253–7269.

Schmidt, K.E., Löwel, S., 2002. Long-range intrinsic connections in cat primary visual cortex. In: Peters, A., Payne, B.R. (Eds.), The cat primary visual cortex. Academic Press, San Diego, CA, pp. 387–426.

Schmidt, K.E., Goebel, R., Löwel, S., Singer, W., 1997a. The perceptual grouping criterion of collinearity is reflected by anisotropies of connections in the primary visual cortex. Eur. J. Neurosci. 9, 1083–1089.

Schmidt, K.E., Kim, D.S., Singer, W., Bonhoeffer, T., Löwel, S., 1997b. Functional specificity of long-range intrinsic and interhemispheric connections in the visual cortex of strabismic cats. J. Neurosci. 17 (14), 5480–5492.

Schmidt, K.E., Lomber, S.G., Innocenti, G.M., 2010. Specificity of neuronal responses in primary visual cortex is modulated by interhemispheric corticocortical input. Cereb. Cortex 20 (12), 2776–2786.

Schmidt, K.E., Lomber, S.G., Payne, B.R., Galuske, R.A.W., 2011. Pattern motion representation in primary visual cortex is mediated by transcortical feedback. NeuroImage 54 (1), 474–484.

Segraves, M.A., Rosenquist, A.C., 1982a. The afferent and efferent callosal connections of retinotopically defined areas in cat cortex. J. Neurosci. 2 (8), 1090–1107.

I. MICROCIRCUITRY

Segraves, M.A., Rosenquist, A.C., 1982b. The distribution of the cells of origin of callosal projections in cat visual cortex. J. Neurosci. 2 (8), 1079–1089.

Shoham, D., Hubener, M., Schulze, S., Grinvald, A., Bonhoeffer, T., 1997. Spatio-temporal frequency domains and their relation to cytochrome oxidase staining in cat visual cortex. Nature 385 (6616), 529–533.

Sincich, L.C., Blasdel, G.G., 2001. Oriented axon projections in primary visual cortex of the monkey. J. Neurosci. 21 (12), 4416–4426.

Sincich, L.C., Horton, J.C., 2002. Divided by cytochrome oxidase: a map of the projections from V1 to V2 in macaques. Science 295 (5560), 1734–1737.

Singer, W., Gray, C.M., 1995. Visual feature integration and the temporal correlation hypothesis. Annu. Rev. Neurosci 18, 555–586.

Skottun, B.C., Grosof, D.H., De Valois, R.L., 1988. Responses of simple and complex cells to random dot patterns: a quantitative comparison. J. Neurophysiol. 59 (6), 1719–1735.

Somogyi, P., Kisvárday, Z.F., Martin, K.A.C., Whitteridge, D., 1983. Synaptic connections of morphologically identified and physiologically characterized large basket cells in the striate cortex of cat. Neuroscience 10 (2), 261–294.

Sompolinsky, H., Shapley, R., 1997. New perspectives on the mechanisms for orientation selectivity. Curr. Opin. Neurobiol 7, 514–522.

Stepanyants, A., Martinez, L.M., Ferecskó, A.S., Kisvárday, Z.F., 2009. The fractions of short- and long-range connections in the visual cortex. Proc. Natl. Acad. Sci. U.S.A. 106 (9), 3555–3560.

Stettler, D.D., Das, A., Bennett, J., Gilbert, C.D., 2002. Lateral connectivity and contextual interactions in macaque primary visual cortex. Neuron 36 (4), 739–750.

Stimberg, M., Wimmer, K., Martin, R., Schwabe, L., Mariño, J., Schummers, J., et al., 2009. The operating regime of local computations in primary visual cortex. Cereb. Cortex 19 (9), 2166–2180.

Stone, J., 1978. The number and distribution of ganglion cells in the cat's retina. J. Comp. Neurol. 180 (4), 753–771.

Sun, J.B., Li, B., Ma, M.H., Diao, Y.C., 1994. Transcallosal circuitry revealed by blocking and disinhibiting callosal input in the cat. Vis. Neurosci. 11, 189–197.

Swadlow, H.A., Kocsis, J.D., Waxman, S.G., 1980. Modulation of impulse conduction along the axonal tree. Annu. Rev. Biophys. Bioeng. 9, 143–179.

Tigges, J., Tigges, M., Anschel, S., Cross, N.A., Letbetter, W.D., McBride, R.L., 1981. Areal and laminar distribution of neurons interconnecting the central visual cortical areas 17, 18, 19, and MT in squirrel monkey (saimiri). J. Comp. Neurol. 202 (4), 539–560.

Tomasi, S., Caminiti, R., Innocenti, G.M., 2012. Areal differences in diameter and length of corticofugal projections. Cereb. Cortex 22 (6), 1463–1472.

Tootell, R.B., Switkes, E., Silverman, M.S., Hamilton, S.L., 1988. Functional anatomy of macaque striate cortex. II. Retinotopic organization. J. Neurosci. 8 (5), 1531–1568.

Treue, S., Martínez Trujillo, J.C., 1999. Feature-based attention influences motion processing gain in macaque visual cortex. Nature 399 (6736), 575–579.

Ts'o, D.Y., Gilbert, C.D., Wiesel, T.N., 1986. Relationships between horizontal interactions and functional architecture in cat striate cortex as revealed by cross-correlation analysis. J. Neurosci. 6 (4), 1160–1170.

Tusa, R.J., Palmer, L.A., Rosenquist, A.C., 1978. The retinotopic organization of area 17 (striate cortex) in the cat. J. Comp. Neurol. 177 (2), 213–235.

Van Hooser, S.D., Heimel, J.A., Chung, S., Nelson, S.B., 2006. Lack of patchy horizontal connectivity in primary visual cortex of a mammal without orientation maps. J. Neurosci. 26 (29), 7680–7692.

Voigt, T.I., LeVay, S., Stamnes, M.A., 1988. Morphological and immunocytochemical observations on the visual callosal projections in the cat. J. Comp. Neurol. 272 (3), 450–460.

Wang, C.-L., Zhang, L., Zhou, Y., Zhou, J., Yang, X.-J., Duan, S., et al., 2007. Activity-dependent development of callosal projections in the somatosensory cortex. J. Neurosci. 27 (42), 11334–11342.

Wang, C., Waleszczyk, W.J., Burke, W., Dreher, B., 2000. Modulatory influence of feedback projections from area 21a on neuronal activities in striate cortex of the cat. Cereb. Cortex 10 (12), 1217–1232.

Wang, G., Ding, S., Yunokuchi, K., 2003. Difference in the representation of cardinal and oblique contours in cat visual cortex. Neurosci. Lett. 338, 77–81.

Wang, G., Nagai, M., Okamura, J., 2011. Orientation dependency of intrinsic optical signal dynamics in cat area 18. NeuroImage 57 (3), 1140–1153.

White, L.E., Bosking, W.H., Williams, S.M., Fitzpatrick, D., 1999. Maps of central visual space in ferret V1 and V2 lack matching inputs from the two eyes. J. Neurosci. 19 (16), 7089–7099.

Wong-Riley, M., 1979. Columnar cortico-cortical interconnections within the visual system of the squirrel and macaque monkeys. Brain. Res. 162 (2), 201–217.

Wörgötter, F., Eysel, U.T., 1989. Axis of preferred motion is a function of bar length in visual cortical receptive fields. Exp. Brain. Res. 76 (2), 307–314.

Wunderle, T., Eriksson, D., Schmidt, K.E., 2013. Multiplicative mechanism of lateral interactions revealed by controlling interhemispheric input. Cereb. Cortex 23 (4), 900–912.

Wunderle, T., Eriksson, D., Peiker, C., Schmidt, K.E., 2015. Input and output gain modulation by the lateral interhemispheric network in early visual cortex. J. Neurosci. 35 (20), 7682–7749.

Yacoub, E., Harel, N., Ugurbil, K., 2008. High-field fMRI unveils orientation columns in humans. Proc. Natl. Acad. Sci. U.S.A. 105 (30), 10607–10612.

Yinon, U., Chen, M., Zamir, S., Gelerstein, S., 1988. Corpus callosum transection reduces binocularity of cells in the visual cortex of adult cats. Neurosci. Lett. 92 (3), 280–284.

Yoshioka, T., Levitt, J.B., Lund, J.S., 1992. Intrinsic lattice connections of macaque monkey visual cortical area V4. J. Neurosci. 12 (7), 2785–2802.

Yoshioka, T., Blasdel, G.G., Levitt, J.B., Lund, J.S., 1996. Relation between patterns of intrinsic lateral connectivity, ocular dominance, and cytochrome oxidase-reactive regions in macaque monkey striate cortex. Cereb. Cortex 6 (2), 297–310.

Yousef, T., Bonhoeffer, T., Kim, D.S., Eysel, U.T., Toth, E., Kisvárday, Z.F., 1999. Orientation topography of layer 4 lateral networks revealed by optical imaging in cat visual cortex (area 18). Eur. J. Neurosci. 11 (12), 4291–4308.

Yousef, T., Tóth, E., Rausch, M., Eysel, U.T., Kisvárday, Z.F., 2001. Topography of orientation centre connections in the primary visual cortex of the cat. Neuroreport 12 (8), 1693–1699.

Zeki, S., Shipp, S., 1989. Modular connections between areas V2 and V4 of macaque monkey visual cortex. Eur. J. Neurosci. 1 (5), 494–506.

9

Neuronal Cell Types in the Neocortex

Rajeevan T. Narayanan[1], Robert Egger[1,2],
Christiaan P.J. de Kock[3] and Marcel Oberlaender[1,4,5]

[1]Computational Neuroanatomy Group, Max Planck Institute for Biological Cybernetics, Tuebingen, Germany [2]Graduate School of Neural Information Processing, University of Tuebingen, Tuebingen, Germany [3]Center for Neurogenomics and Cognitive Research, Neuroscience Campus Amsterdam, VU University Amsterdam, The Netherlands [4]Digital Neuroanatomy, Max Planck Florida Institute for Neuroscience, Jupiter, FL, USA [5]Bernstein Center for Computational Neuroscience, Tuebingen, Germany

9.1 BACKGROUND

The mammalian neocortex is associated with many functions, ranging from sensory perception and motor control, to memory, learning, and cognition. These functions are implemented within highly specialized and complex networks, established by millions of synaptically interconnected neurons.

One of the major challenges in modern neuroscience is thus to identify structural organization principles of the cortical circuitry and to relate these principles to specific cortex functions. However, even within functionally well-defined cortical areas, such as the primary sensory cortices, neurons display highly diverse physiological and morphological properties. Quantitative descriptions of cortical structure and function at cellular levels would thus require tools that allow monitoring signal flow within every neuron at the subcellular resolution and millisecond precision, combined with complete reconstructions of the underlying synaptic wiring diagram. Although the acquisition and analysis of such "dense" cellular structure–function datasets are experimentally impractical, *the concept of neuronal cell types—as the primary building block of neuronal networks—*may instead provide the appropriate level of description for unraveling relationships between cortex structure and function.

As pioneered by Ramon y Cajal more than a century ago, neuroscientists aim to group neurons that share common properties into cell types. Classification schemes to define such cell types remain diverse, ranging from morphological features, such as soma and/or dendrite shape, to biophysical and electrophysiological properties, and ultimately to genetic and molecular profiles (e.g., Ascoli et al., 2008; DeFelipe et al., 2013; Kepecs and Fishell, 2014; Santana et al., 2013; Tripathy et al., 2015). To date, no coherent classification scheme has been established, which unambiguously relates morphological with physiological and molecular/genetic cell types. Hence, one of the "holy grails" of neuroscience research is to provide answers to the questions: *Is it possible to define a set of cell types that is sufficient to describe the organizational principles of cortex and relate structure to function? What are the crucial neuronal features that allow defining such a minimal set of cell types?*

9.2 CELL TYPE CLASSIFICATION BY NEURON MORPHOLOGY

A widely used approach to group neurons into cell types is based on common morphological properties (e.g., Binzegger et al., 2004; Brown and Hestrin, 2009a; Helmstaedter et al., 2013). The assumption is that neurons sharing soma location and dendritic features can be expected to receive similar synaptic input patterns, which may result in common stimulus-evoked activity patterns; comparably, neurons that have similar local and long-range axonal projections might be expected to transmit common stimulus-evoked activity patterns to postsynaptic targets. In the following, we provide a historically motivated overview of how soma, dendrite, and axon morphologies were used to define neuronal cell types.

9.2.1 Somatic Level

One of the classical ways to discriminate neuronal cell types is based on comparing soma sizes and shapes, typically in Nissl-stained coronal brain sections. Along the vertical cortex axis (i.e., from the pial surface toward the white matter), Nissl-stained somata display various shapes, resembling pyramids, ovoids, or spheres, of different sizes. The different soma sizes correlate with systematic changes in soma densities. These vertical soma density gradients gave rise to the *concept of cytoarchitectonic layers* (Brodmann, 1909).

Even though cortical layers are purely defined by vertical gradients in soma size, shape, or density (which are usually ill defined), this concept has been widely accepted to provide a first-order criterion to discriminate between neuronal cell types. Countless physiological, morphological, and molecular studies have grouped data by laminar location and provided layer-specific analyses. However, cortical layers do not have discrete boundaries and are populated by genetically, biophysically, physiologically, and morphologically diverse neurons. For accurate classification on (sub)cellular level, we argue that the preferred approach is to obtain as many parameters as possible including soma location, dendritic, and axonal architecture and ideally physiological characteristics.

9.2.2 Dendritic Level

The first subclassification for cortical neurons involves excitatory (spiny) versus inhibitory (aspiny) neurons. Subclassification of inhibitory neurons is a topic of hot debate and consensus has not been reached (DeFelipe et al., 2013; Kepecs and Fishell, 2014; Rudy et al., 2011). In contrast, subclassification of excitatory neurons is relatively straightforward and the next level of subclassification is grouping excitatory neurons as pyramidal or nonpyramidal neurons. Pyramidal neurons (i.e., pyramid-shaped somata) represent about 70–85% of all cortical neurons (see DeFelipe and Farinas, 1992 for review). They are located throughout all cortical layers except layer 1 and possess dendritic features that discriminate them from neurons with nonpyramidal somata. Nonpyramidal excitatory neurons (spiny stellates) are, for example, found in layer 4 of primary sensory cortices.

Dendrite morphologies of pyramidal and nonpyramidal neurons display a large variety of shapes, some of which represent well-established cell types, such as L5 thick-tufted pyramids, or reflect more sparsely occurring cell types (L5 pyramids with multiple apical dendrites (Hubener et al., 1990), L6 pyramids (see Thomson, 2010 for a review) with a large variety of soma shapes such as ovoid, triangular, or fusiform, inverted and/or bipolar apical dendrites (Bueno-Lopez et al., 1991; de Lima et al., 1990; Peters and Regidor, 1981). Such morphological properties are commonly used to classify excitatory neurons into broad dendritic cell types. However, a nomenclature that is generalizable across cortical areas and/or species has, so far, not been established.

With the development of the patch-clamp technique (Hamill et al., 1981) and improved neuron labeling techniques, such as intracellular labeling with horseradish peroxidase (Horikawa and Armstrong, 1988) or biocytin (Mishra et al., 2010; Pinault, 1996), the large diversity of dendritic shapes within and across layers became a prominent subject in neuroscience research. To date, most of what is known about electrophysiological properties of individual neurons, and how these properties relate to morphological features, originates from recording/labeling approaches in acute brain slices *in vitro*.

The ability to label individual neurons *in vitro*, combined with computer-aided manual tracing systems (such as Neurolucida, Microbrightfield, USA), then allowed reconstructing the 2D (and later 3D) morphology (reviewed in Parekh and Ascoli, 2013; Svoboda, 2011). The digitization of neuron morphologies marks an important step toward more quantitative assessments of morphological cell types. For example, digital reconstructions allowed extracting a variety of morphological features, such as the total dendritic path length, the number of bifurcations, or the length/width of the apical tuft dendrites. Such features were then used to perform supervised or unsupervised cluster analyses to group neurons into objectively determined morphological cell types (e.g., Lubke et al., 2003; Oberlaender et al., 2012a; van Aerde and Feldmeyer, 2013).

Dendrite morphologies, similar to the ones stated above, have been reported in a variety of cortical areas and species. It remains however unknown, which features of the dendrite morphologies are preserved across areas and species, and thus which features will allow defining generalizable dendritic cell classes. Due to the vast amount of *in vitro* labeled dendrite morphologies, neuroinformatics-based approaches thus aim to provide such generalizable principles of dendrite morphologies. The most prominent examples are (i) Neuromorpho.org (Ascoli et al., 2007), an online database containing thousands of *in vitro* labeled dendrite morphologies acquired in different species, brain areas, and across multiple different laboratories, and (ii) the recently launched "BigNeuron community effort" (alleninstitute.org/bigneuron/about/).

Using the Neuromorpho.org dataset, recent approaches thus aimed to provide a general classification scheme of dendritic cell types (Polavaram et al., 2014). However, multiple differences in experimental design across laboratories, such as species, slice thickness, and orientation, result in different degrees of truncating the morphologies. Thus, cut-off dendritic branches and the varying quality of manual tracings result in systematic errors of morphological features that may exceed the biological variability. It remains therefore questionable whether the neuroscience community has generated sufficient complete data to solve the challenge of cell type identification.

9.2.3 Axonal Level

Another widely used approach to classify neurons is based on common postsynaptic targets, i.e., local and long-range axonal projection patterns. This approach is especially influenced by *in vitro* investigations and in particular translaminar, vertical axon profiles (i.e., across layers) are used to distinguish between axonal cell types (Harris and Shepherd, 2015). The focus on vertically projecting intracortical axons is however primarily due to the long-standing dominance of the *columnar concept*. Based on the seminal work of Mountcastle (Mountcastle, 1957; Powell and Mountcastle, 1959), and later on by Hubel and Wiesel (1977), it was shown that vertically aligned cortical neurons respond to common peripheral stimuli, and that these receptive fields change gradually along the horizontal cortex axes. Cortical columns have thus been suggested as the elementary functional building block of cortex (Gilbert and Wiesel, 1979, 1989; Markram, 2008), and the cortical microcircuitry is commonly described by horizontal layers, and vertical, translaminar processing pathways within such columnar units (Feldmeyer et al., 2013; Binzegger et al., 2004; Hill et al., 2012).

However, increasing anatomical and physiological evidence questions the validity and/or completeness of describing cortex by translaminar, vertical pathways. The pioneering work by Gilbert and Wiesel was one of the first to visualize extensive horizontal axons (Gilbert and Wiesel, 1979), which travel lateral distances of multiple millimeters without entry into the white matter. Such pathways were termed long-range *intrinsic pathways* (i.e., extrinsic: via the white matter). It was hence evident—early on—that the elementary functional unit of cortex, the cortical column, is connected both vertically across cytoarchitectonic layers and horizontally to neighboring columns.

For instance, bulk injections of anterograde tracers into a barrel column (Woolsey and Van der Loos, 1970) of the vibrissal part of rodent primary somatosensory cortex revealed horizontal projections that give rise to an hourglass-shaped volume that asymmetrically interconnects columns representing facial whiskers of the same row along the snout (Bernardo et al., 1990). Asymmetries in the distribution of intrinsic, horizontal pathways were also reported in other cortical areas, such as the motor cortex of the monkey (Huntley and Jones, 1991), cat (Keller, 1993; Keller and Asanuma, 1993), and rat (Weiss and Keller, 1994), where horizontal axons project primarily in an anterior–posterior orientation, linking regions related to the activation of related muscle groups.

Detailed understanding of the organizational principles that underlie area-specific intrinsic horizontal projection patterns will thus be key for unraveling common and/or specific mechanisms of cortical stimulus representation and multistimuli integration.

In summary, despite more than a century of investigating organizational principles of cortex at somatic, dendritic, and axonal levels, there is no rigorous scheme for unambiguous morphological cell type classification. The power of predicting dendrite shapes or even intrinsic/extrinsic axonal projections from soma shapes and/or laminar locations remains limited. The primary reason is that to date, as embarrassing as it is, not a single cortical projection neuron has been reconstructed in its entirety (Svoboda, 2011), i.e., including its (i) *exact 3D soma location within cortex*, (ii) *complete 3D dendrite pattern*, (iii) *intrinsic (vertical and horizontal) axon projections*, and (iv) *extrinsic axonal pathways*. However, we argue that this should not prevent us from generating such a scheme. Therefore, in this chapter, we will provide quantitative information about features 1–3 for a dataset of *in vivo* labeled morphologies that comprises about 1% of all excitatory neurons located within a cortical barrel column of rat vibrissal cortex (Narayanan et al., 2015; Oberlaender et al., 2012a). We further discuss strategies to obtain information on the long-range targets of the respective *in vivo* labeled and reconstructed morphologies. Thus, we provide the present state of the art toward generating a coherent morphological classification scheme for excitatory cortical neurons.

9.3 CELL TYPE CLASSIFICATION BY NEURON PHYSIOLOGY

In addition to morphological features, electrophysiological characteristics can also be used to classify neurons (Mainen and Sejnowski, 1995; Tripathy et al., 2015). Electrophysiological characterization focuses around the following themes. First, neuronal responses, as measured at the soma, arise from complex spatiotemporal synaptic input patterns. The general motivation is that functionally comparable neurons have similar responses to the same stimulus (e.g., sensory input), most likely reflecting similar synaptic input patterns. This may suggest a *potential link between soma location and dendritic morphology on the one hand and stimulus-evoked activity on the other hand*. However, somatic responses do not reflect the respective synaptic input patterns in a linear manner. Instead, a large variety of (cell type-specific) passive and active, voltage- and ligand-gated ion channels, located along the dendritic branches, process the synaptic input patterns (for a review, see London and Hausser, 2005). Neuronal responses *in vivo* thus represent the *complex interplay between biophysical mechanisms and spatiotemporal synaptic input patterns*. Physiological classification schemes can thus be based not only on common receptive fields but also on the biophysical properties of individual neurons.

9.3.1 Biophysical Level

The standard approach to assess the biophysical properties of individual cortical neurons is to perform patch-clamp recordings at the soma and/or dendrites of the neuron in acute brain slices *in vitro* (Connors et al., 1982). This allows direct measurement of passive membrane properties and determination of ion channel types along the dendrites. This led to the classification of neurons based on their action potential response to somatic current injections.

Comparable to the major division of morphological types in cortex (excitatory vs. inhibitory), individual neurons in cortex can be segregated into regular-spiking units (RSU) or fast-spiking units (FSU), and this subclassification typically goes hand-in-hand with the two major morphological classes (excitatory = RSU, inhibitory = FSU). However, there are again many exceptions to this general rule and an unambiguous classification scheme based on spiking patterns is also yet to be developed.

A valuable approach was recently developed in an attempt to bridge electrophysiological measurements with neuronal reconstruction by using computational modeling (Druckmann et al, 2007; Hay and Segev, 2014). More specifically, the group of Idan Segev developed a standard procedure that allows converting *in vitro* recorded neurons into biophysically detailed full-compartmental computer models (Druckmann et al., 2007). Simulating the previously measured current injection stimuli within the full-compartmental model, the distribution of the various ion channel types along the dendrites is optimized until the *in vitro* measured responses are recovered in the simulation. This approach promises *to provide one of the missing links to connect dendritic cell types with biophysical and ultimately physiological cell types*.

Nevertheless, it is unlikely that classification of neurons with respect to their action potential responses to somatic current injection *in vitro* is predictive of their responses to complex spatiotemporal synaptic input patterns *in vivo*. In order to understand physiological responses *in vivo*, it is therefore necessary to combine measurements of dendrite morphology and spatial organization of synaptic inputs, with the subcellular distribution of different ion channels resulting from modeling studies as described earlier.

9.3.2 Receptive Field Level

Classification of neurons with respect to their response properties *in vivo* is usually performed by determining their *receptive field*. In sensory cortices, where the description of cortical neuronal receptive fields originates (Mountcastle, 1957), the hypothesis was put forward that *"neurons with similar receptive fields are aligned vertically throughout all layers in a columnar fashion"* (Hubel and Wiesel, 1962; Mountcastle, 1957). As a result, the soma location along the vertical axis is thus by approximation predictive of function and can be used for cell type classification.

Receptive fields of cortical neurons have been thought to arise from thalamocortical inputs, which have specific receptive fields themselves (e.g., Brecht and Sakmann, 2002). However cortical receptive fields in general are broader than receptive fields of thalamocortical neurons (de Kock et al., 2007; Kwegyir-Afful et al., 2005). This may be the result of specific organization of intracortical circuits. In the hierarchy of parameters that can be used for cell type classification, the receptive field of a neuron may be regarded as one of the lowest level parameters, because it is determined by a combination of all the other parameters discussed in this chapter (soma location, dendritic and axonal architecture, synaptic input, and biophysical profiles).

9.4 CELL TYPE CLASSIFICATION BY MOLECULAR/GENETIC PROFILES

Neurons may also be grouped into different cell types based on the *expression of different genetic markers or proteins*. Classically, the basic categories of excitatory and inhibitory neurons were established by the expression of GABA- (an inhibitory neurotransmitter) and glutamate- (an excitatory neurotransmitter) synthesizing proteins (Conti et al., 1987; Jones, 1986; Ribak, 1978).

Recent investigations have proposed that inhibitory interneurons can be grouped into three mutually exclusive classes based on expression of the calcium binding protein *parvalbumin*, the neuropeptide *somatostatin*, and the serotonin receptor *5HT3a* (Rudy et al., 2011). However, these groups do not show clear correlation with cell type classification based on morphology of dendrites and/or axons (Munoz et al., 2014), or with classification based on biophysical profiles or receptive field properties.

Although molecular and/or genetic markers are a potentially powerful tool for investigating and manipulating cortical neurons, it is not yet clear where they belong in the hierarchy of parameters used for determining neuronal cell types. Rather, the appropriate level of cell type-assignment using these methods strongly depends on the scientific question studied, for example, molecular markers that are relevant for development and assembly of cortical networks may differ from molecular markers involved in determining the distribution of different ion channels on dendrites and axons (Hevner et al., 2003).

9.5 CELL TYPE CLASSIFICATION IN RAT BARREL CORTEX

It is still under debate, which of the parameters described above are determinants of cell types. To identify these parameters, it is necessary to integrate data about neurons across all necessary scales, i.e., from 3D soma location, dendrite morphology, intrinsic and extrinsic axon projection patterns, biophysical and receptive field measurements, to molecular profiles and electrical activity across behavioral states.

The rat vibrissal system is an excellent platform for such integration of multimodal data. Individual facial whiskers are represented by stereotypic anatomical reference structures throughout the whisker-to-cortex pathway (for reviews, see Diamond et al., 2008; Feldmeyer et al., 2013). Units located in the brain stem barrelettes (Ma, 1991) send axons to thalamic neurons located in barreloids (Land et al., 1995), which in turn innervate neurons located in barrels (in L4) of primary somatosensory cortex (Welker and Woolsey, 1974). These reference structures follow a somatotopic layout representing the arrangement of the facial whiskers on the animals' snout. Additionally, neurons located in barrellettes, barreloids, and barrels respond preferentially to sensory input from the whisker corresponding to the anatomical reference structure they are located in (Brecht and Sakmann (2002) and Welker (1976)). Finally, the discrete nature of the vibrissal sensory system allows for well-controlled stimulus conditions across different experiments. This correspondence of anatomical and functional somatotopy allows integration of data from individual neurons across various scales and experimental configurations.

To investigate the relationship between synaptic input, functional responses *in vivo*, and axonal output at the single neuron level, we developed a pipeline that allows: (i) recording of sensory-evoked receptive fields of individual cortical neurons, (ii) labeling of the *in vivo* recorded neurons with biocytin, (iii) digital reconstruction of the 3D soma/dendrites/ axon morphologies of the *in vivo* labeled neurons, (iv) quantitative extraction of morphological features, (v) objective classification into morphological cell types, (vi) registration of the morphologies into a precise average model of the barrel cortex geometry, and (vii) determination of the number and subcellular distribution of synaptic inputs impinging onto the *in vivo* recorded/labeled and 3D reconstructed/classified/registered individual neurons.

This approach, which we referred to as *reverse engineering, the structure and function of rat barrel cortex* will be briefly summarized below. Specifically, we will review some of our own recent studies that provided quantitative structural information of individual neurons at somatic, dendritic, axonal and synaptic levels, as well as functional information in terms of spiking probabilities in response to deflections (i.e., passive touch) of individual facial whiskers. Using this multimodal dataset, we define 10 excitatory cell types that share common *input properties* (soma location, dendrite morphology, thalamocortical synapse distributions), *sensory-evoked responses* (spiking probabilities after passive whisker touch), and *output properties* (intrinsic vertical and horizontal axon projection patterns). The present data thus provide an overview of the state of the art toward determining a coherent classification scheme that allows defining the canonical cell types of cortex.

9.5.1 Reconstruction of *In Vivo* Labeled Neurons

Juxtasomal (i.e., loose-patch) recording of individual neurons *in vivo*, combined with labeling the recorded neuron has been a standard approach for decades (Pinault, 1996). However, the success rate for recovering well-filled neurons, in particular with complete axonal morphologies, is usually low (Svoboda, 2011). This has significantly hindered systematic studies that relate the morphology of *in vivo* labeled neurons to their functional responses, e.g., after sensory stimulation. To overcome this present limitation, we recently reported an optimized protocol for cell-attached recordings and biocytin fillings *in vivo* with subsequent histology in order to obtain high-throughput, high-quality axon labeling (Narayanan et al., 2014).

Using this *in vivo* labeling approach, we investigated the structure and function of excitatory neurons located across L2-6 (see original publications: de Kock et al., 2007; de Kock and Sakmann, 2009; Narayanan et al., 2015; Oberlaender et al., 2012a). Neuronal structures were traced manually using Neurolucida software (MicroBrightfield, Williston, VT, USA), or automatically extracted from image stacks using automated tracing software (Oberlaender et al., 2007). Three-dimensional image stacks of up to 5 mm × 5 mm × 0.1 mm were acquired in rat barrel cortex at a resolution of 0.092 μm × 0.092 μm × 0.5 μm per voxel (i.e., at 100 × magnification, numerical aperture 1.4). Manual proof-editing of individual sections and automated alignment across sections were performed using custom-designed software (Dercksen et al., 2014).

9.5.2 Classification into Morphological Excitatory Cell Types

Soma/Dendrite Level (Figure 9.1): To objectively identify morphological excitatory cell types within our sample, we developed a multistep classification approach (Narayanan et al., 2015; Oberlaender et al., 2012a). In the first step, all neurons were grouped according to their mutual distances in a 22-dimensional soma-dendritic feature space using the OPTICS algorithm (Ankerst et al., 1999). This step subdivided our sample into two major soma-dendritic classes: neurons with somata located either above (supragranular L2–3) and within (granular L4) the barrel, or below the barrel (infragranular L5–6). Neurons with somata in L2–4 were termed *[supra-]granular*. The second step revealed *five cell types within the [supra-]granular population, in the following referred to as L2, L3, L4 pyramids (L2py, L3py, L4py), L4 star-pyramids (L4sp), and L4 spiny-stellates (L4ss).* In a final step, we repeated this classification approach for the infragranular neurons and obtained *five additional soma-dendritic cell types, in the following referred to as L5 slender-tufted (L5st), L5 thick-tufted (L5tt), L6 corticocortical (L6cc), L6 corticothalamic (L6ct), and L6 inverted (L6inv) pyramids.*

Axonal Level: We determined axonal parameters for each [supra-]granular neuron and grouped these features by the respective soma-dendritic cell type. One of several features that discriminated well between intrinsic axonal morphologies among cell types was the relative proportion of axon each neuron projected toward supragranular (s),

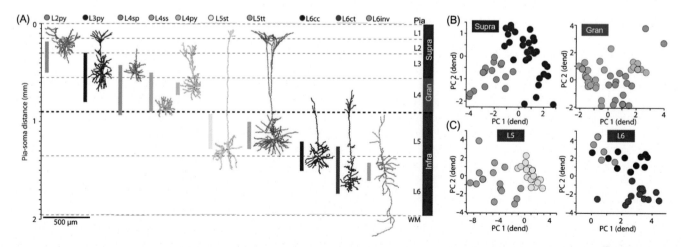

FIGURE 9.1   Soma-dendritic cell types in rat vibrissal cortex. (A) Based on 22 soma-dendritic features, 10 morphological cell types were objectively identified. Colored vertical bars represent respective ranges of soma locations after registration. Note: Cytoarchitectonic layer borders may not coincide with cell type borders. (B) Raster plots of the first two principal components (PCs) of soma-dendritic features that discriminated between [supra-]granular cell types. (C) Raster plots of the first two PCs of soma-dendritic features that discriminated between infragranular cell types. Red outlined circles represent the exemplary neurons in panel (A). *Source: Figure adapted from Narayanan et al. (2015), by permission of Oxford University Press.*

granular (g), and infragranular (i) layers. Plotting the mean and SDs of these indices for the respective soma-dendritic cell types illustrated that [supra-]granular neurons displayed axonal projection patterns that were similar within, but significantly different between soma-dendritic cell types. Similarly, the five infragranular cell types formed disjoint clusters in the soma-dendritic and axonal feature spaces. Thus, because the soma-dendritic morphology predicted the intrinsic axonal projection pattern of the respective neuron, we consider soma-dendritic and axonal cell types as the same and refer to them as *axo-dendritic cell types*. A detailed description of the cell type-specific morphological features is provided below.

9.5.3 Registration of Neuron Morphologies

One of the most important prerequisites to relate cell type-specific soma/dendrite morphologies to putative synaptic input patterns, and cell type-specific axon morphologies to putative synaptic output patterns, is the definition of a standardized 3D reference frame that allows registration of anatomical data, obtained from many animals, to their precise location within the brain. In general, the reference frame describes the 3D geometry of the brain region(s) of interest in terms of anatomical landmarks. Further, it specifies the variability of these landmarks across animals, which serves as a resolution limit to determine the structural overlap between somata/dendrites and axons, and thus to predict synaptic innervation statistically.

We generated such a precise 3D reference frame for rat barrel cortex by reconstructing the pial surface of the entire cortex, the white matter, and the circumferences of 24 barrel columns for 12 animals (Egger et al., 2012). We superimposed the 12 reconstructed barrel cortex geometries using rigid transformations, minimized the distances between the respective center locations of the 24 barrel columns, and calculated the average column center locations, column diameters, and orientations, as well as the average 3D surfaces of the pia and white matter above and below the barrel cortex, respectively. We further determined the variability of these anatomical landmarks across animals. The geometry was remarkably preserved across animals and we defined the resolution limit accordingly as 50 µm.

All reconstructed neurons were registered into this average geometrical reference frame of rat barrel cortex. This allowed determining the precise (i.e., at 50 µm resolution) 3D soma location (depth below the pial surface, depth within the principal column, lateral distance to the center of the principal column), as well as the orientation of the respective neuron within rat barrel cortex, thus compensating for varying angles when sectioning the brains and tissue shrinkage during histology.

9.5.4 Estimation of Putative Synaptic In/Output Patterns

To estimate synaptic connectivity between the *in vivo* recorded/labeled and reconstructed/registered morphologies, we measured structural overlap between axons and dendrites of individual or populations of neurons (Egger et al., 2014). Such statistical approaches are commonly referred to as *Peters' rule* (White, 1979), but the validity of predicting synaptic connectivity by axo-dendritic overlap remains controversial (Mishchenko et al., 2010). The primary reason for this controversy arises from the fact that, to date, a quantitative and coherent framework to measure structural overlap is missing. Specifically, Peters' rule is often misinterpreted in a binary fashion, namely if dendrites and axons of two neurons overlap within a certain volume, it is assumed that they are connected (Brown and Hestrin, 2009b). In contrast, if dendrites and axons do not overlap, there will be no contact, the strongest implication from this approach. However, independent of the spatial scale at which the overlap is measured, within the respective overlap volume, dendrites and axons from other (unstained) neurons will be present and are equally likely to be connected to the stained neurons. Thus, overlap can never be assumed as evidence for a connection but has to be interpreted as a probability for a connection with respect to all neurons present instead.

Consequently, we developed a novel approach, implemented within an interactive software environment called *NeuroNet*, which formulates a coherent framework to measure structural overlap between two neurons, yielding connection probabilities with respect to all neurons present in the overlapping volume. This quantitative version of Peters' rule requires generation of an average dense model of the neuronal circuitry; dense referring to the fact that every neuron within the model of the brain structure of interest: (i) has to be distributed according to measured 3D soma distributions, (ii) is represented by a complete 3D reconstruction of soma/dendrites/axon found at the respective location and (iii) contains information of cell type, as well as subcellular distributions of dendritic spines, diameters and axonal boutons (Egger et al., 2014).

Within the resolution of the reference frame (i.e., 50-µm bins for rat barrel cortex), NeuroNet then allows calculating synaptic innervation between any two neurons in the model, always taking all other neurons within the respective overlap volumes into account. The resultant *dense "statistical" connectome* yields pairwise connection probabilities, numbers of putative synaptic contacts and subcellular synapse distributions for all neurons within an entire brain region, allowing for comparison of these *in silico* measurements with electrophysiological, light- and electron microscopic data (Egger et al., 2014).

We recently illustrated and validated our approach by comparing cell type-specific local and long-range synaptic innervation obtained by NeuroNet with previous results, obtained using paired recordings *in vitro/vivo* (Bruno and Sakmann, 2006; Constantinople and Bruno, 2013; Feldmeyer et al., 2002) or reconstructions of synaptic contacts at electron microscopic levels (Schoonover et al., 2014). Because our *in silico* (i.e., based on the computer model of the barrel cortex) measurements matched the *in vitro/vivo* data, we concluded that our concept of generating an average dense network model and providing a coherent framework to calculate Peters' rule in terms of innervation probabilities is an accurate alternative to any currently available connectivity mapping method. In addition, our approach now opens the possibility to investigate location-specific differences of connectivity within a cell type (Oberlaender et al., 2012a), as well as the presence of higher order (polysynaptic) connectivity patterns within and across cell types (Song et al., 2005).

9.6 INPUT–RESPONSE–OUTPUT EXCITATORY CELL TYPES IN RAT BARREL CORTEX

Using the various tools of the reverse engineering approach introduced above, we were able to compare *input-related* parameters (soma locations, dendritic features, thalamocortical input), *in vivo responses* (ongoing and whisker touch-evoked spiking) and *output-related* parameters (vertical and horizontal intrinsic axonal projections) within and across the 10 canonical cell types of rat barrel cortex.

9.6.1 Cell Type-Specific Input-Related Parameters

Soma Location: Somata of all cell types partially intermingled within and across cytoarchitectonic layer borders (Figure 9.1), as defined by changes in the soma density profile along the vertical axis of rat barrel cortex (Meyer et al., 2010). The majority of excitatory neurons in supragranular L2/3 were classified as L2py or L3py. Somata of L2py were found at depths below the pia surface that ranged from 181 to 506 μm (mean ± SD: 293 ± 97 μm). In contrast, somata of L3py were on average located deeper within L2/3 (510 ± 120 μm), a significant fraction being located within L4 (L3py range: 302–791 μm). There, somata of L3py intermingled with those of the three granular L4 cell types. Somata of L4sp, L4ss, and L4py were found at depths that ranged from 435 to 921 μm (618 ± 125 μm), 526 to 898 μm (758 ± 95 μm), and 611 to 730 μm (677 ± 40 μm), respectively.

Remarkably, in our sample, somata of granular cell types did not intermingle with those of infragranular cell types. Specifically, somata of L5st and L5tt pyramids were found at depths that ranged from 936 to 1278 μm (1098 ± 98 μm) and 1005 to 1273 μm (1165 ± 78 μm), respectively. Thus, somata of the two L5 cell types intermingle, preventing a clear assignment of L5st and L5tt to L5A and L5B, respectively. Furthermore, somata of the L5 cell types partially intermingle with those of L6 cell types. Somata of L6cc, L6ct, and L6inv were found at depths that ranged from 1211 to 1503 μm (1377 ± 86 μm), 1262 to 1722 μm (1492 ± 122 μm) and 1424 to 1610 μm (1505 ± 85 μm), respectively.

As somata of [supra-]granular (L2py, L3py, L4py, L4sp, L4ss) and infragranular (L5st, L5tt, L6cc, L6ct, L6inv) cell types did not overlap, the L4/5 border can be considered as a cytoarchitectonic landmark that divides cortex into two major *cell type strata* (i.e., L2–4 and L5–6), with each stratum containing five axo-dendritic excitatory cell types whose somata partially intermingle. Further, on the basis of this data and the 3D distribution of all excitatory somata within rat barrel cortex (Meyer et al., 2010), we were able to determine the number of neurons per cell type within an average cortical barrel column: L2py: 1095, L3py: 1605, L4py: 459, L4sp: 2570, L4ss: 2752, L5st: 999, L5tt: 1613, L6cc: 1661, L6ct 2226, and L6inv: 788.

Dendrite Morphology: In the following, we briefly highlight the major qualitative differences in terms of dendrite morphology that allow discriminating between the 10 excitatory cell types. (Note: Quantitative discrimination was done by OPTICS-based sorting of 22 features as described above.) The most characteristic feature of L2py neurons are short, tilted apical dendrites (i.e., with respect to the vertical cortex axis) that lack an apical trunk and terminate in widely branched tufts within L1. Dendrites of L3py are on average shorter and less complex (i.e., in terms of branch point numbers) than L2py. Moreover, L3py have typical apical dendrites that project parallel to the vertical cortex axis, terminating in a sparsely branched tuft within L1.

Dendrite morphologies of L4py closely resemble those of L3py. However, their apical dendrite typically terminates within L2, and dendrites of L4py are on average shorter and less complex compared to L3py. L4sp can be easily distinguished from L4py, as they have an apical dendrite that terminates without a tuft in L2 or L3. Finally, L4ss lack an apical dendrite altogether and have small short dendrites.

Somata of excitatory cell types in the infragranular stratum are well separated from those of cell types in the [supra-]granular stratum. However, their dendrites intermingle: apical dendrites of L5 neurons travel through L1–4, and apical tufts of L6 neurons in our sample terminate within L4 and upper L5. The most characteristic feature that discriminates L5st from L5tt is the shape of the apical dendrite, which terminates in sparsely branched tufts within L1. L5tt pyramids have elaborate apical tufts and also more complex basal dendrites and more oblique branches along the apical trunk. This morphology is reflected by average dendritic path lengths and branch point numbers that are more than twice the respective numbers of L5st pyramids.

Dendrite morphologies of L6cc closely resemble those of L4sp, since in our sample, apical dendrites terminate without an apical tuft (i.e., at the L4/5 border). In contrast, L6ct have small apical tuft dendrites, display apical trunks that are densely populated with oblique branches, and have more narrow basal dendrites compared to L6cc. Finally, dendrite morphologies of L6inv are similar to those of L6cc (rendering them as a subclass of L6cc). However, L6inv display either a single apical tuft-like dendrite that is inverted (i.e., points toward the white matter), or two apical dendrites (one terminating within L4 or L5, and one terminating within the white matter).

Thalamocortical Synaptic Innervation: Registering the soma/dendrite morphologies to the average, dense model of rat barrel cortex (as described above) allowed estimating the number of putative synaptic inputs that each *in vivo* recorded/labeled neuron may receive from thalamocortical projection neurons that are located in the thalamic barreloid, which corresponds to the principal column of the respective cortical neuron. This procedure provides quantitative information on synaptic innervation at single neuron and cell type levels across the entire cortical depth (Oberlaender et al., 2012a), distinguishing our approach from layer-specific assessments. For example, thalamocortical axons display one major innervation zone, which defines the vertical and horizontal extents of the L4 barrel. Such dense innervation of L4 is similar across primary sensory cortices and species (Binzegger et al., 2004). Thus, the thalamocortical circuit in rodent S1 and cat V1 is popularly reduced to the thalamus-to-L4 pathway, and L4 is often simplified as the "starting point" of intracortical (canonical) circuits (Douglas and Martin, 2004; Petersen, 2007).

However, thalamocortical axons are not confined to L4. Less dense innervation is also found within L3, L5, and L6, often reaching a second peak at the border between L5 and L6 (Oberlaender et al., 2012b). Further, somata and dendrites of all 10 excitatory cell types partially intermingle within and across cortical layers. Thus, all cortical neurons, not only those in L4, may receive direct monosynaptic input from the thalamus, suggesting that the typical pathway diagrams of *serial (thalamo)cortical processing have to be replaced by parallel pathways* (Constantinople and Bruno, 2013; de Kock et al., 2007; Armstrong-James et al., 1992; Simons, 1978).

According to our analysis of structural overlap between thalamocortical axons and neuronal dendrites (Oberlaender et al., 2012a), virtually every cortical neuron receives direct thalamic input; the number and subcellular distribution of the respective putative synapses being cell type-specific. L2py receive on average 6 ± 12 thalamocortical synapses. In contrast, L3py receive on average 66 ± 63 thalamocortical synapses, the number of inputs being smaller for L3py with somata in L2, compared to those with somata in L4. Furthermore, the vertical soma location impacts the respective subcellular innervation, which is largely restricted to basal dendrites for L3py with somata in L2 and more homogenous within basal and apical oblique dendrites for those with somata in L4.

The three L4 cell types receive most thalamocortical inputs, compared to all other cell types, with L4py, L4sp, and L4ss receiving on average 341 ± 166, 278 ± 159, and 246 ± 123 putative inputs per neuron. Remarkably, these statistically determined numbers match recently reported electron microscopic data in rat barrel cortex (Schoonover et al., 2014) and cat primary visual cortex (da Costa and Martin, 2011). Our measurements further revealed that the number of thalamocortical inputs decreases vertically, from the center of the barrel toward L3 and L5, but also toward the horizontal barrel borders (i.e., toward/in the septum) (Lang et al., 2011; Oberlaender et al., 2012a).

Finally, in our analysis, neurons of all cell types in the infragranular stratum receive substantial direct input from the thalamus. L5st and L5tt receive 165 ± 86 and 195 ± 82 putative thalamic inputs per neuron. L6cc, L6ct, and L6inv were estimated to receive 131 ± 52, 139 ± 84, and 86 ± 31 putative inputs per neuron. Similar to the [supra-]granular cell types, these innervation numbers vary within a respective cell type, depending on the vertical and horizontal location of the soma. For example, L5st have most input when the soma location is close to L4, compared to L5st located deeper within L5. In general, the number of thalamic inputs decreases with increasing radial (horizontal) distance of the soma from the barrel column center.

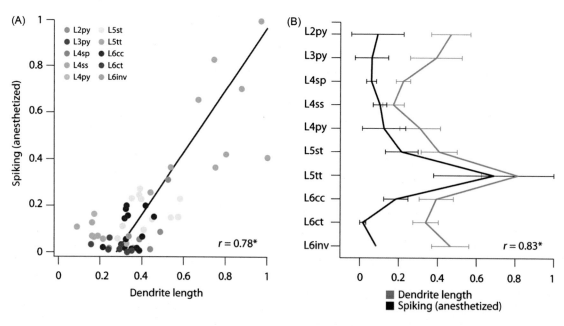

FIGURE 9.2   Structure–function relationship *in vivo*. (A) Significant correlation between dendritic length and spontaneous spiking during anesthetized state at the single neuron level. *p<0.001. (B) Significant correlation between average cell type-specific dendritic length and spontaneous spiking. *p<0.01. *Source: Figure adapted from Oberlaender et al. (2012a), by permission of Oxford University Press.*

9.6.2 Cell Type-Specific Responses *In Vivo*

Despite the fact that somata and dendrites of all excitatory cell types potentially intermingle, our analysis of thalamocortical innervation revealed cell type-specific numbers and subcellular distributions of putative thalamic inputs. Because soma location, dendrite morphology and overlap with thalamocortical axons are first-order determinants of the synaptic input each neuron receives, these input-related parameters may translate into cell type-specific function, such as sensory-evoked spiking responses. Thus, we investigated whether ongoing (i.e., spontaneous spiking) and passive whisker touch-evoked spiking correlate with the morphological cell type, as determined by input-related parameters.

Ongoing Activity: Spontaneous spiking was measured for individual, morphologically identified cortical neurons in anesthetized (de Kock et al., 2007) and awake (de Kock and Sakmann, 2009) rats. Under both experimental conditions, ongoing activity was low across all cell types (see also Barth and Poulet, 2012) and *post hoc* identification of recorded neurons revealed the following organizational principles for spiking activity under urethane anesthesia: L2 = L3, L4py = L4sp = L4ss, L5tt > L5st, L6cc > L6ct. Thus, the analysis of spontaneous spiking activity from individual, morphologically identified neurons revealed cell type-specific differences and recording the spontaneous activity may allow predicting the morphological cell type, even if somata of different cell types intermingle at the depth of the recording pipette.

Next, we explored whether one of the input-related morphological parameters of the 10 cell types correlates directly with cell type-specific ongoing spiking activity (Figure 9.2). The laminar location of the somata did not correlate with the respective spontaneous spike rate (Oberlaender et al., 2012a). In contrast, the dendritic path length was highly correlated with the spontaneous spike rates (Oberlaender et al., 2012a), both at the level of individual neurons as well as at the cell type level. Spontaneous spiking may hence reflect, at least in part, random background activity of the network, where neurons that have longer dendrites—and thus more putative synaptic inputs—have higher ongoing spike rates, compared to neurons with shorter dendrites, even though they are located at the same cortical depth.

In summary, the laminar location of a neuron's soma neither provides conclusive information on the respective dendrite morphology, nor on the respective ongoing activity; but soma location and dendrite morphology combined, allow predicting average spontaneous activity and vice versa. This observation is first evidence that *soma-dendritic cell type* classification schemes can *translate* into *physiological cell type* classification schemes.

Sensory-Evoked Activity: Sensory-evoked spiking was measured for individual, morphologically identified cortical neurons in response to passive whisker touches in anesthetized rats (Oberlaender et al., 2012a). Here, whisker-evoked

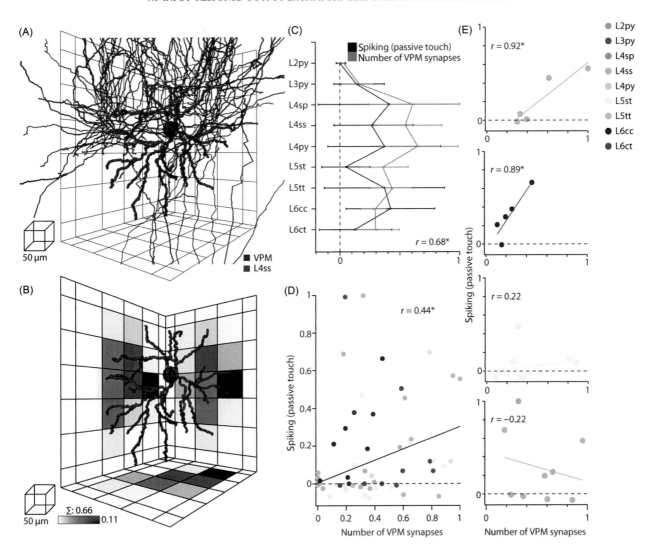

FIGURE 9.3 Relationship between thalamocortical synaptic input and evoked spiking. (A) Structural overlap between thalamocortical axon (blue) and dendrites of L4ss (red). (B) Resultant 3D innervation density at 50 μm resolution (scale represents probability of synaptic contact between the exemplary cells in panel A). (C) Cell type-specific spiking after passive whisker touch correlates with the respective number of statistically determined thalamocortical synapses. (D) Correlation at the single neuron level. (E) Correlation for L4ss and L6cc is linear, which is in general not the case (e.g., L5st and L5tt). *Source: Panels (A) and (B) were adapted from Egger et al. (2014). Panels C–E were adapted from Oberlaender et al. (2012a), by permission of Oxford University Press.*

activity was defined as the average spike rate within 0–50 ms following the stimulus corrected for the average ongoing spike rate of the respective neuron. For whisker-evoked activity, the organizational principles are: L4 > L3 > L2, L5tt > L5st, L6cc > L6ct.

It was previously shown that all excitatory neurons in rat barrel cortex, with the exception of L2py, receive whisker-evoked synaptic input at similar short latencies (Bruno and Sakmann, 2006; Constantinople and Bruno, 2013). This suggests, in line with our anatomical estimates of thalamocortical innervation, that all excitatory neurons receive direct thalamocortical input. Thus, we investigated whether the remaining input-related morphological parameter—thalamocortical innervation—correlates directly with cell type-specific whisker-evoked spiking activity. Even though it is unlikely that synaptic input translates directly into spiking output (e.g., because of non-linear dendritic integration), we found that the number of statistically predicted thalamocortical inputs was significantly correlated with the whisker-evoked spike rate of the respective cell type (Figure 9.3).

At the single neuron level (i.e., within a cell type), the two quantities did not correlate, with the exception of L4ss and L6cc. Two characteristic morphological properties shared by these two cell types are that somata of L4ss and L6cc coincide with the respective innervation peaks (i.e., L4 and L5/6) of the thalamocortical axons, and that

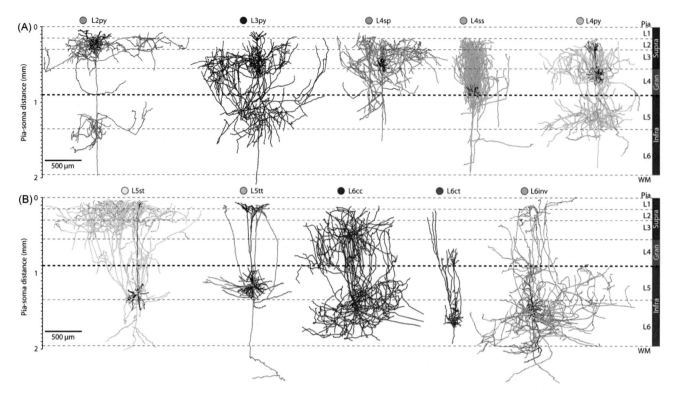

FIGURE 9.4   Axo-dendritic cell types in the rat vibrissal cortex. (A) Semicoronal view of axo-dendritic cell types of the [supra-]granular layers (B) and infragranular layers. Dendrites are in red and axons are in colors. *Source: Figure adapted from Narayanan et al. (2015), by permission of Oxford University Press.*

their dendrite morphologies are relatively simple. Therefore, for these two excitatory cell types, whisker deflection-evoked spiking may directly reflect the relative number of putative thalamocortical synapses.

9.6.3 Cell Type-Specific Output-Related Parameters

The input-related parameters (i.e., soma location, dendritic features, thalamocortical innervation) are correlated with cell type-specific physiological responses *in vivo* (i.e., ongoing and passive whisker touch-evoked spiking). Therefore, we propose that the excitatory cell types of rat barrel cortex, as defined by objective classification of soma-dendritic features, can be regarded as the 10 canonical elements that allow describing sensory-evoked signal flow in terms of input and response. In the following, we investigate whether this *input–response cell type* nomenclature translates to output-related parameters (i.e., intracortical axon projections). If this is the case, we are one major step closer to define a minimal set of cell types that allow investigating structural and functional organizational principles of the cortical circuitry (Figure 9.4).

Axon Morphology: In the following, we describe the major qualitatively differences in terms of intrinsic axon morphology that allow discriminating between the 10 excitatory cell types. (Note: Quantitative discrimination was done by OPTICS-based sorting of 22 soma-dendritic features as described above). To do so, we calculated one-dimensional (1D) axon density profiles along the vertical cortex axis for each *in vivo* recorded/labeled and reconstructed/registered neuron with 50-µm bins (accuracy of the barrel cortex geometry). Further, we discriminated between axonal segments that were located within and outside the principal column (i.e., containing the soma), and grouped the individual profiles by the respective soma-dendritic cell types (Narayanan et al., 2015). The determined vertical axon profiles were highly cell type-specific (Figure 9.5).

Axons of L2py reached average path lengths of 42.73 ± 15.55 mm (214 ± 130 branch points). Most axonal branches of L2py were largely confined to supragranular layers, reaching peak densities within L2. Most axon of L2py projected beyond the dimensions of the principal column, innervating surround columns and septa between them. Axons of L3py were on average longer (path length: 61.24 ± 24.82 mm), but less ramified than those of L2py (branch

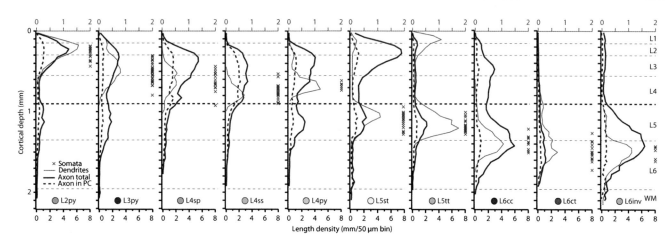

FIGURE 9.5   Cell type-specific vertical projection patterns. Vertical profiles of dendrites (red) and axons (black, dashed profiles: within principal column) in the [supra-]granular layers and infragranular layers. *Source: Figure adapted from Narayanan et al. (2015), by permission of Oxford University Press.*

points: 181 ± 72). In contrast to L2py, L3py collaterals innervated surround columns/septa throughout L1–5, reaching peak densities at the border between L2 and L3, and in the center of L5.

L4sp axons were as elaborate/ramified (68.20 ± 15.20 mm, 185 ± 34) as those of L3py, but largely confined to the [supra-]granular stratum (i.e., L1–4) of the principal column and surround columns, reaching peak densities in L3. Axons of L4py (67.34 ± 21.76 mm, 218 ± 68) resembled those of L3py, but terminated within L2 (i.e., L3py axons reached L1). Discriminating them from the other four [supra-]granular cell types, L4ss axons (49.29 ± 21.38 mm, 177 ± 88) were largely confined to the principal column, where they innervate the [supra-]granular stratum homogeneously between L2 and L4.

L5st axons were more elaborate (86.98 ± 5.71 mm, 216 ± 35) than all [supra-]granular cell types. Axonal branches of L5st densely innervated all surround columns/septa in supragranular layers and remained more confined to the principal column in L5. In contrast to L5st, axonal projections of L5tt were sparse (26.62 ± 15.21 mm, 88 ± 55) and largely confined to L5, where they innervate surround columns/septa. L6cc had the most elaborate axonal projections of all cell types (119.34 ± 64.40 mm, 215 ± 115), innervating all surround columns/septa at all cortical depths, and reaching peak densities at the L5/6 border. In contrast, L6ct had the least amount of intracortical axon of all cell types (15.64 ± 1.49 mm, 48 ± 28), which was mostly confined to L5/6 within the principal column. L6inv axons were as elaborate (111.85 ± 42.25 mm, 232 ± 111) as those of conventional L6cc, but remained largely confined to L5/6.

Transcolumnar Horizontal Pathways: With the exception of L4ss, all cell types projected most of their axon beyond the dimensions of the principal column. We thus calculated the lateral axon extents (with 50-μm bins) for each cell type along the row (i.e., lateral–medial axis) and arc (anterior–posterior axis), respectively (Figure 9.6). Lateral extents were in general asymmetric, with axons spreading on average (across all cell types) 12 ± 6% wider along the row than along the arc. Furthermore, the relative proportion of axon within and outside the principal column deviated for the respective target layers (i.e., L1–3: supragranular, L4: granular, L5–6: infragranular).

To further quantify these cell type- and target layer-specific horizontal extents, we calculated 1D axon density profiles within the dense cortex model along the row and arc for each cell type and target layer (L1–3, L4, L5–6), respectively. Horizontal profiles subdivided into three classes. First, axon density values decreased from the principal column center toward its borders, reaching minima at the centers of the septa (i.e., *up- and- down* horizontal profiles). Significant decreases in axon densities between the columns, when present along the row and arc, were termed *columnar* horizontal profiles. In contrast, significant increases in axon densities between the columns were termed *septal* horizontal profiles.

Second, significant differences between columnar and septal densities were not always present. Depending on the cell type and target layer, septa vanished in the respective axon profiles either along the row, or along the arc. The organizational principles of such *patterned* horizontal axon projections were termed *rowish* (i.e., minima were present only along the arc) and *arcish* (i.e., minima were present only along the row), respectively. Third, septa in some axons profiles were neither present along the row, nor along the arc. Such symmetric horizontal projections

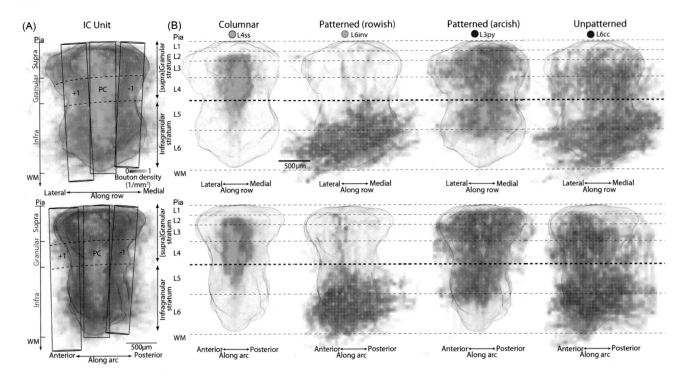

FIGURE 9.6 Cell type-specific horizontal axon projection patterns. (A) 3D bouton density distribution from axons of all reconstructed excitatory neurons located within a cortical column (hourglass-shaped surface represents the volume thresholder at 90% density, representing the intracortical (IC) unit of rat vibrissal cortex). Top panels: view along the row (i.e., lateral–medial); bottom panels: view along the arc (i.e., orthogonal to row). (B) Within the IC unit, horizontal axon projections are cell type-specific and follow three major rules (*columnar, patterned, unpatterned*). *Figure adapted from Narayanan et al. (2015), by permission of Oxford University Press.*

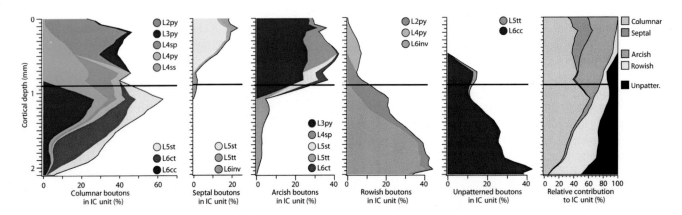

FIGURE 9.7 The structural organizational principles of the intracortical circuitry of rat vibrissal cortex. Deconstructing the IC unit of rat vibrissal cortex into its cell type-specific components revealed that the cortical sheet is divided into two orthogonal horizontal axon strata. *From left to right*: The relative cell type-specific contributions to columnar, septal, patterned (rowish/arcish) and unpatterned pathways within the IC unit. *Figure adapted from Narayanan et al. (2015), by permission of Oxford University Press.*

were termed *unpatterned*. Remarkably, the three rules of transcolumnar horizontal projections—columnar/septal, patterned (rowish/arcish), and unpatterned—were specific for each input–response cell type and target layer.

    To quantify how much axon each cell type contributes to the barrel cortex circuitry, and thus how much of the barrel cortex is organized by the respective horizontal and vertical axonal projections, we scaled our sample up to the total number of neurons located in rat barrel cortex (Meyer et al., 2013). This allowed determining the relative proportions of columnar, septal, arcish, rowish, and unpatterned projections by summing the target layer-specific axon density profiles across all cell types that were assigned to the same organizational principle (Figure 9.7).

Columnar Pathways: Columnar patterns were present throughout the entire cortical depth, representing on average 45% of all axonal projections that originate from excitatory neurons in a barrel column. These intracolumnar vertical pathways can be subdivided into three classes. First, *columnar–local* pathways comprised cell types whose axonal projections patterns remained columnar within the layer containing the respective somata (L2py in supragranular, L4ss and L4sp in granular, L5st and L6ct in infragranular layers). Second, *columnar–translaminar* pathways comprised cell types whose axons remained columnar but projected beyond the layer containing the respective somata (L4ss axons in supragranular, L3py axons in infragranular layers). Third, *septal* pathways represented on average 7% of the intracortical circuitry (i.e., L5st and L5tt). These columnar pathways were confined to the [supra-]granular stratum and may be regarded as a transition from vertical columnar to horizontal transcolumnar projections, in that septal pathways were organized with respect to, but less confined to columnar dimensions.

Patterned Pathways: Arcish horizontal projections (25%) comprised cell types that displayed transcolumnar axons along the arc and columnar ones along the row (primarily L3py in [supra-]granular, L4sp in supragranular, L5st in granular layers). Strikingly, even though arcish pathways originated from cell types with somata in all layers (also L5tt and L6ct), this horizontal organization principle was primarily found within the [supra-]granular stratum. In contrast, rowish axonal projections (13%) were largely confined to infragranular layers. These pathways comprised cell types that were transcolumnar along the row and columnar along the arc (primarily L2py and L6inv).

Unpatterned Pathways: Unpatterned horizontal projections (10%) comprised cell types that were transcolumnar along the row and arc (i.e., primarily L5tt and L6cc). These pathways were found primarily within L5–6, but to a lesser degree also in L4. Consequently, unpatterned pathways intermingle with the patterned ones, but primarily with rowish projections in the infragranular stratum.

9.7 TOWARD A NOMENCLATURE OF CANONICAL CORTICAL CELL TYPES

In this chapter, we have introduced a pipeline to group cortical neurons into cell types that share common morphological and physiological properties. Using this pipeline on the example of rat barrel cortex, we further provided quantitative data that allowed defining 10 excitatory cell types. Remarkably, even though somata and dendrites of these cell types intermingled, the soma-dendritic classification scheme translated into other cell type-specific properties, such as thalamocortical innervation, ongoing/sensory-evoked spiking, and vertical/horizontal intrinsic axonal projections. These cell types can thus be regarded as the 10 fundamental elements of the excitatory circuitry of rat barrel cortex in terms of input–response–output. In the following, we will discuss whether this classification scheme is sufficient to describe the structural and functional organization principles of rat barrel cortex, and how this nomenclature may be generalized to other cell type properties.

9.7.1 Function of Axo-Dendritic Cell Types in Rat Barrel Cortex

Columnar Pathways: The majority of vertically oriented axons (i.e., largely confined to the lateral dimensions of the barrel column) remained local within the layer containing the respective somata. However, two cell types displayed columnar axons that projected translaminar: L4ss→L2/3 and L3py→L5. Because thalamocortical axons are most dense in L4, where they define the lateral extent of a barrel column, the L4ss→L3py→L5 pathway can be regarded as an extension of the columnar Thalamus→L4 pathway, which has been previously referred to as a canonical microcircuit of sensory cortices (Douglas and Martin, 2004; Gilbert and Wiesel, 1979). Our data support this hypothesis. However, we show that this circuit represents only a fraction of the intracortical circuitry (i.e., axon distribution), which is otherwise structured by additional highly specific transcolumnar horizontal pathways.

Unpatterned Pathways: These pathways originated primarily from L6cc, whose axons innervate all columns/septa surrounding the principal column. Because we showed that L6cc display reliable and short-latency whisker-evoked spiking responses (Oberlaender et al., 2012a), we hypothesized that unpatterned pathways give rise to the fast horizontal spread of single whisker-evoked excitation (Petersen et al., 2003), influencing (e.g., broadening) sensory-evoked receptive fields of infragranular neurons (Kwegyir-Afful et al., 2005). In line with this hypothesis, recent findings in mouse primary visual cortex revealed that L6cc encompass a broad spectrum of selectivity to stimulus orientation and are predominantly innervated by L6 neurons (Velez-Fort et al., 2014), suggesting that broadening of cortical receptive fields by unpatterned L6cc projections may be generalizable to other sensory systems and species.

Patterned Pathways: It was suggested that during exploratory whisking (i.e., back and forth movement of the whiskers) coding of spatial (e.g., object shape) and temporal (e.g., object movement) sensory information may be achieved by integrating inputs from multiple whiskers of the same arc and row, respectively (Celikel and Sakmann, 2007).

Providing an anatomical framework for this hypothesis, we found that the majority of horizontal intracortical projections in the [supra-]granular and infragranular stratum interconnected columns that represent whiskers of the same arc and row, respectively. Thus, our data suggest that barrel cortex segregates and differentiates between rowish and arcish multi whisker inputs.

In summary, general functional phenomena of primary sensory cortices, such as broadening of neuronal receptive fields between thalamus and cortex (Kwegyir-Afful et al., 2005), tuning of individual synapses to multiple stimuli (as observed in mouse barrel cortex, Varga et al., 2011; primary visual, Jia et al., 2010; and auditory cortex, Grienberger et al., 2012), and ultimately understanding of complex responses to natural stimuli (Ramirez et al., 2014), will remain elusive without quantification of the pathways by which cortical columns communicate. Consequently, we argue that the concept of cortical columns as the minimal entity of cortical processing has to be extended at the next level of structural representation to "intracortical units" (IC units; i.e., volume comprising all intrinsic axons that project out of a column). The organizational principles of the IC units may in part be similar (e.g., canonical circuit, unpatterned projections in L6), but in general specific for each sensory system (e.g., patches between orientation columns in primary visual cortex, arcish/rowish strata in barrel cortex). Thus, the state of the art of describing cortical circuits by intracolumnar (vertical) layer-specific pathways has to be replaced by cell type-specific intra- and transcolumnar (horizontal) pathways, and we consider our reverse engineering approach as a generalizable roadmap to provide such complete, cell type-specific 3D circuit diagrams for other sensory systems and species.

9.7.2 Behavioral State-Dependence of Functional Axo-Dendritic Cell Types

Our data revealed that sensory-evoked responses correlate with the respective axo-dendritic cell type in rat barrel cortex. However, our example was limited to a very specific stimulus—passive whisker touch—in anesthetized animals. If our cell type nomenclature indeed reveals the canonical elements of cortex, functional responses should remain largely cell type-specific, independent of which kind of sensory stimulus was provided.

For example, despite substantial overlap between somata and dendrites, L5st and L5tt showed largely cell type-specific responses to passive whisker touches. Remarkably, the functional segregation between these two morphologically defined cell types is preserved across behavioral states and stimuli (Oberlaender et al., 2011). Specifically, during voluntary back and forth movement of the whiskers (free whisking in awake animals), L5st increase spiking whereas L5tt remain largely at rest. Given their local axonal projections, L5st pyramids potentially modulate the membrane potential of all dendrites in supragranular layers of multiple barrel columns according to their phase and may lock them to specific parts of the whisking cycle (Crochet and Petersen, 2006).

This example suggests that because soma location and dendrite morphology are first-order determinants of cell type-specific input *and* intrinsic axonal output, it is likely that functional responses to different kinds of whisker stimuli remain cell type-specific. However, under natural conditions, multiple different stimuli will be present simultaneously, for example, when a whisker touches an object during free whisking. Therefore, responses will in general reflect a complex interplay between multiple converging/diverging, synchronous/asynchronous cell type–specific pathways, rendering identification of cell type-specific function more difficult.

9.7.3 Long-Range Targets of Functional Axo-Dendritic Cell Types

So far, we have presented our reasoning that the 10 excitatory cell types of rat barrel cortex can be regarded as primary canonical elements that may elucidate structural and functional organization principles of this particular cortical area. However, one of the challenges that remain is to correlate these cell types with long-range targets in other cortical and subcortical areas. Toward this goal, our cell type nomenclature correlates with defined classes of projection neurons (Harris and Shepherd, 2015); for example, L5st primarily project to the striatum (intratelencephalic neurons), L5tt to the brain stem (pyramidal tract neurons), L6ct to the thalamus (corticothalamic neurons), and L3py to primary motor cortex (M1) and/or secondary somatosensory cortex (S2, intratelencephalic neurons). Further work will need to address whether these broad projection classes segregate further into more area-specific projection cell types and whether these subclasses share common stimulus-specific activity patterns.

For example, recent evidence suggests that L3py in mouse barrel cortex either project to M1 or S2 (Chen et al., 2013). Neurons that increased activity during free whisking were found among S2-projecting but not M1-projecting neurons. Furthermore, a higher fraction of S2-projecting than M1-projecting neurons showed whisker touch-related responses during texture discrimination, whereas a higher fraction of M1-projecting than S2-projecting neurons showed touch-related responses during an object detection task. This suggests cell type-specific transmission of stimulus- and behavior-specific sensory information from barrel cortex via segregated long-range output pathways.

To subdivide the 10 intracortical cell types into such stimulus-specific long-range output pathways, the reverse engineering approach has to be augmented with tools that provide quantitative information about the long-range targets of individual neurons. One potential strategy could be to inject retrograde traces into known long-range targets of rat barrel cortex (Chakrabarti and Alloway, 2006) and to record/label and reconstruct/register these retrogradely labeled neurons, during different sensory stimuli and behavioral states. Targeting individual retrogradely labeled neurons with patch pipettes *in vivo* is however not trivial, but may become feasible given the tools we have introduced above.

9.7.4 Molecular/Genetic Profiles of Functional Axo-Dendritic Cell Types

Up to this point, we have introduced a strategy to define canonical cell types, which interlinks different morphological and functional features in a hierarchical manner. At the highest level were input-related parameters: the precise soma location within cortex and the 3D morphology of the dendrites. These soma-dendritic cell types correlated with functional properties: ongoing and whisker-evoked spiking during anesthetized and awake states; and with intrinsic axonal projections. These cell types may further subdivide into long-range target-specific classes.

Arguably, the definition of molecular/genetic cell types should be based on this input–response–output classification scheme. Extracting the cytosol of each *in vivo* recorded/labeled and reconstructed/registered neuron may allow the identification of genetic correlates of functional axo-dendritic cell types, ideally by a single, but probably by a combination of multiple molecular markers (Sorensen et al., 2015). Only once a quantitative relationship between molecular/genetic markers and input–response–output cell types is established, optogenetic tools will be able to unfold their true potential. At present, genetic cell types are however usually defined without detailed characterization of morphological and/or functional properties. Thus, the meaning of a particular marker with respect to information processing in cortex remains elusive and interpretations from transgenic and/or optogenetic manipulations will remain largely ambiguous.

In conclusion, this chapter reviewed a promising strategy (i.e., reverse engineering) toward generating a nomenclature of the canonical elements of cortex, which will allow describing structural and functional cortical organization in terms of cell types. Specifically, this strategy incorporates recording the spiking activity of individual neurons in response to different sensory stimuli, labeling the recorded neuron *in vivo*, reconstructing complete 3D dendrite and intrinsic axon morphologies, identifying long-range targets, sequencing genetic profiles and finally, registration individual neurons into a dense cortex model to predict its local/long-range synaptic inputs. Proceeding in this way should provide a dataset that allows categorizing the cortical circuitry by its elementary building blocks.

Acknowledgments

We thank the Anatomy Institute of the University of Tuebingen as well as Bert Sakmann for generous support. Funding was provided by the Max Planck Institute for Biological Cybernetics (R.T.N., R.E., M.O.), the Studienstiftung des deutschen Volkes (R.E.), the Bernstein Center for Computational Neuroscience, funded by German Federal Ministry of Education and Research Grant BMBF/FKZ 01GQ1002 (R.T.N., R.E., M.O.), the Werner Reichardt Center for Integrative Neuroscience (M.O.), the Max Planck Florida Institute for Neuroscience (M.O.), and by the VU University, Amsterdam (C.D.K.).

References

Ankerst, M., Breunig, M., Kriegel, H.P., Sander, J., 1999. OPTICS: ordering points to identify the clustering structure. In: Delis, A., Faloutsous, C., Ghandeharizadeh, S. (Eds.), ACM SIGMOD'99 Int. Conf. on Management of Data. ACM Press, Philadelphia, pp. 49–60.

Armstrong-James, M., Fox, K., Das-Gupta, A., 1992. Flow of excitation within rat barrel cortex on striking a single vibrissa. J. Neurophysiol. 68, 1345–1358.

Ascoli, G.A., Donohue, D.E., Halavi, M., 2007. NeuroMorpho.Org: a central resource for neuronal morphologies. J. Neurosci. 27, 9247–9251.

Ascoli, G.A., Alonso-Nanclares, L., Anderson, S.A., Barrionuevo, G., Benavides-Piccione, R., Burkhalter, A., et al., 2008. Petilla terminology: nomenclature of features of GABAergic interneurons of the cerebral cortex. Nat. Rev. Neurosci. 9, 557–568.

Barth, A.L., Poulet, J.F., 2012. Experimental evidence for sparse firing in the neocortex. Trends Neurosci. 35, 345–355.

Bernardo, K.L., McCasland, J.S., Woolsey, T.A., Strominger, R.N., 1990. Local intra- and interlaminar connections in mouse barrel cortex. J. Comp. Neurol. 291, 231–255.

Binzegger, T., Douglas, R.J., Martin, K.A., 2004. A quantitative map of the circuit of cat primary visual cortex. J. Neurosci. 24, 8441–8453.

Brecht, M., Sakmann, B., 2002. Whisker maps of neuronal subclasses of the rat ventral posterior medial thalamus, identified by whole-cell voltage recording and morphological reconstruction. J. Physiol. 538, 495–515.

Brodmann, K., 1909. Vergleichende Lokalisationslehre der Großhirnrinde—in ihren Prinzipien dargestellt auf Grund des Zellenbaues. Barth, Leipzig.

Brown, S.P., Hestrin, S., 2009a. Cell-type identity: a key to unlocking the function of neocortical circuits. Curr. Opin. Neurobiol. 19, 415–421.

Brown, S.P., Hestrin, S., 2009b. Intracortical circuits of pyramidal neurons reflect their long-range axonal targets. Nature 457, 1133–1136.

Bruno, R.M., Sakmann, B., 2006. Cortex is driven by weak but synchronously active thalamocortical synapses. Science 312, 1622–1627.

Bueno-Lopez, J.L., Reblet, C., Lopez-Medina, A., Gomez-Urquijo, S.M., Grandes, P., Gondra, J., et al., 1991. Targets and laminar distribution of projection neurons with 'inverted' morphology in rabbit cortex. Eur. J. Neurosci. 3, 713.

Celikel, T., Sakmann, B., 2007. Sensory integration across space and in time for decision making in the somatosensory system of rodents. Proc. Natl. Acad. Sci. U.S.A. 104, 1395–1400.

Chakrabarti, S., Alloway, K.D., 2006. Differential origin of projections from SI barrel cortex to the whisker representations in SII and MI. J. Comp. Neurol. 498, 624–636.

Chen, J.L., Carta, S., Soldado-Magraner, J., Schneider, B.L., Helmchen, F., 2013. Behaviour-dependent recruitment of long-range projection neurons in somatosensory cortex. Nature 499, 336–340.

Connors, B.W., Gutnick, M.J., Prince, D.A., 1982. Electrophysiological properties of neocortical neurons *in vitro*. J. Neurophysiol. 48, 1302–1320.

Constantinople, C.M., Bruno, R.M., 2013. Deep cortical layers are activated directly by thalamus. Science 340, 1591–1594.

Conti, F., Rustioni, A., Petrusz, P., Towle, A.C., 1987. Glutamate-positive neurons in the somatic sensory cortex of rats and monkeys. J. Neurosci. 7, 1887–1901.

Crochet, S., Petersen, C.C., 2006. Correlating whisker behavior with membrane potential in barrel cortex of awake mice. Nat. Neurosci. 9, 608–610.

da Costa, N.M., Martin, K.A., 2011. How thalamus connects to spiny stellate cells in the cat's visual cortex. J. Neurosci. 31, 2925–2937.

de Kock, C.P., Sakmann, B., 2009. Spiking in primary somatosensory cortex during natural whisking in awake head-restrained rats is cell-type specific. Proc. Natl. Acad. Sci. U.S.A. 106, 16446–16450.

de Kock, C.P., Bruno, R.M., Spors, H., Sakmann, B., 2007. Layer- and cell-type-specific suprathreshold stimulus representation in rat primary somatosensory cortex. J. Physiol. 581, 139–154.

de Lima, A.D., Voigt, T., Morrison, J.H., 1990. Morphology of the cells within the inferior temporal gyrus that project to the prefrontal cortex in the macaque monkey. J. Comp. Neurol. 296, 159–172.

DeFelipe, J., Farinas, I., 1992. The pyramidal neuron of the cerebral cortex: morphological and chemical characteristics of the synaptic inputs. Prog. Neurobiol. 39, 563–607.

DeFelipe, J., Lopez-Cruz, P.L., Benavides-Piccione, R., Bielza, C., Larranaga, P., Anderson, S., et al., 2013. New insights into the classification and nomenclature of cortical GABAergic interneurons. Nat. Rev. Neurosci. 14, 202–216.

Dercksen, V.J., Hege, H.C., Oberlaender, M. (Eds.), 2014. The Filament Editor: an interactive software environment for visualization, proof-editing and analysis of 3D neuron morphology Neuroinformatics 12, 325–339.

Diamond, M.E., von Heimendahl, M., Knutsen, P.M., Kleinfeld, D., Ahissar, E., 2008. 'Where' and 'what' in the whisker sensorimotor system. Nat. Rev. Neurosci. 9, 601–612.

Douglas, R.J., Martin, K.A., 2004. Neuronal circuits of the neocortex. Annu. Rev. Neurosci. 27, 419–451.

Druckmann, S., Banitt, Y., Gidon, A., Schurmann, F., Markram, H., Segev, I., 2007. A novel multiple objective optimization framework for constraining conductance-based neuron models by experimental data. Front. Neurosci. 1, 7–18.

Egger, R., Narayanan, R.T., Helmstaedter, M., de Kock, C.P., Oberlaender, M., 2012. 3D reconstruction and standardization of the rat vibrissal cortex for precise registration of single neuron morphology. PLoS Comput. Biol. 8, e1002837.

Egger, R., Dercksen, V.J., Udvary, D., Hege, H.C., Oberlaender, M., 2014. Generation of dense statistical connectomes from sparse morphological data. Front. Neuroanat. 8, 129.

Feldmeyer, D., Lubke, J., Silver, R.A., Sakmann, B., 2002. Synaptic connections between layer 4 spiny neurone-layer 2/3 pyramidal cell pairs in juvenile rat barrel cortex: physiology and anatomy of interlaminar signalling within a cortical column. J. Physiol. 538, 803–822.

Feldmeyer, D., Brecht, M., Helmchen, F., Petersen, C.C., Poulet, J.F., Staiger, J.F., et al., 2013. Barrel cortex function. Prog. Neurobiol. 103, 3–27.

Gilbert, C.D., Wiesel, T.N., 1979. Morphology and intracortical projections of functionally characterised neurones in the cat visual cortex. Nature 280, 120–125.

Gilbert, C.D., Wiesel, T.N., 1989. Columnar specificity of intrinsic horizontal and corticocortical connections in cat visual cortex. J. Neurosci. 9, 2432–2442.

Grienberger, C., Adelsberger, H., Stroh, A., Milos, R.I., Garaschuk, O., Schierloh, A., et al., 2012. Sound-evoked network calcium transients in mouse auditory cortex *in vivo*. J. Physiol. 590, 899–918.

Hamill, O.P., Marty, A., Neher, E., Sakmann, B., Sigworth, F.J., 1981. Improved patch-clamp techniques for high-resolution current recording from cells and cell-free membrane patches. Pflugers Arch. 391, 85–100.

Harris, K.D., Shepherd, G.M., 2015. The neocortical circuit: themes and variations. Nat. Neurosci. 18, 170–181.

Hay, E., Segev, I., 2014. Dendritic excitability and gain control in recurrent cortical microcircuits. Cereb. Cortex. pii: bhu200. [Epub ahead of print].

Helmstaedter, M., Briggman, K.L., Turaga, S.C., Jain, V., Seung, H.S., Denk, W., 2013. Connectomic reconstruction of the inner plexiform layer in the mouse retina. Nature 500, 168–174.

Hevner, R.F., Daza, R.A., Rubenstein, J.L., Stunnenberg, H., Olavarria, J.F., Englund, C., 2003. Beyond laminar fate: toward a molecular classification of cortical projection/pyramidal neurons. Dev. Neurosci. 25, 139–151.

Hill, S.L., Wang, Y., Riachi, I., Schurmann, F., Markram, H., 2012. Statistical connectivity provides a sufficient foundation for specific functional connectivity in neocortical neural microcircuits. Proc. Natl. Acad. Sci. U.S.A. 109, E2885–E2894.

Horikawa, K., Armstrong, W.E., 1988. A versatile means of intracellular labeling: injection of biocytin and its detection with avidin conjugates. J. Neurosci. Methods 25, 1–11.

Hubel, D.H., Wiesel, T.N., 1962. Receptive fields, binocular interaction and functional architecture in the cat's visual cortex. J. Physiol. 160, 106–154.

Hubel, D.H., Wiesel, T.N., 1977. Ferrier lecture. Functional architecture of macaque monkey visual cortex. Proc. R. Soc. Lond. B. Biol. Sci. 198, 1–59.

Hubener, M., Schwarz, C., Bolz, J., 1990. Morphological types of projection neurons in layer 5 of cat visual cortex. J. Comp. Neurol. 301, 655–674.

Huntley, G.W., Jones, E.G., 1991. Relationship of intrinsic connections to forelimb movement representations in monkey motor cortex: a correlative anatomic and physiological study. J. Neurophysiol. 66, 390–413.

Jia, H., Rochefort, N.L., Chen, X., Konnerth, A., 2010. Dendritic organization of sensory input to cortical neurons *in vivo*. Nature 464, 1307–1312.

Jones, E.G., 1986. Neurotransmitters in the cerebral cortex. J. Neurosurg. 65, 135–153.

I. MICROCIRCUITRY

Keller, A., 1993. Intrinsic synaptic organization of the motor cortex. Cereb. Cortex 3, 430–441.

Keller, A., Asanuma, H., 1993. Synaptic relationships involving local axon collaterals of pyramidal neurons in the cat motor cortex. J. Comp. Neurol. 336, 229–242.

Kepecs, A., Fishell, G., 2014. Interneuron cell types are fit to function. Nature 505, 318–326.

Kwegyir-Afful, E.E., Bruno, R.M., Simons, D.J., Keller, A., 2005. The role of thalamic inputs in surround receptive fields of barrel neurons. J. Neurosci. 25, 5926–5934.

Land, P.W., Buffer Jr., S.A., Yaskosky, J.D., 1995. Barreloids in adult rat thalamus: three-dimensional architecture and relationship to somatosensory cortical barrels. J. Comp. Neurol. 355, 573–588.

Lang, S., Dercksen, V.J., Sakmann, B., Oberlaender, M., 2011. Simulation of signal flow in 3D reconstructions of an anatomically realistic neural network in rat vibrissal cortex. Neural Netw. 24, 998–1011.

London, M., Hausser, M., 2005. Dendritic computation. Annu. Rev. Neurosci. 28, 503–532.

Lubke, J., Roth, A., Feldmeyer, D., Sakmann, B., 2003. Morphometric analysis of the columnar innervation domain of neurons connecting layer 4 and layer 2/3 of juvenile rat barrel cortex. Cereb. Cortex 13, 1051–1063.

Ma, P.M., 1991. The barrelettes—architectonic vibrissal representations in the brainstem trigeminal complex of the mouse. I. Normal structural organization. J. Comp. Neurol. 309, 161–199.

Mainen, Z.F., Sejnowski, T.J., 1995. Reliability of spike timing in neocortical neurons. Science 268, 1503–1506.

Markram, H., 2008. Fixing the location and dimensions of functional neocortical columns. HFSP J. 2, 132–135.

Meyer, H.S., Wimmer, V.C., Oberlaender, M., de Kock, C.P., Sakmann, B., Helmstaedter, M., 2010. Number and laminar distribution of neurons in a thalamocortical projection column of rat vibrissal cortex. Cereb. Cortex 20, 2277–2286.

Meyer, H.S., Egger, R., Guest, J.M., Foerster, R., Reissl, S., Oberlaender, M., 2013. Cellular organization of cortical barrel columns is whisker-specific. Proc. Natl. Acad. Sci. U.S.A. 110, 19113–19118.

Mishchenko, Y., Hu, T., Spacek, J., Mendenhall, J., Harris, K.M., Chklovskii, D.B., 2010. Ultrastructural analysis of hippocampal neuropil from the connectomics perspective. Neuron 67, 1009–1020.

Mishra, A., Dhingra, K., Schuz, A., Logothetis, N.K., Canals, S., 2010. Improved neuronal tract tracing with stable biocytin-derived neuroimaging agents. ACS Chem. Neurosci. 1, 129–138.

Mountcastle, V.B., 1957. Modality and topographic properties of single neurons of cat's somatic sensory cortex. J. Neurophysiol. 20, 408–434.

Munoz, W., Tremblay, R., Rudy, B., 2014. Channelrhodopsin-assisted patching: *in vivo* recording of genetically and morphologically identified neurons throughout the brain. Cell. Rep. 9, 2304–2316.

Narayanan, R.T., Mohan, H., Broersen, R., de Haan, R., Pieneman, A.W., de Kock, C.P., 2014. Juxtasomal biocytin labeling to study the structure–function relationship of individual cortical neurons. J Vis Exp, e51359.

Narayanan, R.T., Egger, R., Johnson, A.S., Mansvelder, H.D., Sakmann, B., de Kock, C.P., et al., 2015. Beyond columnar organization: cell type- and target layer-specific principles of horizontal axon projection patterns in rat vibrissal cortex. Cereb. Cortex Apr 1. pii: bhv053. [Epub ahead of print].

Oberlaender, M., Bruno, R.M., Sakmann, B., Broser, P.J., 2007. Transmitted light brightfield mosaic microscopy for three-dimensional tracing of single neuron morphology. J. Biomed. Opt. 12, 064029.

Oberlaender, M., Boudewijns, Z.S., Kleele, T., Mansvelder, H.D., Sakmann, B., de Kock, C.P., 2011. Three-dimensional axon morphologies of individual layer 5 neurons indicate cell type-specific intracortical pathways for whisker motion and touch. Proc. Natl. Acad. Sci. U.S.A. 108, 4188–4193.

Oberlaender, M., de Kock, C.P., Bruno, R.M., Ramirez, A., Meyer, H.S., Dercksen, V.J., et al., 2012a. Cell type-specific three-dimensional structure of thalamocortical circuits in a column of rat vibrissal cortex. Cereb. Cortex 22, 2375–2391.

Oberlaender, M., Ramirez, A., Bruno, R.M., 2012b. Sensory experience restructures thalamocortical axons during adulthood. Neuron 74, 648–655.

Parekh, R., Ascoli, G.A., 2013. Neuronal morphology goes digital: a research hub for cellular and system neuroscience. Neuron 77, 1017–1038.

Peters, A., Regidor, J., 1981. A reassessment of the forms of nonpyramidal neurons in area 17 of cat visual cortex. J. Comp. Neurol. 203, 685–716.

Petersen, C.C., 2007. The functional organization of the barrel cortex. Neuron 56, 339–355.

Petersen, C.C., Grinvald, A., Sakmann, B., 2003. Spatiotemporal dynamics of sensory responses in layer 2/3 of rat barrel cortex measured *in vivo* by voltage-sensitive dye imaging combined with whole-cell voltage recordings and neuron reconstructions. J. Neurosci. 23, 1298–1309.

Pinault, D., 1996. A novel single-cell staining procedure performed *in vivo* under electrophysiological control: morpho-functional features of juxtacellularly labeled thalamic cells and other central neurons with biocytin or neurobiotin. J. Neurosci. Methods 65, 113–136.

Polavaram, S., Gillette, T.A., Parekh, R., Ascoli, G.A., 2014. Statistical analysis and data mining of digital reconstructions of dendritic morphologies. Front. Neuroanat. 8, 138.

Powell, T.P., Mountcastle, V.B., 1959. Some aspects of the functional organization of the cortex of the postcentral gyrus of the monkey: a correlation of findings obtained in a single unit analysis with cytoarchitecture. Bull. Johns Hopkins Hosp. 105, 133–162.

Ramirez, A., Pnevmatikakis, E.A., Merel, J., Paninski, L., Miller, K.D., Bruno, R.M., 2014. Spatiotemporal receptive fields of barrel cortex revealed by reverse correlation of synaptic input. Nat. Neurosci. 17, 866–875.

Ribak, C.E., 1978. Aspinous and sparsely-spinous stellate neurons in the visual cortex of rats contain glutamic acid decarboxylase. J. Neurocytol. 7, 461–478.

Rudy, B., Fishell, G., Lee, S., Hjerling-Leffler, J., 2011. Three groups of interneurons account for nearly 100% of neocortical GABAergic neurons. Dev. Neurobiol. 71, 45–61.

Santana, R., McGarry, L.M., Bielza, C., Larranaga, P., Yuste, R., 2013. Classification of neocortical interneurons using affinity propagation. Front. Neural. Circuits 7, 185.

Schoonover, C.E., Tapia, J.C., Schilling, V.C., Wimmer, V., Blazeski, R., Zhang, W., et al., 2014. Comparative strength and dendritic organization of thalamocortical and corticocortical synapses onto excitatory layer 4 neurons. J. Neurosci. 34, 6746–6758.

Simons, D.J., 1978. Response properties of vibrissa units in rat SI somatosensory neocortex. J. Neurophysiol. 41, 798–820.

I. MICROCIRCUITRY

Song, S., Sjostrom, P.J., Reigl, M., Nelson, S., Chklovskii, D.B., 2005. Highly nonrandom features of synaptic connectivity in local cortical circuits. PLoS Biol. 3, e68.

Sorensen, S.A., Bernard, A., Menon, V., Royall, J.J., Glattfelder, K.J., Desta, T., et al., 2015. Correlated gene expression and target specificity demonstrate excitatory projection neuron diversity. Cereb. Cortex 25, 433–449.

Svoboda, K., 2011. The past, present, and future of single neuron reconstruction. Neuroinformatics 9, 97–98.

Thomson, A.M., 2010. Neocortical layer 6, a review. Front. Neuroanat. 4, 13.

Tripathy, S.J., Burton, S.D., Geramita, M., Gerkin, R.C., Urban, N.N., 2015. Brain-wide analysis of electrophysiological diversity yields novel categorization of mammalian neuron types. J. Neurophysiol. 113 (10), 3474–3489. http://dx.doi.org/10.1152/jn.00237.2015. Epub 2015 Mar 25.

van Aerde, K.I., Feldmeyer, D., 2013. Morphological and physiological characterization of pyramidal neuron subtypes in rat medial prefrontal cortex. Cereb. Cortex 25 (3), 788–805. http://dx.doi.org/10.1093/cercor/bht278. Epub 2013 Oct 9.

Varga, Z., Jia, H., Sakmann, B., Konnerth, A., 2011. Dendritic coding of multiple sensory inputs in single cortical neurons *in vivo*. Proc. Natl. Acad. Sci. U.S.A. 108, 15420–15425.

Velez-Fort, M., Rousseau, C.V., Niedworok, C.J., Wickersham, I.R., Rancz, E.A., Brown, A.P., et al., 2014. The stimulus selectivity and connectivity of layer six principal cells reveals cortical microcircuits underlying visual processing. Neuron 83, 1431–1443.

Weiss, D.S., Keller, A., 1994. Specific patterns of intrinsic connections between representation zones in the rat motor cortex. Cereb. Cortex 4, 205–214.

Welker, C., 1976. Receptive fields of barrels in the somatosensory neocortex of the rat. J. Comp. Neurol. 166, 173–189.

Welker, C., Woolsey, T.A., 1974. Structure of layer IV in the somatosensory neocortex of the rat: description and comparison with the mouse. J. Comp. Neurol. 158, 437–453.

White, E.L., 1979. Thalamocortical synaptic relations: a review with emphasis on the projections of specific thalamic nuclei to the primary sensory areas of the neocortex. Brain Res. 180, 275–311.

Woolsey, T.A., Van der Loos, H., 1970. The structural organization of layer IV in the somatosensory region (SI) of mouse cerebral cortex. The description of a cortical field composed of discrete cytoarchitectonic units. Brain Res. 17, 205–242.

10

Anterograde Viral Tracer Methods

Kevin Beier

Departments of Biological Sciences and Psychiatry and Behavioral Sciences, Stanford, CA, USA

10.1 AXON TRACT TRACING THROUGH VIRAL-MEDIATED GENE TRANSFER

Classical tracers have proven a powerful method of mapping neuronal connections in the brain. Some, such as the cholera toxin beta subunit (CTB) and fluorogold, are taken up by axon terminals projecting to a given brain region, retrogradely transported, and label the cell bodies of those projecting neurons (Kuypers et al., 1979; Stoeckel and Thoenen, 1975). Others, such as biotinylated dextran amines (BDAs) and *Phaseolus vulgaris* leucoagglutinin (PHA-L), are taken up by cell bodies at the site of injection and anterogradely transported down the axon (Glover et al., 1986; Trojanowski et al., 1981).

However, these methods have limitations. For example, these tracers label both neuronal elements as well as the extracellular space. Also, potentially confounding retrograde uptake by neurons can occur with tracers such as BDA, which are otherwise intended to be anterograde-specific (Reiner et al., 1993; Wang et al., 2014). Classical tracers are frequently taken up by axons of passage, and therefore may label neurons that do not have synaptic terminations in the area injected. Moreover, these tracers will label any of the cell types within the injection site indiscriminately, and these tracers therefore lack the cell-type specificity achievable through modern genetic techniques.

Recombinant viral technology has provided solutions to these problems. Viruses can be engineered to encode a fluorescent protein within the viral genome, such that the fluorescence is contained only within neurons infected by the virus. While most viruses preferentially label the neurons locally within the injected site, specific strategies can be used to infect cells projecting from a distance to the site of injection as will be discussed later in this chapter. While most viruses also appear to infect *en passant* axons, certain viral technologies, such as the monosynaptic rabies virus (RABV) technique, offer a solution to this problem. Viruses can also be engineered to conditionally express genes only in desired cell types. This can be achieved either by the expression of specific promoters encoded in the viral genome, or more commonly, in cells that express a recombinase protein (e.g., Cre or Flp).

Especially with continued developments, viruses offer an extremely powerful tool for axon tract tracing, a so-called "fourth-generation" neuroanatomical tract-tracing technique (Wouterlood et al., 2014). There are many examples in the literature of the use of viruses to map specific pathways in the brain. Here, I will highlight some general methods of how viruses have helped to elucidate the architecture of the brain, with specific discussion of the strengths and weaknesses of different strategies.

10.1.1 Sindbis Virus—High-Resolution Mapping of Small Numbers of Neurons

The Sindbis (SIN) virus is a member of the alphavirus family. It is a negative strand RNA virus and does not have a DNA phase in its life cycle. Genes from SIN vectors are rapidly and highly expressed from viral promoters. This property makes SIN vectors highly useful as anterograde tracers because large amounts of a protein need to be produced in order for that protein to fill the most distal processes of neurons, including axon terminals, which can be located millimeters away from the site of protein synthesis. Just as classical tracers need to effectively fill neurons in order to be visualized, so too do proteins need to be expressed at high levels in order to be visible within those cells.

Axons and Brain Architecture.
DOI: http://dx.doi.org/10.1016/B978-0-12-801393-9.00010-4

However, the virus only achieves this rapid and robust expression by hijacking the endogenous cell transcription and translation machinery and produces its genes at the expense of the host cell, therefore leading to rapid toxicity.

SIN vectors are often injected at a low titer, such that only a few neurons are labeled. This allows the clear identification of each neuron, which is important for reconstruction of axonal arbors.

10.1.2 Example 1: Projections of Individual Olfactory Bulb Input Neurons

As one example of the utility of SIN for axon tracing, recombinant SIN vectors were used to map the projections of specific glomeruli in the olfactory bulb (OB) (Ghosh et al., 2011). The goal was to understand how the mitral and tufted (MT) neurons in specific glomeruli of the OB send information into downstream structures. In the visual and auditory sensory systems in mammals, topographical maps exist in the cortex to encode sensory information. While topographical maps are known to exist in the OB, how this information was transmitted to the olfactory cortex and other downstream structures was unclear.

In a two-glomerulus experiment, recombinant SIN vectors were engineered to express green fluorescent protein (GFP) (green), GFP and tandem dimer tomato (tdTomato) (yellow), or red fluorescent protein (RFP) (red). A small volume of low-titer SIN virus was injected into a single glomerulus in the mouse to label projections from MT neurons in that glomerulus. To investigate the branching patterns of individual MT neurons associated with an individual glomerulus, a single diluted preparation of virus was injected into a targeted glomerulus in order to label a single neuron. This showed that single MT neurons had broad, robust, arborizations and projected to multiple sites; namely, major projections to the anterior olfactory nucleus (AON) and piriform cortex (PC), and less robust projections in more distal sites. Some MT neurons were also observed to innervate a small region of the granule cell layer just proximal to the mirror symmetric glomerulus, suggesting that glomeruli that get input from neurons with the same olfactory receptors may communicate directly with one another.

Next, one color of SIN was injected into one glomerulus, and a different color SIN into a different glomerulus to compare the projection patterns of MT neurons from different glomeruli (Figure 10.1A). While the different colored axons largely remained spatially segregated in one target structure, the AON, a much larger percentage of axons were mixed in another major target, the PC. It therefore appeared that a topographical map was maintained between the OB and AON, but not between the OB and PC. Furthermore, some MT axons diverged from their neighbors (those of the same color) to intermingle with axons of the opposite color. Thus, while a gross topographical map is maintained from the OB to the AON, the representation of information is different in those two regions.

Because of the detailed visualization allowed by the viral vector, it was possible to investigate animal to animal differences by mapping axonal arbors from different animals onto a common reference brain. Axonal projections were mapped onto a common reference brain. Comparing metrics such as numbers of branches in a given region and distance between neuron pairs, the variation in axon configuration within a single animal was found to be as large as that between animals, suggesting the existence of significant variation among axonal arborizations between single MT neurons. To test this hypothesis, different colored viruses were mixed and injected into the same glomerulus, with the intention of labeling a small number of individual neurons with different colors (Figure 10.1B). Homotopic MT neuron pairs (those from the same glomerulus in the same animal) displayed essentially an equivalent amount of variation in branching pattern as heterotypic pairs (different glomeruli in the same animal), and this result was consistent when multiple animals were compared. OB output was therefore shown to be divergent and non-stereotyped, and projections from MT cells to the PC appear random, consistent with other studies (Choi et al., 2011; Miyamichi et al., 2011; Sosulski et al., 2011; Stettler and Axel, 2009).

10.1.3 Example 2: Comparison of Axonal Configurations of Thalamic Neurons

Using a single-neuron, high-resolution analysis, a recent study focused on the projection of neurons in different subdivisions of the motor thalamus in rats. Unique subdivisions of this area are known to differentially receive projections from the basal ganglia and cerebellum (ventral anterior and ventral lateral, respectively). In this study, the authors used a GFP variant that contained a site for palmitoylation, such that the protein was targeted to the cell membrane. As the GFP was expressed robustly from the viral promoter, and the GFP was targeted to the membrane, this SIN vector functioned as a highly sensitive anterograde tracer. Using this virus, the authors showed that thalamic subdivisions had anatomically distinct differences in their axonal projections. One subpopulation projected to both the neostriatum and cortical areas, in layer I. The other projected selectively to cortical areas, in layers II–V, suggesting that these neurons may differentially affect motor circuitry, as, for example, a broad arborization in layer 1 may serve to activate a wide array of L2/3 and L5 pyramidal neurons via their apical tufts. The same group has also used SIN

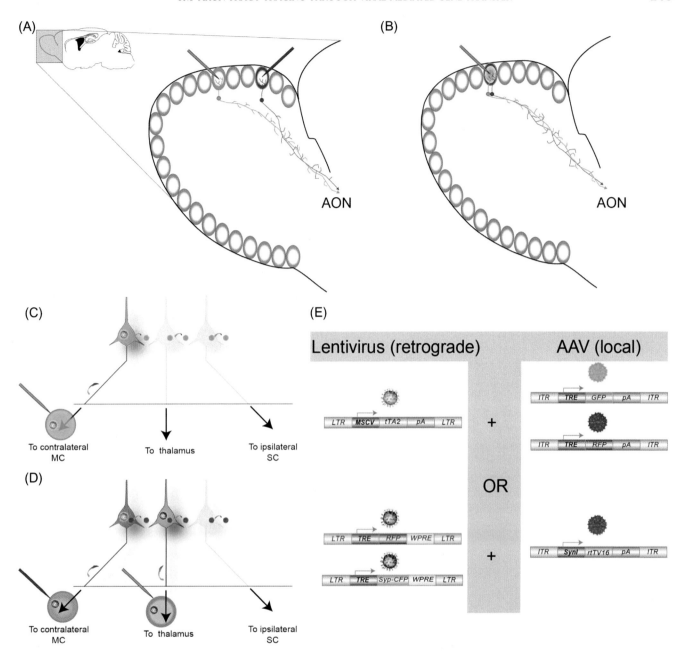

FIGURE 10.1 Strategies for axon tract tracing with viral vectors. (A and B) Strategies used in Ghosh et al. (2011) for mapping axonal output of single olfactory bulb mitral-tufted (MT) neurons. One approach involved injection of one colored virus into one glomerulus, and a different colored virus into a different glomerulus (A), permitting the detailed comparison of MT neurons associated with different glomeruli. The other method involved mixing different colored virions, and injecting into the same glomerulus (B), labeling individual MT cells associated with the same glomerulus. (C–E) Viral strategy published in Watakabe et al. (2014) for labeling axonal projections based on output. AAVs were used as local infection vectors, and lentiviruses were pseudotyped with a glycoprotein that permitted these viruses to be retrogradely transported to neurons that projected to the injection site. The first strategy (C) involved infection with a lentivirus that encoded the tet transactivator (tTA2), and infection by an AAV expressing either GFP or RFP in neurons expressing the tTA2 protein (those infected with lentivirus). The second strategy (D) involved infecting with two lentiviruses, which could be injected into different output sites. Each virus expressed a different fluorophore under the control of the tet transactivator, which was delivered by the local infection vector, AAV. Schematics of the viruses, and the permutations that were used, are shown in (E).

vectors to label axonal arbors in a variety of other structures, where they provide evidence of morphological diversity within a projection system (Honda et al., 2011; Matsuda et al., 2009).

10.1.4 RABV/Vesicular Stomatitis Virus as Rapid Expression Vectors

In addition to SIN, other viral vectors can be used to obtain rapid and high-level fluorescence expression, filling neuronal processes with fluorescent proteins. Therefore, these vectors can also be used as sensitive anterograde tracers. For example, while the native vesicular stomatitis virus (VSV) can transmit among neurons in the brain, the virus can be modified so that it cannot spread to other neurons. Accordingly, researchers can take advantage of the fact that VSV expresses exogenous genes, such as fluorescent proteins, at high levels. For example, in the sea-horse (*Hippocampus erectus*), when VSV was injected into one eye, GFP-labeled axons (those infected by VSV) could be observed crossing the optic chiasm and projecting to the contralateral, but not ipsilateral hemisphere (Mundell et al., 2015). This exclusively contralateral projection differs from the projection present in many organisms, such as the mouse and primate, in which projections are both contralateral and ipsilateral. While no clear evidence of transsynaptic spread was observed, processes were brightly labeled, demonstrating the potential of anterograde tract tracing with VSV.

RABV can also be used as a sensitive anterograde tracer. RABV typically has been used as a transsynaptic tracer, as the wild-type RABV is a neurotropic virus that transmits among neurons in the central nervous system (CNS). While transsynaptic transmission of RABV has been described as *retrograde* in all but one report (Zampieri et al., 2014), by omitting the glycoprotein, the rapid, robust expression of fluorophores from the viral genome permits the filling of axons with fluorescent proteins, and therefore *anterograde* axonal tracing. A combination of *retrograde* transsynaptic tracing (RABV transmitting from one neuron to another) and *anterograde* axon tracing (fluorescent proteins expressed from the RABV genome filling the neuron) for RABV was first demonstrated using a mixture of RABV vectors pseudotyped with either the RABV glycoprotein (retrograde uptake) or the VSV glycoprotein (VSV-G) (primarily local infection) (Wickersham et al., 2013). By injection into the mouse somatosensory thalamus (VPM), thalamic axons projecting into the somatosensory cortex could be observed densely overlapping the apical dendrites of the layer 6 pyramidal neurons that were labeled by fluorescence from the retrogradely trafficked RABV. In addition to these standard vectors expressing a fluorophore, new variants expressing a fluorophore tagged to the presynaptic protein synaptophysin were also developed, potentially allowing the specific visualization of presynaptic termini.

While useful for producing high levels of transgenes in a short period of time, all of these rapid expression vectors, including SIN, RABV, and VSV, are cytotoxic, and therefore, should not be used for long-term gene expression. Adeno-associated viruses (AAVs) are better suited for studies in which robust, long-term expression of genes is required.

10.1.5 AAV—Cell-Type Specificity and Long-Term Expression

One of the strengths of AAVs is their ability to label large numbers of neurons of the same type. This is often accomplished by the use of specific promoters, or by expression of a recombinase, for example, Cre or Flp, in neurons of choice. Recombinases are proteins that can permanently alter the genetic structure of neurons in which they are expressed. This property is often used to either turn off or turn on the expression of a gene, such as a fluorescent protein. For example, if an AAV that conditionally expresses GFP after Cre-mediated recombination is injected into the motor cortex of a mouse that expresses Cre only in layer V cortical pyramidal neurons, then GFP will only be expressed from those layer V pyramidal neurons in the injection site.

10.1.6 Example 1: Projection Specificity of Subpopulations of Serotonin Neurons

Serotonin neurons in the mouse brain, many of which are located in the dorsal raphe (DR), project to most fore-brain regions. However, little was known about the specificity of neurons within the DR, and if there were subtypes of serotonin neurons that projected to different regions of the forebrain. Classical tracers would not have been able to distinguish serotonin neurons from GABA neurons, which are known to also send long-range projections into the forebrain (Bang and Commons, 2012). Whereas classical methods would have labeled all cells in the DR indiscriminately, since the expression of GFP was Cre-dependent, the tracer was only expressed in cells that expressed Cre (i.e., serotonin neurons).

Focal injections targeting subpopulations of serotonin neurons were made by injecting a Cre-conditional AAV into the SERT-Cre mouse line (Muzerelle et al., 2014). Despite the overall diffuse nature of the projections from the raphe nuclei, each subnucleus was shown to have distinct and complementary projections with other nuclei. The serotonin neurons in the DR and median raphe (MR) had largely complementary projections in both the forebrain and hindbrain; and differences were observed within the DR along the rostrocaudal, mediolateral, and dorsoventral axes. As subnuclei of the DR have been shown to respond differently to stressors (Commons, 2008; Hale et al., 2010; Spannuth et al., 2011), an interesting topic of future research will be to address how these differential projections may relate to functional differences.

10.1.7 Example 2: Intersectional Strategy for Investigating Subtype Specificity

AAVs can be used in combination with other viral vectors to specify the neuron types from which axons are traced. They can also be used together with various transgenic mouse lines, for example, those expressing the Cre recombinase in specific projectional subtypes, to define the neurons to be investigated. Figure 10.1C illustrates an elegant experimental design which targets specific neurons based on their somatic location as well as output site. A retrovirus pseudotyped with a variant of the RABV glycoprotein (RABV-G), which is retrogradely trafficked along axons projecting to the injection site (Kato et al., 2011; Mazarakis et al., 2001), was injected into a target (output site) of the cortex; namely, the thalamus or the contralateral cortex. This resulted in labeling of neurons that projected to that site. This virus was then "intersected" by an AAV injection into the region of the cortex whose axons were to be traced.

The method relied on the usage of the tet-ON/tet-OFF system in order to achieve specific, high-level gene expression of fluorescent proteins in the cortical neurons, which were then visualized in axonal processes. Two versions of this technique were employed. In the first version, the tet transactivator was expressed by a lentivirus and one or two separate AAVs contained fluorophores driven by the tet-response element (TRE) (Figures 10.1C and E). In the second version, two lentiviruses contained fluorophores driven by the TRE, and the AAV contained the tet transactivator (Figures 10.1D and E). The first method permitted high levels of fluorescence expression from the AAV vector, which typically yields higher levels of gene expression than the lentivirus. On the other hand, the second method allowed the marking of two separate output sites. The neurons in the cortex that projected to these sites were labeled in different colors, depending on the output site. This method therefore gives a great deal of flexibility regarding the fluorophores and projection subtypes of neurons to be labeled, and also permits the comparison between two or more populations within the same animal.

The power of the intersectional technique is that subtypes of neurons can be defined by two parameters, soma location and target structure. DNA-based viral vectors have the added advantage of being amenable to modification by Cre and Flp recombination and can therefore be used in combination with existing genetic tools in the mouse, a strategy which could further specify the cell type of the labeled neurons (Luo et al., 2008). Other combinations of vectors, such as the retrograde CAV2 (to be discussed later in this chapter) and AAVs can be used to achieve a similar goal (e.g. Schwarz et al., 2015, Beier et al., 2015). One advantage to having multiple parameters defining different neurons in a single experiment is that projections from multiple neuronal types can be compared in the same brain.

10.1.8 Technologies for Whole-Brain Reconstruction of Axons

Most axon tracing experiments rely on the laborious task of manual reconstruction of axonal arbors. However, recent advances in imaging and automation show promise in reducing the labor required in this process. A recent technology, termed fluorescence micro-optical sectioning tomography (fMOST), aims to advance the throughput of axonal tracing by permitting high-resolution automatic reconstruction of axons throughout the brain (Gong et al., 2013). fMOST yields whole-brain images with one-micron voxel resolution. Another technique, termed CLARITY, permits imaging fluorescent signals in intact tissue (Chung et al., 2013). Through removal of the lipids that obscure visualization of thick tissues, CLARITY allows visualization through millimeters of tissue. This effectively circumvents problems inherent in traditional methods using cut tissue, such as tissue warping. Other methods, such as iDISCO (Reiner et al., 2014) have also recently been developed that serve similar purposes.

Another potential advance in axon tracing by virus vectors could be the mapping, in defined neurons, specifically of synaptic terminals rather than whole axons. This would provide a clearer picture of the actual terminations of neurons, rather than the cytoplasmic or membrane-tagged fluorescence often encoded in viruses, which labels axons in passage as well as terminals. A number of reports have used a vesicle marker as a method of preferentially labeling presynaptic terminals (Oh et al., 2014; Watakabe et al., 2014; Xu and Südhof, 2013). Reports in other organisms also highlight the potential for quantification of synaptic terminations from targeted neuronal subtypes (Chen et al., 2014; Mosca and Luo, 2014).

10.2 VIRAL-MEDIATED AXON TRACING TO DELINEATE BRAIN-WIDE CONNECTIVITY

There are at least two major efforts underway to accomplish global mapping of the mouse brain at a mesoscale level. One is the Mouse Brain Architecture (MBA) project, led by a team of researchers at Cold Spring Harbor laboratories. Another is the Mouse Connectivity Atlas, carried out at the Allen Brain Institute (ABI). As the latter has to date published more results, the focus of this segment will be on the ABI project.

Both the MBA and ABI projects are constructing connectivity maps in the mouse brain by using injections to retrogradely fill neurons or anterogradely label their projections with a fluorescent tracer. In both projects, the tracers of choice have been the classical tracer BDA and an AAV expressing GFP (Oh et al., 2014). The bulk of the information analyzed for the ABI project was done using AAVs expressing GFP. CTB and a glycoprotein-deleted RABV are planned for use in the MBA project as a method for achieving retrograde labeling. The idea behind these projects is that a single injection of an AAV expressing GFP into a given region of the brain will permit labeling and reconstruction of axons from neurons located at the injection site. Repeating this for thousands of different sites in the brain, and mapping each set of projections onto a common reference brain, will enable the construction of a brain-wide connectivity map.

To standardize the tissue processing and imaging steps, the ABI project is utilizing a serial two-photon (STP) tomography system (Ragan et al., 2012). Unlike using a traditional cryostat, with which one serially cuts tissue and images each section, STP tomography images the surface of intact tissue (Figure 10.2B and D). After each image, a section of a defined thickness (100 μm for the ABI project) is cut, and the next surface imaged. These images where then processed and registered onto the ABI 3D reference brain (Figure 10.2A and C).

One potential caveat to analyzing axon-based fluorescence in this way is the lack of distinction in identifying axon terminals from axon fibers. Multiple efforts were made to make fluorescence quantification more accurately reflect axon terminations rather than *en passant* axons, such as eliminating major fiber bundles from quantifications, but so far there are no computational algorithms that successfully distinguish passing axons from axon terminals. The ABI project also briefly reported the use of an AAV expressing a synaptophysin-GFP. The use of this fluorescent tag could aid in the specific identification of axonal terminal zones versus *en passant* axons.

A useful feature of the ABI data is that these can be mapped onto a common reference space and therefore the data from multiple experiments can be directly compared, both for injections into the same site as well as injections at different sites (Figure 10.2E and F). The goal is to eliminate the necessity for manual reconstruction and to extrapolate the connections between three nodes in a circuit with only the selection of the middle node. A demonstration of this input–output mapping was published in a recent methods paper from the ABI (Feng et al., 2015), where the authors use the output map and virtual input map to examine input–output relations of two cortical areas, the motor cortex and somatosensory cortex.

10.3 VIRAL DELIVERY OF GENES TO MEASURE OR TO MANIPULATE ACTIVITY IN AXONS

To this point, the techniques that have been described have been used to anatomically label axons by expressing a protein for histochemical or fluorescent visualization. Several recent advances in viral technology have permitted both the functional manipulation and observation of normal axonal activity. These techniques have allowed access to neurons not only on cell body location, but on where these neurons project.

10.3.1 Optogenetics—Manipulation of Axonal Activity Through Light

The ability to manipulate specific populations of neurons was significantly advanced though the development of optogenetics. Traditionally, axon bundles are stimulated with electrodes, a technique which targets all axons nonspecifically. Also, this technique is significantly invasive, requiring implantation of the electrode directly into the brain.

The idea of optogenetics—a minimally invasive, neuron-specific manipulation—was articulated by Francis Crick (1999), during his Kuffler Lectures at the University of California in San Diego in 1999. This dream was first advanced toward realization in 2002, when Gero Miesenböck and colleagues modified neurons so that they could be optically manipulated, using *Drosophila* rhodopsin-containing photoreceptors to optically activate murine neurons in culture (Zemelman et al., 2002). Channelrhodopsin (ChR2), a bacterial photogated ion channel, was discovered by Peter

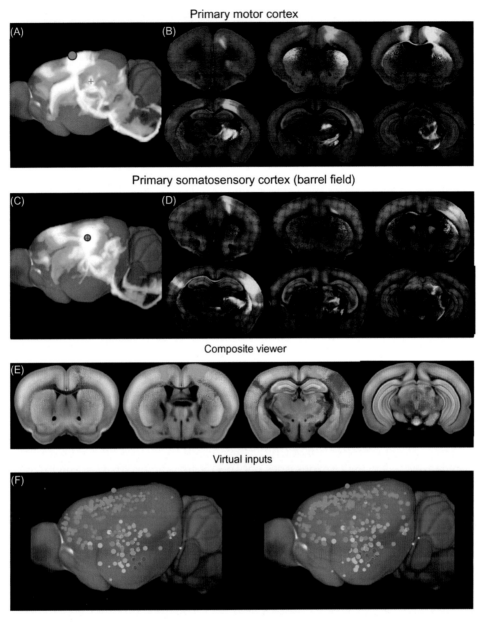

FIGURE 10.2 Examples of data available in the ABI mouse connectivity atlas. (A) A parasagittal view (mouse brain) showing the injection site of an AAV-CMV-GFP in the motor cortex and anterograde projections. Serial coronal sections of this brain are shown in (B). Upper left to lower right image = the rostral–caudal progression. (C and D) The same information as in (A and B), except the AAV-CMV-GFP was injected into the somatosensory (barrel) cortex. (E) These data can be mapped onto a common reference brain and directly compared, as shown in images from the composite viewer. Green, AAV injection into motor cortex; red, AAV injection into somatosensory cortex. (F) Virtual input maps can also be created for defined regions in the brain. For example, the left image is the virtual input map for the primary motor cortex, and the right image, the primary somatosensory cortex. Each dot represents a separate experiment from which anterograde projections were observed in the defined region.

Hegemann and colleagues, also in 2002 (Nagel et al., 2002). The popularity of this technique was established after the practical application of ChR2 in mouse neurons in the lab of Karl Deisseroth (Boyden et al., 2005). In this study, the investigators were able to reliably elicit responses from ChR2-expressing neurons in hippocampal neuronal cultures with application of blue-wavelength light. They were able to achieve millisecond-timescale control of neuronal spiking that was time-locked with light application. In addition, since this technology relied on the expression of a protein (ChR2) within neurons, rather than an injection that labels the extracellular space as well as neurons, the neurons to stimulate could be controlled by the delivery method to specific regions and/or specific cell types.

The literature using ChR2 in different circuits is now extensive and will not be reviewed here. Of specific interest for this chapter is the application of ChR2 to stimulate neurons based on their projection (Figure 10.3A). In this method, an AAV expressing ChR2, either in all cells in an injection site or in cell types defined by Cre expression, is injected into a region of interest (e.g., medial prefrontal cortex). A postsurgical survival period of at least a month is necessary to permit adequate ChR2 expression in the axon terminals to elicit a response to optical stimulation. This long survival time is, in part, because the small single channel conductance of ChR2 requires very high expression of ChR2 in order to reliably elicit action potentials (Bamann et al., 2008). If *in vivo* stimulation is required, then during this incubation time, a fiber optic cable is inserted over a site to which the infected neurons project (e.g., ventral tegmental area (VTA)). Then, at the time of testing, the light can be used to stimulate the axon terminals in the region of interest to elicit neurotransmitter release, with anticipated effect on the postsynaptic neurons. Conversely, axons can be optically silenced with the use of optically activated chloride pumps, like engineered halorhodopins (Zhang et al., 2007) or H+ pumps (Chow et al., 2010). This technique has also been highly effective in acute brain slices, where inputs from specific neurons can be stimulated in a temporally specific fashion. These techniques allow the connectivity between genetically defined populations to be probed with high precision (reviewed in Packer et al., 2013).

While optical stimulation of axonal terminals is a powerful technique for manipulating subpopulations of neurons based on their projection site, one major caveat of this technique is that neurotransmitter release is not necessarily restricted to the site of optical activation. If the optical stimulation is sufficient to evoke an action potential, then antidromic activation—that is, a retrogradely directed action potential—may be elicited (Petreanu et al., 2007). Antidromic activation of terminals has been used to identify ChR2-expressing neurons by physiological recordings in the cell body (Lima et al., 2009). This phenomenon also results in the release of neurotransmitter from collaterals of these neurons. Therefore, while optogenetic activation and inhibition of axons has proven to be a powerful method for manipulating neurons in a cell type-and projection-specific manner, other technologies have arisen to more specifically manipulate localized axon terminals from defined cell types.

10.3.2 Chemogenetic Silencing of Axonal Terminals

One of the weaknesses of optogenetics is that ChR2 is expressed throughout the axon as well as at the axon terminal. Therefore, optogenetic silencing of axons does not specifically and exclusively silence terminals, but is likely to stop axon potentials from progressing through passing axons. This would then inhibit all downstream events, and not specifically block local synaptic transmission (Stuber et al., 2011; Tye et al., 2011). Therefore, in 2014, investigators utilized hM4D, a designer G_i-protein coupled receptor that had previously been used to silence neurons, and made an axon-specific variant that only localizes to neuronal terminals (Stachniak et al., 2014) (Figure 10.3B). When the exogenous small-molecule clozapine-N-oxide (CNO) is administered to neurons, CNO activates hM4D, which then acts to electrically silence neurons by activating the G_i signaling pathway (Armbruster et al., 2007). As CNO/hM4D act to both electrically silence neurons and to inhibit transmitter release, the researchers appended 55 amino acids of the C-terminus of neurexin-1α to the C-terminus of hM4D. This was a clever modification, as axonal distribution can be achieved by two means: direct axonal targeting or selective removal from the somatodendritic compartment. As hM4D already reached axon terminals without modification, the C-terminus of neurexin-1α helped to selectively remove hM4D from the cell body and dendrites, thereby concentrating it at axonal terminals.

What was still needed was to define which neurons were to express the receptor. To do this, the researchers injected an AAV expressing hM4D[nrxn] into a SIM1-Cre mouse, which expresses Cre in a subset of neurons in the paraventricular hypothalamus (PVH). The silencing of these had been shown to inhibit feeding, but the downstream neuron targets that were endogenously being inhibited by these SIM1-Cre-expressing neurons in the PVH remained unclear. CNO was administered in multiple PVH output sites of SIM1-Cre mice injected with AAV-hM4D[nrxn], and the feeding behavior observed. The researchers found certain regions in the periaqueductal grey (PAG)/DR region that preferentially enhanced feeding behavior. From this, it could be inferred that neurotransmitter release from SIM1-Cre neurons in the PAG/DR region is normally directly responsible for inhibiting feeding behavior in mice. Viral expression of the receptor for this ligand-receptor pair therefore provides a powerful method for selectively inhibiting the neuro transmitter release from axons of genetically defined neurons in a spatially defined region.

10.3.3 Fiber Photometry—Measuring Axonal Activity

Advances in genetic technologies have permitted expression of proteins in specific cell types defined by recombinase expression. Expressing ChR2 in axons allows the specific activation of these defined neurons in a temporally

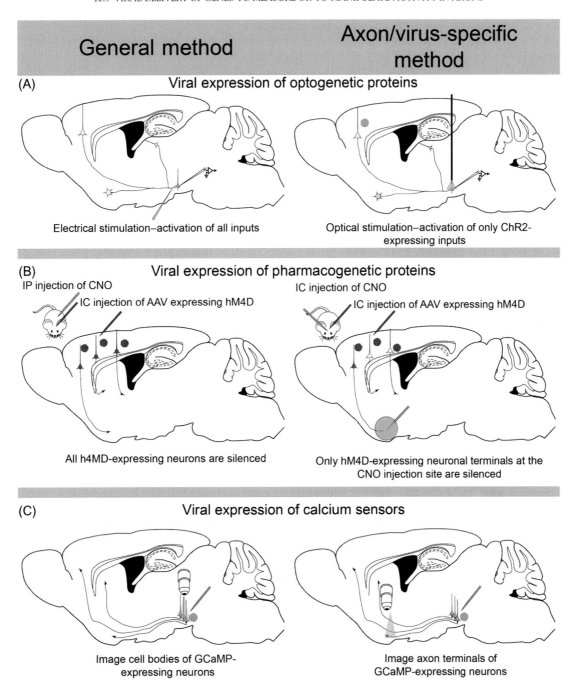

General method	Axon/virus-specific method

(A) Viral expression of optogenetic proteins

Electrical stimulation–activation of all inputs

Optical stimulation–activation of only ChR2-expressing inputs

(B) Viral expression of pharmacogenetic proteins

IP injection of CNO
IC injection of AAV expressing hM4D

IC injection of CNO
IC injection of AAV expressing hM4D

All h4MD-expressing neurons are silenced

Only hM4D-expressing neuronal terminals at the CNO injection site are silenced

(C) Viral expression of calcium sensors

Image cell bodies of GCaMP-expressing neurons

Image axon terminals of GCaMP-expressing neurons

FIGURE 10.3 Viral delivery of functional genes for axonal manipulation. (A) Viruses can deliver ChR2 to neurons in anatomically specific regions in the brain. These neurons can be either somatically stimulated, or axonally stimulated, to cause neurotransmitter release. The major advantage of this strategy for stimulating axonal terminals is that whereas traditional electrode stimulation of inputs would activate axons and terminals indiscriminately (left), optical stimulation will activate only neurons expressing ChR2 (right). (B) Viruses can deliver pharmacogenetic reagents to specific neurons in the brain. By intraperitoneal injection of the hM4D agonist CNO, all neurons expressing hM4D will be silenced (left). If CNO is microinjected directly into the brain, it will silence axon terminals only near the site of CNO infusion (right). (C) Viruses can also deliver genetically encoded calcium sensors (GCaMPs) to neurons in specific regions of the brain. The soma of these neurons can be imaged to record the activity of these neurons (left), or the axons of specific GCaMP-expressing neurons can be imaged by selection of an output site (right).

precise manner. However, optical activation of ChR2 is a nonnatural stimulus and does not necessarily represent the native behavior of these neurons. While many technologies exist to monitor neuronal firing, many of these techniques are appropriate for neuronal firing in the cell bodies and are not sensitive enough to observe activity that occurs in axon terminals. This has been made possible in recent years by the development of increasingly sensitive, protein-based calcium indicators, such as the genetically encodable calcium indicators, GCaMPs (Nakai et al., 2001).

This protein scaffold has been continuously modified and improved, leading to the construction of proteins that can detect single action potentials (Chen et al., 2013). One of these variants, GCaMP5, was used to measure activity in axons from the VTA in the nucleus accumbens (NAc) (Gunaydin et al., 2014). The investigators combined GCaMP technology with a more recent technique using a photon counter in the striatum to observe neuronal activity (Cui et al., 2013). Rather than requiring a microscope to take images, as is common for calcium imaging experiments, this technology quantifies the number of photons absorbed from neurons that were actively fluorescing.

In this study, the investigators injected a Cre-dependent AAV expressing GCaMP5 (AAV-DIO-GCaMP5) in a mouse line (tyrosine hydroxylase, or TH-Cre) in which dopamine (DA) neurons in the express Cre (Lammel et al., 2015) (Figure 10.3C). Through recording calcium signals in the TH-Cre cell bodies, they first showed that VTA neurons expressing TH-Cre had an enhanced calcium signal when the mouse was in a social context as opposed to interacting with a novel object. However, VTA-DA neurons have multiple projection sites, and the investigators wanted to determine if all of the VTA Cre-expressing neurons, or only a subset of them, increased their activity in response to social stimuli. To do this, they first expressed ChR2 in TH-Cre-expressing neurons in the VTA, and optically stimulated axon terminals in the nucleus accumbens medial shell (NAc) or the medial prefrontal cortex (mPFC). They observed that optically stimulating the terminals in the NAc, but not in the mPFC, increased social interaction. However, as ChR2 stimulation is an exogenous stimulus, it was not completely clear if this projection is normally involved in social interaction. In order to address this question, they needed to specifically observe the activity of VTA neurons projecting to the NAc. To do this, they injected AAV-DIO-GCaMP5 into the VTA and measured the GCaMP5 signal in the NAc. Fluorescent signals detected in the NAc could be attributed to GCaMP5 activity in VTA neuron fibers in the NAc. The researchers observed that axons in the NAc showed an increased fluorescence signal during social interaction compared to interaction with a novel object, similar to what was observed when recording calcium signals from the cell body. These results demonstrate the ability to measure calcium signals in axon terminals and open the possibility of not only manipulating neurons based on projection site, but also observing the native activity patterns of these neurons.

10.4 RETROGRADE TRANSMISSION OF VIRUSES—GENETIC ACCESS TO NEURONS BASED ON PROJECTION SITE

Certain viruses are retrogradely trafficked from the axon terminal to the cell body. This has allowed for the labeling of neurons based on their projection site. Herpes virus (HSV), pseudorabies (PRV), and RABV are all examples of viruses used for this purpose. More recently, a number of viruses have been engineered to express transgenes in retrogradely labeled neurons, but not to replicate and spread. An early attempt provided RABV-G to a heterologous vector (lentivirus) and showed that this was sufficient to provide retrograde axonal transport (Mazarakis et al., 2001). This was later done with RABV (Wickersham et al., 2007a), and for other vectors, such as VSV (Beier et al., 2011).

These viruses that can be fitted for retrograde uptake are all enveloped viruses. There are also nonenveloped viruses that display this ability. One example is the canine adenovirus (CAV2) (Kremer et al., 2000). The high efficiency and low toxicity of this virus was exploited in a clever study that used CAV2 to supply a gene specifically to neurons defined by their projection (Hnasko et al., 2006). The *TH* gene, which is necessary for the production of the neuromodulator DA, was made Cre conditional (meaning that it was only expressed in neurons expressing Cre), while norepinephrine-producing neurons, which need TH in the biosynthetic pathway necessary to produce norepinephrine, were engineered to express TH constitutively. CAV2-Cre was injected into the dorsal striatum, where it was taken up by axon terminals of neurons projecting to the injection site, including DA neurons in the substantia nigra pars compacta. Animals that were not injected with CAV2-Cre required injection of L-DOPA for survival (as animals that lack DA do not survive), but, in contrast, CAV2-Cre injection restored a sufficient level of DA production and release to permit survival. Not only could the mice survive and remain viable for over a year without the need for exogenous L-DOPA, but they had largely restored feeding and locomotor behavior, albeit different than in control animals.

There have also been numerous reports of AAVs being retrogradely trafficked. A number of studies have suggested that AAVs are efficiently retrogradely trafficked (Chen et al., 2013; Ciesielska et al., 2011; Kaspar et al., 2002; Masamizu et al., 2011; Passini et al., 2005; Paterna et al., 2004; Salegio et al., 2012; Zhang et al., 2013), while others have suggested that they are not retrogradely trafficked with high efficiency (Burger et al., 2004; Chamberlin et al., 1998; Salegio et al., 2012). A detailed study into this issue suggested that there may not be consistent differences between serotypes, and there may even be stock-to-stock variation between viral preparations (Rothermel et al.,

2013). In addition, the mechanisms by which enveloped and nonenveloped viruses are retrogradely trafficked remain poorly understood.

10.4.1 Retrograde Transsynaptic Tracing—Neuron–Neuron Viral Transmission

The aforementioned techniques described the uptake of viruses by axon terminals after viral injection. This permits the labeling of neuronal populations that project to the site of injection, but do not distinguish neurons with synaptic contacts at the site of injection from those that send passing axons (e.g., see Miyamichi et al., 2011).

Injection damage is a common source of tracer uptake by axons of passage. A simple way to eliminate this problem is to introduce a tracer through means other than injection; for example, achieving retrograde transmission by using viruses that transmit transneuronally in the CNS. The viruses most commonly used for this purpose are HSV and RABV (Figure 10.4A). The capability of some viruses to cross multiple synapses led to circuit mapping based on the timing of transsynaptic spread among neurons (Rathelot and Strick, 2009; Ugolini, 1995). However, timing alone cannot be used to unambiguously delineate neuronal interconnectivity. In addition, these viruses lack cell-type specificity, and therefore the inputs labeled in such experiments represent the inputs to multiple cell types located at the site of injection.

More recently, methods have been developed for the analysis of inputs to defined cell populations. A PRV variant that expressed GFP and the thymidine kinase gene conditionally in cells that expressed Cre (DeFalco et al., 2001) was the first constructed for this purpose. This virus could transmit polysynaptically from cells that expressed Cre, and expressed GFP after Cre-mediated recombination. This study suggested differences in inputs to two populations in the hypothalamus (NPY neurons and leptin receptor-expressing neurons). However, discerning the order of connectivity to each of these populations was very difficult. The nature of those differences, and how they relate to direct connectivity to the Cre-expressing neurons from which infection was initiated, was not clear.

In order to unambiguously identify monosynaptic inputs to "starter neurons" (those from which transsynaptic tracing is initiated), a vector that can infect defined neuronal classes, replicate within those neurons, and spread transsynaptically to input neurons is necessary. Once in the input neurons, the virus would have to replicate (in order to express a gene by which the virus could be detected), but not spread further.

In 2007, a study used a modified RABV to determine neurons making direct, monosynaptic connections onto a starter neuronal population, without labeling polysynaptic input (Figure 10.4C). RABV was modified such that the RABV-G gene was deleted from the genome and replaced by GFP (Wickersham et al., 2007a,b). This modification made the virus unable to spread from cells not expressing the viral glycoprotein. The second necessary alteration to the virus was to pseudotype the virus with the glycoprotein from a different virus, the Avian Sarcoma and Leukosis Virus A (ASLV-A, or EnvA). ASLV is an avian virus that normally only infects avian species. Therefore, this second modification altered the infection properties of the virus: in non-avian species, the EnvA-pseudotyped RABV can only infect cells made to express the EnvA receptor, TVA.

TVA expression is often made to be restricted in neurons that express Cre. Once RABV infects these cells, it can replicate and express GFP, but cannot spread to other neurons, due to the lack of glycoprotein expression in the genome. The virus can be made to transmit transsynaptically to inputs by providing the glycoprotein gene specifically to neurons expressing TVA (often by previous infection of an AAV expressing the glycoprotein gene). By also providing a fluorophore to the starter neurons (those expressing TVA and glycoprotein), the starter neurons and their direct monosynaptic inputs can be discerned.

The monosynaptic RABV method has been employed in many types of studies to address different questions. It has been used as a screening method for identifying inputs (Betley et al., 2013; Haubensak et al., 2010; Krashes et al., 2014), mapping circuit organization and topography (Miyamichi et al., 2011; Stepien et al., 2010; Sun et al., 2014), determining input specificity to regionally distinct or intermingled cell populations (Ogawa et al., 2014; Pollak Dorocic et al., 2014; Wall et al., 2013; Watabe-Uchida et al., 2012; Weissbourd et al., 2014), or investigating timing of circuit formation (Deshpande et al., 2013; Takatoh et al., 2013), among other applications.

In summary, viruses that can label neurons from the injection site (1) provide a method for identifying neurons based on projection site and also (2) provide a way to express transgenes in neurons based on projection. Viruses that can transsynaptically transmit to input neurons provide a method of specifically labeling interconnected populations, with modified viruses permitting the identification of direct, monosynaptic inputs. In addition, other viruses exist that spread among neurons in the opposite direction—that is, not to the *input* neurons, but to the *output* neurons.

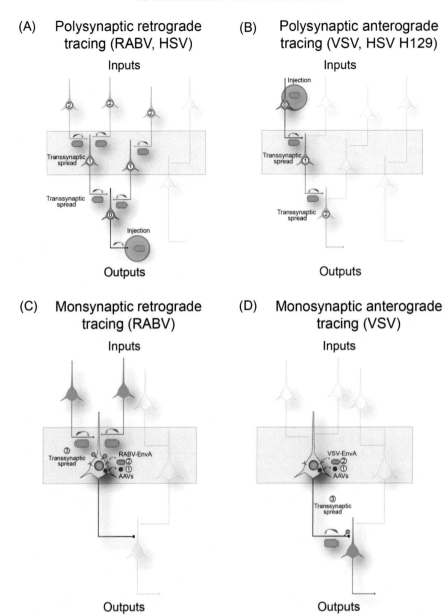

FIGURE 10.4 Examples of strategies to achieve viral anterograde and retrograde transsynaptic transmission. (A) Polysynaptic retrograde tracing involves the labeling of interconnected neurons via multiple synaptic crossings in the retrograde direction—that is, against the normal information flow among neurons. Upon injection, the virus infects neurons locally, as well as those that project to the site of injection. The virus replicates within these cells, exits the dendrites, and infects the presynaptic neurons. The virus then gets transported from the axon terminals to the cell bodies, and then repeats the cycle. "0" marks neurons infected by the initial inoculum ("0th order neurons"), "1" marks 1st order inputs, and "2" marks 2nd order inputs. RABV and HSV are the most common vectors used for this purpose. (B) Polysynaptic anterograde tracing labels multiple neurons in the anterograde direction—with the normal information flow—from the starting site. Here, upon injection, the virus infects only neurons located near the injection site. After infection, the virus must be transported to the axon terminal, where it is released, and infects the postsynaptic neurons, where it replicates and repeats the cycle. The H129 strain and VSV are two known anterograde transsynaptic viruses. (C) Monosynaptically restricted retrograde transsynaptic transmission can be obtained by modification of the viral genome, such that the virus lacks a gene essential for cell-to-cell spread (Wickersham et al., 2007b). This gene can then be re-supplied to the virus by expressing it only in select neurons. Here, (1) AAVs are used to supply two genes to starter neurons—TVA, which can be used to direct specific infection of starter neurons, and RABV-G, the glycoprotein that permits cell-to-cell transmission. Next, (2) RABV pseudotyped with EnvA—to infect only TVA-expressing neurons (here, labeled red)—is injected into the same site, approximately 2 weeks after AAV injections. (3) The virus then replicates within these neurons and expresses GFP (the cell becomes yellow), and because the neurons express RABV-G, the virus then spreads to synaptically connected inputs. The virus can replicate in these inputs and therefore express transgenes encoded in the virus, including a fluorescent protein (green). Since the inputs are not expressing RABV-G, the virus cannot spread further. RABV is the most frequently used monosynaptic retrograde vector. (D) Monosynaptically restricted anterograde transsynaptic transmission works in a similar fashion, but a glycoprotein other than RABV-G is used (Beier et al., 2011). The combination of VSV with glycoproteins such as VSV-G was demonstrated to transmit transsynaptically in the anterograde direction.

10.5 TRANSNEURONAL ANTEROGRADE SPREAD—VIRUSES CAN ALSO DEFINE OUTPUTS

While HSVs had been reported to spread transneuronally both in the anterograde and retrograde directions, the first report of a transsynaptic anterograde-specific virus occurred in 1991 (Zemanick et al., 1991) (Figure 10.4B). In this study, a specific variant of HSV, called H129, was injected into the primate motor cortex. After a few days, virally labeled cell bodies were observed in regions known to receive direct projections from the motor cortex, including the putamen and pontine nuclei (direct, monosynaptic anterograde projections), as well as further downstream regions, such as the globus pallidus and the cerebellar cortex (polysynaptic). This technique was later extended to other organisms, such as the rodent, where the virus infected cells in multiple nodes of the visual system in the brain after intraocular injections, including the lateral geniculate nucleus, suprachiasmatic nucleus, superior colliculus, as well as targets located further downstream, such as the primary visual cortex (V1) (Sun et al., 1996). The virus was recently modified to be Cre-dependent, such that anterograde-specific transsynaptic tracing is initiated from Cre-expressing cells (Lo and Anderson, 2011).

In addition to HSV, VSV has also displayed the ability to spread from one neuron to its outputs. VSV had been reported to spread along the visual pathway in the brain after an intraocular injection (Lundh, 1990), but the viral transmission patterns were not carefully analyzed until many years later (Beier et al., 2011). VSV was injected intraocularly, intranasally, or intracranially, and in each case, labeled interconnected neurons in a pattern consistent with anterograde transsynaptic transmission. VSV vectors have been modified for monosynaptic anterograde transmission (Beier et al., 2011) (Figure 10.4D), and researchers have taken advantage of the broad tropism of VSV for both tract tracing and transsynaptic tracing in a wide variety of organisms (Mundell et al., 2015).

Anterograde transsynaptic technology permits the delivery of a genetic payload to postsynaptic cells *only* if they are synaptically connected to a starter neuron. How does this happen? Take for example, HSV. Many people are naturally infected with HSV, as after initial infection, the virus is not always completely cleared from the body, but rather can become latent within infected cells. Upon reactivation, HSV tends to infect neurons in the periphery, without infecting the CNS (Lyman et al., 2007). In order to avoid infecting neurons in the CNS, upon reactivation, the virus needs to be transmitted from the axon terminal and *not* the cell body or dendrites of the originally infected neuron, as release from the cell body or dendrites would result in infection of descending projections from the brain.

The mechanisms underlying viral anterograde-specific transsynaptic transmission remain unclear. However, clues exist from studies of H129 and PRV, which is a type of Herpes virus. For example, the Bartha strain of PRV, which only transmits transsynaptically retrogradely among neurons, contains a large deletion in three viral genes (Lomniczi et al., 1987). One of these genes is the US9 protein, which has been demonstrated to be responsible for axonal targeting of viral proteins (Tomishima and Enquist, 2001). It is therefore possible that, without US9, viral glycoproteins and capsid proteins do not effectively localize to the axon terminals (Lyman et al., 2007), and therefore virions cannot be released from terminals.

To date, the mechanism by which US9 may be working is incompletely understood. The data suggest, however, that US9, along with the glycoprotein gE, are required for anterograde transsynaptic transmission, and viruses with US9 deletions can transmit among neurons in the retrograde direction (e.g., the Bartha strain of PRV). These data may imply that, at least in the case of HSV, the retrograde direction is the "default state" of the virus, and US9 acts to recruit viral proteins away from the cell body and dendrites and into the axon.

Both HSV and VSV have shown the capability to transmit among neurons either in the anterograde or retrograde directions. (Beier et al., 2011, 2013). For this to be possible, the viral capsid, along with the glycoprotein, must be able to bud from both the dendrites as well as the axon terminals. For VSV, it appears that the membrane-associated glycoprotein is the major determinant of directional transsynaptic viral transmission (Beier et al., 2011). As there are no known specific interactions between the VSV-G and the other viral components in terms of localization within the cell, the glycoprotein likely acts independently to guide the direction of VSV transsynaptic transmission.

10.6 CONCLUSIONS

Viruses provide an extraordinarily powerful set of tools to study the CNS. As viruses evolved different abilities to cope with their unique environments, they have found creative solutions to almost any problem, providing features that can be exploited by neuroscientists. The rapid gene expression of SIN, VSV, and RABV is advantageous for labeling small numbers of neurons in a short time, and the robust expression obtained with these vectors facilitates axonal reconstruction. In contrast, genes from AAVs are expressed much more slowly, but AAVs are nontoxic and can be used for long-term expression. In addition, they are more amenable to work in tandem with modern genetic techniques, for example, by using Cre and Flp to conditionally express genes only in cell types of choice. These vectors are currently the workhorse vectors for the neuroscience community for applications of connectivity mapping.

Beyond these tract tracing vectors, some viruses transmit among neurons in a directionally specific manner and are well-suited for circuitry mapping. Modifications to these viruses now permit restricted infection and transmission in order to more accurately define the connectivity relationships among neurons.

References

Armbruster, B.N., Li, X., Pausch, M.H., Herlitze, S., Roth, B.L., 2007. Evolving the lock to fit the key to create a family of G protein-coupled receptors potently activated by an inert ligand. Proc. Natl. Acad. Sci. U.S.A. 104 (12), 5163–5168.

Bamann, C., Kirsch, T., Nagel, G., Bamberg, E., 2008. Spectral characteristics of the photocycle of channelrhodopsin-2 and its implication for channel function. J. Mol. Biol. 375, 686–694.

Bang, S.J., Commons, K.G., 2012. Forebrain GABAergic projections from the dorsal raphe nucleus identified by using GAD67-GFP knock-in mice. J. Comp. Neurol. 520, 4157–4167.

Beier, K.T., Saunders, A., Oldenburg, I. A, Miyamichi, K., Akhtar, N., Luo, L., et al., 2011. Anterograde or retrograde transsynaptic labeling of CNS neurons with vesicular stomatitis virus vectors. Proc. Natl. Acad. Sci. U.S.A. 108 (37), 15414–15419.

Beier, K., Saunders, A., Oldenburg, I., Sabatini, B., Cepko, C., 2013. Vesicular stomatitis virus with the rabies virus glycoprotein directs retrograde transsynaptic transport among neurons in vivo. Front. Neural. Circuits 7, 11.

Beier, K.T., Steinberg, E.E., DeLoach, K.E., Xie, S., Gao, X.J., Miyamichi, K., et al., 2015. Circuit architecture of midbrain dopamine neurons revealed by systematic input -output mapping. Cell 162, 622–634.

Betley, J.N., Fang, Z., Cao, H., Ritola, K.D., Sternson, S.M., 2013. Parallel, redundant circuit organization for homeostatic control of feeding behavior. Cell 155 (6), 1337–1350.

Boyden, E.S., Zhang, F., Bamberg, E., Nagel, G., Deisseroth, K., 2005. Millisecond-timescale, genetically targeted optical control of neural activity. Nat. Neurosci. 8 (9), 1263–1268.

Burger, C., Gorbatyuk, O.S., Velardo, M.J., Peden, C.S., Williams, P., Zolotukhin, S., et al., 2004. Recombinant AAV viral vectors pseudotyped with viral capsids from serotypes 1, 2, and 5 display differential efficiency and cell tropism after delivery to different regions of the central nervous system. Mol. Ther. 10 (2), 302–317.

Chamberlin, N.L., Du, B., De Lacalle, S., Saper, C.B., 1998. Recombinant adeno-associated virus vector: use for transgene expression and anterograde tract tracing in the CNS. Brain Res. 793, 169–175.

Chen, J.L., Carta, S., Soldado-Magraner, J., Schneider, B.L., Helmchen, F., 2013. Behaviour-dependent recruitment of long-range projection neurons in somatosensory cortex. Nature 499, 336–340.

Chen, Y., Akin, O., Nern, A., Tsui, C.Y.K., Pecot, M.Y., Zipursky, S.L., 2014. Cell-type-specific labeling of synapses in vivo through synaptic tagging with recombination. Neuron 81 (2), 280–293.

Choi, G.B., Stettler, D.D., Kallman, B.R., Bhaskar, S.T., Fleischmann, A., Axel, R., 2011. Driving opposing behaviors with ensembles of piriform neurons. Cell 146 (6), 1004–1015.

Chow, B.Y., Han, X., Dobry, A.S., Qian, X., Chuong, A.S., Li, M., et al., 2010. High-performance genetically targetable optical neural silencing by light-driven proton pumps. Nature 463 (7277), 98–102.

Chung, K., Wallace, J., Kim, S.-Y., Kalyanasundaram, S., Andalman, A.S., Davidson, T.J., et al., 2013. Structural and molecular interrogation of intact biological systems. Nature 497, 332–337.

Ciesielska, A., Mittermeyer, G., Hadaczek, P., Kells, A.P., Forsayeth, J., Bankiewicz, K.S., 2011. Anterograde axonal transport of AAV2-GDNF in rat basal ganglia. Mol. Ther. 19 (5), 922–927.

Commons, K.G., 2008. Evidence for topographically organized endogenous 5-HT-1A receptor-dependent feedback inhibition of the ascending serotonin system. Eur. J. Neurosci. 27 (August), 2611–2618.

Crick, F., 1999. The impact of molecular biology on neuroscience. Philos. Trans. R. Soc. Lond. B. Biol. Sci. 354, 2021–2025.

Cui, G., Jun, S.B., Jin, X., Pham, M.D., Vogel, S.S., Lovinger, D.M., et al., 2013. Concurrent activation of striatal direct and indirect pathways during action initiation. Nature 494 (7436), 238–242.

DeFalco, J., Tomishima, M., Liu, H., Zhao, C., Cai, X., Marth, J.D., et al., 2001. Virus-assisted mapping of neural inputs to a feeding center in the hypothalamus. Science 291 (March), 2608–2613.

Deshpande, A., Bergami, M., Ghanem, A., Conzelmann, K.-K., Lepier, A., Götz, M., et al., 2013. Retrograde monosynaptic tracing reveals the temporal evolution of inputs onto new neurons in the adult dentate gyrus and olfactory bulb. Proc. Natl. Acad. Sci. U.S.A. 110 (12), E1152–E1161.

Feng, D., Lau, C., Ng, L., Li, Y., Kuan, L., Sunkin, S.M., et al., 2015. Exploration and visualization of connectivity in the adult mouse brain. Methods 73, 90–97.

Ghosh, S., Larson, S.D., Hefzi, H., Marnoy, Z., Cutforth, T., Dokka, K., et al., 2011. Sensory maps in the olfactory cortex defined by long-range viral tracing of single neurons. Nature 472 (7342), 217–220.

Glover, J.C., Petursdottir, G., Jansen, J.K., 1986. Fluorescent dextran-amines used as axonal tracers in the nervous system of the chicken embryo. J. Neurosci. Methods 18 (3), 243–254.

Gong, H., Zeng, S., Yan, C., Lv, X., Yang, Z., Xu, T., et al., 2013. Continuously tracing brain-wide long-distance axonal projections in mice at a one-micron voxel resolution. NeuroImage 74, 87–98.

Gunaydin, L.A., Grosenick, L., Finkelstein, J.C., Kauvar, I.V., Fenno, L.E., Adhikari, A., et al., 2014. Natural neural projection dynamics underlying social behavior. Cell 157 (7), 1535–1551.

Hale, M.W., Johnson, P.L., Westerman, A.M., Abrams, J.K., Shekhar, A., Lowry, C.A., 2010. Multiple anxiogenic drugs recruit a parvalbumin-containing subpopulation of GABAergic interneurons in the basolateral amygdala. Prog. Neuropsychopharmacol. Biol. Psychiatry 34 (7), 1285–1293.

Haubensak, W., Kunwar, P.S., Cai, H., Ciocchi, S., Wall, N.R., Ponnusamy, R., et al., 2010. Genetic dissection of an amygdala microcircuit that gates conditioned fear. Nature 468 (7321), 270–276.

Hnasko, T.S., Perez, F.A., Scouras, A.D., Stoll, E.A., Gale, S.D., Luquet, S., et al., 2006. Cre recombinase-mediated restoration of nigrostriatal dopamine in dopamine-deficient mice reverses hypophagia and bradykinesia. Proc. Natl. Acad. Sci. U.S.A. 103, 8858–8863.

Honda, Y., Furuta, T., Kaneko, T., Shibata, H., Sasaki, H., 2011. Patterns of axonal collateralization of single layer V cortical projection neurons in the rat presubiculum. J. Comp. Neurol. 519, 1395–1412.

Kaspar, B.K., Erickson, D., Schaffer, D., Hinh, L., Gage, F.H., Peterson, D.A., 2002. Targeted retrograde gene delivery for neuronal protection. Mol. Ther. 5 (1), 50–56.

Kato, S., Kuramochi, M., Takasumi, K., Kobayashi, K., Inoue, K., Takahara, D., et al., 2011. Neuron-specific gene transfer through retrograde transport of lentiviral vector pseudotyped with a novel type of fusion envelope glycoprotein. Hum. Gene. Ther. 22 (December), 1511–1523.

Krashes, M.J., Shah, B.P., Madara, J.C., Olson, D.P., Strochlic, D.E., Garfield, A.S., et al., 2014. An excitatory paraventricular nucleus to AgRP neuron circuit that drives hunger. Nature 507 (7491), 238–242.

Kremer, E.J., Boutin, S., Chillon, M., 2000. Canine adenovirus vectors: an alternative for adenovirus-mediated gene transfer. J. Virol. 74, 1.

Kuypers, H.G., Bentivoglio, M., van der Kooy, D., Catsman-Berrevoets, C.E., 1979. Retrograde transport of bisbenzimide and propidium iodide through axons to their parent cell bodies. Neurosci. Lett. 12 (1), 1–7.

Lammel, S., Steinberg, E.E., Földy, C., Wall, N.R., Beier, K., Luo, L., et al., 2015. Diversity of transgenic mouse models for selective targeting of midbrain dopamine neurons. Neuron 85 (2), 429–438. http://dx.doi.org/10.1016/j.neuron.2014.12.036.

Lima, S.Q., Hromádka, T., Znamenskiy, P., Zador, A.M., 2009. PINP: a new method of tagging neuronal populations for identification during in vivo electrophysiological recording. PLoS One 4 (7), e6099.

Lo, L., Anderson, D.J., 2011. A {Cre}-dependent, anterograde transsynaptic viral tracer for mapping output pathways of genetically marked neurons. Neuron 72 (6), 938–950.

Lomniczi, B., Watanabe, S., Ben-Porat, T., Kaplan, A.S., 1987. Genome location and identification of functions defective in the Bartha vaccine strain of pseudorabies virus. J. Virol. 61 (3), 796–801.

Lundh, B., 1990. Spread of vesicular stomatitis virus along the visual pathways after retinal infection in the mouse. Acta Neuropathol. 79 (4), 395–401.

Luo, L., Callaway, E.M., Svoboda, K., 2008. Genetic dissection of neural circuits. Neuron 57 (5), 634–660.

Lyman, M.G., Feierbach, B., Curanovic, D., Bisher, M., Enquist, L.W., 2007. Pseudorabies virus Us9 directs axonal sorting of viral capsids. J. Virol. 81 (20), 11363–11371.

Masamizu, Y., Okada, T., Kawasaki, K., Ishibashi, H., Yuasa, S., Takeda, S., et al., 2011. Local and retrograde gene transfer into primate neuronal pathways via adeno-associated virus serotype 8 and 9. Neuroscience 193, 249–258.

Matsuda, W., Furuta, T., Nakamura, K.C., Hioki, H., Fujiyama, F., Arai, R., et al., 2009. Single nigrostriatal dopaminergic neurons form widely spread and highly dense axonal arborizations in the neostriatum. J. Neurosci. 29 (2), 444–453.

Mazarakis, N.D., Azzouz, M., Rohll, J.B., Ellard, F.M., Wilkes, F.J., Olsen, A.L., et al., 2001. Rabies virus glycoprotein pseudotyping of lentiviral vectors enables retrograde axonal transport and access to the nervous system after peripheral delivery. Hum. Mol. Genet. 10 (19), 2109–2121.

Miyamichi, K., Amat, F., Moussavi, F., Wang, C., Wickersham, I., Wall, N.R., et al., 2011. Cortical representations of olfactory input by trans-synaptic tracing. Nature 472 (7342), 191–196.

Mosca, T.J., Luo, L., 2014. Synaptic organization of the Drosophila antennal lobe and its regulation by the teneurins. eLife 3, 1–29.

Mundell, N., Beier, K.T., Pan, Y., Lapan, S., Göz Aytürk, D., Berezovskii, V.K., et al., 2015. Vesicular stomatitis virus enables gene transfer and transsynaptic tracing in a wide range of organisms. J. Comp. Neurol. 523, 1639–1663.

Muzerelle, A., Scotto-Lomassese, S., Bernard, J.F., Soiza-Reilly, M., Gaspar, P., 2014. Conditional anterograde tracing reveals distinct targeting of individual serotonin cell groups (B5–B9) to the forebrain and brainstem. Brain Struct. Funct. http://dx.doi.org/10.1007/s00429-014-0924-4.

Nagel, G., Ollig, D., Fuhrmann, M., Kateriya, S., Musti, A.M., Bamberg, E., et al., 2002. Channelrhodopsin-1: a light-gated proton channel in green algae. Science 296, 2395–2398.

Nakai, J., Ohkura, M., Imoto, K., 2001. A high signal-to-noise Ca(2+) probe composed of a single green fluorescent protein. Nat. Biotechnol. 19, 137–141.

Ogawa, S.K., Cohen, J.Y., Hwang, D., Uchida, N., Watabe-Uchida, M., 2014. Organization of monosynaptic inputs to the serotonin and dopamine neuromodulatory systems. Cell Rep. 8 (4), 1105–1118.

Oh, S.W., Harris, J.A., Ng, L., Winslow, B., Cain, N., Mihalas, S., et al., 2014. A mesoscale connectome of the mouse brain. Nature 508 (7495), 207–214.

Packer, A.M., Roska, B., Häusser, M., 2013. Targeting neurons and photons for optogenetics. Nat. Neurosci. 16 (7), 805–815.

Passini, M.A., Macauley, S.L., Huff, M.R., Taksir, T.V., Bu, J., Wu, I.H., et al., 2005. AAV vector-mediated correction of brain pathology in a mouse model of Niemann-Pick A disease. Mol. Ther. 11 (5), 754–762.

Paterna, J., Feldon, J., Büeler, H., Bu, H., 2004. Transduction profiles of recombinant adeno-associated virus vectors derived from serotypes 2 and 5 in the nigrostriatal system of rats. J. Virol. 78 (13), 6808–6817.

Petreanu, L., Huber, D., Sobczyk, A., Svoboda, K., 2007. Channelrhodopsin-2-assisted circuit mapping of long-range callosal projections. Nat. Neurosci. 10 (5), 663–668.

Pollak Dorocic, I., Fürth, D., Xuan, Y., Johansson, Y., Pozzi, L., Silberberg, G., et al., 2014. A whole-brain atlas of inputs to serotonergic neurons of the dorsal and median raphe nuclei. Neuron 83 (3), 663–678.

Ragan, T., Kadiri, L.R., Venkataraju, K.U., Bahlmann, K., Sutin, J., Taranda, J., et al., 2012. Serial two-photon tomography for automated ex vivo mouse brain imaging. Nat. Methods 9 (3), 255–258.

Rathelot, J.-A., Strick, P.L., 2009. Subdivisions of primary motor cortex based on cortico-motoneuronal cells. Proc. Natl. Acad. Sci. U. S. A. 106 (3), 918–923.

Reiner, A., Veenman, C., Honig, M., 1993. Anterograde tracing using biotinylated dextran amine. Neurosci. Prot. 93 (section 050-14).

Reiner, N., Wu, Z., Simon, D.J., Yang, J., Ariel, P., Tessier-Lavigne, M., 2014. iDISCO: a simple, rapid method to immunolabel large tissue samples for volume imaging. Cell 159 (4), 896–910.

Rothermel, M., Brunert, D., Zabawa, C., Díaz-Quesada, M., Wachowiak, M., 2013. Transgene expression in target-defined neuron populations mediated by retrograde infection with adeno-associated viral vectors. J. Neurosci. 33 (38), 15195–15206.

Salegio, E.A., Samaranch, L., Kells, A.P., Forsayeth, J., Bankiewicz, K., 2012. Guided delivery of adeno-associated viral vectors into the primate brain. Adv. Drug Deliv. Rev. 64 (7), 598–604.

Schwarz, L.A., Miyamichi, K., Gao, X.J., Beier, K.T., Weissbourd, B., DeLoach, K.E., et al., 2015. Viral-Genetic Tracing of the Input -Output Organization of A Central Norepinephrine Circuit. Nature 524 (7563), 88–92.

Sosulski, D.L., Bloom, M.L., Cutforth, T., Axel, R., Datta, S.R., 2011. Distinct representations of olfactory information in different cortical centres. Nature 472 (7342), 213–216.

Spannuth, B.M., Hale, M.W., Evans, A.K., Lukkes, J.L., Campeau, S., Lowry, C.A., 2011. Investigation of a central nucleus of the amygdala/dorsal raphe nucleus serotonergic circuit implicated in fear-potentiated startle. Neuroscience 179, 104–119.

Stachniak, T.J., Ghosh, A., Sternson, S.M., 2014. Chemogenetic synaptic silencing of neural circuits localizes a hypothalamus→midbrain pathway for feeding behavior. Neuron 82 (4), 797–808.

Stepien, A.E., Tripodi, M., Arber, S., 2010. Monosynaptic rabies virus reveals premotor network organization and synaptic specificity of cholinergic partition cells. Neuron 68 (3), 456–472.

Stettler, D.D., Axel, R., 2009. Representations of odor in the piriform cortex. Neuron 63 (6), 854–864.

Stoeckel, K., Thoenen, H., 1975. Retrograde axonal transport of nerve growth factor: specificity and biological importance. Brain Res. 85 (2), 337–341.

Stuber, G.D., Sparta, D.R., Stamatakis, A.M., van Leeuwen, W. a, Hardjoprajitno, J.E., Cho, S., et al., 2011. Excitatory transmission from the amygdala to nucleus accumbens facilitates reward seeking. Nature 475 (7356), 377–380.

Sun, N., Cassell, M.D., Perlman, S., 1996. Anterograde, transneuronal transport of herpes simplex virus type 1 strain {H}129 in the murine visual system. J. Virol. 70 (8), 5405–5413.

Sun, Y., Nguyen, A.Q., Nguyen, J.P., Le, L., Saur, D., Choi, J., et al., 2014. Cell-type-specific circuit connectivity of hippocampal CA1 revealed through Cre-dependent rabies tracing. Cell Rep. 7 (1), 269–280.

Takatoh, J., Nelson, A., Zhou, X., Bolton, M.M., Ehlers, M.D., Arenkiel, B.R., et al., 2013. Emergence of exploratory whisking. Neuron 77(2), 346–360.

Tomishima, M.J., Enquist, L.W., 2001. A conserved alpha-herpesvirus protein necessary for axonal localization of viral membrane proteins. J. Cell Biol. 154, 741–752.

Trojanowski, J.Q., Gonatas, J.O., Gonatas, N.K., 1981. Conjugates of horseradish peroxidase (HRP) with cholera toxin and wheat germ agglutinin are superior to free HRP as orthogradely transported markers. Brain Res. 223 (2), 381–385.

Tye, K.M., Prakash, R., Kim, S.-Y., Fenno, L.E., Grosenick, L., Zarabi, H., et al., 2011. Amygdala circuitry mediating reversible and bidirectional control of anxiety. Nature 471 (7338), 358–362.

Ugolini, G., 1995. Specificity of rabies virus as a transneuronal tracer of motor networks: transfer from hypoglossal motoneurons to connected second-order and higher order central nervous system cell groups. J. Comp. Neurol. 356 (3), 457–480.

Wall, N., DeLaParra, M., Callaway, E., Kreitzer, A., 2013. Differential innervation of direct- and indirect-pathway striatal projection neurons. Neuron 79 (2), 347–360.

Wang, Q., Henry, A.M., Harris, J.A., Oh, S.W., Joines, K.M., Nyhus, J., et al., 2014. Systematic comparison of adeno-associated virus and biotinylated dextran amine reveals equivalent sensitivity between tracers and novel projection targets in the mouse brain. J. Comp. Neurol. 522, 1989–2012.

Watabe-Uchida, M., Zhu, L., Ogawa, S.K., Vamanrao, A., Uchida, N., 2012. Whole-brain mapping of direct inputs to midbrain dopamine neurons. Neuron 74 (5), 858–873.

Watakabe, A., Takaji, M., Kato, S., Kobayashi, K., Mizukami, H., Ozawa, K., et al., 2014. Simultaneous visualization of extrinsic and intrinsic axon collaterals in Golgi-like detail for mouse corticothalamic and corticocortical cells: a double viral infection method. Front. Neural Circuits 8 (September), 1–18.

Weissbourd, B., Ren, J., DeLoach, K.E., Guenthner, C.J., Miyamichi, K., Luo, L., 2014. Presynaptic partners of dorsal raphe serotonergic and GABAergic neurons. Neuron 83 (3), 645–662.

Wickersham, I.R., Finke, S., Conzelmann, K.-K., Callaway, E.M., 2007a. Retrograde neuronal tracing with a deletion-mutant rabies virus. Nat. Methods 4 (1), 47–49.

Wickersham, I.R., Lyon, D.C., Barnard, R.J.O., Mori, T., Finke, S., Conzelmann, K., et al., 2007b. Monosynaptic restriction of transsynaptic tracing from single, genetically targeted neurons. Neuron 53 (3) 639–647.

Wickersham, I.R., Sullivan, H. a, Seung, H.S., 2013. Axonal and subcellular labelling using modified rabies viral vectors. Nat. Commun. 4, 2332.

Wouterlood, F.G., Bloem, B., Mansvelder, H.D., Luchicchi, A., Deisseroth, K., 2014. A fourth generation of neuroanatomical tracing techniques: exploiting the offspring of genetic engineering. J. Neurosci. Methods 235, 1–18.

Xu, W., Südhof, T.C., 2013. A neural circuit for memory specificity and generalization. Science 339 (March), 1290–1295.

Zampieri, N., Jessell, T.M., Murray, A.J., 2014. Mapping sensory circuits by anterograde transsynaptic transfer of recombinant rabies virus. Neuron 81 (4), 766–778.

Zemanick, M.C., Strick, P.L., Dix, R.D., 1991. Direction of transneuronal transport of herpes simplex virus 1 in the primate motor system is strain-dependent. Proc. Natl. Acad. Sci. U. S. A. 88 (18), 8048–8051.

Zemelman, B.V., Lee, G.A., Ng, M., Miesenböck, G., 2002. Selective photostimulation of genetically chARGed neurons. Neuron 33, 15–22.

Zhang, F., Wang, L.-P., Brauner, M., Liewald, J.F., Kay, K., Watzke, N., et al., 2007. Multimodal fast optical interrogation of neural circuitry. Nature 446 (April), 633–639.

Zhang, S.-J., Ye, J., Miao, C., Tsao, A., Cerniauskas, I., Ledergerber, D., et al., 2013. Optogenetic dissection of entorhinal-hippocampal functional connectivity. Science 340 (April), 1232627-1–1234627-14.

I. MICROCIRCUITRY

AXON DYNAMICS

Overview

Neuroanatomical mapping of axon configuration, as we saw in Section I, can serve as almost a Rosetta Stone of brain architecture and neural communication. Principles of spatial divergence, quasihierarchical area organization, and synaptic efficacy are read out or at least inferred. In an important aspect, however, the analogy does not hold; that is, axons are not static, not stonelike. They are active and changing, both anatomically and physiologically.

In Chapter 11, Vincenzo De Paola and colleagues discuss various methods for live imaging of axons, and review key results that axons engage in substantial remodeling both throughout life and, as best we can establish from animal models, under different pathological conditions. In Chapter 12, Dirk Bucher reminds us that, contrary to traditional views, axons are not faithful transmission lines of the neural code, but rather engage in complex physiological events, including failures in impulse propagation or elicitation of ectopic action potentials.

Both authors comment, in their conclusions, on the limited nature of our current views. "Our understanding of the physiological settings and biochemical pathways regulating when and how synapses form and break is in its infancy" (Grillo et al.). "Conceptually, to appreciate the role that axons may play … would require an understanding of coding at both the neuron and network level that simply is not advanced enough in most systems" (Bucher). But the implications are positive and exciting: lots of work to do, rapidly improving techniques, and possibly totally new paradigms ahead.

11

In Vivo Visualization of Single Axons and Synaptic Remodeling in Normal and Pathological Conditions

Federico W. Grillo[1], Alison J. Canty[2],
Peter Bloomfield[3] and Vincenzo De Paola[3]

[1]MRC Centre for Developmental Neurobiology, Institute of Psychiatry, Psychology and Neuroscience, Kings College London, London, UK [2]Wicking Dementia Research and Education Centre, School of Medicine, University of Tasmania, Hobart, Australia [3]MRC Clinical Sciences Centre, Imperial College London, London, UK

11.1 NEURONAL LABELING METHODS

The scientific quest of understanding the structure, function, development, and pathology of the central nervous system (CNS) has greatly relied on our ability to image neurons and their subcellular compartments. Famously *la reazione nera* developed by Camillo Golgi enabled Santiago Ramón y Cajal to observe individual neurons and initiate the "neuron doctrine" over 100 years ago (DeFelipe, 2010). Since then, the introduction of fluorescence-based microscopy techniques has represented one of the major technological leaps in the study of brain anatomy. More recently, developments in genetic labeling tools and deep tissue imaging in the late 90s have allowed optical access of the living mammalian brain. In this chapter, we will examine the methods used to visualize axons in the brain and what they have taught us about axon development and plasticity throughout the lifespan.

Conventional light microscopy of living tissue suffers the degradation of resolution and contrast caused by light scattering, making deep tissue imaging challenging (Denk and Svoboda, 1997). The introduction of confocal laser scanning microscopy (CLSM) and its deployment in the study of fluorescently labeled biological specimens has revolutionized modern microscopy (Amos et al., 1987). CLSM allows the selective detection of emitted light from the focal plane (using pinholes to reject out-of-focus photons), enabling greater resolution below the sample's surface. Concurrently, excitation of the sample is not only achieved in the focal plane but throughout the depth of illuminated tissue causing the signal-to-noise ratio to decrease rapidly at increasing depths (Denk and Svoboda, 1997). Advances in laser technology have seen the introduction of mode-locked lasers with pulses below 1 ps in duration with high frequency repetitions, in the order of 100 MHz. Such lasers make 2-photon laser scanning microscopy (2PLSM) applicable to live tissue imaging (Denk et al., 1990). 2PLSM makes use of the ability of fluorophores to absorb two photons of infrared light in a single quantum event to generate a phenomenon known as "localization of excitation." As the probability of a two-photon-absorption-event increases quadratically with the excitation intensity, the excitation of the fluorophore is extremely rare and effectively only happens at the focal volume. Since in 2PLSM, all photons (even the scattered ones) are collected and contribute to the image, and with infrared light giving better penetration in scattering tissue, 2PLSM is now the technique of choice for the longitudinal live imaging of cells and subcellular elements deep under the brain surface.

Axons and Brain Architecture.
DOI: http://dx.doi.org/10.1016/B978-0-12-801393-9.00011-6

In parallel with the improvement of optical microscopy, the range of commercially available tools to identify living cells has increased. For example, a micropipette injection of biocytin or conjugates of the enzyme Horse Radish Peroxidase into single neurons achieves whole-cell labeling (Horikawa and Armstrong, 1988). Horse Radish Peroxidase produces an opaque permanent staining of the cell following a reaction with a substrate in postmortem tissue. The resulting product may be imaged with bright field or electron microscopy (EM). Biocytin injection also requires subsequent fixation but can be coupled with either bright field or fluorescent probes. A number of biochemical dyes can be used to optically image neurons and their compartments. Luciferase Yellow produces transient fluorescent labeling that can be imaged in live cells but with potential toxic effects. Furthermore, fluorescent dextrans have been successfully employed for the anterograde or retrograde tracing of axons (Glover et al., 1986). Carbocyanine dyes, such as DiO and DiI, have also been widely used to trace axonal arbors and reconstruct the somas and dendrites of the projecting neurons. These fluorophores are lipophilic, travel passively via the plasma membrane, and produce bright labeling of the entire neuron (Honig and Hume, 1989). More recently the family of Alexa dyes offers a wide range of fluorescent probes that are excitable at different wavelengths and are produced in membrane permeable and membrane impermeable versions (Panchuk-Voloshina et al., 1999), making them useful for labeling neuronal populations or single cells in patch clamp experiments. The first images of neurons in an intact mammalian brain were obtained using the injection of a calcium sensitive dye (calcium green-1) in the rat somatosensory cortex (Svoboda et al., 1997), and the first reported observation of living axons was achieved using viral vector-mediated labeling of neurons in the barrel cortex of the rat (Lendvai et al., 2000). Another early *in vivo* imaging study describes imaging of Climbing Fibers in the cerebellum using dextran dyes injected in the inferior olive (Nishiyama et al., 2007).

One of the great revolutions of modern cell biology came from the introduction of genetically encoded fluorescent probes, providing experimenters with an ever-increasing palette of fluorescent proteins that can be expressed under cell type-specific promoters. One of the first to be used was green fluorescent protein (GFP), isolated from the jellyfish and introduced as a fluorescent marker for cells over 20 years ago (Chalfie et al., 1994). Since then it has proved to be an invaluable tool in countless studies due to its photostability, brightness, and virtually absent toxicity. GFP is easily excitable with single photon and 2-photon modalities (Potter et al., 1996). The popularity of GFP spurred researchers to look for equally practical fluorescent proteins with different spectral properties. Many different fluorescent proteins (XFPs) exist today, covering the entire visible light spectrum and beyond in the near infrared, offering the experimenter the chance of simultaneous multicolor labeling at a cellular and subcellular level (Chudakov et al., 2010). The introduction of XFPs inside living neurons has been pivotal to our current understanding of the cellular and molecular mechanisms governing the formation and function of the CNS. Today, expression of exogenous XFPs is generally achieved with one of three methods: the generation and maintenance of transgenic animal lines, the local injection of viral vectors, or through electroporation of DNA plasmids. *In vivo* imaging studies have made wide use of all of these methods.

Viral vectors have been employed in studies that focused on experience-dependent remodeling of layer 2/3 axons in the mouse barrel cortex (Marik et al., 2010b), to assess regeneration of the climbing fibers in the cerebellum (Allegra Mascaro et al., 2013) and to look at baseline dynamic changes of axon branches and presynaptic boutons in the visual cortex of monkeys (Stettler et al., 2006). The advantages of using viral vectors include localized labeling, fast expression, the ability to restrict expression to a genetically defined subset of neurons (e.g., via specific promoters or CRE recombinase lines), the possibility of infecting neurons of different species and, to a certain degree, controlling density of labeling. Anterograde or retrograde trans-synaptic labeling of CNS neurons can also be achieved using viral constructs (Beier et al., 2011), although they have not been used yet to study axons in longitudinal imaging experiments due to their toxicity. Transgenic animal lines on the other hand have provided scientists with a powerful tool to study neurons within the intact brain with minimal external perturbation that may cause toxicity (van den Pol and Ghosh, 1998).

The introduction of the *Thy1*-GFP mouse lines represented a major breakthrough in the *in vivo* imaging field. In these lines, cytoplasmic (e.g., *Thy1*-GFP-M; Feng et al., 2000), membrane- or synaptically targeted (e.g., *Thy1*-mGFP-L15; De Paola et al., 2003; Livet et al., 2002) expression of XFPs is driven by the murine *Thy1* promoter and part of the regulatory regions of the *Thy1* gene itself. XFPs are expressed in specific patterns that are consistent in successive generations in the different lines and in many cases producing an intravital Golgi like staining of excitatory neurons. Only a small percentage of neurons express the XFPs in the *Thy1* lines, enabling the discrimination of individual axons.

In particular, the *Thy1*-GFP-M and *Thy1*-mGFP-L15 lines have been used to visualize axons in 2-photon *in vivo* imaging experiments in the mouse somatosensory cortex (Canty et al., 2013a; De Paola et al., 2006; Grillo et al., 2013; Mostany et al., 2013) (Figure 11.1).

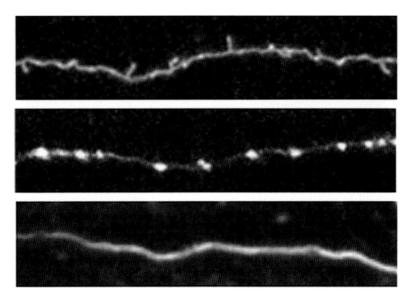

FIGURE 11.1   *In vivo* imaging of single axons. Examples of individual cortical layer 1 axons imaged *in vivo* using *Thy1*-mGFP-L15 line (membrane-targeted GFP, top and bottom) and *Thy1*-GFP-M lines (middle). The field of view spans 15 microns in the *x*-axis for all the panels.

The Brainbow mouse lines make use of the expression pattern of the *Thy1* promoter system to drive the expression of three XFPs (cyan, yellow, and red) in varying concentrations and random combinations in individual neurons. This strategy can label excitatory neurons with up to 166 different color hues in the same brain, allowing discrimination of neighboring neurons and their projecting axons (Livet, 2007). Although so far this technique has had limited use for the longitudinal *in vivo* imaging of axonal projections, new improvements may soon provide a powerful tool for axonal tracing *in vivo* (Cai et al., 2013). GFP has also been introduced specifically in inhibitory interneurons; the GAD65-GFP mouse line displays strong GFP expression in the majority of GABAergic neurons in the brain. Expression is driven by the promoter and accessory regulatory regions of the glutamic acid decarboxylase (GAD) enzyme, responsible for GABA production at nerve terminals (Lopez-Bendito et al., 2004). The GAD67-GFP (somatic GAD) lines also label most interneurons in the brain, each line with slightly differing expression patterns (Tamamaki et al., 2003). This is similar to the *Thy1* lines, with different labeling patterns being inherited by the offspring and resulting stable but unique interneuron classes labeled (Ma et al., 2006). GABAergic axons have been imaged *in vivo* using transgenic mouse lines (Keck et al., 2011; Lee et al., 2006). Further refinement of the population of neurons labeled has made use of Cre recombinase technology to drive XFP expression in specific neuronal populations under strong promoters (Huang et al., 2002; Taniguchi et al., 2011).

In utero electroporation at different embryonic stages is another widely used neuronal labeling technique that consists of injecting plasmid DNA in the ventricle of mouse or rat embryos. The injected exogenous DNA is then forced into neuronal progenitor cells, found lining the ventricular surface, with an electric field pulse. DNA, a negatively charged molecule, is attracted to the positive side of the field so that the transfection can be directed to the area of interest (e.g., hippocampus, somatosensory, or visual cortices). Depending on the developmental stage different neuronal populations can be transfected, as progenitors of different layers are found depending on the embryonic stage (Fukuchi-Shimogori and Grove, 2001; LoTurco et al., 2009; Saito and Nakatsuji, 2001; Tabata and Nakajima, 2001).

In order to visualize neuronal structures, fluorescent neurons need to be optically accessible in the intact brain. Two surgical techniques are now routinely used to image neurons in the brain *in situ* using 2PLSM in mouse and rat models. One approach consists of thinning the skull bone down to 30 µm thickness allowing transcranial imaging through the bone. This approach has the advantage of not exposing the brain directly therefore causing limited damage. The downside is that the imaging area is small (200–300 µm in diameter) and, in the case of chronic experiments, the procedure needs to be repeated for successive time points (Yu and Zuo, 2014). This technique is most commonly reported in studies that focus on dendrites and dendritic spines, rather than axonal processes, mainly because of poorer resolution and the restricted imaging areas, which is not optimal to track thin and long axonal segments. The preferred technique to image axons and their presynaptic boutons is using cranial window implantation (Holtmaat et al., 2009). Here a craniotomy (removal of a skull bone flap) is performed over the brain area of interest, the underlying dura membrane is preserved, and the window is then sealed by a glass coverslip and dental cement. The result is a large imaging field (up to 3–4 mm wide) that allows the use of sparsely labeled mouse lines and importantly optical access for repeated imaging experiments of the same neurons over extended periods of time and up to 1 year (Mostany et al., 2013). The wider imaging area together with the employment of sparse labeling enables the tracing of long axonal stretches that can contain hundreds of synaptic boutons (Grillo et al., 2013).

Imaging efforts have focused not only on the structure of neurites, dendrites, and axons but also importantly on synaptic structures such as dendritic spines and axonal boutons (Holtmaat and Svoboda, 2009). Such studies demonstrate that these elements maintain a plastic potential throughout the life of the rodent and are not restricted to early postnatal development. Importantly, electron microscopy has confirmed in a variety of conditions that the structures observed *in vivo*, on the basis of morphology, are indeed in most cases structurally complete synaptic connections between neurons (Holtmaat and Svoboda, 2009). Synaptic sites can also be reliably imaged using fluorescent markers of synaptic proteins. Postsynaptic density-95 (PSD-95) has been successfully employed to image synaptic sites and their dynamics on dendritic spines *in vivo* in the mammalian cortex (Gray et al., 2006; Cane et al., 2014). A similar approach for a presynaptic counterpart, overexpression of synaptophysin-GFP, has only been shown in preliminary results (Holtmaat and Svoboda, 2009). Future experiments will likely avoid the issues of overexpression by using genetically encoded antibody-like strategies, such as the FingR intrabodies (Gross et al., 2013) or conditional strategies (Fortin et al., 2014).

11.2 AXONAL DYNAMICS DURING DEVELOPMENT: FROM SPECIFICATION TO MAINTENANCE

The dynamic behavior of axons throughout development and adulthood reflects the need of the developing brain to establish fully functional neuronal networks, refining connections between network components and allowing plastic events to shape the circuitry in response to experience or neuronal damage. We can define five developmental stages that axons go through: (i) axon specification, (ii) axon growth, (iii) axon branching and presynaptic elaboration (iv) axon pruning, and (v) axon maintenance during adulthood and aging (Figure 11.2).

Newly born neurons start migrating from the periventricular region towards their cortical layer of competence. Migrating neurons usually polarize their processes, with the trailing process generally becoming the axon. The specified axon can then extend and be guided by external cues to its final target while along its growth path it can stem interstitial elongating branches to secondary target areas. When it reaches eventual innervation sites, the

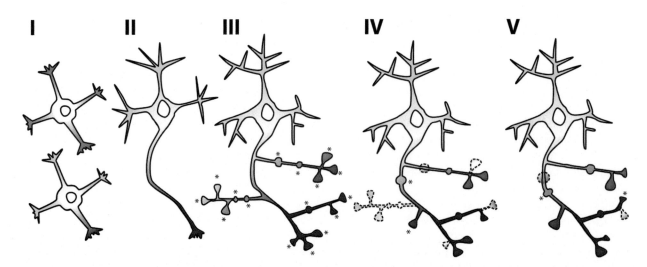

FIGURE 11.2   Five stages of axon development. (I) *Axon specification.* In neural progenitors, one process is selected to develop into an axon while the remainder become dendrites. (II) *Axon growth.* The specified axon elongates towards its targets using environmental cues. (III) *Axon branching and presynaptic elaboration.* Sprouts arise from the main axonal shaft once it reaches its target regions. At the same time, synaptic contacts are established in the form of *TBs* and *EPBs*. (IV) *Axon pruning.* Redundant connections are eliminated at the final stages of the juvenile period. (V) *Axon maintenance.* Axons and their synapses are stably maintained in adulthood with minimal branch changes and low synaptic turnover. Overall, the density of presynaptic terminals stays constant with boutons being added and eliminated in equal numbers.

axon undergoes extensive branching and forms synapses with an exuberant number of targets (Lewis et al., 2013). Synaptic competition between overabundant pre- and postsynaptic partners shapes the developing circuitry by inducing a net elimination of contacts, a phenomenon known as synaptic pruning. Both small- and large-scale axon retractions are triggered during this phase establishing the final mature network by selecting and maintaining fewer synapses compared to the developmental stage (Luo and O'Leary, 2005). In physiological conditions, axon dynamics carry on throughout adult life, albeit at a reduced and balanced pace with no net difference in branch extension/ retraction and synapse formation/elimination (Holtmaat and Svoboda, 2009). It appears that even in the later stages of life, in aged mice, axon branches retain their dynamic properties yet synaptic formation and elimination rates are surprisingly increased (Grillo et al., 2013), implying age-related instability of connectivity.

Neuronal polarization is the initiating event that leads to axon formation and specification, making all other processes future dendrites. Most of the studies done to investigate the mechanisms that regulate the initial stages of axon formation have been done *in vitro*. It is not clear whether axon specification is determined in a cell-autonomous or nonautonomous manner, due to the difficulty of recreating the complex environmental cues present in the intact brain. It is also not clear if neurons inherit a polarization directly from their progenitor cells. What is clear is the importance of the neuronal cytoskeleton including microtubules for axonal extension and the actin cytoskeleton that acts as a molecular filter at the axon initial segment (Neukirchen and Bradke, 2011).

The main axonal driving growth takes place at the growth cone, a structure also responsible for following guidance cues to the final target. The growth cone contains a core of microtubules, highly organized asymmetric protein structures with a minus end and a plus end. In axons, microtubules arrange themselves in a parallel orientation to form filaments with the plus end always pointing distally and the minus end directed towards the soma. At the growth cone the microtubules form a central domain, which interacts with the peripheral domain formed by an actin network that invades the growth cone lamellipodia and filopodia. The dynamic behavior of this complex, dependent on axonal transport, generates the forces that cause axon elongation (Suter and Miller, 2011). Besides the extension of the distal tip, axons also undergo tensions over their entire length due to overall tissue enlargement during development. Indirect evidence of an axon stretching phenomenon comes from studies in peripheral rat axons that measure the distance between nodes of Ranvier. During development the internodal distance can double with no reduction of axon diameter (Abe et al., 2004). Component addition along the axon has been confirmed in drosophila models of axon growth (O'Toole et al., 2008). The strategies that neurons adopt to extend and enlarge their axons during development are likely to be cell type-specific. For instance, *in vivo* imaging of developing mouse cortex has shown how the growth cones of Cajal-Retzius cells (cortical layer 1 inhibitory interneuron) and thalamocortical axons differ in their behavior (Portera-Cailliau et al., 2005). Cajal-Retzius growth cones are large and complex with numerous filopodia, they tend to follow a tortuous path often splitting in two branches as if exploring wider target areas before retracting to focus in a single direction. Thalamocortical axons, on the contrary, display smaller and simpler growth cones with a lower number of filopodia and tend to follow straight trajectories. In both cases, the prevalence of growth cones is significantly decreased after postnatal day 9 (Portera-Cailliau et al., 2005).

Axonal branching appears to be a separate process to axon elongation. For instance, cortical neurons of superficial layers extend their axons across the corpus callosum in a fasciculate fashion, avoiding branching before being myelinated. Axons are therefore actively prevented from branching in white matter tracts but form branches once they reach the contralateral target areas (Lewis et al., 2013). Evidence from *in vivo* studies suggests that the formation of the terminal branches and consequently of synaptic contacts is dependent on neuronal activity of the presynaptic cell (Mizuno et al., 2007; Wang et al., 2007), and at a later stage on activity of the postsynaptic targets (Mizuno et al., 2010). As development progresses in the early postnatal life, redundant circuitry is gradually refined through genetically guided mechanisms and spontaneous activity (Gogolla et al., 2007). In general axons that have a more efficient transmission will displace competing terminals at the same synaptic site (Buffelli et al., 2003), though homeostatic mechanisms exist in order to limit the size of axonal terminal arbors (Kasthuri and Lichtman, 2003). *In vivo* imaging of the developing mouse somatosensory cortex has shown that axonal branch extension and retraction, as well as synapse formation and elimination, are simultaneous events that have an overall bias towards branch growth and synapse formation (Portera-Cailliau et al., 2005). At the end of the developmental process, neural circuits are prominently shaped by sensory experience and learning during region specific time windows known as critical periods, when the network ability to structurally respond to external stimuli peaks (Gogolla et al., 2007). Sensory manipulations such as monocular deprivation prior to the critical period are well known to dramatically, and permanently, alter the anatomical distribution of thalamocortical and intracortical terminals in the visual cortex (Antonini and Stryker, 1993; Trachtenberg and Stryker, 2001). During the critical period learning can leave an underlying anatomical trace that can be reactivated later in life when the animal is presented to the same learning condition (Linkenhoker et al., 2005). *In vivo* time lapse imaging experiments found a lasting physical trace of juvenile experience in the dendritic spines that formed as a consequence of the closure of one eye. Such spines persisted into adulthood and were enlarged when the same eye was shut again later in life (Hofer et al., 2009). Experiments investigating similar cellular mechanisms on the presynaptic side have yet to be reported. Interestingly, it has been described that degradation of perineuronal nets, extracellular matrix components, can reactivate structural plasticity to levels resembling those seen during critical periods (Pizzorusso et al., 2002). Some evidence suggests that these extracellular protein nets protect strong memories to persist through the adult life by preventing learning-induced extinction (Gogolla et al., 2009).

One of the great achievements of chronic *in vivo* 2PLSM has been to show that the adult nervous system of mammals is far from being a structurally crystallized environment where neuronal connections at rest are fixed. Together with new transgenic tools to label subsets of living neurons, *in vivo* imaging techniques enabled researchers to observe synaptic turnover in real time. While a large proportion of synapses are generally stable for potentially a lifetime, a small group of postsynaptic elements are constantly added and removed without altering the overall number (Grutzendler et al., 2002; Trachtenberg et al., 2002). Synaptic remodeling is also evident at presynaptic terminals and axon branches as shown by long-term *in vivo* 2PLSM. It appears that the dynamic properties of axonal arbors depend on the neurons of origin. The overall branching architecture is fairly stable in the mouse cortex (layer 1 of the barrel field) but a consistent 60% of axonal branches of layer 6 axons remodel, compared to 25% of branches originating from thalamocortical axons, over the period of a few weeks. These remodeling events included retractions and elongations up to 150 μm in length (De Paola et al., 2006). The same study established that different axon types have different rates of synaptic turnover. *Terminaux boutons* (TBs), presynaptic terminals found on axons with cell bodies that reside in layer 6 of the cortex, displayed high plasticity rates with around half the population being turned over after 24 days of imaging. In contrast, presynaptic varicosities on the axonal shaft, known as *En Passant boutons* (EPBs) were reported mainly on axons with cell bodies residing in layer 2/3 and 5 axons as well as thalamocortical axons. EPBs were in general less plastic than TBs with only around 20% being replaced over the same period of time (24 days), suggesting that a large proportion of EPBs may be very stable, possibly surviving an entire lifetime. These dynamic behaviors of different circuits in the adult mouse brain have been confirmed in the same cortical region by other investigators (Grillo et al., 2013) and similar dynamic properties have been observed in the visual cortex of the macaque with approximately 7% weekly turnover rate for EPBs and 14% for TBs (Stettler et al., 2006). Apart from the reported high rates of TB turnover, axonal boutons are thought to be generally more stable than dendritic spines. This view has been confirmed in a study that imaged spines and boutons in the same animals where spines were found to be more dynamic than boutons in somatosensory, visual, and auditory cortex (Majewska et al., 2006). This mismatch of dynamic properties may be explained by the presence of multisynapse boutons, where a single presynaptic terminal has more than one active zone contacting a number of separate spine heads. These multiple synapse boutons have been found to be preferentially contacted by newly formed spines in the mouse somatosensory cortex (Knott et al., 2006), allowing for additions of dendritic spines in the absence of a change in total number of presynaptic terminals.

Another brain area that has received attention from the *in vivo* imaging community is the cerebellum due to its accessible location on the brain surface. Axons have been imaged also in this highly organized brain structure, where similar plastic properties to the cortical areas have been described (Nishiyama et al., 2007). Climbing fibers in the cerebellum are composed of a main ascending branch that innervates Purkinje cells and transverse branches that contact interneurons. *In vivo* time lapse imaging has shown that over a short time frame (4 h) the main ascending branches are extremely stable while 36% of the transverse branches are dynamic, demonstrating that structural plasticity can be regulated in very specific subdomains of the same neuron (Nishiyama et al., 2007). Another study focused on the parallel fibers and their synaptic dynamics during baseline and motor training (Carrillo et al., 2013). Similar to other axonal populations, EPBs detected on the parallel fibers were found to be remarkably stable, around 10% of the EPBs are replaced over a 2-week period. Subsequently, the same mice were trained to perform a range of acrobatic climbing skills for the following 8 weeks. Imaging the parallel fibers over the training period highlighted a striking loss of dynamic EPBs that dropped to around 3% every 2 weeks, with the remainder being prevalently EPB losses rather than gains pointing to a reduction of overall synaptic contacts in response to motor learning (Carrillo et al., 2013). To date this study is the only to address the changes in excitatory axonal bouton dynamics in response to learning a specific behavioral task (see also Chen et al., 2015 and related news and views by Grillo et al., 2015).

11.3 AGING

During the aging process, the body gradually enters a phase of decline that affects the entire organism, including the brain. Cognitive, motor, and sensory degradation are common aspects shared by all higher organisms including humans. Great effort has gone into studying the process of aging. Recently two separate studies have looked at the effects of aging on synaptic dynamics in the somatosensory cortex (Grillo et al., 2013; Mostany et al., 2013). One of the studies focused on dendritic spines imaged on layer 5 pyramidal neurons and on EPB rich axons in *Thy1*-GFP-M mice across four periods in the lifespan. Juvenile mice (1 month old) had the highest spine and bouton density and the highest turnover rates. Young adults (3–5 months old) exhibited reduced number of synaptic elements and reduced overall turnover rate both for dendritic spines and axonal boutons, in accordance with the phenomenon of synaptic pruning. Interestingly, in contrast with the general idea that aging reduces the number of synapses, the number of synaptic contacts increased in a mature group (8–15 months old) and in an aged group (20 months old) progressively. Dendritic spines also displayed a marked increase in turnover that was not significant in the case of axonal boutons (Mostany et al., 2013).

Another study that focused exclusively on presynaptic boutons and axon dynamics described a conserved density of synaptic boutons during aging but also a strong increase in synaptic dynamics in specific cortical circuits (Grillo et al., 2013, Figure 11.3). Two age groups were imaged: young adults (4–6 months old) and aged (22–24 months old). Branch dynamics were found to involve the same proportion of processes and the events of retraction and elongation were comparable in both age groups. Overall, both in TB rich and EPB rich axons the number of synaptic boutons was comparable between age groups. TBs were found to have comparable turnover rates during aging and adult life, suggesting that this class of bouton is plastic throughout life. Surprisingly, EPB rich axons had increased rates of synaptic turnover in aged animals with 15% of EPBs being replaced every 4 days compared to just 8% in young adults. While young adults appeared to have a very stable population of boutons, aged animals were characterized by higher destabilization rates that might support the idea of aging circuits unfit for long-term memory consolidation. Furthermore, large synaptic boutons were stable on young axons, hardly ever disappearing, but were unstable in the aged brain (Grillo et al., 2013). The picture that emerges from these studies is that the aging brain, contrary to

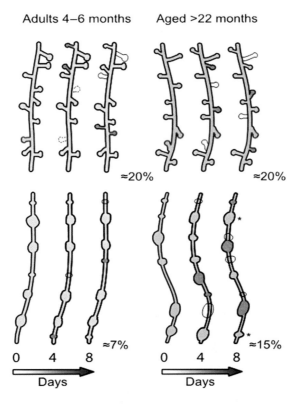

FIGURE 11.3   Schematic representation of the circuit-specific effect of aging on cortical synapses. (Top) Based on recently published work (Grillo et al., 2013): synaptic turnover is comparable with TB rich axons (i.e., from layer 6) in young adult (green) and aged (purple) mice. TBs are gained and lost at similar rates in both age groups over 4 days. (Bottom) EPB rich axons (i.e., from layer 2/3/5 and thalamus) have larger rates of synaptic change in the aged compared to the young adult cortex. EPBs are gained and lost more frequently on aged axons. The size of EPBs on aged axons is subject to larger changes between sessions (*). Large EPBs are very stable on young adult axons while they are turned over more frequently in the aged brain. Percentages in each panel represent an approximation of bouton turnover over 4 days for each condition. Dashed outline represents a lost bouton while new boutons are shown in orange.

the idea of synaptic reduction, maintains comparable synaptic densities but is characterized by increased synaptic dynamics, which may have a detrimental effect on the function of specific neural networks.

11.3.1 Axonal and Synaptic Dynamics II: Neurodevelopmental Diseases

For the brain to function in a healthy manner, neuronal ensembles must interconnect to create a large synaptic network. Quantification of synapses in both the pre- (axonal) and post- (dendritic) synaptic domains offers assessment of connectome complexity. As we have seen in the first section of this chapter, *in vivo* microscopy provides a method of quantification of synaptic densities and dynamic changes at high resolution, however, imaging of synapses in the human brain relies on postmortem tissue assessment. In this section, we will investigate how the methods used to investigate synaptic alterations in the context of human pathology can be related to animal investigation of neurodevelopmental disorders. Patient data from a number of neurodevelopmental disorders, including autism spectrum disorders (ASDs), Rett syndrome and Fragile X in terms of symptom presentation and postmortem synapse morphology, will be discussed. We will then investigate genetic models of these disorders and see how behavioral and synaptic detriments in animals relate to the human conditions. This approach enables investigation of synaptic dynamics in neurodevelopmental disorders, which cannot easily be assessed in humans *in vivo*.

Historically, neuronal integrity in postmortem human brain tissue is observed by Golgi staining. This enables visualization of the gross architecture of neurons and synapses at the endpoint of disease and can help determine the final anatomical rearrangements in neurodevelopmental and cognitive disorders. The main hindrance for these types of investigation is the temporal interpretation. These studies are limited to endpoints of disease, preventing investigation of the development or progression of axonal pathology. Consequently, deficits or delays in synapse formation and elimination are difficult to detect.

Following gene identification in human genome wide association studies (GWAS), mouse models of neurodevelopmental disorders can be generated to investigate the temporal development of axonal pathology.

Through careful and rigorous experimental design, synaptic integrity can be quantified in healthy and pathological tissues. In postmortem tissue, Golgi staining provides a binary staining of the neuron and its processes, with resolution sufficient to include dendritic spines and in some cases axonal boutons. However, *in vivo* imaging of dendritic spines and axonal boutons as they are formed and eliminated, as well as quantification of synapses in fixed tissue, can produce a more comprehensive picture of the disordered circuitry.

Neurodevelopmental disorders can be severe and disabling, impairing growth and function of the CNS. They affect a significant number of individuals and many do not currently have effective medications or treatments. Many neurodevelopmental disorders have known gene mutations contributing to their symptomatic expression, however, the effect of these genetic alterations on the function and connectivity of circuits *in vivo* is unclear (Schubert et al., 2014). In this section, we look at some well-characterized genetic mutations and see how changes in synapse dynamics may be causing functional impairment in these disorders, in the absence of dendritic spine density alterations.

11.4 AUTISM

Autism Spectrum Disorders (ASD) are disorders of social interaction, emotional and cognitive function. They are associated with a vast range of phenotypes, which are apparent at many stages of development and adulthood. Initially, ASD was described by Leo Kanner as an infantile disorder of impaired social reciprocity and later developed by Hans Asperger to describe a wider "autistic psychopathy" (Zoghbi, 2003). It is estimated that 1 in 68 children in the United States are affected by ASD according to estimates from the Center for Disease Control's Autism and Developmental Disabilities Monitoring Network, with similar prevalence across socioeconomic status and ethnic groups. Specific treatment and medications for ASD are not available and the symptoms are largely heterogeneous with current treatment targeting symptomatic features such as energy levels, ability to focus, seizures, and depression (http://www.cdc.gov/ncbddd/autism/data.html).

Postmortem investigation of ASD patients has revealed limbic and cerebellar abnormalities, as well as altered morphology in hippocampal neurons and white matter pathology (Ellegood et al., 2012; Wolff et al., 2015). In prefrontal networks, axonal alterations included a decrease in the largest axons that communicate over long distances and an excessive number of thin axons that linked neighboring areas, particularly around the anterior cingulate cortex

(Zikopoulos and Barbas, 2010). Such phenotypes suggest altered axon growth and guidance in ASD. Postmortem investigation of spine densities has produced conflicting data, where early data suggested decreased spine numbers (Williams et al., 1980), and more recent investigations observing increased spine density on projection neurons (Hutsler and Zhang, 2010). While postmortem investigation of spines demonstrated elevated spine density in the latter study, it is difficult to say whether this is present from an early stage or progresses as a result of the disease process. A number of gene mutations associated with synaptic proteins in ASD have been identified from patient GWAS and can be used to model features of the disease in animals. These ASD genetic mutations produce behavioral and tissue pathologies similar to those seen in patients with autism. Genetic models of ASD are some of the most comprehensively assessed in terms of behavior and neuroanatomy. Below are the most thoroughly characterized genes associated with synaptic alterations in ASD, with emphasis on axonal phenotypes.

Neurexins are presynaptic adhesion molecules, which connect with postsynaptic neuroligins. There are three neurexin isoforms, each with an alpha and beta subtype and neurexins are expressed across all neuronal types. Neurexin1α−/− mice have impaired neuronal activity alongside deficits in prepulse inhibition (PPI) responses, elevated grooming activity, poor performance in nest building, however, these mice also demonstrated improved motor learning as compared to wild type animals. In this investigation, no social interaction or spatial memory-associated impairments were observed (Etherton et al., 2009). In contrast, a more recent investigation showed neurexin1α−/− alters social behavior, with reduced social investigation in −/− animals and reduced locomotor activity in novel environments (Grayton et al., 2013). Animals with a neurexin 2α−/− demonstrated deficits in sociability and social memory. In open field exploration, grooming was increased, reflecting greater levels of anxiety. Deficits in presynaptic transmitter release proteins were present following tissue analysis, however, no data on spine density or dynamics were presented (Dachtler et al., 2014).

Integrin β3 (ITB3) has been linked to ASD, particularly as a genetic risk factor (Napolioni et al., 2011; Weiss et al., 2006). The gene, located on chromosome 17, has also been implicated in serotonergic signaling as well as being associated with the ASD susceptibility gene SLC6A5, a serotonin transporter. ITB3 functions include control of cell adhesion, platelet function, and cell signaling. ITB3 knockout (−/−) mice have distinct behavioral phenotypes, which are similar to the behavior observed in patients with ASD. A number of studies have demonstrated a lack of social preference in ITB3−/− mice (Carter et al., 2011; Nadler et al., 2004; Yang et al., 2011), as well as elevations of grooming behavior, suggestive of anxiety (Carter et al., 2011; Ellegood et al., 2012). ITB3−/− mice have an 11% reduction in whole brain volume, with significant deficits in regions including the hippocampus, striatum, thalamus and frontal, parieto-temporal cortex, occipital and entorhinal cortices. A prominent white matter pathology has been found compared to wild type animals (Ellegood et al., 2012), suggesting incomplete axonal connectivity. To date, no *in vivo* synaptic evaluation has been conducted with this model.

Neuroligins are postsynaptic cell adhesion molecules, which interact with neurexins on the presynaptic domain. There are four isoforms of neuroligin (NL1–4) which act as ligands for neurexins (Ichtchenko et al., 1995). The presence of each isoform is dependent on the form of activity at the synapse (NL1 is excitatory, NL2 is inhibitory, NL3 is both inhibitory and excitatory, and NL4 is purely found in glycinergic synapses). More than 30 mutations of NL genes have been implicated in ASD. NL3 is affected by a point mutation, in some cases by a whole gene deletion, and has attracted the most experimental investigation (Chadman et al., 2008; Isshiki et al., 2014; Jamain et al., 2003; Radyushkin et al., 2009; Sanders et al., 2011; Tabuchi et al., 2007; Varoqueaux et al., 2006). The behavioral consequences of neuroligin knockout vary, however, deletion of each isoform produces a distinct phenotype related to ASD. NL1 null animals have impaired spatial memory and increased repetitive behaviors (Blundell et al., 2010), while NL2−/− demonstrates elevated levels of anxiety (Blundell et al., 2009). NL3 null animals have been investigated multiple times with conflicting behavioral observations, the first study demonstrated an increase in inhibitory synaptic transmission (Tabuchi et al., 2007), a later study reported minimal behavioral deficits (Chadman et al., 2008) and most recently, olfactory deficits have been observed (Radyushkin et al., 2009). A recent NL4−/− investigation produced a mouse with repetitive behavior, impaired social interaction and reduced amounts of ultrasonic communication (Ju et al., 2014). Assessment of tissue from triple knockout of NL1-3 suggests that functional impairment does not arise from an alteration in synaptic number, rather by changes in transmitter properties at the synapses (Varoqueaux et al., 2006). A more recent NL3−/− investigation revealed, similar to the investigation by Varoqueaux et al., that spine density was not different from wild type animals, however, the turnover rate of spines was greatly elevated (Isshiki et al., 2014).

Contactin-associated protein 2 is a neurexin-related cell adhesion molecule which has been associated with ASD. Knockdown of Contactin-associated protein 2 produces phenotypes including epilepsy and seizures, reduced communication and social behavior, reduced numbers of GABAergic interneurons, and impaired neuronal activity synchronicity (Peñagarikano et al., 2011). Assessment of spine densities in fixed tissue revealed a reduced number of spines in knockout mice (Anderson et al., 2012).

SH3 and multiple ankyrin repeat domains 3 (Shank3) are thought to be the cause of the neurodevelopmental disorder Phelan-McDermid syndrome, an ASD. GWAS mutant screens have demonstrated a shank3 deletion in ASD outside of the Phelan-McDermid syndrome. Shank3, located on chromosome 22, is part of the postsynaptic scaffolding and signaling complex in glutamatergic synapses. Shank3 mutants display repetitive grooming behavior and impaired social interaction and neuronal activity perturbations. Tissue assessment of Shank3−/− animals revealed reduced spine densities on medium spiny neurons and widespread striatal hypertrophy (Peça et al., 2011).

11.5 RETT SYNDROME

Rett syndrome is a neurodevelopmental disorder, which primarily affects cognition and social function in a progressive manner. Rett syndrome is considered part of ASD and is one of the most severe and disabling forms of ASD. It is an X-linked mutation of the *MECP2* gene, thus affects almost exclusively females. Symptoms include loss of speech and motor control, diminished compulsivity, impaired sympathetic function, anxiety, seizures, and Parkinsonian tremors. The manifestation of an *MECP2* mutation can range from a normal phenotype to severe disability. One in 10,000 females in the United States are estimated to have Rett Syndrome. Other than supportive measures, such as feeding tubes, surgery, physical support, and seizure medication, there is currently no treatment for Rett Syndrome (Zoghbi, 2003). Postmortem tissue investigation of Rett syndrome has revealed a lower spine density on CA1 pyramidal neurons than age-matched controls (Belichenko et al., 1994; Chapleau et al., 2009), as well as reduced dendrite length (Armstrong et al., 1995).

Generation of *MECP2* null mice produces animals with neurological symptoms mimicking Rett syndrome (Guy et al., 2001). Both whole animal and brain-specific *MECP2* knockout produces "Rett syndrome-like" behavior and phenotypes including reduced bodyweight and impaired hind limb clasping. In open field assessment, *MECP2*−/− mice demonstrated less exploratory behavior and rearing, as well as spending more time immobile (Guy et al., 2001). The synaptic morphology of Rett/*MECP2*−/− mice is complex. A reduced spine density has recently been observed in fixed tissue (Xu et al., 2014), however, using multiple time points has revealed dysregulation of spines at a presymptomatic stage, but a restored number at later time points (Chapleau et al., 2012). An *in vivo* imaging study of *MECP2*−/− mice demonstrated a reduced spine density as well as rapidly changing numbers, sizes, and motilities of spines (Landi et al., 2011).

11.6 FRAGILE X SYNDROME

Fragile X syndrome is an inherited monogenic form of mental retardation. The disorder affects 1 in 4000 males and 1 in 8000 females, symptoms include difficulties in social interaction, learning, linguistic ability, impaired attention, and poor emotional interaction. Nearly half of those diagnosed with fragile X will meet diagnostic criteria for ASD. An expansion of the CGG trinucleotide repeat affecting the Fragile X mental retardation 1 (*FMR1*) gene results in a deficit of fragile X mental retardation protein production, which is responsible for neural development. Similar to Rett syndrome and ASDs, there is no medication currently available for Fragile X. In postmortem human studies, a higher density of spines is observed on dendrites, proposed to be a result of reduced levels of synapse elimination (Irwin et al., 2000). These observations from an early age are also carried through to later life (Irwin et al., 2001). Earlier studies observed differences in spines in Fragile X patients, however, spine densities were not quantified (Hinton et al., 1991; Rudelli et al., 1985).

The *FMR1*−/− mouse has been well characterized from a behavioral and pathological perspective. In wild type animals, learning-induced fragile X mental retardation protein production and neurotransmitter-induced protein synthesis were normal. However in *FMR1*−/− mice, these processes were disrupted (Greenough et al., 2001). In *FMR1*−/− mice, anxiety and aggression were unaffected, but spatial learning in the radial maze was reduced and in open field testing, greater levels of locomotion were observed (Mineur et al., 2002). *FMR1*−/− mice also have impaired motor learning skills with a forearm reaching task, reduced synapse strength and impaired LTP (Padmashri et al., 2013). The earliest *in vivo* time lapse imaging of an *FMR1*−/− demonstrated a delay in stabilization of spines and an increased density of dendritic spines on cortical pyramidal neurons. *FMR1* null mice also demonstrated a developmental delay in the down regulation of spine turnover seen in wild type animals (Cruz-Martín et al., 2010). More recent *in vivo* imaging of dendritic spines revealed that *FMR1−/−* mice have a similar density of spines to wild type, with a higher rate of turnover, with more spines forming and eliminating in the time course. Additionally, motor learning-induced spine formation in cortical layer 1 was impaired in *FMR1−/−* mice (Padmashri et al., 2013).

Typically, neurodevelopmental disorders have similar clinical symptoms with diagnosis often overlapping. It is important to note that existing mouse models replicate only some aspects of the diseases and are not fully representative of the human syndromes. *In vivo* synaptic dynamics associated with these features can be studied in rodent models and correlated with postmortem patient and control tissue. Both human and animal studies demonstrate how spine dynamics are involved in the pathophysiology of neurodevelopmental disorders. It is increasingly clear that many cases of disordered cognition arise not purely from a reduced or increased number of synapses but also from an altered rate of synaptic turnover. This is similar to the cognitive impairment observed in aged mice (Grillo et al., 2013) suggesting a significant component of cognitive competence is maintenance of appropriate synapses in the connectome.

The studies of developmental disorders discussed here demonstrate how synaptic alterations can be assessed in genetic mouse models using *in vivo* spine imaging and how they correlate with postmortem tissue from humans. Behavioral phenotypes, similar to those observed in human patients, can be reproduced in animal models of these disorders. This has allowed investigators to model the synaptic pathology underlying the disorders associated with specific behavioral phenotypes. Where dysfunction arises in patients, with a seemingly normal spine density, it would seem that the turnover rate of spines is higher than that of controls, which could explain part of the cognitive deficit.

While dendritic spine dynamics reveal distinct alterations associated with genes underlying severe neurodevelopmental disorders, little is known about axonal alterations in the disorders covered here. The spine densities seen in postmortem human tissue correspond with animal models, where impaired spine dynamics prove to be a significant component of disorder. To date there are no studies investigating presynaptic changes arising from the progression of neurodevelopmental disorders but we expect a large number of studies to be conducted investigating presynaptic alterations in these models of disease in the future.

11.6.1 Axonal and Synaptic Dynamics III: Lesions and Neurological Diseases

How mature axons respond to disease states has been of interest since the dawn of neuroscience research as neuroscientists try to couple overt disease symptoms with specific pathology in the nervous system. Indeed Ramon y Cajal spent countless hours peering down the lens of his microscope to produce illustrations of silver stained axonal pathology evident after traumatic injury. His observations of reactive axonal changes and degenerative processes maintain their currency in the twenty-first century. For the majority of time since then, neuroscientists have either collected postmortem human tissue for histological analysis or generated animal models of disease to observe changes that occur during disease progression. Axons have the crucial role of directly communicating with other neurons by means of electrical and chemical neurotransmission, hence understanding how they respond to trauma and disease, before resulting loss in function or frank degeneration, becomes of vital importance to reveal the mechanisms of nervous system adaptation. Axonal projections can be up to 20,000 longer than the cell body, exposing them as particularly vulnerable elements in many diseases.

Clinically, a number of imaging modalities are in use including computerized axial tomography, magnetic resonance imaging and positron emission tomography which contribute to understanding the disease process when coupled with other functional and cognitive testing and, where possible, postmortem histological analysis. Diffusion tensor imaging is the most recent imaging modality that has been applied in clinical research and produces images of cortical tract integrity based on the diffusion of water molecules. The axial and radial diffusivities derived from diffusion tensor imaging have shown promise as noninvasive surrogate markers of axonal damage and demyelination, respectively (Budde et al., 2008). However, none of these noninvasive imaging techniques offer the resolution to study the intact nervous system at the level of single axons and synapses.

Using animal models to infer understanding of complex disease states in humans allows a more controlled, reproducible, and invasive approach to understand axon dynamics. As described previously in this chapter, axons have historically been observed by their morphology and neurochemistry in fixed tissue slices using silver and Golgi staining methods. With the advent of live imaging, including intrinsic imaging and 2PLSM, we can see superficial circuitry in the peripheral or central nervous system, and in some cases deep inside the brain. As such, we are now beginning to understand more about how these vulnerable processes are differentially affected in various states of disease that were previously only discerned through static imaging using classical postmortem anatomical techniques.

Using these imaging techniques, there has been a recent explosion of studies describing reactive changes in the neuropil in the injured and disease states (Brown et al., 2010; Grutzendler and Gan, 2006; Liebscher et al., 2013; Murmu et al., 2013; Ochs et al., 2014). A common approach that has been particularly informative in understanding how axons respond in disease states has already been described in this chapter—the use of XFP gene insertions under the control of neuron-specific promoters such as *Thy1* (Caroni, 1997), which results in labeling of 1–2% of excitatory neurons throughout the nervous system. When imaged with a near infrared laser through a cranial window in animal models of disease, the same circuitry can be observed throughout the disease process. By and large, users of this approach have studied dendritic arbors and plasticity owing to their easily identified bright processes, high density of dendritic spines and the ease at which their branched morphologies can be reconstructed in deep stacks of layered images. Axon trajectories and their dynamic properties are more problematic to visualize. White matter tracts are usually situated deep within the brain, making the penetration of imaging lasers virtually impossible. The outer layers of the brain, specifically the upper layers of the cerebral cortex, are more amenable to imaging in the living brain. In these upper cortical regions, the axons form a veritable melting pot of sub populations of local, intracortical, corticocallosal, and subcortical projections all intermingled with each other in a rich neuropil. Based on their branching patterns and synaptic makeup, axonal projections coming from neurons in layers 2, 3, 5, and 6 can be identified morphologically, as can those originating in the thalamus (De Paola et al., 2006). Observing real-time changes in axonal connectivity in the murine cortex over the trajectory of neurodegenerative disease or following trauma is now possible, albeit with a large array of technical challenges and limitations (Brown et al., 2011; Canty and De Paola, 2011).

11.7 TRAUMA

Whether induced by general aging, disease or trauma, subsequent axon pathologies can be remarkably similar. The usual process of axonal degeneration is a disruption in the axonal cytoskeleton and/or in axon transport with the ensuing functional dysfunction leading to degeneration of the axonal process.

Traumatic brain injury is a dynamic process resulting in alterations in structure and function of virtually all elements of the brain that may continue for long periods following injury. A variety of injury paradigms are applied to animal models—such as needlestick injury, fluid percussion injury, vascular lesions, and diffuse axonal injury—all of which model traumatic brain injuries in humans to varying degrees. Historically, mature axons in the mammalian CNS that have been transected rarely show signs of functional recovery. The lack of significant regenerative growth is likely due to a combination of intrinsic properties of the neurons and the glial environment surrounding the injury site. In recent years, significant regenerative attempts have been documented, largely through spontaneous (and generally abortive) regenerative sprouting, as well as via collateral branching and remodeling of surviving axonal projections. Indeed some populations of axons injured by laser ablation through a cranial window do show significant regrowth and restoration of synaptic connectivity, albeit with different trajectories and connectivity to their original projections (Allegra Mascaro et al., 2013; Canty et al., 2013a).

Following laser-mediated transection (Jackson et al., 2015), axons typically undergo a period of retraction from the lesion site, followed by stabilization of the surviving axonal stump. In some cases, axons show signs of sprouting or regeneration from the severed axonal shaft. Canty and colleagues describe a specific population of cortical axons rich in TBs with cell bodies residing in layer 6 of the cortex, that mount a significant regenerative attempt within 2 weeks of injury (Canty et al., 2013a). Axons typically grew with altered trajectories, but were able to establish synaptic contacts with new targets that showed comparable synaptic dynamics (density, turnover, stability) to the prelesion state. For those axons not showing signs of any significant regrowth, synaptic dynamics on injured axons tend to be subpopulation specific. For example, EPB rich axons undergo a period of retraction over several days, however, synaptic density and turnover remain relatively stable and are comparable to prelesion dynamics. Conversely, TB rich axons show a rapid (within 6 h) 10% reduction in synaptic density that is stable for at least 30 days with comparable stability of synapses following the initial loss (Canty et al., 2013b) (Figure 11.4).

In a similar model of laser ablation of cerebellar axons, labeled via lentivirus encoded GFP injections into the inferior olive, climbing fibers also show signs of new axonal sprouts which contain varicosities in close opposition to dendrites suggesting some level of functional restoration (Allegra Mascaro et al., 2013).

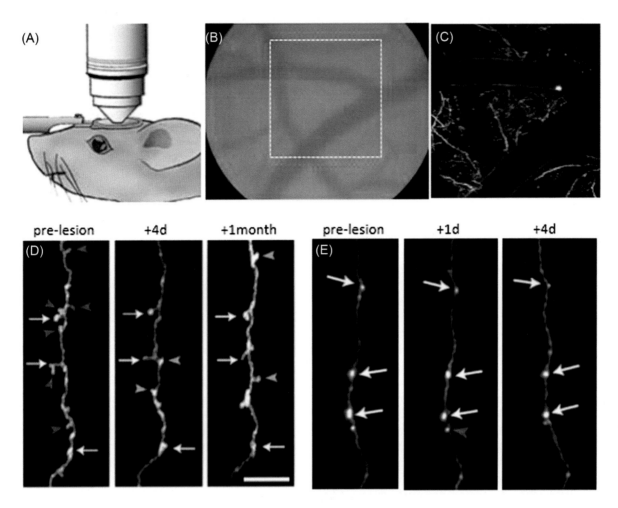

FIGURE 11.4 *In vivo* 2-photon imaging of cortical axons. (A) Cranial window insertion overlying the somatosensory cortex. (B) Vasculature. (C) Same area as in B imaged with 2-photon microscopy. (D) Synaptic dynamics in naive or lesioned cortical layer 6 axons. (E) Synaptic dynamics in naive or lesioned cortical layer 2/3 axons. Arrows indicate synapses that are stable (white), gained (green), or lost (red). Scale bar in D and E: 10μm (Canty et al., 2013b).

11.8 SENSORY MANIPULATIONS

In the first part of this chapter, axon development and plasticity in mature circuits was described. Having established that axons retain a certain degree of structural plasticity in adult animals in the absence of any particular perturbation, experimenters turned their attention to experience dependent plastic phenomena. Peripheral lesions of sensory inputs to the cerebral cortex offer valuable opportunities to investigate the response of different axon populations in both deprived and spared cortical regions, in the absence of overt trauma and bleeding. Retinal and mystacial whisker innervation to the cortex is topographically organized making these accessible regions of the cortex ideal hotspots to study axon plasticity. Without directly lesioning these cortical neuron populations, these sensory deprivation models allow the investigation of otherwise intact neurons and their axonal projections. A growing body of work in both of these systems reveals differences in axonal remodeling within and adjacent to the projection zones (reviewed by Gilbert and Li, 2012; Jones, 2000).

In the macaque, retinal lesions induced axon growth and increased axon branch dynamics in the corresponding deprived area of the macaque visual cortex with higher rates of bouton plasticity being observed. In addition, axonal sprouting was also boosted in areas immediately surrounding the deprived cortex (Yamahachi et al., 2009). Cortical interneurons in V1 of the macaque, labeled with viral vectors, showed that prior to deafferentation inhibitory projections have stable axonal arbors and bouton turnover of 10% over 7 days (Marik et al., 2014). Following retinal lesion, rapid and extensive growth and pruning of inhibitory processes was evident both within and adjacent to the lesion projection zone. The reciprocal pattern of sprouting and retraction of excitatory and inhibitory axons in somatosensory regions of the cortex may serve to maintain a balance of excitatory/inhibitory input to the reorganized cortex. In nondeprived areas of the macaque V1 cortex, there were minimal changes in total arborization, but elevated synaptic dynamics. Elevated bouton turnover was observed in all inhibitory neurons within and beyond the deprived regions, with an overall decrease in total density only occurring in the deprived regions of the cortex (Marik et al., 2014). Remapping of excitatory and inhibitory axonal populations occurs almost immediately and continues to progress in the weeks following the peripheral lesions.

Similar experiments have been carried out in the mouse barrel cortex, the somatosensory cortical area responsible for whisker representation, following the plucking of two rows of whiskers. Here, excitatory long range horizontal projections in the upper layers of the cortex were reported to invade nondeprived barrels, a phase which was preceded by inhibitory axons growing into the deprived areas, with subsequent retraction of preexisting collaterals (Marik et al., 2010a). A 3.6-fold net increase in axonal projections was observed after 14 days of sensory deprivation, much of which extended into newly deprived regions of the cortex. Synaptic bouton density and turnover increased in these projections. Using viral vector injections to label inhibitory projections, massive reorganization and 2.5-fold expansion in axonal arbors were evident within the deprived cortical regions after only 2 days of sensory deprivation.

Importantly inhibitory neurons show a different dynamic behavior compared to excitatory cells. In the visual cortex of mice layer 2/3 GABAergic interneurons have a relatively low baseline plasticity rate, with between 2% and 3% EPBs being turned over in normal conditions every 4 days. Following monocular deprivation the same interneurons, which innervate layer 5 pyramidal cells, had an increased rate of eliminated EPBs (7%) and a decreased rate of EPBs being added (0.7%) (Chen et al., 2011). Interestingly, in response to changes in activity the reaction of the postsynaptic inputs and the presynaptic outputs of spiny interneurons in the mouse visual cortex has a different time onset. Following a retinal lesion, synaptic input to the deprived visual cortex is ablated. The spines on the inhibitory cells in the lesion projection zone are rapidly destabilized and eliminated at faster rates compared to baseline turnover; the decrease in spine density is detected by 6 h after the lesion. The decrease in synaptic bouton density follows soon after, within 24 h postlesion, but nevertheless with a slower onset compared to spine elimination (Keck et al., 2011; van Versendaal et al., 2012). These findings on inhibitory circuits support a model where the loss of sensory input causes a loss of excitatory drive of the network which in turn results in a loss of inhibitory input to the deprived areas. Importantly these studies in the visual cortex did not detect axon branching phenotypes following deprivation. Such findings are in contrast with what is seen in the mouse barrel field and monkey visual cortex, though the different experimental approaches may account for some of this discrepancy.

11.9 DEGENERATION

Axon degeneration is a characteristic event that occurs in many neurodegenerative conditions including glaucoma, stroke, traumatic brain injury, and motor neuropathies. Axonal degeneration occurs in at least three phases—an acute and rapid degeneration phase on both sides of the lesion (Kerschensteiner et al., 2005; Knoferle et al., 2010), followed by a period of quiescence/latency, then rapid cytoskeletal disassembly, fragmentation and granular degeneration of the axon distal to the injury site. This latter phase of distinct morphological and molecular changes is known as Wallerian degeneration (Coleman and Freeman, 2010). These phases of degeneration occur to variable extents in different regions and axonal populations within the nervous system, and can vary significantly according to the trauma paradigm. In the spinal cord, *Thy1*-GFP dorsal column axons injured by a fine needle undergo acute axonal degeneration occurring on both sides of the lesion up to several hundred micrometers from the lesion site, whereas in laser ablated membrane-bound GFP cortical neurons rapid axonal degeneration only occurs within a few micrometers either side of the lesion site (Canty et al., 2013b), before stabilization of the axon on both sides is achieved, albeit temporarily. In this latter paradigm, owing to the shorter distances involved, the fragmentation kinetics of the degenerating axons can be monitored in their entirety. Our unpublished observations indicate that the kinetics of fragmentation are quite unique compared to those described for in other regions of the CNS (Canty et al., submitted).

Diffuse axonal injury typically occurs following rapid acceleration/deceleration and is reproduced in experimental settings using impact acceleration models, inertial acceleration models, and fluid percussion models. Axons are stretched without undergoing immediate transection, and all models result in characteristic cytoskeletal rearrangements and eventual secondary axotomy depending on the severity of the trauma. Much of our understanding of secondary axotomy comes from *in vitro* models of injured cortical axons with direct transection (Chuckowree and Vickers, 2003; Dickson et al., 2000), mechanical stretching (Iwata et al., 2004), physical oscillation of neurons (Nakayama et al., 2001), and transient pressurized fluid deflection of axons (Chung et al., 2005). Directly observing such changes in *in vivo* models of diffuse axonal injury is more problematic with resulting vascular pathology often making the visualization of single axons difficult. Diffusion tensor imaging has been successfully used to detect white matter deterioration in both mild and moderate–severe closed-skull brain injury models in the mouse (Bennett et al., 2012). Changes in axial and mean diffusivity are observed which correlate with postmortem histological analysis—silver staining in mild injury, and APP accumulations in axonal swellings following moderate–severe injury. Similarly, computed topography and magnetic resonance imaging of human patients after head injury cannot show axonal pathology in any detail.

11.10 DEMENTIA

Alzheimer's disease (AD), the most prevalent form of dementia, is characterized by the deposition of extracellular beta-amyloid plaques and intracellular neurofibrillary pathology. This neuronal pathology selectively affects different regions of the brain, primarily the temporal and frontal lobes of the cerebrum and the hippocampus, within which excitatory synapses and the processes of projection neurons are particularly vulnerable. In other types of dementia, pathological changes can originate in other cortical areas, with widespread degeneration occurring throughout the cortex and hippocampus in the latter stages in almost all forms of dementia. It was widely accepted that the presence of plaques precedes axonal pathology, however, the correlative or causative links between the two remain under intense discussion in the literature (Jacobsen et al., 2006; Spires and Hyman, 2004). Reactive axonal changes in the vicinity of plaques are most likely both regenerative and degenerative and result in a wide range of axonal pathologies. Axonal swellings typically contain abnormal accumulations of microtubule-associated proteins, phosphorylated neurofilaments, tau, organelles, and molecular motor proteins. Together, these disrupt axonal transport and contribute to axonal pathology (Stokin et al., 2005). Axon pathologies include dystrophic neurites, neurofibrillary tangles (in axons), and neuropil threads (in dendrites, reviewed by Vickers et al., 2009). In addition, these dystrophic neurites are often associated with areas of demyelination around the plaques (Mitew et al., 2010), which would further compromise any remaining functional capacity. Aggregates of a range of proteins including Tau, TAR DNA-binding protein 43 (TDP-43), Lewy bodies (in Parkinson's disease), and polyglutamine (in Huntington's disease) are reported in cortical axons (Vickers et al., 2009). Signs of both dendritic (Adlard and Vickers, 2002; Spires et al., 2005) and axonal deflection (Woodhouse et al., 2006) and withering in the vicinity of plaques have also been described.

The disruption of the axonal cytoskeleton is evidenced by changes in phosphorylation of the intermediate filaments such as the neurofilament triplet protein, alpha internexin and is described in a large body of literature (Selkoe, 2001). These structural changes would likely lead to altered axonal transport, with a number of transgenic models of AD describing axon transport disruptions concomitant with axonopathy (Adalbert et al., 2009; Christensen et al., 2014; Jawhar et al., 2012; Salehi et al., 2006; Stokin et al., 2005; Wirths et al., 2006). Nonpyramidal neurons, such as inhibitory cortical interneurons, seem preferentially unaffected and show no overt cell loss or reactive changes (Mitew et al., 2013; Sampson et al., 1997) within the vicinity of the plaques, or further afield.

Synapse loss is consistently observed in AD and correlates well with cognitive impairment (Knafo et al., 2009; Masliah et al., 2001; Scheff and Price, 2006; Terry et al., 1991). Using a variety of experimental techniques including synaptosome preparations and immune labeling of tissue sections, reductions in markers of excitatory synapses (synaptophysin, vGLUT) are evident in preclinical AD and end stage AD in humans, and in transgenic mouse models of AD, particularly within and immediately surrounding amyloid plaques (APP/PS-1: Mitew et al., 2013; Tg2576: Spires et al., 2005). Interestingly, markers of inhibitory synapses (vesicular GABA transporter and GAD65) within and adjacent to plaques are largely unaffected, suggesting that some axons subtypes are protected from synaptic loss (Mitew et al., 2013).

Using live imaging of an amyloid precursor protein (APP) expressing model of AD (Tg2576) coupled with GFP-labeled neurons, a profound decrease in dendritic spine density has been reported both within (approximately 50%) and beyond plaque containing regions (approaching 25%) compared to age-matched controls (Jung and Herms, 2012; Spires et al., 2005). Imaging axons in the same experimental model has proven more technically challenging as axons are thinner making them more difficult to localize to any depth in the cortical neuropil. In addition axons generally have lower densities of synaptic structures in the upper cortical layers, meaning that longer stretches of individual axons need to be imaged to reach statistical significance. As a result, there is currently a scarcity of live axonal imaging studies in animal models of dementia.

11.11 STROKE

For those that survive cerebrovascular accidents, a myriad of brain pathologies have been described from necrotic infarctions, to regions of the brain that experience significant prolonged, or transient, decreases in blood flow. Those axons in the center of infarct degenerate very rapidly, along with all of the other resident cells in the areas. Axons which remain intact in the penumbra regions surrounding cerebral infarctions are of considerable interest for their potential to maintain and restore any loss of function. Using rodent models of middle cerebral artery occlusion, cranial windows can be inserted overlying the medio-lateral cortical areas that include the territory of the middle cerebral artery and adjacent areas. Such a preparation creates an ideal experimental model to study axon dynamics in the infarcted regions, the penumbra where blood flow has been compromised and the adjacent regions of healthy tissue. Dendritic plasticity has been described in some detail in these regions, with characteristic rearrangements occurring in the surrounding neuropil (Brown et al., 2008, 2010).

Elongation of nodes of Ranvier, which are comprised of regular gaps in the myelin sheath vital for efficient propagation of action potentials along the axon, is reported to occur in various disease states including vascular disease and ischemic accidents (reviewed by Arancibia-Carcamo and Attwell, 2014). With nodal elongation, often accompanied by reshuffling of sodium and potassium channels within the axon, action potential propagation and synchronicity of neuronal firing are likely to be affected. In axons adjacent to infarcts resulting from microvascular disease, nodal and paranodal lengths increase by approximately 20% and 80% respectively, without marked changes in axonal architecture as described by neurofilament integrity (Hinman et al., 2015).

11.12 MULTIPLE SCLEROSIS

Multiple sclerosis (MS) is an autoimmune disease whereby peripheral immune cells infiltrate the CNS with resultant "attacks" on myelin integrity, and thereby act to expose these long and vulnerable projection axons. Myelinated axons appear throughout the CNS and are concentrated in deep white matter tracks as well as regions in the cerebral cortex. Axon pathologies, described by axonal swellings, end bulbs of transected axons and altered cytoskeletal organization are considered end state pathologies in sclerotic regions of the nervous system in MS brains (reviewed by Vickers et al., 2009). A distinguishing feature of axon pathology in MS, and in mouse models such as experimental autoimmune encephalomyelitis (EAE), is the presence of ovoid bodies, hypothesized to represent disruption of fast axonal transport (Koo et al., 1990; Sherriff et al., 1994) or alterations of axonal microtubules that lead to Wallerian degeneration (Erturk et al., 2007; Imitola et al., 2011). These areas of degeneration can be detected using MRI, but to date postmortem tissue analysis remains the major route of pathological investigation (Spence et al., 2014).

11.13 CONCLUDING REMARKS AND FUTURE PROSPECTS

Axon and synaptic impairment are linked to several, currently incurable disorders of cognition (e.g., autism), metabolism (e.g., diabetes) and aging (e.g., AD), which represent global health and economic burdens for our society. Understanding the mechanisms regulating the formation, elimination and regeneration of synapses, i.e., the synaptic life cycle, is therefore a major theme in neurobiology and medicine. Since synaptic dysfunction is a key process in neurodegenerative and neuropsychiatric disorders, and usually occurs at an early stage of disease progression, targeting the events leading to synaptic impairment has potential therapeutic value. For example, rescuing synaptic deficits in a mouse model of prion disease increases its survival (Moreno et al., 2012) and synaptic treatments can ameliorate symptoms in mouse models of complex neurodevelopmental disorders such as Rett syndrome and autism, paving the way for human trials (reviewed by Delorme et al., 2013).

However, our understanding of the physiological settings and biochemical pathways regulating when and how synapses form and break is in its infancy, precluding the identification of much needed additional potential targets for intervention. As we discussed in the first part of this chapter, synapses are highly dynamic submicrometer structures, and it is important to study them in real time, using high-resolution techniques, ideally in living organisms. Much progress has been made to elucidate synaptic physiology by focusing on the postsynapse, but relatively less is known about how remodeling of axons and presynaptic boutons contributes to cognition. In addition, because of their reduced complexity and higher accessibility, axons in the peripheral nervous system have been studied more extensively, leaving gaps in our understanding of axonal dynamics in the brain.

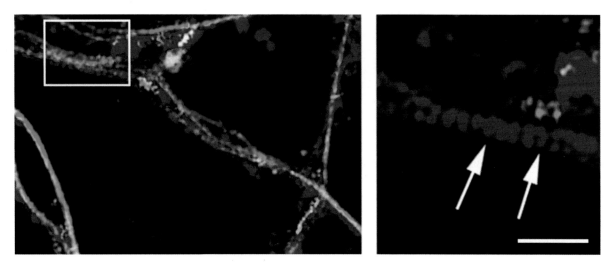

FIGURE 11.5 Super-resolved images of axons *in vitro*. Red: phalloidin; green: beta-III tubulin; blue: synapsin. White arrows point to actin rings labeled with phalloidin (Xu et al., 2013). Scale bar: 600 nm. *Image courtesy of Lucien West.*

In this chapter, we have highlighted the potential of dissecting the signaling pathways that control the basal levels of cellular (axonal and synaptic) dynamics, to uncover new ways to manipulate these processes for clinical applications. With this new knowledge, it may be possible to design pharmaceutical approaches to fine-tune (increase or decrease) the rate of axonal and synaptic reorganization for therapeutic intervention.

We identify three optical-based technologies as main areas for future development, which are likely to have a transformative impact on the study of axon biology in health and disease. Firstly, molecular imaging *in vivo* could potentially lead to the discovery of new principles in axon cell biology (e.g., in how organelles and synaptic proteins are transported; Cane et al., 2014; Fortin et al., 2014; Gross et al., 2013; Kleele et al., 2014), as most of what we currently know comes from *in vitro* or postmortem studies. In particular, super-resolution microscopy (Figure 11.5), which enables imaging at the nanoscale, offers the ideal tool to study synapse architecture and function. Size ranges of key presynaptic components such as active zones, synaptic vesicles and the synaptic cleft are below the resolution of conventional microscopy (diameter 150–450 nm, diameter 35–50 nm, width 20 nm, respectively; Dani et al., 2010; Ribrault et al., 2011). In addition, the presence of thousands of different molecules makes the synapse an incredibly heterogeneous subcellular compartment in terms of protein composition (O'Rourke et al., 2012). By breaking the diffraction limit, super-resolution microscopy will undoubtedly provide unprecedented *in vivo* insights (Berning et al., 2012) into axonal structure, development and plasticity in the not so distant future.

Secondly, combining structural and functional imaging studies will ultimately inform us on how electrical activity controls axon and presynaptic remodeling in intact cortical circuits. For example, sensory experience is a major determinant of cortical neural firing and synaptic stability, however, as we have discussed, most *in vivo* imaging studies have identified structural dynamics in a small fraction (5–10%) of dendritic spines as a prominent cellular mechanism for experience-dependent functional rewiring of cortical circuits (Holtmaat and Svoboda, 2009). Unanswered questions include (i) whether in addition to axonal bouton formation and elimination (Marik et al., 2010a), further changes in the size of persistent boutons contribute to experience dependent plasticity *in vivo*, (ii) whether synapses from all cortical layers remodel in response to sensory experience manipulations, and (iii) whether large-scale axonal remodeling occurs in cells other than L2/3 neurons (Marik et al., 2010a). These questions can be addressed in the adult rodent somatosensory cortex, which offers several advantages for the study of experience dependent plasticity, including topography and ease of input manipulations (Holtmaat and Svoboda, 2009) as evidenced by similar studies such as Grillo et al. (2013) and Holtmaat et al. (2009). Structural markers (such as tdTomato) and activity sensors such as genetically encoded calcium sensors can now be simultaneously used to investigate the functional implications of structural remodeling. Calcium transients from presynaptic boutons in response to as few as 3–5 action potentials have been demonstrated using the less sensitive GCaMP3 (Petreanu et al., 2012). The ultrasensitive GCaMP6 (Chen et al., 2013) could then be used to measure fluorescence changes in intracortical and thalamocortical neurons projecting to cortical layer 1 in response to whisker stimulation. Inducible GCaMP6 floxed alleles can be crossed with cell type-specific Cre-driver lines (available from The Jackson Laboratory) to introduce the calcium sensor in layer 2/3, 5, and 6 neurons. Thalamocortical neurons could be labeled by injections of a viral mixture in the posterior medial nucleus of the thalamus. This approach could allow the study of axonal bouton activity changes in defined populations of cortical and thalamocortical neurons in response to sensory experience manipulations. Using optogenetics and calcium imaging it may be possible to simultaneously activate axons and visualize axonal and presynaptic dynamics in layer 2/3, 5, and 6 pyramidal neurons by crossing layer-specific Cre-expressing mouse lines with floxed Channelrhodopsin-2 (ChR2) or GCaMP6 reporter mouse lines, which coexpress a cell morphology marker (Madisen et al., 2012) (e.g., tdTomato or EYFP). This strategy could also be used to simultaneously activate/ inhibit and visualize presynaptic dynamics in selected pyramidal neuron populations using Designer Receptors Exclusively Activated by Designer Drugs (Pienaar et al. 2014; Rogan and Roth, 2011) and hyperpolarizing opto/ pharmacogenetic tools. Thus, similar to the idea discussed above of dissecting the signaling mechanisms underlying axonal and presynaptic dynamics to design drugs targeting synapses for therapeutic interventions, by learning more about the functional circuitry of ensembles of neurons from *in vivo* calcium imaging, one could modify individual components (e.g., layer-specific excitatory neurons or defined classes of interneurons) with optogenetics, or relevant brain regions with deep brain stimulation, to restore function.

Finally, imaging of human-induced pluripotent stem cell-derived neurons in the mouse brain *in vivo* will enable further physiological assessments of the role of the environment (e.g., stress) on neural circuit development and dynamics than currently possible in a dish (Espuny-Camacho et al., 2013). Longitudinal *in vivo* studies using patient-derived neurons grafted in the mouse brain will represent a new paradigm, which will open up novel routes of investigation into human neurocognitive disorders. Since the vast majority of work linking synaptic dysfunction and cognitive impairment has focused on postsynaptic spine defects (Penzes et al., 2011) and early developmental stages, a focus on axons and the presynaptic side of the synapse in adult cortical circuits provides an opportunity to discover new mechanisms of human disease. We expect this work will deliver new fundamental insights into how synaptic connectivity and repair are controlled in both the healthy and diseased brain.

Abbreviations

AD Alzheimer's disease
APP amyloid precursor protein
ASD autism spectrum disorders
CNS central nervous system
CLSM confocal laser scanning microscopy
EPBs *En Passant* boutons
FMR1 Fragile X mental retardation 1
GAD glutamic acid decarboxylase
GFP green fluorescent protein
GWAS genome wide association studies
ITB3 integrin β3
MS multiple sclerosis
NL neuroligin
PSD postsynaptic density-95
TBs *Terminaux* boutons
XFPs different fluorescent proteins
2PLSM 2-photon laser scanning microscopy

References

Abe, I., Ochiai, N., Ichimura, H., Tsujino, A., Sun, J., Hara, Y., 2004. Internodes can nearly double in length with gradual elongation of the adult rat sciatic nerve. J. Orthop. Res. 22, 571–577.

Adalbert, R., Nogradi, A., Babetto, E., Janeckova, L., Walker, S.A., Kerschensteiner, M., et al., 2009. Severely dystrophic axons at amyloid plaques remain continuous and connected to viable cell bodies. Brain 132, 402–416.

Adlard, P.A., Vickers, J.C., 2002. Morphologically distinct plaque types differentially affect dendritic structure and organisation in the early and late stages of Alzheimer's disease. Acta Neuropathol. 103, 377–383.

Allegra Mascaro, A.L., Cesare, P., Sacconi, L., Grasselli, G., Mandolesi, G., Maco, B., et al., 2013. *In vivo* single branch axotomy induces GAP-43-dependent sprouting and synaptic remodeling in cerebellar cortex. Proc. Natl. Acad. Sci. U.S.A. 110, 10824–10829.

Amos, W.B., White, J.G., Fordham, M., 1987. Use of confocal imaging in the study of biological structures. Appl. Opt. 26, 3239–3243.

Anderson, G.R., Galfin, T., Xu, W., Aoto, J., Malenka, R.C., Südhof, T.C., 2012. Candidate autism gene screen identifies critical role for cell-adhesion molecule CASPR2 in dendritic arborization and spine development. Proc. Natl. Acad. Sci. U.S.A. 109, 18120–18125.

Antonini, A., Stryker, M.P., 1993. Rapid remodeling of axonal arbors in the visual-cortex. Science 260, 1819–1821.

Arancibia-Carcamo, I.L., Attwell, D., 2014. The node of Ranvier in CNS pathology. Acta Neuropathol. 128, 161–175.

Armstrong, D., Dunn, J.K., Antalffy, B., Trivedi, R., 1995. Selective dendritic alterations in the cortex of Rett syndrome. J. Neuropathol. Exp. Neurol. 54, 195–201.

Beier, K.T., Saunders, A., Oldenburg, I.A., Miyamichi, K., Akhtar, N., Luo, L., et al., 2011. Anterograde or retrograde transsynaptic labeling of CNS neurons with vesicular stomatitis virus vectors. Proc. Natl. Acad. Sci. U.S.A. 108, 15414–15419.

Belichenko, P., Oldfors, A., Hagberg, B., Dahlstrom, A., 1994. Rett syndrome: 3-D confocal microscopy of cortical pyramidal dendrites and afferents. NeuroReport 5, 1509–1513.

Bennett, R.E., Mac Donald, C.L., Brody, D.L., 2012. Diffusion tensor imaging detects axonal injury in a mouse model of repetitive closed-skull traumatic brain injury. Neurosci. Lett. 513, 160–165.

Berning, S., Willig, K.I., Steffens, H., Dibaj, P., Hell, S.W., 2012. Nanoscopy in a living mouse brain. Science 335, 551.

Blundell, J., Tabuchi, K., Bolliger, M.F., Blaiss, C.A., Brose, N., Liu, X., et al., 2009. Increased anxiety-like behavior in mice lacking the inhibitory synapse cell adhesion molecule neuroligin 2. Genes Brain Behav. 8, 114–126.

Blundell, J., Blaiss, C.A., Etherton, M.R., Espinosa, F., Tabuchi, K., Walz, C., et al., 2010. Neuroligin 1 deletion results in impaired spatial memory and increased repetitive behavior. J. Neurosci. 30, 2115–2129.

Brown, C.E., Wong, C., Murphy, T.H., 2008. Rapid morphologic plasticity of peri-infarct dendritic spines after focal ischemic stroke. Stroke 39, 1286–1291.

Brown, C.E., Boyd, J.D., Murphy, T.H., 2010. Longitudinal *in vivo* imaging reveals balanced and branch-specific remodeling of mature cortical pyramidal dendritic arbors after stroke. J. Cereb. Blood Flow Metab. 30, 783–791.

Brown, K.M., Barrionuevo, G., Canty, A.J., De Paola, V., Hirsch, J.A., Jefferis, G.S., et al., 2011. The DIADEM data sets: representative light microscopy images of neuronal morphology to advance automation of digital reconstructions. Neuroinformatics 9 (2–3), 143–157.

Budde, M.D., Kim, J.H., Liang, H.F., Russell, J.H., Cross, A.H., Song, S.K., 2008. Axonal injury detected by *in vivo* diffusion tensor imaging correlates with neurological disability in a mouse model of multiple sclerosis. NMR Biomed. 21, 589–597.

Buffelli, M., Burgess, R.W., Feng, G.P., Lobe, C.G., Lichtman, J.W., Sanes, J.R., 2003. Genetic evidence that relative synaptic efficacy biases the outcome of synaptic competition. Nature 424, 430–434.

Cai, D., Cohen, K.B., Luo, T., Lichtman, J.W., Sanes, J.R., 2013. Improved tools for the Brainbow toolbox. Nat. Methods 10, 540–547.

Cane, M., Maco, B., Knott, G., Holtmaat, A., 2014. The relationship between PSD-95 clustering and spine stability *in vivo*. J. Neurosci. 34, 2075–2086.

Canty, A.J., De Paola, V., 2011. Axonal reconstructions going live. Neuroinformatics 9, 129–131.

Canty, A.J., Huang, L., Jackson, J.S., Little, G.E., Knott, G., Maco, B., et al., 2013a. In-vivo single neuron axotomy triggers axon regeneration to restore synaptic density in specific cortical circuits. Nat. Commun. 4, 2038.

Canty, A.J., Teles-Grilo Ruivo, L.M., Nesarajah, C., Song, S., Jackson, J.S., Little, G.E., et al., 2013b. Synaptic elimination and protection after minimal injury depend on cell type and their prelesion structural dynamics in the adult cerebral cortex. J. Neurosci. 33, 10374–10383.

Caroni, P., 1997. Overexpression of growth-associated proteins in the neurons of adult transgenic mice. J. Neurosci. Methods 71, 3–9.

Carrillo, J., Cheng, S.Y., Ko, K.W., Jones, T.A., Nishiyama, H., 2013. The long-term structural plasticity of cerebellar parallel fiber axons and its modulation by motor learning. J. Neurosci. 33, 8301–8307.

Carter, M.D., Shah, C.R., Muller, C.L., Crawley, J.N., Carneiro, A.M.D., Veenstra-VanderWeele, J., 2011. Absence of preference for social novelty and increased grooming in integrin β3 knockout mice: initial studies and future directions. Autism Res. 4, 57–67.

Chadman, K.K., Gong, S., Scattoni, M.L., Boltuck, S.E., Gandhy, S.U., Heintz, N., et al., 2008. Minimal aberrant behavioral phenotypes of neuroligin-3 R451C knockin mice. Autism Res. 1, 147–158.

Chalfie, M., Tu, Y., Euskirchen, G., Ward, W.W., Prasher, D.C., 1994. Green Fluorescent protein as a marker for gene expression. Science 263, 802–805.

Chapleau, C.A., Calfa, G.D., Lane, M.C., Albertson, A.J., Larimore, J.L., Kudo, S., et al., 2009. Dendritic spine pathologies in hippocampal pyramidal neurons from Rett syndrome brain and after expression of Rett-associated MECP2 mutations. Neurobiol. Dis. 35, 219–233.

Chapleau, C.A., Boggio, E.M., Calfa, G., Percy, A.K., Giustetto, M., Pozzo-Miller, L., 2012. Hippocampal CA1 pyramidal neurons of Mecp2 mutant mice show a dendritic spine phenotype only in the presymptomatic stage. Neural Plast. 2012, 9.

Chen, J.L., Lin, W.C., Cha, J.W., So, P.T., Kubota, Y., Nedivi, E., 2011. Structural basis for the role of inhibition in facilitating adult brain plasticity. Nat. Neurosci. 14, 587–594.

Chen, S.X., Kim, A.N., Peters, A.J., Komiyama, T., 2015. Subtype-specific plasticity of inhibitory circuits in motor cortex during motor learning. Nat. Neurosci. 18 (8), 1109–1115.

Chen, T.W., Wardill, T.J., Sun, Y., Pulver, S.R., Renninger, S.L., Baohan, A., et al., 2013. Ultrasensitive fluorescent proteins for imaging neuronal activity. Nature 499, 295–300.

Christensen, D.Z., Huettenrauch, M., Mitkovski, M., Pradier, L., Wirths, O., 2014. Axonal degeneration in an Alzheimer mouse model is PS1 gene dose dependent and linked to intraneuronal Abeta accumulation. Front. Aging Neurosci. 6, 139.

Chuckowree, J.A., Vickers, J.C., 2003. Cytoskeletal and morphological alterations underlying axonal sprouting after localized transection of cortical neuron axons *in vitro*. J. Neurosci. 23, 3715–3725.

Chudakov, D.M., Matz, M.V., Lukyanov, S., Lukyanov, K.A., 2010. Fluorescent Proteins and their applications in imaging living cells and tissues. Physiol. Rev. 90, 1103–1163.

Chung, R.S., Staal, J.A., McCormack, G.H., Dickson, T.C., Cozens, M.A., Chuckowree, J.A., et al., 2005. Mild axonal stretch injury in vitro induces a progressive series of neurofilament alterations ultimately leading to delayed axotomy. J. Neurotrauma 22, 1081–1091.

Coleman, M.P., Freeman, M.R., 2010. Wallerian degeneration, wld(s), and nmnat. Annu. Rev. Neurosci. 33, 245–267.

Cruz-Martín, A., Crespo, M., Portera-Cailliau, C., 2010. Delayed stabilization of dendritic spines in fragile X mice. J. Neurosci. 30, 7793–7803.

Dachtler, J., Glasper, J., Cohen, R.N., Ivorra, J.L., Swiffen, D.J., Jackson, A.J., et al., 2014. Deletion of [alpha]-neurexin II results in autism-related behaviors in mice. Transl. Psychiatr. 4, e484.

Dani, A., Huang, B., Bergan, J., Dulac, C., Zhuang, X., 2010. Superresolution imaging of chemical synapses in the brain. Neuron 68, 843–856.

DeFelipe, J., 2010. From the connectome to the synaptome: an epic love story. Science 330, 1198.

De Paola, V., Arber, S., Caroni, P., 2003. AMPA receptors regulate dynamic equilibrium of presynaptic terminals in mature hippocampal networks. Nat. Neurosci. 6, 491–500.

De Paola, V., Holtmaat, A., Knott, G., Song, S., Wilbrecht, L., Caroni, P., et al., 2006. Cell type-specific structural plasticity of axonal branches and boutons in the adult neocortex. Neuron 49, 861–875.

II. AXON DYNAMICS

Delorme, R., Ey, E., Toro, R., Leboyer, M., Gillberg, C., Bourgeron, T., 2013. Progress toward treatments for synaptic defects in autism. Nat. Med. 19, 685–694.

Denk, W., Svoboda, K., 1997. Photon upmanship: why multiphoton imaging is more than a gimmick. Neuron 18, 351–357.

Denk, W., Strickler, J.H., Webb, W.W., 1990. Two-photon laser scanning fluorescence microscopy. Science 248, 73–76.

Dickson, T.C., Adlard, P.A., Vickers, J.C., 2000. Sequence of cellular changes following localized axotomy to cortical neurons in glia-free culture. J. Neurotrauma 17, 1095–1103.

Ellegood, J., Henkelman, R.M., Lerch, J.P., 2012. Neuroanatomical assessment of the integrin β3 mouse model related to autism and the serotonin system using high resolution MRI. Front. Psychiatr. 3, 37.

Erturk, A., Hellal, F., Enes, J., Bradke, F., 2007. Disorganized microtubules underlie the formation of retraction bulbs and the failure of axonal regeneration. J. Neurosci. 27, 9169–9180.

Espuny-Camacho, I., Michelsen, K.A., Gall, D., Linaro, D., Hasche, A., Bonnefont, J., et al., 2013. Pyramidal neurons derived from human pluripotent stem cells integrate efficiently into mouse brain circuits *in vivo*. Neuron 77, 440–456.

Etherton, M.R., Blaiss, C.A., Powell, C.M., Südhof, T.C., 2009. Mouse neurexin-1α deletion causes correlated electrophysiological and behavioral changes consistent with cognitive impairments. Proc. Natl. Acad. Sci. U.S.A. 106, 17998–18003.

Feng, G.P., Mellor, R.H., Bernstein, M., Keller-Peck, C., Nguyen, Q.T., Wallace, M., et al., 2000. Imaging neuronal subsets in transgenic mice expressing multiple spectral variants of GFP. Neuron 28, 41–51.

Fortin, D.A., Tillo, S.E., Yang, G., Rah, J.C., Melander, J.B., Bai, S., et al., 2014. Live imaging of endogenous PSD-95 using ENABLED: a conditional strategy to fluorescently label endogenous proteins. J. Neurosci. 34, 16698–16712.

Fukuchi-Shimogori, T., Grove, E.A., 2001. Neocortex patterning by the secreted signaling molecule FGF8. Science 294, 1071–1074.

Gilbert, C.D., Li, W., 2012. Adult visual cortical plasticity. Neuron 75, 250–264.

Glover, J.C., Petursdottir, G., Jansen, J.K.S., 1986. Fluorescent dextran-amines used as axonal tracers in the nervous-system of the chicken-embryo. J. Neurosci. Meth. 18, 243–254.

Gogolla, N., Galimberti, I., Caroni, P., 2007. Structural plasticity of axon terminals in the adult. Curr. Opin. Neurobiol. 17, 516–524.

Gogolla, N., Caroni, P., Luthi, A., Herry, C., 2009. Perineuronal nets protect fear memories from erasure. Science 325, 1258–1261.

Gray, N.W., Weimer, R.M., Bureau, I., Svoboda, K., 2006. Rapid redistribution of synaptic PSD-95 in the neocortex in vivo. PLoS Biol. 4 (11), e370.

Grayton, H.M., Missler, M., Collier, D.A., Fernandes, C., 2013. Altered social behaviours in neurexin 1α knockout mice resemble core symptoms in neurodevelopmental disorders. PLoS ONE 8, e67114.

Greenough, W.T., Klintsova, A.Y., Irwin, S.A., Galvez, R., Bates, K.E., Weiler, I.J., 2001. Synaptic regulation of protein synthesis and the fragile X protein. Proc. Natl. Acad. Sci. U.S.A. 98, 7101–7106.

Grillo, F.W., Song, S., Teles-Grilo Ruivo, L.M., Huang, L., Gao, G., Knott, G.W., et al., 2013. Increased axonal bouton dynamics in the aging mouse cortex. Proc. Natl. Acad. Sci. U.S.A. 110, E1514–E1523.

Grillo, W.F., West, L., De Paola, V., 2015. Removing synaptic brakes on learning. Nat. Neurosci. 18, 1062–1064.

Gross, G.G., Junge, J.A., Mora, R.J., Kwon, H.B., Olson, C.A., Takahashi, T.T., et al., 2013. Recombinant probes for visualizing endogenous synaptic proteins in living neurons. Neuron 78, 971–985.

Grutzendler, J., Gan, W.B., 2006. Two-photon imaging of synaptic plasticity and pathology in the living mouse brain. NeuroRx 3, 489–496.

Grutzendler, J., Kasthuri, N., Gan, W.B., 2002. Long-term dendritic spine stability in the adult cortex. Nature 420, 812–816.

Guy, J., Hendrich, B., Holmes, M., Martin, J.E., Bird, A., 2001. A mouse Mecp2-null mutation causes neurological symptoms that mimic Rett syndrome. Nat. Genet. 27, 322–326.

Hinman, J.D., Lee, M.D., Tung, S., Vinters, H.V., Carmichael, S.T., 2015. Molecular disorganization of axons adjacent to human lacunar infarcts. Brain 138, 736–745.

Hinton, V.J., Brown, W.T., Wisniewski, K., Rudelli, R.D., 1991. Analysis of neocortex in three males with the fragile X syndrome. Am. J. Med. Genet. 41, 289–294.

Hofer, S.B., Mrsic-Flogel, T.D., Bonhoeffer, T., Hubener, M., 2009. Experience leaves a lasting structural trace in cortical circuits. Nature 457, 313–317.

Holtmaat, A., Svoboda, K., 2009. Experience-dependent structural synaptic plasticity in the mammalian brain. Nat. Rev. Neurosci. 10, 647–658.

Holtmaat, A., Bonhoeffer, T., Chow, D.K., Chuckowree, J., De Paola, V., Hofer, S.B., et al., 2009. Long-term, high-resolution imaging in the mouse neocortex through a chronic cranial window. Nat. Protoc. 4, 1128–1144.

Honig, M.G., Hume, R.I., 1989. DiI and DiO—Versatile fluorescent dyes for neuronal labeling and pathway tracing. Trends. Neurosci. 12 333-&.

Horikawa, K., Armstrong, W.E., 1988. A versatile means of intracellular labeling: injection of biocytin and its detection with avidin conjugates. J. Neurosci. Methods 25, 1–11.

Huang, Z.J., Yu, W., Lovett, C., Tonegawa, S., 2002. Cre/loxP recombination-activated neuronal markers in mouse neocortex and hippocampus. Genesis 32, 209–217.

Hutsler, J.J., Zhang, H., 2010. Increased dendritic spine densities on cortical projection neurons in autism spectrum disorders. Brain Res. 1309, 83–94.

Ichtchenko, K., Hata, Y., Nguyen, T., Ullrich, B., Missler, M., Moomaw, C., et al., 1995. Neuroligin 1: a splice site-specific ligand for beta-neurexins. Cell 81, 435–443.

Imitola, J., Cote, D., Rasmussen, S., Xie, X.S., Liu, Y., Chitnis, T., et al., 2011. Multimodal coherent anti-Stokes Raman scattering microscopy reveals microglia-associated myelin and axonal dysfunction in multiple sclerosis-like lesions in mice. J. Biomed. Opt. 16, 021109.

Irwin, S.A., Galvez, R., Greenough, W.T., 2000. Dendritic spine structural anomalies in Fragile-X mental retardation syndrome. Cereb. Cortex 10, 1038–1044.

Irwin, S.A., Patel, B., Idupulapati, M., Harris, J.B., Crisostomo, R.A., Larsen, B.P., et al., 2001. Abnormal dendritic spine characteristics in the temporal and visual cortices of patients with fragile-X syndrome: a quantitative examination. Am. J. Med. Genet. 98, 161–167.

Isshiki, M., Tanaka, S., Kuriu, T., Tabuchi, K., Takumi, T., Okabe, S., 2014. Enhanced synapse remodelling as a common phenotype in mouse models of autism. Nat. Commun., 5.

Iwata, A., Stys, P.K., Wolf, J.A., Chen, X.H., Taylor, A.G., Meaney, D.F., et al., 2004. Traumatic axonal injury induces proteolytic cleavage of the voltage-gated sodium channels modulated by tetrodotoxin and protease inhibitors. J. Neurosci. 24, 4605–4613.

Jackson, J., Canty, A.J., Huang, L., De Paola, V., 2015. Laser mediated scar-free microlesions in the mouse neocortex to investigate neuronal degeneration and regeneration. Curr. Protoc. Neurosci.

Jacobsen, J.S., Wu, C.C., Redwine, J.M., Comery, T.A., Arias, R., Bowlby, M., et al., 2006. Early-onset behavioral and synaptic deficits in a mouse model of Alzheimer's disease. Proc. Natl. Acad. Sci. U.S.A. 103, 5161–5166.

Jamain, S., Quach, H., Betancur, C., Råstam, M., Colineaux, C., Gillberg, I.C., et al., 2003. Mutations of the X-linked genes encoding neuroligins NLGN3 and NLGN4 are associated with autism. Nat. Genet. 34, 27–29.

Jawhar, S., Trawicka, A., Jenneckens, C., Bayer, T.A., Wirths, O., 2012. Motor deficits, neuron loss, and reduced anxiety coinciding with axonal degeneration and intraneuronal Abeta aggregation in the 5XFAD mouse model of Alzheimer's disease. Neurobiol Aging 33, 196e129–140.

Jones, E.G., 2000. Cortical and subcortical contributions to activity-dependent plasticity in primate somatosensory cortex. Annu. Rev. Neurosci. 23, 1–37.

Ju, A., Hammerschmidt, K., Tantra, M., Krueger, D., Brose, N., Ehrenreich, H., 2014. Juvenile manifestation of ultrasound communication deficits in the neuroligin-4 null mutant mouse model of autism. Behav. Brain Res. 270, 159–164.

Jung, C.K., Herms, J., 2012. Role of APP for dendritic spine formation and stability. Exp. Brain Res. 217, 463–470.

Kasthuri, N., Lichtman, J.W., 2003. The role of neuronal identity in synaptic competition. Nature 424, 426–430.

Keck, T., Scheuss, V., Jacobsen, R.I., Wierenga, C.J., Eysel, U.T., Bonhoeffer, T., et al., 2011. Loss of sensory input causes rapid structural changes of inhibitory neurons in adult mouse visual cortex. Neuron 71, 869–882.

Kerschensteiner, M., Schwab, M.E., Lichtman, J.W., Misgeld, T., 2005. *In vivo* imaging of axonal degeneration and regeneration in the injured spinal cord. Nat. Med. 11, 572–577.

Kleele, T., Marinkovic, P., Williams, P.R., Stern, S., Weigand, E.E., Engerer, P., et al., 2014. An assay to image neuronal microtubule dynamics in mice. Nat. Commun. 5, 4827.

Knafo, S., Alonso-Nanclares, L., Gonzalez-Soriano, J., Merino-Serrais, P., Fernaud-Espinosa, I., Ferrer, I., et al., 2009. Widespread changes in dendritic spines in a model of Alzheimer's disease. Cereb. Cortex 19, 586–592.

Knoferle, J., Koch, J.C., Ostendorf, T., Michel, U., Planchamp, V., Vutova, P., et al., 2010. Mechanisms of acute axonal degeneration in the optic nerve *in vivo*. Proc. Natl. Acad. Sci. U.S.A. 107, 6064–6069.

Knott, G.W., Holtmaat, A., Wilbrecht, L., Welker, E., Svoboda, K., 2006. Spine growth precedes synapse formation in the adult neocortex *in vivo*. Nat. Neurosci. 9, 1117–1124.

Koo, E.H., Sisodia, S.S., Archer, D.R., Martin, L.J., Weidemann, A., Beyreuther, K., et al., 1990. Precursor of amyloid protein in Alzheimer disease undergoes fast anterograde axonal transport. Proc. Natl. Acad. Sci. USA 87, 1561–1565.

Landi, S., Putignano, E., Boggio, E.M., Giustetto, M., Pizzorusso, T., Ratto, G.M., 2011. The short-time structural plasticity of dendritic spines is altered in a model of Rett syndrome. Sci. Rep., 1.

Lee, W.C., Huang, H., Feng, G., Sanes, J.R., Brown, E.N., So, P.T., et al., 2006. Dynamic remodeling of dendritic arbors in GABAergic interneurons of adult visual cortex. PLoS Biol. 4, e29.

Lendvai, B., Stern, E.A., Chen, B., Svoboda, K., 2000. Experience-dependent plasticity of dendritic spines in the developing rat barrel cortex *in vivo*. Nature 404, 876–881.

Lewis Jr., T.L., Courchet, J., Polleux, F., 2013. Cell biology in neuroscience: cellular and molecular mechanisms underlying axon formation, growth, and branching. J. Cell Biol. 202, 837–848.

Liebscher, S., Page, R.M., Kafer, K., Winkler, E., Quinn, K., Goldbach, E., et al., 2013. Chronic gamma-secretase inhibition reduces amyloid plaque-associated instability of pre- and postsynaptic structures. Mol. Psychiatr.

Linkenhoker, B.A., von der Ohe, C.G., Knudsen, E.I., 2005. Anatomical traces of juvenile learning in the auditory system of adult barn owls. Nat. Neurosci. 8, 93–98.

Livet, J., 2007. The brain in color: transgenic "Brainbow" mice for visualizing neuronal circuits. Med. Sci. (Paris) 23, 1173–1176.

Livet, J., Sigrist, M., Stroebel, S., De Paola, V., Price, S.R., Henderson, C.E., et al., 2002. ETS gene Pea3 controls the central position and terminal arborization of specific motor neuron pools. Neuron 35, 877–892.

Lopez-Bendito, G., Sturgess, K., Erdelyi, F., Szabo, G., Molnar, Z., Paulsen, O., 2004. Preferential origin and layer destination of GAD65-GFP cortical interneurons. Cereb. Cortex 14, 1122–1133.

LoTurco, J., Manent, J.B., Sidiqi, F., 2009. New and improved tools for *in utero* electroporation studies of developing cerebral cortex. Cereb. Cortex 19, I120–I125.

Luo, L., O'Leary, D.D., 2005. Axon retraction and degeneration in development and disease. Annu. Rev. Neurosci. 28, 127–156.

Ma, Y., Hu, H., Berrebi, A.S., Mathers, P.H., Agmon, A., 2006. Distinct subtypes of somatostatin-containing neocortical interneurons revealed in transgenic mice. J. Neurosci. 26, 5069–5082.

Madisen, L., Mao, T., Koch, H., Zhuo, J.M., Berenyi, A., Fujisawa, S., et al., 2012. A toolbox of Cre-dependent optogenetic transgenic mice for light-induced activation and silencing. Nat. Neurosci. 15, 793–802.

Majewska, A.K., Newton, J.R., Sur, M., 2006. Remodeling of synaptic structure in sensory cortical areas *in vivo*. J. Neurosci. 26, 3021–3029.

Marik, S.A., Yamahachi, H., McManus, J.N., Szabo, G., Gilbert, C.D., 2010a. Axonal dynamics of excitatory and inhibitory neurons in somatosensory cortex. PLoS Biol. 8, e1000395.

Marik, S.A., Yamahachi, H., McManus, J.N.J., Szabo, G., Gilbert, C.D., 2010b. Axonal dynamics of excitatory and inhibitory neurons in somatosensory cortex. PLoS Biol., 8.

Marik, S.A., Yamahachi, H., Meyer zum Alten Borgloh, S., Gilbert, C.D., 2014. Large-scale axonal reorganization of inhibitory neurons following retinal lesions. J. Neurosci. 34, 1625–1632.

Masliah, E., Mallory, M., Alford, M., DeTeresa, R., Hansen, L.A., McKeel Jr., D.W., et al., 2001. Altered expression of synaptic proteins occurs early during progression of Alzheimer's disease. Neurology 56, 127–129.

Mineur, Y.S., Sluyter, F., de Wit, S., Oostra, B.A., Crusio, W.E., 2002. Behavioral and neuroanatomical characterization of the Fmr1 knockout mouse. Hippocampus 12, 39–46.

Mitew, S., Kirkcaldie, M.T., Halliday, G.M., Shepherd, C.E., Vickers, J.C., Dickson, T.C., 2010. Focal demyelination in Alzheimer's disease and transgenic mouse models. Acta Neuropathol. 119, 567–577.

II. AXON DYNAMICS

Mitew, S., Kirkcaldie, M.T., Dickson, T.C., Vickers, J.C., 2013. Altered synapses and gliotransmission in Alzheimer's disease and AD model mice. Neurobiol. Aging 34, 2341–2351.

Mizuno, H., Hirano, T., Tagawa, Y., 2007. Evidence for activity-dependent cortical wiring: formation of interhemispheric connections in neonatal mouse visual cortex requires projection neuron activity. J. Neurosci. 27, 6760–6770.

Mizuno, H., Hirano, T., Tagawa, Y., 2010. Pre-synaptic and post-synaptic neuronal activity supports the axon development of callosal projection neurons during different post-natal periods in the mouse cerebral cortex. Eur. J. Neurosci. 31, 410–424.

Moreno, J.A., Radford, H., Peretti, D., Steinert, J.R., Verity, N., Martin, M.G., et al., 2012. Sustained translational repression by eIF2alpha-P mediates prion neurodegeneration. Nature 485, 507–511.

Mostany, R., Anstey, J.E., Crump, K.L., Maco, B., Knott, G., Portera-Cailliau, C., 2013. Altered synaptic dynamics during normal brain aging. J. Neurosci. 33, 4094–4104.

Murmu, R.P., Li, W., Holtmaat, A., Li, J.Y., 2013. Dendritic spine instability leads to progressive neocortical spine loss in a mouse model of Huntington's disease. J. Neurosci. 33, 12997–13009.

Nadler, J.J., Moy, S.S., Dold, G., Simmons, N., Perez, A., Young, N.B., et al., 2004. Automated apparatus for quantitation of social approach behaviors in mice. Genes, Brain Behav. 3, 303–314.

Nakayama, Y., Aoki, Y., Niitsu, H., 2001. Studies on the mechanisms responsible for the formation of focal swellings on neuronal processes using a novel in vitro model of axonal injury. J. Neurotrauma 18, 545–554.

Napolioni, V., Lombardi, F., Sacco, R., Curatolo, P., Manzi, B., Alessandrelli, R., et al., 2011. Family-based association study of ITGB3 in autism spectrum disorder and its endophenotypes. Eur. J. Hum. Genet. 19, 353–359.

Neukirchen, D., Bradke, F., 2011. Neuronal polarization and the cytoskeleton. Semin. Cell Dev. Biol. 22, 825–833.

Nishiyama, H., Fukaya, M., Watanabe, M., Linden, D.J., 2007. Axonal motility and its modulation by activity are branch-type specific in the intact adult cerebellum. Neuron 56, 472–487.

O'Rourke, N.A., Weiler, N.C., Micheva, K.D., Smith, S.J., 2012. Deep molecular diversity of mammalian synapses: why it matters and how to measure it. Nat. Rev. Neurosci. 13, 365–379.

O'Toole, M., Latham, R., Baqri, R.M., Miller, K.E., 2008. Modeling mitochondrial dynamics during *in vivo* axonal elongation. J. Theor. Biol. 255, 369–377.

Ochs, S.M., Dorostkar, M.M., Aramuni, G., Schon, C., Filser, S., Poschl, J., et al., 2014. Loss of neuronal GSK3beta reduces dendritic spine stability and attenuates excitatory synaptic transmission via beta-catenin. Mol. Psychiatr.

Padmashri, R., Reiner, B.C., Suresh, A., Spartz, E., Dunaevsky, A., 2013. Altered structural and functional synaptic plasticity with motor skill learning in a mouse model of Fragile X syndrome. J. Neurosci. 33, 19715–19723.

Panchuk-Voloshina, N., Haugland, R.P., Bishop-Stewart, J., Bhalgat, M.K., Millard, P.J., Mao, F., et al., 1999. Alexa dyes, a series of new fluorescent dyes that yield exceptionally bright, photostable conjugates. J. Histochem. Cytochem. 47, 1179–1188.

Peça, J., Feliciano, C., Ting, J.T., Wang, W., Wells, M.F., Venkatraman, T.N., et al., 2011. Shank3 mutant mice display autistic-like behaviours and striatal dysfunction. Nature 472, 437–442.

Peñagarikano, O., Abrahams, B.S., Herman, E.I., Winden, K.C., Gdalyahu, A., Dong, H., et al., 2011. Absence of CNTNAP2 leads to epilepsy, neuronal migration abnormalities and core autism-related deficits. Cell 147, 235–246.

Penzes, P., Cahill, M.E., Jones, K.A., VanLeeuwen, J.E., Woolfrey, K.M., 2011. Dendritic spine pathology in neuropsychiatric disorders. Nat. Neurosci. 14, 285–293.

Petreanu, L., Gutnisky, D.A., Huber, D., Xu, N.L., O'Connor, D.H., Tian, L., et al., 2012. Activity in motor-sensory projections reveals distributed coding in somatosensation. Nature 489, 299–303.

Pienaar, I.S., Sharma, P., Elson, J.L., De Paola, V., Dexter, D.T., 2014. A DREADD approach for acute stimulation of cholinergic neurons of the pedunculopontine nucleus reverses Parkinsonism in the lactacystin model of Parkinson's disease [abstract]. Mov. Disord. 29 (Suppl. 1), 381.

Pizzorusso, T., Medini, P., Berardi, N., Chierzi, S., Fawcett, J.W., Maffei, L., 2002. Reactivation of ocular dominance plasticity in the adult visual cortex. Science 298, 1248–1251.

Portera-Cailliau, C., Weimer, R.M., De Paola, V., Caroni, P., Svoboda, K., 2005. Diverse modes of axon elaboration in the developing neocortex. PLoS Biol. 3, 1473–1487.

Potter, S.M., Wang, C.M., Garrity, P.A., Fraser, S.E., 1996. Intravital imaging of green fluorescent protein using two-photon laser-scanning microscopy. Gene 173, 25–31.

Radyushkin, K., Hammerschmidt, K., Boretius, S., Varoqueaux, F., El-Kordi, A., Ronnenberg, A., et al., 2009. Neuroligin-3-deficient mice: model of a monogenic heritable form of autism with an olfactory deficit. Genes Brain Behav. 8, 416–425.

Ribrault, C., Sekimoto, K., Triller, A., 2011. From the stochasticity of molecular processes to the variability of synaptic transmission. Nat. Rev. Neurosci. 12, 375–387.

Rogan, S.C., Roth, B.L., 2011. Remote control of neuronal signaling. Pharmacol. Rev. 63, 291–315.

Rudelli, R.D., Brown, W.T., Wisniewski, K., Jenkins, E.C., Laure-Kamionowska, M., Connell, F., et al., 1985. Adult fragile X syndrome. Acta Neuropathol. 67, 289–295.

Saito, T., Nakatsuji, N., 2001. Efficient gene transfer into the embryonic mouse brain using *in vivo* electroporation. Dev. Biol. 240, 237–246.

Salehi, A., Delcroix, J.D., Belichenko, P.V., Zhan, K., Wu, C., Valletta, J.S., et al., 2006. Increased App expression in a mouse model of Down's syndrome disrupts NGF transport and causes cholinergic neuron degeneration. Neuron 51, 29–42.

Sampson, V.L., Morrison, J.H., Vickers, J.C., 1997. The cellular basis for the relative resistance of parvalbumin and calretinin immunoreactive neocortical neurons to the pathology of Alzheimer's disease. Exp. Neurol. 145, 295–302.

Sanders, S.J., Ercan-Sencicek, A.G., Hus, V., Luo, R., Murtha, M.T., Moreno-De-Luca, D., et al., 2011. Multiple recurrent *de novo* copy number variations (CNVs), including duplications of the 7q11.23 Williams–Beuren syndrome region, are strongly associated with autism. Neuron 70, 863–885.

Scheff, S.W., Price, D.A., 2006. Alzheimer's disease-related alterations in synaptic density: neocortex and hippocampus. J. Alzheimers Dis. 9, 101–115.

Schubert, D., Martens, G.J., Kolk, S.M., 2014. Molecular underpinnings of prefrontal cortex development in rodents provide insights into the etiology of neurodevelopmental disorders. Mol. Psychiatr.

Selkoe, D.J., 2001. Alzheimer's disease: genes, proteins, and therapy. Physiol. Rev. 81, 741–766.

Sherriff, F.E., Bridges, L.R., Gentleman, S.M., et al., 1994. Markers of axonal injury in post-mortem human brain. Acta Neuropathol. 88, 433–439.

Spence, R.D., Kurth, F., Itoh, N., Mongerson, C.R., Wailes, S.H., Peng, M.S., et al., 2014. Bringing CLARITY to gray matter atrophy. Neuroimage 101, 625–632.

Spires, T.L., Hyman, B.T., 2004. Neuronal structure is altered by amyloid plaques. Rev. Neurosci. 15, 267–278.

Spires, T.L., Meyer-Luehmann, M., Stern, E.A., McLean, P.J., Skoch, J., Nguyen, P.T., et al., 2005. Dendritic spine abnormalities in amyloid precursor protein transgenic mice demonstrated by gene transfer and intravital multiphoton microscopy. J. Neurosci. 25, 7278–7287.

Stettler, D.D., Yamahachi, H., Li, W., Denk, W., Gilbert, C.D., 2006. Axons and synaptic boutons are highly dynamic in adult visual cortex. Neuron 49, 877–887.

Stokin, G.B., Lillo, C., Falzone, T.L., Brusch, R.G., Rockenstein, E., Mount, S.L., et al., 2005. Axonopathy and transport deficits early in the pathogenesis of Alzheimer's disease. Science 307, 1282–1288.

Suter, D.M., Miller, K.E., 2011. The emerging role of forces in axonal elongation. Prog. Neurobiol. 94, 91–101.

Svoboda, K., Denk, W., Kleinfeld, D., Tank, D.W., 1997. In vivo dendritic calcium dynamics in neocortical pyramidal neurons. Nature 385, 161–165.

Tabata, H., Nakajima, K., 2001. Efficient in utero gene transfer system to the developing mouse brain using electroporation: visualization of neuronal migration in the developing cortex. Neuroscience 103, 865–872.

Tabuchi, K., Blundell, J., Etherton, M.R., Hammer, R.E., Liu, X., Powell, C.M., et al., 2007. A neuroligin-3 mutation implicated in autism increases inhibitory synaptic transmission in mice. Science 318, 71–76.

Tamamaki, N., Yanagawa, Y., Tomioka, R., Miyazaki, J., Obata, K., Kaneko, T., 2003. Green fluorescent protein expression and colocalization with calretinin, parvalbumin, and somatostatin in the GAD67-GFP knock-in mouse. J. Comp. Neurol. 467, 60–79.

Taniguchi, H., He, M., Wu, P., Kim, S., Paik, R., Sugino, K., et al., 2011. A resource of Cre driver lines for genetic targeting of GABAergic neurons in cerebral cortex. Neuron 71, 995–1013.

Terry, R.D., Masliah, E., Salmon, D.P., Butters, N., DeTeresa, R., Hill, R., et al., 1991. Physical basis of cognitive alterations in Alzheimer's disease: synapse loss is the major correlate of cognitive impairment. Ann. Neurol. 30, 572–580.

Trachtenberg, J.T., Stryker, M.P., 2001. Rapid anatomical plasticity of horizontal connections in the developing visual cortex. J. Neurosci. 21, 3476–3482.

Trachtenberg, J.T., Chen, B.E., Knott, G.W., Feng, G., Sanes, J.R., Welker, E., et al., 2002. Long-term in vivo imaging of experience-dependent synaptic plasticity in adult cortex. Nature 420, 788–794.

van den Pol, A.N., Ghosh, P.K., 1998. Selective neuronal expression of green fluorescent protein with cytomegalovirus promoter reveals entire neuronal arbor in transgenic mice. J. Neurosci. 18, 10640–10651.

van Versendaal, D., Rajendran, R., Saiepour, M.H., Klooster, J., Smit-Rigter, L., Sommeijer, J.P., et al., 2012. Elimination of inhibitory synapses is a major component of adult ocular dominance plasticity. Neuron 74, 374–383.

Varoqueaux, F., Aramuni, G., Rawson, R.L., Mohrmann, R., Missler, M., Gottmann, K., et al., 2006. Neuroligins determine synapse maturation and function. Neuron 51, 741–754.

Vickers, J.C., King, A.E., Woodhouse, A., Kirkcaldie, M.T., Staal, J.A., McCormack, G.H., et al., 2009. Axonopathy and cytoskeletal disruption in degenerative diseases of the central nervous system. Brain Res. Bull. 80, 217–223.

Wang, C.L., Zhang, L., Zhou, Y., Zhou, J., Yang, X.J., Duan, S.M., et al., 2007. Activity-dependent development of callosal projections in the somatosensory cortex. J. Neurosci. 27, 11334–11342.

Weiss, L.A., Kosova, G., Delahanty, R.J., Jiang, L., Cook Jr., E.H., Ober Jr., C., et al., 2006. Variation in ITGB3 is associated with whole-blood serotonin level and autism susceptibility. Eur. J. Hum. Genet. 14, 923–931.

Williams, R.S., Hauser, S.L., Purpura, D.P., DeLong, G., Swisher, C.N., 1980. Autism and mental retardation: neuropathologic studies performed in four retarded persons with autistic behavior. Arch. Neurol. 37, 749–753.

Wirths, O., Weis, J., Szczygielski, J., Multhaup, G., Bayer, T.A., 2006. Axonopathy in an APP/PS1 transgenic mouse model of Alzheimer's disease. Acta Neuropathol. 111, 312–319.

Wolff, J.J., Gerig, G., Lewis, J.D., Soda, T., Styner, M.A., Vachet, C., et al., 2015. Altered corpus callosum morphology associated with autism over the first 2 years of life. Brain.

Woodhouse, A., Vickers, J.C., Dickson, T.C., 2006. Cytoplasmic cytochrome c immunolabelling in dystrophic neurites in Alzheimer's disease. Acta Neuropathol. 112, 429–437.

Xu, K., Zhong, G., Zhuang, X., 2013. Actin, spectrin, and associated proteins form a periodic cytoskeletal structure in axons. Science 339, 452–456.

Xu, X., Miller, E.C., Pozzo-Miller, L., 2014. Dendritic spine dysgenesis in Rett syndrome. Front. Neuroanatomy 8.

Yamahachi, H., Marik, S.A., McManus, J.N.J., Denk, W., Gilbert, C.D., 2009. Rapid axonal sprouting and pruning accompany functional reorganization in primary visual cortex. Neuron 64, 719–729.

Yang, M., Silverman, J.L., Crawley, J.N., 2011. Automated three-chambered social approach task for mice. Curr Protoc Neurosci., Chapter 8:Unit 8.26.

Yu, X.Z., Zuo, Y., 2014. Two-photon in vivo imaging of dendritic spines in the mouse cortex using a thinned-skull preparation. Jove-J. Vis. Exp.

Zikopoulos, B., Barbas, H., 2010. Changes in prefrontal axons may disrupt the network in autism. J. Neurosci. 30, 14595–14609.

Zoghbi, H.Y., 2003. Postnatal neurodevelopmental disorders: meeting at the synapse? Science 302, 826–830.

12

Contribution of Axons to Short-Term Dynamics of Neuronal Communication

Dirk Bucher

Federated Department of Biological Sciences, New Jersey Institute of Technology and Rutgers University, University Heights, Newark, NJ, USA

12.1 INTRODUCTION

In most (if not all) neurons, electrical activity is not just the result of linear summation of synaptic inputs to a fixed threshold for action potential (spike) initiation, but rather the outcome of complex nonlinear operations. At chemical synapses, the transmitter release machinery as well as the binding and gating properties of postsynaptic receptors render most communication between neurons dependent on prior activity, facilitating and/or depressing synaptic strength and efficacy (Regehr and Stevens, 2001; Zucker and Regehr, 2002). In addition, relatively complex complements of voltage- and calcium-dependent ion channels with different gating properties affect subthreshold behavior and excitability, and therefore the neuron's input–output function (Hille, 2001; Taylor et al., 2009). Such dynamics of neural signaling occur over time scales of milliseconds to tens of minutes and are "short term" in the sense that no persistent changes in the signaling machinery are required, such as changes in the properties or abundance of membrane proteins. In addition, both synaptic properties and membrane excitability can be changed relatively rapidly by neuromodulation. Neuromodulators are often released diffusely and either activate G protein-coupled receptors, which leads to phosphorylation of target proteins like ion channels or other receptors, or they activate ionotropic receptors, which changes membrane potential and input resistance (Bucher and Marder, 2013; Nadim and Bucher, 2014).

At the level of single neurons, the rates and temporal patterns of spikes represent the neural code. It is still a commonly held perception that specific spike firing patterns are shaped exclusively through the complex dynamics of both synapses and integrative properties of somatodendritic structures and the proximal axon. The propagating part of the axon (the "axon proper") is thought to more or less faithfully transmit the pattern of spikes encoded in the proximal neuronal compartments to electrotonically distant presynaptic sites. This view is inconsistent with what are relatively old findings of complex activity-dependent changes in spike conduction velocity and even spike failures, particularly in peripheral or long centrally projecting axons (Swadlow et al., 1980).

Over the last few decades, additional evidence has been collected for a potentially important contribution of axons to the short-term dynamics of neural communication in a variety of systems (Bucher and Goaillard, 2011; Debanne, 2004; Debanne et al., 2011; Sasaki, 2013). Some common themes have emerged that underlie axonal dynamics in most of these cases. First, axonal excitability, i.e., the complement of ionic mechanisms that govern membrane behavior, in most neurons is substantially more complex than often credited. A variety of different voltage- or calcium-gated ion channels, as well as ion pumps, are found along axons. Because of the wide range of voltage- and time dependencies of the resulting currents, axonal properties can depend on prior activation in a much wider time range than can be explained by simple refractory effects described in the classic literature. As spike propagation critically depends on excitability, propagation speed can change substantially with repeated activation. Therefore, the time intervals between subsequent spikes can change during propagation from initiation site to presynaptic sites, potentially altering the temporal code. In some axons, even the spike rate can be altered, as changes in excitability can lead to ectopic axonal spike initiation or changes in the likelihood of propagation failures.

Second, the notion that spikes are "all-or-none" events and that neuronal communication with spikes is therefore purely digital, is not true in many cases. Activity-dependent changes in axonal excitability can substantially alter the shape of the spike waveform and the resting membrane potential, and in some neurons, even subthreshold signals are propagated much further than originally thought possible and influence the voltage trajectory of propagating spikes. As the activation of calcium influx at presynaptic sites is voltage dependent, activity-dependent changes in spike shape can be associated with changes in transmitter release.

Third, in a variety of cases, neuromodulator receptors (of both metabotropic and ionotropic types) can be expressed in extrasynaptic axonal membrane, where they can affect axonal excitability, spike shape, and propagation, dependent on neuromodulatory state.

In this chapter, I will highlight some of these findings and discuss their potential role for neural computations. The examples given illustrate general principles governing electrical behavior of axons, and similar properties can probably be found in many axons in many systems. However, when a given aspect of axonal properties is examined in detail, the diversity may be as large as for other neuron properties, often not even adhering to commonly used classifications of neuron types (Nusser, 2009). At this level, categorizing axon types into invertebrate/vertebrate, mammalian/nonmammalian, myelinated/unmyelinated, central/peripheral, or sensory/motor may not be useful, as a given property may be found only in a subset of axons of one category, but also in axons of another category (Bucher and Goaillard, 2011).

12.2 COMPLEX AXONAL EXCITABILITY

The ionic mechanisms underlying spike generation and propagation were first described quantitatively by Hodgkin and Huxley, based on experiments performed on the giant axon of the squid (Hodgkin and Huxley, 1952a,b). Their model fairly accurately described the behavior of this axon on the basis of only two voltage-gated currents, a fast and inactivating sodium current responsible for the regenerative nature of the fast depolarizing phase of the spike, and a slower and noninactivating potassium current, the "delayed rectifier" responsible for repolarization to resting membrane potential. This relatively simple model has dominated the textbook understanding of axon function for more than half a century, even though it was found relatively early that axons can express other currents. In fact, a quantitative model of the third voltage-dependent current ever described, the inactivating A-type potassium current, has first been achieved in an axon (Connor, 1975).

Since then, abundant evidence has accumulated showing that ionic mechanisms in axons, even in the propagating part, can be almost as rich and complex as those associated with signal integration and spike initiation (Figure 12.1A). Different sodium currents (including persistent ones), many different potassium currents with a wide range of gating properties, hyperpolarization-activated inward currents, calcium currents, and a range of noncanonical ionic mechanisms have all been found (Bucher and Goaillard, 2011; Krishnan et al., 2009). As in proximal compartments, the complement of ionic mechanisms may be very cell type specific, and there is still a dearth of information about membrane properties in axons of most neuron types (Nusser, 2009). Large peripheral axons have been amenable to electrophysiological and pharmacological studies for quite some time (Waxman and Ritchie, 1993), but there is a distinct lack of data about central axons, mostly owing to the fact that they are fairly small structures ($<5.0\,\mu m$ or even $\leq1.0\,\mu m$) and often inaccessible to standard electrophysiology techniques. Some indirect information has been derived from advancements in immunohistochemical localization of ion channels (Vacher et al., 2008), but only recently have patch clamp recordings from swellings of cut ends of axons in slices ("blebs") allowed detailed biophysical and physiological studies of small central axons (Hu and Shu, 2012).

I do not intend to give a detailed account here of the types of ion channels found in axons or their specific properties. However, it is important to note that ionic conductances beyond those minimally required for spike propagation are in many cases expressed at levels that clearly indicate a substantial contribution to excitability. In addition, in many cases, ion channels are targeted precisely and with high specificity to the axon, and are not just "sloppily" expressed throughout the neuron. This is particularly evident in the highly spatially organized channel distribution in myelinated axons, and from the findings that a given type of current can be based on a different channel gene or isoforms than are similar currents in other compartments of the same neuron (Bucher and Goaillard, 2011; Vacher et al., 2008). Most importantly, whatever the actual complement of ionic mechanisms in a given axon may be, relatively complex combinations of currents—gated with a range of different voltage dependencies and time constants

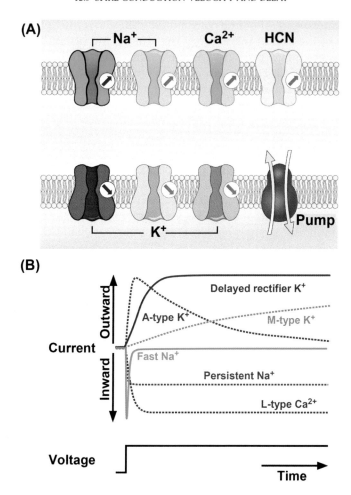

FIGURE 12.1 Diversity of ionic mechanisms in axons. (A) The original quantitative description of spike initiation and propagation, based on the giant axon of the squid, included only a fast and inactivating sodium channel (green), and a delayed rectifier potassium channel (red). Most axons contain additional channels, which can include channels giving rise to other inward currents, for example, persistent sodium, calcium, and hyperpolarization-activated (HCN) channels. Many axons also contain multiple potassium currents, with a range of different gating properties. In addition, other ionic mechanisms, like ion pumps, can have important effects on axonal properties. (B) Schematic current responses to a depolarizing voltage step shown to illustrate the diversity of voltage- and time dependencies of currents through some of these ion channels.

associated with activation and inactivation (Figure 12.1B)—will give rise to highly nonlinear behavior, particularly during repetitive activity. Depending on the range of time constants underlying such dynamics, spike shape and propagation can become highly dependent on prior activity, minimally on the immediately preceding interval, but potentially on spike history over many minutes.

12.3 SPIKE CONDUCTION VELOCITY AND DELAY

Before considering the potential impact of changes in axonal excitability on spike propagation, it is important to understand what determines spike velocity and delay. The speed of spike propagation in an axon, i.e., the conduction velocity, depends on the specific properties of the axonal membrane and cytosol, and on the diameter of the axon. These relationships were defined relatively early on (Del Castillo and Moore, 1959; Hodgkin, 1954; Hodgkin and Huxley, 1952a,b; Waxman, 1975). Once a spike is initiated, passive spread of current depolarizes the membrane ahead of the active site. The larger the distance over which this depolarization is sufficient to bring the membrane voltage to threshold, the faster the spike propagates along the axon. This distance is determined by the axon's passive properties, namely the length constant and time constant (Hodgkin and Rushton, 1946; Rall, 1969). The length constant depends on diameter and specific membrane and axial resistances, and describes the distance over which steady-state voltage changes can spread. The time constant describes how much this change in membrane potential is delayed by the capacitance of the axonal membrane. The latter is important because for fast transient signals it prevents reaching the full voltage change predicted by the length constant, therefore slowing spike propagation.

In unmyelinated axons, propagation is continuous and velocity is essentially limited by the parameters described above. Under the assumption of identical specific membrane and cytosolic properties, conduction velocity in unmyelinated axons is proportional to the square root of the diameter (Hodgkin, 1954). The giant axon of the squid, with a diameter of 1 mm, can propagate spikes at ~25 m/s (Rosenthal and Bezanilla, 2000). Small diameter unmyelinated axons like hippocampal mossy fibers or slow cutaneous C-fibers show conduction velocities far below 1 m/s (Kress et al., 2008; Weidner et al., 1999).

Not only in vertebrates but also in some annelids and crustaceans (Hartline and Colman, 2007; Roots, 2008), many axons projecting over larger distances are myelinated. The mode of spike propagation in myelinated axons is discontinous, which is referred to as saltatory conduction (Huxley and Stampfli, 1949). Voltage-gated channels are mostly restricted to the regions in and directly adjacent to the Nodes of Ranvier (i.e., interruptions in the myelin sheath formed by glia cells tightly wrapped around the axon). The low abundance of ion channels and the myelin sheath of the internodal axon serve to significantly increase electrical resistance and therefore the length constant, and to reduce capacitance and therefore the time constant. Consequently, near-instantaneous passive spread of potentials is substantially increased, and impulses jump from node to node, the only regions where channel gating processes significantly take up time. Under the assumption of otherwise identical properties, conduction velocity in myelinated axons increases linearly with diameter, and for larger diameters is much faster than in unmyelinated axons of the same size. However, conduction velocity in myelinated axons does not just depend on diameter but also on myelin thickness and internodal distances (Waxman, 1980). The largest classes of mammalian peripheral sensory and motor axons reach about 20 μm in diameter and show conduction velocities of 120 m/s (Manzano et al., 2008), similar to those found in some large cortico-spinal axons (Evarts, 1965). Conduction velocity in myelinated axons of shrimp can even reach 200 m/s (Xu and Terakawa, 1999).

Conduction velocity on its own says very little about functional aspects of neuronal communication. What really matters is conduction delay, which of course depends on both spike conduction velocity and axonal length. Very long axons like vertebrate peripheral motor axons, or invertebrate axons involved in escape responses, have evolved to propagate spikes rapidly and minimize delay out of behavioral necessity, either through myelination or large diameters (Hartline and Colman, 2007). However, delay in itself can have important functional implications for neural coding, for example, in the computation of time differences representing spatial information in sensory systems (Konishi, 2003). Both the summation and activity-dependent plasticity of synaptic signals critically depend on spike intervals or synchrony of converging inputs, and are therefore vulnerable to changes in axonal conduction delays (Izhikevich, 2006). In consequence, it is important to note that the changes in axonal excitability discussed in the following only matter in terms of changes to the temporal patterns that they ultimately cause. Even large percentage changes in conduction velocities may have little effect if the average conduction delay is very small. In this sense, one can speculate that the evolution of fast spike propagation should be viewed not only in the context of minimizing delay but also in the context of increasing temporal precision (Wang, 2008).

12.4 THE RECOVERY CYCLE OF AXONAL EXCITABILITY

With minimal sets of axonal ionic currents, fairly accurate predictions of conduction velocity of single spikes in unmyelinated axons have been achieved on the basis of morphology, traveling wave considerations, and ionic conductances (George et al., 2015; Muratov, 2000; Tasaki, 2004). However, it is not clear if such descriptions apply to axons with more complex excitability and, in particular, to spike propagation during repetitive activity. In general, repetitive axonal firing changes conduction velocity because spikes travel through regions of altered membrane excitability in the wake of preceding activity.

The most widely appreciated form of excitability changes are refractory effects explained by the ionic theory of spike initiation and propagation based on the fast sodium and delayed rectifier potassium conductances of the squid giant axon (Hodgkin and Huxley, 1952a,b). During the spike, the sodium conductance responsible for the depolarization rapidly inactivates, and the potassium conductance repolarizes the membrane. Following the spike, sodium channels have to recover from inactivation and potassium channels have to close for the membrane to regain excitability. Immediately following the spike is the *absolute* refractory period in which no other spike can be initiated or propagated, ensuring unidirectional propagation. During the subsequent *relative* refractory period, excitability is still reduced, associated with an increased spike threshold and slowed conduction velocity. In most axons, depending on the specific gating properties of the channels present, refractory effects last for only a few milliseconds. This means

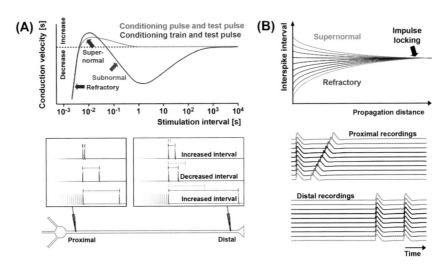

FIGURE 12.2   The recovery cycle of axonal excitability and conduction velocity and its consequences for spike intervals during propagation. (A) Schematic of the dependence of conduction velocity of a test pulse following either a single conditioning spike (green line) or a train of spikes (purple line) on the stimulation interval. At small intervals, the axonal membrane is in the relative refractory period (blue arrow) and conduction velocity is decreased. With increasing intervals, conduction velocity increases and exceeds that of the response to the conditioning pulse for a range of intervals (supernormal period, red arrow). When trains of action potentials precede the test pulse, conduction velocity can be reduced for a large range of intervals (subnormal period, brown arrow). The lower panels show the consequences for spike intervals recorded close to the initiation site (proximal) and close to distal presynaptic sites. Refractory and subnormal excitability increase intervals, whereas supernormal excitability decreases them. (B) In very long or slowly propagating axons, pairs of spikes initiated at intervals corresponding to the range from relative refractory (red) to supernormal (green) period eventually all converge onto the same interval after propagation to a sufficient distance (impulse locking). Spikes traveling in the refractory excitability period left in the wake of the preceding spike are slowed and the interval is increasing, but only to a value that corresponds to the transition point between refractory and supernormal period. At that point, excitability and conduction velocity for both spikes is equal. Spikes traveling in the supernormal excitability left in the wake of the preceding spikes travel faster, and the interval is decreasing. Again, they only do so until the interval reaches the value at which excitability and conduction velocity for both spikes are equal.

that they affect axonal spike propagation only at fairly high frequencies. However, it was known decades before any quantitative description of the ionic mechanisms underlying the spike, that the recovery process of axons is complex, including both increases and decreases of excitability, dependent on the stimulation interval (Adrian, 1920).

Later studies carefully quantified these activity-dependent changes in spike propagation, showing that axonal excitability can be altered far beyond simple refractory effects (Figure 12.2A). For example, Bullock used a conditioning and a test stimulus at different intervals to show that in earthworm giant fibers and frog sciatic A-fibers, the few milliseconds of refractory period were followed by a "supernormal" period during which excitability and conduction velocity were increased (Bullock, 1951). This effect was detectable at intervals up to 100 ms or more. In frog myelinated sciatic fibers, Raymond later used conditioning trains instead of a single stimulus, and showed that in this case the supernormal period was followed by a subnormal period, an extended period of reduced excitability and conduction velocity (Raymond, 1979). The duration and magnitude of both supernormal and subnormal excitability was dependent on the conditioning stimulus frequency and duration, and the total duration of changed excitability reached tens of minutes or more with long conditioning stimulus regimes. The oscillation between reduced and enhanced excitability is collectively referred to as the "recovery cycle," and is an important diagnostic tool for axonal excitability changes associated with peripheral neuropathies (Burke et al., 2001; Krishnan et al., 2009). This behavior may play an important role for temporal coding, as the changes in conduction velocity mean that spike intervals are changing from proximal initiation sites to distal presynaptic sites (Figure 12.2A, lower panel). It is important to note that the time scale of the recovery cycle is shorter in fast propagating axons, but does not appear to be qualitatively different between myelinated and unmyelinated axons (Swadlow et al., 1980).

A curious consequence of the interval dependence of axonal conduction velocity can be seen in some axons that have large conduction delays because they are either very long or propagate spikes slowly (Figure 12.2B). If a spike substantially alters the excitability of an axon and therefore the conduction velocity of a following spike, the interval between these two spikes changes gradually during propagation along the axon. Given enough time and distance, the interval will be changed to a value at which the excitability change caused by the first spike is significantly less severe (Moradmand and Goldfinger, 1995). A spike traveling in the relative refractory period left in the wake of a preceding spike will travel more slowly, but only until the interval has increased to the value at which the membrane is not refractory anymore. Past that point, both spikes will be propagated with the same velocity at a fixed interval that corresponds to the end of the refractory period in the recovery cycle. A spike traveling in the supernormal excitability left in the wake of a preceding spike will travel faster than the first one only to the distance at which the interval has decreased to a value at which excitability is not supernormal anymore. For both cases, this interval is the same, the transition point between refractory and supernormal period. Therefore, spikes initiated with a fairly broad range of intervals converge to the same interval value after long enough propagation. Such equalization of intervals has for example been observed in rabbit efferent visual cortical neurons in the corpus callosum, and was termed "impulse entrainment" or "impulse locking" (Kocsis et al., 1979).

Except for the refractory period, the mechanisms underlying the recovery cycle are poorly understood and may be very diverse across different neuron types (Bucher and Goaillard, 2011). In many cases, altered excitability is at least somewhat associated with changes in the spike voltage trajectory and after-potentials. Spikes in the refractory period often have a reduced amplitude, as inactivation decreases the number of available sodium channels (Moradmand and Goldfinger, 1995). Supernormal conduction can be associated with a depolarizing after-potential following the preceding spike. This after-depolarization can arise from delayed extracellular potassium clearance, which shifts the potassium equilibrium potential to less negative values (Bowe et al., 1987); passive capacitive charging, particularly of the internodal membrane in myelinated axons (David et al., 1995); or persistent or resurgent voltage-gated sodium currents (McIntyre et al., 2002). Subnormal conduction is often associated with slow after-hyperpolarization. This activity-dependent hyperpolarization can be due to cumulative activation of slow potassium currents during repetitive activation (Burke et al., 2001; Schwarz et al., 2006). However, slow changes in excitability and hyperpolarization lasting many minutes in many axons are likely due to the electrogenic nature of the sodium/potassium-ATPase (Moldovan and Krarup, 2006; Scuri et al., 2007). This pump is essential for generating and maintaining a negative membrane potential, but as it is activated by increases in intracellular sodium concentration, the massive sodium influx during highly repetitive spiking can lead to long-lasting hyperpolarization.

It should be noted that the relationship between changes in voltage trajectory and changes in excitability is often only tentative or even ambiguous. Consequently, the recovery cycle can often not be predicted from changes in membrane potential. For example, substantial changes in conduction delay during repetitive activity in a crustacean axon are not clearly correlated with measures of spike shape or baseline membrane potential (Ballo et al., 2012). One reason for this ambiguity is the voltage dependence of the ion channel gating mechanisms; that is, activation of ion channels affects membrane potential, which in turn affects the activation or inactivation of other channels (Figure 12.3). For example, the most obvious prediction of the effect of membrane potential changes on axonal excitability and spike conduction velocity is that depolarization increases excitability and hyperpolarization decreases it. Depolarization moves the membrane potential closer to spike threshold and reduces the amount of current needed for spiking, while hyperpolarization moves the membrane potential further away from threshold and increases the amount of current needed. However, more sustained changes can have the opposite effect. Depolarization can inactivate sodium channels and therefore reduce the number of channels available for spiking, while hyperpolarization removes inactivation and therefore has the opposite effect. In such cases, lack of hyperpolarizing ionic mechanisms would actually decrease conduction delay. For example, sodium channel inactivation slows spike conduction in axons innervating the rat cranial meninges, and block of the sodium/potassium-ATPase increases this effect (De Col et al., 2008). In rat hippocampal axons, block of slow potassium channels increases sodium channel inactivation and slows conduction (Vervaeke et al., 2006). The presence of fast and inactivating (A-type) potassium currents in many axons (Bucher and Goaillard, 2011; Debanne et al., 2011; Krishnan et al., 2009) means that the level of outward currents that functionally oppose the sodium currents can also depend on membrane potential, as availability of such channels decreases with depolarization and increases with hyperpolarization (Figure 12.3A). The inactivation state of both sodium and A-type potassium channels is reflected in changing spike shapes during repetitive spiking at different membrane potentials (Figure 12.3B). Intracellular recordings of crustacean motor axons during bursting activity show substantial changes in the voltage trajectory of successive spikes, dependent on depolarization or hyperpolarization (Ballo and Bucher, 2009). Reduction in spike amplitudes over successive spikes is more severe at depolarized membrane potentials, consistent with enhanced sodium channel inactivation. A-type potassium channels appear to be mostly inactivated at these levels of depolarization, and spikes are therefore broad but duration does not change over successive spikes. At hyperpolarized potentials, sodium channels appear to be able to recover from inactivation in between successive spikes, and spike amplitudes are therefore stable. However, A-type channels recover more slowly, leading to an increase in spike durations over successive spikes.

A second reason for a lack of clear correlation between membrane potential and excitability is that concomitant or at least temporally overlapping activation of opposing inward and outward currents can substantially alter total membrane

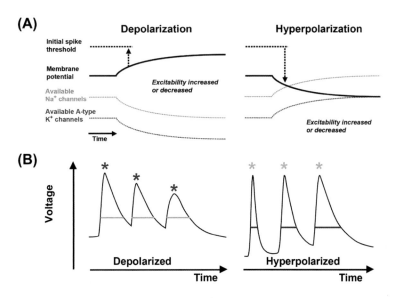

FIGURE 12.3 The ambiguous effects of changes in membrane potential on axonal excitability and conduction velocity. (A) Depolarization brings the membrane potential closer to spike threshold, that is, less current would be required to reach threshold and therefore excitability and conduction velocity should be increased. Hyperpolarization increases the voltage to threshold and therefore should decrease excitability. However, depolarization can inactivate a portion of sodium channels, and hyperpolarization can remove inactivation. A decrease in available channels reduces excitability, while an increase does the opposite. If the axon also contains a fast and inactivating (A-type) potassium current, depolarization can increase excitability because the inactivation of A-type channels removes functional opposition to sodium channel activation. Removal of inactivation has the opposite effect. Both depolarization and hyperpolarization can therefore either increase or decrease excitability and conduction velocity, dependent on voltage, time course of membrane potential changes, and voltage dependencies and relative magnitudes of different currents. (B) The effect of depolarization and hyperpolarization on the availability of inactivating currents can be reflected in the spike shapes during repetitive activity. In this hypothetical case, sodium channels are rapidly inactivating with successive spikes when the membrane depolarizes, so spike amplitudes decrease substantially (blue asterisks). A-type potassium channels are already mostly inactivated. Therefore, even the initial spike takes fairly long to repolarize, and the successive spike width does not change much (gray lines). At hyperpolarized membrane potentials, removal of sodium channel inactivation is fast between successive spikes, so spike amplitude does not change much (gray asterisks). A-type potassium channels are initially mostly deinactivated, so the first spike repolarizes rapidly. However, removal of inactivation between spikes is slow and therefore incomplete, so spike width increases substantially (blue lines).

conductance (and therefore resistance) without being reflected in large changes in membrane potential. Fairly similar spike shapes can be produced by vastly different underlying current amplitudes, dependent on how much temporal overlap or separation exists between sodium and potassium currents (Sengupta et al., 2010). Another example is the fairly common balance of activity-dependent axonal hyperpolarization by H-currents, i.e., hyperpolarization-activated inward currents. These currents limit hyperpolarization and spike slowing during highly repetitive activity (Baginskas et al., 2009; Ballo et al., 2012). However, if hyperpolarizing currents and inward rectification are well balanced, their opposing signs mean that total membrane currents can be very different at quite similar membrane potentials, with potentially important consequences for input resistance and therefore conduction velocity (Bucher and Goaillard, 2011).

Axonal excitability can also change in response to activity in neighboring axons or whole brain regions, as a result of altered and inhomogeneous field potentials and/or extracellular ion concentrations (Anastassiou et al., 2010). Such "ephaptic" interactions can promote synchronous propagation in neighboring axons (Katz and Schmitt, 1940). For myelinated axons, the tight wrapping by glia cells was thought to prevent such interactions. However, oligodendrocytes can depolarize in response to axonal activity, and as they myelinate multiple axons, neuroglia interactions may represent a powerful way to synchronize bundles of myelinated axons (Fields, 2008a; Yamazaki et al., 2007).

12.5 CHANGES IN SPIKE INTERVAL STRUCTURE DURING PROPAGATION

Analyzing the recovery cycle with either simple paired pulses or trains and test pulses is useful to obtain some measure of the dependence of excitability and conduction velocity on spiking activity. However, it does not necessarily allow predicting how complex temporal patterns of spikes generated at proximal neuronal compartments are transformed during propagation to distal presynaptic sites. In order to gauge how much the recovery cycle matters for neural coding, two considerations are necessary. First, it is important to have some understanding of the range of naturally occurring spike patterns in a neuron. If a neuron only fires occasionally, for example, in the context of sparse coding of sensory information in the cortex (Olshausen and Field, 2004), changes in excitability do not matter as long as the normal spike intervals are larger than the recovery time of the axonal membrane. Second, it has to be considered how much the coding scheme utilized in a signaling pathway actually depends on temporal precision. If information is contained in just the rate of spikes, changes in delay and interval structure will not matter. In contrast, temporal coding schemes in which information is contained in spike intervals, latencies, or firing phase (Panzeri et al., 2010) may be exquisitely sensitive to axonal excitability changes. Other than in the context of primary sensory signaling, coding schemes are rarely well enough understood at the neuron and network level to make such an assessment. However, there is plenty of evidence from cellular neurophysiological studies showing that synaptic communication often is very sensitive to spike intervals and timing (Feldman, 2012; Zucker and Regehr, 2002).

Ultimately, one has to determine for each given neuron or network if dynamics of axonal propagation contribute to information processing or merely constitute noise. As an analogy, imagine using an old landline telephone connection. At the microphone, sound waves are encoded as electrical currents, propagated through cables, and decoded back into sound waves at a receiver. The connection may be noisy or limit the transmitted frequency band, but the pertinent information is still transmitted if the person on the receiving end understands what you are saying. However, if the transmission process did not just degrade the signal but helped shape it, the message received would differ in some significant way from what you said. Relating back to axonal spike propagation, in cases in which the temporal code is significantly and meaningfully altered, the changes during propagation are part of the process of encoding information, and not just noise.

As stated above, in most systems the consequences of propagation dynamics for coding are unknown, but in some cases spike patterns are so severely altered that it is hard to imagine that the activation of postsynaptic neurons could be unaffected. The ionic mechanisms discussed above can exert their effects on axonal excitability at time scales far exceeding single intervals. Therefore, propagation can become dependent not just on single intervals, but on the overall activity level, i.e., the history of spiking. As the recovery cycle can change as a function of highly repetitive prior activity, transformation of temporal patterns during propagation at a given time can be highly dependent on the mean rate of activity (Figure 12.4A). One example is the change in intraburst spike frequency during propagation

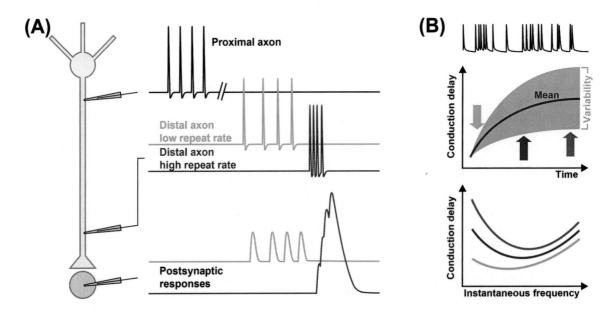

FIGURE 12.4   History-dependent transformations of temporal patterns during propagation. (A) The interval changes induced by propagation can be highly dependent on overall prior activity levels. In the hypothetical case depicted, similar to the behavior of human C-fibers (Weidner et al., 2002), bursts of spikes initiated proximally with the same intraburst interval structure can be transformed in opposite ways during propagation, dependent on the repeat rate. Postsynaptic responses to these changed patterns are shown to illustrate that even without any synaptic dynamics, summation can lead to vastly different synaptic communication. (B) Schematic of the effect of long "unpatterned" (*Poisson*-like) stimulation of an axon (upper panel) to illustrate further that conduction delay is dependent on activity at multiple time scales. The mean conduction delay increases with stimulation time, and so does the variability of delay (middle panel). The lower panel shows a hypothetical relationship between instantaneous frequency (the inverse of the spike interval) and conduction delay, which changes between three different times during the stimulation (arrows in middle panel). This behavior is similar to findings in a crustacean stomatogastric motor axon (Ballo et al., 2012).

in human C-fibers (Weidner et al., 2002). When stimulated at the foot with repeating trains (bursts), overall conduction delay to a recording site at knee level progressively slowed. However, the actual intraburst interval structure changed substantially depended on the burst repeat rate, i.e., the overall activity level. When a train of stimuli was repeated at a fairly low rate, intraburst intervals slowly increased, i.e., spike frequency decreased within each burst. In contrast, when the same train, i.e., the same number of stimuli with identical intervals, was repeated at a much higher rate, it caused a several fold increase in intraburst spike frequency. It is possible that this effect serves as a kind of contrast enhancement. Overall bursting activity in C-fibers can change substantially, for example, in response to inflammatory mediators. Postsynaptic neurons would then be activated not just with an overall increased rate but also with an increased frequency at a faster time scale.

In the lobster stomatogastric nervous system, the several centimeters long motor axons of the pyloric dilator (PD) neurons substantially alter the interval structure of motor bursts generated proximally (Ballo and Bucher, 2009; Ballo et al., 2012). Within each single burst, conduction delay over the several centimeters of peripheral nerves can vary ~30% across individual spikes, increasing and decreasing in a way that is not clearly correlated with spike intervals. When centrally generated bursting activity is blocked, the axon slowly depolarizes. When the motor nerve is then stimulated electrically, the axon slowly hyperpolarizes again, with a time constant of several minutes. If this is done with a *Poisson*-like stimulation instead of bursts, it becomes clear that not only the mean conduction delay increases but also the variability of delay (Figure 12.4B). The relationship between delay and instantaneous stimulus frequency is nonmonotonic and changes substantially over the course of several minutes of stimulation. Therefore, conduction delay is dependent on prior activity at two time scales, on the preceding spike interval and the history of spiking over the preceding few minutes.

12.6 NONUNIFORM AXONAL PROPERTIES AND SPIKE FAILURES

The changes in spike propagation discussed so far depend solely on activity-dependent changes in excitability and will therefore be expressed even when excitability and diameter are uniform along the axon, or uniformly spatially repeated in the case of myelinated axons. In contrast, changes in properties along the axon due to differences in diameter, or in the complement and density of ion channels, can affect propagation even for single spikes, and can interact with activity-dependent mechanisms to produce changes in spike patterns.

In many axons, the diameter changes along the path of projection, most notably so at branch points and varicosities or boutons. According to cable theory, such nonuniformities result in altered conduction velocity if they are associated with an impedance mismatch (Goldstein and Rall, 1974; Swadlow et al., 1980). Input impedance depends on diameter, and when a spike travels towards a region of changed diameter, the difference in impedance means that the current ahead of the spike will either be faster or take longer in bringing the membrane to threshold. This will slow down or speed up propagation. As discussed, axons with larger diameters propagate spikes faster. However, if a spike approaches a region of an axon at which the diameter increases, it will first slow down, as this region has lower input impedance and the current ahead of the spike will therefore cause a smaller depolarization.

At branch points, the combined impedance of the two daughter branches has to be considered (Rall, 1959). Propagation through branch points depends on the geometric ratio between the diameters of parent and daughter branches. If the mean diameter of daughter branches is about 63% of the diameter of the parent branch, propagation is unperturbed. If the daughter branches are thinner, conduction velocity is increased. If they are thicker, conduction velocity is decreased, and with larger impedance mismatches, propagation can even fail.

In axonal trees, the interactions between multiple morphological inhomogeneities can be complex (Manor et al., 1991). For example, electrotonically close successive branch points, with daughter branches thicker than needed for unperturbed propagation, will decrease conduction velocity more than is explained by the sum of the effects of each individual branch point. Furthermore, decrease in conduction velocity at an increase in diameter is larger than the increase in velocity at the equivalent decrease in diameter. This means that spike propagation through varicosities and boutons always increases conduction delay.

The effects of morphological inhomogeneities on spike propagation are particularly interesting when they interact with activity-dependent changes in excitability. Obviously, an axonal branch point with a large impedance mismatch that always causes spike failures does not make a lot of sense. However, if spike failures are conditional on prior activity, branch points can act as a frequency filter (Figure 12.5A). At a crustacean motor axon branch point, higher frequencies of spiking lead to spike failures, presumably due to extracellular potassium accumulation and subsequent depolarization, and sodium channel inactivation that lowers the safety factor for propagation (Grossman et al., 1979). In this case, the branch point acts as a low-pass filter and failures constitute a nonsynaptic locus of depression seen at the downstream neuromuscular junction. Repetitive activity can also lead to the *relief* of spike

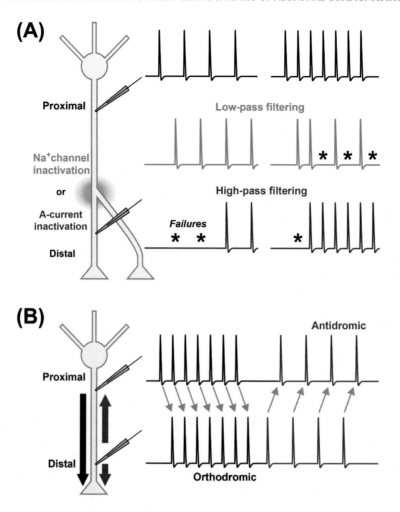

FIGURE 12.5   Spike failures and ectopic spike initiation. (A) Axonal branch points can be vulnerable to spike failures because of impedance mismatch between parent and daughter branches, and inactivating currents can contribute. If the branch point shows an increased likelihood of spike failures with higher stimulation frequencies (green traces), it acts as a low-pass filter for spike propagation. This can for example be the case if sodium channels inactivate substantially at higher spike frequencies. If the branch points shows a decreased likelihood of spike failures with higher stimulation frequencies (red traces), it acts as a high-pass filter for spike propagation. This can be the case if A-type potassium currents prohibit the propagation of single or low-frequency spikes at the branch points, but get inactivated at higher frequencies. (B) Axons can initiate spikes "ectopically," that is, at some distance from the usual proximal spike initiation site, for example, as a result of excitability changes in response to normal, orthodromically propagated spikes. Ectopic spikes can then travel both orthodromically towards terminals, and antidromically back to the proximal part of the neuron, where they can interfere with normal spike initiation.

failures at axonal branch points. In some hippocampal pyramidal cells, single spikes can have a high probability of propagation failure at branch points. However, during repetitive activity, later spikes are usually propagated successfully (Debanne et al., 1997). The crucial ionic mechanism underlying this behavior is thought to be the presence of a fast inactivating (A-type) potassium current. This transient outward current prevents sufficient depolarization at the branch point for a single spike, but its subsequent partial inactivation allows propagation of following spikes. Therefore, in this case branch points act as high-pass filters and constitute a nonsynaptic locus of facilitation for downstream synaptic outputs.

Spike propagation through varicosities can also be highly affected by repetitive activity. In the neurohypophysis, slowing of spike propagation at varicosities is exacerbated by local ionic mechanisms (Muschol et al., 2003). Varicosities are sites of calcium-dependent vesicle release, and spike-mediated increase in local calcium concentration slowly activates calcium-dependent potassium currents that reduce excitability and further slow propagation. This has been called "stuttering conduction," as progressive slowing of spikes only takes place at varicosities but not the rest of the propagating axonal sections.

Not all nonuniform axonal properties are associated with morphological inhomogeneities. In many cases, pathological dysregulation of excitability affects spike propagation, for example, in the case of activity-dependent conduction failures in demyelinating neuropathies (Krishnan et al., 2009). However, there are also some examples from the healthy nervous system. In some neurons, the spike initiation site at soma or proximal axons may contain very different ionic mechanisms than the rest of the axon. In Purkinje neurons, there is a relatively high rate of axonal propagation failures during repetitive spiking recorded in the soma (Khaliq and Raman, 2005). In very thin axons, the relatively small numbers of ion channels can make excitability subject to thermal and gating noise and degrade reliability of propagation (Faisal and Laughlin, 2007). Spontaneous opening of ion channel clusters induced by noise can make conduction "microsaltatory," as an approaching spike can jump ahead in discrete steps.

12.7 ECTOPIC AXONAL SPIKE INITIATION

Spike initiation is usually thought to occur only at proximal compartments of a neuron, as a result of synaptic integration and/or spontaneous membrane potential fluctuations. Under some circumstances, however, spikes can also be initiated in the axon, sometimes at substantial electrotonic distance from proximal compartments. These spikes can travel both orthodromically to synaptic targets, and antidromically to the proximal compartments, where they can interfere with normal spike initiation (Cattaert and Bevengut, 2002; Pinault, 1995). In peripheral neuropathies, such ectopic spike initiation is commonly due to changes in myelination and/or dysregulation of ion channels (Krishnan et al., 2009; Poliak and Peles, 2003). Ectopic spiking may also make up a substantial part of eleptiform activity in the brain (Pinault, 1995). In the healthy nervous system, there are a number of possible causes for ectopic spike initiation, including activity-dependent changes in excitability, axo-axonal electrical coupling, and axonal neuromodulation. In many cases, the underlying mechanisms for ectopic spike initiation are not understood. However, it has been argued that in some neurons distal axons constitute semi-independent sites of integration, both of prior activity and environmental signals (Pinault, 1995).

Early work in peripheral axons showed that repetitive stimulation can be followed by "reflex discharge," that is, prolonged firing generated at some distance and traveling antidromically back to the stimulation site (Figure 12.5B). This was true both in sensory (Toennies, 1938) and motor axons (Standaert, 1963). In the latter case, ectopic spiking contributed to muscle potentiation (Standaert, 1964), which means that distally generated spikes detected as antidromic signals in proximal recording sites can also travel orthodromically towards terminals. In human sensory and motor axons, the underlying mechanism for ectopic spike initiation in response to high frequency activity is thought to be hyperexcitability due to extracellular potassium accumulation (Bostock and Bergmans, 1994; Kiernan et al., 1997).

A recent resurgence of interest in ectopic spike initiation is due to a number of studies showing that ectopic spike initiation may play an important role in the mammalian brain, particularly in the hippocampus. In some hippocampal and cortical interneurons, long-lasting spiking activity is integrated in the axon over many minutes and leads to persistent ectopic firing (Sheffield et al., 2011). This activity is dependent on spike propagation from the proximal part of the neuron, but not on somatic depolarization. Persistent firing is dependent on calcium channels but not on chemical transmission, and it can spread to neighboring axons, presumably through gap-junctional coupling (Sheffield et al., 2013). In cortical interneurons *in vivo*, such persistent firing can cause long-lasting inhibition of pyramidal cells and is therefore thought to play a role in limiting the spread of epileptiform activity or other hyperexcitability states (Suzuki et al., 2014).

Axonal electrical coupling appears to also play an important role in oscillatory brain states. Such coupling is apparent in small but rapid depolarizations ("spikelets") caused by spikes in neighboring axons, and has the obvious advantage of being a very rapid mechanism to promote synchrony of firing. For example, fast cerebellar oscillations rely on synchronization through gap junctions between Purkinje cell axons (Middleton et al., 2008). In the hippocampus, network-entrained spiking in CA1 pyramidal cells during high frequency network oscillations (sharp wave ripple complexes) is exclusively generated in the axons and depends on axo-axonal coupling (Bahner et al., 2011). It also depends on suppression of centrally generated activity through perisomatic inhibition. Axonal spike initiation is enhanced by depolarization through the presence of low tonic concentrations of GABA (see Section 12.9). Interestingly, these ectopically generated spikes backpropagate to the soma and can cause long-lasting synaptic depression (Bukalo et al., 2013). In CA3 pyramidal cells, high frequency ectopic spikes are generated during gamma oscillations (Dugladze et al., 2012). Here, backpropagation into the soma is largely prevented by inhibitory axo-axonal connections to the axon initial segment, and this inhibition has only minor impact on orthodromic spike initiation. Therefore, axonal and somatic activity is functionally segregated and the normal polarity of information flow is maintained. Most of these findings come from slice preparations, but there is evidence that axo-axonal electrical coupling plays a highly significant role in the intact hippocampus. Recordings from CA1 pyramidal cells in freely moving rats during spatial exploration tasks show that about 30% of all spikes are triggered by spikelets and not proximal chemical synaptic inputs (Epsztein et al., 2010).

There are also a number of examples of ectopic axonal spike initiation from invertebrate preparations. In leech motor neurons, suppression of centrally generated activity can result in bursting activity generated in peripheral nerves (Maranto and Calabrese, 1983b). Interestingly, this activity may also depend on electrical coupling between peripheral axons (Maranto and Calabrese, 1983a). In a crustacean stomatogastric motor axon, ectopic spiking or even bursting can also occur in response to suppression of centrally generated activity. It relies on relief from activity-dependent hyperpolarization and is aided by inward currents that usually balance hyperpolarization (Ballo and Bucher, 2009; Bucher et al., 2003). Ectopically generated spikes backpropagate and can also suppress centrally generated activity. Therefore, the two spike initiation sites show mutual inhibition through propagated spikes. A number of axons in the stomatogastric nervous system produce ectopic spiking in response to biogenic amines (Ballo et al., 2009; Daur et al., 2009; Goaillard et al., 2004; Meyrand et al., 1992). These examples are discussed in Section 12.9.

In leech sensory neurons, another mechanism of ectopic spike initiation has been described in detail, namely spike reflection at branch points (Baccus, 1998; Baccus et al., 2000). If impedance mismatch at a branch point results in a spike being close to failure, the imposed propagation delay can exceed the refractory period, so that a second spike gets initiated and travels antidromically. In this sensory neuron, the antidromic spike can invade upstream collaterals and enhance synaptic output there.

12.8 SPIKE SHAPE AND ANALOG SIGNALING

Signaling and activation of synapses with spikes is usually viewed as digital, in other words as a discrete, discontinuous, "all-or-none" representation of information. At any given time, a neuron either spikes or it does not, equivalent to the 0s and 1s in binary code. The evolutionary argument for why such a mode is necessary is that communication over longer distances has to rely on regenerative potentials because the high axial resistance and leaky membrane make neuronal processes poor conductors and limit passive spatial spread of potentials. Hodgkin famously stated that a long thin axon has about the same axial resistance as a thick copper wire over several times the distance between Earth and Saturn (Hodgkin, 1964). However, some small local neurons do not use spikes, and communication with postsynaptic neurons can be analog, as transmitter release is a graded function of membrane potential (Roberts and Bush, 1981). From an information theoretical point of view, this analog signaling allows for much higher bit rates to be transmitted (de Ruyter van Steveninck and Laughlin, 1996), but is costly in terms of energy consumption (Debanne et al., 2013). There are also some invertebrate neurons, e.g., in the crustacean stomatogastric ganglion, that use both spike-mediated and graded forms of transmission (Graubard et al., 1983). Spike-mediated and graded transmission in these neurons function in parallel and can be experimentally separated (Ayali et al., 1998).

In other cases, spike signaling itself can exceed the all-or-none digital mode, mainly through changes in spike shape and interactions with graded potentials at distal synaptic sites. The synapse itself can make use of the temporal pattern of spikes, and decode temporal information not just on the basis of postsynaptic summation, but because transmitter release often shows short-term dynamics (Regehr and Stevens, 2001; Zucker and Regehr, 2002). Classical cases are short-term facilitation based on presynaptic accumulation of residual calcium and subsequent enhanced transmitter release, and short-term depression based on depletion of vesicles. However, activity-dependent changes in presynaptic spike voltage trajectory can also contribute to synaptic dynamics (Figure 12.6A). The shape of presynaptic depolarization impacts the amount of calcium influx at the terminal, and synaptic currents can be very sensitive to even small changes in spike shape (Sabatini and Regehr, 1997). Postsynaptic responses can show depression because sodium channel inactivation during repetitive spiking leads to a reduction in spike amplitudes, as shown for example in cultured hippocampal neurons (Brody and Yue, 2000). Fast inactivation of potassium currents can lead to spike broadening and subsequent facilitation of synaptic responses, for example, at hippocampal mossy fiber boutons (Geiger and Jonas, 2000).

Changes in spike shape may play a more significant role in short-term synaptic dynamics than is generally appreciated and mean that spikes do not activate presynaptic sites entirely in an all-or-none manner. However, as presynaptic terminals have long been known to be sites of integration, for example, for presynaptic chemical inputs (Frank and Fuortes, 1957), the contribution of changes in spike shape to synaptic dynamics does not really represent a paradigm shift in our understanding of axonal function, at least if these changes happen locally at the terminals.

In contrast, many neurons may integrate analog and digital components into synaptic communication in a related but different manner that depends on subthreshold somatic events that can spread to presynaptic sites. Effects of

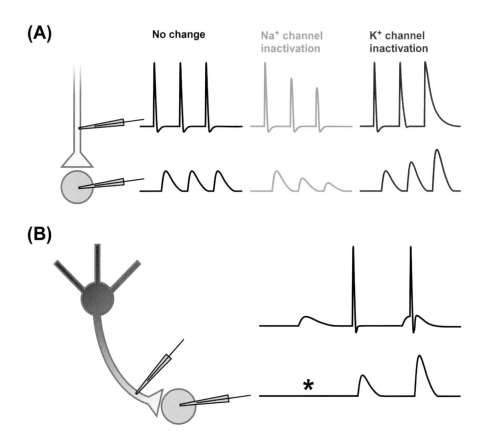

FIGURE 12.6 Analog components of spike-mediated synaptic transmission. (A) Activity-dependent changes in spike shape can contribute to short-term synaptic plasticity. Reduction of spike amplitude due to sodium channel inactivation during repetitive spiking can decrease presynaptic calcium influx, and therefore reduce transmitter release. In this case, the synapse can show short-term depression, even if there is no shortage of synaptic vesicles. Increase of spike duration due to inactivation of transient potassium currents can enhance presynaptic calcium influx, and therefore increase transmitter release. In this case, the synapse can show short-term facilitation, even if there is no accumulation of residual calcium during repetitive activation. (B) Neurons with electrotonically short axons can show a mixture of analog and digital signaling. If the electrotonic length of the axon allows subthreshold events like EPSPs to spread all the way to presynaptic sites, these changes in membrane potential can alter spike-mediated synaptic transmission. Subthreshold events alone are not sufficient to elicit postsynaptic responses (asterisk), but coincidence of subthreshold events and spikes result in postsynaptic responses of different amplitude compared to those elicited by spikes alone.

membrane potential fluctuations at the soma on transmitter release at axonal presynaptic sites have originally been described in invertebrates, for example in the *Aplysia* L10 neuron (Shapiro et al., 1980). More recently, it was discovered that many mammalian central synapses are also susceptible to modulation by subthreshold somatic potentials (Alle and Geiger, 2008; Debanne et al., 2013; Sasaki, 2013). It turns out that in many cell types even relatively thin axons have fairly large length constants. Therefore, subthreshold events can spread over several hundred micrometers and at least reach more proximal presynaptic sites.

Graded depolarizing potentials spreading from the soma are not sufficient to induce transmitter release on their own, as is the case at graded synapses, but can increase synaptic responses to spikes (Figure 12.6B). In some synapses, this facilitation is due to inactivation of potassium currents and subsequent spike broadening, for example, D-type currents in cortical axons that are strongly inactivated by even small depolarizations (Shu et al., 2007). Alternatively, subthreshold depolarization can increase basal calcium levels through activation of voltage-gated calcium channels at presynaptic sites and therefore facilitate spike-mediated transmitter release, for example, in cerebellar interneurons (Christie et al., 2011). Spread of subthreshold depolarizations into the axon may also affect spike failures. The relief of propagation failures dependent on A-currents in hippocampal pyramidal cells described in Section 12.6 is also susceptible to changes in somatic membrane potential (Debanne et al., 1997). These hybrid forms of analog and digital signaling may play an important role for local circuit operation in many brain areas (Debanne et al., 2013).

12.9 NEUROMODULATION OF AXONS

Perhaps the strongest argument that many of the axonal properties described above are actually important for neuron and network function comes from studies showing that they can be modified by neuromodulation, i.e., chemical signaling from other neurons (Bucher and Goaillard, 2011). Neuromodulation is a somewhat loosely defined term, and I use it here mostly for diffuse release of small transmitter molecules that then act on axonal receptors independent of postsynaptic specializations. Broadly, such neuromodulatory effects can be divided into two categories, those mediated by ligand-gated ion channels (ionotropic receptors), and those mediated by G protein-coupled (metabotropic) receptors (Nadim and Bucher, 2014). Activation of ion channels can change axonal membrane potential and input resistance, and therefore affect both the gating state of voltage-gated ion channels, as well as spike shape and propagation. G protein activation can result in second messenger-mediated changes in the gating properties of ion channels, and thus also affect excitability, spike shape, and propagation.

Ionotropic effects on axons are usually mediated by classic transmitters like GABA or acetylcholine (ACh). Ionotropic $GABA_A$ receptors are the major receptors in fast inhibitory synapses and mediate fast presynaptic inhibition at many axon terminals. Interestingly, presynaptic potentials mediated by $GABA_A$ receptors are often depolarizing, as the specific types of axonal transporters located in these axons cause a high intracellular chloride concentration (Rocha-Gonzalez et al., 2008). These potentials are still mostly inhibitory because they cause shunting, that is, a decrease in input resistance that can exert graded control over the shape of the action potential and also prevent active invasion of spikes into the terminal (Jackson and Zhang, 1995). However, they can also cause spike ectopic initiation (Cattaert et al., 1999). In addition, $GABA_A$ receptors can be expressed at substantial distance to synapses in some peripheral axons, both myelinated and unmyelinated ones (Kocsis and Sakatani, 1995). Recently, the modulatory role of $GABA_A$ receptors in the mammalian brain has received a lot of attention. Low concentrations of ambient GABA, for example, from synaptic spillover, are sufficient to activate high affinity $GABA_A$ receptors, often in extrasynaptic membrane (Belelli et al., 2009). Ectopic spiking, for example in the case of the CA1 axonal plexus that generates synchronous activity during fast oscillations (Bahner et al., 2011), is thought to be mainly due to activation of such receptors. In contrast, $GABA_A$ receptors may cause hyperpolarization in cortical axons (Xia et al., 2014). This hyperpolarization affects spike waveform and reduces calcium transients at downstream synapses.

Similar effects on extrasynaptic membrane potential, spike shape, and propagation may be mediated by ACh through nicotinic receptors (nAChRs), and by glutamate acting through AMPA receptors. In the central nervous system, nAChRs are mainly expressed presynaptically, but apparently mostly directly at axonal presynaptic sites or in their immediate vicinity (Mansvelder et al., 2009). Therefore, it is not yet clear to which degree ACh may affect propagation of spike patterns. However, in C-fibers of the peripheral nervous system, nAChRs can substantially affect excitability and spike propagation (Lang et al., 2003). AMPA receptors, the ubiquitous mediators of fast glutamatergic transmission at postsynaptic sites, have only recently been shown to also act on axonal properties. In CA3 pyramidal cells, glutamate release from astrocytes activates AMPA receptors in long stretches of the axon (Sasaki et al., 2011). The resulting depolarization leads to inactivation of potassium channels, broadens spikes, and consequently increases synaptic transmission. AMPA receptors also mediate activity-dependent changes in spike propagation in hippocampal mossy fibers (Chida et al., 2015).

Modulation of axons can also be due to biogenic amines. In C-fibers, serotonin (5-HT) substantially alters the recovery cycle in a way that dampens the change in spike intervals during propagation (Lang et al., 2006). This effect is mediated by ionotropic receptors, whereas most other examples of aminergic axonal modulation likely involve metabotropic actions. In the crustacean stomatogastric nervous system, octopamine elicits ectopic spikes in a descending modulatory axon (Goaillard et al., 2004), and in a sensory axon (Daur et al., 2009). In both cases, the additional spikes significantly alter the rhythmic motor patterns generated by the target circuits. Ectopic spiking has also been reported in stomatogastric motor axons in peripheral nerves. The axon of the lateral gastric neuron responds to 5-HT by prolonging centrally generated bursts, with a significant effect on downstream muscle activation (Meyrand et al., 1992). While in this case ectopic spike initiation is dependent on prior propagation of centrally generated activity, ectopic spiking in the axon of the PD neuron shows the opposite dependency. Ectopic spiking can be elicited by low concentrations of dopamine (DA), but only in the absence of centrally generated bursting (Bucher et al., 2003). The underlying depolarization is caused by a cAMP-mediated increase of an H-current produced by hyperpolarization-activated, cyclic nucleotide-gated channels. This depolarization is overcome during centrally generated activity by a substantial activity-dependent hyperpolarization (Ballo et al., 2010). However, the balance between hyperpolarization and H-current plays an important role during centrally generated rhythmic bursting. The large increase in mean conduction delay and variability of delay during rhythmic bursting over several minutes (Figure 12.4B) is exacerbated when the H-current is blocked, but almost completely eliminated when it is enhanced by DA (Ballo et al., 2012). Therefore, temporal fidelity is substantially improved by DA modulation. Such modulatory effects are not restricted to ion channels. In leech sensory axons, 5-HT inhibits the sodium/potassium-ATPase, and the reduced activity-dependent hyperpolarization facilitates synaptic outputs (Scuri et al., 2007).

In the vertebrate nervous system, contributions of metabotropic receptors to axonal modulation are less well characterized, but there are some examples. GABA$_B$ receptors regulate calcium transients in optic nerve axons of the neonatal rat (Sun and Chiu, 1999). Recently, DA was suggested to control the shape of the propagating spike in cortical pyramidal axons through modulation of inactivating potassium currents (Yang et al., 2013). Interestingly, DA acted through both D1- and D2-like receptors mechanisms. D1-like receptors caused a decrease in potassium currents, whereas D2-like receptors had the opposite effect. The sparseness of information about axonal neuromodulation in the mammalian central nervous system may be more a reflection of technical limitations rather than lack of significance. It appears likely that a growing appreciation of the possibility of axonal neuromodulation and the improved techniques for biophysical characterization of axonal membranes will lead to many more discoveries of this kind in the near future.

12.10 LONG-TERM REGULATION OF AXON PHYSIOLOGY

The main topic of this chapter is the contribution of axons to short-term dynamics of neural signaling. However, there is now also evidence that axonal properties can be subject to activity-dependent long-term regulatory changes. I will briefly discuss some of these findings because in many cases, long-term changes in morphology and protein expression are highly likely to also affect short-term dynamics. Long-term changes in axonal excitability and reliability of spike propagation are of particular interest and widely appreciated in the context of human peripheral neuropathies associated with ischemia, diabetes, demyelination, amyotrophic lateral sclerosis, epilepsy, and spinal cord injury (Krishnan et al., 2009; Nashmi and Fehlings, 2001; Poliak and Peles, 2003). In fact, analysis of axonal excitability and dynamics, particularly recovery cycle measurements, is a widely used diagnostic tool for such diseases (Bostock et al., 1998; Krishnan et al., 2009). However, long-term regulatory processes may also play an important role in changing or maintaining axonal function in the healthy nervous system, both in the context of regulating mean conduction delay and in the context of activity-dependent changes in excitability.

Long-term recordings of both mean conduction delay and recovery cycle in callosal axons showed somewhat inconsistent results in that not all neurons expressed changes in mean delay, and activity-dependent excitability changes were much more stable than mean delays (Swadlow, 1982, 1985). Cultured cortical neurons show activity-dependent long-term changes in conduction velocity (Bakkum et al., 2008), and both cultured hippocampal neurons and avian brainstem auditory neurons show activity-dependent changes in proximal axonal excitability and the position of the spike initiation site (Grubb and Burrone, 2010; Kuba et al., 2010); but to which degree channel expression and ionic mechanisms may be regulated in an activity-dependent in the axon proper is not clear.

Another line of evidence for long-term regulation comes from observed changes in axon morphology. Higher order axonal branches can undergo substantial morphological changes in an activity-dependent manner (Holtmaat and Svoboda, 2009), although it is not clear to which degree those reflect altered connectivity rather than changes in axonal function. A clearer case is the compensation of axonal path length by different conduction velocities in some systems. This compensation ensures similar delays across different axonal branches of one neuron or across axons of different individual neurons, both in the case of unmyelinated and myelinated axons (Seidl, 2014). In myelinated axons, such compensatory regulation appears to be at least partly due to axon interactions with myelinating glia cells, as differences in conduction velocity are correlated with regional differences in myelin thickness and internodal distance (Salami et al., 2003; Seidl, 2014). Dynamics of myelination is not restricted to a developmental time scale, but can occur relatively rapidly throughout adulthood in an activity-dependent manner. For example, changes in neural activity in peripheral nerves of adult rats, induced by load changes, affect myelin thickness and internodal distances (Canu et al., 2009). Such changes may play an important role in regulating conduction velocity and temporal precision (Wang and Young, 2014), which may be a crucial component of timing-dependent processes in the brain, for example, in the context of oscillations and synchrony underlying cognitive function (Fields, 2008b; Pajevic et al., 2014), or finely tuned motor tasks like piano playing (Bengtsson et al., 2005).

12.11 CONCLUDING REMARKS

The concepts and examples discussed here make a strong point that axons, including the propagating part, can play an important role in the short-term dynamics of neural communication. While particularly in the peripheral nervous system, many general aspects of axonal signaling beyond faithful conduction have been established a long time ago, and their potential implications discussed in detail (Swadlow et al., 1980), some obvious challenges have precluded broad recognition of axonal dynamics and their inclusion in models of neural function. Conceptually, to appreciate the role that axons may play in shaping patterns of spiking activity and activation of synapses would require an understanding of coding at both the neuron and network level that simply is not advanced enough in most systems. In the absence of such understanding, it will be hard to gauge if the changes that axonal propagation can induce in the patterns of spiking activity merely constitute noise or actually contribute to the encoding of information (Bucher and Goaillard, 2011). The answer may depend not just on the particular neuron studied but also differ between physiological states.

Technically, studying axons in most cases is a challenge simply because of their often diminutive size and extended length. It is to be expected that relatively new methods, like patch clamp recordings from axonal blebs, will in the nearer future significantly advance the appreciation of complex axonal excitability even in small central axons (Hu and Shu, 2012). In addition, optical methods will serve this goal, at least in cases where use of electrodes is not feasible (Bradley et al., 2009; Fleidervish et al., 2010). However, these techniques will be mostly employed in reduced preparations like brain slices. To gain insights into axonal contributions to actual neural functions, such methods will have to be combined with advances in recording axonal spikes *in vivo* (Barry, 2015).

Acknowledgments

This work was supported in part by National Institute of Neurological Disorders and Stroke Grant NS083319. I thank Drs. Farzan Nadim and Nelly Daur for many helpful discussions about axonal dynamics.

References

Adrian, E.D., 1920. The recovery process of excitable tissues: part I. J. Physiol. 54 (1–2), 1–31.

Alle, H., Geiger, J.R., 2008. Analog signalling in mammalian cortical axons. Curr. Opin. Neurobiol. 18 (3), 314–320.

Anastassiou, C.A., Montgomery, S.M., Barahona, M., Buzsaki, G., Koch, C., 2010. The effect of spatially inhomogeneous extracellular electric fields on neurons. J. Neurosci. 30 (5), 1925–1936.

Ayali, A., Johnson, B.R., Harris-Warrick, R.M., 1998. Dopamine modulates graded and spike-evoked synaptic inhibition independently at single synapses in pyloric network of lobster. J. Neurophysiol. 79 (4), 2063–2069.

Baccus, S.A., 1998. Synaptic facilitation by reflected action potentials: enhancement of transmission when nerve impulses reverse direction at axon branch points. Proc. Natl. Acad. Sci. U.S.A. 95 (14), 8345–8350.

Baccus, S.A., Burrell, B.D., Sahley, C.L., Muller, K.J., 2000. Action potential reflection and failure at axon branch points cause stepwise changes in EPSPs in a neuron essential for learning. J. Neurophysiol. 83 (3), 1693–1700.

Baginskas, A., Palani, D., Chiu, K., Raastad, M., 2009. The H-current secures action potential transmission at high frequencies in rat cerebellar parallel fibers. Eur. J. Neurosci. 29 (1), 87–96.

Bahner, F., Weiss, E.K., Birke, G., Maier, N., Schmitz, D., Rudolph, U., et al., 2011. Cellular correlate of assembly formation in oscillating hippocampal networks *in vitro*. Proc. Natl. Acad. Sci. U.S.A. 108 (35), E607–616.

Bakkum, D.J., Chao, Z.C., Potter, S.M., 2008. Long-term activity-dependent plasticity of action potential propagation delay and amplitude in cortical networks. PLoS ONE 3 (5), e2088.

Ballo, A.W., Bucher, D., 2009. Complex intrinsic membrane properties and dopamine shape spiking activity in a motor axon. J. Neurosci. 29 (16), 5062–5074.

Ballo, A.W., Keene, J.C., Goeritz, M.L., Bucher, D., 2009. Dopamine modulates hyperpolarization-activated inward current in a peripheral motor axon. Abstract Viewer/Itinerary Planner, Society for Neuroscience, Program No. 178.7.

Ballo, A.W., Keene, J.C., Troy, P.J., Goeritz, M.L., Nadim, F., Bucher, D., 2010. Dopamine modulates Ih in a motor axon. J. Neurosci. 30 (25), 8425–8434.

Ballo, A.W., Nadim, F., Bucher, D., 2012. Dopamine modulation of ih improves temporal fidelity of spike propagation in an unmyelinated axon. J. Neurosci. 32 (15), 5106–5119.

Barry, J.M., 2015. Axonal activity *in vivo*: technical considerations and implications for the exploration of neural circuits in freely moving animals. Front. Neurosci. 9, 1–12.

Belelli, D., Harrison, N.L., Maguire, J., Macdonald, R.L., Walker, M.C., Cope, D.W., 2009. Extrasynaptic GABAA receptors: form, pharmacology, and function. J. Neurosci. 29 (41), 12757–12763.

Bengtsson, S.L., Nagy, Z., Skare, S., Forsman, L., Forssberg, H., Ullen, F., 2005. Extensive piano practicing has regionally specific effects on white matter development. Nat. Neurosci. 8 (9), 1148–1150.

Bostock, H., Bergmans, J., 1994. Post-tetanic excitability changes and ectopic discharges in a human motor axon. Brain 117 (Pt 5), 913–928.

Bostock, H., Cikurel, K., Burke, D., 1998. Threshold tracking techniques in the study of human peripheral nerve. Muscle Nerve 21 (2), 137–158.

Bowe, C.M., Kocsis, J.D., Waxman, S.G., 1987. The association of the supernormal period and the depolarizing afterpotential in myelinated frog and rat sciatic nerve. Neuroscience 21 (2), 585–593.

Bradley, J., Luo, R., Otis, T.S., DiGregorio, D.A., 2009. Submillisecond optical reporting of membrane potential *in situ* using a neuronal tracer dye. J. Neurosci. 29 (29), 9197–9209.

Brody, D.L., Yue, D.T., 2000. Release-independent short-term synaptic depression in cultured hippocampal neurons. J. Neurosci. 20 (7), 2480–2494.

Bucher, D., Goaillard, J.M., 2011. Beyond faithful conduction: short-term dynamics, neuromodulation, and long-term regulation of spike propagation in the axon. Prog. Neurobiol. 94 (4), 307–346.

Bucher, D., Marder, E., 2013. SnapShot: Neuromodulation. Cell 155 (2) 482–482 e481.

Bucher, D., Thirumalai, V., Marder, E., 2003. Axonal dopamine receptors activate peripheral spike initiation in a stomatogastric motor neuron. J. Neurosci. 23 (17), 6866–6875.

Bukalo, O., Campanac, E., Hoffman, D.A., Fields, R.D., 2013. Synaptic plasticity by antidromic firing during hippocampal network oscillations. Proc. Natl. Acad. Sci. U.S.A. 110 (13), 5175–5180.

Bullock, T.H., 1951. Facilitation of conduction rate in nerve fibers. J. Physiol. 114 (1–2), 89–97.

Burke, D., Kiernan, M.C., Bostock, H., 2001. Excitability of human axons. Clin. Neurophysiol. 112 (9), 1575–1585.

Canu, M.H., Carnaud, M., Picquet, F., Goutebroze, L., 2009. Activity-dependent regulation of myelin maintenance in the adult rat. Brain Res. 1252, 45–51.

Cattaert, D., Bevengut, M., 2002. Effects of antidromic discharges in crayfish primary afferents. J. Neurophysiol. 88 (4), 1753–1765.

Cattaert, D., El Manira, A., Bevengut, M., 1999. Presynaptic inhibition and antidromic discharges in crayfish primary afferents. J. Physiol. Paris 93 (4), 349–358.

Chida, K., Kaneko, K., Fujii, S., Yamazaki, Y., 2015. Activity-dependent modulation of the axonal conduction of action potentials along rat hippocampal mossy fibers. Eur. J. Neurosci. 41 (1), 45–54.

Christie, J.M., Chiu, D.N., Jahr, C.E., 2011. Ca(2+)-dependent enhancement of release by subthreshold somatic depolarization. Nat. Neurosci. 14 (1), 62–68.

Connor, J.A., 1975. Neural repetitive firing: a comparative study of membrane properties of crustacean walking leg axons. J. Neurophysiol. 38 (4), 922–932.

Daur, N., Nadim, F., Stein, W., 2009. Regulation of motor patterns by the central spike-initiation zone of a sensory neuron. Eur. J. Neurosci. 30 (5), 808–822.

David, G., Modney, B., Scappaticci, K.A., Barrett, J.N., Barrett, E.F., 1995. Electrical and morphological factors influencing the depolarizing afterpotential in rat and lizard myelinated axons. J. Physiol. 489 (Pt 1), 141–157.

De Col, R., Messlinger, K., Carr, R.W., 2008. Conduction velocity is regulated by sodium channel inactivation in unmyelinated axons innervating the rat cranial meninges. J. Physiol. 586 (4), 1089–1103.

de Ruyter van Steveninck, R.R., Laughlin, S.B., 1996. The rate of information transfer at graded-potential synapses. Nature 379 (6566), 642–645.

Debanne, D., 2004. Information processing in the axon. Nat. Rev. Neurosci. 5 (4), 304–316.

Debanne, D., Guerineau, N.C., Gahwiler, B.H., Thompson, S.M., 1997. Action-potential propagation gated by an axonal I(A)-like K+ conductance in hippocampus. Nature 389 (6648), 286–289.

Debanne, D., Campanac, E., Bialowas, A., Carlier, E., Alcaraz, G., 2011. Axon physiology. Physiol. Rev. 91 (2), 555–602.

Debanne, D., Bialowas, A., Rama, S., 2013. What are the mechanisms for analogue and digital signalling in the brain? Nat. Rev. Neurosci. 14 (1), 63–69.

Del Castillo, J., Moore, J.W., 1959. On increasing the velocity of a nerve impulse. J. Physiol. 148, 665–670.

Dugladze, T., Schmitz, D., Whittington, M.A., Vida, I., Gloveli, T., 2012. Segregation of axonal and somatic activity during fast network oscillations. Science 336 (6087), 1458–1461.

Epsztein, J., Lee, A.K., Chorev, E., Brecht, M., 2010. Impact of spikelets on hippocampal CA1 pyramidal cell activity during spatial exploration. Science 327 (5964), 474–477.

Evarts, E.V., 1965. Relation of discharge frequency to conduction velocity in pyramidal tract neurons. J. Neurophysiol. 28, 216–228.

Faisal, A.A., Laughlin, S.B., 2007. Stochastic simulations on the reliability of action potential propagation in thin axons. PLoS Comput. Biol. 3 (5), e79.

Feldman, D.E., 2012. The spike-timing dependence of plasticity. Neuron 75 (4), 556–571.

Fields, R.D., 2008a. Oligodendrocytes changing the rules: action potentials in glia and oligodendrocytes controlling action potentials. Neuroscientist 14 (6), 540–543.

Fields, R.D., 2008b. White matter in learning, cognition and psychiatric disorders. Trends Neurosci. 31 (7), 361–370.

Fleidervish, I.A., Lasser-Ross, N., Gutnick, M.J., Ross, W.N., 2010. Na+ imaging reveals little difference in action potential-evoked Na+ influx between axon and soma. Nat. Neurosci. 13 (7), 852–860.

Frank, K., Fuortes, M., 1957. Presynaptic and postsynaptic inhibition of monosynaptic reflexes. Paper presented at the Federation Proceedings.

Geiger, J.R., Jonas, P., 2000. Dynamic control of presynaptic Ca(2+) inflow by fast-inactivating K(+) channels in hippocampal mossy fiber boutons. Neuron 28 (3), 927–939.

George, S., Foster, J.M., Richardson, G., 2015. Modelling *in vivo* action potential propagation along a giant axon. J. Math. Biol. 70 (1-2), 237–263.

Goaillard, J.M., Schulz, D.J., Kilman, V.L., Marder, E., 2004. Octopamine modulates the axons of modulatory projection neurons. J. Neurosci. 24 (32), 7063–7073.

Goldstein, S.S., Rall, W., 1974. Changes of action potential shape and velocity for changing core conductor geometry. Biophys. J. 14 (10), 731–757.

Graubard, K., Raper, J.A., Hartline, D.K., 1983. Graded synaptic transmission between identified spiking neurons. J. Neurophysiol. 50 (2), 508–521.

Grossman, Y., Parnas, I., Spira, M.E., 1979. Mechanisms involved in differential conduction of potentials at high frequency in a branching axon. J. Physiol. 295, 307–322.

Grubb, M.S., Burrone, J., 2010. Activity-dependent relocation of the axon initial segment fine-tunes neuronal excitability. Nature 465 (7301), 1070–1074.

Hartline, D.K., Colman, D.R., 2007. Rapid conduction and the evolution of giant axons and myelinated fibers. Curr. Biol. 17 (1), R29–35.

Hille, B., 2001. *Ion Channels of Excitable Membranes.*, third ed. Sinauer, Sunderland, MA.

Hodgkin, A.L., 1954. A note on conduction velocity. J. Physiol. 125 (1), 221–224.

Hodgkin, A.L., 1964. The conduction of the nervous impulse. C. C. Thomas, Springfield, IL.

Hodgkin, A.L., Huxley, A.F., 1952a. Propagation of electrical signals along giant nerve fibers. Proc. R. Soc. Lond. B. Biol. Sci. 140 (899), 177–183.

Hodgkin, A.L., Huxley, A.F., 1952b. A quantitative description of membrane current and its application to conduction and excitation in nerve. J. Physiol. 117, 500–544.

Hodgkin, A.L., Rushton, W.A., 1946. The electrical constants of a crustacean nerve fibre. Proc. R. Soc. Med. 134 (873), 444–479.

Holtmaat, A., Svoboda, K., 2009. Experience-dependent structural synaptic plasticity in the mammalian brain. Nat. Rev. Neurosci. 10 (9), 647–658.

Hu, W., Shu, Y., 2012. Axonal bleb recording. Neurosci. Bull. 28 (4), 342–350.

Huxley, A.F., Stampfli, R., 1949. Evidence for saltatory conduction in peripheral myelinated nerve fibres. J. Physiol. 108 (3), 315–339.

Izhikevich, E.M., 2006. Polychronization: computation with spikes. Neural Comput. 18 (2), 245–282.

Jackson, M.B., Zhang, S.J., 1995. Action potential propagation and propagation block by GABA in rat posterior pituitary nerve terminals. J. Physiol. 483 (Pt 3), 597–611.

Katz, B., Schmitt, O.H., 1940. Electric interaction between two adjacent nerve fibres. J. Physiol. 97 (4), 471–488.

Khaliq, Z.M., Raman, I.M., 2005. Axonal propagation of simple and complex spikes in cerebellar Purkinje neurons. J. Neurosci. 25 (2), 454–463.

Kiernan, M.C., Mogyoros, I., Hales, J.P., Gracies, J.M., Burke, D., 1997. Excitability changes in human cutaneous afferents induced by prolonged repetitive axonal activity. J. Physiol. 500 (Pt 1), 255–264.

Kocsis, J.D., Sakatani, K., 1995. Modulation of axonal excitability by neurotransmitter receptors. In: Waxman, S.G., Kocsis, J.D., Stys, P.K. (Eds.), The Axon. Structure, Function and Pathophysiology Oxford University Press, New York, NY.

Kocsis, J.D., Swadlow, H.A., Waxman, S.G., Brill, M.H., 1979. Variation in conduction velocity during the relative refractory and supernormal periods: a mechanism for impulse entrainment in central axons. Exp. Neurol 65 (1), 230–236.

Konishi, M., 2003. Coding of auditory space. Annu. Rev. Neurosci. 26, 31–55.

Kress, G.J., Dowling, M.J., Meeks, J.P., Mennerick, S., 2008. High threshold, proximal initiation, and slow conduction velocity of action potentials in dentate granule neuron mossy fibers. J. Neurophysiol. 100 (1), 281–291.

Krishnan, A.V., Lin, C.S., Park, S.B., Kiernan, M.C., 2009. Axonal ion channels from bench to bedside: a translational neuroscience perspective. Prog. Neurobiol. 89 (3), 288–313.

Kuba, H., Oichi, Y., Ohmori, H., 2010. Presynaptic activity regulates Na(+) channel distribution at the axon initial segment. Nature 465 (7301), 1075–1078.

Lang, P.M., Burgstahler, R., Sippel, W., Irnich, D., Schlotter-Weigel, B., Grafe, P., 2003. Characterization of neuronal nicotinic acetylcholine receptors in the membrane of unmyelinated human C-fiber axons by *in vitro* studies. J. Neurophysiol. 90 (5), 3295–3303.

Lang, P.M., Moalem-Taylor, G., Tracey, D., Bostock, H., Grafe, P., 2006. Activity-dependent modulation of axonal excitability in unmyelinated peripheral rat nerve fibers by the 5-HT(3) serotonin receptor. J. Neurophysiol.

Manor, Y., Koch, C., Segev, I., 1991. Effect of geometrical irregularities on propagation delay in axonal trees. Biophys. J. 60 (6), 1424–1437.

Mansvelder, H.D., Mertz, M., Role, L.W., 2009. Nicotinic modulation of synaptic transmission and plasticity in cortico-limbic circuits. Semin. Cell Dev. Biol. 20 (4), 432–440.

Manzano, G.M., Giuliano, L.M., Nobrega, J.A., 2008. A brief historical note on the classification of nerve fibers. Arq. Neuropsiquiatr. 66 (1), 117–119.

Maranto, A.R., Calabrese, R.L., 1983a. Neural control of the hearts in the leech, *Hirudo medicinales*. I. Anatomy, electrical coupling, and innervation of the hearts. J. Comp. Physiol. A 154, 367–380.

Maranto, A.R., Calabrese, R.L., 1983b. Neural control of the hearts in the leech, *Hirudo medicinales*. II. Myogenic activity and its control by heart motoneurons. J. Comp. Physiol. A 154, 381–391.

McIntyre, C.C., Richardson, A.G., Grill, W.M., 2002. Modeling the excitability of mammalian nerve fibers: influence of afterpotentials on the recovery cycle. J. Neurophysiol. 87 (2), 995–1006.

Meyrand, P., Weimann, J.M., Marder, E., 1992. Multiple axonal spike initiation zones in a motor neuron: serotonin activation. J. Neurosci. 12 (7), 2803–2812.

Middleton, S.J., Racca, C., Cunningham, M.O., Traub, R.D., Monyer, H., Knopfel, T., et al., 2008. High-frequency network oscillations in cerebellar cortex. Neuron 58 (5), 763–774.

Moldovan, M., Krarup, C., 2006. Evaluation of Na^+/K^+ pump function following repetitive activity in mouse peripheral nerve. J. Neurosci. Methods 155 (2), 161–171.

Moradmand, K., Goldfinger, M.D., 1995. Computation of long-distance propagation of impulses elicited by Poisson-process stimulation. J. Neurophysiol. 74 (6), 2415–2426.

Muratov, C.B., 2000. A quantitative approximation scheme for the traveling wave solutions in the Hodgkin–Huxley model. Biophys. J. 79 (6), 2893–2901.

Muschol, M., Kosterin, P., Ichikawa, M., Salzberg, B.M., 2003. Activity-dependent depression of excitability and calcium transients in the neurohypophysis suggests a model of "stuttering conduction". J. Neurosci. 23 (36), 11352–11362.

Nadim, F., Bucher, D., 2014. Neuromodulation of neurons and synapses. Curr. Opin. Neurobiol. 29C, 48–56.

Nashmi, R., Fehlings, M.G., 2001. Mechanisms of axonal dysfunction after spinal cord injury: with an emphasis on the role of voltage-gated potassium channels. Brain Res. Brain Res. Rev. 38 (1-2), 165–191.

Nusser, Z., 2009. Variability in the subcellular distribution of ion channels increases neuronal diversity. Trends Neurosci 32 (5), 267–274.

Olshausen, B.A., Field, D.J., 2004. Sparse coding of sensory inputs. Curr. Opin. Neurobiol. 14 (4), 481–487.

Pajevic, S., Basser, P.J., Fields, R.D., 2014. Role of myelin plasticity in oscillations and synchrony of neuronal activity. Neuroscience 276, 135–147.

Panzeri, S., Brunel, N., Logothetis, N.K., Kayser, C., 2010. Sensory neural codes using multiplexed temporal scales. Trends Neurosci. 33 (3), 111–120.

Pinault, D., 1995. Backpropagation of action potentials generated at ectopic axonal loci: hypothesis that axon terminals integrate local environmental signals. Brain Res. Brain Res. Rev. 21 (1), 42–92.

Poliak, S., Peles, E., 2003. The local differentiation of myelinated axons at nodes of Ranvier. Nat. Rev. Neurosci. 4 (12), 968–980.

Rall, W., 1959. Branching dendritic trees and motoneuron membrane resistivity. Exp. Neurol. 1, 491–527.

Rall, W., 1969. Time constants and electrotonic length of membrane cylinders and neurons. Biophys. J. 9, 1483–1508.

Raymond, S.A., 1979. Effects of nerve impulses on threshold of frog sciatic nerve fibres. J. Physiol. 290 (2), 273–303.

Regehr, W.G., Stevens, C.F., 2001. Physiology of synaptic transmission and short-term plasticity. In: Cowan, W.M., Sudhof, T.C., Stevens, C.F. (Eds.), Synapses The Johns Hopkins University Press, Baltimore, MD, pp. 135–176.

Roberts, A., Bush, B.M.H., 1981. Neurons Without Impulses. Cambridge University Press, Cambridge, UK.

Rocha-Gonzalez, H.I., Mao, S., Alvarez-Leefmans, F.J., 2008. Na+, K+, 2Cl− cotransport and intracellular chloride regulation in rat primary sensory neurons: thermodynamic and kinetic aspects. J. Neurophysiol. 100 (1), 169–184.

Roots, B.I., 2008. The phylogeny of invertebrates and the evolution of myelin. Neuron. Glia. Biol. 4 (2), 101–109.

Rosenthal, J.J., Bezanilla, F., 2000. Seasonal variation in conduction velocity of action potentials in squid giant axons. Biol. Bull. 199 (2), 135–143.

Sabatini, B.L., Regehr, W.G., 1997. Control of neurotransmitter release by presynaptic waveform at the granule cell to Purkinje cell synapse. J. Neurosci. 17 (10), 3425–3435.

Salami, M., Itami, C., Tsumoto, T., Kimura, F., 2003. Change of conduction velocity by regional myelination yields constant latency irrespective of distance between thalamus and cortex. Proc. Natl. Acad. Sci. U.S.A. 100 (10), 6174–6179.

Sasaki, T., 2013. The axon as a unique computational unit in neurons. Neurosci. Res. 75 (2), 83–88.

Sasaki, T., Matsuki, N., Ikegaya, Y., 2011. Action-potential modulation during axonal conduction. Science 331 (6017), 599–601.

Schwarz, J.R., Glassmeier, G., Cooper, E.C., Kao, T.C., Nodera, H., Tabuena, D., et al., 2006. KCNQ channels mediate IKs, a slow K+ current regulating excitability in the rat node of Ranvier. J. Physiol. 573 (Pt 1), 17–34.

Scuri, R., Lombardo, P., Cataldo, E., Ristori, C., Brunelli, M., 2007. Inhibition of Na+/K+ ATPase potentiates synaptic transmission in tactile sensory neurons of the leech. Eur. J. Neurosci. 25 (1), 159–167.

Seidl, A.H., 2014. Regulation of conduction time along axons. Neuroscience 276, 126–134.

Sengupta, B., Stemmler, M., Laughlin, S.B., Niven, J.E., 2010. Action potential energy efficiency varies among neuron types in vertebrates and invertebrates. PLoS Comput. Biol. 6, e1000840.

Shapiro, E., Castellucci, V.F., Kandel, E.R., 1980. Presynaptic membrane potential affects transmitter release in an identified neuron in Aplysia by modulating the Ca2+ and K+ currents. Proc. Natl. Acad. Sci. U.S.A. 77 (1), 629–633.

Sheffield, M.E., Best, T.K., Mensh, B.D., Kath, W.L., Spruston, N., 2011. Slow integration leads to persistent action potential firing in distal axons of coupled interneurons. Nat. Neurosci. 14 (2), 200–207.

Sheffield, M.E., Edgerton, G.B., Heuermann, R.J., Deemyad, T., Mensh, B.D., Spruston, N., 2013. Mechanisms of retroaxonal barrage firing in hippocampal interneurons. J. Physiol. 591 (Pt 19), 4793–4805.

Shu, Y., Yu, Y., Yang, J., McCormick, D.A., 2007. Selective control of cortical axonal spikes by a slowly inactivating K+ current. Proc. Natl. Acad. Sci. U.S.A. 104 (27), 11453–11458.

Standaert, F.G., 1963. Post-tetanic repetitive activity in the cat soleus nerve: its origin, course, and mechanism of generation. J. Gen. Physiol. 47, 53–70.

Standaert, F.G., 1964. The mechanisms of post-tetanic potentiation in cat soleus and gastrocnemius muscles. J. Gen. Physiol. 47, 987–1001.

Sun, B.B., Chiu, S.Y., 1999. N-type calcium channels and their regulation by GABAB receptors in axons of neonatal rat optic nerve. J. Neurosci. 19 (13), 5185–5194.

Suzuki, N., Tang, C.S., Bekkers, J.M., 2014. Persistent barrage firing in cortical interneurons can be induced *in vivo* and may be important for the suppression of epileptiform activity. Front. Cell Neurosci. 8, 76.

Swadlow, H.A., 1982. Impulse conduction in the mammalian brain: physiological properties of individual axons monitored for several months. Science 218 (4575), 911–913.

Swadlow, H.A., 1985. Physiological properties of individual cerebral axons studied *in vivo* for as long as one year. J. Neurophysiol. 54 (5), 1346–1362.

Swadlow, H.A., Kocsis, J.D., Waxman, S.G., 1980. Modulation of impulse conduction along the axonal tree. Annu. Rev. Biophys. Bioeng. 9, 143–179.

Tasaki, I., 2004. On the conduction velocity of nonmyelinated nerve fibers. J. Integr. Neurosci. 3 (2), 115–124.

Taylor, A.L., Goaillard, J.M., Marder, E., 2009. How multiple conductances determine electrophysiological properties in a multicompartment model. J. Neurosc.i 29 (17), 5573–5586.

Toennies, J.F., 1938. Reflex discharge from the spinal cord over the dorsal roots. J. Neurophysiol. 1, 378–390.

Vacher, H., Mohapatra, D.P., Trimmer, J.S., 2008. Localization and targeting of voltage-dependent ion channels in mammalian central neurons. Physiol. Rev. 88 (4), 1407–1447.

Vervaeke, K., Gu, N., Agdestein, C., Hu, H., Storm, J.F., 2006. Kv7/KCNQ/M-channels in rat glutamatergic hippocampal axons and their role in regulation of excitability and transmitter release. J. Physiol. 576 (Pt 1), 235–256.

Wang, S., Young, K.M., 2014. White matter plasticity in adulthood. Neuroscience 276, 148–160.

Wang, S.S., 2008. Functional tradeoffs in axonal scaling: implications for brain function. Brain Behav. Evol. 72 (2), 159–167.

Waxman, S.G., 1975. Integrative properties and design principles of axons. Int. Rev. Neurobiol. 18, 1–40.

Waxman, S.G., 1980. Determinants of conduction velocity in myelinated nerve fibers. Muscle Nerve 3 (2), 141–150.

Waxman, S.G., Ritchie, J.M., 1993. Molecular dissection of the myelinated axon. Ann. Neurol. 33 (2), 121–136.

Weidner, C., Schmelz, M., Schmidt, R., Hansson, B., Handwerker, H.O., Torebjork, H.E., 1999. Functional attributes discriminating mechano-insensitive and mechano-responsive C nociceptors in human skin. J. Neurosci. 19 (22), 10184–10190.

Weidner, C., Schmelz, M., Schmidt, R., Hammarberg, B., Orstavik, K., Hilliges, M., et al., 2002. Neural signal processing: the underestimated contribution of peripheral human C-fibers. J. Neurosci. 22 (15), 6704–6712.

Xia, Y., Zhao, Y., Yang, M., Zeng, S., Shu, Y., 2014. Regulation of action potential waveforms by axonal GABAA receptors in cortical pyramidal neurons. PLoS ONE 9 (6), e100968.

Xu, K., Terakawa, S., 1999. Fenestration nodes and the wide submyelinic space form the basis for the unusually fast impulse conduction of shrimp myelinated axons. J. Exp. Biol. 202 (Pt 15), 1979–1989.

Yamazaki, Y., Hozumi, Y., Kaneko, K., Sugihara, T., Fujii, S., Goto, K., et al., 2007. Modulatory effects of oligodendrocytes on the conduction velocity of action potentials along axons in the alveus of the rat hippocampal CA1 region. Neuron Glia. Biol. 3 (4), 325–334.

Yang, J., Ye, M., Tian, C., Yang, M., Wang, Y., Shu, Y., 2013. Dopaminergic modulation of axonal potassium channels and action potential waveform in pyramidal neurons of prefrontal cortex. J. Physiol. 591 (Pt 13), 3233–3251.

Zucker, R.S., Regehr, W.G., 2002. Short-term synaptic plasticity. Annu. Rev. Physiol. 64, 355–405.

II. AXON DYNAMICS

AXON GUIDANCE

Overview

Axons mapped in their entirety, as set forth in several chapters in Section I, have complex and even convoluted trajectories. Inevitably, the question arises of how long-distance axons make their way to one or more target structures and how specific branch points are determined. Related to this, the two chapters in this section review several examples of white matter tract formation and the associated mechanisms.

Chapter 13, by Austen Sitko and Carol Mason, discusses re-sorting of axons in the retinofugal, olfactory, auditory, somatosensory, and corticocortical pathways with attention to the identified mechanisms influencing pretarget axon organization. Three general categories of agent mechanisms are distinguished; namely, sorting by topography, chronotopy, and typography.

Chapter 14, by Linda Richards and colleagues, extensively reviews background data on cell identity and axon guidance, and then focuses on the corpus callosum and its agenesis as a model of axon guidance and tract formation. The authors discuss key players in projection cell identity and axon guidance among identified transcription factors and signaling molecules, and discuss several mouse models helpful for deciphering the mechanisms impacting on callosal agenesis in specific and axon guidance in general.

Both chapters comment on the obvious relevance of tract development and organization to diffusion tensor imaging studies in the normal and diseased brain. This issue will be revisited extensively in Section IV.

13

Organization of Axons in Their Tracts

Austen A. Sitko[1] and Carol A. Mason[1,2]

[1]Department of Neuroscience, College of Physicians and Surgeons, Columbia University, New York, NY, USA [2]Departments of Pathology & Cell Biology, and Ophthalmology, College of Physicians and Surgeons, Columbia University, New York, NY, USA

13.1 INTRODUCTION

Since the introduction of the chemoaffinity hypothesis (Sperry, 1963), developmental neurobiologists have identified a host of molecular gradients and cues in intermediate and final target regions that help establish orderly connections between projection neurons and their targets. With this emphasis on source–target correspondence, the field has focused less attention on axons within developing tracts, leaving several questions: what, if any, defined axon arrangements exist prior to target entry; through what molecular environments do axons navigate in their tracts; and what interactions occur between axons and the tract environment, and between cohorts of axons? Furthermore, is pretarget axon organization a crucial step in the creation of accurate synaptic connections in the final target, or are cues within the target sufficient to establish accurate circuitry, leaving tract order largely incidental?

The idea that axons are pre-sorted in their tracts is certainly not new. Though his contributions to the field have been credited for the emphasis on chemoaffinity-driven axon–target interactions, Sperry himself suggested that axons are organized, or at least guided, by chemical cues along the pathway: "Not only the details of synaptic association within terminal centers, but also the routes by which the fibers reach their synaptic zones would seem to be subject to regulation during growth by differential chemical affinities" (Attardi and Sperry, 1963). Sperry's contemporaries, too, hinted at the possibility of pretarget axon guidance cues in the tract, asserting that "mechanical factors may play an insignificant role in determining how and where a growing nerve fiber shall end," but "more subtle forces are at work, forces whose basic mechanisms are almost totally unknown at the present time" (Barnard and Woolsey, 1956). In the more than half-century since, researchers have identified numerous chemoattractive and chemorepulsive sources and gradients that guide axons within their targets or through intermediate choice points, such as at the central nervous system midline in the optic chiasm or spinal cord floorplate. Concurrent with research on intermediate and final targets, some investigators also explored the organization of axons within developing tracts. Reports from the late 1970s identified pretarget axon order in the developing visual pathway, with Scholes even declaring that "common sense" dictated the importance of fiber–fiber interactions and pretarget sorting in the formation of developing pathways (Cook and Horder, 1977; Scholes, 1979).

The field has since revealed much about the order of retinal axons along the optic nerve and tract, and identified axon organization in other sensory and central tracts. The question of how axons are arranged in their tracts has gained more traction, and recent studies have utilized advanced molecular genetic and selective labeling techniques to directly probe whether such pretarget axon order is required for accurate innervation of targets. In this chapter, we first review what is known about the modes and degrees of pretarget axon organization across sensory, motor, and central neural circuits, before surveying the molecular mechanisms at play as axons course through their tracts. We then explore recent research on the relationship between tract organization and the specificity of synaptic connectivity within final target regions.

13.2 EVIDENCE OF TRACT ORDER: TOPOGRAPHY, CHRONOTOPY, AND TYPOGRAPHY

13.2.1 Visual System

Sperry's pioneering work underlying the chemoaffinity hypothesis was conducted on retinal ganglion cell (RGC) axon terminals as they map onto the optic tectum (Sperry, 1951, 1963). Though topographic order of axon terminations in their targets has been characterized in many other systems since the 1960s, the retinofugal pathway has remained one of the dominant classic model systems. As such, many of the first steps toward understanding pretarget axon sorting were taken in the visual system, with a multitude of studies characterizing axon order in the optic nerve and tract.

RGCs are the only projection neurons that exit the retina and connect to central brain targets, and the hallmark of the visual system is its retinotopy, the topographic correspondence between RGC location in the retina and terminal targets in the brain (Lund et al., 1974). During development, RGCs extend axons out of the retina and into the optic nerve to the optic chiasm, the midline choice point or intermediate target region of the retinofugal pathway. RGC axons then grow into the optic tract, extending to the primary image-processing visual targets in the thalamus and midbrain, the dorsal lateral geniculate nucleus (dLGN) and superior colliculus (SC). While subsets of RGC axons extend to other nonimage-processing targets (e.g., the suprachiasmatic nucleus and ventral lateral geniculate nucleus), tracts projecting to these regions and retinotopy therein have been less well characterized. Instead, pretarget axon sorting has been examined along two segments of the retinothalamic pathway: the prechiasmatic optic nerve and the postchiasmatic optic tract.

Early hypotheses on the relationship between tract organization and targeting fall along a continuum between two extremes. At one extreme, retinotopy is thought to arise from maintenance of neighbor relationships between RGC axons from the retina to the brain targets; whereas at the other extreme, simple chemoaffinity within the target is thought to be sufficient to organize incoming axons, leaving axon order in the tract either nonexistent or inconsequential. In support of the first of these hypotheses, RGC axons were observed to bundle into fascicles as they navigate toward the optic disc within the developing chick retina (Goldberg and Coulombre, 1972). This preponderance of fasciculating fibers prior to exiting the retina was suggested to help maintain order along the remainder of the developing visual pathway. When technical advances allowed for closer examination, however, a remarkable extent of splitting and crossing-over of axons between fascicles was observed in both the retina (Simon and O'Leary, 1991) and optic stalk (Simon and O'Leary, 1991; Williams and Rakic, 1985). One-to-one neighbor relationships are not preserved between RGC axons upon entry into the optic nerve, and thus, as suggested by Barnard and Woolsey in 1956, simple mechanical maintenance of neighbor relationships must be insufficient to explain retinotopic mapping in central targets.

Further studies using classical tracing methods and electrophysiology revealed that as RGC axons exit the eye and enter the optic nerve, they maintain a rough retinotopic arrangement that coarsely reflects their neighbor relationships within the retina, but gradually become less ordered as they near the optic chiasm. This progressive scattering of retinotopic order along the length of the nerve has been described in frog (Montgomery et al., 1998), chick (Ehrlich and Mark, 1984), mouse (Chan and Chung, 1999; Plas et al., 2005), rat (Chan and Guillery, 1994; Simon and O'Leary, 1991), ferret (Reese and Baker, 1993), cat (Horton et al., 1979; Naito, 1986), and monkey (Naito, 1989). These studies of retinotopic axon order in the optic nerve again demonstrate that one-to-one neighbor relationships are not maintained. However, despite finding that axon order in the cat optic nerve is not a simple reflection of retinal coordinates, Horton et al. "do not conclude that the fibres are arranged haphazardly." Instead, the authors speculate: "once the fibres diverge quickly in the initial portion of the nerve, they tend to maintain a more constant disposition, hinting that the scatter is somehow controlled" (Horton et al., 1979). Thus, though it may still be unclear how, axon organization in the nerve might serve as a functional prelude to successful navigation of later guidance cues or targeting steps.

Retinofugal axons next navigate the optic chiasm, the midline choice point in binocular visual systems, where ipsi- and contralaterally projecting RGC axons from each eye diverge into the optic tract of the appropriate hemisphere. Although degree of binocularity varies across species, axon reorganization at the chiasm is a key step in the development of the visual circuit and raises the question of how ipsi- and contralateral RGC axons are organized in the optic nerve prior to reaching the chiasm. By backlabeling retinal axons from the optic tract, studies in rodents concluded that, similar to retinotopy in the nerve, some order exists early in the nerve between future ipsi- and contralaterally projecting RGC axons, but is progressively lost as the axons near the chiasm (Baker and Jeffery, 1989; Colello and Guillery, 1990). More recent approaches utilizing a genetic marker of ipsilateral identity confirm the segregation of ipsilateral axons from their contralateral neighbors in the nerve but suggest that a greater degree of order is retained in the distal optic nerve than previously thought (Sitko and Mason, unpublished).

At the chiasm itself, several guidance factors have been identified on radial glia and a specialized neuronal population that mediate the decision of retinal axons to project ipsilaterally (reviewed in Petros et al., 2008, and discussed below). While efforts are ongoing to identify the full symphony of decussation cues in the chiasm, details and mechanisms of topographic axon sorting therein are considerably less clear than decussation cues in the chiasm. Topographic order appears to be largely, but not completely, lost as axons unbundle from nerve fascicles, spread out across and through the depth of the chiasm, and then re-sort into orderly arrangements as they enter the optic tract (Chan and Chung, 1999; Chan and Guillery, 1994). Indeed, the fascicular composition of RGC axons changes entirely within the optic chiasm, as axons from one eye split into many separate fascicles that cross contralateral axon bundles from the opposite eye at right angles, forming a braid-like configuration at the midline (Colello and Guillery, 1998).

After navigating the complex array of cues at the optic chiasm, RGC axons enter into the optic tract. Though reports vary in the details and extent of order, a coarse topography of axons has been reported in the optic tract of many species (Chan and Chung, 1999; Chan and Guillery, 1994; Plas et al., 2005; Reese and Baker, 1993; Reh et al., 1983; Torrealba et al., 1982). The minor variations between these studies can be attributed to different axon tracing methodologies, ranging from degeneration studies to various retrograde and anterograde axonal tracers, and to disagreement between physiological and anatomical data. Additionally, it is sometimes difficult to find agreement among published histological analyses because many studies used unconventional planes of section in attempts to take true cross-sections of the tract, a fiber bundle that hugs the curves of the perimeter of the thalamus. Regardless of apparent inconsistencies across these studies, three broad conclusions surface from this work, which are expanded upon in the following paragraphs and summarized in Figure 13.1. First, there is a coarse topographic order among retinal axons in the optic tract (Figure 13.1A). Second, evidence supports an age-related, or chronotopic, order of retinal axons in the tract (Figure 13.1B), and finally, several studies argue for order among functional subclasses of RGC axons, including between ipsi- and contralateral axons (Figure 13.1C).

Anterograde labeling of retinal quadrants reveals that dorsal RGC axons course through the medial optic tract and ventral RGC axons through the lateral tract (Chan and Chung, 1999; Chan and Guillery, 1994; Plas et al., 2005; Reese and Baker, 1993; Reese and Cowey, 1990; Reh et al., 1983). However, in the distal optic tract of the mouse, just

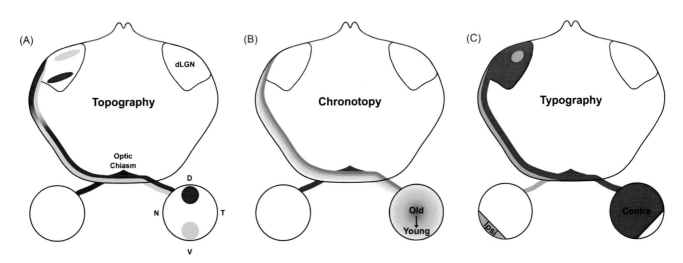

FIGURE 13.1 Axons in developing tracts are organized by three common modes: topography (A), chronotopy (B), and typography (C), as illustrated here in frontal sections of the mouse retinogeniculate pathway. (A) RGC axons are ordered topographically in the mouse optic nerve and tract, with dorsal and ventral RGC axons switching positions near the dLGN. Dorsal RGC axons (blue) innervate the ventromedial aspect of the dLGN, while ventral RGC axons (yellow) innervate dorsolateral dLGN. The dLGN is a three-dimensional structure, but the anterior–posterior axis is collapsed in this schema into one two-dimensional plane for simplicity; however, note that dorsal axons terminate more posteriorly than ventral axons. Nasal and temporal RGC axons, not shown here, are less segregated relative to dorsal/ventral axons. (B) In chronotopic order, younger axons (light blue) enter the superficial aspect of the tract and course along the pia, displacing older axons (dark blue) deeper into the tract. (C) Typography is represented by eye-specific axon order (i.e., ipsi- and contralateral RGC axons). Ipsilateral axons (green) course more superficially (i.e., medially, along the pia) in the tract relative to contralateral axons (red). D, dorsal; V, ventral; N, nasal; T, temporal; dLGN, dorsal lateral geniculate nucleus.

prior to axon entry into the dLGN, dorsal RGC axons are found laterally and ventral RGC axons medially, opposite to their position in the proximal tract (Plas et al., 2005; Figure 13.1A). This may be a reflection of a twist in the optic tract as a whole, which is evident in wholemount views of the surface of the thalamus, and also corresponds to the eventual dorsal–ventral mapping of RGCs along the lateral–medial axis in the SC (Plas et al., 2005). The chemical or mechanical factors that create this positional shift of dorsal and ventral axons in the tract are unknown. It is noteworthy that in studies that directly examined the topographic order of retinal axons in the tract, in both mammalian (Chan and Guillery, 1994; Plas et al., 2005; Reese and Baker, 1993) and nonmammalian species (Ehrlich and Mark, 1984; Montgomery et al., 1998; Reh et al., 1983; Thanos and Bonhoeffer, 1983), most conclude that the segregation between dorsal and ventral retinal axons is more distinct compared to that between nasal and temporal retinal axons, which are often described as being positioned adjacent, mixed, or otherwise overlapping in the tract.

Chronotopy, or age-related order, was first suggested as a mode of axon organization in the frog, where back-labeling a portion of the optic tract results in annular retrograde labeling in the retina, denoting a cohort of RGCs born at the same time (Fawcett et al., 1984; Reh et al., 1983). Axons are topographically ordered within age-related bundles, where dorsal and ventral RGCs extend axons laterally and medially, respectively, while nasal and temporal RGC axons are mixed together across much of the tract (Reh et al., 1983). Note that this arrangement differs from topographic order in mammalian tracts. The authors propose that active reordering mechanisms play a part in the topographic arrangement of fibers, while chronotopic order likely stems from passive mechanisms related to space constraints on incoming axons exiting the optic chiasm. Furthermore, Reh et al. hypothesize the presence of attractive cues along the dorsal neuraxis that could preferentially act upon ventral RGC axons, providing a tract-specific mechanism of active axon organization that would occur independently of target-derived cues (Reh et al., 1983). Mammalian optic tracts also display chronotopic axon order (Figure 13.1B), with younger axons running along the glial endfeet-lined pial surface, displacing older axons progressively deeper in the tract (Colello and Guillery, 1992; Reese, 1987; Reese and Cowey, 1990; Reese et al., 1997; Walsh and Guillery, 1985). Interestingly, age-related axon order is largely lost within the chiasm itself, though it reappears in the tract (Colello and Guillery, 1998).

Chronotopic order of axons in the optic tract may reflect segregation of functional subtypes of RGCs. Work in the cortex has shown that timing of neuronal differentiation is critical for establishing neuronal subtype identity and laminar position (Molyneaux et al., 2007), a principle that appears to be true of RGCs in the retinofugal system as well. RGCs of differing soma diameter, an early proxy for identifying RGC subtypes, are generated at different times in the retina (Rapaport et al., 1996; Reese et al., 1994). Furthermore, timing is important for the appropriate differentiation of ipsilateral RGCs (Bhansali et al., 2014), and birthdate of RGC subtypes correlates to target matching strategies in the developing SC and dLGN (Osterhout et al., 2014). These recent studies provide support for earlier findings, which used axon diameter and cell morphology to show that functional subsets of RGC axons are segregated in the optic tracts of rodents (Reese, 1987), cats (Guillery et al., 1982; Torrealba et al., 1982), and monkeys (Reese and Cowey, 1990).

The development of genetic tools and identification of RGC subtype-specific markers (e.g., Blackshaw et al., 2004; Dhande and Huberman, 2014) will afford even more precise analyses of the axon organization of molecularly defined RGC subtypes. In the meantime, two subtypes of RGC subtypes can be defined based on their trajectory at the chiasm of binocular animals: those projecting ipsi- or contralaterally. A moderate degree of order between ipsi- and contralateral axon cohorts, or "eye-specific" order, has been described in the optic tract of cats (Torrealba et al., 1982). In neonatal mice, contralateral axons are spread across the entire optic tract, with ipsilateral axons constrained to the anterior-lateral edge (Godement et al., 1984; Sitko and Mason, unpublished) (Figure 13.1C). The earliest-born RGCs in the mouse extend axons from the dorsocentral retina and project either ipsi- or contralaterally. Even though the early ipsilateral RGCs arising from this region are transient, their axons segregate in the optic tract from those of their contralaterally projecting counterparts, underscoring the presence of eye-specific order in the tract (Soares and Mason, 2015).

A major question arising from this body of work is whether pretarget axon organization is functionally important in the formation of the retinofugal pathway. Though most of the studies in the optic nerve and tract to date are descriptive, they argue that pretarget axon organization is governed by a consistent logic and is likely an important step in the formation of the visual circuit. For instance, in the cichlid fish optic tract, topographic order is maintained among retinal axon fascicles until very near the optic tectum, at which point the fascicles rearrange as they enter the appropriate tectal region (Scholes, 1979). Scholes, like Reh et al. (1983) and Horton et al. (1979), infers that the consistent maintenance of axon order along the length of the tract is indicative of target-independent cues in the tract; that is, short-range fiber–fiber interactions may provide organization in the tract prior to reaching target-derived guidance cues (Scholes, 1979). Furthermore, retinotopy in the frog optic nerve and tract is largely normal following surgical removal of the optic tectum early in development (Reh et al., 1983). Axon order in these tectum-less frogs provides even stronger evidence that fiber–fiber interactions and/or environmental cues in the tract organize axons *en route* to the target.

In summary, there are at least three different types of axon order in the developing optic tract: coarse topography, chronotopy (which may reflect or occur independently of functional specificity), and eye-specificity (which may be considered *typo*graphic order based on cohort identity) (Figure 13.1). These different modes of order are not easy to fully synthesize together, as they do not always appear in register with one another. However, early reports of a lack of fiber order in the optic tract (Horton et al., 1979) can be reconciled with later findings of axon order by an overlap of different modes of axon order. One mode (e.g., retinotopy) may be obscured by another (e.g., chronotopy), depending on the methodology used to identify axon order. In other words, failure to find a certain type of axon order at points along the visual pathway does not necessarily demonstrate a lack of order. Future studies may identify further contradictions between modes of axon order in the optic tract or even new modes altogether, which may, for instance, involve genetically defined subsets of RGCs.

13.2.2 Olfactory System

Whereas the visual system relays topographically related sensory information between sense organ and brain target, the olfactory system transduces chemical stimuli, which do not require a one-to-one spatial map to represent the sensory information. While there are accordingly fundamental differences in the relationship between sense organ and target in the visual and olfactory systems, they both exhibit order among axons in their tracts. Olfactory sensory neurons (OSNs) lining the olfactory epithelium (OE) of the nose send their axons into the brain target, the olfactory bulb (OB). In the OB, OSN axons are organized by a "typographic" principle, where OSNs of the same type, i.e., expressing the same olfactory receptor (OR), coalesce into distinct target glomeruli via presumed homotypic axon–axon interactions based on OR expression (Feinstein and Mombaerts, 2004). As retinotopy is the hallmark of visual system targeting, OR-based typography is the hallmark of olfactory system targeting. However, there is a coarse spatial pattern of expression of ORs among OSNs in the OE, as well as some topographic correspondence between the OE and OB. Thus, while ultimate axon sorting decisions based on OR expression occur within the glomerular layer of the OB, pretarget ordering of axons in the olfactory nerve may serve a preliminary role in establishing the olfactory sensory circuit (Miller et al., 2010).

Topographic correspondence between the OE and OB was first identified over a half century ago, using degeneration studies in the rabbit olfactory system (Clark, 1951). The findings were confirmed by subsequent anatomical and physiological studies, which describe a correspondence of the dorsal–ventral and mediolateral axes in the OE to the same axes in the OB, but little to no rostral–caudal correspondence between the two structures (Costanzo and O'Connell, 1978; Land, 1973; Saucier and Astic, 1986). With the development of more advanced genetic tools, a series of studies showed first that there is zonal patterning in the OE, with OSNs expressing certain ORs restricted to particular zones (Ressler et al., 1993; Vassar et al., 1993); and second, that all OSNs expressing one type of OR project to specific, bilaterally symmetric glomeruli in the OB, creating what Ressler et al. term an "epitope map" of olfactory stimuli in the OB (Ressler et al., 1994; Vassar et al., 1994). Importantly, this glomerular map in the OB is remarkably consistent across animals, demonstrating a highly stereotyped typographic map of ORs in the OB (Ressler et al., 1994; Vassar et al., 1994). None of these studies directly examined axon arrangements in the olfactory nerve, but the degree of order collectively described between the OE and OB provided the first suggestions that organization of axons entering into the target may be important for wiring the circuit.

Further exploration of the olfactory nerve found that OSN axons undergo organizational events as they grow toward the OB. Two subpopulations of OSNs were identified in *Xenopus* tadpoles based on their relative expression levels of Neuropilin (Nrp) and Plexin, both receptors to Semaphorins (Semas) (Satoda et al., 1995). The two cohorts of axons are intermingled early in the olfactory nerve but become increasingly sorted as they project to the OB. In the distal olfactory nerve, there is a sharp border between the two bundles of axons, the segregation of which reflects their final segregation in the OB (Satoda et al., 1995). The position of additional axon cohorts in the olfactory nerve is defined by expression of other molecular markers, with a similar degree of order described in all cases (Imai et al., 2009; Miller et al., 2010). Regardless of the marker examined, OSN axon cohorts exit the OE in overlapping swathes of axons and become more orderly along the nerve. There is a very clear order of axon bundles prior to entry into the inner nerve layer of the OB, where axons undergo further OR-based sorting into individual glomeruli (Imai et al., 2009; Miller et al., 2010).

Similar to investigations in the visual system, studies in the olfactory system aimed to determine if this pretarget axon order arises due to target-derived cues or is driven independently by active organizational mechanisms occurring within the olfactory nerve. The fact that pretarget sorting of axons in the olfactory nerve is apparent by E12 in mice, at least 72 h prior to onset of synaptogenesis in the OB, argues in favor of target-independent organizational mechanisms (Miller et al., 2010). Even stronger evidence that axon sorting in the olfactory nerve occurs independently of OB cues comes from a series of experiments that surgically (Graziadei et al., 1978) or genetically ablated the OB (St John et al., 2003) or subsets of target cells in the OB (Bulfone et al., 1998). In all such experiments, OSN axons sort normally in the nerve and form OR-specific glomerulus-like bundles, which invade ectopic brain regions in the absence of a true OB. Similar to optic tectum removal experiments in frog (Reh et al., 1983), these target-ablating experiments provide compelling evidence of active and independent pre-sorting of axons in the olfactory nerve.

Studies examining higher order olfactory targets beyond the OB have noted that central olfactory regions fail to maintain topographic correspondence to the spatial organization of glomeruli in the OB (Luskin and Price, 1982; Sosulski et al., 2011). Topographic order of OB efferents can only be found in the earliest portion of the lateral olfactory tract (LOT), and axon order reflecting spatial information of the OB is presumably lost along the length of the LOT (Price and Sprich, 1975). A more recent study found that OB efferents are in fact arranged chronotopically in the LOT, with younger axons coursing nearer the pial surface and older axons more deeply in the tract (Yamatani et al., 2004), notably similar to chronotopic order in the optic tract. This finding again highlights the fact that lack of evidence of one mode of axon organization does not necessarily mean a lack of organization overall.

13.2.3 Auditory System

Relative to the other sensory systems, axon guidance in the auditory system is particularly poorly understood. Tonotopy—the organization of sound frequencies along an axis—is evident in the basilar membrane of the cochlea and all brainstem and central auditory centers (reviewed in Appler and Goodrich, 2011). Tonotopic organization at each auditory brain target develops precisely before the onset of hearing and processing of acoustic information (Appler and Goodrich, 2011; Rubel and Fritzsch, 2002). Though it has yet to be directly tested, preservation of tonotopy among axons within the auditory nerve and tracts connecting auditory brain centers could be a means of establishing precise tonotopy in each auditory target early in development.

At least one portion of the auditory pathway maintains an incredibly precise tonotopic map through the tract. Axons in the crossed dorsal cochlear tract (XDCT) connecting the nucleus magnocellularis (NM) and nucleus laminaris (NL) in the chick auditory brainstem are arranged in strict correspondence with their positions in both NM and NL (Kashima et al., 2013). Multicellular labeling of regions in the NM by electroporation of dextran dyes in two areas along the tonotopic axis shows a clear correspondence between the position of groups of neurons in the NM and their axons in the XDCT. The authors also followed axons from individually labeled NM cells and found a strikingly linear relationship between the position of the soma in the NM and their respective axons in the XDCT (Kashima et al., 2013). To date, this study is one of the highest resolution in any system, tracing individual axons as they project through the tract to their synaptic targets.

The degree of tonotopy in this brainstem auditory tract appears far more precise than any axon order so far described in the retinotectal or olfactory systems, in which axons are more coarsely organized. A compelling question raised by this work is whether the higher degree of precision found in this auditory tract corresponds to a more precise initial mapping of axon terminations in the target relative to other systems. In other words, does the NL rely less upon successive activity-driven refinement of axon terminals than does, for instance, the dLGN in the visual system? The relationship between degree of pretarget axon order and accuracy of initial target innervation has been discussed in the visual system, where, relative to mammals, fish and frogs display greater order in the nerve and tract and greater initial accuracy of terminals in the target (Simon and O'Leary, 1991). Based on this correlation in the visual system, it is reasonable to expect that the auditory system exhibits a greater degree of precision in its early circuit formation than do other sensory systems. Why some systems may deploy different developmental strategies of circuit formation is unclear and is an area ripe for comparative evolutionary study.

13.2.4 Spinal Sensorimotor System

The sensorimotor tracts in the spinal cord are another system in which topography of projections, similar to retinotopy in the visual system, spatially reflects the information being transmitted. In this case, somatotopy, the representation of the body map, must be contained in both motor efferents projecting to specific muscle fibers and sensory afferents conveying touch and pain information from the periphery. The somatotopic correspondence of the periphery, neuronal nuclei in the spinal cord, and the sensory and motor cortices has long been known. Indeed much of the research on the development of the sensorimotor system has focused on understanding mechanisms of neuronal diversification and migration to their correct locations at these discrete points in the system (reviewed in Kania, 2014). However, details of motor and sensory axon tract organization within this system remain less clear.

Early tract tracing experiments revealed an orderly arrangement of ascending and descending axon tracts in the chick spinal cord, where newer axons grow along older ones (Nornes et al., 1980), similar to chronotopic layering of axons in the optic tract and LOT. The authors concluded that positional information of neurons in the spinal cord "can be used as a mechanism for imparting order to the presynaptic components" of the pathway, hypothesizing that contact-mediated guidance and selective fasciculation underlie such pretarget axon order (Nornes et al., 1980). Another early study in the chick showed that motor neuron axons destined for a specific muscle course together in distinct bundles but do not maintain strict neighbor relationships along the path to the muscle (Lance-Jones and Landmesser, 1981). Instead, axons rearrange along the path and sort out within the limb plexus into discrete fascicles bound for specific muscles. This finding led the authors to suggest that active mechanisms of pretarget axon sorting form the circuit rather than either passive maintenance of neighbor relationships or exuberant, disorganized growth followed by pruning (Lance-Jones and Landmesser, 1981). More recent retrograde labeling studies tracked motor axon fascicles through the peripheral sciatic-tibial nerve in rats and found somatotopy preserved along the length of the nerve (Badia et al., 2010). Further details of topographic organization of axon bundles in the spinal cord and periphery remain largely undefined.

Other studies have found clear subtype organization (i.e., typographic order) of axons in the spinal cord prior to target innervation. For instance, motor neuron axons innervating fast and slow muscle fibers in the chick hindlimb fasciculate separately from each other starting in the proximal regions of the limb, well before reaching their target muscles (Milner et al., 1998). However, motor axons projecting to different parts of fast muscle fibers remain intermingled *en route* to the muscle, implying that a second step of axon sorting must occur within the target to organize this subset of inputs (Milner et al., 1998). Additionally, mechanosensory and proprioceptive sensory afferents segregate along the medial–lateral axis in the dorsal column of the mouse spinal cord. Notably, axons are arranged somatotopically within this typographic order (Niu et al., 2013), revealing again the layering of modes of axon organization. Given that these afferents terminate in a modality-specific manner in the medulla, pretarget typographic order of the axons may be functionally important in establishing modality-based targeting patterns (Niu et al., 2013).

One of the better-understood aspects of axon organization in the periphery is the relationship between motor and sensory axon fascicles, which run alongside each other in peripheral nerves (Honig et al., 1998). During the earliest phases of axon extension, motor neuron axons enter peripheral nerve tracts, displaying a high degree of specificity in their initial trajectory, followed by sensory afferents, which use motor axon fascicles as a tracking system to lead them to their respective targets (Huettl et al., 2011; Landmesser and Honig, 1986; Wang and Marquardt, 2013; Wang et al., 2014). Motor and sensory axons are closely associated but segregated—indeed their segregation can be recapitulated *in vitro* (Gallarda et al., 2008)—and have a hierarchical dependence on each other, such that sympathetic efferents rely on sensory afferents, which in turn rely on motor efferents to project to their respective targets (Wang et al., 2014). In sum, the sensory and motor projections from and to the periphery again demonstrate the common principles of topographic, typographic, and chronotopic order, with a necessary layering of axon order modalities within tracts.

13.2.5 Central Axon Tracts: To, From, and Within the Cortex

While the sensory systems, especially the visual system and spinal circuitry, have dominated axon guidance study for the last century, the last few decades have seen increased interest in examining the development of central neural circuits. The thalamocortical projection in particular has proved a useful model for understanding complex axon guidance and axon organizational principles. This section will survey work describing the axon tracts entering into (thalamocortical), exiting from (corticothalamic), and coursing within (corpus callosum) the cortex. While there are other corticofugal tracts, the corticothalamic tract is the best studied with regard to axon trajectories and organization, and as such, the discussion in this section will reflect that focus.

In both rodents and primates, there is a topographic correspondence between primary thalamic nuclei and target regions in the cortex (Caviness and Frost, 1980; Hohl-Abrahao and Creutzfeldt, 1991). An early study of the thalamocortical tract used anterograde tract labeling in *ex vivo* slice preparations of mouse brain to examine the entire projection from the ventrobasal complex in the thalamus to the primary sensory cortex (Bernardo and Woolsey, 1987). Though the order of specific subsets of axons within the tract was not examined, the authors found that axons diverged from their neighbors and rejoined other fascicles at numerous points along their path. The patterns of axon rearrangement were consistent from animal to animal, indicating that while neighbor relationships between thalamocortical axons (TCAs) are not maintained, there is an underlying orderliness in the fiber bundling patterns along the thalamocortical tract. One particular aspect of the tract described by Bernardo and Woolsey is a 180° rotation of the entire bundle of axons in the internal capsule (IC), which accounts for a rotated topographic map between the thalamus and sensory cortex (Bernardo and Woolsey, 1987). The 180° rotation in the mouse optic tract, discussed in Section 13.2.1, could similarly account for the transposition of the topographic map between retina and SC (Plas et al., 2005).

Subsequent studies sought to describe the axon organization along the thalamocortical pathway in more detail and discover the underlying mechanisms of tract order and target specificity. Explants of thalamic tissue innervate cortical slices promiscuously *in vitro*, regardless of whether the cortical slice is from the appropriate region for the thalamic explant to target (Molnar and Blakemore, 1991), suggesting that chemotropic cues in the cortex are insufficient to organize TCAs. Later studies from the same group confirmed the suspicion that TCAs are topographically ordered as they course through the IC and approach the pallial/subpallial boundary (PSPB), before entering the cortex (Molnar et al., 1998). Several cues in this subpallial region have been identified that orient subsets of TCAs, arranging them topographically inside the corridor (Dufour et al., 2003; Powell et al., 2008; Seibt et al., 2003).

In addition to the maintenance of topographic order of TCAs in the thalamocortical tract, there are distinct spatial and temporal relationships between the reciprocal projections of TCAs and corticothalamic axons (CTAs) along this pathway. Anterograde labeling of small populations of cortical and thalamic projection neurons shows axons from each population running alongside each other in the same fascicles, rather than two separate tracts of TCA and CTA bundles (Molnar et al., 1998). Thus, ascending and descending axons in this tract are both topographically ordered and course in close association with each other. Additionally, the meeting of CTAs and TCAs at the PSPB is temporally regulated, as CTAs extend into the PSPB first and then wait for the arrival of TCAs (reviewed in Leyva-Diaz and Lopez-Bendito, 2013). This interaction is required for the proper guidance of both cohorts and was dubbed the "handshake hypothesis" (Molnar and Blakemore, 1995; discussed in more detail below). Similar to the temporal regulation of motor and sensory axons in the spinal cord, the chronological sequence of axon extension is a key element guiding the development of the reciprocal connections between cortex and thalamus.

Reciprocal contralaterally projecting corticocortical axons make up another major white matter tract in the cortex, the corpus callosum (CC). Structural studies identified an anterior–posterior distribution of fibers within the human CC based on axon diameter (Aboitiz et al., 1992) and broadly topographic correspondence between a neuron's regional location in the cortex and its axon position in the CC of mice (Ozaki and Wahlsten, 1992), cats (Nakamura and Kanaseki, 1989), and humans (de Lacoste et al., 1985). Diffusion tensor imaging in human subjects has confirmed the topographic correspondence of cortical brain regions across the anterior–posterior axis of the CC in greater detail (Hofer and Frahm, 2006). Histological analyses in both macaques and human subjects confirm diffusion tensor imaging findings of anterior–posterior order of axon bundles corresponding to different cortical regions (Caminiti et al., 2013). Evidence from anterograde and retrograde labeling studies in the cat visual cortex shows further that subregion topographic maps are contained within the gross topography along the rostral–caudal axis of the CC (Payne and Siwek, 1991).

The development of genetic tools and advanced labeling techniques has provided higher resolution views of axon order in the CC. Axons of layer II/III sensory and motor cortical neurons in the mouse are topographically ordered as they traverse the CC to innervate the contralateral cortex (Zhou et al., 2013). Zhou et al. examined whether the topographic relationships between the neurons were preserved in their axons and, further, whether such order might affect their homotopic targeting in the contralateral cortex. The authors found a clear correspondence between neuron position in one hemisphere, axon position in the CC, and targeting position in the contralateral hemisphere: specifically, the medial–lateral axis is mirrored between the two hemispheres and maps onto the dorsal–ventral axis of the CC. This is true of axons arising from separate sensory and motor cortical areas as well as those axons arising from discrete medial–lateral positions within each cortical area (Zhou et al., 2013).

The multiple tracts coursing into, away from, and within the cortex demonstrate the principles of axon organization that have been identified in other tracts discussed so far. Similar to tracts in the spinal, visual, and auditory pathways, topography is evident in all of the cortical tracts described. Like sensory and motor axons in the spinal cord, CTAs and TCAs in the thalamocortical tract engage in heterotypic interactions that are integral to the formation of their reciprocal connections, while maintaining segregation of axon cohorts according to their origin. The reciprocal CTA and TCA projections are also temporally regulated in a similar manner to sensory and motor axons in the spinal cord. Because sensory modalities are organized into topographically defined subregions of the cortex, typographic order overlays topographic arrangements in this system.

Despite the differences evident across the systems discussed in this section, three major themes of axon organization are evident: topography, chronotopy or temporal regulation, and subtype organization or typography (schematized in the visual system in Figure 13.1). Additionally, in all but the auditory system, axons in tracts make numerous rearrangements along their paths, which, while disrupting neighbor–neighbor relationships, appear orderly and consistent. Pretarget axon order is likely governed by many common mechanisms across systems in order to achieve the combinatorial axon order of the three general organizational modes discussed in Section 13.2.

13.3 MOLECULES AND MECHANISMS GUIDING PRETARGET AXON ORGANIZATION

Despite the gaps remaining in our knowledge of the modes and degree of axon pre-sorting in all white matter tracts, mechanisms guiding such order have begun to be identified. This section explores known and hypothesized mechanisms of pretarget axon organization in developing circuits, divided into two broad categories: first, cues extrinsic to the axons, e.g., glia and choice points or intermediate targets, and second, axon-intrinsic (i.e., axon–axon) mechanisms, such as fasciculation, repulsion, and other axon–axon interactions (summarized in Figure 13.2).

Curiously, the majority of progress made in unraveling tract-specific mechanisms of axon organization has occurred in tracts that have been less well described than the retinofugal pathway. This is because while research in the visual system led the way in anatomical characterization of axon organization in developing tracts, it has lagged

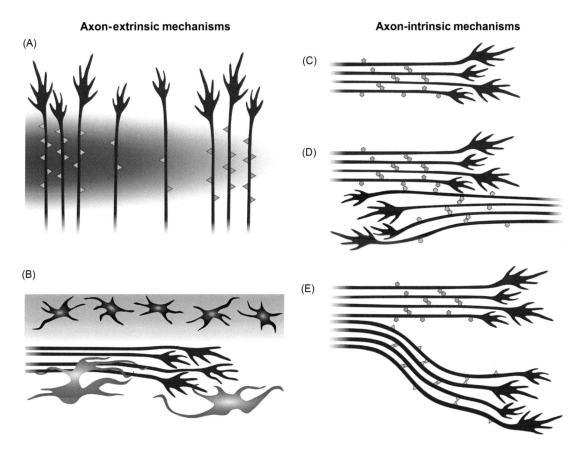

FIGURE 13.2 Axon-extrinsic (A and B) and axon-intrinsic (C and E) mechanisms influence the organization of axons in developing tracts. (A) Intermediate choice points, e.g., the IC in the thalamocortical pathway, express gradients of attractive and repellant guidance molecules. Axons are ordered according to their relative expression levels of receptors. Here, yellow receptors mediate attraction to the molecular gradient (in purple), and green receptors mediate other inhibitory cues and are segregated away via repulsion. (B) ECM, astroglia, and microglia also influence the organization of axons in tracts. Molecular gradients can exist in the ECM (gray), and different types of glia (red and orange) express guidance molecules and can extend cytoplasmic processes between axon fascicles. Axons fasciculate via homotypic adhesion (C), and are also organized by heterotypic axon–axon interactions (D and E). (D) Grossly heterotypic axon–axon interactions occur between reciprocally projecting axon cohorts, such as CTAs and TCAs, or sensory and motor axons in the periphery. Reciprocally projecting cohorts (red and blue axons) cofasciculate while maintaining appropriate segregation between each other. (E) Finely heterotypic axon–axon interactions occur between axon subtypes originating from the same source and projecting in the same direction, e.g., ILc and MLc axons in the marginal zone of the spinal cord. Cohorts of axons from different neuronal subtypes selectively defasciculate from each other at appropriate points in their pathway, while axons within each subfascicle homotypically adhere with each other.

behind other systems in terms of studying the molecular mechanisms guiding that organization. The exception to this rule is the midline choice point, the optic chiasm, which has been well characterized. As such, we begin our discussion in this section with the role of intermediate targets and choice points in organizing axon tracts, starting with the optic chiasm. The remainder of the organizational mechanisms discussed throughout this section will come from studies conducted in other systems.

13.3.1 Axon-Extrinsic Cues

13.3.1.1 Intermediate Targets, Choice Points, and Guidepost Cells

As discussed in the first section, RGC axons in the developing binocular visual system are orderly in the proximal optic nerve, unravel and undergo rearrangements near and within the midline choice point (the optic chiasm), and then regain order in the optic tract. Midline radial glia and a raft of early-born neurons caudal to the midline glial palisade provide guidance cues at the optic chiasm (reviewed in Erskine and Herrera, 2014; Petros et al., 2008) that direct the crossing of contralateral RGC axons (Charron et al., 2003; Erskine et al., 2011; Kuwajima et al., 2010, 2012; Williams et al., 2006) and the repulsion of ipsilateral RGC axons (Petros et al., 2009, 2010; Williams et al., 2003). Similarly, a transient neuronal population at the cortical midline called the subcallosal sling is important for guiding axons across the callosal midline during the CC development (Suarez et al., 2014). Subcallosal sling neurons express Sema3C, which attracts Nrp1-expressing pioneer cingulate cortex neurons (Niquille et al., 2009; Piper et al., 2009). These two guidance molecules have lately been found to play an important role in establishing the dorsal–ventral organization of callosal axons (Zhou et al., 2013).

In another midline system, the developing ventral spinal cord, a host of molecular mechanisms have been elucidated that guide axons to, through, and out of the floor plate, which contains radial neuroepithelial cells that function as glia (reviewed in Neuhaus-Follini and Bashaw, 2015). Early work in the chick floorplate revealed that crossing the midline leads to a change in surface cell adhesion molecule (CAM) expression from TAG-1 to L1 on commissural and motor neuron axons (Dodd et al., 1988). This change in surface protein expression on axons and their growth cones could be a common mechanism guiding midline-crossing axons through and beyond the choice point into the continuation of their tracts. It is possible that such midline-induced changes in surface expression could lead to specific bundling and organizational behaviors of postmidline axons in their developing tracts. This has yet to be tested in the visual system, but a change in surface CAM expression on ipsilateral and/or contralateral RGC axons could affect interactions between the two axon cohorts or between one group of axons and the tract environment.

In systems where axon tracts navigate across a midline choice point, it is challenging to disentangle the molecular mechanisms at the choice point from later organizational mechanisms at play within the tract or target. In other words, are postdecussation organizational events tract-specific or consequences of events occurring at the choice point (e.g., putative changes in CAM expression on axons)? This question is experimentally challenging, given the difficulty in isolating molecular events in the tract from those at the midline or at the target, especially in systems where the same guidance molecules are present at multiple steps in the pathway.

Recent work in the auditory system has shed light on this issue by showing that aberrant navigation of the midline decussation point leads to subtle but critical functional defects of subsequent synaptic connections, apparently independently of target-derived cues (Michalski et al., 2013). Conditional knockout of Robo3 from contralaterally projecting ventral cochlear nucleus neurons causes their axons to misroute ipsilaterally. The misrouted axons innervate their appropriate target, the medial nucleus of the trapezoid body, but with incorrect laterality. While axon targeting appeared grossly normal, closer structural and physiological examination revealed failures in synaptic maturation. It is unlikely that Robo3 is directly involved in synaptic development in this case, as it is downregulated immediately after the axons decussate in wild-type (wt) mice. Further, temporally controlled conditional knockdown of Robo3 after decussation showed no effect on synaptic development (Michalski et al., 2013). The authors conclude that the act of decussation at the midline conditions the axons of ventral cochlear nucleus neurons for appropriate synaptic development. Hence, while this study does not examine axon order in the postdecussation tract, it underscores the idea that individual events along a pathway, such as interaction with the choice point or intermediate target, can have subtle and surprising effects on downstream steps in the developing circuit (e.g., targeting or, theoretically, tract order).

Work in the thalamocortical system has demonstrated how a non-midline intermediate target region can influence axon order in a tract (Figure 13.2A). As described earlier, TCAs are topographically ordered as they pass through the developing striatum and IC (Garel and Lopez-Bendito, 2014), where a population of migrating guidepost cells creates a permissive corridor for extending TCAs in an otherwise growth-prohibitive environment (Bielle et al., 2011;

Lopez-Bendito et al., 2006). Perturbing the development of the corridor region leads to disordered TCA topography in the corridor, causing a subset of TCAs to fail to reach the cortex (Garel et al., 2002). This evidence that topography of incoming TCAs relies on cues in the corridor region spurred a series of studies aimed at delineating the cues present along the thalamocortical pathway in the subpallium.

Gradient expression of Slit1 by corridor cells contributes to the organization of intermediate and rostral TCAs in the subpallium (Bielle et al., 2011). In addition to Slit1-expressing corridor cells, there are several gradients of guidance cues across the rostral–caudal axis of the striatum at the time of TCA extension through the region. Subsets of thalamic projection neurons express specific combinations of guidance receptors and CAMs that correspondingly interact with gradients of guidance cues in the subpallium, leading to topographic organization of TCAs (reviewed in Garel and Lopez-Bendito, 2014; Molnar et al., 2012). Differential expression of EphA3, 4, and 7 on TCAs mediates their interaction with a rostral-low to caudal-high ephrin-A5 gradient in the corridor and subpallium (Dufour et al., 2003); TCAs expressing EphB1 and B2 receptors are repelled from the ventrolateral telencephalon by ephrin-B1 (Robichaux et al., 2014); responsiveness to the high-rostral low-caudal netrin-1 gradient in the developing striatum is mediated by differential TCA expression of DCC and Unc5 (Powell et al., 2008), Robo1, and its coreceptor FLRT3 (Leyva-Diaz et al., 2014); Neurogenin2 guides TCA responsiveness to subpallium cues (Seibt et al., 2003); and NrCAM mediates responsiveness to Sema3F in the subpallium (Demyanenko et al., 2011).

The CTA waiting period in the subpallium, a necessary step in the "handshake" with TCAs, required for mutual guidance, is regulated by transient expression of PlexinD1 on pioneer CTAs. PlexinD1 binds to Sema3E on radial glia near the PSPB in the intermediate target region of the pathway, mediating a growth-inhibition response (Deck et al., 2013). This waiting period is essential for the correct projection of CTAs because it temporally positions them to engage in necessary interactions with reciprocal TCAs (discussed in more detail below, Section 13.3.2.2). A waiting period has also been described for TCAs, which cluster and pause below the cortical plate before entering the cortex and navigating to the appropriate cortical regions (Ghosh and Shatz, 1992; Leyva-Diaz and Lopez-Bendito, 2013; Lund and Mustari, 1977). The mechanisms and rationale for the TCA waiting period below the cortex remain unclear, but recent evidence suggests that ascending TCAs require descending CTAs in order to cross the PSPB and innervate the cortex (Chen et al., 2012).

13.3.1.2 Glia and Extracellular Matrix Cues in and Along Tracts

Glia and extracellular matrix (ECM) elements have been described in many axon tracts, and evidence for their role in axon organization within tracts is growing. At midline choice points, glia provide cues for decussation and other axon segregation events, and they, along with transient neuronal populations, also create physical and chemorepellent barriers that are essential for the overall formation of the midline tracts (reviewed in Raper and Mason, 2010; Suarez et al., 2014). For instance, as discussed earlier, radial glia at the optic chiasm direct midline decussation of RGCs while a transient neuronal population provides a barrier behind the chiasm (Petros et al., 2008; Raper and Mason, 2010). It is unclear whether and if so, how the early-born midline neurons contribute to axon order at the chiasm. Likewise, glia are essential in the development of the CC, where a glial wedge provides a physical and chemorepellent boundary at the ventromedial boundary, indusium griseum glia form a dorsomedial boundary, and midline zipper glia provide a ventromedial boundary (reviewed in Suarez et al., 2014). Although glial structures and the transient neuronal populations in the subcallosal sling of the CC are important in the formation of the midline tracts and the guidance of axons across the midline, it is not known whether and how each of these glial populations contributes to axon organization within the CC.

Outside of midline choice points, glia and ECM elements support early axon outgrowth. They specify boundaries by surrounding developing tracts (Marcus and Easter, 1995) and provide growth-permissive substrates essential for pioneer axon extension (Raper and Mason, 2010). For example, expression of growth cone-collapsing glycosaminoglycans in the chick hindlimb delineates growth-permissive corridors for axons extending into the limb (Tosney and Landmesser, 1985). Beyond these roles in tract formation, however, there is also some evidence that molecules expressed by glia or in the ECM are integral for organizing axons within tracts. In the ferret visual system, for instance, radial glia-associated chondroitin sulfate proteoglycans (CSPGs) are distributed asymmetrically in the optic tract. CSPGs are more densely localized to the deep part of the tract, opposite the superficial aspect where the youngest RGC axons chronotopically extend (Reese et al., 1997). Enzymatic removal of CSPGs abolishes chronotopy, as young axons, no longer inhibited by the CSPG border, run throughout the entire width of the tract (Leung et al., 2003). Heparan sulfate proteoglycans (HSPGs) also contribute extraaxonal cues to organize axons in their tracts. In the zebrafish retinotectal system, HSPGs act non-cell-autonomously on RGC axons to correct naturally occurring pathfinding errors prior to axon entry into the tectum (Poulain and Chien, 2013).

Other evidence supporting a role for glia in organizing axons within tracts remains largely circumstantial. For example, in ferrets, interfascicular glia predominate in the optic nerve, radial glia in the optic chiasm, and in the postchiasm optic tract, glia cluster along the pia (Colello and Guillery, 1992; Guillery and Walsh, 1987). The changing glial profiles are presumed to be involved in the different modes of axon organization at each step of the pathway. Guillery and Walsh (1987) suggest that age-related order in the mouse optic tract arises from radial glia at the chiasm guiding growth cones to the pial surface, whereas the relative lack of age-related axon order in the optic nerve may be due to the preponderance of interfascicular glia found there. Similarly, in fish, the optic nerve/optic tract boundary is demarcated by a sharp change in glia morphology and molecular profile: glia in the nerve express a vimentin-like protein and those in the tract express glial fibrillary acidic protein (GFAP). This change in glial profile spatially coincides with a marked re-sorting of retinal axons into their final topographic order, again providing indirect evidence that the change in glial environment has involved these axon rearrangements (Maggs and Scholes, 1986).

Additionally, work in the antennal lobe of the moth *Manduca sexta* has demonstrated a role for glia in organizing OSN axons. Direct effects of the sorting zone glia on OSN axons have been demonstrated *in vitro* (Tucker et al., 2004), and electron microscopy analysis shows a close association between OSN axon growth cones and glia in the glial-rich sorting region where axons rearrange prior to glomerulus entry (Oland et al., 1998). Subsequent experiments demonstrated the necessity of glia in this sorting region, as OSN axons in glia-deficient moths fail to correctly fasciculate or target glomeruli appropriately (Rossler et al., 1999).

In the vertebrate olfactory system, olfactory ensheathing glia (OEG) are positioned favorably to play a similar organizational role as glia do in the insect antennal lobe (reviewed in Tolbert et al., 2004) (Figure 13.2B). OEG surround OSN axons and extend cytoplasmic processes in between axon fascicles, heterogeneously expressing several guidance molecules, including ephrin-B2, NCAM, GAP-43, laminin, HSPGs, Sema3A, and a host of growth factors (reviewed in Chuah and West, 2002). Furthermore, OEG precede OSN axon fascicles growing into the OB, and could thus provide guidance or organizational cues to leading OSN growth cones (Chuah and West, 2002; Tolbert et al., 2004). Loss-of-function studies in Sema3A mutant mice specifically implicate OEG in pretarget organization of olfactory axons. Sema3A is expressed by OEG in the developing mouse olfactory system, and Nrp1-expressing OSN axons misroute and mistarget in Sema3A knockout mice (Schwarting et al., 2000). Axon–axon interactions are implicated in this scenario too, as Nrp1-negative axons, i.e., those that do not respond to Sema3A, also suffer misrouting defects in the OB in Sema3A mutants (Schwarting et al., 2000).

Recently, attention has turned to potential noncanonical roles of microglia in tract formation. Perturbing microglia activity leads to marked defasciculation of callosal axons and subsequent disturbance in the dorsal–ventral order in the CC (Pont-Lezica et al., 2014). Strikingly, microglia in the embryonic mouse brain at E14.5 cluster at decision points of axon tracts, displaying a distinctly different distribution than at E12.5 or postnatally, when microglia are more evenly spread throughout the brain (Squarzoni et al., 2014). Eliminating or immune-inactivating microglia in the embryonic brain produces wiring defects in dopaminergic axon outgrowth and positioning of neocortical interneurons (Squarzoni et al., 2014). These findings provide compelling evidence for a novel role of microglia in tract formation.

13.3.2 Axon-Intrinsic Cues

13.3.2.1 Fasciculation and Homotypic Axon–Axon Interactions

Despite the fact that fasciculation has long been considered a fundamental mechanism of neural circuit development (reveiwed in Raper and Mason, 2010; Wang and Marquardt, 2013), our understanding of the mechanisms underlying axon–axon interactions and how they contribute to sorting of axons in tracts remains limited. In their review of axon–axon interactions, Wang and Marquardt (2013) offer an insightful distinction between homotypic and heterotypic axon fasciculation, which we will similarly use to guide the present review of axon–axon interactions involved in pretarget axon organization (Figure 13.2C–E). We first consider homotypic fasciculation situations where like axons, those of the same source or subset, are joined together by adhesive forces or corralled together by surround repulsion.

Fasciculation of homotypic axons (Figure 13.2C) has been studied most commonly in the pioneer–follower model of early tract formation. Pioneer axons provide a path along which younger axons follow, but while loss of pioneer axons often disturbs tract formation, follower axons in some systems can successfully navigate to their targets in the absence of pioneers (Raper and Mason, 2010). An elegant set of experiments in the developing zebrafish retinotectal system provides a case in which pioneer RGC axons are both necessary and sufficient for leading later-born

RGC axons along the correct path to the optic tectum (Pittman et al., 2008). The results of this study also imply that pioneer–follower RGC axon–axon interactions trump extraaxonal cues along the tract. Transplanting mutant pioneer RGCs from zebrafish lacking the Slit receptor Robo2 into wt retinae caused wt axons to make routing errors alongside transplanted mutant pioneers. Conversely, transplanting wt pioneer RGCs into mutant retinae rescued many host axons from making pathfinding errors (Pittman et al., 2008). These results advocate for the importance of direct axon–axon interactions in generating organized vertebrate axon tracts. Slit and Robo are also involved in motor axon fasciculation in the mouse, where Slit2 directly promotes fasciculation via interactions with Robo1 and Robo2 by an autocrine/juxtaparacrine mechanism (Jaworski and Tessier-Lavigne, 2012).

Surround repulsion can also lead to homotypic fasciculation by corralling like axons together away from external repulsive signals. Despite the fact that such surround-repulsion scenarios result from extraaxonal cues, we consider them in this section, as an increase in homotypic fasciculation is the end result. For instance, motor axons expressing Nrp1 fasciculate via surround repulsion from Sema3A in the mouse forelimb. In the absence of Sema3A, motor axons defasciculate and subsequently make errors in dorsal–ventral pathfinding in the limb plexus region (Huber et al., 2005). On the other hand, Sema3F–Nrp2 interactions do not affect fasciculation of motor axons in the limb (Huber et al., 2005). They do, however, drive fasciculation of vomeronasal axons in the murine accessory olfactory system, although they are not involved in pathfinding in this system (Cloutier et al., 2002).

The Eph:ephrin family of cell surface proteins is also involved in mediating selective homotypic fasciculation of axon cohorts via selective surround repulsion. EphB2-expressing sensory and motor axons fasciculate via surround repulsion from ephrin-B1 produced in the limb bud mesenchyme. Ephrin-B1 expression on sensory axons indicates that a more complicated interaction may be involved in addition to the surround-repulsion from the mesenchyme (Luxey et al., 2013). Loss of ephrin-B1 in mice leads to defasciculation of axon tracts, especially the oculomotor nerve (Davy et al., 2004); and in double knockout mutants lacking both EphB2 and EphB3, axons extending from the habenular nuclei extensively defasciculate. Fasciculation is normal, however, in each single knockout mutant (Orioli et al., 1996). Whether appropriate maintenance of homotypic fasciculation is necessary for appropriate innervation and synaptogenesis in the target remains to be elucidated.

13.3.2.2 Heterotypic Axon–Axon Interactions

Axon cohorts in tracts can be either grossly or finely heterotypic (Figure 13.2D–E). This section will first consider grossly heterotypic axon–axon interactions, prime examples of which include TCAs and CTAs in the thalamocortical tract, or sensory and motor spinal axons in the periphery, where two cohorts from different sources run segregated from each other in a tract containing reciprocal projections (Figure 13.2D). Finely heterotypic axon–axon interactions are those occurring between subsets of axons arising from the same source but representing different molecularly defined neuronal subtypes; these cases will be considered at the end of this section. Common guidance molecules participate in both types of heterotypic axon–axon interactions, including the Eph:ephrin family, Semas, Plexins, and Neuropilins.

In the peripheral nerve tracts, motor axons extend before sensory axons, which in turn rely on motor axons to form correct patterns of connections (Wang and Marquardt, 2013). While these two heterotypic axon cohorts run closely together, sensorimotor function requires their appropriate segregation. In the axial nerves, this segregation has been found to depend on repulsive transaxonal signaling between motor axon EphA3/EphA4 receptors and sensory axon ephrin-A ligands. Perturbing ephrinA–EphA forward signaling accordingly results in severe wiring defects in the peripheral nerves (Gallarda et al., 2008). However, EphA3/4 and ephrin-As are also required for sensory axons to track along preformed motor axon pathways in the limb (Wang et al., 2011). This curious dual action of EphAs/ephrin-As relies on the forward and reverse signaling capabilities of the binding partners; namely, segregation of motor and sensory axons results from forward signaling and sensory axon tracking along motor axons utilizes reverse signaling (Wang et al., 2011). The reliance on motor axons for sensory axon pathfinding was further demonstrated by assessing the sensory axon routes taken in two mouse strains that display motor axon pathfinding errors (EphA4 and Ret mutants). In both cases, as well as in analogous zebrafish mutants, sensory axons mimicked the pathfinding errors made by motor axons (Wang et al., 2014).

The reciprocal projections comprising the thalamocortical pathway are the other prime example of grossly heterotypic axon–axon interactions in developing tracts. Recent work demonstrates a mutual reliance of TCAs and CTAs on each other for successful tract organization and navigation to their respective targets (e.g., Chen et al., 2012; Deck et al., 2013), and some of the molecules involved in these axon–axon interactions have been identified. The CTA waiting period, as discussed in Section 13.3.1.1 is governed by Sema3E/Plexin-D1. If Sema3E/Plexin-D1 signaling is perturbed, all corticofugal axons pass the PSPB prematurely, bypassing their encounter with TCAs. They subsequently take a corticosubcerebral trajectory, rather than dividing into the appropriate corticothalamic and corticosubcerebral paths (Deck et al., 2013). This finding highlights the necessity of the subpallial "handshake" between thalamic and cortical axons. In the converse scenario, when CTA outgrowth is prevented, TCAs are rendered incapable of navigating across the PSPB and into the cortex (Chen et al., 2012). This mutual reliance is suggestive of heterotypic axon–axon interactions or fasciculation occurring in the PSPB and along the tract. Curiously, TCAs and CTAs collapse upon contact with each other *in vitro*, and thus are not intrinsically inclined to cofasciculate (Bagnard et al., 2001). From this, the "handshake" between TCAs and CTAs is likely mediated by a combination of axon-extrinsic cues in the surrounding ventral telencephalon, which may in turn affect direct axon–axon interactions between the two axon populations. The intimate association between thalamic and cortical axons in the IC and along the tract (Molnar et al., 1998) implicates cofasciculation between these heterotypic axon cohorts. Likewise, their codependency on each other for guidance, and the fact that a subset of CTAs track along misrouted TCAs in EphB1/2 double knockout mice (Robichaux et al., 2014) provides further support to a model in which direct axon–axon interactions are involved in organization and guidance of these axonal cohorts.

The Eph:ephrin family may be involved in such an interaction. In the primary olfactory nerve of the moth *Manduca sexta*, Eph receptors and ephrins are differentially expressed among glomeruli and are candidates for mediating OSN axon sorting prior to glomerulus innervation (Kaneko and Nighorn, 2003). Importantly, there is no detectable Eph or ephrin expression intrinsic to the OSN axons' target, the antennal lobe, signifying that Eph–ephrin repulsive signaling involved in sorting axons would necessarily occur via direct axon–axon interactions (Kaneko and Nighorn, 2003).

Eph–ephrin-mediated sorting of moth OSN axons involves a different kind of heterotypy than has been considered thus far. As opposed to the grossly heterotypic interactions of reciprocal projections in the spinal cord or thalamus, in finely heterotypic axon–axon interactions, axons originating from the same source must segregate into appropriate subfascicles representing different neuronal subtypes prior to target innervation. Sorting of axons into subfascicles in a tract is thought to occur by selective inhibition of fasciculation, such that individual cohorts can separate from a larger group and select the correct trajectory to the appropriate target (Figure 13.2E). An example comes from the chick spinal cord marginal zone, where intermediate and medial longitudinal commissural (ILc and MLc) axons run parallel in the lateral and ventral funiculi, respectively. ILc axons express Robo, which is thought to inhibit N-cadherin-driven fasciculation in order to separate ILc axons away from the MLc fascicle. Accordingly, loss of Robo function leads to ILc axons aberrantly coursing alongside MLc axons in the ventral funiculus and subsequently mistargeting in the brain (Sakai et al., 2012).

Semas and their Nrp and Plexin receptors are also involved in sorting axons within tracts. In the mouse OE, different populations of OSN axons express Sema3A and Nrp1 in a roughly regionally defined manner. These OSN axons are likewise segregated from each other in the olfactory nerve, and selectively deleting either Nrp1 or Sema3A in OSNs leads to a blurring of pretarget axon order and shifts in glomerular position in the OB (Imai et al., 2009). Because these effects are found in conditional mutants that only affect subsets of OSNs, they occur independently of Sema3A surround expression by OEG. Given this result, the authors conclude that axon–axon interactions are critical for establishing pretarget axon organization in the olfactory nerve (Imai et al., 2009).

Sema3A and Nrp1 are also implicated in establishing the dorsal–ventral segregation of axons in the CC (Zhou et al., 2013). This order reflects a medial-high to lateral-low gradient of Nrp1 expression in the cortex and a reciprocal, though broader expression of Sema3A, which is slightly higher in the lateral cortex. Disrupting these molecular gradients perturbs both axon position in the CC and the position of axon terminations in the contralateral cortex. The authors propose that both axon-extrinsic and axon–axon interactions are at play (Zhou et al., 2013). It is intriguing to speculate that Sema/Nrp interactions may be a conserved mechanism in mediating axon sorting across other systems as well.

13.3.2.3 Activity-Driven Axon Interactions

Neural activity, both spontaneous and activity-dependent, is a known factor in synaptic maturation and refinement of axonal projections in their targets (reviewed in Zhang and Poo, 2001), but it may also play a role in the development and organization of axon tracts. *In vitro* experiments demonstrate that neuronal firing patterns regulate expression levels of CAMs in dorsal root ganglion neurons (Itoh et al., 1997). In the olfactory system, single-cell microarrays of OSNs showed that OR-driven cyclic adenosine monophosphate (cAMP) levels complementarily regulate Nrp1 and Sema3A expression, two key players in modulating OSN axon organization in the olfactory nerve (discussed above, Imai et al., 2009). Furthermore, both pretarget axon order in the olfactory nerve and glomerular patterning in the OB are perturbed in mice lacking adenylyl cyclase 3, a key enzyme involved in the production of cAMP (Miller et al., 2010).

A series of studies in the chick spinal cord demonstrated the relationship between spontaneous bursting activity and both pathfinding decisions and axon fasciculation (reviewed in Hanson et al., 2008). Pharmacological blockade of neurotransmitters governing early bursting activity leads to dorsal–ventral pathfinding errors of spinal motor axons, while slight increases in bursting rate disrupt motor neuron pool specific fasciculation and targeting choices. Again, these effects are likely mediated by activity-dependent changes in guidance molecule expression patterns (Hanson et al., 2008). While blocking evoked synaptic transmission during development affects neither the pretarget axon topography nor subsequent targeting of TCAs (Molnar et al., 2002), spontaneous activity in thalamic neurons mediates TCA growth rates via transcriptional regulation of Robo1 (Mire et al., 2012). Thus, different types of neuronal activity may have highly specific effects on the expression of guidance molecules and thereby influence the organization of axon tracts.

13.3.3 Tracts and Targeting

Many studies, including OB (St John et al., 2003) or optic tectum removal experiments (Reh et al., 1983), indicate that pretarget axon organization occurs largely independently of target-derived cues. However, the functional relevance of pretarget tract organization in circuit formation is still unclear, specifically whether or not tract order is necessary for appropriately specific synaptic connections of terminals in their targets. Some of the studies discussed above examine the relationship between axon order in tracts and subsequent order in the targets and provide hints that pre-sorting of axons in tracts is in fact necessary for accurate target innervation.

In the olfactory system, selectively perturbing Nrp1 or Sema3A expression in a subset of OSNs disrupts axon order in consistent directions along the central-peripheral axis of the olfactory nerve and produces concomitant shifts in glomerular position in the OB (Imai et al., 2009). Similarly, in the CC, constitutive knockout of Sema3A or selective knockout of Nrp1 in the motor cortex both lead to a blurring of dorsal–ventral axon segregation in the CC (Zhou et al., 2013). More strikingly, there is a linear relationship between dorsal–ventral position of an axon in the CC and its subsequent medial–lateral position in the target contralateral cortex. This linear relationship between tract position and targeting is true both in the wt context as well as in the experimental manipulations of CC axon order, leading the authors to conclude that axon position in the tract is a determining factor in establishing correct terminations within the target. Importantly, postnatal refinement is largely unable to correct mistargeted axons in the contralateral cortex, underscoring the significance of early axon targeting decisions even in systems with known refinement mechanisms at play (Zhou et al., 2013).

Findings from other systems also support the importance of pretarget axon order for fidelity of axon targeting. Blocking Robo1 signaling in chick spinocerebellar neurons causes their axons to aberrantly fasciculate with other ascending spinal axons in the medial rather than lateral funiculus of the spinal cord marginal zone. This incorrect positioning in the tract results in the axons bypassing the cerebellum and misprojecting instead to other hindbrain targets (Sakai et al., 2012). In the chick retinotectal system, blocking NCAM expression in focal subsets of RGCs causes affected axons to run in ectopically diffuse positions in the optic nerve and tract. While roughly 15% of these ectopic axons are able to make dramatic course corrections to reach the appropriate location once in the tectum, the rest project incorrectly (Thanos et al., 1984). More direct evidence comes from the thalamocortical tract, where many cues within the subpallium organize ascending TCAs. Selectively perturbing subpallium development blurs the topography of extending TCAs, leading to similarly blurred topography in the somatosensory cortex. An elegant set of control experiments substantiates the conclusion that the targeting defects are a direct result of defective pretarget axon order, rather than aberrant cues in either the thalamus or cortex (Lokmane et al., 2013).

On the other hand, not all studies corroborate the importance of axon pre-sorting for accuracy of target innervation. For one, plasticity is a hallmark of developing neural circuits, so pretarget errors may be correctable once axons invade the target. Disrupted HSPG synthesis in the zebrafish optic tract, for instance, leads to axon disorganization in the optic tract, but RGC axons still appear to innervate the optic tectum in topographically correct positions (Lee et al., 2004). In the mouse periphery, loss of ephrin-B1 leads to defasciculation of motor and sensory axons, but motor axons innervate the limb in grossly normal patterns (Luxey et al., 2013). Further, while much evidence in the thalamocortical system supports a role for axon pre-sorting in establishing accurate connections, cues within the cortex also drive guidance and targeting behavior (reviewed in Garel and Lopez-Bendito, 2014). When the arealization of the cortex is perturbed, for example, TCAs retain a normal topography in the subpallium but undergo extensive rearrangements within the cortex to innervate the appropriate target region (Shimogori and Grove, 2005). It is worth noting, however, that this study assessed gross targeting outcomes but not maturation of synaptic connections. As discussed previously, Michalski et al. (2013) demonstrated in the auditory system the degree of subtlety in downstream effects of pathfinding perturbations. Thus, while the TCAs innervating a disorganized cortex grossly find their correct targets, it is possible that subtle defects persist at the synaptic level. The field is beginning to piece together the puzzle of how each step of circuit development affects subsequent steps along the pathway.

13.4 CONCLUSIONS

In 1956, very early in the history of considering pretarget axon sorting, Barnard and Woolsey reflected upon the trajectory of the field, noting a shift from focusing on "origins and terminations of the major fiber tracts," toward examining "intratract localization of one sort or another." They contested the false dichotomy that axons must either maintain highly specific neighbor relationships in order to innervate a target appropriately or else they course along their pathways haphazardly until arriving at their destination (Barnard and Woolsey, 1956). The last several decades of research highlight the complexity and subtlety of many developmental and guidance mechanisms in growing neural circuits. Taking stock of the current understanding of intratract axon organization, we find that (i) topographic, chronotopic, and typographic order of axons are common modes of pretarget axon sorting (Figure 13.1) in white matter tracts, and (ii) molecular mechanisms mediating these modes of order fall generally into axon-extrinsic and axon-intrinsic categories (Figure 13.2). Glia, transient or migrating neuronal populations, and the ECM provide organizational cues to axons within the tracts. Homotypic and heterotypic axon–axon interactions are proving to be important components in organizing developing tracts, though the detailed mechanisms of these interactions are still somewhat elusive. Finally, current evidence strongly supports a role for pretarget axon order in creating accurate connections in targets, though plasticity in developing circuits is capable of mending early wiring errors to some degree.

13.4.1 Challenges and Perspectives

The axon guidance field has historically divided neural circuits into experimentally manageable segments, focusing on the transcriptional regulation of the source neurons, axon guidance decisions at choice points or intermediate targets, and terminal guidance and refinement events within targets. As our understanding of each segment of the neural pathways improves, so too have the available experimental tools, opening up new avenues of inquiry into the guidance mechanisms within tracts coursing between the segments so far studied. These new avenues of inquiry of course uncover greater overall complexity in developing neural circuits. Determining whether pretarget axon organization is a mere consequence of either a neuron's transcriptional identity or events at choice points, as opposed to an active step in organizing the overall circuit, remains the greatest challenge. In other words, are guidance events and axon–axon interactions within the tract instructive for subsequent targeting events, or are local guidance cues within the target sufficient for ordering incoming axons, so long as they reach the target? Additionally, it is unclear if defasciculation in a tract necessarily leads to tract and target disruptions, or if other organizational mechanisms can compensate for disordered fasciculation within tracts.

One particular challenge in answering these questions is the fact that the same molecules may be acting at intermediate targets, between axons (fasciculation and repulsion), and in the target (between axons and target cells). New approaches will be needed in order to fully understand how each element of the pathway contributes to final targeting events. In this regard, novel *in vitro* assays may assist in teasing apart the relative contributions of guidance molecules at choice points and within tracts. In tracts with less-well-understood pretarget axon organization, advanced labeling techniques like Brainbow (Lu et al., 2009) or electroporation of fluorescent proteins (Saito and Nakatsuji, 2001) along with tissue clearing methods to enhance visualization (e.g., Erturk et al., 2012; Kuwajima et al., 2013; Tomer et al., 2014) will be particularly useful in more fully detailing modes of order within tracts. Additionally, modern molecular genetic techniques now allow for the selective perturbation of gene expression in specific neuronal subpopulations or at particular points along an axon's course, which will allow us to better comprehend how each part of the pathway contributes to formation of the whole circuit.

Understanding the subtleties of white matter tract formation by studying pretarget axon order in normal development is a necessary step toward unraveling the basis of neurodevelopmental disorders. Diffusion tensor imaging studies have found white matter tract abnormalities in subjects with autism spectrum disorder (Wolff et al., 2012) and schizophrenia (Kubicki et al., 2007). A better grasp on the principles of tract development and organization, and the relationship between tracts and targets, is necessary to more effectively study and treat developmental disorders of neural circuitry. Given the multistepwise process of circuit formation, perturbations in pretarget order may not necessarily produce drastic shifts in targeting, and as such could appear relatively inconsequential to the overall order of the circuit. However, the devil is likely to be in the details, as it were, and the effects such perturbations have on targeting are more likely to be subtle and possibly even undetectable except at the level of individual synapses. Examining developing axon tracts with a keen eye toward such subtle aberrations in circuit formation and function is central to unlocking underlying mechanisms of complex developmental brain disorders.

Abbreviations

cAMP Cyclic adenosine monophosphate
CC Corpus callosum
CSPG Chondroitin sulfate proteoglycan
CTA Corticothalamic axon
dLGN Dorsal lateral geniculate nucleus
ECM Extracellular matrix
HSPG Heparan sulfate proteoglycan
IC Internal capsule
ILc Intermediate longitudinal commissural
LOT Lateral olfactory tract
MLc Medial longitudinal commissural
NL Nucleus laminaris
NM Nucleus magnocellularis
Nrp Neuropilin
OB Olfactory bulb
OE Olfactory epithelium
OEG Olfactory ensheathing glia
OR Olfactory receptor
OSN Olfactory sensory neuron
PSPB Pallial/subpallial boundary
RGC Retinal ganglion cell
SC Superior colliculus
Sema Semaphorin
TCA Thalamocortical axons
wt Wild-type
XDCT Crossed dorsal cochlear tract

References

Aboitiz, F., Scheibel, A.B., Fisher, R.S., Zaidel, E., 1992. Fiber composition of the human corpus callosum. Brain Res. 598 (1–2), 143–153.

Appler, J.M., Goodrich, L.V., 2011. Connecting the ear to the brain: molecular mechanisms of auditory circuit assembly. Prog. Neurobiol. 93 (4), 488–508. http://dx.doi.org/10.1016/j.pneurobio.2011.01.004

Attardi, D.G., Sperry, R.W., 1963. Preferential selection of central pathways by regenerating optic fibers. Exp. Neurol. 7, 46–64.

Badia, J., Pascual-Font, A., Vivo, M., Udina, E., Navarro, X., 2010. Topographical distribution of motor fascicles in the sciatic-tibial nerve of the rat. Muscle Nerve 42 (2), 192–201. http://dx.doi.org/10.1002/mus.21652

Bagnard, D., Chounlamountri, N., Puschel, A.W., Bolz, J., 2001. Axonal surface molecules act in combination with semaphorin 3a during the establishment of corticothalamic projections. Cereb. Cortex 11 (3), 278–285.

Baker, G.E., Jeffery, G., 1989. Distribution of uncrossed axons along the course of the optic nerve and chiasm of rodents. J. Comp. Neurol. 289 (3), 455–461. http://dx.doi.org/10.1002/cne.902890309

Barnard, J.W., Woolsey, C.N., 1956. A study of localization in the cortico-spinal tracts of monkey and rat. J. Comp. Neurol. 105 (1), 25–50.

Bernardo, K.L., Woolsey, T.A., 1987. Axonal trajectories between mouse somatosensory thalamus and cortex. J. Comp. Neurol. 258 (4), 542–564. http://dx.doi.org/10.1002/cne.902580406

Bhansali, P., Rayport, I., Rebsam, A., Mason, C., 2014. Delayed neurogenesis leads to altered specification of ventrotemporal retinal ganglion cells in albino mice. Neural. Dev. 9, 11. http://dx.doi.org/10.1186/1749-8104-9-11

Bielle, F., Marcos-Mondejar, P., Leyva-Diaz, E., Lokmane, L., Mire, E., Mailhes, C., et al., 2011. Emergent growth cone responses to combinations of Slit1 and Netrin 1 in thalamocortical axon topography. Curr. Biol. 21 (20), 1748–1755. http://dx.doi.org/10.1016/j.cub.2011.09.008

Blackshaw, S., Harpavat, S., Trimarchi, J., Cai, L., Huang, H., Kuo, W.P., et al., 2004. Genomic analysis of mouse retinal development. PLoS Biol. 2 (9), E247. http://dx.doi.org/10.1371/journal.pbio.0020247

Bulfone, A., Wang, F., Hevner, R., Anderson, S., Cutforth, T., Chen, S., et al., 1998. An olfactory sensory map develops in the absence of normal projection neurons or GABAergic interneurons. Neuron 21 (6), 1273–1282.

Caminiti, R., Carducci, F., Piervincenzi, C., Battaglia-Mayer, A., Confalone, G., Visco-Comandini, F., et al., 2013. Diameter, length, speed, and conduction delay of callosal axons in macaque monkeys and humans: comparing data from histology and magnetic resonance imaging diffusion tractography. J. Neurosci. 33 (36), 14501–14511. http://dx.doi.org/10.1523/JNEUROSCI.0761-13.2013

Caviness Jr., V.S., Frost, D.O., 1980. Tangential organization of thalamic projections to the neocortex in the mouse. J. Comp. Neurol. 194 (2), 335–367. http://dx.doi.org/10.1002/cne.901940205

Chan, S.O., Chung, K.Y., 1999. Changes in axon arrangement in the retinofugal [correction of retinofungal] pathway of mouse embryos: confocal microscopy study using single- and double-dye label. J. Comp. Neurol. 406 (2), 251–262.

Chan, S.O., Guillery, R.W., 1994. Changes in fiber order in the optic nerve and tract of rat embryos. J. Comp. Neurol. 344 (1), 20–32. http://dx.doi.org/10.1002/cne.903440103

Charron, F., Stein, E., Jeong, J., McMahon, A.P., Tessier-Lavigne, M., 2003. The morphogen sonic hedgehog is an axonal chemoattractant that collaborates with netrin-1 in midline axon guidance. Cell 113 (1), 11–23.

Chen, Y., Magnani, D., Theil, T., Pratt, T., Price, D.J., 2012. Evidence that descending cortical axons are essential for thalamocortical axons to cross the pallial-subpallial boundary in the embryonic forebrain. PLoS One 7 (3), e33105. http://dx.doi.org/10.1371/journal.pone.0033105

Chuah, M.I., West, A.K., 2002. Cellular and molecular biology of ensheathing cells. Microsc. Res. Tech. 58 (3), 216–227. http://dx.doi.org/10.1002/jemt.10151

Clark, W.E., 1951. The projection of the olfactory epithelium on the olfactory bulb in the rabbit. J. Neurol. Neurosurg. Psychiatry 14 (1), 1–10.

Cloutier, J.F., Giger, R.J., Koentges, G., Dulac, C., Kolodkin, A.L., Ginty, D.D., 2002. Neuropilin-2 mediates axonal fasciculation, zonal segregation, but not axonal convergence, of primary accessory olfactory neurons. Neuron 33 (6), 877–892.

Colello, R.J., Guillery, R.W., 1990. The early development of retinal ganglion cells with uncrossed axons in the mouse: retinal position and axonal course. Development 108 (3), 515–523.

Colello, R.J., Guillery, R.W., 1992. Observations on the early development of the optic nerve and tract of the mouse. J. Comp. Neurol. 317 (4), 357–378. http://dx.doi.org/10.1002/cne.903170404

Colello, S.J., Guillery, R.W., 1998. The changing pattern of fibre bundles that pass through the optic chiasm of mice. Eur. J. Neurosci. 10 (12), 3653–3663.

Cook, J.E., Horder, T.J., 1977. The multiple factors determining retinotopic order in the growth of optic fibres into the optic tectum. Philos. Trans. R. Soc. Lond., B, Biol. Sci. 278 (961), 261–276.

Costanzo, R.M., O'Connell, R.J., 1978. Spatially organized projections of hamster olfactory nerves. Brain Res. 139 (2), 327–332.

Davy, A., Aubin, J., Soriano, P., 2004. Ephrin-B1 forward and reverse signaling are required during mouse development. Genes Dev. 18 (5), 572–583. http://dx.doi.org/10.1101/gad.1171704

de Lacoste, M.C., Kirkpatrick, J.B., Ross, E.D., 1985. Topography of the human corpus callosum. J. Neuropathol. Exp. Neurol. 44 (6), 578–591.

Deck, M., Lokmane, L., Chauvet, S., Mailhes, C., Keita, M., Niquille, M., et al., 2013. Pathfinding of corticothalamic axons relies on a rendezvous with thalamic projections. Neuron 77 (3), 472–484. http://dx.doi.org/10.1016/j.neuron.2012.11.031

Demyanenko, G.P., Riday, T.T., Tran, T.S., Dalal, J., Darnell, E.P., Brennaman, L.H., et al., 2011. NrCAM deletion causes topographic mistargeting of thalamocortical axons to the visual cortex and disrupts visual acuity. J. Neurosci. 31 (4), 1545–1558. http://dx.doi.org/10.1523/JNEUROSCI.4467-10.2011

Dhande, O.S., Huberman, A.D., 2014. Retinal ganglion cell maps in the brain: implications for visual processing. Curr. Opin. Neurobiol. 24 (1), 133–142. http://dx.doi.org/10.1016/j.conb.2013.08.006

Dodd, J., Morton, S.B., Karagogeos, D., Yamamoto, M., Jessell, T.M., 1988. Spatial regulation of axonal glycoprotein expression on subsets of embryonic spinal neurons. Neuron 1 (2), 105–116.

Dufour, A., Seibt, J., Passante, L., Depaepe, V., Ciossek, T., Frisen, J., et al., 2003. Area specificity and topography of thalamocortical projections are controlled by ephrin/Eph genes. Neuron 39 (3), 453–465.

Ehrlich, D., Mark, R., 1984. The course of axons of retinal ganglion cells within the optic nerve and tract of the chick (*Gallus gallus*). J. Comp. Neurol. 223 (4), 583–591. http://dx.doi.org/10.1002/cne.902230409

Erskine, L., Herrera, E., 2014. Connecting the retina to the brain. ASN Neuro 6 (6). http://dx.doi.org/10.1177/1759091414562107

Erskine, L., Reijntjes, S., Pratt, T., Denti, L., Schwarz, Q., Vieira, J.M., et al., 2011. VEGF signaling through neuropilin 1 guides commissural axon crossing at the optic chiasm. Neuron 70 (5), 951–965. http://dx.doi.org/10.1016/j.neuron.2011.02.052

Erturk, A., Becker, K., Jahrling, N., Mauch, C.P., Hojer, C.D., Egen, J.G., et al., 2012. Three-dimensional imaging of solvent-cleared organs using 3DISCO. Nat. Protoc. 7 (11), 1983–1995. http://dx.doi.org/10.1038/nprot.2012.119

Fawcett, J.W., Taylor, J.S., Gaze, R.M., Grant, P., Hirst, E., 1984. Fibre order in the normal *Xenopus* optic tract, near the chiasma. J. Embryol. Exp. Morphol. 83, 1–14.

Feinstein, P., Mombaerts, P., 2004. A contextual model for axonal sorting into glomeruli in the mouse olfactory system. Cell 117 (6), 817–831. http://dx.doi.org/10.1016/j.cell.2004.05.011

Gallarda, B.W., Bonanomi, D., Muller, D., Brown, A., Alaynick, W.A., Andrews, S.E., et al., 2008. Segregation of axial motor and sensory pathways via heterotypic trans-axonal signaling. Science 320 (5873), 233–236. http://dx.doi.org/10.1126/science.1153758

Garel, S., Lopez-Bendito, G., 2014. Inputs from the thalamocortical system on axon pathfinding mechanisms. Curr. Opin. Neurobiol. 27, 143–150. http://dx.doi.org/10.1016/j.conb.2014.03.013

Garel, S., Yun, K., Grosschedl, R., Rubenstein, J.L., 2002. The early topography of thalamocortical projections is shifted in Ebf1 and Dlx1/2 mutant mice. Development 129 (24), 5621–5634.

Ghosh, A., Shatz, C.J., 1992. Pathfinding and target selection by developing geniculocortical axons. J. Neurosci. 12 (1), 39–55.

Godement, P., Salaun, J., Imbert, M., 1984. Prenatal and postnatal development of retinogeniculate and retinocollicular projections in the mouse. J. Comp. Neurol. 230 (4), 552–575. http://dx.doi.org/10.1002/cne.902300406

Goldberg, S., Coulombre, A.J., 1972. Topographical development of the ganglion cell fiber layer in the chick retina. A whole mount study. J. Comp. Neurol. 146 (4), 507–518. http://dx.doi.org/10.1002/cne.901460406

Graziadei, P.P., Levine, R.R., Graziadei, G.A., 1978. Regeneration of olfactory axons and synapse formation in the forebrain after bulbectomy in neonatal mice. Proc. Natl. Acad. Sci. U.S.A. 75 (10), 5230–5234.

Guillery, R.W., Walsh, C., 1987. Changing glial organization relates to changing fiber order in the developing optic nerve of ferrets. J. Comp. Neurol. 265 (2), 203–217. http://dx.doi.org/10.1002/cne.902650205

Guillery, R.W., Polley, E.H., Torrealba, F., 1982. The arrangement of axons according to fiber diameter in the optic tract of the cat. J. Neurosci. 2 (6), 714–721.

Hanson, M.G., Milner, L.D., Landmesser, L.T., 2008. Spontaneous rhythmic activity in early chick spinal cord influences distinct motor axon pathfinding decisions. Brain Res. Rev. 57 (1), 77–85. http://dx.doi.org/10.1016/j.brainresrev.2007.06.021

Hofer, S., Frahm, J., 2006. Topography of the human corpus callosum revisited—comprehensive fiber tractography using diffusion tensor magnetic resonance imaging. Neuroimage 32 (3), 989–994. http://dx.doi.org/10.1016/j.neuroimage.2006.05.044

Hohl-Abrahao, J.C., Creutzfeldt, O.D., 1991. Topographical mapping of the thalamocortical projections in rodents and comparison with that in primates. Exp. Brain Res. 87 (2), 283–294.

Honig, M.G., Frase, P.A., Camilli, S.J., 1998. The spatial relationships among cutaneous, muscle sensory and motoneuron axons during development of the chick hindlimb. Development 125 (6), 995–1004.

Horton, J.C., Greenwood, M.M., Hubel, D.H., 1979. Non-retinotopic arrangement of fibres in cat optic nerve. Nature 282 (5740), 720–722.

Huber, A.B., Kania, A., Tran, T.S., Gu, C., De Marco Garcia, N., Lieberam, I., et al., 2005. Distinct roles for secreted semaphorin signaling in spinal motor axon guidance. Neuron 48 (6), 949–964. http://dx.doi.org/10.1016/j.neuron.2005.12.003

Huettl, R.E., Soellner, H., Bianchi, E., Novitch, B.G., Huber, A.B., 2011. Npn-1 contributes to axon–axon interactions that differentially control sensory and motor innervation of the limb. PLoS Biol 9 (2), e1001020. http://dx.doi.org/10.1371/journal.pbio.1001020

Imai, T., Yamazaki, T., Kobayakawa, R., Kobayakawa, K., Abe, T., Suzuki, M., et al., 2009. Pre-target axon sorting establishes the neural map topography. Science 325 (5940), 585–590. http://dx.doi.org/10.1126/science.1173596

Itoh, K., Ozaki, M., Stevens, B., Fields, R.D., 1997. Activity-dependent regulation of N-cadherin in DRG neurons: differential regulation of N-cadherin, NCAM, and L1 by distinct patterns of action potentials. J. Neurobiol. 33 (6), 735–748.

Jaworski, A., Tessier-Lavigne, M., 2012. Autocrine/juxtaparacrine regulation of axon fasciculation by Slit-Robo signaling. Nat. Neurosci. 15 (3), 367–369. http://www.nature.com/neuro/journal/v15/n3/abs/nn.3037.html - supplementary-information.

Kaneko, M., Nighorn, A., 2003. Interaxonal Eph-ephrin signaling may mediate sorting of olfactory sensory axons in *Manduca sexta*. J. Neurosci. 23 (37), 11523–11538.

Kania, A., 2014. Spinal motor neuron migration and the significance of topographic organization in the nervous system. Adv. Exp. Med. Biol. 800, 133–148. http://dx.doi.org/10.1007/978-94-007-7687-6_8

Kashima, D.T., Rubel, E.W., Seidl, A.H., 2013. Pre-target axon sorting in the avian auditory brainstem. J. Comp. Neurol. 521 (10), 2310–2320. http://dx.doi.org/10.1002/cne.23287

Kubicki, M., McCarley, R., Westin, C.F., Park, H.J., Maier, S., Kikinis, R., et al., 2007. A review of diffusion tensor imaging studies in schizophrenia. J. Psychiatr. Res. 41 (1-2), 15–30. http://dx.doi.org/10.1016/j.jpsychires.2005.05.005

Kuwajima, T., Rebsam, A., Bhansali, P., Yoshida, Y., Takegahara, N., Kumanogoh, A., et al., 2010. Nr-CAM, Plexin-A1, and Semaphorin6D are important for contralateral retinal axon guidance through the optic chiasm and in the dLGN. In: Paper Presented at the Society for Neuroscience, San Diego, CA.

Kuwajima, T., Yoshida, Y., Takegahara, N., Petros, T., Kumanogoh, A., Jessell, T.M., et al., 2012. Optic chiasm presentation of Semaphorin6D in the context of Plexin-A1 and Nr-CAM promotes retinal axon midline crossing. Neuron 74 (4), 676–690. doi: http://dx.doi.org/10.1016/j.neuron.2012.03.025.

Kuwajima, T., Sitko, A.A., Bhansali, P., Jurgens, C., Guido, W., Mason, C., 2013. ClearT: a detergent- and solvent-free clearing method for neuronal and non-neuronal tissue. Development 140 (6), 1364–1368. http://dx.doi.org/10.1242/dev.091844

Lance-Jones, C., Landmesser, L., 1981. Pathway selection by chick lumbosacral motoneurons during normal development. Proc. R. Soc. Lond. B, Biol. Sci. 214 (1194), 1–18.

Land, L.J., 1973. Localized projection of olfactory nerves to rabbit olfactory bulb. Brain Res. 63, 153–166.

Landmesser, L., Honig, M.G., 1986. Altered sensory projections in the chick hind limb following the early removal of motoneurons. Dev. Biol. 118 (2), 511–531.

Lee, J.S., von der Hardt, S., Rusch, M.A., Stringer, S.E., Stickney, H.L., Talbot, W.S., et al., 2004. Axon sorting in the optic tract requires HSPG synthesis by ext2 (dackel) and extl3 (boxer). Neuron 44 (6), 947–960. http://dx.doi.org/10.1016/j.neuron.2004.11.029

Leung, K.M., Taylor, J.S., Chan, S.O., 2003. Enzymatic removal of chondroitin sulphates abolishes the age-related axon order in the optic tract of mouse embryos. Eur. J. Neurosci. 17 (9), 1755–1767.

Leyva-Diaz, E., Lopez-Bendito, G., 2013. In and out from the cortex: development of major forebrain connections. Neuroscience 254, 26–44. http://dx.doi.org/10.1016/j.neuroscience.2013.08.070

Leyva-Diaz, E., del Toro, D., Menal, M.J., Cambray, S., Susin, R., Tessier-Lavigne, M., et al., 2014. FLRT3 is a Robo1-interacting protein that determines Netrin-1 attraction in developing axons. Curr. Biol. 24 (5), 494–508. http://dx.doi.org/10.1016/j.cub.2014.01.042

Lokmane, L., Proville, R., Narboux-Neme, N., Gyory, I., Keita, M., Mailhes, C., et al., 2013. Sensory map transfer to the neocortex relies on pre-target ordering of thalamic axons. Curr. Biol. 23 (9), 810–816. http://dx.doi.org/10.1016/j.cub.2013.03.062

Lopez-Bendito, G., Cautinat, A., Sanchez, J.A., Bielle, F., Flames, N., Garratt, A.N., et al., 2006. Tangential neuronal migration controls axon guidance: a role for neuregulin-1 in thalamocortical axon navigation. Cell 125 (1), 127–142. http://dx.doi.org/10.1016/j.cell.2006.01.042

Lu, J., Tapia, J.C., White, O.L., Lichtman, J.W., 2009. The interscutularis muscle connectome. PLoS Biol. 7 (2), e32. http://dx.doi.org/10.1371/journal.pbio.1000032

Lund, R.D., Mustari, M.J., 1977. Development of the geniculocortical pathway in rats. J. Comp. Neurol. 173 (2), 289–306. http://dx.doi.org/10.1002/cne.901730206

Lund, R.D., Lund, J.S., Wise, R.P., 1974. The organization of the retinal projection to the dorsal lateral geniculate nucleus in pigmented and albino rats. J. Comp. Neurol. 158 (4), 383–403. http://dx.doi.org/10.1002/cne.901580403

Luskin, M.B., Price, J.L., 1982. The distribution of axon collaterals from the olfactory bulb and the nucleus of the horizontal limb of the diagonal band to the olfactory cortex, demonstrated by double retrograde labeling techniques. J. Comp. Neurol. 209 (3), 249–263. http://dx.doi.org/10.1002/cne.902090304

Luxey, M., Jungas, T., Laussu, J., Audouard, C., Garces, A., Davy, A., 2013. Eph:ephrin-B1 forward signaling controls fasciculation of sensory and motor axons. Dev. Biol. 383 (2), 264–274. http://dx.doi.org/10.1016/j.ydbio.2013.09.010

Maggs, A., Scholes, J., 1986. Glial domains and nerve fiber patterns in the fish retinotectal pathway. J. Neurosci. 6 (2), 424–438.

Marcus, R.C., Easter Jr., S.S., 1995. Expression of glial fibrillary acidic protein and its relation to tract formation in embryonic zebrafish (*Danio rerio*). J. Comp. Neurol. 359 (3), 365–381. http://dx.doi.org/10.1002/cne.903590302

Michalski, N., Babai, N., Renier, N., Perkel, D.J., Chedotal, A., Schneggenburger, R., 2013. Robo3-driven axon midline crossing conditions functional maturation of a large commissural synapse. Neuron 78 (5), 855–868. http://dx.doi.org/10.1016/j.neuron.2013.04.006

Miller, A.M., Maurer, L.R., Zou, D.J., Firestein, S., Greer, C.A., 2010. Axon fasciculation in the developing olfactory nerve. Neural Dev. 5, 20. http://dx.doi.org/10.1186/1749-8104-5-20

Milner, L.D., Rafuse, V.F., Landmesser, L.T., 1998. Selective fasciculation and divergent pathfinding decisions of embryonic chick motor axons projecting to fast and slow muscle regions. J. Neurosci. 18 (9), 3297–3313.

Mire, E., Mezzera, C., Leyva-Diaz, E., Paternain, A.V., Squarzoni, P., Bluy, L., et al., 2012. Spontaneous activity regulates Robo1 transcription to mediate a switch in thalamocortical axon growth. Nat. Neurosci. 15 (8), 1134–1143. http://dx.doi.org/10.1038/nn.3160

Molnar, Z., Blakemore, C., 1991. Lack of regional specificity for connections formed between thalamus and cortex in coculture. Nature 351 (6326), 475–477. http://dx.doi.org/10.1038/351475a0

Molnar, Z., Blakemore, C., 1995. How do thalamic axons find their way to the cortex? Trends Neurosci. 18 (9), 389–397.

Molnar, Z., Adams, R., Blakemore, C., 1998. Mechanisms underlying the early establishment of thalamocortical connections in the rat. J. Neurosci. 18 (15), 5723–5745.

Molnar, Z., Lopez-Bendito, G., Small, J., Partridge, L.D., Blakemore, C., Wilson, M.C., 2002. Normal development of embryonic thalamocortical connectivity in the absence of evoked synaptic activity. J. Neurosci. 22 (23), 10313–10323.

Molnar, Z., Garel, S., Lopez-Bendito, G., Maness, P., Price, D.J., 2012. Mechanisms controlling the guidance of thalamocortical axons through the embryonic forebrain. Eur. J. Neurosci. 35 (10), 1573–1585. http://dx.doi.org/10.1111/j.1460-9568.2012.08119.x

Molyneaux, B.J., Arlotta, P., Menezes, J.R., Macklis, J.D., 2007. Neuronal subtype specification in the cerebral cortex. Nat. Rev. Neurosci. 8 (6), 427–437. http://dx.doi.org/10.1038/nrn2151

Montgomery, N.M., Tyler, C., Fite, K.V., 1998. Organization of retinal axons within the optic nerve, optic chiasm, and the innervation of multiple central nervous system targets Rana pipiens. J. Comp. Neurol. 402 (2), 222–237.

Naito, J., 1986. Course of retinogeniculate projection fibers in the cat optic nerve. J. Comp. Neurol. 251 (3), 376–387. http://dx.doi.org/10.1002/cne.902510308

Naito, J., 1989. Retinogeniculate projection fibers in the monkey optic nerve: a demonstration of the fiber pathways by retrograde axonal transport of WGA-HRP. J. Comp. Neurol. 284 (2), 174–186. http://dx.doi.org/10.1002/cne.902840203

Nakamura, H., Kanaseki, T., 1989. Topography of the corpus callosum in the cat. Brain Res. 485 (1), 171–175.

Neuhaus-Follini, A., Bashaw, G.J., 2015. Crossing the embryonic midline: molecular mechanisms regulating axon responsiveness at an intermediate target. Wiley Interdiscip Rev. Dev. Biol. http://dx.doi.org/10.1002/wdev.185

Niquille, M., Garel, S., Mann, F., Hornung, J.P., Otsmane, B., Chevalley, S., et al., 2009. Transient neuronal populations are required to guide callosal axons: a role for semaphorin 3C. PLoS Biol. 7 (10), e1000230. http://dx.doi.org/10.1371/journal.pbio.1000230

Niu, J., Ding, L., Li, J.J., Kim, H., Liu, J., Li, H., et al., 2013. Modality-based organization of ascending somatosensory axons in the direct dorsal column pathway. J. Neurosci. 33 (45), 17691–17709. http://dx.doi.org/10.1523/JNEUROSCI.3429-13.2013

Nornes, H.O., Hart, H., Carry, M., 1980. Pattern of development of ascending and descending fibers in embryonic spinal cord of chick. I. Role of position information. J. Comp. Neurol. 192 (1), 119–132. http://dx.doi.org/10.1002/cne.901920108

Oland, L.A., Pott, W.M., Higgins, M.R., Tolbert, L.P., 1998. Targeted ingrowth and glial relationships of olfactory receptor axons in the primary olfactory pathway of an insect. J. Comp. Neurol. 398 (1), 119–138.

Orioli, D., Henkemeyer, M., Lemke, G., Klein, R., Pawson, T., 1996. Sek4 and Nuk receptors cooperate in guidance of commissural axons and in palate formation. EMBO J. 15 (22), 6035–6049.

Osterhout, J.A., El-Danaf, R.N., Nguyen, P.L., Huberman, A.D., 2014. Birthdate and outgrowth timing predict cellular mechanisms of axon target matching in the developing visual pathway. Cell Rep. 8 (4), 1006–1017. http://dx.doi.org/10.1016/j.celrep.2014.06.063

Ozaki, H.S., Wahlsten, D., 1992. Prenatal formation of the normal mouse corpus callosum: a quantitative study with carbocyanine dyes. J. Comp. Neurol. 323 (1), 81–90. http://dx.doi.org/10.1002/cne.903230107

Payne, B.R., Siwek, D.F., 1991. The visual map in the corpus callosum of the cat. Cereb. Cortex 1 (2), 173–188.

Petros, T.J., Rebsam, A., Mason, C.A., 2008. Retinal axon growth at the optic chiasm: to cross or not to cross. Annu. Rev. Neurosci. 31, 295–315. http://dx.doi.org/10.1146/annurev.neuro.31.060407.125609

Petros, T.J., Shrestha, B.R., Mason, C., 2009. Specificity and sufficiency of EphB1 in driving the ipsilateral retinal projection. J Neurosci 29 (11), 3463–3474. http://dx.doi.org/10.1523/JNEUROSCI.5655-08.2009

Petros, T.J., Bryson, J.B., Mason, C., 2010. Ephrin-B2 elicits differential growth cone collapse and axon retraction in retinal ganglion cells from distinct retinal regions. Dev. Neurobiol. 70 (11), 781–794. http://dx.doi.org/10.1002/dneu.20821

Piper, M., Plachez, C., Zalucki, O., Fothergill, T., Goudreau, G., Erzurumlu, R., et al., 2009. Neuropilin 1-Sema signaling regulates crossing of cingulate pioneering axons during development of the corpus callosum. Cereb. Cortex 19 (Suppl. 1), i11–21. http://dx.doi.org/10.1093/cercor/bhp027

Pittman, A.J., Law, M.Y., Chien, C.B., 2008. Pathfinding in a large vertebrate axon tract: isotypic interactions guide retinotectal axons at multiple choice points. Development 135 (17), 2865–2871. http://dx.doi.org/10.1242/dev.025049

Plas, D.T., Lopez, J.E., Crair, M.C., 2005. Pretarget sorting of retinocollicular axons in the mouse. J. Comp. Neurol. 491 (4), 305–319. http://dx.doi.org/10.1002/cne.20694

Pont-Lezica, L., Beumer, W., Colasse, S., Drexhage, H., Versnel, M., Bessis, A., 2014. Microglia shape corpus callosum axon tract fasciculation: functional impact of prenatal inflammation. Eur. J. Neurosci. 39 (10), 1551–1557. http://dx.doi.org/10.1111/ejn.12508

Poulain, F.E., Chien, C.B., 2013. Proteoglycan-mediated axon degeneration corrects pretarget topographic sorting errors. Neuron 78 (1), 49–56. http://dx.doi.org/10.1016/j.neuron.2013.02.005

Powell, A.W., Sassa, T., Wu, Y., Tessier-Lavigne, M., Polleux, F., 2008. Topography of thalamic projections requires attractive and repulsive functions of Netrin-1 in the ventral telencephalon. PLoS Biol. 6 (5), e116. http://dx.doi.org/10.1371/journal.pbio.0060116

Price, J.L., Sprich, W.W., 1975. Observations on the lateral olfactory tract of the rat. J. Comp. Neurol. 162 (3), 321–336. http://dx.doi.org/10.1002/cne.901620304

Rapaport, D.H., Rakic, P., LaVail, M.M., 1996. Spatiotemporal gradients of cell genesis in the primate retina. Perspect. Dev. Neurobiol. 3 (3), 147–159.

Raper, J., Mason, C., 2010. Cellular strategies of axonal pathfinding. Cold Spring Harb. Perspect. Biol. 2 (9). http://dx.doi.org/10.1101/cshperspect.a001933

Reese, B.E., 1987. The distribution of axons according to diameter in the optic nerve and optic tract of the rat. Neuroscience 22 (3), 1015–1024.

Reese, B.E., Baker, G.E., 1993. The re-establishment of the representation of the dorso-ventral retinal axis in the chiasmatic region of the ferret. Vis. Neurosci. 10 (5), 957–968.

Reese, B.E., Cowey, A., 1990. Fibre organization of the monkey's optic tract: I. Segregation of functionally distinct optic axons. J. Comp. Neurol. 295 (3), 385–400. http://dx.doi.org/10.1002/cne.902950304

III. AXON GUIDANCE

Reese, B.E., Thompson, W.F., Peduzzi, J.D., 1994. Birthdates of neurons in the retinal ganglion cell layer of the ferret. J. Comp. Neurol. 341 (4), 464–475. http://dx.doi.org/10.1002/cne.903410404

Reese, B.E., Johnson, P.T., Hocking, D.R., Bolles, A.B., 1997. Chronotopic fiber reordering and the distribution of cell adhesion and extracellular matrix molecules in the optic pathway of fetal ferrets. J. Comp. Neurol. 380 (3), 355–372.

Reh, T.A., Pitts, E., Constantine-Paton, M., 1983. The organization of the fibers in the optic nerve of normal and tectum-less *Rana pipiens*. J. Comp. Neurol. 218 (3), 282–296. http://dx.doi.org/10.1002/cne.902180305

Ressler, K.J., Sullivan, S.L., Buck, L.B., 1993. A zonal organization of odorant receptor gene expression in the olfactory epithelium. Cell 73 (3), 597–609.

Ressler, K.J., Sullivan, S.L., Buck, L.B., 1994. Information coding in the olfactory system: evidence for a stereotyped and highly organized epitope map in the olfactory bulb. Cell 79 (7), 1245–1255.

Robichaux, M.A., Chenaux, G., Ho, H.Y., Soskis, M.J., Dravis, C., Kwan, K.Y., et al., 2014. EphB receptor forward signaling regulates area-specific reciprocal thalamic and cortical axon pathfinding. Proc. Natl. Acad. Sci. U.S.A. 111 (6), 2188–2193. http://dx.doi.org/10.1073/pnas.1324215111

Rossler, W., Oland, L.A., Higgins, M.R., Hildebrand, J.G., Tolbert, L.P., 1999. Development of a glia-rich axon-sorting zone in the olfactory pathway of the moth *Manduca sexta*. J. Neurosci. 19 (22), 9865–9877.

Rubel, E.W., Fritzsch, B., 2002. Auditory system development: primary auditory neurons and their targets. Annu. Rev. Neurosci. 25, 51–101. http://dx.doi.org/10.1146/annurev.neuro.25.112701.142849

Saito, T., Nakatsuji, N., 2001. Efficient gene transfer into the embryonic mouse brain using *in vivo* electroporation. Dev. Biol. 240 (1), 237–246. http://dx.doi.org/10.1006/dbio.2001.0439

Sakai, N., Insolera, R., Sillitoe, R.V., Shi, S.H., Kaprielian, Z., 2012. Axon sorting within the spinal cord marginal zone via Robo-mediated inhibition of N-cadherin controls spinocerebellar tract formation. J. Neurosci. 32 (44), 15377–15387. http://dx.doi.org/10.1523/JNEUROSCI.2225-12.2012

Satoda, M., Takagi, S., Ohta, K., Hirata, T., Fujisawa, H., 1995. Differential expression of two cell surface proteins, neuropilin and plexin, in *Xenopus* olfactory axon subclasses. J. Neurosci. 15 (1 Pt 2), 942–955.

Saucier, D., Astic, L., 1986. Analysis of the topographical organization of olfactory epithelium projections in the rat. Brain Res. Bull. 16 (4), 455–462.

Scholes, J.H., 1979. Nerve fibre topography in the retinal projection to the tectum. Nature 278 (5705), 620–624.

Schwarting, G.A., Kostek, C., Ahmad, N., Dibble, C., Pays, L., Puschel, A.W., 2000. Semaphorin 3A is required for guidance of olfactory axons in mice. J. Neurosci. 20 (20), 7691–7697.

Seibt, J., Schuurmans, C., Gradwhol, G., Dehay, C., Vanderhaeghen, P., Guillemot, F., et al., 2003. Neurogenin2 specifies the connectivity of thalamic neurons by controlling axon responsiveness to intermediate target cues. Neuron 39 (3), 439–452.

Shimogori, T., Grove, E.A., 2005. Fibroblast growth factor 8 regulates neocortical guidance of area-specific thalamic innervation. J. Neurosci. 25 (28), 6550–6560. http://dx.doi.org/10.1523/JNEUROSCI.0453-05.2005

Simon, D.K., O'Leary, D.D., 1991. Relationship of retinotopic ordering of axons in the optic pathway to the formation of visual maps in central targets. J. Comp. Neurol. 307 (3), 393–404. http://dx.doi.org/10.1002/cne.903070305

Soares, C.A., Mason, C.A., 2015. Transient ipsilateral retinal ganglion cell projections to the brain: extent, targeting, and disappearance. Dev. Neurobiol. http://dx.doi.org/10.1002/dneu.22291

Sosulski, D.L., Bloom, M.L., Cutforth, T., Axel, R., Datta, S.R., 2011. Distinct representations of olfactory information in different cortical centres. Nature 472 (7342), 213–216. http://dx.doi.org/10.1038/nature09868

Sperry, R.W., 1951. Regulative factors in the orderly growth of neural circuits. Growth Suppl. 10, 63–87.

Sperry, R.W., 1963. Chemoaffinity in the orderly growth of nerve fiber patterns and connections. Proc. Natl. Acad. Sci. U.S.A. 50, 703–710.

Squarzoni, P., Oller, G., Hoeffel, G., Pont-Lezica, L., Rostaing, P., Low, D., et al., 2014. Microglia modulate wiring of the embryonic forebrain. Cell Rep. 8 (5), 1271–1279. http://dx.doi.org/10.1016/j.celrep.2014.07.042

St John, J.A., Clarris, H.J., McKeown, S., Royal, S., Key, B., 2003. Sorting and convergence of primary olfactory axons are independent of the olfactory bulb. J. Comp. Neurol. 464 (2), 131–140. http://dx.doi.org/10.1002/cne.10777

Suarez, R., Gobius, I., Richards, L.J., 2014. Evolution and development of interhemispheric connections in the vertebrate forebrain. Front. Hum. Neurosci. 8, 497. http://dx.doi.org/10.3389/fnhum.2014.00497

Thanos, S., Bonhoeffer, F., 1983. Investigations on the development and topographic order of retinotectal axons: anterograde and retrograde staining of axons and perikarya with rhodamine *in vivo*. J. Comp. Neurol. 219 (4), 420–430. http://dx.doi.org/10.1002/cne.902190404

Thanos, S., Bonhoeffer, F., Rutishauser, U., 1984. Fiber-fiber interaction and tectal cues influence the development of the chicken retinotectal projection. Proc. Natl. Acad. Sci. U.S.A. 81 (6), 1906–1910.

Tolbert, L.P., Oland, L.A., Tucker, E.S., Gibson, N.J., Higgins, M.R., Lipscomb, B.W., 2004. Bidirectional influences between neurons and glial cells in the developing olfactory system. Prog. Neurobiol. 73 (2), 73–105. http://dx.doi.org/10.1016/j.pneurobio.2004.04.004

Tomer, R., Ye, L., Hsueh, B., Deisseroth, K., 2014. Advanced CLARITY for rapid and high-resolution imaging of intact tissues. Nat. Protoc. 9 (7), 1682–1697. http://dx.doi.org/10.1038/nprot.2014.123

Torrealba, F., Guillery, R.W., Eysel, U., Polley, E.H., Mason, C.A., 1982. Studies of retinal representations within the cat's optic tract. J. Comp. Neurol. 211 (4), 377–396. http://dx.doi.org/10.1002/cne.902110405

Tosney, K.W., Landmesser, L.T., 1985. Development of the major pathways for neurite outgrowth in the chick hindlimb. Dev. Biol. 109 (1), 193–214.

Tucker, E.S., Oland, L.A., Tolbert, L.P., 2004. *In vitro* analyses of interactions between olfactory receptor growth cones and glial cells that mediate axon sorting and glomerulus formation. J. Comp. Neurol. 472 (4), 478–495. http://dx.doi.org/10.1002/cne.20058

Vassar, R., Ngai, J., Axel, R., 1993. Spatial segregation of odorant receptor expression in the mammalian olfactory epithelium. Cell 74 (2), 309–318.

Vassar, R., Chao, S.K., Sitcheran, R., Nunez, J.M., Vosshall, L.B., Axel, R., 1994. Topographic organization of sensory projections to the olfactory bulb. Cell 79 (6), 981–991.

Walsh, C., Guillery, R.W., 1985. Age-related fiber order in the optic tract of the ferret. J. Neurosci. 5 (11), 3061–3069.

Wang, L., Marquardt, T., 2013. What axons tell each other: axon-axon signaling in nerve and circuit assembly. Curr. Opin. Neurobiol. 23 (6), 974–982. http://dx.doi.org/10.1016/j.conb.2013.08.004

Wang, L., Klein, R., Zheng, B., Marquardt, T., 2011. Anatomical coupling of sensory and motor nerve trajectory via axon tracking. Neuron 71 (2), 263–277. http://dx.doi.org/10.1016/j.neuron.2011.06.021

III. AXON GUIDANCE

Wang, L., Mongera, A., Bonanomi, D., Cyganek, L., Pfaff, S.L., Nusslein-Volhard, C., et al., 2014. A conserved axon type hierarchy governing peripheral nerve assembly. Development 141 (9), 1875–1883. http://dx.doi.org/10.1242/dev.106211

Williams, R.W., Rakic, P., 1985. Dispersion of growing axons within the optic nerve of the embryonic monkey. Proc. Natl. Acad. Sci. U.S.A. 82 (11), 3906–3910.

Williams, S.E., Mann, F., Erskine, L., Sakurai, T., Wei, S., Rossi, D.J., et al., 2003. Ephrin-B2 and EphB1 mediate retinal axon divergence at the optic chiasm. Neuron 39 (6), 919–935.

Williams, S.E., Grumet, M., Colman, D.R., Henkemeyer, M., Mason, C.A., Sakurai, T., 2006. A role for Nr-CAM in the patterning of binocular visual pathways. Neuron 50 (4), 535–547. http://dx.doi.org/10.1016/j.neuron.2006.03.037

Wolff, J.J., Gu, H., Gerig, G., Elison, J.T., Styner, M., Gouttard, S., et al., 2012. Differences in white matter fiber tract development present from 6 to 24 months in infants with autism. Am. J. Psychiatry. 169 (6), 589–600. http://dx.doi.org/10.1176/appi.ajp.2011.11091447

Yamatani, H., Sato, Y., Fujisawa, H., Hirata, T., 2004. Chronotopic organization of olfactory bulb axons in the lateral olfactory tract. J. Comp. Neurol. 475 (2), 247–260. http://dx.doi.org/10.1002/cne.20155

Zhang, L.I., Poo, M.M., 2001. Electrical activity and development of neural circuits. Nat. Neurosci. 4 (Suppl), 1207–1214. http://dx.doi.org/10.1038/nn753

Zhou, J., Wen, Y., She, L., Sui, Y.N., Liu, L., Richards, L.J., et al., 2013. Axon position within the corpus callosum determines contralateral cortical projection. Proc. Natl. Acad. Sci. U.S.A. 110 (29), E2714–2723. http://dx.doi.org/10.1073/pnas.1310233110

14

Cortical Architecture, Midline Guidance, and Tractography of 3D White Matter Tracts

Laura R. Morcom[1], Timothy J. Edwards[1] and Linda J. Richards[1,2]

[1]Queensland Brain Institute, The University of Queensland, Brisbane, Australia [2]School of Biomedical Sciences, The University of Queensland, Brisbane, Australia

14.1 INTRODUCTION

The circuitry of the brain underlies its ability to function. These circuits are formed and refined during development under the influence of a combination of genetic/molecular and activity-dependent mechanisms. Connections between neurons form integrated circuits that receive, process, produce, and store information. The output component of a neuron is its axon, which connects to the dendrites of the neurons that receive its information. The circuitry of the cerebral cortex encompasses both short-range and long-range axonal connections between neurons. Short-range connections form local circuits whereas long-range connections form between different brain regions in the same or opposing cerebral hemisphere. Long-range axonal connections cluster into large white matter tracts containing both myelinated and unmyelinated axons. The correct targeting of axons during brain development is therefore of fundamental importance in establishing the functional circuits of the brain.

The anatomical connections of the nervous system, including the cerebral cortex, have been studied using histological techniques encompassing immunohistochemical staining and tract tracing. More recently, these studies have been complemented by magnetic resonance imaging (MRI) techniques that utilize diffusion MRI (dMRI) as the basis for tractography. Combining these techniques has revolutionized our understanding of the anatomical connectivity of the human brain in both normal and pathological states. It has also enabled more rapid crossover of studies of brain connectivity in human pathologies that can be investigated in animal models. Finally, dMRI can be performed in living human subjects, providing an opportunity to conduct a longitudinal examination of brain connectivity. These studies are providing unprecedented knowledge about brain development, normal brain function, and brain degeneration.

In this chapter, we focus on the development of cerebral connectivity and provide a review of how cortical axonal connections are formed and how they can be studied using the complementary technologies of MRI and histology.

14.2 DEVELOPMENT OF CIRCUITS IN THE BRAIN

14.2.1 Formation of the Neocortex

The neocortex is a highly organized structure present in all mammals that processes sensory, motor, language, emotional, and associative information. It is the largest component of the cerebral cortex and comprises six layers of neurons that are grouped according to their primary input or output circuitry. Two of the most fundamental

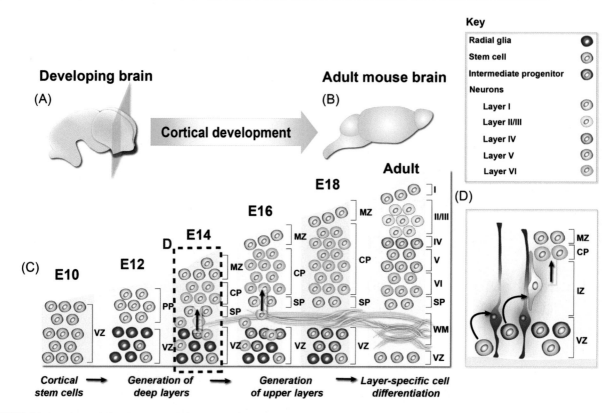

FIGURE 14.1   Prenatal development of the mammalian neocortex. Within the developing mouse embryo (A), the neocortex develops from a thin collection of cortical stem cells around embryonic day (E)10, to form a complex structure composed of six layers of mature neurons in the adult (B). Cortical stem cells give rise to radial glia (dark blue cells in C and D), the progenitor cells of all major cell types in the mature brain that provide the scaffold for immature neurons to migrate away from the ventricular zone (VZ) and into the cortical plate (CP). Radial glia generate intermediate progenitor cells (red cells in C and D), which divide and generate the majority of neurons in the brain. The first neurons (pink cells in C) to be born in the cortex form the preplate (PP) around E12, but are split into the marginal zone (MZ) and the subplate (SP) by later born neurons (green cells in C) migrating radially into the CP from E14. Neurons are generated in an inside out fashion, whereby later born neurons (such as those in layer II/III) must migrate past those born earlier. Neurons differentiate into many mature projection neuron subclasses that grow an axon and form local or distant connections. Neurons leaving the VZ for the CP migrate through the intermediate zone (IZ), a region that will contain the axons of cortical neurons forming the white matter (WM). Production of neurons depletes the progenitor cell pool in the mature cortex and very few stem cells remain in the adult.

questions in neuroscience are: how does the complexity of the mature neocortex take shape during development? and how are connections between often distant cortical areas correctly established and maintained? To begin to answer these questions, we must first consider the earliest events in neurodevelopment: the establishment of the cortical plate and its architecture, and subsequent patterning of a primordial cortex of neurons into structurally distinct areas specialized to deal with specific types of inputs and outputs. Disruption to any of these developmental events can have a profound impact on brain wiring and connectivity.

The neocortex emerges from the rostral-most brain vesicle of the neural tube (the prosencephalon), around embryonic day (E)10 in mice and gestational week (GW)5 in humans (Müller and O'Rahilly, 1988). At this stage of development, the neocortex is a thin sheet of pseudostratified neuroepithelial cells that will shortly transform into radial glia: the major progenitors of neurons and glia (Figure 14.1A–C; Aaku-Saraste et al., 1996; Hartfuss et al., 2001; Heins et al., 2002; Noctor et al., 2002). Radial glia are bipolar cells that reside in the ventricular zone of the developing brain; they have both an apical attachment to the underlying ventricular surface and a basal attachment to the pial surface of the brain. Radial glia can undergo either symmetric divisions to produce more radial glia or asymmetric divisions to give rise to a radial glia and an intermediate progenitor or neuron (Figure 14.1D; Chenn and McConnell, 1995; Noctor et al., 2004). Intermediate progenitors undergo mainly symmetric division to produce two daughter neurons, thereby generating the majority of the pyramidal neurons in the brain (Haubensak et al., 2004; Miyata et al., 2004; Noctor et al., 2004; Wu et al., 2005; Sessa et al., 2008; Kowalczyk et al., 2009). In contrast, it is well established that in rodents interneurons are generated outside the cortex in the ganglionic eminences and preoptic area, and migrate tangentially through the marginal zone and intermediate zone to populate the cortical plate (Anderson et al., 2001, 2002; Gelman et al., 2009). At the end of neurogenesis, radial glia differentiate into ependymal cells or astrocytes (Voigt, 1989; Spassky et al., 2005). This terminal differentiation event depletes the pool of progenitors in the ventricular zone, and very few stem cells remain in the lateral ventricular zone in the adult (Zhao et al., 2008).

The neocortex is a six-layered structure, with layer I closest to the pial surface and layer VI closest to the ventricle during development, and to the white matter in the adult (Figure 14.1C). The first layer of postmitotic neurons in the cortex is called the preplate. The preplate is derived from the early divisions of neuroepithelial cells and is split into the subplate and marginal zone by migrating neurons born in the ventricular zone that form the cortical plate around E13 in mice (Allendoerfer and Shatz, 1994; Marín-Padilla, 1998). Neurons destined for the cortical plate are generated in an inside out fashion, whereby neurons of the deeper layers (V–VI) are born first (from E12 in mice) and migrate away from the ventricular zone along the processes of radial glial cells to reach the correct layer (Figure 14.1; Angevine and Sidman, 1961; Caviness, 1982). Upper layer neurons (born on or after E15 in mice) must migrate past deeper layer neurons to form layers II and III of the neocortex (Caviness, 1982; Tissir and Goffinet, 2003). Neurons in the cortical plate then differentiate into a variety of projection neuron subtypes, which differ in terms of their molecular profile, morphology, electrophysiological properties, and connections (Migliore and Shepherd, 2005; Molyneaux et al., 2007). Differentiation of neurons into subtypes is driven by transcription factor programs that are initiated within progenitor cells or within the immature neurons, and are fundamental for establishing the blueprint of connectivity within the neocortex (reviewed by Molyneaux et al., 2007; Leone et al., 2008; Fame et al., 2011).

14.2.2 dMRI as a Means of Studying Neocortical Development

One means of studying neocortical development is through MRI. MRI is the measurement of the response of protons (most of which are contained in water molecules) to static and oscillating magnetic fields (Stejskal and Tanner, 1965). dMRI is an extension of the MRI method that models the microstructural properties of the brain based on the directionality of the diffusion of hydrogen atoms, on a macroscale (Taylor and Bushell, 1985; Le Bihan et al., 1986; Turner et al., 1990; Le Bihan, 1995). dMRI as a research imaging modality has traditionally been concerned with modeling white matter tracts in the mature brain. The same measurements of water diffusion, however, can be combined with histological techniques to offer a great deal of detail about how brain structure on a microscale or mesoscale correlates with the dynamic events within the developing neocortex. These findings will be discussed throughout the chapter, but the basic concepts underpinning dMRI must first be briefly introduced.

In the brain, the diffusion of water molecules is restricted to varying degrees in different compartments of tissue. Diffusion is unrestricted and random (or "Brownian") within compartments containing cerebrospinal fluid, as they contain no barriers to diffusion (Figure 14.2A and B). Water molecules in the grey matter reside within or between the cell bodies of neurons and glia, and their diffusion is therefore somewhat restricted by barriers such as cellular membranes, although it is still mostly random (Figure 14.2C). This type of diffusion, that is not selectively restricted in specific directions, is termed isotropic diffusion. In situations where cell membranes restrict the direction in which water molecules can move (as is the case within the white matter), the diffusion of water is anisotropic—i.e., it is highly restricted in some directions, but not in others (Figure 14.2D; Thomsen et al., 1987; Chenevert et al., 1990; Moseley et al., 1990, 1991; Le Bihan et al., 1993; Beaulieu, 2002). This physical restriction of the Brownian motion of water molecules underpins dMRI, as the movement of protons between applications of radiofrequency (RF) gradients results in a change in signal that can be measured within the MRI scanner and reconstructed. The critical insight of the early founders of this dMRI technique was that determining the extent of diffusion along a number of directions of an RF gradient (which can vary widely and can influence accurate resolution of diffusion direction) can be used as a noninvasive and indirect measure of the structure and orientation of white matter tracts in the brain.

In a dMRI, a single voxel can be considered a functional unit of protons, and within this voxel (typically $1–2.5\,mm^3$), the three-dimensional diffusion of water (Figure 14.2) can be modeled as an ellipsoidal tensor defined by the direction of its principal axes (eigenvectors ε_1, ε_2, and ε_3) and their magnitude (eigenvalues; λ_1, λ_2, and λ_3). Fractional anisotropy (FA) describes the shape of the ellipsoid. The isotropic diffusion that occurs in the grey matter is best modeled by a spherical ellipsoid where eigenvalues are close to equal (e.g., $\lambda_1 {\sim} \lambda_2$, ${\sim} \lambda_3$) and FA is close to zero. Anisotropic diffusion is modeled by a narrow ellipsoid where the major diffusion eigenvector is near-parallel to the direction of the white matter, there is a significant difference between eigenvalues (e.g., $\lambda_1 >> \lambda_2, >> \lambda_3$), and FA is high (close to 1). Various other scalar values, such as mean diffusivity (the average of the three eigenvalues), axial diffusivity (the principal eigenvalue λ_1), and radial diffusivity (an average of the second and third eigenvalues $[\lambda_2, + \lambda_3]/2$) among others, can be used to describe the structure of tissue within a voxel; these are outlined in Figure 14.2 for various tensor shapes. Diffusion properties are influenced by numerous tissue properties at the microstructural level, including axonal thickness, degree of myelination, cytoskeletal components, orientation and density, as well as dMRI artifacts such as temperature and head movement (Beaulieu, 2002). As a result, FA is very sensitive to microstructural changes, but is unable to determine the nature of these changes; indeed, it follows that an infinite number of combinations of eigenvectors can result in the same FA value, with each reflecting a different underlying microstructure. FA is regularly implemented as a surrogate measure of the organization of white matter, but should be interpreted with care, and where possible in conjunction with other measures of the tensor model. These concepts will be revisited throughout the chapter and will be built upon to describe how white matter connections, which arise from a complex sequence of developmental events can be modeled noninvasively to describe how connectivity is established.

Water diffusion in the grey matter of the postnatal human brain can be considered restricted but isotropic (Figure 14.2); however, this is not the case in the prenatal neocortex (Mori et al., 2001; McKinstry et al., 2002; Huang et al., 2006, 2009; Ball et al., 2013; Xu et al., 2014). During the second and third trimesters in humans, the FA of the cortex increases to coincide with the radial migration of immature bipolar neurons along the radial glial scaffold that runs perpendicular to the pial surface (Gupta et al., 2005). As the neocortex matures, neuronal morphologies become far more complex, and thalamocortical and corticocortical afferents begin to innervate the cortical plate, giving rise to complex cytoarchitecture that is no longer radially oriented (Gadisseux et al., 1992; Xu et al., 2014). Therefore, somewhat paradoxically, the normal maturation of the neocortex is associated with a decrease in anisotropy and mean diffusivity secondary to a reduction in the principal eigenvalue (λ_1, the axial component of diffusivity).

These microstructural changes are demonstrated in more detail by the refinement of techniques such as dMR microscopy, which allows for the reconstruction of changes in tissue microstructure on the scale of tens of microns in mouse models (Aggarwal et al., 2010, 2014). Progress in these methods is blurring the distinction between the micro- and mesoscale, and the histological correlates of the radial organization of diffusivity in the neocortex (mostly accounted for by radial glial fibers) are intuitively similar to dMRI-based tractography results of the developing cortical plate (Aggarwal et al., 2014; Xu et al., 2014). Using these techniques, the developing cortical plate can be delineated into three distinct layers based on differences in FA: the ventricular zone containing radially organized progenitors, the intermediate zone comprising tangentially oriented early axonal projections, and the radially oriented developing cortical plate (McKinstry et al., 2002; Huang et al., 2009; Xu et al., 2014). As GABAergic interneurons migrate tangentially into the developing cortex, together with the maturation of radial glia into astrocytes and the arrival of afferent axonal projections, the cortex becomes more complex and the clear relationship between microstructure and mesoscale representations of the diffusion signal breaks down. At this point in development, tractography reveals a reorganization of fibers towards tangential and corticocortical (fibers connecting different cortical regions) orientations (Xu et al., 2014). This change coincides with the specification of projection neurons and the sprouting of long-range projections, which will mature into the long-range white matter tracts that allow for information transfer between distant cortical areas in the postnatal brain.

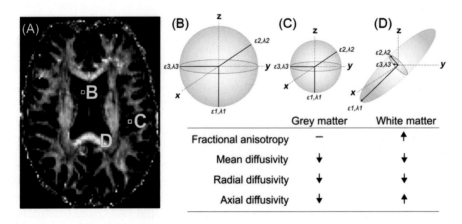

FIGURE 14.2 **Biological basis of the tensor model.** In a dMRI (A, horizontal view of an adult human brain depicted), diffusion within the cerebrospinal fluid (CSF) is unrestricted and isotropic, and is modeled by a spherical tensor (B, depicted right). All eigenvalues (λ1, λ2, and λ3) are approximately equal, such that there is a low FA, high mean diffusivity, and high radial diffusivity. In the grey matter (C), diffusion is nonselectively restricted and can be modeled by a similarly shaped but smaller sphere compared to the CSF, reflecting comparable decreases in λ1, λ2, and λ3. Relative to the CSF, all measures of diffusivity are decreased and the FA remains mostly unchanged. (D) In the white matter, diffusion can be modeled as an ellipsoid with a principal eigenvector (ε1) parallel to the direction of white matter. The FA is increased relative to the CSF; mean diffusivity and radial diffusivity are decreased, reflecting a smaller ε2 and ε3, and axial diffusivity is increased, reflecting a larger λ1 relative to the λ2 and λ3 eigenvalues.

14.2.3 Arealization

A crucial aspect of functional wiring of the brain is that areas that process similar information are interconnected. In part, this is driven by the early patterning of the neocortex into different functional areas. These areas share homology in their development but vary in terms of their final connectivity (inputs and outputs), cytoarchitecture, and repertoire of neural subclasses. The patterning of the cortex into different functional areas is under strict intrinsic control but can later be modified by sensory input (O'Leary and Sahara, 2008; Alfano and Studer, 2013).

The very early secretion of molecules from various signaling centers in the forebrain patterns the neocortex into functionally distinct areas from around E10.5 in mice (Figure 14.3). The best characterized secreted morphogens involved in this initial arealization are members of the fibroblast growth factor (FGF) family of ligands from the rostral patterning center, wingless-related MMTV integrations site (WNT) and bone morphogenic protein (BMP) ligands from the cortical hem dorsally, and sonic hedgehog (SHH) from the prechordal plate ventrally (Figure 14.3A; Munoz-Sanjuan et al., 1999; Grove and Fukuchi-Shimogori, 2003; O'Leary et al., 2007). These ligands diffuse across the cortex, bind to their corresponding receptors on proliferating progenitor cells, and induce regionalized gene programs that specify brain regions by influencing cell proliferation and cell death (Ohkubo et al., 2002; Storm et al., 2006; Fernandes et al., 2007). Each progenitor cell or neuron is subject to a specific combination of identity cues depending on its macroscale location, and it is the interpretation of these cues (on the cellular microscale) that results in the demarcation of cortical areas (Figure 14.3C and D). Four transcription factors are expressed in complementary gradients across the cortex and establish boundaries between functional areas (Figure 14.3B; O'Leary et al., 2007). In addition to their induction by morphogens, reciprocal regulation of these transcription factors also sets up their gradients (Figure 14.3B).

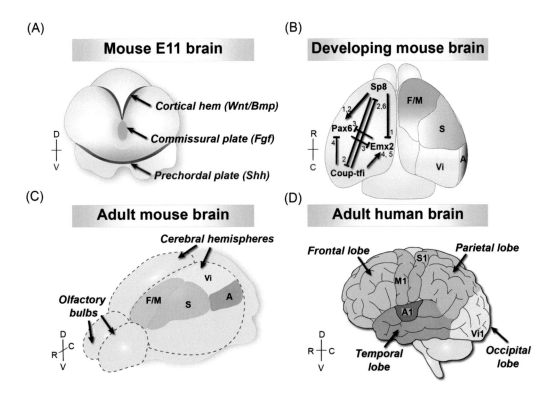

FIGURE 14.3 **Arealization of the developing neocortex.** (A) Early expression of diffusible ligands from three major patterning centers in the forebrain establishes areal identity. *Wnts* and *Bmps* are expressed from the cortical hem, *Fgfs* are expressed from the commissural plate (also known as the rostral patterning center), and *Shh* is expressed from the prechordal plate. Gradients of these morphogens influence the expression of transcription factors that establish boundaries between functionally distinct areas (B). *Sp8* and *Pax6* are expressed in high rostral to low caudal gradients and promote frontal/rostral identity. *Coup-tfi* is expressed in a high caudolateral to low rostromedial gradient and inhibits rostral identity by repressing *Sp8* and *Pax6* expression. *Emx2* is expressed in a high caudal to low rostral gradient and promotes caudal identity. These areal identities have a clear functional distinction in the adult mouse (C) and human (D) brain and form the motor (M, M1), somatosensory (S, S1), auditory (A, A1), and visual (Vi, Vi1) cortices. C=caudal, D=dorsal, F/M=frontal, R=rostral, and V=ventral. Numbers in B refer to the following references: [1]Zembrzycki et al. (2007), [2]Borello et al. (2013), [3]Muzio et al. (2002), [4]Faedo et al. (2008), [5]Hamasaki et al. (2004), [6]Alfano et al. (2014).

Paired box gene 6 (*Pax6*) and Specificity protein 8 (*Sp8*) are expressed in a high rostral to low caudal gradient and are required to establish rostral frontal/motor areas. Mutant mice that are deficient in PAX6 or SP8 display a rostral shift in arealization, such that the frontal/motor and somatosensory areas are diminished, whereas the visual and auditory areas are expanded rostrally (Bishop et al., 2000; Muzio et al., 2002; Sahara et al., 2007; Zembrzycki et al., 2007). *Pax6* conditional knockout mice also show significant reductions in interhemispheric connectivity by dMRI. This is evidenced by a reduction in tractography-generated interhemispheric streamlines associated with (i) decreases in FA and axial diffusivity and (ii) an increase in radial diffusivity, suggesting a decrease in axonal number (Boretius et al., 2009). This change is associated with a reduction in the size of cortical areas that normally give rise to the majority of interhemispheric connections.

Chicken ovalbumin upstream promoter transcription factor 1 (COUP-TF1, also known as nuclear receptor subfamily 2, group F, member 1 or NR2F1) is expressed in a high caudolateral to low rostromedial gradient and represses frontal/motor area identity. *Coup-tf1* conditional knockout mice have a remarkably expanded frontal/motor area across most of the cortex, at the expense of somatosensory, visual, and auditory areas (Armentano et al., 2007). Empty spiracles homologue 2 (EMX2) is expressed in a similar high caudomedial to low rostrolateral gradient. In contrast with COUP-TF1, EMX2 is thought to actively promote caudal area identity, as *Emx2* knockout mice have expanded frontal/motor and somatosensory areas and smaller visual and auditory regions (Bishop et al., 2000).

The mechanisms of regional fate specification downstream of these broadly expressed transcription factors are as yet poorly understood; however, three transcription factors have been identified that may play a role: T box brain 1 (*Tbr1*), Basic helix loop helix protein 5 (*Bhlh5*), and *Coup-tf1*. *Tbr1* specifies frontal identity by promoting the expression of the frontal cortex marker gene Autism susceptibility candidate 2 (*Auts2*; Bedogni et al., 2010). *Tbr1* knockout mice show a gene expression pattern consistent with a respecification of frontal areas into more caudal identity. *Bhlh5* specifies the identity of the somatosensory and caudal motor regions (Joshi et al., 2008), and *Bhlh5* null mice have a larger primary visual area and somatosensory barrel field associated with changes in area-specific gene expression, but no change in the graded expression of transcription factors (PAX6, COUP-TF1, EMX2, and SP8 are all normally expressed). Another study found that PAX6 regulates BHLH5 (Holm et al., 2007) suggesting that BHLH5 could be a mediator of arealization downstream of graded transcription factors. The postmitotic expression of *Coup-tf1* has also been shown to be more important for specifying regional identity than *Coup-tf1* expression in progenitor cells. Re-expression of COUP-TF1 in postmitotic neurons of *Coup-tf1* null mice largely rescues the arealization phenotype (Alfano et al., 2014). How these and other genetic pathways downstream of the transcription factors PAX6, COUP-TF1, EMX2, and SP8 result in the acquisition of area-specific characteristics and discrete areal boundaries remains to be elucidated. This is important as subtle changes in arealization, and therefore the connectivity of one or more cortical areas may result from disruption to signaling pathways downstream of early patterning and transcription factor programs.

Arealization is initially set up intrinsically, but can later be modified by sensory input. Sensory input to the cortex comes primarily from thalamocortical neurons. Thalamocortical input can influence the development of a cortical area in terms of its gene expression, cytoarchitecture, and cell morphology (Li et al., 2013). Following sensory deprivation in the visual or auditory system, the corresponding cortical area diminishes in size and is often innervated by sensory afferents from different modalities (Schneider, 1973; Frost, 1982; Sur et al., 1988; Roe et al., 1990; Dehay et al., 1991; Rakic et al., 1991; Roe et al., 1993; Sharma et al., 2000; Hunt et al., 2006). These rewiring events appear to instruct functional changes in areas receiving these ectopic sensory afferents (Frost et al., 2000; Ptito et al., 2001), expanding the processing capacity for that modality, and ultimately resulting in improved performance outcomes (Métin and Frost, 1989; Roe et al., 1990; Sadato et al., 1996; Kujala et al., 1997; Kupers and Ptito, 2014).

Subtler changes in thalamocortical connectivity have been observed in several human disorders. The stabilization of projections between the thalamus and cortex occurs mainly in the third trimester; this critical window of development is therefore disrupted in prematurely born infants. Tractography studies in neonates who have been born prematurely show markedly reduced structural connectivity between the thalamus and neocortex (Ball et al., 2013); changes in this connectivity around birth are associated with poorer cognitive outcome in the ensuing years of development (Ball et al., 2015).

14.2.4 Cortical Connectivity

Excitatory projection neurons in the mature cortex can be subdivided into two major classes based on their axonal projections (Harris and Shepherd, 2015). These are intratelencephalic projection neurons and corticofugal (cortico-subcortical) projection neurons. Intratelencephalic projection neurons project to other regions within the cortex are found in all cortical layers and can be further subdivided into interhemispheric or intrahemispheric projection neurons. Intrahemispheric projection neurons can have single or multiple axonal connections to neurons within different layers of the same cortical column, to other cortical columns within the same functional area (intrinsic or intra-areal), or to other functional areas within the cortex (inter-areal). Axons that form intrahemispheric connections do so by navigating through the grey matter or by contributing to white matter tracts (Watakabe et al., 2014). Interhemispheric connections are formed by long-range projection neurons that form commissures across the interhemispheric midline to innervate both homotopic and heterotopic contralateral cortical areas (Boyd et al., 1971; Yorke and Caviness, 1975; Hedreen and Yin, 1981; Segraves and Rosenquist, 1982; Krubitzer et al., 1998; Catania, 2001; Zhou et al., 2013). The major forebrain commissures connecting the cortical hemispheres in mammals are the corpus callosum, hippocampal commissure, and anterior commissure (depicted for mouse in Figure 14.4A). Callosal neurons are primarily found in layers II–III and V and a proportion are bifurcating, i.e., they project to the contralateral cortex as well as structures in the ipsilateral hemisphere such as the striatum (Wilson, 1987; Koester and O'Leary, 1993; Reiner et al., 2003; Veinante and Deschênes, 2003; Mitchell and Macklis, 2005; Garcez et al., 2007; Hattox and Nelson, 2007; Lickiss et al., 2012; Watakabe et al., 2014). Corticofugal projection neurons are mainly found in the deep layers of the cortex (V–VI); some of the major descending projections are the corticothalamic, corticocollicular, corticopontine, and corticospinal tracts.

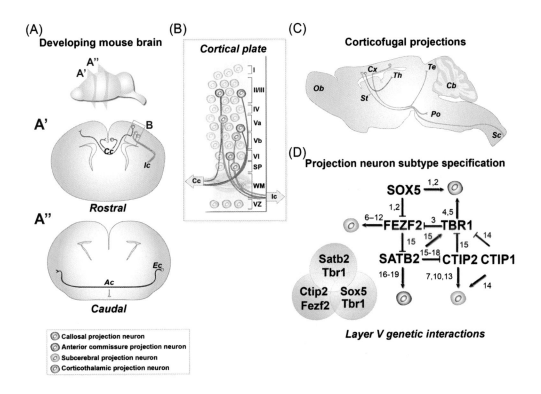

FIGURE 14.4 **Specification of projection neuron subtypes.** (A) Neocortical projection neurons can form connections either within the telencephalon or outside the telencephalon (a corticofugal projection) via the internal capsule (Ic; A') projecting to either the thalamus (corticothalamic projection neurons, blue cells) or subcerebral targets (subcerebral projection neuron, green cells) (B). A subset of intratelencephalic neurons projects across the telencephalic midline via commissures such as the corpus callosum (Cc; A', purple cells) or anterior commissure (Ac; A'', red cells) and connect with either homotopic or heterotopic targets in the contralateral hemisphere. (B) Most callosally projecting and subcerebrally projecting neurons originate from layers II/III and V, whereas corticothalamic projection neurons reside in layer VI (B and C). (D) Projection neuron subtype is specified by a complex network of interactions between transcription factors. *Ctip1*, *Ctip2*, and *Fezf2* are major determinants of subcerebral projection neuron identity, whereas *Satb2* downregulates *Ctip2* expression and is a major determinant of callosal neuron identity. Corticothalamic projection neuron identity is determined by *Sox5* expression secondary to direct repression of *Fezf2*. Cb= cerebellum, Cx= cortex, Ob= olfactory bulb, Po=pons, Sc=spinal cord, St=striatum, Te=tectum, Th=thalamus. Numbers in D refer to the following references: [1]Lai et al. (2008), [2]Shim et al. (2012), [3]Han et al. (2011), [4]O'Leary and Koester (1993), [5]McKenna et al. (2011), [6]Chen et al. (2005), [7]Chen et al. (2008), [8]Molyneaux et al. (2005), [9]Rouaux and Arlotta (2010), [10]Rouaux and Arlotta (2013), [11]De la Rossa et al. (2013), [12]Lodato et al. (2014), [13]Arlotta (2005), [14]Cánovas et al. (2015), [15]Srinivasan et al. (2012), [16]Alcamo et al. (2008), [17]Britanova et al. (2008), [18]Srivatsa et al. (2014), [19]Leone (2014).

14.2.5 The Specification of Neocortical Projection Neurons

The first layers to be generated in the cortex are the subplate and layer VI which contain neurons that form subcortical connections to the thalamus and the spinal cord, and serve as pioneers of the corticothalamic and corticospinal tracts (Figure 14.4B and C; McConnell et al., 1989; Ghosh et al., 1990; De Carlos and O'leary, 1992; Ghosh and Shatz, 1993; McConnell et al., 1994; Clancy et al., 2009; Kanold, 2009; Kanold and Luhmann, 2010; Molnár et al., 2010). The transcription factor Sex determining region Y box 5 (SOX5) is required for the correct generation of these connections. SOX5 is expressed throughout the developing cortical plate at E13.5 but becomes restricted to layers V–VI and the subplate postnatally. It promotes the development of corticothalamic projection neurons by repressing a corticospinal fate through direct repression of *Fezf2* (enhancer of FEZ family zinc finger 2; Lai et al., 2008; Shim et al., 2012). Neurons born at E11.5 that normally generate the subplate and layer VI are disorganized in Sox5 knockout mice, upregulate the transcription factor *Ctip2* (chicken ovalbumin upstream promoter transcription factor-interacting protein 2), and project their axons towards the corticospinal tract (Lai et al., 2008). TBR1 also specifies corticothalamic fate in layer VI projection neurons. Almost all corticothalamic axons highly express TBR1, which is required for timely corticothalamic innervation by birth. TBR1 overexpression in deep layers (at E12.5 and E13.5) is sufficient to redirect corticospinal neurons toward the thalamus (McKenna et al., 2011) because it directly represses the corticospinal-specifying transcription factor FEZF2 (Han et al., 2011). SOX5 and TBR1 are therefore crucial for the development of corticothalamic neurons by suppressing the later born corticospinal fate through repression of *Fezf2* (Figure 14.4D).

Layer Va predominantly contains medially projecting callosal neurons and layer Vb predominantly contains corticofugal neurons, a large majority of which form the corticospinal tract (Figure 14.4C; Harwell et al., 2012). These projection neuron subtypes are generated on the same birth date and are largely intermingled during embryonic development; however, corticofugal projection neurons extend axons 2 days earlier than interhemispheric projection neurons (Richards et al., 1997). Furthermore, subclasses of corticofugal neurons target different subcortical structures (Figure 14.4C). This leads to the question of how such a diverse mix of projection neuron subtypes with the same birth date is specified. FEZF2 is normally expressed in layers V and VI, and specifies corticospinal motor neuron identity by upregulating a suite of genes that promote corticospinal identity and repress callosal and other corticofugal fates (Molyneaux et al., 2005; Rouaux and Arlotta, 2010; la Rossa De et al., 2013; Lodato et al., 2014). Corticocollicular and corticospinal neurons are greatly reduced in *Fezf2* knockout mice (Chen et al., 2005; Molyneaux et al., 2005) and these projection neurons instead project axons through the corpus callosum (Chen et al., 2005, 2008; Molyneaux et al., 2005). This phenotype can be partially rescued by CTIP2 overexpression in cells born at E13.5 (Chen et al., 2008).

Ctip2 knockout mice display a disorganized corticospinal tract that does not project past the pons (Arlotta et al., 2005); therefore CTIP2 is important for the last stages of corticospinal tract guidance. CTIP2 and FEZF2 overexpression in upper layer progenitors at E14.5 induces rewiring of upper layer neurons to a corticospinal projection neuron identity (Chen et al., 2008; Rouaux and Arlotta, 2013). However, FEZF2 does not directly regulate *Ctip2*, and upper layer neurons that are rewired in response to overexpression of FEZF2 cannot project past the pons, as they fail to express CTIP2 (Chen et al., 2008). Similar to CTIP2, CTIP1 directly represses *Tbr1* in layer V and is required for the projection of the corticospinal tract past the pons (Cánovas et al., 2015). *Fezf2*, *Ctip1*, and *Ctip2* are therefore master regulator genes that are required for the correct specification and formation of the corticospinal tract (Figure 14.4D).

Special AT-rich sequence-binding protein 2 (SATB2) is a transcription factor that is highly expressed in layers II–III and V, and is required for the development of callosal neurons from layer V. FEZF2 represses the callosal trajectory of layer V neurons by repressing *Satb2* (Srinivasan et al., 2012). In FEZF2-negative cells, however, SATB2 directly represses *Ctip2* to inhibit the corticofugal fate and initiate a gene expression program that facilitates the acquisition of a callosal fate (Alcamo et al., 2008; Britanova et al., 2008; Srinivasan et al., 2012; Srivatsa et al., 2014). In *Satb2* knockout mice, cells that normally express *Satb2* in layers II–III and V are molecularly respecified to a corticofugal projection neuron type (Alcamo et al., 2008; Britanova et al., 2008). Importantly, cells in both deep and upper layers that normally express SATB2 upregulate CTIP2. However, only deeper layer neurons are able to contribute axons to the corticofugal tract suggesting that upper layer neurons cannot be completely respecified (Leone et al., 2014). Interestingly, early expression of SATB2 is required for the correct formation of the corticospinal tract. When *Satb2* is conditionally knocked out in the cortex (E15.5 or earlier), the corticospinal tract fails to extend past the cerebral peduncle or enter the spinal cord by P15. This phenotype is alleviated when *Satb2* is conditionally knocked out after E16.5 (Leone et al., 2014).

Although these transcription factors regulate many genes that are likely to be involved in the molecular identity of the different projection neuron subtypes, two axon guidance receptors Unc5 homolog c (UNC5C) and Deleted in colorectal cancer (DCC), and their ligand Netrin1 (NTN1), have been shown to facilitate the downstream guidance of corticofugal versus callosal axons in layer V. DCC and UNC5C mediate an attractive and repulsive response to NTN1, respectively (Keino-Masu et al., 1996; Serafini et al., 1996; Hong et al., 1999; Finger et al., 2002; Srivatsa et al., 2014). Srivatsa et al. (2014) found that SATB2 positively regulates *Unc5c* and inhibits *Dcc*, whereas CTIP2 inhibits *Unc5c*. They showed that overexpressing SATB2, or UNC5C or knocking down DCC in deep layers in the *Satb2* knockout mice led to the redirection of these axons across the corpus callosum, thereby partially rescuing the phenotype. This suggests that *Unc5c* and *Dcc* are axon guidance genes downstream of SATB2 that mediate the callosal versus corticofugal fate of layer V projection neurons. As NTN1 is expressed in the basal ganglia, surrounding the internal capsule (Métin et al., 1997), it is likely that *Unc5c* expression downstream of SATB2 and repression of *Dcc* by SATB2 results in the repulsion of SATB2-expressing callosal axons away from the source of NTN1 where corticofugal neurons must navigate. In the absence of SATB2 and the presence of CTIP2, however, *Unc5c* is repressed and DCC is able to mediate a positive response to NTN1, directing axons toward subcortical targets.

The specification of callosal neurons in upper layers is less well studied. As described above, upper layer neurons in *Satb2* knockout mice, despite molecularly expressing Ctip2, are unable to project corticofugally or callosally (Leone et al., 2014). Instead, their axons project aberrantly to intrahemispheric targets. It should be noted that cells that do not normally express SATB2 continue to cross the midline and form a small corpus callosum in *Satb2* knockout mice. In addition, not all retrogradely labeled callosal neurons express SATB2 (Alcamo et al., 2008; Lickiss et al., 2012), suggesting that a subset of callosal neurons do not require SATB2 to form callosal connections. Furthermore, upper layer neurons continue to project their axons across the corpus callosum when *Satb2* is knocked down only in upper layer progenitors (Zhang et al., 2012), suggesting that *Satb2* may play different roles in callosal axon specification depending on the layer of origin. These cells have defects in axonal growth, somal spacing, and dendritic arborization. Interestingly, the repression of *Ctip2* by SATB2 depends on the coexpression of the proto-oncogene *Ski* (Baranek et al., 2012). SATB2 binds the DNA sequence upstream of *Ctip2*, assembling the Nucleosome remodeling deacetylase (NURD) complex for deacytelation (inactivation) of CTIP2 (Britanova et al., 2008). SKI, however, is required for the recruitment of Histone deacetylase 1 (HDAC1) to this protein complex, and is therefore indispensible for SATB2 repression of *Ctip2* (Baranek et al., 2012). SKI and SATB2 are highly coexpressed in the upper layers, where they repress the expression of *Ctip2* (Baranek et al., 2012). Evidence of this is that *Ski* and *Satb2* knockout mice both show expanded expression of *Ctip2* into the upper layers of the cortex. Interestingly, *Ski* and *Ctip2* are coexpressed in layer V, where *Satb2* and *Ski* are almost never coexpressed (Baranek et al., 2012). Whether SATB2 directly represses *Ctip2* in layer V (without SKI) remains to be investigated; however, layer V neurons regain the ability to cross the corpus callosum in *Satb2/Ctip2* double knockout mice, suggesting that CTIP2 is at least partially responsible for the loss of callosal axons in *Satb2* knockout mice (Srivatsa et al., 2014).

These papers represent a decade of progress toward understanding the transcriptional regulation of projection neuron fate specification. Although several studies have identified the suite of genes downstream of these transcription factors, how these gene programs regulate projection neurons to acquire mature characteristics and make functional connections in the brain remains to be elucidated. Axonal guidance genes, which encode families of ligands and receptors, likely act downstream of these transcription factors to regulate the development of connections within the complex cellular and molecular environment of the brain.

14.2.6 The Corpus Callosum as a Model System for Studying Developing White Matter Tracts

The corpus callosum is the largest white matter tract in eutherian mammals, and is particularly prominent in the human brain, carrying approximately 190 million axons from one cortical hemisphere across the midline of the forebrain and into the opposite cortical hemisphere (Tomasch, 1954; Gazzaniga, 2005). Congenital absence (or agenesis) of the corpus callosum (AgCC) is a relatively common human brain malformation (1:4000 live births; Hetts et al., 2006) that is associated with a large number of neurodevelopmental disorders (Edwards et al., 2014) and results in a wide spectrum of neurological deficits (Paul et al., 2007).

The corpus callosum in AgCC patients can be completely absent (complete AgCC), partially absent (partial AgCC), or reduced in thickness (hypoplasia of the CC). The formation of Probst bundles is a common feature of AgCC. These axons do not cross the midline but instead run anteroposteriorly along the midline (Probst, 1901). All of these phenotypes have been recapitulated in mouse models of AgCC (Ren et al., 2007), which have begun to reveal some of the cellular processes of corpus callosum development that may lead to their emergence. Normal corpus callosum formation depends on a multitude of developmental events, including the correct generation of midline glial cells, and the specification of projection neurons to a callosal identity, as well as the guidance of axons out of the cortical plate toward the midline of the brain, then across the midline to their targets in the contralateral hemisphere. Corpus callosum development and AgCC provide an excellent system for translational studies that aim to investigate the genetics and anatomy of white matter disorders in humans, using animal models that recapitulate the phenotype for experimental manipulation.

14.2.6.1 Intermediate Zone Guidance

Once callosal neurons have been specified in the cortex, they project axons into the white matter underlying the neocortex and turn medially toward the midline. At this stage, distinct axon guidance receptors downstream of transcription factors are able to guide the axon to turn medially toward the midline and across the corpus callosum, or inferiorly to enter the internal capsule. Callosal axons respond to repulsive cues from the lateral cortex (Zhao et al., 2011), one of which is probably NTN1 in mice. As described in Section 14.2.4, UNC5C expression on callosal axons downstream of SATB2 facilitates the chemorepulsive response to NTN1 expressed in ventral–lateral regions of the mouse forebrain, repelling them from making a lateral turn. T-cadherin (also known as cadherin 13) also regulates this decision. T-cadherin is normally expressed in subcortically projecting neurons of the deeper layers. Suppressing its activity causes subcortical projection neurons to incorrectly project medially toward the midline, and overexpressing T-cadherin in upper layer callosal neurons causes them to project laterally (Hayano et al., 2014). Neurogenin2 (NGN2) plays a complementary role, promoting the medial turning of callosal axons. *Ngn2* knockout mice show partial AgCC without Probst bundles. Interestingly knockdown of *Ngn2* in upper layer neurons results in lateral turning of affected axons but no change in their transcriptional profiles (they continue to express upper layer markers rather than deep layer markers such as *Ctip2*), suggesting that NGN2 is downstream of fate specification (Hand and Polleux, 2011). Another mouse model with a similar phenotype (AgCC without Probst bundles) lacks both the Neuronal differentiation (*Neurod*) 2 and 6 genes. Axons in double knockout mice are able to turn medially out of the cortex toward the midline but fail to grow toward it, instead defasciculating and stalling in the subventricular zone of the cingulate cortex (Bormuth et al., 2013). These examples demonstrate that guidance decisions must be made well before callosal axons reach the midline.

14.2.6.2 Guidepost Cells

Guidepost cells are populations of neurons or glia that express guidance cues at different stages of white matter tract formation. They enable step-by-step navigation of the brain environment, facilitating the guidance of axons across long distances. The corpus callosum crosses the midline rostrally at the boundary between the cingulate cortex and septum (cortico-septal boundary). This region of the midline comprises a heterogeneous mix of both neuronal cells and glia that act as guidepost cells to attract callosal axons toward the midline, channel them through the midline, and repel them into the contralateral hemisphere (Figure 14.5A and B). The dorsal boundary of the corpus callosum is the indusium griseum, which comprises both neurons and glia (the indusium griseum glia; Shu and Richards, 2001). During development, radial glia reside ventral and lateral to the corpus callosum comprising the glial wedge (Shu and Richards, 2001; Shu et al., 2003a). The subcallosal sling is a predominantly neuronal population that forms a characteristic U-shape underneath the corpus callosum (Silver et al., 1982; Shu et al., 2003b). In addition, a population of GABAergic neurons populates the middle and lateral regions of the corpus callosum during midline crossing (Niquille et al., 2009, 2013). All of these guidepost cell populations have been shown to have a functional influence on corpus callosum formation by secreting guidance factors (Table 14.1) or making cellular contacts with the callosal axons.

For the corpus callosum to form correctly, distinct midline glial populations must be generated in the correct proportions and position in order for the midline crossing of callosal axons to occur (Figure 14.5B; Gobius and Richards, 2011). The glial wedge and indusium griseum glia must first correctly differentiate from radial glia within the ventricular zone of the medial cortex and then mature morphologically. These glial populations are some of the first glia to differentiate in the forebrain, as the majority of astrocytes are generated postnatally in the cerebral cortex. Several morphogens, transcription factors, and enzymes are required to produce the glial populations of the cerebral midline, and these are briefly reviewed below.

Nuclear factor one (NFI) transcription factors A and B are important for the transition of the radial glia progenitors into the mature astroglial cells that make up the glial wedge and indusium griseum glia. NFIA and NFIB are expressed in the indusium griseum glia, glial wedge, and subcallosal sling as well as in cortical neurons (Shu et al., 2003a; Piper et al., 2009a). *Nfia* and *Nfib* knockout mice have delayed maturation of these cell populations, which remain radially oriented and lack expression of mature glial markers during embryonic development. Consequently, these knockout mice fail to form a corpus callosum by birth (Neves Das et al., 1999; Shu et al., 2003a; Steele-Perkins et al., 2005; Piper et al., 2009a).

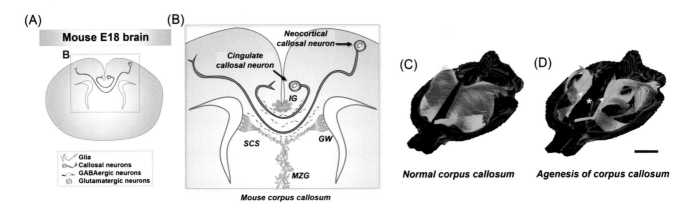

FIGURE 14.5 **Guidepost cells regulate midline axon guidance of the corpus callosum.** (A) The axons that form the corpus callosum originate from neurons within the cingulate cortex and the neocortex. (B) Callosal axon guidance across the midline is facilitated by multiple glial and neuronal populations. The indusium griseum (IG) is positioned dorsal to the corpus callosum and comprises glia and neurons. The glial wedge (GW) consists of glia that arise from the medial ventricular zones and extend processes toward the midline, providing the ventral boundary to the corpus callosum. The subcallosal sling (SCS) is a predominantly neuronal population containing some glia that forms a characteristic U-shape ventral to the developing corpus callosum. The midline is also populated with a population of GABAergic neurons during midline crossing. Gradients of diffusible axon guidance molecules and cell contact guidance cues that are expressed by these cellular populations must first attract/repel callosal axons toward the midline, and then into the contralateral cortex after midline crossing. Midline zipper glia (MZG) are positioned ventral to the corpus callosum and are involved in midline development prior to corpus callosum formation. (C and D). The corpus callosum can be visualized as a large bundle of mediolaterally oriented (colored red) tractography-generated streamlines crossing the telencephalic midline (C). In a mouse with agenesis of the corpus callosum (asterisks in D), Probst bundles are visualized as streamlines running anteroposteriorly (colored green) on either side of the telencephalic midline (arrowheads in D). Scale bar= 2mm in panels C and D.

Glial wedge cells normally retain a medially oriented polarity and cluster their processes to form a wedge shape. The indusium griseum glia, however, must translocate their cell body to the pial surface and retract their processes from the ventricular zone. This translocation event is regulated by FGF signaling. Glial-specific knockout of the Fgf receptor 1 (*Fgfr1*) results in complete and partial AgCC that is associated with the absence of the indusium griseum glia due to incomplete radial glial translocation to the pial surface. Cells destined to become indusium griseum glia remain attached to the ventricular surface and their soma lies close to the ventricle in *Fgfr1* knockout mice and following cell-specific knockdown of *Fgfr1* (Smith et al., 2006). The heparin sulfotransferases HS6ST1 and HS2ST interact with FGF signaling to inhibit this process, probably normally acting to prevent excessive somal translocation and loss of the glial wedge. In *Hs6st1* and *Hs2st1* knockout mice, AgCC (complete and partial) is associated with an increase in the number of indusium griseum glia at the midline at the expense of the glial wedge (Conway et al., 2011; Clegg et al., 2014). The phosphorylated form of the FGF intracellular signaling effector, extracellular signal-related kinase (ERK) 1/2 is significantly increased at the midline in *Hs6st1* and *Hs2st1* knockout mice, particularly within the nuclei of cells within the medial ventricular zone (Clegg et al., 2014). This callosal phenotype can be rescued in a subset of *Hs6st1* knockout mice that are also heterozygous for *Fgf8* or have been administered an inhibitor targeting the ERK-activating kinase MEK (Clegg et al., 2014), suggesting that downstream activation of FGF signaling is at least partially responsible for the phenotype. An inhibitor of FGF signaling, Sprouty (SPRY) 1/2 also keeps ERK signaling in check. *Spry1/2* double knockout mice have AgCC with Probst bundles, associated with increased phosphorylated ERK1/2, and midline glia that span the entire length of the cingulate cortex from the ventricular zone to the pial surface, thereby ectopically transecting the path of callosal axons (Magnani et al., 2012).

Altered FGF signaling (increased levels FGF8 and phosphorylated ERK1/2) downstream of GLI family zinc finger 3 (GLI3) also results in glial fibrillary acidic protein (GFAP)-positive glia spanning the entire cingulate cortex, and absence of the indusium griseum glia. *Gli3* mutant mice also show expanded expression of the repulsive axon guidance ligand Slit homolog (*Slit*) 2 into the dorsomedial cortex and ectopically in the septum (Magnani et al., 2012; Amaniti et al., 2013, 2015). Similarly, conditional loss of the tumor suppressor Neurofibromatosis 2 (*Nf2*) in the cortex results in a similar glial phenotype, with upregulation of SLIT2 and an increase in glia that extend their processes to the pia and cross the callosal axon path (Lavado et al., 2014). Deleting one copy of *Slit2* restores callosal formation in *Nf2* mouse mutants, suggesting that ectopic SLIT2 expression is preventing callosal crossing in these mice. These findings highlight the importance of correct development of guidepost neurons and disruption to genes involved in glial development associated with AgCC in humans (Edwards et al., 2014).

TABLE 14.1   Expression of Axon Guidance Molecules at the Telencephalic Midline

Callosal axons	IG	GW	MZG	Septum	Subcallosal sling	Guidepost neurons	
						GABAergic interneurons	Glutamatergic neurons
DCC[b]	DRAXIN[i]	DRAXIN[i]	EPHA4[c]	EPHB2[c]	DRAXIN[i]	EPHB2[l]	SEMA3C[l]
EPHB1[c]	EPHA4[c]	EPHA4[c]	EPHB2[c]	EFNB1[c]	EPHA4[c]	EPHB3[a,l]	
EPHB2[c]	EPHB2[c]	EPHB2[c]	EFNB1[c]	EFNB3[c]	SEMA3A[l]	EPHA4[a,l]	
EPHB3[c]	EPHB2[c]	EFNB1[c]	EFNB3[c]	NTN1[b,j]	SEMA3C[d]	EFNA1[l]	
EPHB4[c]	EFNB1[c]	EFNB2[c]		SLIT1[e]	ROBO1[k]	EFNA4[l]	
EFNB1[c]	EFNB3[c]	EFNB3[c]		SLIT3[e]		EFNB1[l]	
EFNB2[c]	NTN1[b,j]	ROBO1[e]		ROBO3[b]		EFNB2[l]	
NRP1[d]	ROBO1[e]	SLIT1[e]				NRP1a[l]	
ROBO1[a,e–g]	ROBO2[e]	SLIT2[e]				NRP2[l]	
ROBO2[a,f–g]	SEMA3C[d]	SLIT3[e]				SEMA3A[l]	
RYK[h]	SLIT1[e]	WNT5A[h]					
	SLIT2[e]						
	SLIT3[e]						
	WNT5A[h]						

GW, glial wedge; IG, indusium griseum glia; MZG, midline zipper glia.
[a]*Very low levels of expression.*
[b]*Fothergill et al. (2014).*
[c]*Mendes et al. (2006).*
[d]*Piper et al. (2009).*
[e]*Unni et al. (2012).*
[f]*Shu and Richards (2001).*
[g]*Andrews et al. (2006).*
[h]*Keeble et al. (2006).*
[i]*Islam et al. (2009).*
[j]*Serafini et al. (1996).*
[k]*Shu et al. (2003b).*
[l]*Niquelle et al. (2011).*

14.2.6.3 Attraction to the Midline

Guidepost cells at the telencephalic midline secrete a wide variety of soluble axon guidance proteins, which signal to surface receptors on callosal axons to mediate corpus callosum formation (Table 14.1). These cues direct the attraction of callosal axons toward and across the midline and their repulsion into the contralateral hemisphere. Many axon guidance genes have been implicated in the development of the corpus callosum, as mice lacking the expression of these genes display AgCC (Table 14.2).

Class 3 semaphorins (SEMA3) and their receptor Neuropilin 1 (NRP1), which is expressed on callosal axons of the cingulate cortex (Piper et al., 2009b), have been shown to regulate the pathfinding of these axons, which are the first population of callosal axons to cross the midline. *Nrp1*^Sema- mice (which express NRP1 that lacks the semaphorin-binding domain) and *Sema3c* knockout mice show partial and complete AgCC associated with aberrant guidance of cingulate pioneering axons (Table 14.2; Gu et al., 2003; Niquille et al., 2009; Piper et al., 2009b). SEMA3C is normally expressed in the indusium griseum and subcallosal sling, and can bind and attract callosal axons *in vivo* (Niquille et al., 2009). SEMA3C binding to NRP1 elicits both positive outgrowth and guidance in cingulate cortical explants (Piper et al., 2009b). *Sema3a* is also expressed in the subcallosal sling and in GABAergic interneurons, and has been shown to repel both cingulate cortical and neocortical explants *in vitro*. SEMA3C is therefore likely to attract NRP1-expressing cingulate callosal axons toward the midline, whereas SEMA3A expression from the subcallosal sling is likely involved in preventing the overgrowth of these cingulate axons into the septum, and channeling them across the midline.

The secreted protein, NTN1, is an important attractive guidance cue for commissural axons of the spinal cord (Kennedy et al., 1994; Serafini et al 1996). In contrast, in the cerebral cortex, NTN1 differentially acts as an attractant

TABLE 14.2 Malformation of the Corpus Callosum in Mice Lacking Axon Guidance Genes

Mouse mutant	AgCC	pAgCC	Refs.
Dcc	✓		Fazeli et al. (1997), Ren et al. (2007)
Dcckanga	✓		Fothergill et al. (2014)
Draxin	✓	✓	Islam et al. (2009), Ahmed et al. (2011)
Epha5		✓	Mendes et al. (2006)
Ephb1	✓	✓	Mendes et al. (2006)
Ephb2	✓	✓	Mendes et al. (2006)
EfnB3	✓	✓	Mendes et al. (2006)
Ephb1/b2	✓	✓	Mendes et al. (2006)
Ephb1/b3	✓	✓	Mendes et al. (2006)
Ephb1/Ephb4	✓	✓	Mendes et al. (2006)
Ephb2/b3	✓	✓	Mendes et al. (2006)
Efnb3/Ephb1	✓	✓	Mendes et al. (2006)
Efnb3/Ephb2	✓	✓	Mendes et al. (2006)
Efnb3/Epha4	✓[a]	✓[a]	Mendes et al. (2006)
Fzd3	✓		Wang et al. (2002, 2006)
Ntn1	✓		Fothergill et al. (2014), Serafini et al. (1996)
Nrp1(Sema)	✓	✓	Gu et al. (2003), Piper et al. (2009)
Robo1		✓	Andrews et al. (2006)
Robo1/Robo2		✓	López-Bendito et al. (2007)
Ryk		✓	Keeble et al. (2006)
Slit2		✓	Bagri et al. (2002), Shu et al. (2003c), Unni et al. (2012)
Slit3		✓	Unni et al. (2012)
Slit1/Slit3		✓	Unni et al. (2012)
Slit1/Slit2	✓		Unni et al. (2012)
Sema3c	✓	✓	Niquille et al. (2009)
Efnb1	✓		Bush and Soriano (2009)
Efnb1ΔV[a]	✓		Bush and Soriano (2009)

AgCC, Complete agenesis of the corpus callosum; pAgCC, partial agenesis of the corpus callosum.
[a]Mice express EFNB1 that lacks a valine required for PDZ reverse signaling.

depending on the neuronal population (Fothergill et al., 2014). For example, NTN1 attracts cingulate callosal axons to the midline. It does not generally act as an attractant for neocortical callosal axons, but rather modulates the repellent actions of another ligand, SLIT2 (see later).

14.2.6.4 Midline Crossing

Eph receptors (EPH) and their ephrin ligands (EFN) are axon guidance genes that influence growth and guidance through cell–cell contact, as both the receptor and the ligands are anchored to the membrane. Eph receptor tyrosine kinases are divided into two subclasses, A and B, which differ in their binding affinities for ephrins. Ephrins are also divided into an A subclass with a glycosyl-phosphatidylinositol (GPI) anchor to the plasma membrane and a B subclass that possesses transmembrane domains to embed the ligand within the membrane (Egea and Klein, 2007). Ephs and ephrins are widely expressed at the telencephalic midline on both callosal axons and glial guidepost cells (Table 14.1; Mendes et al., 2006). Generally, EFNA class ligands can bind EPHA receptors and EFNB class ligands can bind EPHB receptors, with the exception that EPHA4 binds both EFNA and EFNB ligands (Pasquale, 2005).

Eph-ephrin binding can elicit both forward signaling within the EPH-expressing cell but also reverse signaling within the ephrin-expressing cell, and Ephs and ephrins can also interact on the membrane when expressed in the same cell (Egea and Klein, 2007). For these reasons, the functional role of Eph/ephrin signaling in corpus callosum development remains elusive. *Ephb1*, *Ephb2*, *Epha5*, and *Efnb3* appear to be most important for corpus callosum formation as knockout mice for these genes show partial or complete AgCC with Probst bundles; however, EPHB3 appears to also be involved as *Ephb2/b3* double knockout mice show increased penetrance compared to *Ephb2* single knockout mice (Table 14.2; Hu et al., 2003; Mendes et al., 2006). *In vitro* studies suggest substrate-embedded EFNB1, EFNB2, and EFNA5 stimulate neurite outgrowth of cortical axons (Hu et al., 2003; Mendes et al., 2006), suggesting that ephrins may act as growth-promoting cues during midline crossing.

All three *Slit* family members are expressed in the telencephalic midline and their receptors, roundabout homolog (ROBO)1 and 2 but not 3, are expressed on callosal axons (Table 14.1). Analyses of single and double mutant knockout mice have demonstrated that corpus callosum formation absolutely requires SLIT2 expression, but SLIT1, SLIT3, ROBO1, and ROBO2 are also involved (Table 14.2; Bagri et al., 2002; Shu et al., 2003c; Unni et al., 2012). *Slit2*, *Slit3*, *Robo1*, *Slit1/3*, and *Robo1/2* knockout mice have partial AgCC with Probst bundles, and *Slit1/2* double knockout mice show complete AgCC with Probst bundles (Table 14.2; Unni et al., 2012). DMRI and carbocyanine dye tracing experiments have revealed that *Robo1*, *Robo1/2*, and *Slit2* knockout mice display ipsilateral misguidance of callosal axons ventrally into the septum (López-Bendito et al., 2007; Unni et al., 2012). SLIT2, expressed by the indusium griseum glia and the glial wedge cells, acts to repel callosal axons across the midline and prevent them from entering the septum. This is based on the observations that (i) SLIT2 repels both cingulate and neocortical explants *in vitro* (Shu and Richards, 2001; Unni et al., 2012; Fothergill, Donahoo et al., 2014), (ii) the glial wedge repels callosal axons *in vitro*, an effect which is abolished when ROBO1 and ROBO2 blocking peptides are added to the culture (Shu et al., 2003c), (iii) repression of SLIT2 signaling in the glial wedge on one side of the brain causes axons to be misguided and project aberrantly into the septum, and (iv) overexpression of SLIT2 at the cortico-septal boundary on one side of the brain causes axon misguidance and the formation of Probst bundles on the side ipsilateral to the manipulation (Lavado et al., 2014).

A recent study suggests that axon guidance families interact to generate the appropriate axonal response to different cues. DCC is a transmembrane receptor that is expressed on cingulate callosal axons, while its ligand, NTN1, is expressed in the ventral septum and in the indusium griseum. *Dcc* knockout mice and *Ntn1*-deficient (hypomorphic mutation) mice have fully penetrant complete AgCC with Probst bundles (Table 14.2; Serafini et al., 1996; Fazeli et al., 1997; Ren et al., 2007; Fothergill et al., 2014). The interaction of ROBO1 and DCC receptors in response to simultaneous NTN1 and SLIT2 cues is important for the silencing of SLIT2-mediated repulsion of callosal axons before they reach the midline. In the spinal cord, DCC and ROBO1 proteins interact through their CC1 and P3 intracellular domains, respectively (Stein and Tessier-Lavigne, 2001). In E17 neocortical explants (precrossing), Fothergill et al. (2014) demonstrated that SLIT2 was unable to repel neocortical axons in the presence of NTN1, an effect that was dependent on both ROBO1 and DCC. They showed that the level of DCC protein decreases *in vivo* from E17 until birth, at which point SLIT2 is able to repel neocortical axons. This study was the first to report how callosal axons might respond to multiple guidance cues while navigating the complex midline environment.

14.2.6.5 Exit from the Midline

After callosal axons cross the midline, they must stop being attracted to midline guidepost cells and navigate away from the midline into the contralateral hemisphere. As described above, SLIT2 plays an important role in the initial repulsion of axons from the midline environment (Shu et al., 2003c). The axon guidance receptor Receptor-like tyrosine kinase (RYK) has also been implicated in the regulation of guidance away from the midline. *Ryk* knockout mice have a unique callosal phenotype. Carbocyanine dye tracing in these mice reveals callosal axons crossing the midline and forming Probst bundles in the contralateral hemisphere (Keeble et al., 2006). E18 (postcrossing) but not E16 or E17 neocortical explants are repelled by WNT5A in a RYK- and calcium/calmodulin-dependent protein kinase (CAMK2A)-dependent fashion (Keeble et al., 2006; Hutchins et al., 2011). Furthermore, *Ryk* and *Camk2a* knockdown results in reduced outgrowth and postcrossing misguidance of callosal axons *in vivo* (Hutchins et al., 2011). WNT/RYK signaling to downstream CAMK2A modulates calcium activity within the axon, initiating repulsive growth and guidance away from the midline in mice.

14.2.6.6 Contralateral Targeting

The mechanisms by which callosal axons innervate appropriate heterotopic and homotopic regions in the contralateral cortex, after midline crossing, are less well studied but these processes may be disrupted in a number of neurological and mental disorders (Fenlon and Richards, 2015). However, two general rules governing this guidance have been identified. The first is that axon position within the callosal tract during midline crossing is important for targeting in the contralateral hemisphere, and the second is that activity matching between cortical hemispheres regulates the targeting of a subset of callosal projections.

Callosal axons from different cortical areas are segregated within the corpus callosum during midline crossing. For example, cingulate axons cross dorsal to neocortical callosal axons, and axons from the medial neocortex (primary motor cortex) cross the midline dorsal to axons from lateral regions of the neocortex (primary somatosensory cortex; Piper et al., 2009b; Zhou et al., 2013). SEMA3A/NRP1 signaling regulates this topography. SEMA3A is expressed within the cortical plate in a high caudal and lateral to low rostral and medial gradient, and NRP1 is highly expressed on callosal axons derived from caudal and medial regions. Cell-autonomous disruption of *Nrp1* expression within the medial primary motor cortex causes mixing of callosal fibers derived from the primary somatosensory and primary motor cortex within the callosal tract at the midline and aberrant targeting of the primary motor cortex to the contralateral somatosensory cortex. Interestingly, *Sema3a* knockout mice display normal midline crossing; however, primary motor and primary somatosensory callosal axons mix at the midline and target the incorrect cortical region in the opposite hemisphere rather than remaining segregated and innervating their homotopic regions (Zhou et al., 2013).

Mouse studies of contralateral callosal targeting have identified that callosal axons derived from the primary somatosensory, motor, and visual areas sparsely innervate homotopic regions and densely innervate the border region between primary and secondary areas. This dense targeting to the border region is highly sensitive to manipulations in activity in both the somatosensory and visual systems (Olavarria and Van Sluyters, 1995; Mizuno et al., 2007; Wang et al., 2007; Ageta-Ishihara et al., 2009; Huang et al., 2013; Suárez et al., 2014). Furthermore, a balance of both intrinsic (endogenous cortical activity) and extrinsic (from the sensory periphery) activity between ipsilateral and contralateral areas appears to be particularly important for targeting (Mizuno et al., 2010; Huang et al., 2013; Suárez et al., 2014). Unilaterally disrupting sensory activity from the periphery or endogenous cortical activity in one hemisphere also disrupts the major projection from the primary somatosensory cortex which targets the border region between the primary and secondary somatosensory regions in the contralateral hemisphere (Mizuno et al., 2007; Wang et al., 2007; Suárez et al., 2014). Furthermore, bilaterally disrupting sensory input or endogenous cortical activity rescues the targeting to the border region (Suárez et al., 2014). The activity-dependent molecular mechanisms that regulate this axon targeting remain to be elucidated.

14.3 dMRI-BASED IMAGING OF BRAIN CONNECTIVITY

14.3.1 Modeling the Development of Normal and Abnormal White Matter Tracts

In addition to investigating the cellular and molecular mechanisms that regulate the development of white matter tracts in the brain, sophisticated neuroimaging techniques allow the investigation of anatomical connectivity in living and postmortem human and animal brains. This provides a means for translational and comparative studies of brain development and plasticity. dMRI generates detailed information regarding axonal white matter tracts based on the diffusion of water molecules within the brain (Mori and Zhang, 2006) as described in Section 14.2.2 in the context of the developing cortical plate. However, the dynamic structural changes of the developing brain are not limited to early cortical development. During preterm and neonatal life, the white matter undergoes significant changes in microstructure and organization. At this time, most long-range white matter tracts are still being established and stabilized, and the mean diffusivity of the white matter decreases while the FA tends to increase (Hüppi et al., 1998; Neil et al., 1998; Mukherjee et al., 2002; Huang et al., 2006). These changes are correlated with the progressive myelination of maturing white matter tracts, as well as earlier changes in the cytoskeletal complexity of axons, changes in axon diameter and packing, population of these tracts with oligodendrocytes, and concomitant decreases in the water content of the tissue (Hüppi et al., 1998; Neil et al., 1998; Huang et al., 2006; Hüppi and Dubois, 2006; Huang et al., 2009). The exact contribution of each of these processes to the changes in the dMRI signal is not entirely clear; however, by further investigating changes in specific components of the dMRI tensor, a clearer picture can be drawn about how different changes in microstructure are reflected in the tensor model. For example, the diffusion parallel to a white matter tract (the principal eigenvalue of a tensor) is unlikely to be affected by an increase in myelination; it does, however, change in response to differences in water content, and barriers to diffusion such as the cytoskeleton (Mukherjee et al., 2002). Decreases in the minor eigenvalues reflect a decrease in the diffusion of water perpendicular to the white matter tract, and so would be expected to correlate with processes such as myelination that occur in early neonatal life (Mukherjee et al., 2002). A significant amount of diffusion is also likely to occur extracellularly within these tissues. In this case, changes in the two minor eigenvalues may also reflect decreased diffusion between axons as the tissue water content decreases and barriers to extracellular diffusion (such as more axons) increase (Norris, 2001).

dMRI can involve modeling of diffusion in over 100 directions. Under these circumstances, the simple tensor model is inadequate and it becomes more worthwhile to model connectivity between specific brain regions by means of tractography. An initial approach to white matter tractography was to generate streamlines starting from a specified location (the "seed") by interpreting the major eigenvalue of each voxel as the approximate direction of the white matter and propagating a continuous line from voxel to voxel by linking the major eigenvectors into a coherent virtual tract (Conturo et al., 1999; Mori et al., 1999; Mori and van Zijl, 2002). However, the value of this approach is limited because it fails to resolve distinct populations of axons contributing to crossing white matter fibers and different directions of anisotropies within the same voxel (Alexander et al., 2001; Frank, 2001; Alexander et al., 2002; Tuch et al., 2002). This is not ideal, as the complexity of the mammalian brain necessitates that essentially every axonal projection must encounter other axonal projections with different orientations. To overcome this limitation, various algorithms have been proposed to model complex white matter architectures, including diffusion-spectrum imaging (Wedeen et al., 2005), high angular resolution diffusion imaging, q-ball imaging, and constrained spherical deconvolution (Tournier et al., 2004, 2007; Jbabdi and Johansen-Berg, 2011). These more advanced interpretations of the MR signal are based on the assumption that each fiber direction independently contributes to the final tensor model, and they are being further developed to resolve multiple fiber populations, including situations in which crossing fibers meet when axons of different orientations enter the same tract. Using these approaches, it is possible to virtually dissect white matter tracts within the brain, identify novel white matter structures that result from neurodevelopmental disorders, and determine changes in scalar values (such as FA) along white matter tracts.

14.3.2 Corpus Callosum Dysgenesis and Visualizing Axonal Plasticity During Brain Development

Corpus callosum dysgenesis in human or animal brains can be identified by a reduction or complete absence of midline crossing fibers as evident in structural T1- or T2-weighted images or, alternatively, can be identified from modeling with whole-brain tractography or seeded from regions of interest at the midline or in the cortex (Figures 14.5C,D and 14.6). As noted above, AgCC in both human and mouse brains is commonly associated with the formation of Probst bundles, which occur when callosal axons fail to cross the midline and instead form longitudinal intrahemispheric bundles situated in both hemispheres. These bundles run anterior to posterior and can be easily visualized by dMRI and tractography (Figures 14.5C,D and 14.6). As described in Section, mouse models of AgCC demonstrate that the formation of Probst bundles is associated with failure of callosal axons to cross the midline, rather than a defect in axonal generation, specification or navigation toward the midline. The development of Probst bundles has not been extensively investigated but they are electrically active in rodents (Lefkowitz et al., 1991), supporting the possibility that functional connections may exist within the bundle.

In mice, corpus callosum dysgenesis is associated with loss of axon guidance genes (Table 14.2), loss of genes that regulate midline formation as well as guidepost cell differentiation and migration (described in Section). Using *Ntn1* and *Dcc* knockout mice as a model, the anatomy of the Probst bundle has been described using dMRI with tractography combined with carbocyanine dye tracing (Ren et al., 2007). It was demonstrated that axons within the dorsal Probst bundle are more organized, running in a rostral to caudal direction along the midline. In contrast, ventral parts of the Probst bundle appeared more disorganized and had random trajectories. In partial AgCC mouse models, MRI combined with carbocyanine dye tracing can determine which populations of callosal axons contribute to the formation of a Probst bundle. In *Robo1* and *Slit2* knockout mice, tractography modeling of interhemispheric projections has revealed that interhemispheric projections from the visual cortex appear to be reduced, whereas axons from the motor and parietal cortices aberrantly enter both the ipsilateral and contralateral septum (Unni et al., 2012).

The use of dMRI has recently provided a platform for the discovery of the remarkable plasticity in the development of axonal tracts in the brain of human subjects with corpus callosum dysgenesis. dMRI coupled with tractography has identified a series of novel interhemispheric connections, originating from cortical areas, that display an impressive degree of developmental plasticity. Of particular note, ectopic tracts projecting through the anterior and posterior commissures have been identified in individuals with AgCC that do not exist in normal individuals (Tovar-Moll et al., 2014). These novel connections appear to consistently connect homotopic regions of the parietal cortex, and therefore could partially compensate for the lack of interhemispheric connectivity normally mediated by the corpus callosum. In individuals with partial AgCC, where a small remnant of callosal axons still cross the midline

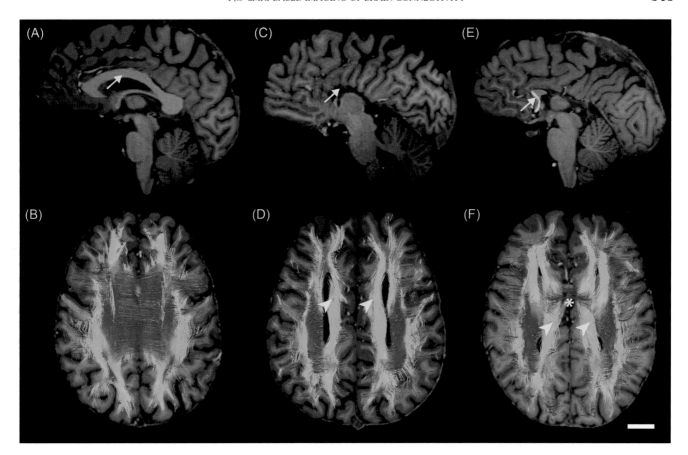

FIGURE 14.6 **Anatomical features of agenesis of corpus callosum in humans revealed by T1-weighted MRI and-based tractography.** dMRI The normal corpus callosum is a prominent white matter structure that can be easily visualized in a sagittal view of a T1-weighted MR image (A, yellow arrow). dMRI-based tractography reconstructs the corpus callosum as a large band of fibers running mediolaterally (red) connecting the hemispheres (B, horizontal view). In individuals with agenesis of the corpus callosum, the corpus callosum can be completely absent (C, yellow arrow), or a partial remnant may remain (partial agenesis of the corpus callosum, yellow arrow in E). In both cases, callosal axons that would normally cross the midline can be identified by tractography as large, longitudinal Probst bundles (arrowheads in D and F). In partial agenesis, a small remnant of callosal connections can be identified by tractography (asterisk in F). Scale bar= 2 cm.

(Figure 14.6E and F), an additional asymmetric ectopic connection has been identified by tractography connecting the frontal cortex with the contralateral parieto-occipital region (Tovar-Moll et al., 2007; Wahl et al., 2009; Bénézit et al., 2014; Tovar-Moll et al., 2014). This ectopic connection, which has been termed the "sigmoid bundle," represents a dramatic change in brain connectivity compared to the mostly homotopic organization of callosal fibers seen in the normal corpus callosum (Hofer and Frahm, 2006). These ectopic cortical projections appear to be functionally connected as the "strength" of structural connectivity (determined by quantifying the number of streamlines between the two cortical areas) positively correlates with the degree of functional connectivity (correlation in blood-oxygen-level dependent (BOLD) signal) between the two parietal areas (Tovar-Moll et al., 2014).

14.3.3 Validation of dMRI Methods to Model Brain Connectivity

dMRI and tractography allow brain connectivity to be modeled noninvasively in three-dimensional space, and are therefore important tools in human and animal studies. However, dMRI-based tractography is an indirect method based on a measure of water diffusion that can be influenced by multiple microstructural properties leading to several important limitations (Jbabdi and Johansen-Berg, 2011). These include:

1. The direction of a connection cannot be inferred from dMRI and tractography. This is a fundamental limitation of dMRI as the diffusion of water molecules is symmetrical along axial and radial axes.

2. Inability to establish precise cortical origins and terminations (in terms of both identifying cortical layers and determining where a connection first enters the cortex). This is primarily due to high isotropic diffusion within the grey matter that does not allow for accurate fiber tracking along directions of principal diffusion.
3. Ambiguity in determining multiple intravoxel axon trajectories from diffusion measurements. This is particularly important as even advanced algorithms for modeling white matter structure within a voxel can lead to ambiguous interpretations of axonal structure and direction. Consequently, tractography can be highly susceptible to false positives (reconstructing tracts that do not exist biologically) and false negatives (failing to reconstruct tracts that do exist).

It is therefore important to validate tractography findings with some form of "ground truth" of connectivity. Traditionally, this has been achieved *in vivo* with carbocyanine dye tract tracing. This approach has been particularly useful in modeling ectopic white matter tracts in animal models of altered brain connectivity, such as the structure of Probst bundles in AgCC mice (Ren et al., 2007). A major limitation of traditional tract tracing techniques, however, is that only a small population of axons can be studied, corresponding to where the tracer is injected (usually a single injection site in the brain). When multiple injection sites are used, it is difficult to reconstruct each labeled tract in three-dimensional space. One of the most comprehensive projects for establishing a gold standard for mammalian brain connectivity has been the Allen Mouse Brain Connectivity Atlas (Oh et al., 2014; http://connectivity.brain-map.org/). It comprises a mesoscale connectome of the mouse brain based on the digitization and quantification of large numbers of viral tracings of axons. This can overcome the traditional limitations of viewing the three-dimensional structure of labeled axons, but has also presented new challenges in that these data still contain false positive signals and cannot easily distinguish the connection strength or termination sites of axons (Oh et al., 2014). Given the difficulties in tract tracing, it is useful to employ complementary MRI or other imaging techniques, in addition to tracing, to correlate with structural changes determined by tractography. The most important tools for this are functional MRI, which measures changes in the BOLD signal as an indirect measure of brain activity, and magnetoencephalography, which detects changes in magnetic fields that occur secondarily to the electrical activity that occurs normally in the brain.

14.3.4 Modeling Whole-Brain Connectivity

A significant advantage of using dMRI to investigate structural connectivity in the brain is that it can be used to study whole-brain connectivity in three dimensions in a way that is not possible using histological tract tracing. Given the complexity of how the brain is structurally wired, it is not always helpful to consider individual white matter tracts in isolation. Instead, dMRI can reveal complex structural networks between distant brain areas. This idea has been extended by considering the brain as a simple graph, with each anatomical area represented as a node, and each connection as an edge between nodes (Zalesky et al., 2010; Bullmore and Bassett, 2011). Describing connectivity in the human brain using a graph theory approach allows connectional data derived from either dMRI or functional MRI to be described as a macroscale connectome (Sporns, 2011).

14.4 CONCLUSIONS

The anatomical and functional connectivity of the brain underpins its capacity for information processing and storage. As the largest fiber tract in the human brain, the corpus callosum provides an excellent model for understanding the mechanisms that regulate the development of brain wiring. Developmental dysgenesis of the corpus callosum is a common neurodevelopmental defect, the etiology of which can be better understood by applying our knowledge of the variety of developmental events required for the correct formation of this axon tract. A large number of genes have been implicated in mouse callosal development (Donahoo and Richards, 2009) and these are being applied to understanding human callosal disorders (Edwards et al., 2014b).

In addition to genetics, neuroimaging is a critical platform for translational discoveries based on human and mouse studies by providing a wealth of spatial and temporal information about anatomical and functional connectivity. It can be applied longitudinally in living subjects to examine the development and plasticity of axonal wiring in normal and pathological brains, as well as monitoring the effect of treatment. Furthermore, studies that relate cognitive abilities with brain wiring can test hypotheses about the degree to which connectivity underpins brain function. Currently, limitations include an inability to unambiguously resolve crossing fibers and microstructural details and a limited ability to validate tractography findings. In brain development, new advances are required to improve human fetal imaging *in utero* to provide faster scanning with higher resolution and minimal movement artifacts in a safe manner. Additional methods for neuroimaging anatomical and gene expression changes in smaller groups of neurons, or even individual neurons forming circuits, are urgently required as these will allow real-time changes in brain connectivity and plasticity, and the mechanisms regulating these changes, to be discovered. It is an extremely exciting time for this field of science, as the fruits of decades of basic science research can now be applied to identifying the causes of human disorders of brain wiring and finding treatments that will restore function.

References

Aaku-Saraste, E., Hellwig, A., Huttner, W.B., 1996. Loss of occludin and functional tight junctions, but not ZO-1, during neural tube closure—remodeling of the neuroepithelium prior to neurogenesis. Dev. Biol. 180, 664–679.

Ageta-Ishihara, N., Takemoto-Kimura, S., Nonaka, M., Adachi-Morishima, A., Suzuki, K., Kamijo, S., et al., 2009. Control of cortical axon elongation by a GABA-driven Ca^{2+}/Calmodulin-dependent protein kinase cascade. J. Neurosci. 29, 13720–13729.

Aggarwal, M., Mori, S., Shimogori, T., Blackshaw, S., Zhang, J., 2010. Three-dimensional diffusion tensor microimaging for anatomical characterization of the mouse brain. Magn. Reson. Med. 64, 249–261.

Aggarwal, M., Gobius, I., Richards, L.J., Mori, S., 2014. Diffusion MR microscopy of cortical development in the mouse embryo. Cereb. Cortex 25, 1970–1980.

Ahmed, G., Shinmyo, Y., Ohta, K., Islam, S.M., Hossain, M., Naser, I.B., et al., 2011. Draxin inhibits axonal outgrowth through the netrin receptor DCC. J. Neurosci. 31, 14018–14023.

Alcamo, E.A., Chirivella, L., Dautzenberg, M., Dobreva, G., Fariñas, I., Grosschedl, R., et al., 2008. Satb2 regulates callosal projection neuron identity in the developing cerebral cortex. Neuron 57, 364–377.

Alexander, A.L., Hasan, K.M., Lazar, M., Tsuruda, J.S., Parker, D.L., 2001. Analysis of partial volume effects in diffusion-tensor MRI. Magn. Reson. Med. 45, 770–780.

Alexander, D.C., Barker, G.J., Arridge, S.R., 2002. Detection and modeling of non-Gaussian apparent diffusion coefficient profiles in human brain data. Magn. Reson. Med. 48, 331–340.

Alfano, C., Studer, M., 2013. Neocortical arealization: evolution, mechanisms, and open questions. Dev. Neurobiol. 73, 411–447.

Alfano, C., Magrinelli, E., Harb, K., Hevner, R.F., Studer, M., 2014. Postmitotic control of sensory area specification during neocortical development. Nat. Commun. 5, 5632.

Allendoerfer, K.L., Shatz, C.J., 1994. The subplate, a transient neocortical structure: its role in the development of connections between thalamus and cortex. Ann. Rev. Neurosci. 17, 185–218.

Amaniti, E.-M., Hasenpusch-Theil, K., Li, Z., Magnani, D., Kessaris, N., Mason, J.O., et al., 2013. Gli3 is required in Emx1+ progenitors for the development of the corpus callosum. Dev. Biol. 376, 113–124.

Amaniti, E.-M., Fu, C., Lewis, S., Saisana, M., Magnani, D., Mason, J.O., et al., 2015. Expansion of the piriform cortex contributes to corticothalamic pathfinding defects in Gli3 conditional mutants. Cereb. Cortex 25, 460–471.

Anderson, S.A., Marin, O., Horn, C., Jennings, K., Rubenstein, J.L., 2001. Distinct cortical migrations from the medial and lateral ganglionic eminences. Development 128, 353–363.

Anderson, S.A., Kaznowski, C.E., Horn, C., Rubenstein, J.L., McConnell, S.K., 2002. Distinct origins of neocortical projection neurons and interneurons *in vivo*. Cereb. Cortex 12, 702–709.

Andrews, W., Liapi, A., Plachez, C., Camurri, L., Zhang, J., Mori, S., et al., 2006. Robo1 regulates the development of major axon tracts and interneuron migration in the forebrain. Development 133, 2243–2252.

Angevine, J.B., Sidman, R.L., 1961. Autoradiographic study of cell migration during histogenesis of cerebral cortex in the mouse. Nature 192, 766–768.

Arlotta, P., Molyneaux, B.J., Chen, J., Inoue, J., Kominami, R., Macklis, J.D., 2005. Neuronal subtype-specific genes that control corticospinal motor neuron development *in vivo*. Neuron 45, 207–221.

Armentano, M., Chou, S.-J., Tomassy, G.S., Leingärtner, A., O'Leary, D.D., Studer, M., 2007. COUP-TFI regulates the balance of cortical patterning between frontal/motor and sensory areas. Nat. Neurosci. 10, 1277–1286.

Bagri, A., Marín, O., Plump, A.S., Mak, J., Pleasure, S.J., Rubenstein, J.L.R., et al., 2002. Slit proteins prevent midline crossing and determine the dorsoventral position of major axonal pathways in the mammalian forebrain. Neuron 33, 233–248.

Ball, G., Srinivasan, L., Aljabar, P., Counsell, S.J., Durighel, G., Hajnal, J.V., et al., 2013. Development of cortical microstructure in the preterm human brain. Proc. Natl. Acad. Sci. U.S.A. 110, 9541–9546.

Ball, G., Pazderova, L., Chew, A., Tusor, N., Merchant, N., Arichi, T., et al., 2015. Thalamocortical connectivity predicts cognition in children born preterm. Cereb. Cortex [Epub ahead of print].

Baranek, C., Dittrich, M., Parthasarathy, S., Bonnon, C.G., Britanova, O., Lanshakov, D., et al., 2012. Protooncogene Ski cooperates with the chromatin-remodeling factor Satb2 in specifying callosal neurons. Proc. Natl. Acad. Sci. U.S.A. 109, 3546–3551.

Beaulieu, C., 2002. The basis of anisotropic water diffusion in the nervous system—a technical review. NMR Biomed. 15, 435–455.

Bedogni, F., Hodge, R.D., Elsen, G.E., Nelson, B.R., Daza, R.A., Beyer, R.P., et al., 2010. Tbr1 regulates regional and laminar identity of postmitotic neurons in developing neocortex. Proc. Natl. Acad. Sci. U.S.A. 107, 13129–13134.

Bénézit, A., Hertz-Pannier, L., Dehaene-Lambertz, G., Monzalvo, K., Germanaud, D., Duclap, D., et al., 2014. Organising white matter in a brain without corpus callosum fibres. Cortex 63C, 155–171.

Bishop, K.M., Goudreau, G., O'Leary, D.D., 2000. Regulation of area identity in the mammalian neocortex by Emx2 and Pax6. Science 288, 344–349.

Boretius, S., Michaelis, T., Tammer, R., Ashery-Padan, R., Frahm, J., Stoykova, A., 2009. *In vivo* MRI of altered brain anatomy and fiber connectivity in adult pax6 deficient mice. Cereb. Cortex 19, 2838–2847.

Bormuth, I., Yan, K., Yonemasu, T., Gummert, M., Zhang, M., Wichert, S., et al., 2013. Neuronal basic helix–loop–helix proteins neurod2/6 regulate cortical commissure formation before midline interactions. J. Neurosci. 33, 641–651.

Boyd, E.H., Pandya, D.N., Bignall, K.E., 1971. Homotopic and nonhomotopic interhemispheric cortical projections in the squirrel monkey. Exp. Neurol. 32, 256–274.

Britanova, O., de Juan Romero, C., Cheung, A., Kwan, K.Y., Schwark, M., Gyorgy, A., et al., 2008. Satb2 is a postmitotic determinant for upper-layer neuron specification in the neocortex. Neuron 57, 378–392.

Bullmore, E.T., Bassett, D.S., 2011. Brain graphs: graphical models of the human brain connectome. Annu. Rev. Clin. Psychol. 7, 113–140.

Bush, J.O., Soriano, P., 2009. Ephrin-B1 regulates axon guidance by reverse signaling through a PDZ-dependent mechanism. Genes Dev. 23, 1586–1599.

Catania, K.C., 2001. Areal and callosal connections in the somatosensory cortex of the star-nosed mole. Somatosens. Mot. Res. 18, 303–311.

Caviness, V.S., 1982. Neocortical histogenesis in normal and reeler mice: a developmental study based upon [3 H] thymidine autoradiography. Brain Res. 4, 293–302.

Cánovas, J., Berndt, F.A., Sepúlveda, H., Aguilar, R., Veloso, F.A., Montecino, M., et al., 2015. The specification of cortical subcerebral projection neurons depends on the direct repression of TBR1 by CTIP1/BCL11a. J. Neurosci. 35, 7552–7564.

Chen, B., Schaevitz, L.R., McConnell, S.K., 2005. Fezl regulates the differentiation and axon targeting of layer 5 subcortical projection neurons in cerebral cortex. Proc. Natl. Acad. Sci. U.S.A. 102, 17184–17189.

Chen, B., Wang, S.S., Hattox, A.M., Rayburn, H., Nelson, S.B., McConnell, S.K., 2008. The Fezf2–Ctip2 genetic pathway regulates the fate choice of subcortical projection neurons in the developing cerebral cortex. Proc. Natl. Acad. Sci. U.S.A. 105, 11382–11387.

Chenevert, T.L., Brunberg, J.A., Pipe, J.G., 1990. Anisotropic diffusion in human white matter: demonstration with MR techniques *in vivo*. Radiology 177, 401–405.

Chenn, A., McConnell, S.K., 1995. Cleavage orientation and the asymmetric inheritance of Notchl immunoreactivity in mammalian neurogenesis. Cell 82, 631–641.

Clancy, B., Teague-Ross, T.J., Nagarajan, R., 2009. Cross-species analyses of the cortical GABAergic and subplate neural populations. Front. Neuroanat. 3, 20.

Clegg, J.M., Conway, C.D., Howe, K.M., Price, D.J., Mason, J.O., Turnbull, J.E., et al., 2014. Heparan sulfotransferases Hs6st1 and Hs2st keep Erk in check for mouse corpus callosum development. J. Neurosci. 34, 2389–2401.

Conturo, T.E., Lori, N.F., Cull, T.S., Akbudak, E., Snyder, A.Z., Shimony, J.S., et al., 1999. Tracking neuronal fiber pathways in the living human brain. Proc. Natl. Acad. Sci. U.S.A. 96, 10422–10427.

Conway, C.D., Howe, K.M., Nettleton, N.K., Price, D.J., Mason, J.O., Pratt, T., 2011. Heparan sulfate sugar modifications mediate the functions of slits and other factors needed for mouse forebrain commissure development. J. Neurosci. 31, 1955–1970.

De Carlos, J.A., O'leary, D.D., 1992. Growth and targeting of subplate axons and establishment of major cortical pathways [published erratum appears in J Neurosci 1993 Mar; 13(3): following table of contents]. J. Neurosci. 12, 1194–1211.

Dehay, C., Horsburgh, G., Berland, M., Killackey, H., Kennedy, H., 1991. The effects of bilateral enucleation in the primate fetus on the parcellation of visual cortex. Brain Res. 62, 137–141.

Donahoo, A.-L.S., Richards, L.J., 2009. Understanding the mechanisms of callosal development through the use of transgenic mouse models. Semin. Pediatr. Neurol. 16, 127–142.

Edwards, T.J., Sherr, E.H., Barkovich, A.J., Richards, L.J., 2014. Clinical, genetic and imaging findings identify new causes for corpus callosum development syndromes. Brain 137, 1579–1613.

Egea, J., Klein, R., 2007. Bidirectional Eph–ephrin signaling during axon guidance. Trends Cell Biol. 17, 230–238.

Fame, R.M., MacDonald, J.L., Macklis, J.D., 2011. Development, specification, and diversity of callosal projection neurons. Trends Neurosci. 34, 41–50.

Fazeli, A., Rayburn, H., Simons, J., Bronson, R.T., Gordon, J.I., Tessier-Lavigne, M., et al., 1997. Phenotype of mice lacking functional deleted in colorectal cancer (Dcc) gene. Nature 386, 796–804.

Fenlon, L.R., Richards, L.J., 2015. Contralateral targeting of the corpus callosum in normal and pathological brain function. Trends Neurosci. 38, 264–272.

Fernandes, M., Gutin, G., Alcorn, H., McConnell, S.K., Hébert, J.M., 2007. Mutations in the BMP pathway in mice support the existence of two molecular classes of holoprosencephaly. Development 134, 3789–3794.

Finger, J.H., Bronson, R.T., Harris, B., Johnson, K., Przyborski, S.A., Ackerman, S.L., 2002. The netrin 1 receptors Unc5h3 and Dcc are necessary at multiple choice points for the guidance of corticospinal tract axons. J. Neurosci. 22, 10346–10356.

Fothergill, T., Donahoo, A.L., Douglass, A., Zalucki, O., Yuan, J., Shu, T., et al., 2014. Netrin-DCC signaling regulates corpus callosum formation through attraction of pioneering axons and by modulating Slit2-mediated repulsion. Cereb. Cortex 24, 1138–1151.

Frank, L.R., 2001. Anisotropy in high angular resolution diffusion-weighted MRI. Magn. Reson. Med. 45, 935–939.

Frost, D.O., 1982. Anomalous visual connections to somatosensory and auditory systems following brain lesions in early life. Brain Res. 3, 627–635.

Frost, D.O., Boire, D., Gingras, G., Ptito, M., 2000. Surgically created neural pathways mediate visual pattern discrimination. Proc. Natl. Acad. Sci. U.S.A. 97, 11068–11073.

Gadisseux, J.F., Evrard, P., Mission, J.P., Caviness, V.S., 1992. Dynamic changes in the density of radial glial fibers of the developing murine cerebral wall: a quantitative immunohistological analysis. J. Comp. Neurol. 322, 246–254.

Garcez, P.P., Henrique, N.P., Furtado, D.A., Bolz, J., Lent, R., Uziel, D., 2007. Axons of callosal neurons bifurcate transiently at the white matter before consolidating an interhemispheric projection. Eur. J. Neurosci. 25, 1384–1394.

Gazzaniga, M.S., 2005. Forty-five years of split-brain research and still going strong. Nat. Rev. Neurosci. 6, 653–659.

Gelman, D.M., Martini, F.J., Nóbrega-Pereira, S., Pierani, A., Kessaris, N., Marin, O., 2009. The embryonic preoptic area is a novel source of cortical GABAergic interneurons. J. Neurosci. 29, 9380–9389.

Ghosh, A., Shatz, C.J., 1993. A role for subplate neurons in the patterning of connections from thalamus to neocortex. Development 117, 1031–1047.

Ghosh, A., Antonini, A., McConnell, S., Shatz, C., 1990. Requirement for subplate neurons in the formation of thalamocortical connections. Nature 347, 179.

Gobius, I., Richards, L., 2011. Creating Connections in the Developing Brain. Morgan & Clay pool Publishers.

Grove, E.A., Fukuchi-Shimogori, T., 2003. Generating the cerebral cortical area map. Ann. Rev. Neurosci. 26, 355–380.

Gu, C., Rodriguez, E.R., Reimert, D.V., Shu, T., Fritzsch, B., Richards, L.J., et al., 2003. Neuropilin-1 conveys semaphorin and VEGF signaling during neural and cardiovascular development. Dev. Cell 5, 45–57.

Gupta, R.K., Hasan, K.M., Trivedi, R., Pradhan, M., Das, V., Parikh, N.A., et al., 2005. Diffusion tensor imaging of the developing human cerebrum. J. Neurosci. Res. 81, 172–178.

Han, W., Kwan, K.Y., Shim, S., Lam, M.M., Shin, Y., Xu, X., et al., 2011. TBR1 directly represses Fezf2 to control the laminar origin and development of the corticospinal tract. Proc. Natl. Acad. Sci. U.S.A. 108, 3041–3046.

Hand, R., Polleux, F., 2011. Neurogenin2 regulates the initial axon guidance of cortical pyramidal neurons projecting medially to the corpus callosum. Neural Dev. 6, 1–16.

Harris, K.D., Shepherd, G.M., 2015. The neocortical circuit: themes and variations. Nat. Neurosci. 18, 170–181.

Hartfuss, E., Galli, R., Heins, N., Götz, M., 2001. Characterization of CNS precursor subtypes and radial glia. Dev. Biol. 229, 15–30.

Harwell, C.C., Parker, P.R., Gee, S.M., Okada, A., McConnell, S.K., Kreitzer, A.C., et al., 2012. Sonic hedgehog expression in corticofugal projection neurons directs cortical microcircuit formation. Neuron 73, 1116–1126.

Hattox, A.M., Nelson, S.B., 2007. Layer V neurons in mouse cortex projecting to different targets have distinct physiological properties. J. Neurophysiol. 98, 3330–3340.

Haubensak, W., Attardo, A., Denk, W., Huttner, W.B., 2004. Neurons arise in the basal neuroepithelium of the early mammalian telencephalon: a major site of neurogenesis. Proc. Natl. Acad. Sci. U.S.A. 101, 3196–3201.

Hayano, Y., Zhao, H., Kobayashi, H., Takeuchi, K., Norioka, S., Yamamoto, N., 2014. The role of T-cadherin in axonal pathway formation in neocortical circuits. Development 141, 4784–4793.

Hedreen, J.C., Yin, T.C., 1981. Homotopic and heterotopic callosal afferents of caudal inferior parietal lobule in *Macaca mulatta*. J. Comp. Neurol. 197, 605–621.

Heins, N., Malatesta, P., Cecconi, F., Nakafuku, M., Tucker, K.L., Hack, M.A., et al., 2002. Glial cells generate neurons: the role of the transcription factor Pax6. Nat. Neurosci. 5, 308–315.

Hetts, S.W., Sherr, E.H., Chao, S., Gobuty, S., Barkovich, A.J., 2006. Anomalies of the corpus callosum: an MR analysis of the phenotypic spectrum of associated malformations. Am. J. Roentgenol. 187, 1343–1348.

Hofer, S., Frahm, J., 2006. Topography of the human corpus callosum revisited—comprehensive fiber tractography using diffusion tensor magnetic resonance imaging. Neuroimage 32, 989–994.

Holm, P.C., Mader, M.T., Haubst, N., Wizenmann, A., Sigvardsson, M., Götz, M., 2007. Loss-and gain-of-function analyses reveal targets of Pax6 in the developing mouse telencephalon. Mol. Cell. Neurosci. 34, 99–119.

Hong, K., Hinck, L., Nishiyama, M., Poo, M.-M., Tessier-Lavigne, M., Stein, E., 1999. A ligand-gated association between cytoplasmic domains of UNC5 and DCC family receptors converts netrin-induced growth cone attraction to repulsion. Cell 97, 927–941.

Hu, Z., Yue, X., Shi, G., Yue, Y., Crockett, D.P., Blair-Flynn, J., et al., 2003. Corpus callosum deficiency in transgenic mice expressing a truncated ephrin-A receptor. J. Neurosci. 23, 10963–10970.

Huang, H., Zhang, J., Wakana, S., Zhang, W., Ren, T., Richards, L.J., et al., 2006. White and gray matter development in human fetal, newborn and pediatric brains. Neuroimage 33, 27–38.

Huang, H., Xue, R., Zhang, J., Ren, T., Richards, L.J., Yarowsky, P., et al., 2009. Anatomical characterization of human fetal brain development with diffusion tensor magnetic resonance imaging. J. Neurosci. 29, 4263–4273.

Huang, Y., Song, N.-N., Lan, W., Zhang, Q., Zhang, L., Zhang, L., et al., 2013. Sensory input is required for callosal axon targeting in the somatosensory cortex. Mol. Brain 6, 53.

Hunt, D.L., Yamoah, E.N., Krubitzer, L., 2006. Multisensory plasticity in congenitally deaf mice: how are cortical areas functionally specified? Neuroscience 139, 1507–1524.

Hutchins, B.I., Li, L., Kalil, K., 2011. Wnt/calcium signaling mediates axon growth and guidance in the developing corpus callosum. Dev. Neurobiol. 71, 269–283.

Hüppi, P.S., Dubois, J., 2006. Diffusion tensor imaging of brain development. Semin. Fetal Neonatal Med. 11, 489–497.

Hüppi, P.S., Maier, S.E., Peled, S., Zientara, G.P., Barnes, P.D., Jolesz, F.A., et al., 1998. Microstructural development of human newborn cerebral white matter assessed *in vivo* by diffusion tensor magnetic resonance imaging. Pediatr. Res. 44, 584–590.

Islam, S.M., Shinmyo, Y., Okafuji, T., Su, Y., Naser, I.B., Ahmed, G., et al., 2009. Draxin, a repulsive guidance protein for spinal cord and forebrain commissures. Science 323, 388–393.

Jbabdi, S., Johansen-Berg, H., 2011. Tractography: where do we go from here? Brain Connect. 1, 169–183.

Joshi, P.S., Molyneaux, B.J., Feng, L., Xie, X., Macklis, J.D., Gan, L., 2008. Bhlhb5 regulates the postmitotic acquisition of area identities in layers II–V of the developing neocortex. Neuron 60, 258–272.

Kanold, P.O., 2009. Subplate neurons: crucial regulators of cortical development and plasticity. Front. Neuroanat. 3, 16.

Kanold, P.O., Luhmann, H.J., 2010. The subplate and early cortical circuits. Ann. Rev. Neurosci. 33, 23–48.

Keeble, T.R., Halford, M.M., Seaman, C., Kee, N., Macheda, M., Anderson, R.B., et al., 2006. The Wnt receptor Ryk is required for Wnt5a-mediated axon guidance on the contralateral side of the corpus callosum. J. Neurosci. 26, 5840–5848.

Keino-Masu, K., Masu, M., Hinck, L., Leonardo, E.D., Chan, S.S.-Y., Culotti, J.G., et al., 1996. Deleted in colorectal cancer (DCC) encodes a netrin receptor. Cell 87, 175–185.

Koester, S.E., O'Leary, D.D., 1993. Connectional distinction between callosal and subcortically projecting cortical neurons is determined prior to axon extension. Dev. Biol. 160, 1–14.

Kowalczyk, T., Pontious, A., Englund, C., Daza, R.A.M., Bedogni, F., Hodge, R., et al., 2009. Intermediate neuronal progenitors (basal progenitors) produce pyramidal–projection neurons for all layers of cerebral cortex. Cereb. Cortex 19, 2439–2450.

Krubitzer, L., Clarey, J.C., Calford, M.B., 1998. Interhemispheric connections of somatosensory cortex in the flying fox. J Comp Neurol 402, 538–559.

Kujala, T., Alho, K., Huotilainen, M., Ilmoniemi, R.J., Lehtokoski, A., Leinonen, A., et al., 1997. Electrophysiological evidence for cross-modal plasticity in humans with early-and late-onset blindness. Psychophysiology 34, 213–216.

Kupers, R., Ptito, M., 2014. Structural, metabolic and functional changes in the congenitally blind brain. Int. J. Psychophysiol. 94, 152.

la Rossa De, A., Bellone, C., Golding, B., Vitali, I., Moss, J., Toni, N., et al., 2013. *In vivo* reprogramming of circuit connectivity in postmitotic neocortical neurons. Nat. Neurosci. 16, 193–200.

Lai, T., Jabaudon, D., Molyneaux, B.J., Azim, E., Arlotta, P., Menezes, J.R., et al., 2008. SOX5 controls the sequential generation of distinct corticofugal neuron subtypes. Neuron 57, 232–247.

Lavado, A., Ware, M., Paré, J., Cao, X., 2014. The tumor suppressor Nf2 regulates corpus callosum development by inhibiting the transcriptional coactivator Yap. Development 141, 4182–4193.

Le Bihan, D., 1995. Molecular diffusion, tissue microdynamics and microstructure. NMR Biomed. 8, 375–386.

Le Bihan, D., Breton, E., Lallemand, D., Grenier, P., Cabanis, E., Laval-Jeantet, M., 1986. MR imaging of intravoxel incoherent motions: application to diffusion and perfusion in neurologic disorders. Radiology 161, 401–407.

Le Bihan, D., Turner, R., Douek, P., 1993. Is water diffusion restricted in human brain white matter? An echo-planar NMR imaging study. Neuroreport 4, 887–890.

Lefkowitz, M., Durand, D., Smith, G., Silver, J., 1991. Electrical properties of axons within probst bundles of acallosal mice and callosi that have reformed upon glial-coated polymer implants. Exp. Neurol. 113, 306–313.

Leone, D.P., Srinivasan, K., Chen, B., Alcamo, E., McConnell, S.K., 2008. The determination of projection neuron identity in the developing cerebral cortex. Curr. Opin. Neurobiol. 18, 28–35.

Leone, D.P., Heavner, W.E., Ferenczi, E.A., Dobreva, G., Huguenard, J.R., Grosschedl, R., et al., 2014. Satb2 regulates the differentiation of both callosal and subcerebral projection neurons in the developing cerebral cortex. Cereb. Cortex [Epub ahead of print].

Li, H., Fertuzinhos, S., Mohns, E., Hnasko, T.S., Verhage, M., Edwards, R., et al., 2013. Laminar and columnar development of barrel cortex relies on thalamocortical neurotransmission. Neuron 79, 970–986.

Lickiss, T., Cheung, A.F., Hutchinson, C.E., Taylor, J.S., Molnár, Z., 2012. Examining the relationship between early axon growth and transcription factor expression in the developing cerebral cortex. J. Anat. 220, 201–211.

Lodato, S., Molyneaux, B.J., Zuccaro, E., Goff, L.A., Chen, H.-H., Yuan, W., et al., 2014. Gene co-regulation by Fezf2 selects neurotransmitter identity and connectivity of corticospinal neurons. Nat. Neurosci. 17, 1046–1054.

López-Bendito, G., Flames, N., Ma, L., Fouquet, C., Di Meglio, T., Chédotal, A., et al., 2007. Robo1 and Robo2 cooperate to control the guidance of major axonal tracts in the mammalian forebrain. J. Neurosci. 27, 3395–3407.

Magnani, D., Hasenpusch-Theil, K., Benadiba, C., Yu, T., Basson, M.A., Price, D.J., et al., 2012. Gli3 controls corpus callosum formation by positioning midline guideposts during telencephalic patterning. Cereb. Cortex 24, 186–198.

Marín-Padilla, M., 1998. Cajal–Retzius cells and the development of the neocortex. Trends Neurosci. 21, 64–71.

McConnell, S.K., Ghosh, A., Shatz, C.J., 1989. Subplate neurons pioneer the first axon pathway from the cerebral cortex. Science 245, 978–982.

McConnell, S.K., Ghosh, A., Shatz, C.J., 1994. Subplate pioneers and the formation of descending connections from cerebral cortex. J. Neurosci. 14, 1892–1907.

McKenna, W.L., Betancourt, J., Larkin, K.A., Abrams, B., Guo, C., Rubenstein, J.L., et al., 2011. Tbr1 and Fezf2 regulate alternate corticofugal neuronal identities during neocortical development. J. Neurosci. 31, 549–564.

McKinstry, R.C., Mathur, A., Miller, J.H., Ozcan, A., Snyder, A.Z., Schefft, G.L., et al., 2002. Radial organization of developing preterm human cerebral cortex revealed by non-invasive water diffusion anisotropy MRI. Cereb. Cortex 12, 1237–1243.

Mendes, S.W., Henkemeyer, M., Liebl, D.J., 2006. Multiple Eph receptors and B-class ephrins regulate midline crossing of corpus callosum fibers in the developing mouse forebrain. J. Neurosci. 26, 882–892.

Métin, C., Frost, D.O., 1989. Visual responses of neurons in somatosensory cortex of hamsters with experimentally induced retinal projections to somatosensory thalamus. Proc. Natl. Acad. Sci. U.S.A. 86, 357–361.

Métin, C., Deléglise, D., Serafini, T., Kennedy, T.E., Tessier-Lavigne, M., 1997. A role for netrin-1 in the guidance of cortical efferents. Development 124, 5063–5074.

Migliore, M., Shepherd, G.M., 2005. An integrated approach to classifying neuronal phenotypes. Nat. Rev. Neurosci. 6, 810–818.

Mitchell, B.D., Macklis, J.D., 2005. Large-scale maintenance of dual projections by callosal and frontal cortical projection neurons in adult mice. J. Comp. Neurol. 482, 17–32.

Miyata, T., Kawaguchi, A., Saito, K., Kawano, M., Muto, T., Ogawa, M., 2004. Asymmetric production of surface-dividing and non-surface-dividing cortical progenitor cells. Development 131, 3133–3145.

Mizuno, H., Hirano, T., Tagawa, Y., 2007. Evidence for activity-dependent cortical wiring: formation of interhemispheric connections in neonatal mouse visual cortex requires projection neuron activity. J. Neurosci. 27, 6760–6770.

Mizuno, H., Hirano, T., Tagawa, Y., 2010. Pre-synaptic and post-synaptic neuronal activity supports the axon development of callosal projection neurons during different post-natal periods in the mouse cerebral cortex. Eur. J. Neurosci. 31, 410–424.

Molnár, Z., Wang, W.Z., Piñon, M.C., Blakey, D., Kondo, S., Oeschger, F., et al., 2010. Subplate and the formation of the earliest cerebral cortical circuits New Aspects of Axonal Structure and Function. Springer US, Boston, MA, pp. 19–31.

Molyneaux, B.J., Arlotta, P., Hirata, T., Hibi, M., Macklis, J.D., 2005. Fezl is required for the birth and specification of corticospinal motor neurons. Neuron 47, 817–831.

Molyneaux, B.J., Arlotta, P., Menezes, J.R.L., Macklis, J.D., 2007. Neuronal subtype specification in the cerebral cortex. Nat. Rev. Neurosci. 8, 427–437.

Mori, S., van Zijl, P.C.M., 2002. Fiber tracking: principles and strategies—a technical review. NMR Biomed. 15, 468–480.

Mori, S., Zhang, J., 2006. Principles of diffusion tensor imaging and its applications to basic neuroscience research. Neuron 51, 527–539.

Mori, S., Crain, B.J., Chacko, V.P., van Zijl, P.C., 1999. Three-dimensional tracking of axonal projections in the brain by magnetic resonance imaging. Ann. Neurol. 45, 265–269.

Mori, S., Itoh, R., Zhang, J., Kaufmann, W.E., van Zijl, P.C., Solaiyappan, M., et al., 2001. Diffusion tensor imaging of the developing mouse brain. Magn. Reson. Med. 46, 18–23.

Moseley, M.E., Cohen, Y., Kucharczyk, J., Mintorovitch, J., Asgari, H.S., Wendland, M.F., et al., 1990. Diffusion-weighted MR imaging of anisotropic water diffusion in cat central nervous system. Radiology 176, 439–445.

III. AXON GUIDANCE

Moseley, M.E., Kucharczyk, J., Asgari, H.S., Norman, D., 1991. Anisotropy in diffusion-weighted MRI. Magn. Reson. Med. 19, 321–326.

Mukherjee, P., Miller, J.H., Shimony, J.S., Philip, J.V., Nehra, D., Snyder, A.Z., et al., 2002. Diffusion-tensor MR imaging of gray and white matter development during normal human brain maturation. Am. J. Neuroradiol. 23, 1445–1456.

Munoz-Sanjuan, I., Simandl, B.K., Fallon, J.F., Nathans, J., 1999. Expression of chicken fibroblast growth factor homologous factor (FHF)-1 and of differentially spliced isoforms of FHF-2 during development and involvement of FHF-2 in chicken limb development. Development 126, 409–421.

Muzio, L., Di Benedetto, B., Stoykova, A., Boncinelli, E., Gruss, P., Mallamaci, A., 2002. Emx2 and Pax6 control regionalization of the pre-neuronogenic cortical primordium. Cereb. Cortex 12, 129–139.

Müller, F., O'Rahilly, R., 1988. The development of the human brain from a closed neural tube at stage 13. Anat. Embryol. 177, 203–224.

Neil, J.J., Shiran, S.I., McKinstry, R.C., Schefft, G.L., Snyder, A.Z., Almli, C.R., et al., 1998. Normal brain in human newborns: apparent diffusion coefficient and diffusion anisotropy measured by using diffusion tensor MR imaging. Radiology 209, 57–66.

Neves Das, L., Duchala, C.S., Godinho, F., Haxhiu, M.A., Colmenares, C., Macklin, W.B., et al., 1999. Disruption of the murine nuclear factor IA gene (Nfia) results in perinatal lethality, hydrocephalus, and agenesis of the corpus callosum. Proc. Natl. Acad. Sci. U.S.A. 96, 11946–11951.

Niquille, M., Garel, S., Mann, F., Hornung, J.-P., Otsmane, B., Chevalley, S., et al., 2009. Transient neuronal populations are required to guide callosal axons: a role for semaphorin 3C. PLoS Biol. 7, e1000230.

Niquille, M., Minocha, S., Hornung, J.-P., Rufer, N., Valloton, D., Kessaris, N., et al., 2013. Two specific populations of GABAergic neurons originating from the medial and the caudal ganglionic eminences aid in proper navigation of callosal axons. Dev. Neurobiol. 73, 647–672.

Noctor, S.C., Flint, A.C., Weissman, T.A., Wong, W.S., Clinton, B.K., Kriegstein, A.R., 2002. Dividing precursor cells of the embryonic cortical ventricular zone have morphological and molecular characteristics of radial glia. J. Neurosci. 22, 3161–3173.

Noctor, S.C., Martínez-Cerdeño, V., Ivic, L., Kriegstein, A.R., 2004. Cortical neurons arise in symmetric and asymmetric division zones and migrate through specific phases. Nat. Neurosci. 7, 136–144.

Norris, D.G., 2001. The effects of microscopic tissue parameters on the diffusion weighted magnetic resonance imaging experiment. NMR Biomed. 14, 77–93.

O'Leary, D.D., Chou, S.-J., Sahara, S., 2007. Area patterning of the mammalian cortex. Neuron 56, 252–269.

Oh, S.W., et al., 2014. A mesoscale connectome of the mouse brain. Nature 508, 207–214.

Ohkubo, Y., Chiang, C., Rubenstein, J., 2002. Coordinate regulation and synergistic actions of BMP4, SHH and FGF8 in the rostral prosencephalon regulate morphogenesis of the telencephalic and optic vesicles. Neuroscience 111, 1–17.

Olavarria, J.F., Van Sluyters, R.C., 1995. Overall pattern of callosal connections in visual cortex of normal and enucleated cats. J. Comp. Neurol. 363, 161–176.

O'Leary, D.D., Sahara, S., 2008. Genetic regulation of arealization of the neocortex. Curr. Opin. Neurobiol. 18, 90–100.

Pasquale, E.B., 2005. Eph receptor signalling casts a wide net on cell behaviour. Nat. Rev. Mol. Cell. Biol. 6, 462–475.

Paul, L.K., Brown, W.S., Adolphs, R., Tyszka, J.M., Richards, L.J., Mukherjee, P., et al., 2007. Agenesis of the corpus callosum: genetic, developmental and functional aspects of connectivity. Nat. Rev. Neurosci. 8, 287–299.

Piper, M., Moldrich, R.X., Lindwall, C., Little, E., Barry, G., Mason, S., et al., 2009a. Multiple non-cell-autonomous defects underlie neocortical callosal dysgenesis in Nfib-deficient mice. Neural. Dev. 4, 43.

Piper, M., Plachez, C., Zalucki, O., Fothergill, T., Goudreau, G., Erzurumlu, R., et al., 2009b. Neuropilin 1-Sema signaling regulates crossing of cingulate pioneering axons during development of the corpus callosum. Cereb. Cortex 19, i11–i21.

Probst, M., 1901. The structure of complete dissolving corpus callosum of the cerebrum and also the microgyry and heterotropy of the grey substance. Arch. Psychiatr. Nervenkr. 34, 709–786.

Ptito, M., Giguère, J.-F., Boire, D., Frost, D.O., Casanova, C., 2001. When the auditory cortex turns visual. Prog. Brain Res. 134, 447–458.

Rakic, P., Suner, I., Williams, R.W., 1991. A novel cytoarchitectonic area induced experimentally within the primate visual cortex. Proc. Natl. Acad. Sci. U.S.A. 88, 2083–2087.

Reiner, A., Jiao, Y., Del Mar, N., Laverghetta, A.V., Lei, W.L., 2003. Differential morphology of pyramidal tract-type and intratelencephalically projecting-type corticostriatal neurons and their intrastriatal terminals in rats. J. Comp. Neurol. 457, 420–440.

Ren, T., Zhang, J., Plachez, C., Mori, S., Richards, L.J., 2007. Diffusion tensor magnetic resonance imaging and tract-tracing analysis of Probst bundle structure in Netrin1- and DCC-deficient mice. J. Neurosci. 27, 10345–10349.

Richards, L.J., Koester, S.E., Tuttle, R., O'leary, D.D., 1997. Directed growth of early cortical axons is influenced by a chemoattractant released from an intermediate target. J. Neurosci. 17, 2445–2458.

Roe, A.W., Pallas, S.L., Hahm, J.-O., Sur, M., 1990. A map of visual space induced in primary auditory cortex. Science 250, 818–820.

Roe, A.W., Garraghty, P.E., Esguerra, M., Sur, M., 1993. Experimentally induced visual projections to the auditory thalamus in ferrets: evidence for a W cell pathway. J. Comp. Neurol. 334, 263–280.

Rouaux, C., Arlotta, P., 2010. Fezf2 directs the differentiation of corticofugal neurons from striatal progenitors *in vivo*. Nat. Neurosci. 13, 1345–1347.

Rouaux, C., Arlotta, P., 2013. Direct lineage reprogramming of post-mitotic callosal neurons into corticofugal neurons *in vivo*. Nat. Cell Biol. 15, 214–221.

Sadato, N., Pascual-Leone, A., Grafman, J., Ibañez, V., Deiber, M.-P., Dold, G., et al., 1996. Activation of the primary visual cortex by Braille reading in blind subjects. Nature 380, 526–528.

Sahara, S., Kawakami, Y., Izpisua Belmonte, J.C., O'Leary, D.D., 2007. Sp8 exhibits reciprocal induction with Fgf8 but has an opposing effect on anterior-posterior cortical area patterning. Neural Dev. 2, 10.

Schneider, G.E., 1973. Early lesions of superior colliculus: factors affecting the formation of abnormal retinal projections. Brain Behav. Evol. 8 (91–109).

Segraves, M.A., Rosenquist, A.C., 1982. The afferent and efferent callosal connections of retinotopically defined areas in cat cortex. J. Neurosci. 2, 1090–1107.

Serafini, T., Colamarino, S.A., Leonardo, E.D., Wang, H., Beddington, R., Skarnes, W.C., et al., 1996. Netrin-1 is required for commissural axon guidance in the developing vertebrate nervous system. Cell 87, 1001–1014.

Sessa, A., Mao, C.-A., Hadjantonakis, A.-K., Klein, W.H., Broccoli, V., 2008. Tbr2 directs conversion of radial glia into basal precursors and guides neuronal amplification by indirect neurogenesis in the developing neocortex. Neuron 60, 56–69.

Sharma, J., Angelucci, A., Sur, M., 2000. Induction of visual orientation modules in auditory cortex. Nature 404, 841–847.

Shim, S., Kwan, K.Y., Li, M., Lefebvre, V., Šestan, N., 2012. Cis-regulatory control of corticospinal system development and evolution. Nature 486, 74–79.

Shu, T., Richards, L.J., 2001. Cortical axon guidance by the glial wedge during the development of the corpus callosum. J. Neurosci. 21, 2749–2758.

Shu, T., Butz, K.G., Plachez, C., Gronostajski, R.M., Richards, L.J., 2003a. Abnormal development of forebrain midline glia and commissural projections in Nfia knock-out mice. J. Neurosci. 23, 203–212.

Shu, T., Li, Y., Keller, A., Richards, L.J., 2003b. The glial sling is a migratory population of developing neurons. Development 130, 2929–2937.

Shu, T., Sundaresan, V., McCarthy, M.M., Richards, L.J., 2003c. Slit2 guides both precrossing and postcrossing callosal axons at the midline *in vivo*. J. Neurosci. 23, 8176–8184.

Silver, J., Lorenz, S.E., Wahlsten, D., Coughlin, J., 1982. Axonal guidance during development of the great cerebral commissures: descriptive and experimental studies, *in vivo*, on the role of preformed glial pathways. J. Comp. Neurol. 210, 10–29.

Smith, K.M., Ohkubo, Y., Maragnoli, M.E., Rašin, M.-R., Schwartz, M.L., Šestan, N., et al., 2006. Midline radial glia translocation and corpus callosum formation require FGF signaling. Nat. Neurosci. 9, 787–797.

Spassky, N., Merkle, F.T., Flames, N., Tramontin, A.D., García-Verdugo, J.M., Alvarez-Buylla, A., 2005. Adult ependymal cells are postmitotic and are derived from radial glial cells during embryogenesis. J. Neurosci. 25, 10–18.

Sporns, O., 2011. The human connectome: a complex network. Ann. N. Y. Acad. Sci. 1224, 109–125.

Srinivasan, K., Leone, D.P., Bateson, R.K., Dobreva, G., Kohwi, Y., Kohwi-Shigematsu, T., et al., 2012. A network of genetic repression and derepression specifies projection fates in the developing neocortex. Proc. Natl. Acad. Sci. U.S.A. 109, 19071–19078.

Srivatsa, S., Parthasarathy, S., Britanova, O., Bormuth, I., Donahoo, A.-L., Ackerman, S.L., et al., 2014. Unc5C and DCC act downstream of Ctip2 and Satb2 and contribute to corpus callosum formation. Nat. Commun. 5, 3708.

Steele-Perkins, G., Plachez, C., Butz, K.G., Yang, G., Bachurski, C.J., Kinsman, S.L., et al., 2005. The transcription factor gene Nfib is essential for both lung maturation and brain development. Mol. Cell. Biol. 25, 685–698.

Stein, E., Tessier-Lavigne, M., 2001. Hierarchical organization of guidance receptors: silencing of netrin attraction by slit through a Robo/DCC receptor complex. Science 291, 1928–1938.

Stejskal, E.O., Tanner, J.E., 1965. Spin diffusion measurements: spin echoes in the presence of a time-dependent field gradient. J. Chem. Phys. 42, 286–288.

Storm, E.E., Garel, S., Borello, U., Hebert, J.M., Martinez, S., McConnell, S.K., et al., 2006. Dose-dependent functions of Fgf8 in regulating telencephalic patterning centers. Development 133, 1831–1844.

Suárez, R., Fenlon, L.R., Marek, R., Avitan, L., Sah, P., Goodhill, G.J., et al., 2014. Balanced interhemispheric cortical activity is required for correct targeting of the corpus callosum. Neuron 82, 1289–1298.

Sur, M., Garraghty, P.E., Roe, A.W., 1988. Experimentally induced visual projections into auditory thalamus and cortex. Science 242, 1437–1441.

Taylor, D.G., Bushell, M.C., 1985. The spatial mapping of translational diffusion coefficients by the NMR imaging technique. Phys. Med. Biol. 30, 345–349.

Thomsen, C., Henriksen, O., Ring, P., 1987. *In vivo* measurement of water self diffusion in the human brain by magnetic resonance imaging. Acta Radiol. 28, 353–361.

Tissir, F., Goffinet, A.M., 2003. Reelin and brain development. Nat. Rev. Neurosci. 4, 496–505.

Tomasch, J., 1954. Size, distribution, and number of fibres in the human corpus callosum. Anat. Rec. 119, 119–135.

Tournier, J.-D., Calamante, F., Gadian, D.G., Connelly, A., 2004. Direct estimation of the fiber orientation density function from diffusion-weighted MRI data using spherical deconvolution. Neuroimage 23, 1176–1185.

Tournier, J.-D., Calamante, F., Connelly, A., 2007. Robust determination of the fibre orientation distribution in diffusion MRI: non-negativity constrained super-resolved spherical deconvolution. Neuroimage 35, 1459–1472.

Tovar-Moll, F., Moll, J., de Oliveira-Souza, R., Bramati, I., Andreiuolo, P.A., Lent, R., 2007. Neuroplasticity in human callosal dysgenesis: a diffusion tensor imaging study. Cereb. Cortex 17, 531–541.

Tovar-Moll, F., Monteiro, M., Andrade, J., Bramati, I.E., Vianna-Barbosa, R., Marins, T., et al., 2014. Structural and functional brain rewiring clarifies preserved interhemispheric transfer in humans born without the corpus callosum. Proc. Natl. Acad. Sci. U.S.A. 111, 7843–7848.

Tuch, D.S., Reese, T.G., Wiegell, M.R., Makris, N., Belliveau, J.W., Wedeen, V.J., 2002. High angular resolution diffusion imaging reveals intravoxel white matter fiber heterogeneity. Magn. Reson. Med. 48, 577–582.

Turner, R., Le Bihan, D., Maier, J., Vavrek, R., Hedges, L.K., Pekar, J., 1990. Echo-planar imaging of intravoxel incoherent motion. Radiology 177, 407–414.

Unni, D.K., Piper, M., Moldrich, R.X., Gobius, I., Liu, S., Fothergill, T., et al., 2012. Multiple slits regulate the development of midline glial populations and the corpus callosum. Dev. Biol. 365, 36–49.

Veinante, P., Deschênes, M., 2003. Single-cell study of motor cortex projections to the barrel field in rats. J. Comp. Neurol. 464, 98–103.

Voigt, T., 1989. Development of glial cells in the cerebral wall of ferrets: direct tracing of their transformation from radial glia into astrocytes. J. Comp. Neurol. 289, 74–88.

Wahl, M., Strominger, Z., Jeremy, R.J., Barkovich, A.J., Wakahiro, M., Sherr, E.H., et al., 2009. Variability of homotopic and heterotopic callosal connectivity in partial agenesis of the corpus callosum: a 3T diffusion tensor imaging and Q-ball tractography study. Am. J. Neuroradiol. 30, 282–289.

Wang, Y., Thekdi, N., Smallwood, P.M., Macke, J.P., Nathans, J., 2002. Frizzled-3 is required for the development of major fiber tracts in the rostral CNS. J. Neurosci. 22, 8563–8573.

Wang, Y., Zhang, J., Mori, S., Nathans, J., 2006. Axonal growth and guidance defects in Frizzled3 knock-out mice: a comparison of diffusion tensor magnetic resonance imaging, neurofilament staining, and genetically directed cell labeling. J. Neurosci. 26, 355–364.

Wang, C.-L., Zhang, L., Zhou, Y., Zhou, J., Yang, X.-J., Duan, S.-M., et al., 2007. Activity-dependent development of callosal projections in the somatosensory cortex. J. Neurosci. 27, 11334–11342.

Watakabe, A., Takaji, M., Kato, S., Kobayashi, K., Mizukami, H., Ozawa, K., et al., 2014. Simultaneous visualization of extrinsic and intrinsic axon collaterals in Golgi-like detail for mouse corticothalamic and corticocortical cells: a double viral infection method. Front. Neural Circuits 8, 110.

Wedeen, V.J., Hagmann, P., Tseng, W.-Y.I., Reese, T.G., Weisskoff, R.M., 2005. Mapping complex tissue architecture with diffusion spectrum magnetic resonance imaging. Magn. Reson. Med. 54, 1377–1386.

Wilson, C.J., 1987. Morphology and synaptic connections of crossed corticostriatal neurons in the rat. J. Comp. Neurol. 263, 567–580.

Wu, S.-X., Goebbels, S., Nakamura, K., Nakamura, K., Kometani, K., Minato, N., et al., 2005. Pyramidal neurons of upper cortical layers generated by NEX-positive progenitor cells in the subventricular zone. Proc. Natl. Acad. Sci. U.S.A. 102, 17172–17177.

Xu, G., Takahashi, E., Folkerth, R.D., Haynes, R.L., Volpe, J.J., Grant, P.E., et al., 2014. Radial coherence of diffusion tractography in the cerebral white matter of the human fetus: neuroanatomic insights. Cereb. Cortex 24, 579–592.

Yorke, C.H., Caviness, V.S., 1975. Interhemispheric neocortical connections of the corpus callosum in the normal mouse: a study based on anterograde and retrograde methods. J. Comp. Neurol. 164, 233–245.

Zalesky, A., Fornito, A., Bullmore, E.T., 2010. Network-based statistic: identifying differences in brain networks. Neuroimage 53, 1197–1207.

Zembrzycki, A., Griesel, G., Stoykova, A., Mansouri, A., 2007. Genetic interplay between the transcription factors Sp8 and Emx2 in the patterning of the forebrain. Neural. Dev. 2, 8.

Zhang, L., Song, N.-N., Chen, J.-Y., Huang, Y., Li, H., Ding, Y.-Q., 2012. Satb2 is required for dendritic arborization and soma spacing in mouse cerebral cortex. Cereb. Cortex 22, 1510–1519.

Zhao, C., Deng, W., Gage, F.H., 2008. Mechanisms and functional implications of adult neurogenesis. Cell 132, 645–660.

Zhao, H., Maruyama, T., Hattori, Y., Sugo, N., Takamatsu, H., Kumanogoh, A., et al., 2011. A molecular mechanism that regulates medially oriented axonal growth of upper layer neurons in the developing neocortex. J. Comp. Neurol. 519, 834–848.

Zhou, J., Wen, Y., She, L., Sui, Y.-N., Liu, L., Richards, L.J., et al., 2013. Axon position within the corpus callosum determines contralateral cortical projection. Proc. Natl. Acad. Sci. U.S.A. 110, E2714–E2723.

TRACTOGRAPHY

Overview

This volume opened with chapters on single axon configuration and microcircuitry and now draws to a close with four chapters relating to axons tracts and imaging these *in vivo* in the human brain. Diffusion imaging is one of a small number of noninvasive techniques that can begin to address interconnectivity *in vivo*, including from the perspective of ongoing behaviors and longitudinally across time. Resolution, however, is limited, and final results depend closely on modeling algorithms. Intense interest is necessarily focused on validation and improvement of currently available algorithms. Most often, as set forth by all the chapters in this section, this involves histological validation, either in postmortem human tissue or by tracer injections in nonhuman primates. Other themes developed in previous chapters, such as computational architecture and axonal trajectories, are revisited here in the context of diffusion imaging and the new technique of polarized light imaging (PLI) (Chapter 18).

In Chapter 15, Giorgio Innocenti and colleagues extensively review neuroanatomical data on axon diameter in nonhuman primates, and critically consider the several imaging models used to derive comparable data in humans *in vivo*. An important focus in this chapter is on the relevance of axon diameter to neural computations and on the consequences of pathway-specific differences in axon caliber and length. These are considered in the context of preferentially faster or slower intercommunication in an evolutionary framework. Increased brain volume, as in humans, is proposed to have resulted in interneuronal communication that is slower and more temporally dispersed.

Chapter 16, by Ralf Galuske and colleagues, presents a critical review of various diffusion imaging procedures and discusses how these might be validated by histological investigations in human postmortem material. This includes thoughtful commentary on the strengths and limitations of the various available techniques, as well as a frank assessment of the need for better understanding of structure–function relationships and the associated need for better techniques, especially those that can elucidate the more dynamic phenomena of learning and plasticity in the brain.

Of considerable direct concern to diffusion imaging are issues of evaluation and independent validation. One source for this is from higher resolution neuroanatomical data in postmortem brains, as treated in Chapters 15 and 16. Chapter 17, by Farzad Mortazavi and Douglas Rosene, addresses several histological modifications for optimal visualization of individual axons and populations of axons in nonhuman primate brains. These include customized plane of section, improved antibody penetration via microwave treatment and tissue clearing techniques, and steps toward better visualization of neuropil and white matter organization in thick tissue slabs. The authors present examples of right-angle axon branches in several different regions and discuss implications for geometric organization and substructure within the classical fiber bundles.

Another important source of validation for the models used in diffusion imaging is from independent technical approaches. Chapter 18, by Karl Zilles and colleagues, gives a detailed narrative of the relatively new technique of PLI. This produces ultra-high-resolution visualization of nerve fibers in postmortem brains (human, nonhuman primates, and rodents) by exploiting the birefringent property of myelin. Image acquisition by polarimeter has an in-plane sampling resolution of $64 \times 64\,\mu m^2$; and acquisition by a polarization microscope provides higher spatial resolution of $1.3 \times 1.3\,\mu m^3$ in-plane. Crossing of fibers can be directly visualized at these resolutions, and an added capability is that fiber orientation is encoded by different colors. PLI also is applied to higher resolution myeloarchitecture mapping of cortical regions.

Many will agree with these Authors that diffusion imaging can rightly be considered a major methodological revolution, already allowing noninvasive identification of axonal tracts in animals and humans and promising further rapid advances; for example, the characterization of axon diameters within tracts. This, together with high-resolution imaging by PLI and by improved histological techniques, refocuses attention on white matter organization and intracortical myeloarchitectonics. These domains have been relatively neglected for decades, but, as continuing work will certainly demonstrate, hold important insights to brain organization, under normal and pathological conditions.

15

The Diameters of Cortical Axons and Their Relevance to Neural Computing

Giorgio M. Innocenti[1,2], Marie Carlén[1] and Tim B. Dyrby[3]

[1]Department of Neuroscience, Karolinska Institutet, Stockholm, Sweden [2]Brain and Mind Institute, EPFL, Lausanne, Switzerland [3]Danish Research Centre for Magnetic Resonance, Centre for Functional and Diagnostic Imaging and Research, Copenhagen University Hospital Hvidovre, Hvidovre, Denmark

The study of brain connections underwent four recent methodological revolutions. The first was the implementation of retrograde axonal transport methods initiated by the work of Kristensson and Olsson (1971) and followed by a host of other methodologies (e.g., Bentivoglio et al., 1979; Katz et al., 1984). The second was the introduction of anterograde transport methods, which allowed the detailed visualization of axonal arbors (King et al., 1989). The third revolution is underway and it concerns the application of water diffusion methods allowing the identification of axonal tracts, noninvasively, in animals and humans, and possibly the characterization of the spectrum of axon diameters involved in each tract (below). The fourth revolution involves the usage of transsynaptic tracing in the study of neural pathways (below). This chapter is about the implications of the above methods for the characterization of brain networks beyond the limits of the current, admirable, connectome enterprise. A number of recent, excellent reviews complement several aspects of the present chapter (Bucher and Goaillard, 2011; Debanne et al., 2011; Nave, 2010; Taveggia et al., 2010; Zatorre et al., 2012).

15.1 COMPUTATIONAL NEUROANATOMY AND THE COMPUTATIONAL PROPERTIES OF AXONS

The complexity of axonal arbors prompted one of us to suggest that the axon, rather than transferring information faithfully between two brain sites, participates in the processing of information, i.e., it has computational properties (Innocenti et al., 1994). Computational neuroanatomy has now become a subfield of its own and includes both the operations performed by neurons on the basis of their morphological properties and the computational methods used to acquire structural information by measurements of neurons in histological preparations, or *ex vivo*, or *in vivo*. These approaches are important if one considers the close link between brain structure and function, and the modeling that this allows. The concept of axonal computation was further elaborated by proposing (Innocenti, 1999; Figure 15.1) that the axon performs computation in at least three domains: (i) mapping, i.e., transforming the topographical position of the cell body of origin into the position of its terminal boutons; (ii) amplification, i.e., distributing different numbers of boutons to its different targets, and (iii) temporal transformations, i.e., reaching the different targets with different delays, due to the conduction properties of the terminal branches of each axon and to the different conduction properties of different axons. To these, Debanne et al. (2011) have added three functional computations, that is, (i) activity-dependent shaping of action potential; (ii) signal amplification along the axons; and (iii) axonal integration.

Among the different computations performed by the axons, the temporal transformations have come into focus recently since the discovery that different cortical connections use axons of different diameters (Caminiti et al., 2009).

Axons unlike cables are computational devices performing transformations in:

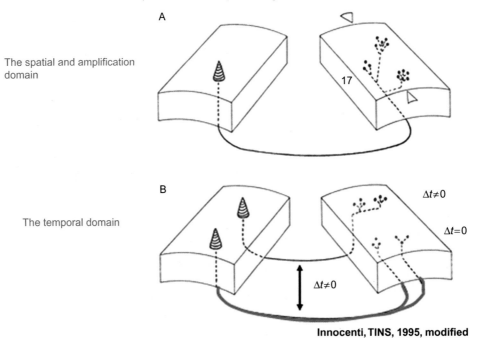

Innocenti, TINS, 1995, modified

FIGURE 15.1   The computational properties of axons. Particularly relevant to this chapter are the transformations axons perform in the temporal domain due to their geometry, i.e., thickness and length. Axons of different diameter, conduct action potential at different speed and this creates temporal delays in the information they carry to targets. On the other hand, while the geometry of the terminal arbors often causes synchronous activation of targets, some axonal arbors operate like delay lines. See also text. *From Innocenti (1995), modified.*

The finding was heralded by the report that different sectors of the primate corpus callosum (CC) contain axons of different diameters (Aboitiz et al., 1992; LaMantia and Rakic, 1990). As different CC sectors also contain axons of different cortical origin, the inference that different areas may communicate with different diameter axons was rather attractive. Indeed, that was validated with injections of anterograde tracers in different areas in the monkey (Caminiti et al., 2009; Tomasi et al., 2012). This established a hierarchy with thinner axons originating from prefrontal, parietal, and temporal association areas and thicker axons originating from primary motor, somatosensory, and visual areas (Figure 15.2). From the diameter of the axons, their conduction velocity could be computed and from conduction velocity and pathway length, one could calculate the delays generated by the axons from their cortical origin to the midline of the CC and to the contralateral cortex. This established a hierarchy predicting faster conduction from premotor, motor, and somatosensory areas and slower conduction from the temporal areas (Figure 15.3). One potentially important aspect of this is that for each area axonal diameters and conduction velocities are dispersed within a spectrum. Presumably the conduction delays are as well, although this could not be precisely calculated in the absence of information on the length of the individual axons. We will discuss this issue further.

Obviously, similar concepts could be extended to other cortical connections and this was done by considering short intracortical connections, cortical connections to the thalamus, to the basal ganglia as well as connections from parietal cortex to frontal and prefrontal areas (Innocenti et al., 2014). These findings are presented in Figure 15.4 where the different axonal systems are characterized from their median axon diameter, pathway length, and conduction delays. The general messages which can be gathered from those data are that: (i) axonal diameters depend not only on the area of origin of a projection but also on the target; that is, the same area will send axons of different diameters to its different targets. For example, prefrontal area 9 will send thicker axons to the CC, thinner ones to the striatum, thalamus, or internal capsule. (ii) Unlike the long connections, the short intracortical connections seem to be similar across areas. (iii) The axon diameter of the projections and the pathway lengths can be expected to generate shorter conduction delays for the premotor, motor, and somatosensory connections. We have speculated that the latter finding might indicate that motor and somatosensory areas will process more information per unit time and that this could establish a reference baseline or temporal frame for the information pertaining to other sensory modalities or resulting from operations of intracortical association. More specifically, we have proposed

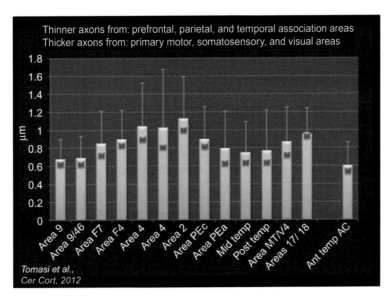

FIGURE 15.2 Mean (columns; whiskers are standard deviations) and median (diamonds) diameters of interhemispheric axons from different areas of cortex. Notice that primary motor, somatosensory, and visual areas communicate with thicker axons than "association" areas. *From Tomasi et al. (2012), modified.*

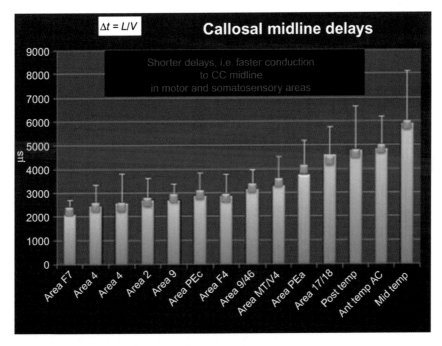

FIGURE 15.3 Temporal delays to the callosal midline caused by axonal diameters and lengths. Notice the hierarchy of delays, favoring motor and somatosensory areas. *From Tomasi et al. (2012), modified.*

that the motor and somatosensory processing speed primacy may be an important component of the mechanisms responsible for body ownership (Innocenti et al., 2014).

The recent documentation of unmyelinated segments along cortical axons (Tomassy et al., 2014) raises interesting problems on the conduction of CNS axons. A similar finding was reported for axons in the process of myelination (Remahl and Hilderbrand, 1990) in line with recent observations and models of myelination showing that Ranvier nodes are formed by a lateral growth of oligodendrocytes lamellae along the axon (Snaidero et al., 2014). This raises the question of how action potentials are transported in developing axons. Presumably at uneven speed and

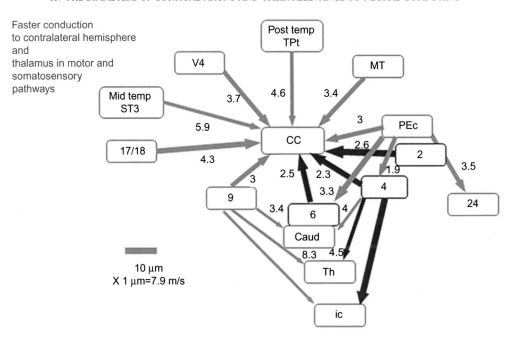

FIGURE 15.4 Schematic representation of diameters and estimated conduction delays for connections studied, in Tomasi et al. (2012) and Innocenti et al. (2014). Thickness of arrow segments is proportional to median diameter; their length is proportional to pathway length; projections from area 4 are the means of two animals. Delays are calculated on conduction velocities and pathway lengths. Red arrows emphasize the faster conduction of pathways originating in somatosensory and motor areas. *From Innocenti et al. (2014), modified.*

with low security. This opens the possibility that the gradual formation of Ranvier nodes might be the mechanism whereby the conduction properties of the axon are adjusted to functional demand in development.

It should be asked if the estimates of conduction velocities and delays computed from anatomical measurement would fit the electrophysiological estimates. This is the case, although the electrophysiological data remain scanty (Tomasi et al., 2012). One problem was encountered with corticospinal projection. We had previously noticed that for this projection the conduction velocities estimated from electrophysiological data were higher than those computed from structural data (Innocenti et al., 2014). This discrepancy was confirmed (Firmin et al., 2015). One reason for the discrepancy seems to be that the electrophysiological data obtained by antidromic invasion of corticospinal projecting neurons are strongly biased toward large cell bodies (Humphrey and Corrie, 1978), which also have the thickest axons (Tomasi et al., 2012).

15.2 METHODOLOGICAL PERSPECTIVES

15.2.1 Transsynaptic Tracing

The morphological characterization of axons has thus far encountered two limitations. The first has been the difficulty of tracing connections across synapses. Therefore, it has been impossible or extremely laborious to identify the target neurons of a given projection and hence to determine if axons of different diameter target different neuronal populations. The use of rabies viruses in genetically modified rodents is a promising way to overcome this difficulty. Rabies virus infects neurons and spreads strictly retrogradely to upstream input neurons, where new viral particles are produced causing further retrograde spread (Ugolini, 2007). A recent and very important development in the use of rabies virus for retrograde circuit tracing has been the introduction of genetic modifications in the viral genome restricting the retrograde spread to monosynaptic connections, i.e., inhibiting spread beyond the presynaptic input neurons (Wickersham et al., 2014). It is thus possible to select a neuronal population as a starter population, and to label and trace selectively the presynaptic neurons giving inputs to the starter population. Most often loxP-Cre recombinase strategies with a combination of mouse Cre-lines and dual viral systems are used for retrograde rabies tracing of input neurons to genetically defined starter populations (Pollak Dorocic et al., 2014; Wall et al., 2013; Watabe-Uchida et al., 2012; Wickersham et al., 2014). The use of rabies tracing for determination of axonal parameters awaits testing.

15.2.2 Axon Diffusion Weighted Imaging

15.2.2.1 Diffusion Weighted Imaging

The second limitation has been the impossibility to characterize axon diameters in humans using noninvasive techniques. Changes in white matter structure probably underlie a host of neurological and psychiatric conditions, ranging from developmental dyslexia (Klingberg et al., 2000) to schizophrenia (Innocenti et al., 2003). They might also be the hallmark of special talents, where they be musical or others. Recently, techniques based on water diffusion estimate are aiming at overcoming this limitation.

Diffusion magnetic resonance imaging (MRI) is an imaging technique that uses the physical process of the always-ongoing motion of water molecules to measure geometrical features of the microstructural environment. The image resolution of MRI is typically in cubic millimeters, whereas the microstructural environment consists of an intra- and extracellular space separated by the membranes of cell bodies, dendrites, and axons at the micrometer length scale. The water diffusion signal measured in each voxel of an MR image is therefore the sum of signals generated by water diffusion within each of these microstructural tissue compartments. If we know how molecules diffuse within these different compartments, specifically the intra- versus extracellular spaces, we can design biophysical models that allow to estimate specific geometrical features of the microstructural environment such as axon diameters, their distribution and density, as well as the exchange of water molecules across cell membranes. By varying the scan parameters of a diffusion MR sequence we generally control two parameters, namely (i) the period of diffusion time (t_d) within which we observe water diffusion, and (ii) the sensitive range of maximal molecular displacement of water molecules that can take place within the selected diffusion time, i.e., the q-value (Callaghan et al., 1991) and for review (Cohen and Assaf, 2002). It should be noted however, that most often and especially on clinical MRI systems the q-value and diffusion time are merged for simplicity into the broadly known parameter the b-value ($= q^{2*}t_d$). Diffusion MRI is a fast growing imaging field in clinical research thanks to the introduction of the widely used diffusion tensor imaging (DTI), which has shown high sensitivity to microstructural alterations in brain tissue (Basser et al., 1994). However, DTI is nonspecific, and to obtain sensitivity and specificity to geometries such as axon diameters and densities one therefore has to select an appropriate biophysical model and MRI setup.

15.2.2.2 Biophysical Models of Axonal Geometries in Neuronal Tissue

Neuronal tissue is complex and many different cellular compartments contribute to the diffusion MRI signal. One can generate models of the individual geometrical cellular structures believed to appear in brain tissue and use these to predict the fraction of water molecules scattering belonging to each of these compartments. Panagiotaki et al. (2012) presented a taxonomic collection of compartmental diffusion models representative for various intra- and extracellular spaces that can be combined to design a biophysical model of, for example, axon diameters and axon diameter distribution. Stanisz et al. (1997) introduced the first specific biophysical model for estimating axon diameters from diffusion MRI, validated on fixed bovine optic nerve. Their biophysical model used elongated ellipsoids as a model of axons and their diameter, while surrounding cell bodies were modeled as spheres. Water exchange across cell membranes and surface/volume ratios was also considered. To obtain reliable estimates using such a relative complex model, they used a comprehensive acquisition MRI protocol including many q-values for different diffusion times. Their results were compared with EM and clearly supported the feasibility of estimating axon diameters from MRI (Stanisz et al., 1997).

In later work by Assaf et al. (2008) and Alexander et al. (2010), introducing, respectively, the AxCaliber and ActiveAx frameworks, the biophysical models were further simplified (Alexander, 2008; Alexander et al., 2010; Assaf and Basser, 2005; Assaf et al., 2008). The MRI protocol complexity was optimized and applied in the living human brain (Alexander et al., 2010). In the following, we will focus on the work from the AxCaliber and the ActiveAx frameworks, but the considerations and challenges underlying these are believed to be general for microstructural imaging.

The basic biophysical model of the microstructural environment in white matter and its believed contributing compartments for estimating ADD are illustrated in Figure 15.5 It is assumed that axons are organized as straight nontouching tubes, and due to the isolating myelin and low density of neuronal or glial cell bodies, it is assumed that in the measured MRI signal, the fraction of water molecules exchanging between the intra- and extracellular space is small and is not accounted for in the biophysical model. In *the extracellular space*, signal contributions are assumed to originate both from hindered and free diffusion. Hindered diffusion is water molecules that can displace most freely along axons but across they are hindered due to a tortuosity effect related to axonal density and can be modeled as an anisotropic diffusion tensor (Basser et al., 1994; Szafer and Zhong, 1995). Free diffusion is modeled as an isotropic diffusion tensor and handles, for example, partial volume effects in a voxel containing both

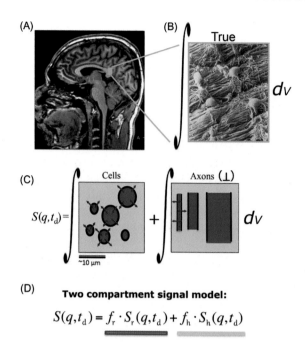

FIGURE 15.5 The basic biophysical principle of microstructural imaging for estimating axon diameters noninvasively with diffusion MRI. (A) The voxel intensities in an MRI, i.e., the green box here shown on a T1-weighted MRI represents in diffusion MRI the sum or integration of all signal contributions of molecular diffusion hindrance and restrictions within that voxel. (B) Shows the real histological elements within an axon tract visualized by scanning EM. (C) The intensity or signal (S) is a function of q-value and diffusion time (t_d) and is assumed to represent a fraction of intra- and extracellular spaces (visualized as blue vs. green). Exchange of water molecules across cell membranes also takes place (red arrows) but often this signal contribution is ignored. To measure the fraction of the intracellular space pertaining to axons, hence information of their size, the direction of the diffusion weighting gradient must be applied perpendicular to the axons. If the gradient is applied along the axons, no restrictions will be observed. (D) Based on an assumption of the geometries in the cellular spaces, one can express S in a biophysical model as resulting from fractions (f) of intra and extracellular space. It is the intracellular space that includes the model of axon diameters obtained when measured perpendicular to the axons or with shells of b-values. The scanning 3D EM of axons and oligodendrocytes in (B) is obtained from https://neurophilosophy.wordpress.com/2006/11/22/time-lapse-movies-of-oligodendrocyte-progenitor-migration/.

cerebrospinal fluid and white matter (Barazany et al., 2009). For the *intracellular space*, water molecules are restricted within axons and so their signal contribution reflects information of axonal geometrics with an assumed shape, i.e., modeled as straight parallel nonpermeable tubes. These restricted intracellular models typically use as input various basic scan parameters related to the q-value and diffusion time (e.g.,Callaghan, 1995; Neuman, 1974; van Gelderen et al., 1994). However, within a voxel the axons do not have a single diameter but an ADD.

From EM and LM studies, the axonal histogram appears to follow a Gamma distribution with a given mean and having a tail of low probabilities representing a small population of larger axons (Aboitiz et al., 1992; Caminiti et al., 2009; LaMantia and Rakic, 1990; Riise and Pakkenberg, 2011). The Gamma distribution therefore is widely used in generating biologically plausible synthetic substrates of axons with known mean ADD and densities. These can be used as "ground truth" for verifying the MRI sequence and biophysical model in a simulation setup (Alexander et al., 2010; Dyrby et al., 2013; Hall and Alexander, 2009). Figure 15.6 shows an example of an EM and synthetic generated substrate for the same region in CC. AxCaliber that is based on the composite hindered and restricted model of diffusion outputs a Gamma distributed ADD that optimally fits the MRI signal; from its alpha and beta parameters (that describe a Gamma distribution), the mean ADD can be calculated (Assaf and Basser, 2005; Assaf et al., 2008). ActiveAx uses the minimal model of white mater diffusion and does not assume a given underlying ADD. Instead it outputs an axonal index describing a mean axon diameter of the underlying ADD (Alexander, 2008; Alexander et al., 2010). Biophysical models for estimating axon diameter are based on highly simplified assumptions of the microstructural environment to reduce model complexity and fitting problems, and therefore include only significantly contributing compartments. However, new knowledge of the various compartments' models are continuously added to better represent the true axonal environment, in particular, the extracellular space (Lam et al., 2014) as well as for handling fiber dispersion effects (see Section 15.2.2.5.2) (Zhang et al., 2011b). Similarly, when using perfusion-fixed tissue for acquiring high-quality diffusion MRI data sets (Dyrby et al., 2011) in long-scan sessions, an additional but small fraction of restricted compartment mimicking trapped (not moving) molecules

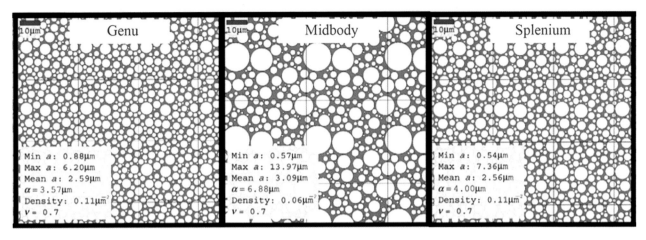

FIGURE 15.6 Virtual white matter environments, i.e., substrates of various ADD can be generated based on histograms of ADD obtained from, e.g., EM studies and used in Monte Carlo simulations to verify estimates of a biophysical model (Alexander et al., 2010; Dyrby et al., 2013; Hall and Alexander, 2009). The example show three substrates generated with different ADD's inspired by EM results in the human brain (Aboitiz et al., 1992) and monkey (LaMantia and Rakic, 1990) in callosal regions representing the genu, midbody, and splenium. For each substrate, the minimum and maximum axon diameter is given (a), the mean as well as the volume weighted mean (α) of the ADD, axonal density, and intraaxonal volume fraction (ν) of 70%. *Adaption of figure 3 in Alexander et al. (2010); with permission by Elsevier.*

must be added to the biophysical model for ensuring an optimal match to the data, which is not needed for *in vivo* estimates (Alexander et al., 2010; Dyrby et al., 2013). This minor signal fraction could be related to a change in membrane permeability due to fixation (Sønderby et al., 2013) and water molecules trapped in glial cells (Stanisz et al., 1997) or maybe due to a kind of fixation artifact where cells exposed to formaldehyde can have small blebs on the membrane (Fox et al., 1985).

15.2.2.3 Tuning MRI Parameters for Sensitivity to Axons

It is essential that the acquired MRI data set provide the correct anatomical contrasts from which model parameters can be estimated. The key is choosing a representative range of MRI parameters, i.e., a combination of q-values and diffusion times (or b-values), and the type of MRI sequence. However, often *ad hoc* sample strategies that include a comprehensive imaging protocol with a wide range of different q-values and diffusion times are used. Typically, such *ad hoc* protocols require experimental MRI scanners with strong diffusion weighting gradient hardware to achieve high q-values that are currently not available in clinical scanners, as well as hours and even days of scanning which challenges *in vivo* usage. As a result they are often constrained to a single direction perpendicular to the axons along which diameter information can be obtained. This poses a further challenge as it requires prior knowledge of the orientation of the axonal bundle to be investigated. The CC is among the few WM structures where the orientation of axon bundles is roughly perpendicular to the cross-sectional plane, i.e., the sagittal plane. Furthermore, ADD and density vary in anteroposterior axonal compartments (Aboitiz et al., 1992; LaMantia and Rakic, 1990) corresponding to different cortical origins of callosal axons (Caminiti et al., 2009; Tomasi et al., 2012), making CC a good structure for studying the performance of biophysical models.

For human *in vivo* microstructural imaging, scan time is limited to less than an hour, and therefore the *ad hoc* protocols must be significantly reduced in number of measurements. One can empirically optimize an *ad hoc* protocol by using design optimization strategies. Alexander (2008) introduced such an optimization framework and named it *ActiveImaging* (AI).

The idea here is to use *a priori knowledge* of the whole imaging framework. This includes a biophysical model of the underlying microstructure and assumptions on the models parameters such as the diffusion coefficients as well as the range of axon diameters for which optimal sensitivity is wanted. Additionally, MR properties of tissue such as the noise level and T1- and T2-relaxation times are taken into account, together with scanner hardware properties such as the gradient strength and a maximal scan time. When including such *a priori* information, ActiveAx demonstrated that as few as three unique b-values (combination of q-values and t_d) could be an optimal choice for estimating axon diameters. AI indicates that more b-values might not lead to further improvements as shown in Figure 15.7. For example, in Alexander et al. (2010), four unique b-values were used, but interestingly the sequence parameters of two of the unique b-values was similar (Alexander et al., 2010; Dyrby et al., 2013). The AI is a powerful general

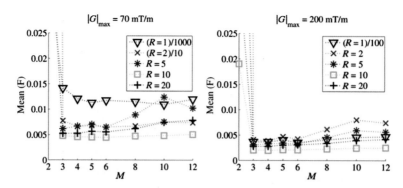

FIGURE 15.7   An experimental framework based on *a priori* knowledge of the microstructural environment can be used to optimize the MRI sequence parameters needed for axon diameters estimates. *A priori* knowledge includes the expected T1- and T2-relaxations times, maximal gradient strength (G_{max}) provided by the scanner hardware, the noise level and the range of single axonal diameters or radii (R) (not ADD). The *y*-axis shows the cost function results based on the Cramér–Rao lower bound function for a given M-number of unique *b*-values shown on the *x*-axis. For a range of single axon radii (R), minimal cost function is found for a relative few number of unique *b*-values. Increasing G_{max} to 200 mT/m clearly only $M=3$ or 4 *b*-values is needed for the estimation of axon diameters using the given setup. *From Alexander (2008), figure 1; with permission by Wiley InterScience.*

framework and can be used to optimize not only gradient waveforms (Drobnjak et al., 2010, 2013) but also a range of different MR sequences, including, for example, magnetization transfer imaging (Cercignani and Alexander, 2006).

To enable rotational invariant estimates of axon diameters (that is, independence from the specific orientation of the WM structure to be measured) each unique *b*-value in the imaging protocol must be acquired in many noncollinear different directions on the unit sphere (Jones et al., 1999), i.e., as a shell in the same way as for DTI (Jones, 2004). Increasing angular resolution (determined by the number of directions in the shell) also increases the number of measurements and hence the total scan time. In clinical settings, acquiring three *b*-values as shells with 90 directions each (in total 360 measurements) can be achieved in about 1 h (Alexander et al., 2010).

Using the optimized ActiveAx protocol for clinical scanners, Alexander et al. (2010) demonstrated the possibility of estimating rotational invariant axon diameters in living human subjects. For comparison, they also estimated axon diameters from high-quality MRI obtained on *ex vivo* monkey brain, and demonstrated for both human and monkey that the gradient of varying mean ADD and density across the CC is in agreement with the literature of EM and LM data as shown in Figure 15.8. Angular resolution was not part of AI in this first study, but later the authors included it as part of AI optimization. No clear difference was found between the three different shells (Dyrby et al., 2013).

15.2.2.4 Fitting the Biophysical Model to MRI

With a biophysical model and an acquired MRI data set in hand, the model is then fitted to the measured signal for estimating axon diameters, density, etc. Avoiding local minima in the fitting procedure and ensuring robust results are often a challenge in microstructural imaging. It increases with model complexity when more and finer compartmental parameters are estimated. A further challenge is the sparsity of the data set as caused by reducing the number of measurements to enable clinical use and/or when there is limited gradient strength. Advanced nonlinear optimization/fitting methods such as the widely known Levenberg–Marquardt algorithm, or a sequence of these nonlinear methods as in ActiveAx, are often used, to ensure robustness in the fitting. Constraints are added to the model parameters as well as to the order of their optimization. These nonlinear optimization strategies are very complex and time consuming, such that even on todays computer clusters it can take days to estimate axon diameters from a whole-brain data set.

Accelerated Microstructure Imaging via Convex Optimization (AMICO), presented by Daducci et al. (2015), is a new, extremely fast, and robust model-fitting framework for advanced biophysical models. The approach is to reorganize the mathematical formulation of the biophysical models from a nonlinear to a linear convex optimization problem. Linear optimization strategies are much faster than the nonlinear ones, and the AMICO was shown to speed up data processing up to four orders of magnitude: hence from days to hours and, importantly, with accuracy similar to that of the nonlinear methods. The authors expect the AMICO framework to become a general mathematical formulation that can be applied to most biophysical models. Indeed, one can design advanced biophysical models and now efficiently fit these to acquired MRI signal; but this of course still requires that the measured MRI signal include the information of the microstructural environment needed to estimate axon diameter.

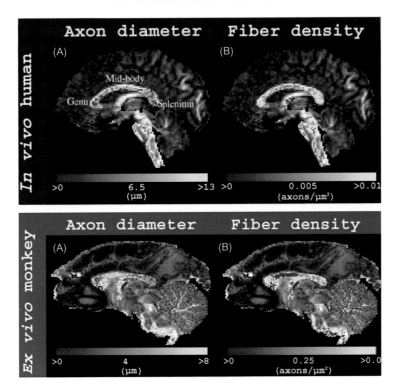

FIGURE 15.8 Axon diameter estimates and densities have been demonstrated in living humans using standard clinical MRI scanners and on fixed monkey brains using a preclinical MRI scanner with the strongest gradient hardware available. For both human and *ex vivo* monkey brains, the ActiveAx axon diameter estimates across the corpus callosum from genu, midbody, and splenium follows a low-higher-lower trend. This is similar albeit not identical to what was found in EM and LM studies (Aboitiz et al., 1992; Caminiti et al., 2009; LaMantia and Rakic, 1990; Caminiti et al., 2013). One difference regards the relatively thin axons in the splenium, where thicker axons were seen with LM and EM. Two factors combined may cause this difference. First, the splenium contains more transversally cut axons and straight axons than the genu hence less macroscopic effects to bias axon diameter to higher values with MRI. Second, due to the less macroscopic effects in splenium, it is possible that the sizes of the axons now falls below the lower-bound where they cannot correctly be determined and just will appear as small. In fixed monkey brain, the regional differences are sharper due to the stronger gradient hardware, i.e., 70 mT/m for clinical MRI versus 140 mT/m. *Modified from figures 7 and 12 from Alexander et al. (2010); with permission by Elsevier.*

15.2.2.5 Why Axonal Diameter Overestimation When Using MRI?

Axon diameter estimation from diffusion MRI is a noninvasive measure compared with LM and EM analyses. Nevertheless, when imaging tissue, biophysical models tend to overestimate axon diameters as compared with EM and LM results. Trendwise estimates of ADD from MRI do follow those of EM/LM, indicating that MRI has the sensitivity required to capture the true ADD. Comparable results are shown for simulations using ADD estimated from EM-based substrates and simulations are therefore essential for identifying and understanding methodological challenges and limitations (Drobnjak et al., 2015; Dyrby et al., 2013). There are several known underlying factors that can lead to difference between MRI and EM/LM, but these cannot yet fully explain all mismatches observed, and important questions remain to be investigated.

15.2.2.5.1 The Diffusion Process Using MRI Is Within a Volume

As the diffusion MRI is a measure within a volume (voxel) and not within a plane (i.e., a slice like in EM and LM), the axon diameter estimates from MRI are actually weighted by the volume of the individual axon within the ADD (Alexander et al., 2010). Hence, the larger axons in the ADD will contribute more to the signal than the smaller ones and thus bias the axon diameter estimate from MRI toward the larger diameters within the ADD.

15.2.2.5.2 Macroscopic Effects—A Yet Underestimated Confounding Effect

Diffusion MRI is made sensitive to geometrical features at the microstructural level, but since the MRI signal is acquired in low resolution with voxel sizes in cubic millimeters, it is sensitive to macroscopic confounds. Macroscopic confounds originate from locally crossing, bending, fanning, and undulating axons, or from any nonstraight axon projections, and are in some contradiction with the basic assumption underlying the biophysical models (Nilsson et al., 2011; Zhang et al., 2011b). Macroscopic effects lead to increased axon diameter estimates from MRI simply because the cross-sectional area may not be measured perfectly perpendicular to the ADD. Hence, some axons will appear larger than they actually are.

It has been suggested that some macroscopic confounds in white matter can be handled if included in the biophysical model. Zhang et al. (2011b) incorporated a fiber dispersion model in ActiveAx and demonstrated how the axon diameter decreased as expected in regions assumed to include minimal dispersion, i.e., consisting of highly aligned axons as in mid-sagittal CC (Zhang et al., 2011b). Also, the crossing of axonal bundles can be handled if incorporated in the model (Zhang et al., 2011a), but much work still remains to both validate and disentangle the ADD from macroscopic confounds along nonstraight tract systems. It should be noted that the above model for handling macroscopic confounds requires that each b-value in the MRI data set is acquired as a shell as done with ActiveAx.

15.2.2.5.3 The Sensitivity Profile of MR Sequences to Distinguish Axon Diameters

Sensitivity of the diffusion MRI signal to axon diameter (i.e., ability to correctly distinguish between ranges of axons diameters) depends on the anatomical contrasts provided by the MRI sequence chosen. The LM and EM literature indicates that the range of axon diameters in the CNS of mammalian brains ranges from >0.1 μm (being typically unmyelinated, short intracortical connections) to giant axons of about 10 μm (LaMantia and Rakic, 1990). The ADD correlates to some extent with brain size but generally the mean ADD is <1 μm. Ideally, MRI sequences should provide sensitivity profiles within this range of diameters, but practically this is rarely the case. Dyrby and colleagues demonstrated, for the widely used PGSE sequence, that the sequence has an axon diameter sensitivity profile that determines the min–max range of axons that can be correctly distinguished (Dyrby et al., 2013). Outside this range, diameters might be too small (the minimum range) to be correctly distinguished or too large for the water molecules to robustly detect the restriction (Dyrby et al., 2013). The sensitivity also includes accuracy of diameter measurement, i.e., the ability to correctly differentiate between (for example) 4 and 5 μm versus 4.5 and 5 μm. This a topic still to be investigated.

The main factor constraining sensitivity to the smallest axon diameters is related to the hardware of the MR scanner, i.e., the limited maximal gradient strength and its slew rate, whereas for the largest axons it is more the tissue T1- and T2-relaxation times and hence a compromise with signal-to-noise ratio (SNR). The use of different MRI sequences can potentially expand the range of sensitivity profiles for axon diameters on both ends of the spectrum. As axon diameter estimates depend upon the chosen MRI sequence, as well as the scanner hardware, we could argue that axon diameter estimates from MRI are not true diameters but more an axonal diameter index (Alexander et al., 2010, Dyrby et al., 2013).

The widely used sequence for microstructure imaging is the pulsed-gradient-spin-echo (PGSE) (Tanner and Stejskal, 1968) sequence. This is a standard diffusion MRI sequence installed on clinical systems, and has a specificity profile to mean ADD ranging from about the smallest (being about 2 μm, lower bound) to the largest geometries <10 μm. This was demonstrated with simulations by Dyrby et al. (2013) for ActiveAx. Results suggested that the lower bound of about 2 μm for the mean of a gamma ADD is typically constrained by the hardware of the MR scanner limited gradients strength (G_{max}) over SNR, which limits contrast to the shortest displacement of water molecules. However, the largest diameters are compromised by limited diffusion time due to relative short T2-relaxation and SNR (Dyrby et al., 2013). There exists a high correlation between "ground truth" and estimated axon diameters in simulations. However and interestingly, although the sizes of ADD from EM/LM falls within the lower bound when measured in tissue, the mean ADD profile measured across the 10 subregions of CC estimated from MRI are overestimated *above* the lower bound and show similar axonal contrast as EM/LM for any maximal gradient strength used in the AI optimization, i.e., 60, 140, 200, and 300 mT/m (Dyrby et al., 2013). Macroscopic effects as described in Section 15.2.2.5.2 cannot alone explain such a large overestimation as observed in tissue.

Various authors suggest that oscillating gradient spin-echo (OGSE) sequence offers benefits over the simple PGSE for probing smaller structures (Drobnjak et al., 2010; Gore et al., 2010; Xu et al., 2014). OGSE sequence is simply a PGSE sequence where the gradient pulse is replaced by an oscillating train of pulses. OGSE sequence hence enables shorter diffusion times not possible with PGSE, and hence it can probe smaller length scales (Parsons et al., 2006). For example, high frequency allows the effective diffusion time to be so short that it is possible to get estimates of free water diffusivity (intrinsic diffusivity), or to obtain a surface-to-volume ratio. This property is used in a method called temporal diffusion spectroscopy (TDS) where tissue is probed by a range of OGSE frequencies in order to create a diffusion spectrum. Different tissues have different microstructural properties and hence give different diffusion spectrums revealing differences at small length scales (Aggarwal et al., 2012; Lundell et al., 2014b). TDS is very useful in situations when the biophysical model is not known, as it provides a way of characterizing different tissues. However, it is difficult to use to investigate sensitivity to individual microstructural parameters.

In situations when the biophysical model is known, like in ActiveAx or AxCaliber approaches, sensitivity to axon diameter can be investigated directly. Recent simulations show that, for the simple case of gradients perfectly perpendicular to straight parallel fibers, PGSE always gives maximum sensitivity to axon diameter compared to other MR sequences. However, in real-world scenarios where fibers have unknown and dispersed orientation, low-frequency OGSE provides higher sensitivity to axon diameter (Drobnjak et al., 2015). This happens because when diffusion gradient and fibers are not perpendicular, OGSE achieves a similar sensitivity to axon diameter as PGSE but with much lower b-value. This provides higher total signal and hence better overall sensitivity. It is not straightforward to translate this knowledge into finding an optimal sequence for measuring axon diameter as model complexity also increase, and AI optimization techniques become necessary. For example, in case both axon diameter and diffusion coefficient are estimated, the optimal protocol contains PGSE and OGSE of frequency correlated to the size of the axons (Drobnjak et al., 2010).

EM and LM studies have demonstrated the existence of giant axons of about 10 μm but these are expected to be very few (LaMantia and Rakic, 1990). Although few in number, there is actually an increased chance of detecting these giant axons with MRI because axon estimates are volume weighted, as previously mentioned. Sensitivity to such large diameters (>10 μm) demands long diffusion times for the water molecules to detect the geometrical boundary. Due to a relative short T2-relaxation compared with diffusion time needed to probe larger diameters, then both PGSE and OGSE might not be an optimal sequence of choice. However, the STimulated Echo Acquisition Mode (STEAM) sequence (Merboldt et al., 1991; Tanner, 1970) allows the use of the much longer T1-relaxation over the T2 and enables long diffusion times. Care must be taken when using the STEAM sequence because the gradient crushers used to handle imperfect 180 RF pulses and the imaging gradients together here defined as butterfly gradients can significantly contribute an unwanted diffusion weighting during the mixing period. This can bias the directionalities of the experimental design and, concurrently, axon diameter estimates. A solution is to avoid the use of crushers for diffusion weighting image volumes or to use a simple compensation to the directional scheme file uploaded to the scanner that we recently have introduced (Lundell et al., 2014a).

15.3 EVOLUTIONARY TRENDS

A fundamental question in brain science is: how does brain connectivity scale to brain size? This question has been approached by using gross anatomical parameters, e.g., by correlating brain volumes to white matter volumes, or to volumes of specific white matter compartments, in particular the CC. However, the gross anatomical estimates are difficult to relate to the computational elements of the CNS, i.e., neurons or axons. Indeed they are affected by volumes of blood vessels, of glial compartments, of extracellular space, etc. A better approach is to estimate the degree of connectivity as function of brain size. This can be done by counting axons in specific white matter tracts of by trying to estimate streamlines computed by noninvasive approaches based on diffusion MRI (Hänggi et al., 2014). These approaches, though, overlook uncertainties on the relation between axon numbers and connectional strength expressed as number and density of synapses or size of synapses. Additional difficulties to the task of extrapolating connectional strength from morphological data were raised (Glickfeld et al., 2014). These include the possibility that a fraction of the projection may have inhibitory effects, by contacting interneurons, or that the different axons may convey different activation properties of their parent cell body.

In this chapter, we will restrict ourselves to the question: how does brain size scale to conduction delays generated by axonal morphologies? The answer to this question is also difficult. In a previous paper, Innocenti et al. (1995) have taken the conduction delays of CC axons computed from antidromic invasion of neuronal cell bodies in homologous areas (V1) as the most reliable estimate of conduction delays in animals of different brain size. The perusal of the available literature did not provide evidence in favor of a systematic relationship between brain size and interhemispheric conduction delays. Conduction delays were rather preserved in spite of differences in brain size but with species-specific differences favoring carnivores (cats) versus rodents, or even more rabbits. Unfortunately, as discussed above, conduction measurements based on such an approach are biased by microelectrode sampling, not necessarily the same for different electrodes, or different laboratories.

More recently, we have estimated axon diameter from myelin-stained axons in macaques, chimpanzee, and humans (Caminiti et al., 2009; Figure 15.9). A similar approach was attempted by Olivares et al. (2001) by measuring axon diameters in EM preparations of rat, rabbit, cat, dog caw, and horse. The findings in the two studies are very

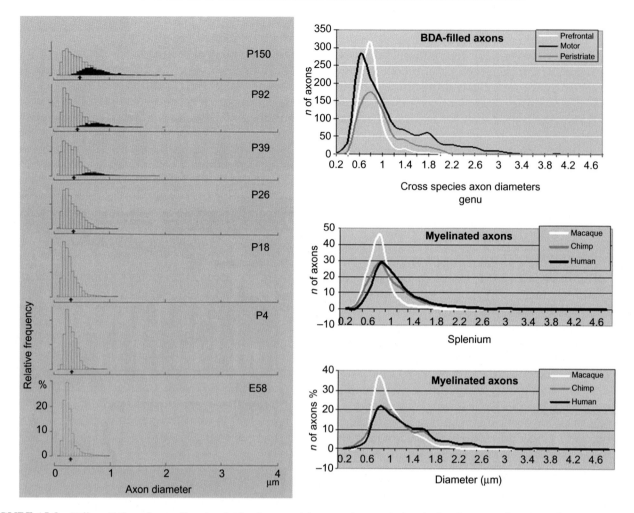

FIGURE 15.9 Differentiation of axon diameters in development (electron microscopic data in development of cat CC (left panel) and evolution (right panels) (Innocenti, 2011). Top: Differences in the spectrum of diameters of BDA traced callosal axon originating in different areas, in the macaque monkey; bottom: differences in the spectrum of myelinated axon diameters in compartments of the corpus callosum, genu, and splenium, containing axons originating in prefrontal and visual areas. Notice that as discussed elsewhere (Innocenti, 2011) in both development and evolution, i.e., over very different time scales, the differentiation consists in dispersion of axon diameters due to the emergence of large diameter axons. *Reproduced from Innocenti (2011).*

similar: "The most common fiber diameter is within the same range in all species studied (0.11–0.2 μm), although the different species vary in the size of this peak. Generally in larger brain species (dog, cow, and horse), the peak is smaller and the rightward tail of the histogram is higher and longer than in smaller-brained species, who have higher peaks and smaller rightward tails (Olivares et al., 2001). "Most striking was that the dispersion of axon diameters increases substantially in both chimpanzees and humans compared with macaques because of an increased number of axons greater than 1–1.5 μm" (Caminiti et al., 2009; Figure 15.9). The addition of a limited number of large axons in bigger brains was also noticed in a comparison of callosal axons in monkey and man (Schüz and Preissl, 1996; Wang et al., 2008). This solution might be the only possible in evolution. If the diameter of all axons were to be increased proportionally to brain size, brain volume would proportionally increase, requiring still larger axons and so on toward an explosive brain enlargement. Nevertheless, it seems clear that brain size is not the only determinant of axonal diameter as (i) cats and dogs have a greater proportion of large axons in the splenium, than species with a larger brain (Olivares et al., 2001), confirming some kind of carnivore advantage as discussed above and (ii) axonal size apparently did not increase further between chimpanzee and human, in spite of a more than threefold increase in brain volume (Caminiti et al., 2009).

The neuroanatomical data seem to be compelling. An increase in conduction delay from macaques to chimpanzee to humans can be inferred from Figure 15.10.

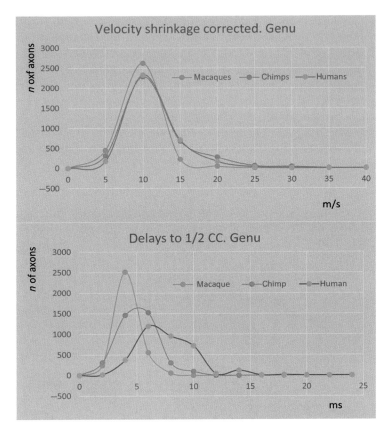

FIGURE 15.10   Example of computed changes in conduction velocity and conduction delays of axons in evolution. Axons in the genu of the corpus callosum, presumably interconnecting prefrontal areas. Top panel, distribution of conduction velocities; axon diameters are corrected for a 30% shrinkage. Notice the similarly increased number of axons with velocities between 15 and 25 m/s in chimpanzee and humans. Bottom panel, the increased conduction velocities in chimpanzee and human do not compensate for the increased brain volumes; this causes interhemispheric communication to slow down and to become temporally dispersed. *Data from Caminiti et al. (2009).*

The conclusions of the two papers quoted above were different. Olivares et al. (2001) felt that the data confirmed Ringo's et al. hypothesis (1994) that in evolution brain enlargement led to slower communication between the hemispheres, hence favoring the emergence of functional hemispheric specialization. Our own work suggested that evolution has increased variability of conduction delays, hence enriching cortical dynamics (Caminiti et al., 2009) as also suggested by theoretical work (Nunez et al., 2014 and below). An additional exciting possibility is that firing rate of axons might increase with diameter (Perge et al., 2012).

15.3.1 Variability

As mentioned above, axon diameters varied characteristically in each of the projections studied. The range of variation increases with axon diameter. Therefore, the range in the macaque is smallest in the projections to the striatum and maximal in motor and somatosensory projections to CC, as well as in motor cortex projections into the internal capsule and the pyramidal tract (Innocenti et al., 2014). Obviously, it is expected that the spectrum of conduction velocities shall expand in parallel. This variability in axon diameters and conduction velocity is an interesting aspect of the morphological variability which characterizes neurons. What could its meaning be? As we have suggested previously (Caminiti et al., 2009), the differences in axon diameter and the related conduction velocity could mean that the transmission of information originating in cortex (and probably anywhere else in the brain) arrives at its targets dispersed in time. Furthermore, if the distribution of axon diameters remains more or less stable with the increase of brain size, e.g., between chimpanzee and humans, the increased conduction distances should increase the temporal dispersion of information. Figure 15.10 documents the increased variability of conduction velocities and conduction delays for callosal axons of the genu of CC, presumably interconnecting the prefrontal areas. In chimpanzee and humans, axons with conduction velocity between 15 and 25 m/s increase in similar proportions (These values are corrected against estimated 30% histological shrinkage.). The combined effect of dispersion of axon diameters and increased brain volume expand the calculated range of conduction delays to the CC midline from about 6ms in the macaque to 8.5 in the chimpanzee and 18ms in humans.

We speculated that the range of conduction delays expanded dramatically in the transition between the australopithecine and the emergence of *Homo sapiens* when brain volume increased from about 400 to 1500 cm^3. These suggestions must be complemented by the following considerations. First, the data available derive from the computation of conduction delays caused by the distribution of axon diameters (hence conduction velocities) along conduction distance computed on the trajectory of axons in the bulk of the tract (see also Caminiti et al., 2014). In other words, they do not represent the real dispersion of delays along the single, individual axons, which seems impossible to evaluate by current anatomical techniques. As the individual axons do not have identical trajectories, and therefore conduction distances, such an estimate would be useful to have. Second, axons with different diameters may well contact different target neurons. In particular, work on the interhemispheric connection in the ferret visual cortex has suggested that some of the fastest axons may contact inhibitory interneurons (Makarov et al., 2008). In the mouse visual system, the projections to different areas convey different neuronal activation properties of V1 neurons (Glickfeld et al., 2014), and it would not be surprising that the axonal properties of such projections were also found to be different. Finally, the relations between axon diameter and conduction velocity of myelinated and unmyelinated axons seem well established on the basis of work by pioneers of electrophysiology. However, some of these concepts might have to be sharpened in the future, for example, by considering the changes in conduction velocity dependent on variations in the *g*-ratio, the previous state of the axon (Swadlow, 1985), or other modulations due to interaction with extracellular ionic concentration, axon–glia interaction, etc. (e.g. Adriano et al., 2011; see Bucher and Goaillard, 2011; Debanne et al., 2011 for reviews).

15.4 MODELING CONDUCTION DELAYS

Innocenti and collaborators used a simplified model of interhemispheric connections to explore the consequences of conduction delays (Caminiti et al., 2009). The model consisted of naturally bursting neurons of the Hindmarsch–Rose type: one excitatory and one inhibitory neuron in each hemisphere. The excitatory neurons were reciprocally connected (the dee connection). The inhibitory neuron received connections from the excitatory neurons of the other hemisphere (the dei connection) and projected back to the excitatory neuron of the same hemisphere. In the range of dei delays calculated for the macaque monkey, the burst length increased linearly with the dei, and for all values of dee delays. Above that it produced bifurcations, i.e., the burst length increased or decreased, depending on the value of the dee delay. In other words, the combined effect of the dee and dei delays was the regulation of the oscillatory frequency of the net.

The importance of delays on the behavior of artificial neural networks has not escaped the attention of theoreticians. More references to their work can be found in the selected papers quoted below. Roxin et al. (2005) investigated the role of delays in pattern formation in randomly connected neural networks. Coombes and Laing (2009) analyzed the role of delays on the dynamics of a Wilson–Cowan neural network. Pérez et al. (2011) reported that with certain network topologies, delays can set the phase differences between the oscillatory regimes of remote neural populations and also determine whether the interconnected cells can set in any coherent firing at all. Similarly, Qian (2014) reported synchronization transitions, induced by delays in Newman–Watts small world neural networks. Veltz and Faugeras (2011) analyzed how the introduction of conduction delays between neurons changes the stationary states of Wilson–Cowan–Amari type networks. The collective message from this work probably still is (Coombes and Laing, 2009):"There remains much to be discovered about the role of delays in more realistic neural models" although attempts in this direction exist (see also Deco et al., 2011 and references therein).

15.5 DEVELOPMENT AND PLASTICITY

Innocenti (2011) has recently discussed the development of axonal diameters as a process of neuronal differentiation (i.e., neurons becoming different from each other) and stressed the similarity between axonal differentiation in development and evolution. Figure 15.9 illustrates the factual basis for that statement, and the argument will not be reiterated here.

Early investigations of the role of neural activity in myelination in development provided ambiguous results. Gyllensten and Malmfors (1963) were probably the first to report a decrease in the number of myelinated axons and a 60% decrease in the number of axons larger than 1 μm in the optic nerve of mice dark-reared from birth up to day 20 or 30. In dark-reared kittens, myelination of the optic tract seemed to be delayed at 3 weeks but no longer at 4 weeks and there was a slight increase in the mean axon size (Moore et al., 1976). Conversely, artificial eye opening accelerated myelination in the optic nerve of the rabbit (Tauber et al., 1980). In contrast to the results of Gyllensten and Malmfors (above), an increase in the number of myelinated axons and in the diameter of axons was found in the optic nerve of dark-reared rats (Fukui and Bedi, 1991). If species or other factors cause the differences remains unclear.

The role of experience in the development of white mater tracts was rekindled by the finding by Bengtsson et al. (2005) that fractional anisotropy (FA) obtained from diffusion MRI (Basser et al., 1994) increased in the internal capsule, CC, and arcuate fasciculus of adult pianists (aged 32.6 years, on average) proportionally to the time spent in practicing during childhood. Also, in rats, FA in white matter pathways has been shown to correlate with the learning rate of a novel motor skill (Sampaio-Baptista et al., 2013). Unfortunately, changes in FA cannot be safely interpreted as changes in myelination; indeed no relationship between diffusivity changes and myelin basic protein (MBP) could be found by Sampaio-Baptista et al. (2013). Difficulties of interpretation also arise with the decrease in cross-sectional surface area of the posterior CC in macaques raised in "nursery" conditions versus controls (Sánchez et al., 1998). As reported in rodents (Markham et al., 2009), macroscopic changes could be due to an increase in glial cell processes or unmyelinated axons. Differences in the axonal composition of the CC was found in rats raised from P 23-25 until P 53-55 in isolation versus enriched condition, but the differences depended also on sex (Juraska and Kopcik, 1988).

Deprivation of mother–pup interaction was found to alter the expression of (MBP) in male but not in female pups (Kikusui et al., 2007). Two weeks of social isolation after weaning decreased the expression of MBP and myelin-associated glycoprotein (MAG) and increased the g-ratio in the prefrontal cortex of mice (Makinodan et al., 2012). The effects were restricted to a critical period which did not extend beyond P35. However, a decreased myelination was reported in the prefrontal cortex of mice deprived of social interactions as late as at 8 weeks (Liu et al., 2012).

On the whole, the results quoted above suggest that myelination in development might be regulated by experience. Although this seems reasonable given the precise timing of neuronal interactions required in adult neural circuits (Buzsáki et al., 2013), the available evidence does not rule out the possibility that altered selection of axons (Innocenti and Price, 2005) might have also played a role in the results reported. An additional question regards the role of experience in myelination in adulthood. Complex environmental housing—which, as mentioned above, modifies the axonal composition of the splenium—was reported to have no effect after four months of age (Markham et al., 2009).

15.5.1 Plasticity in Adulthood

Neuronal activity is inherently a consequence of experiencing the world. Neuronal activity might affect myelination, not only in development but also in the adult brain, although technological limitations have hampered direct testing of the hypothesis. Further, it needs to be established whether any change in myelination in turn affects the circuitry, e.g., axonal parameters determining conduction speed and ultimately behavior. To address these questions, it is central to evoke activity without causing tissue injury or inflammation, which could affect the glial compartment, including the cells involved in myelin production. Optogenetics (Yizhar et al., 2011) provide the opportunity to drive neuronal activity with light, circumventing tissue damage related to the use of electrodes for electrical stimulation of activity. In a recent study, optogenetic stimulation was employed to unilaterally activate cortical layer V neurons in premotor cortex of adult mice (Gibson et al., 2014). The activation, as expected, induced not only circling ambulation of the animals but also proliferation of oligodendrocyte precursors (OPCs) and increased number of differentiated myelin-producing oligodendrocytes in the stimulated premotor cortex and the underlying gray matter/white matter border, a compartment where corticospinal, callosal, other cortico-cortical and cortical afferent axons cross. The ontogenetic induction of motor activity was found to be accompanied by thickening of the myelin and by a small decrease in the g-ratio within the stimulated premotor circuit. This suggests enhanced conduction velocities in the axons, which could affect motor function. Therefore, the animals' motor performance was assayed. The authors found the forelimb swing speed during normal gait to be improved in the forelimb driven by the stimulated hemisphere, supporting the notion that neuronal activity drives adaptations in myelination which in turn influence the underlying neural circuits and behavior.

The functional role of adult myelin adaptations found further support in a recent study showing that adult-born oligodendrocytes are required for learning novel motor skills. Using genetically modified mice, McKenzie et al. (2014) could show that knockout of the myelin regulatory factor MYRF in oligodendrocyte progenitor cells results in severe loss of adult-born oligodendrocytes and less newly formed myelinating sheaths in the CC. These changes were accompanied by a failure to master a complex motor skill, implying the necessity of adult oligodendrogenesis and myelin production in learning novel motor skills.

However, many questions remain to be answered. The changes in myelination are probably the only reliable evidence of the role of activity or experience on myelination. Oligodendrocyte proliferation is not, since there is no reason *a priori* to expect that in adult, experimental animals they will be involved in myelination. Furthermore, oligodendrocyte proliferation may not be prominently involved in myelin turn over in humans (Yeung et al., 2014).

In conclusion, the importance of the axon in the proliferation of OPCs and in oligodendrocyte survival seems well established *in vitro* (reviewed by Barres and Raff, 1999), possibly mediated by activity (Demerens et al., 1996) and/or via astrocytes (Ishibashi et al., 2006). Conclusive *in vivo* evidence is still missing, particularly in adult animals. Major questions to be tackled regard the relationship between the proliferation of OPCs, their differentiation, and myelin formation. The relations between myelin formation, axonal diameter, and conduction properties need to be investigated as does the possibility that in development myelin may be labile for some time, before it is stabilized into adult myelin, as happens to the axons themselves (Innocenti and Price, 2005).

The molecular mechanisms of axonal myelination have been reviewed recently (e.g., Nave, 2010; Taveggia et al., 2010) and fall beyond the scopes of this chapter.

15.5.2 The Mysterious Unmyelinated Axons

The fraction of unmyelinated axons in intensively studied structures, such as the CC, varies very much among different studies. The estimate of 30% unmyelinated axons is sometimes quoted, disregarding the more complete study by Aboitiz et al. (1992) which reported 16% unmyelinated fibers in the genu and about 5% elsewhere, in line with a comparable distribution in the macaque monkey (LaMantia and Rakic, 1990). The existence of these fibers in long pathways is puzzling as they would result in very long conduction delays. One hypothesis is that these axons may have tonic, rather than phasic influence on their targets (LaMantia and Rakic, 1990). Another is that they may contribute modulate to effects mediated by the faster myelinated axons (Aboitiz et al., 1992).

We propose that unmyelinated axons might represent a pool of undifferentiated, poorly connected axons, to be recruited under exceptional circumstances such as those imposed by increased functional demands. Alternatively, they might be understimulated axons which have never acquired, or have lost, myelin sheath in development.

15.6 PERSPECTIVES

White matter abnormalities have been often identified in neurological and psychiatric conditions ranging from developmental dyslexia (Klingberg et al., 2000) to schizophrenia (Innocenti et al., 2003; reviewed by Fields, 2008). The physiopathogenetic mechanism is unclear and it could refer to loss of axons, maintenance of exuberant axons which are normally eliminated in development (Innocenti and Price, 2005) and/or to structural changes affecting individual axons and their conduction properties. While structural axonal abnormalities seem to be beyond doubt in some conditions, in particular in demyelinating diseases such as MS, more subtle, pathogenetic changes in axonal structures can be expected. This seems to be an avenue for further investigations in which the noninvasive estimates of microscopic axonal morphology by diffusion MRI, as mentioned above, are bound to play a major role.

References

Aboitiz, F., Scheibel, A.B., Fisher, R.S., Zaidel, E., 1992. Fiber composition of the human corpus callosum. Brain. Res. 598, 143–153.

Adriano, E., Perasso, L., Panfoli, I., Ravera, S., Gandolfo, C., Mancardi, G., et al., 2011. A novel hypothesis about mechanisms affecting conduction velocity of central myelinated fibers. Neurochem. Res. 36, 1732–1739.

Aggarwal, M., Jones, M.V., Calabresi, P.A., Mori, S., Zhang, J., 2012. Probing mouse brain microstructure using oscillating gradient diffusion MRI. Magn. Reson. Med. 67, 98–109.

Alexander, D.C., 2008. A general framework for experiment design in diffusion MRI and its application in measuring direct tissue-microstructure features. Magn. Reson. Med. 60, 439–448.

Alexander, D.C., Hubbard, P.L., Hall, M.G., Moore, E.A., Ptito, M., Parker, G.J.M., et al., 2010. Orientationally invariant indices of axon diameter and density from diffusion MRI. Neuroimage 52, 1374–1389.

Assaf, Y., Basser, P.J., 2005. Composite hindered and restricted model of diffusion (CHARMED) MR imaging of the human brain. Neuroimage 27, 48–58.

Assaf, Y., Blumenfeld-Katzir, T., Yovel, Y., Basser, P.J., 2008. AxCaliber: a method for measuring axon diameter distribution from diffusion MRI. Magn. Reson. Med. 59, 1347–1354.

Barazany, D., Basser, P.J., Assaf, Y., 2009. *In vivo* measurement of axon diameter distribution in the corpus callosum of rat brain. Brain 132, 1210–1220.

Barres, B. a., Raff, M.C., 1999. Axonal control of oligodendrocyte development. J. Cell Biol. 147, 1123–1128.

Basser, P.J., Mattiello, J., LeBihan, D., 1994. Mr diffusion tensor spectroscopy and imaging. Biophys. J. 66, 259–267.

Bengtsson, S.L., Nagy, Z., Skare, S., Forsman, L., Forssberg, H., Ullén, F., 2005. Extensive piano practicing has regionally specific effects on white matter development. Nat. Neurosci. 8, 1148–1150.

Bentivoglio, M., Kuypers, H.G.J.M., Catsman-Berrevoets, C.E., Dann, O., 1979. Fluorescent retrograde neuronal labeling in rat by means of substances binding specifically to adenine. Neurosci. Lett. 12, 235–240.

Bucher, D., Goaillard, J.M., 2011. Beyond faithful conduction: short-term dynamics, neuromodulation, and long-term regulation of spike propagation in the axon. Prog. Neurobiol. 94, 307–346.

Buzsáki, G., Logothetis, N., Singer, W., 2013. Scaling brain size, keeping timing: evolutionary preservation of brain rhythms. Neuron 80, 751–764.

Callaghan, P.T., 1995. Pulsed-gradient spin-echo NMR for planar, cylindrical, and spherical pores under conditions of wall relaxation. J. Magn. Reson. 113, 53–59.

Callaghan, P.T., Coy, A., MacGowan, D., Packer, K.J., Zelaya, F.O., 1991. Diffraction-like effects in NMR diffusion studies of fluids in porous solids. Nature 351, 467–469.

Caminiti, R., Ghaziri, H., Galuske, R., Hof, P.R., Innocenti, G.M., 2009. Evolution amplified processing with temporally dispersed slow neuronal connectivity in primates. Proc. Natl. Acad. Sci. U.S.A. 106, 19551–19556.

Caminiti, R., Carducci, F., Piervincenzi, C., Battaglia-Mayer, A., Confalone, G., Visco-Comandini, F., et al., 2013. Diameter, length, speed and conduction delay of callosal axons in macaque monkeys and humans. Comparing data from histology and MRI diffusion tractography. J. Neurosci. 33, 14501–14511.

Cercignani, M., Alexander, D.C., 2006. Optimal acquisition schemes for *in vivo* quantitative magnetization transfer MRI. Magn. Reson. Med. 56, 803–810.

Cohen, Y., Assaf, Y., 2002. High *b*-value *q*-space analyzed diffusion-weighted MRS and MRI in neuronal tissues—a technical review. NMR Biomed. 15, 516–542.

Coombes, S., Laing, C., 2009. Delays in activity-based neural networks. Philos. Trans. A, Math. Phys. Eng. Sci. 367, 1117–1129.

Daducci, A., Canales-Rodríguez, E.J., Zhang, H., Dyrby, T.B., Alexander, D.C., Thiran, J.-P., 2015. Accelerated microstructure imaging via convex optimization (AMICO) from diffusion MRI data. Neuroimage 105, 32–44.

Debanne, D., Campanac, E., Bialowas, A., Carlier, E., Alcaraz, G., 2011. Axon physiology. Physiol. Rev. 91, 555–602.

Deco, G., Jirsa, V.K., McIntosh, A.R., 2011. Emerging concepts for the dynamical organization of resting-state activity in the brain. Nat. Rev. Neurosci. 12, 43–56.

Demerens, C., Stankoff, B., Logak, M., Anglade, P., Allinquant, B., Couraud, F., et al., 1996. Induction of myelination in the central nervous system by electrical activity. Proc. Natl. Acad. Sci. U.S.A. 93, 9887–9892.

Drobnjak, I., Siow, B., Alexander, D.C., 2010. Optimizing gradient waveforms for microstructure sensitivity in diffusion-weighted MR. J. Magn. Reson. 206, 41–51.

Drobnjak, I., Cruz, G., Alexander, D.C., 2013. Optimising oscillating waveform-shape for pore size sensitivity in diffusion-weighted MR. Microporous Mesoporous Mater. 178, 11–14.

Drobnjak, I., Zhang, H., Ianus, A., Caden, E., Alexander, D.C., 2015. PGSE, OGSE and Sensitivity to axon diameter in diffusion MRI: insight from a simulation study. Magn. Reson. Med. http://dx.doi.org/10.1002/mrm.25631.

Dyrby, T.B., Baaré, W.F.C., Alexander, D.C., Jelsing, J., Garde, E., Søgaard, L.V., 2011. An *ex vivo* imaging pipeline for producing high-quality and high-resolution diffusion-weighted imaging data sets. Hum. Brain. Mapp. 32, 544–563.

Dyrby, T.B., Søgaard, L.V., Hall, M.G., Ptito, M., Alexander, D.C., 2013. Contrast and stability of the axon diameter index from microstructure imaging with diffusion MRI. Magn. Reson. Med. 70, 711–721.

Fields, R.D., 2008. White matter in learning, cognition and psychiatric disorders. Trends. Neurosci. 31, 361–370.

Firmin, L., Field, P., Maier, M.A., Kraskov, A., Kirkwood, P.A., Nakajima, K., et al., 2015. Axon diameters and conduction velocities in the macaque pyramidal tract axon diameters and conduction velocities in the macaque pyramidal tract. J. Neurophysiol. 112, 1229–1240.

Fox, C.H., Johnson, F.B., Whiting, J., Roller, P.P., 1985. Formaldehyde fixation. J. Histochem. Cytochem. 33, 845–853.

Fukui, Y., Bedi, K.S., 1991. Quantitative study of the development of neurons and synapses in rats reared in the dark during early postnatal life. 1. Superior colliculus. J. Anat. 174, 49–60.

Gibson, E.M., Purger, D., Mount, C.W., Goldstein, A.K., Lin, G.L., Wood, L.S., et al., 2014. Neuronal activity promotes oligodendrogenesis and adaptive myelination in the mammalian brain. Science 344, 1252304. http://dx.doi.org/10.1126/science.1252304.

Glickfeld, L.L., Reid, R.C., Andermann, M.L., 2014. A mouse model of higher visual cortical function. Curr. Opin. Neurobiol. 24, 28–33.

Gore, J.C., Xu, J., Colvin, D.C., Yankeelov, T.E., Parsons, E.C., Does, M.D., 2010. Characterization of tissue structure at varying length scales using temporal diffusion spectroscopy. NMR. Biomed. 23, 745–756.

Gyllensten, L., Malmfors, T., 1963. Myelinization of the optic nerve and its dependence on visual function—a quantitative investigation in mice. J. Embryol. Exp. Morphol. 11, 255–266.

Hall, M.G., Alexander, D.C., 2009. Convergence and parameter choice for Monte-Carlo simulations of diffusion MRI. IEEE. Trans. Med. Imaging. 28, 1354–1364.

Hänggi, J., Fövenyi, L., Liem, F., Meyer, M., Jäncke, L., 2014. The hypothesis of neuronal interconnectivity as a function of brain size—a connectomal organisation principle? Front. Hum. Neurosci. 8, 915. http://dx.doi.org/10.3389/fnhum.2014.00915.

Humphrey, D.R., Corrie, W.S., 1978. Properties of pyramidal tract neuron system within a functionally defined subregion of primate motor cortex. J. Neurophysiol. 41, 216–243.

Innocenti, G.M., 1995. Exuberant development of connections, and its possible permissive role in cortical evolution. Trends. Neurosci. 18, 397–402.

Innocenti, G.M., 2011. Development and evolution: two determinants of cortical connectivity Braddick, O. Atkinson, J. Innocenti, G.M. (Eds.), Progress in Brain Research, vol. 189 Academic Press, Burlington, CA, pp. 65–75.

Innocenti, G.M., Price, D.J., 2005. Exuberance in the development of cortical networks. Nat. Rev. Neurosci. 6, 955–965.

Innocenti, G.M., Lehmann, P., Houzel, J.C., 1994. Computational structure of visual callosal axons. Eur. J. Neurosci. 6, 918–935.

Innocenti, G.M., Aggoun-Zouaoui, D., Lehmann, P., 1995. Cellular aspects of callosal connections and their development. Neuropsychologia 33, 961–987.

Innocenti, G.M., Ansermet, F., Parnas, J., 2003. Schizophrenia, neurodevelopment and corpus callosum. Mol. Psychiatry. 8, 261–274.

Innocenti, G.M., Vercelli, A., Caminiti, R., 2014. The diameter of cortical axons depends both on the area of origin and target. Cereb. Cortex. 24, 2178–2188.

Ishibashi, T., Dakin, K. a., Stevens, B., Lee, P.R., Kozlov, S.V., Stewart, C.L., et al., 2006. Astrocytes promote myelination in response to electrical impulses. Neuron 49, 823–832.

Jones, D.K., 2004. The effect of gradient sampling schemes on measures derived from diffusion tensor MRI: a Monte Carlo study. Magn. Reson. Med. 51, 807–815.

Jones, D.K., Horsfield, M.A., Simmons, A., 1999. Optimal strategies for measuring diffusion in anisotropic systems by magnetic resonance imaging. Magn. Reson. Med. 42, 515–525.

Juraska, J.M., Kopcik, J.R., 1988. Sex and environmental influences on the size and ultrastructure of the rat corpus callosum. Brain. Res. 450, 1–8.

Katz, L.C., Burkhalter, a, Dreyer, W.J., 1984. Fluorescent latex microspheres as a retrograde neuronal marker for *in vivo* and in vitro studies of visual cortex. Nature 310, 498–500.

Kikusui, T., Kiyokawa, Y., Mori, Y., 2007. Deprivation of mother–pup interaction by early weaning alters myelin formation in male, but not female, ICR mice. Brain. Res. 1133, 115–122.

King, M. a, Louis, P.M., Hunter, B.E., Walker, D.W., 1989. Biocytin: a versatile anterograde neuroanatomical tract-tracing alternative. Brain. Res. 497, 361–367.

Klingberg, T., Hedehus, M., Temple, E., Salz, T., Gabrieli, J.D., Moseley, M.E., et al., 2000. Microstructure of temporo-parietal white matter as a basis for reading ability: evidence from diffusion tensor magnetic resonance imaging. Neuron 25, 493–500.

Kristensson, K., Olsson, Y., 1971. Retrograde axonal transport of protein. Brain Res. 29, 363–365.

Lam, W.W., Jbabdi, S., Miller, K., 2014. A model for extra-axonal diffusion spectra with frequency-dependent restriction. Magn. Reson. Med. http://dx.doi.org/10.1002/mrm.25363

LaMantia, A.S., Rakic, P., 1990. Cytological and quantitative characteristics of four cerebral commissures in the rhesus monkey. J. Comp. Neurol. 291, 520–537.

Liewald, D., Miller, R., Logothetis, N., Wagner, H.-J., Schüz, A., 2014. Distribution of axon diameters in cortical white matter: an electron-microscopic study on three human brains and a macaque. Biol. Cybern. 108, 541–555.

Liu, J., Dietz, K., DeLoyht, J.M., Pedre, X., Kelkar, D., Kaur, J., et al., 2012. Impaired adult myelination in the prefrontal cortex of socially isolated mice. Nat. Neurosci. 15, 1621–1623.

Lundell, H., Alexander, D.C., Dyrby, T.B., 2014a. High angular resolution diffusion imaging with stimulated echoes: compensation and correction in experiment design and analysis. NMR. Biomed. 27, 918–925.

Lundell, H., Sønderby, C.K., Dyrby, T.B., 2014b. Diffusion weighted imaging with circularly polarized oscillating gradients. Magn. Reson. Med. 73, 1171–1176.

Makarov, V.A., Schmidt, K.E., Castellanos, N.P., Lopez-Aguado, L., Innocenti, G.M., 2008. Stimulus-dependent interaction between the visual areas 17 and 18 of the 2 hemispheres of the ferret (*Mustela putorius*). Cereb. Cortex. 18, 1951–1960.

Makinodan, M., Rosen, K., Corgas, G., 2012. A critical period for social experience. Science 337, 1357–1360.

Markham, J.A., Herting, M.M., Luszpak, A.E., Juraska, J.M., Greenough, W.T., 2009. Myelination of the corpus callosum in male and female rats following complex environment housing during adulthood. Brain. Res. 1288, 9–17.

McKenzie, I.A., Ohayon, D., Li, H., de Faria, J.P., Emery, B., Tohyama, K., et al., 2014. Motor skill learning requires active central myelination. Science 346, 318–322.

Merboldt, K.D., Hanicke, W., Frahm, J., 1991. Diffusion imaging using stimulated echoes. Magn. Reson. Med. 19, 233–239.

Moore, C.L., Kalil, R., Richards, W., 1976. Development of myelination in optic tract of the cat. J. Comp. Neurol. 165, 125–136.

Nave, K.-A., 2010. Myelination and support of axonal integrity by glia. Nature 468, 244–252.

Neuman, C.H., 1974. Spin echo of spins diffusing in a bounded medium. J. Chem. Phys. 60, 4508–4511.

Nilsson, M., Lätt, J., Ståhlberg, F., Westen, D., Hagslätt, H., 2011. The importance of axonal undulation in diffusion Mr measurements: a Monte Carlo simulation study. NMR. Biomed. 25, 795–805.

Nunez, P.L., Srinivasan, R., Fields, R.D., 2014. EEG functional connectivity, axon delays and white matter disease. Clin. Neurophysiol. 126, 110–120.

Olivares, R., Montiel, J., Aboitiz, F., 2001. Species differences and similarities in the fine structure of the mammalian corpus callosum. Brain. Behav. Evol. 57, 98–105.

Panagiotaki, E., Schneider, T., Siow, B., Hall, M.G., Lythgoe, M.F., Alexander, D.C., 2012. Compartment models of the diffusion Mr signal in brain white matter: a taxonomy and comparison. Neuroimage 59, 2241–2254.

Parsons, E.C., Does, M.D., Gore, J.C., 2006. Temporal diffusion spectroscopy: theory and implementation in restricted systems using oscillating gradients. Magn. Reson. Med. 55, 75–84.

Pérez, T., Garcia, G.C., Eguíluz, V.M., Vicente, R., Pipa, G., Mirasso, C., 2011. Effect of the topology and delayed interactions in neuronal networks synchronization. PLoS. ONE. 6, 1–9.

Perge, J.A., Niven, J.E., Mugnaini, E., Balasubramanian, V., Peter Sterling, P., 2012. Why do axons differ in caliber? J. Neurosci.. 32, 626–638.

Pollak Dorocic, I., Fürth, D., Xuan, Y., Johansson, Y., Pozzi, L., Silberberg, G., et al., 2014. Whole-brain atlas of inputs to serotonergic neurons of the dorsal and median raphe nuclei. Neuron 83, 663–678.

Qian, Y., 2014. Time delay and long-range connection induced synchronization transitions in Newman–Watts small-world neuronal networks. PLoS. ONE. 9, 1–9.

Remahl, S., Hilderbrand, C., 1990. Relations between axons and oligodendroglial cells during initial myelination. II. The individual axon. J. Neurocytol. 19, 883–898.

Riise, J., Pakkenberg, B., 2011. Stereological estimation of the total number of myelinated callosal fibers in human subjects. J. Anat. 218, 277–284.

Ringo, J.L., Doty, R.W., Demeter, S., Simard, P.Y., 1994. Time is of the essence: a conjecture that hemispheric specialization arises from interhemispheric conduction delay. Cerebral. Cortex 4, 331–343.

Roxin, A., Brunel, N., Hansel, D., 2005. Role of delays in shaping spatiotemporal dynamics of neuronal activity in large networks. Phys. Rev. Lett. 94, 1–4.

Sampaio-Baptista, C., Khrapitchev, A. a, Foxley, S., Schlagheck, T., Scholz, J., Jbabdi, S., et al., 2013. Motor skill learning induces changes in white matter microstructure and myelination. J. Neurosci. 33, 19499–19503.

Sánchez, M.M., Hearn, E.F., Do, D., Rilling, J.K., Herndon, J.G., 1998. Differential rearing affects corpus callosum size and cognitive function of rhesus monkeys. Brain. Res. 812, 38–49.

Schüz, A., Preissl, H., 1996. Basic connectivity of the cerebral cortex and some considerations on the corpus callosum. Neurosci. Biobehav. Rev. 20, 567–570.

Seidl, A.H., 2013. Regulation of conduction time along axons. Neuroscience 276, 126–134.

Snaidero, N., Möbius, W., Czopka, T., Hekking, L.H.P., Mathisen, C., Verkleij, D., et al., 2014. Myelin membrane wrapping of CNS axons by PI(3,4,5)P3-dependent polarized growth at the inner tongue. Cell 156, 277–290.

Sønderby, C.K., Lundell, H.M., Søgaard, L.V., Dyrby, T.B., 2013. Apparent exchange rate imaging in anisotropic systems. Magn. Reson. Med. 72, 756–762.

Stanisz, G.J., Szafer, A., Wright, G.A., Henkelman, R.M., 1997. An analytical model of restricted diffusion in bovine optic nerve. Magn. Reson. Med. 37, 103–111.

Swadlow, H.A., 1985. Physiological properties of individual cerebral axons studied *in vivo* for as long as one year. J. Neurophysiol. 54, 1346–1362.

Szafer, A., Zhong, J., 1995. Theoretical model for water diffusion in tissues. Magn. Reson. Med. 33, 697–712.

Tanner, J.E., 1970. Use of the stimulated echo in NMR diffusion measurements. J. Chem. Phys. 52, 2523–2526.

Tanner, J.E., Stejskal, E.O., 1968. Restricted self-diffusion of protons in colloidal systems by the pulsed-gradient spin-echo method. J. Chem. Phys. 49, 1768–1777.

Tauber, H., Waehneldt, T.V., Neuhoff, V., 1980. Myelination in rabbit optic nerves is accelerated by artificial eye opening. Neurosci. Lett. 16, 235–238.

Taveggia, C., Feltri, M.L., Wrabetz, L., 2010. Signals to promote myelin formation and repair. Nat. Rev. Neurosci. 6, 276–287.

Tomasi, S., Caminiti, R., Innocenti, G.M., 2012. Areal differences in diameter and length of corticofugal projections. Cereb. Cortex. 22, 1463–1472.

Tomassy, G.S., Berger, D.R., Chen, H.-H., Kasthuri, N., Hayworth, K.J., Vercelli, A., et al., 2014. Distinct profiles of myelin distribution along single axons of pyramidal neurons in the neocortex. Science 344, 319–324.

Ugolini, G., 2007. Rabies virus as a transneuronal tracer of neuronal connections. Adv. Virus. Res. 79, 165–202.

van Gelderen, P., DesPres, D., van Zijl, P.C., Moonen, C.T., 1994. Evaluation of restricted diffusion in cylinders. Phosphocreatine in rabbit leg muscle. J. Magn. Reson. B 103, 255–260.

Veltz, R., Faugeras, O., 2011. Stability of the stationary solutions of neural field equations with propagation delays. J. Math. Neurosci. 1, 1.

Wall, N.R., De La Parra, M., Callaway, E.M., Kreitzer, A.C., 2013. Differential innervation of direct- and indirect-pathway striatal projection neurons. Neuron 79, 347–360.

Wang, S.-H., Shultz, J.R., Burish, M.J., Kimberly, H., Harrison, K.H., Hof, P.R., et al., 2008. Shaping of white matter composition by biophysical scaling constraints. J. Neurosci. 28, 4047–4056.

Watabe-Uchida, M., Zhu, L., Ogawa, S.K., Vamanrao, A., Uchida, N., 2012. Whole-brain mapping of direct inputs to midbrain dopamine neurons. Neuron 74, 858–873.

Wickersham, I.R., Lyon, D.C., Barnard, R.J., Mori, T., Finke, S., Conzelmann, K.K., et al., 2014. Monosynaptic restriction of transsynaptic tracing from single, genetically targeted neurons. Neuron 53, 639–647.

Xu, J., Li, H., Harkins, K.D., Jiang, X., Xie, J., Kang, H., et al., 2014. Mapping mean axon diameter and axonal volume fraction by MRI using temporal diffusion spectroscopy. Neuroimage 103, 10–19.

Yeung, M.S.Y., Zdunek, S., Bergmann, O., Bernard, S., Salehpour, M., Alkass, K., et al., 2014. Dynamics of oligodendrocyte generation and myelination in the human brain. Cell 159, 766–774.

Yizhar, O., Fenno, L.E., Davidson, T.J., Mogri, M., Deisseroth, K., 2011. Optogenetics in neural systems. Neuron 71, 9–34.

Zatorre, R.J., Fields, R.D., Johansen-Berg, H., 2012. Plasticity in gray and white: neuroimaging changes in brain structure during learning. Nat. Neurosci. 15, 528–536.

Zhang, H., Dyrby, T.B., Alexander, D.C., 2011a. Axon diameter mapping in crossing fibers with diffusion MRI. Med. Image Comput. Comput. Assist. Interv. 14, 82–89.

Zhang, H., Hubbard, P.L., Parker, G.J.M., Alexander, D.C., 2011b. Axon diameter mapping in the presence of orientation dispersion with diffusion MRI. Neuroimage 56, 1301–1315.

16

Critical Review and Comparison of Axonal Structures in MRI/DTI and Histology

Ralf Galuske[1], Arne Seehaus[2] and Alard Roebroeck[2]

[1]Systems Neurophysiology, Department of Biology, TU Darmstadt, Darmstadt, Germany
[2]Faculty of Psychology and Neuroscience, Maastricht University, Maastricht, The Netherlands

16.1 INTRODUCTION

The analysis of the connectivity in the human brain forms an important basis for our understanding of brain function. As pointed out in previous chapters of this book, the majority of tracing techniques which allow for the visualization of connectivity in the brain rely on active transport processes and invasive approaches and are, thus, not applicable to the human brain. Until the last decade of the last century, the direct study of connectivity in the human was limited to coarse postmortem approaches such as the investigation of the global direction of large fiber tracts or degeneration studies after brain lesions. However, a targeted understanding of the fine structure of connections could only be gained by analogies to findings in other species (often primates).

Over the past two decades, this situation has substantially changed. On the one hand, the development of magnetic resonance techniques, in particular diffusion imaging approaches, proved to be very powerful for both *in vivo* and postmortem visualization of fiber tracts in the human brain on a macro- and mesoscale level (centimeters to tenths of millimeters). These will be discussed in Section 16.1.2 and are already outlined in Chapter 15 of this book. On the other hand, techniques for fine axonal tracing became available for postmortem application in the human brain allowing for a detailed visualization of axonal connectivity on a micro- and mesoscopic level (millimeters to micrometers). These issues will be outlined in the following part of this contribution, and more perspectives will be given in the contribution by Zilles et al. in the following chapter.

Nevertheless, the macro- and the microscopic level of research on the human connectome have to be linked which culminates in the interpretation of data on the mesoscopic level. Here, we need strong cross-validation approaches, which will mutually support each other toward a realistic understanding of the human connectome. Besides all these considerations, we also need to realize that the description of anatomical connectivity on any of these scales does not tell us much about its functional role and impact. More work is needed using a combination of approaches, some of which are presented here, as well as new studies on the functional impact of brain connectivity such as causality analyses. For the time being, the overall framework still needs to be interpreted within the anatomical network.

16.2 HISTOLOGY AND TRACER STUDIES AND THEIR USE FOR STUDYING CONNECTIVITY IN THE HUMAN BRAIN

Articulating clear statements about the connectivity in the human brain is historically a difficult issue. Until late twentieth century, for example, we only bluntly knew that there is a connection between eye and visual cortex from electrophysiology, from following white matter tracts, and from closer investigations in animals. But how about the fine structure of visual cortex, as one of the best examined regions of the mammalian brain? Well, we had learned something from the cat and the macaque and should not underestimate old Golgi studies on cortical circuits, but for the actual situation in the human we knew very little for a long time. Only in 1989 did Burkhalter et al. present first data on targeted studies on axons in the human brain traversing the border between V1 and V2 and their termination patterns (Burkhalter and Bernardo, 1989). In follow-up studies, they examined the postnatal development of local and interareal connections in this region (Burkhalter, 1993; Burkhalter et al., 1993). In both cases, they used a very promising approach of tracing axons with lipophilic dyes which so far had been used predominantly in cell culture (Honig and Hume, 1989) and prenatally developing nervous tissues (Godement et al., 1987). The results of the studies by the Burkhalter group led to a validation of previous data gained, for example, in the macaque. This opened a new window to the study of micro-anatomical structures in the human brain. After the validation that in the mammalian brain the dyes would not spread trans-synaptically (Lübke and Albus, 1992; Galuske and Singer, 1996), these dyes offered a great potential for the study of connectivity in the human brain, postmortem. There are limitations to this approach, unfortunately: it can only be applied to very fresh postmortem tissue within the first few days of fixation. Furthermore, we are dealing here with a diffusion process with time constants of at least 3 months to more than a year, which precludes the analysis of very long fiber tracts, for example, from one cerebral lobe to the other.

Given all these confounds, it was nevertheless possible to identify intrinsic connections in areas of the human brain for which no clear nonhuman primate analogue exists, such as Wernicke's area (Galuske et al., 2000) and Broca's area (Hutsler and Galuske, 2003; Tardif et al., 2007) (Figure 16.1).

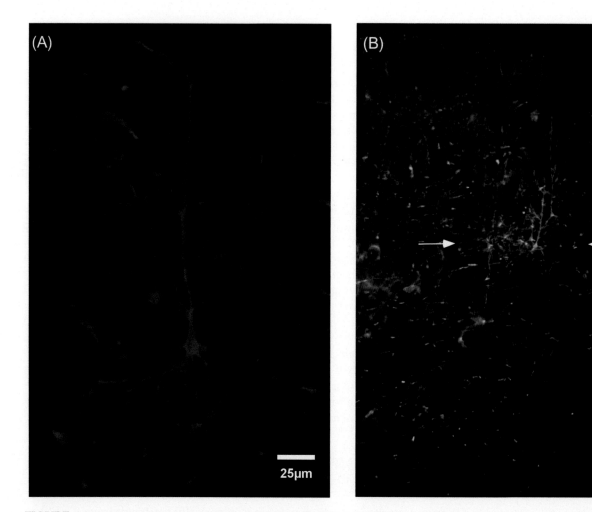

FIGURE 16.1   Long-range intrinsic connections in area 22 of the human brain visualized using the carbocyanine dyes DiI and DiA. (A) Individual pyramidal neuron in layer III at a distance of about 5 mm from the site of DiI injection. (B) Cluster of intrinsically projecting neurons and axon terminations in supragranular layers, about 4 mm distant to the site of DiA injection.

These studies revealed that the long-range intrinsic connectivity exhibited a patchy and reciprocal connectivity pattern in these regions as had been described for the visual cortex in different mammalian species. This is an important finding as the patterns of intrinsic connectivity are closely linked to the columnar functional architecture of a given cortical area, thought to preferentially link modules of similar functional specialization (Gilbert and Wiesel, 1989; Schmidt et al., 1997, but see also Martin et al., 2014). Even though the functional properties of these modules may not yet be well understood in regions such as Wernicke's area, this topographical information will contribute to a better interpretation of functional MR data gained at ultrahigh resolution (Formisano et al., 2003; Yacoub et al., 2008; Olman et al., 2012; Moerel et al., 2013). Beyond local circuits—ranging from tens of micrometers to a few millimeters—these techniques have also been applied to the study of connectivity patterns between cortical areas, in both occipital (Burkhalter and Bernardo, 1989) and temporal cortices (Galuske et al., 1999). Also here, in a limited range of distances of up to 2 cm, interareal connections could be visualized between V1 and V2 or between the different auditory regions. Preliminary analyses of the properties of these connections revealed a hierarchical pattern of reciprocally connected cortical areas and suggested a separation of these connections into different processing streams (Galuske et al., 1999) (Figure 16.2).

Even though the availability of such data sounds very promising, one has to keep in mind the limitation of this technique and conclude that this approach alone will only very partially reveal the human connectome. However, it offers a great window to bridge the gap between macro- and micro techniques such as Golgi impregnation or postmortem visualization of individual neurons (Buhl and Schlote, 1987; Galuske et al., 1993) and is a useful approach

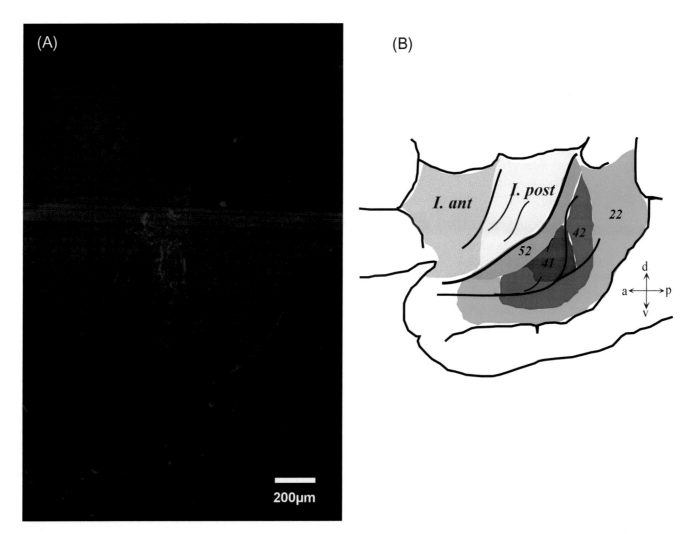

FIGURE 16.2   Interareal connections in the human temporal cortex. (A) Bundle of axons arising from a DiI injection into area 41 and terminating in supragranular layers of area 42. (B) Schematic representation of the different areas (arabic numbers) in the upper part of the human temporal cortex according to Brodmann's parcellization scheme. I. ant, anterior insula; I. post, posterior insula; d, dorsal; p, posterior; v, ventral; a, anterior. *Source: Modified from Economo and Koskinas (1925).*

for the validation of techniques which operate on a much larger scale. Hopefully, this technique can be combined with other approaches which reveal the projection pathways of axons in white and grey matter; and as will be pointed out below, both targeted tract tracing and high-resolution fiber tracing have proven to be an important tool to better interpret MR diffusion findings in many respects.

16.3 dMRI AND TRACTOGRAPHY AND ITS USE FOR THE STUDY OF CONNECTIVITY

Diffusion MRI (dMRI) of the human brain is an important medical imaging technique that has found application in clinical radiology practice as well as in fundamental human neuroscience research (Jones, 2008; Le Bihan and Johansen-Berg, 2012). In dMRI, signal is sensitized to the microscopic thermal motion (or self-diffusion) of water on a voxel-by-voxel (volume element) basis. At diffusion times on the order of milliseconds, this signal can be employed to measure micrometer scale tissue properties in millimeter scale voxels. Noninvasive access to such properties *in vivo* is of crucial importance. For instance, edema, cell swelling, or tissue deterioration can leave detectable traces in the dMRI signal leading to diagnostic potential for acute conditions. In addition, dMRI is often used to characterize white matter axon bundles that connect remote cortical gray matter areas in the brain, so-called tractography (Basser et al., 2000; Conturo et al., 1999; Mori et al., 1999).

The most used model for dMRI data is the diffusion tensor (Basser et al., 1994; Jones and Leemans, 2011; Le Bihan et al., 2001; Tournier et al., 2011) which gives diffusion tensor imaging (DTI) its name. The diffusion tensor describes the diffusion of water molecules in a voxel as a 3D Gaussian distribution. The tensor's eigenvectors and corresponding eigenvalues say something about the local direction of water diffusion and local tissue properties. Particularly, the mean diffusivity (MD, the mean of the eigenvalues) and the fractional anisotropy (FA, the normalized variance of the eigenvalues) have been intensively used as sensitive markers of microstructure.

However, the diffusion tensor model suffers from at least two important limitations (Alexander et al., 2010; Assaf and Basser, 2005; Jones, 2010; Jones and Cercignani, 2010). First, the diffusion tensor and quantities computed from it are physiologically unspecific: they reflect an unknown mixture of multiple physiological parameters of interest. For instance, FA measured in a single fiber bundle is a complex mixture of fiber orientation dispersion, axonal density (the intra-axonal volume fraction), and, to a lesser degree, diameter and myelination of the axons. Modeling methods to alleviate this limitation focus on biophysical compartment models which consider the dMRI signal as a sum of contributions from multiple microstructural compartments, such as intra-axonal and extra-axonal (see Chapter 15). Second, the diffusion tensor poorly models partial volume effects in which different tissue types (e.g., white matter, grey matter, and cerebrospinal fluid) or multiple white matter tracts contribute to the signal in a single voxel. This is particularly harmful in the case of multiple differently oriented white matter tracts, such as crossing fiber pathways, which the diffusion tensor is incapable of representing them accurately. This has stimulated investigation into new acquisition protocols, such as high-angular resolution diffusion imaging (Frank, 2001; Tuch et al., 2002), and new analysis strategies that can estimate multiple fiber orientations in the same voxel.

Diffusion spectrum imaging (DSI; Wedeen et al., 2005) and Q-ball imaging (QBI; Tuch, 2004; Tuch et al., 2003) were among the first to estimate a multimodal (i.e., non-Gaussian) water displacement probability density function without making any prior assumptions about its shape. A crucial abstraction made in QBI was to model only the 2D orientation distribution function (ODF) on the sphere, with its local maxima representing fiber orientations. The modeling of more than one diffusion orientation in a voxel enables superior fiber tractography through regions with multiple populations of fibers with different orientations (Wedeen et al., 2008). Subsequently, many more methods have been proposed to identify multiple fiber orientations per voxel, including constrained spherical deconvolution (Tournier et al., 2004, 2007), the extended ball and stick model (Behrens et al., 2007), the diffusion orientation transform (Ozarslan et al., 2006), the composite hindered and restricted model of diffusion (Assaf and Basser, 2005), generalized diffusion tensors (Liu et al., 2004; Ozarslan and Mareci, 2003), and generalized q-sample imaging (Yeh et al., 2010).

Once the fiber orientations have been estimated for each voxel, tractography algorithms aim to reconstruct the axonal fiber bundles connecting different brain areas. The first tractography approaches that have been developed connect discrete steps along the preferential diffusion directions (Basser et al., 2000; Conturo et al., 1999; Mori et al., 1999), the so-called deterministic "streamline" approach. Tensor deflection or tensor projection was introduced to use more than just the primary eigenvector of the local diffusion tensor model (Lazar et al., 2003; Westin et al., 2002). The streamline approach was soon generalized to multi-direction fiber models (Tuch et al., 2003). Probabilistic tractography approaches characterize the variability of tractography results due to the uncertainty of the estimated local fiber directions (Behrens et al., 2003; Parker et al., 2003). This sets them apart from deterministic approaches which give a single deterministic answer for the connection of a given region-of-interest to any part of the brain. More precisely, the result of a probabilistic tractography algorithm for every single well-defined seed point is a 3D map of visitation counts for tracks through a voxel. The same noisy streamline or diffusing particle principles have been applied to multi-direction models to create probabilistic multi-direction approaches (Behrens et al., 2007; Parker and Alexander, 2005).

A final recent development step is a move from local step-wise reconstruction of fiber trajectories to a global goodness-of-fit of the entire candidate fiber. Here, the measure of fit quantifies the joint likelihood of the fiber given all voxel data it passes through (Jbabdi et al., 2007; Sherbondy et al., 2008, 2009; Tuch et al., 2002; Zalesky and Fornito, 2009). The global fit measure makes tractography less sensitive to modeling errors caused by local noise (Jbabdi et al., 2007). Recently, a graph-based tractography algorithm (Iturria-Medina et al., 2007) and its extension to a multiple direction fiber model (Sotiropoulos et al., 2010) have been proposed. These algorithms re-conceptualize the global tractography problem as a shortest path search in a graph, in which nodes are represented by the center of each white matter voxel. Since graph weights are then defined as the probability of voxel-center connections given the local ODFs, a shortest path from one point to another in this graph constitutes a globally optimized fiber. Since in a shortest path search all possible nodes are visited and the path lengths recorded, the $n\%$ shortest paths then correspond to the $n\%$ most likely paths in the probabilistic tractography sense. Thus, graph-based methods—and in fact, global methods in general—are naturally used as probabilistic methods.

A number of studies have recently used dMRI and tractography to investigate human brain connectivity post-mortem on the mesoscale in order to gain deeper insight into the fine connectivity pattern of the human brain and in order to illustrate the power of diffusion methods for anatomical investigations. For neuroanatomical studies the use of postmortem tissue has a number of advantages, most importantly the absence of motion artifacts, the possibility to use small radiofrequency coils which are located very close to the brain and to scan the tissue probe for very long times up to days. This usually outweighs the disadvantages such as a reduced diffusivity, resulting in a higher achievable spatial resolution as compared *to in vivo* tissue. Aggarwal et al. (2013) created a 3D atlas of high-resolution (125–255 μm isotropic) DTI results at 11.7 T for the human brain stem. They reconstructed fiber pathways with high fidelity using DTI streamline tractography for several structures including the decussation of the pyramidal tract. They also used ADC maps to delineate the nuclei in this tissue at a similar precision level as with myelin-stained sections of the same specimen. In an even more impressive study, Dell'Aqua et al. (2013) combined dMRI of the human cerebellum down to 100 μm resolution at 7 T with immunohistochemical labeling of different neuronal and glial elements to distinguish cerebellar cortical layers and model the trajectories of most of its circuits. The histological results proved that diffusion characteristics for each cerebellar cortical layer reflect the respective anatomical patterns and structures. Leuze et al. (2014) modeled layer-specific intracortical connectivity in human V1 by tracking intracortical fiber pathways running tangentially and radially within specific cortical layers. They validated these dMRI-based findings (acquired at 242 μm^3 at 9.4 T) with cell and myelin stains and polarized light imaging of sections from the same tissue blocks. Kleinnijenhuis et al. (2013a) could similarly distinguish cortical layers and the V1/V2 border based on dMRI signal properties acquired at a resolution of 300 μm at 11.7 T. They showed that reduced FA derived from DTI analysis varied over the cortical depth, in particular at the stria of Gennari, the inner band of Baillarger, and the deepest layer of the cortex. From these results (Kleinnijenhuis et al., 2013b), by examining neurite orientation dispersion and density imaging (NODDI; Zhang et al., 2012), they further elaborated a more sophisticated multi-compartment dMRI signal model.

16.4 dMRI AND TRACTOGRAPHY VALIDATION STUDIES USING HISTOLOGY AND TRACING TECHNIQUES

An important combined application of dMRI-based and histological postmortem connectivity analyses is the validation of dMRI results. Even though dMRI is widely accepted and used in a vast range of clinical and basic neuroscience applications, one has to keep in mind that it is not a direct approach to visualize neuroanatomical structures. As pointed out above, the interpretation of dMRI with respect to connectivity and structural visualization relies on different mathematical models which relate the diffusivity information in each voxel to microstructural features. For this reason, it is important to validate these models by comparison to other sources of information about the underlying structures.

One option is to obtain this information in the MRI space already. For example, phantoms can be built that feature artificial "fiber pathways" (e.g., Pullens et al., 2010). From these phantoms, dMRI volumes are obtained and one can evaluate how closely the modeled microarchitecture and connectivity match reality. This approach has the advantage to provide ground truth data about the scanned object—it is fully known what results to expect from the dMR image. On the downside, a phantom can only provide a very simplified model of the actual situation in the brain. To get more detailed information, it is desirable to compare dMR images to the actual neuroanatomy of the scanned object, which can be investigated with a variety of histological techniques, some of them having been described above. This principally allows for an analysis of the true anatomical structures underlying a dMR image at microscopic resolution and is therefore often considered a gold standard in validating dMRI.

There are, however, some challenges associated with the histological validation approach. For one, histological analyses are only possible *ex vivo* which precludes performing such validation studies on living subjects. Therefore, these studies have to be performed on samples of fixed brain tissue, on which first dMRI is acquired, and subsequently histological analyses are conducted. One challenge here is that fixed postmortem tissue has different properties with respect to dMRI than *in vivo* tissue, in particular a reduced diffusivity. As stated above, this is not a problem in terms of achievable spatial resolution. However, one has to be cautious in generalizing results such as optimized parameters to the *in vivo* case where many characteristic values of dMRI are different. Second, histological analyses at least up to now usually require to slice the tissue. As a consequence, one has to compare a 3D data set acquired in the dMRI scan to a series of 2D data sets obtained from the stained sections. Restoring the 3D information from these sections is, depending on the staining method, very costly or even impossible. However, there are ongoing attempts to solve this challenge (e.g., Amunts et al., 2013). Lastly, the innocuously sounding aim to highlight all nerve fibers in a piece of tissue is not easily achievable with current histological methods (see Chapter 17 of this book for more details along this line). Instead, there are various methods all of which highlight different anatomical features and stain different subsets of the available fiber material. On the upside, this makes histological analyses very flexible and allows answering very specific questions. On the downside, there is no one study that answers all the questions regarding dMRI characteristics. Instead, one has to intelligently combine different histological approaches that validate dMRI on different conceptual levels in order to get comprehensive insights.

Focusing on DTI as the simplest of all dMRI models, we will outline in the following, two different validation approaches that have recently been performed.

16.5 VALIDATING DIFFUSION TENSORS

The basis of a DTI scan on the voxel level is the diffusion tensor. DTI is a unidirectional model of fiber orientation: It is supposed to describe with its largest diffusion axis the major fiber direction in a voxel and with the two minor axes indicate the prominence of this direction. As described above, multiple fiber orientations in one voxel cannot be described in this model. Theoretically, they should result in spherical or disc-shaped diffusion tensors and should be indistinguishable from isotropic diffusion or noise. Due to this simplification however, DTI is a very robust model for unidirectional fiber pathways which it can mathematically correctly describe. A validation study on the level of diffusion tensors should therefore provide, on the level of individual voxels, information about the true local fiber orientation distribution (FOD). This would allow us, first, to investigate the behavior of diffusion tensors in different areas and see how it meets the above expectations. Second, it would allow us to correlate the scalar measures FA, MD, etc. to anatomical characteristics and properties of the FODs to see what is represented by these values.

One example for this approach is a study by Leergaard et al. (2010) who coregistered dMRI volumes of rat brains to tissue sections obtained from the same probes and stained for myelin. They manually registered fiber orientations in two regions of interest (one unidirectional in the corpus callosum, one with crossing fibers in the superior colliculus) to determine histological FODs. These were compared to DTI and a multidirectional dMRI model (DSI) to show how DTI correctly models only the unidirectional, while DSI correctly models both situations. Although this method is a very precise way to obtain histological FODs, the manual registration of fiber directions is very effortful and can only be conducted for relatively small parts of a tissue sample. An automated technique to obtain FODs was more recently proposed by Budde and Frank (2012) and Budde and Annese (2013): Using structure tensors, it is possible to describe image orientedness locally in each pixel of a digitized stained tissue section. Pooling this information over image compartments that correspond to the voxel size in an imaging experiment, one can obtain histological FODs at arbitrary "voxel sizes" from any kind of fiber staining. Seehaus et al. (2015) applied this voxel-based approach to a large sample of human brain tissue. The authors coregistered a DTI volume obtained from a piece of human premotor cortex at high spatial resolution (0.34 mm) to a series of myelin-stained sections from the same tissue. In this data set, diffusion tensor orientations and scalar measures were compared to their structure tensor analogues by means of orientation differences and correlations (Figure 16.3).

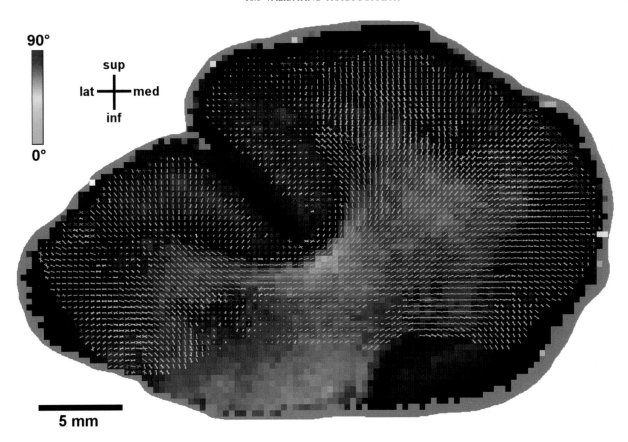

FIGURE 16.3 Illustration of diffusion tensor orientations and corresponding histological structure tensor orientations displayed on an FA map. The color indicates the level of mismatch. One can observe that DTI is an accurate model of the orientation of (myelinated) fibers particularly in white matter where FA is high. In gray matter, the match becomes worse at some but not all locations. Bands of orthogonal orientations of diffusion tensors and myelinated fibers were observed in cortical layers III/IV. This is presumably due to bundles of apical dendrites which are invisible to the myelin staining but likely to contribute to the dMRI signal. The latter shows the importance of combining different histological techniques in order to fully understand dMRI models. *Source: Taken from Seehaus et al. (2015) Histological validation of high-resolution DTI in human post mortem tissue. Front. Neuroanatomy 9:98. doi: 10.3389/fnana.2015.00098.*

This study confirms FA and MD as measures of the (non-)uniformity of the FOD and justifies the usage of FA as an indicator of the goodness-of-fit of DTI results. Moreover, the supposed behavior of diffusion tensors in the presence of unidirectional (good model) and multidirectional (diffusion tensor averages of different orientations) fiber architecture was verified. Lastly, the authors investigated whether DTI could serve as a model of gray matter fiber structure. It was revealed that this was indeed the case in the deeper cortical layers where diffusion tensors were almost as strongly aligned to fiber directions as in white matter, albeit at lower FA values. Even in the middle layers, the diffusion tensors still showed relative robust orientation patterns, pointing radially toward the cortical surface. However, these orientations were not in alignment with the myelin stain but presumably with bundles of unstained apical dendrites. Only in the most superficial layers, the diffusion tensors appeared to be rather meaninglessly oriented. This was a very interesting finding since it is another indication that with spatial resolutions higher than currently used for *in vivo* studies, the scope of DTI can potentially be extended from white matter to a significant portion of gray matter, even if not all.

16.6 VALIDATING TRACTOGRAPHY

Another important level of dMRI validation is to compare the results obtained from tractography algorithms to actual connectivity patterns. Here, the crucial question is first how sensitive and specific dMRI tractography is with respect to the true connections. In other words: How likely is it that a particular fiber connection is represented by tractography (sensitivity) and how likely is it that a tractography pathway shows an existing fiber connection

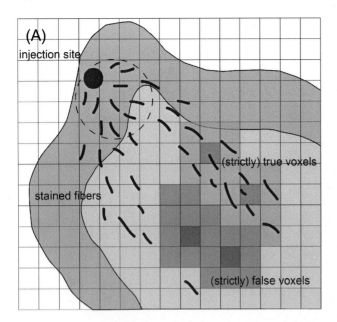 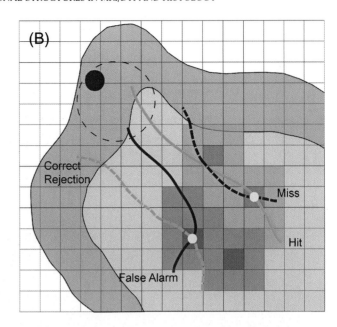

FIGURE 16.4   Schematic of signal detection theory used for tractography validation. (A) A dMRI volume and data from a tracer injection (red snippets) are coregistered to a common coordinate system. dMRI voxels are identified as being connected to the injection site (true voxel) or not (false voxel). (B) If a modeled pathway originating in a true voxel indeed projects to the injection site, it is considered a hit. Likewise, misses, false alarms, and correct rejections are defined. From the numbers of hits, misses, etc., sensitivity (hits/(hits + misses)) and specificity (correct rejections/(correct rejections + false alarms)) can be derived. *Source: Taken from Seehaus et al. (2013).*

(specificity)? Second, one is interested in relating this information to the choice of model parameters such as the algorithm's stopping criterion which determines the length of modeled streamlines or fiber tracks.

As done by Catani et al. (2002), the connectivity information in such a validation study can come from anatomy atlases. This works very well when only interested in macroscale connections (i.e., whether or not certain areas are connected). If, however, the aim is to investigate the validity of individual modeled fiber tracks, histological data which are obtained on the same tissue probe and highlight fiber connections are necessary. Unfortunately, a simple myelin staining like the Gallyas protocol often used in the diffusion tensor validation studies is not suitable to yield this kind of information: Here, fibers are stained very densely, which, particularly in white matter, precludes the separation of individual fibers. Instead, tracer studies like the ones outlined in the first section of this chapter seem better suited for the problem at hand because they provide precise information about fibers that arise from or project to a designated location (i.e., the injection site). In this respect the information about connectivity gained from dMRI tractography and from histological tracers is very different in its nature: While dMRI models the major contributors to the overall diffusivity in each voxel, tracers stain only a very small and specific portion of the whole fiber population in each voxel. Therefore, the goodness-of-fit of modeled fiber pathways cannot be estimated as directly as in the studies described above in validating diffusion tensors. However, one follows a more abstract approach and only asks whether certain pairs of points in the brain are connected in one or the other set of fiber paths. This situation is mathematically described by signal detection theory. In this statistical approach, two sets of binary data are compared, one is interpreted as the ground truth (signal) and the other as a model of it. Classifying and counting the cases where the model correctly or incorrectly reproduces the true situation leads to an estimate of the model's sensitivity and specificity.

This approach was employed in a study by Seehaus et al. (2013), where carbocyanine tracing was used postmortem in a sample of human temporal lobe to validate and optimize DTI tractography (Figure 16.4). To this end the tissue probe was coregistered to a DTI volume and voxels were classified based on whether they contained stained fiber material (indicating a connection to the injection site) or not. Then, tractography was conducted starting at the various voxels and it was analyzed where the modeled tracks proceeded to with respect to the injection site. In this way, it was shown how sensitivity and specificity of DTI tractography depends on the modeled pathway length and the FA stopping criterion. The latter is an important parameter in streamline tractography algorithms. It determines the minimum amount of anisotropy of diffusion tensors along a modeled fiber pathway, making the algorithm more strict or lenient in finding connections. Accordingly, high FA thresholds made the results more specific, low thresholds more sensitive. Finding the point of optimal balance between achieved sensitivity and specificity provides the grounds for an optimal value for conducting tractography in postmortem tissue.

16.7 FUTURE DIRECTIONS FOR HISTOLOGICAL VALIDATION

The validation studies presented here had to cope with the confound of 2D sectioning in one way or the other. If it were possible to obtain and analyze 3D histological volumes, this would open up a vast amount of new validation approaches and would help to make the existing techniques more complete and precise. One approach to obtain such 3D volumes is tissue clearing which, despite being known for almost a century, started to gain attention in neuroanatomy only in the last decade (Dodt et al., 2007; Chung and Deisseroth, 2013). The principle behind this is to make a piece of tissue transparent by immersing it in a medium that matches the refractive index of its proteins. With appropriate protocols, the tissue remains physically stable and permeable to external macromolecules, allowing for the appliance of different staining methods which are suited to visualize axonal structures and connectivity patterns (more on this in Chapters 10 and 17 of this book). Tissue treated this way can then be analyzed using 3D microscopy techniques such as confocal microscopy in combination with long working distance objectives. This approach is of course again limited by the ability of the respective microscopical technique to optically penetrate the tissue. Nevertheless, it opens a great window to conduct 3D structure tensor analyses in the near future and to directly apply tractography algorithms to histological volumes.

16.7.1 The Future of Studies on the Human Connectome

As pointed out above, already on a purely structural level the connectome can be addressed on different scales from several millimeters to only a few micrometers. DMRI techniques offer a very valuable tool which contributes reliable data on the mesoscopic level. But in the end, it comes to its limits with its spatial resolution in the range of 100 μm or hopefully a bit less in the future. Another important confound is that dMRI is not yet very well able to identify the connectivity structure in the gray matter. However, in this respect there are promising studies that already show great potential (e.g., Dell'Acqua et al., 2013; Leuze et al., 2014, just to mention a small sample) and there are many more studies currently on the way. Only when we are able to safely reconstruct fiber tracts from the gray matter of one region to the gray matter of another region, can we make firm statements about these connections. Nevertheless, one has to keep in mind that even then one has to be cautious because diffusion may show the presence of a connection but not its direction. Here, targeted tracer studies are still somewhat of a gold standard. But as we tried to point out, they also have their limitation, be it the spatial range of tracing using carbocyanine dyes or be it the optical penetration of the tissue when using clearing protocols. Thus, only a thoughtful and clever combination of microscopic and MRI approaches will be our best chance to gain optimal insight into the connectome on a structural level.

Last but not least, structural connectivity may provide the framework for interactions in the brain, but we should not forget that understanding interaction in the brain also needs a functional analysis to better understand the impact of the respective connection. Thus, we have to include the importance of structure–function relationships with its mutual dependencies in mind: Structure determines function by providing the anatomical basis of all neuronal signal transmission. On the other hand, function shapes structure by means of learning, or more generally, plasticity. Based on these considerations, functional data, in particular from EEG, MEG, and ultrahigh-field fMRI, need to be taken into account, each with their own set of limitations. In the end, all of these methods have to be combined in an insightful and critical way to complement our view on the human connectome.

References

Aggarwal, M., Zhang, J., Pletnikova, O., Crain, B., Troncoso, J., Mori, S., 2013. Feasibility of creating a high-resolution 3D diffusion tensor imaging based atlas of the human brainstem: a case study at 11.7 T. NeuroImage 74, 117–127.

Alexander, D.C., Hubbard, P.L., Hall, M.G., Moore, E.A., Ptito, M., Parker, G.J., et al., 2010. Orientationally invariant indices of axon diameter and density from diffusion MRI. Neuroimage 52 (4), 1374–1389.

Amunts, K., Lepage, C., Borgeat, L., Mohlberg, H., Dickscheid, T., Rousseau, M.É., et al., 2013. BigBrain: an ultrahigh-resolution 3D human brain model. Science 340 (6139), 1472–1475.

Assaf, Y., Basser, P.J., 2005. Composite hindered and restricted model of diffusion (CHARMED) MR imaging of the human brain. Neuroimage 27 (1), 48–58.

Basser, P.J., Mattiello, J., LeBihan, D., 1994. Estimation of the effective self-diffusion tensor from the NMR spin echo. J. Magn. Reson. B 103 (3), 247–254.

Basser, P.J., Pajevic, S., Pierpaoli, C., Duda, J., Aldroubi, A., 2000. In vivo fiber tractography using DT-MRI data. Magn. Reson. Med. 44 (4), 625–632.

Bastiani, M., Oros-Peusquens, A., Brenner, D., Moellenhoff, K., Seehaus, A.K., Celik, A., et al. (2013). "Cortical fiber insertions and automated layer classification in human motor cortex from 9.4T Diffusion MRI" in ISMRM 21st Annual Meeting and Exhibition (Salt Lake City, UT), 2124.

Behrens, T.E.J., Berg, H.J., Jbabdi, S., Rushworth, M.F.S., Woolrich, M.W., 2007. Probabilistic diffusion tractography with multiple fibre orientations: What can we gain? Neuroimage 34 (1), 144–155.

Behrens, T.E.J., Woolrich, M.W., Jenkinson, M., Johansen-Berg, H., Nunes, R.G., Clare, S., et al., 2003. Characterization and propagation of uncertainty in diffusion-weighted MR imaging. Magn. Reson. Med. 50 (5), 1077–1088.

Budde, M.D., Annese, J., 2013. Quantification of anisotropy and fiber orientation in human brain histological sections. Front. Integr. Neurosci. 7, 3.

Budde, M.D., Frank, J.A., 2012. Examining brain microstructure using structure tensor analysis of histological sections. NeuroImage 63, 1–10.

Buhl, E.H., Schlote, W., 1987. Intracellular lucifer yellow staining and electron microscopy of neurones in slices of fixed epitumourous human cortical tissue. Acta Neuropathol. 75, 140–146.

Burkhalter, A., Bernardo, K.L., 1989. Organization of corticocortical connections in human visual cortex. Proc. Natl. Acad. Sci. U.S.A. 86, 1071–1075.

Burkhalter, A., 1993. Development of forward and feedback connections between areas V1 and V2 of human visual cortex. Cereb. Cortex 3, 476–487.

Burkhalter, A., Bernardo, K.L., Charles, V., 1993. Development of local circuits in human visual cortex. J. Neurosci. 13, 1916–1931.

Catani, M., Howard, R.J., Pajevic, S., Jones, D.K., 2002. Virtual in vivo interactive dissection of white matter fasciculi in the human brain. NeuroImage 17, 77–94.

Chung, K., Deisseroth, K., 2013. CLARITY for mapping the nervous system. Nat. Methods 10, 508–513.

Conturo, T.E., Lori, N.F., Cull, T.S., Akbudak, E., Snyder, A.Z., Shimony, J.S., et al., 1999. Tracking neuronal fiber pathways in the living human brain. Proc. Natl. Acad. Sci. 96 (18), 10422–10427.

Dell'Acqua, F., Bodi, I., Slater, D., Catani, M., Modo, M., 2013. MR diffusion histology and micro-tractography reveal mesoscale features of the human cerebellum. Cerebellum 12, 923–931.

Dodt, H.U., Leischner, U., Schierloh, A., Jährling, N., Mauch, C.P., Deininger, K., et al., 2007. Ultramicroscopy: three-dimensional visualization of neuronal networks in the whole mouse brain. Nat. Methods 4, 331–336.

Economo, C., Koskinas, G.N., 1925. Die Cytoarchitektonik der Hirnrinde des erwachsenen Menschen. Springer Verlag, Vienna.

Formisano, E., Kim, D.S., Salle, F., Di, Moortele, P.F., van de, Ugurbil, K., Goebel, R., 2003. Mirror-symmetric tonotopic maps in human primary auditory cortex. Neuron 40, 859–869.

Frank, L.R., 2001. Anisotropy in high angular resolution diffusion-weighted MRI. Magn. Reson. Med. 45 (6), 935–939.

Galuske, R.A., Singer, W., 1996. The origin and topography of long-range intrinsic projections in cat visual cortex: a developmental study. Cereb. Cortex 6, 417–430.

Galuske, R.A., Delius, J.A., Singer, W., 1993. Lucifer yellow filling of immunohistochemically pre-labeled neurons: a new method to characterize neuronal subpopulations. J. Histochem. Cytochem. 41, 1043–1052.

Galuske, R.A., Schuhmann, A., Bratzke, H., Schlote, W., 1999. Cortico-cortical interconnectivity between auditory and language areas in the human temporal cortex. Soc. Neurosci. Abstr. 25 (654), 5.

Galuske, R.A., Schlote, W., Bratzke, H., Singer, W., 2000. Interhemispheric asymmetries of the modular structure in human temporal cortex. Science 289, 1946–1949.

Gilbert, C.D., Wiesel, T.N., 1989. Columnar specificity of intrinsic horizontal and corticocortical connections in cat visual cortex. J. Neurosci. 9 (7), 2432–2442.

Godement, P., Vanselow, J., Thanos, S., Bonhoeffer, F., 1987. A study in developing visual systems with a new method of staining neurons and their processes in fixed tissue. Development 101, 697–713.

Honig, M.G., Hume, R.I., 1989. Carbocyanine dyes. Novel markers for labelling neurons. Trends Neurosci. 12, 336–338.

Hutsler, J., Galuske, R.A.W., 2003. Hemispheric asymmetries in cerebral cortical networks. Trends Neurosci. 26, 429–435.

Iturria-Medina, Y., Canales-Rodriguez, E.J., Melie-Garcia, L., Valdes-Hernandez, P.A., Martinez-Montes, E., Alemán-Gómez, Y., et al., 2007. Characterizing brain anatomical connections using diffusion weighted MRI and graph theory. Neuroimage 36 (3), 645–660.

Jbabdi, S., Woolrich, M.W., Andersson, J.L.R., Behrens, T.E.J., 2007. A Bayesian framework for global tractography. Neuroimage 37 (1), 116–129.

Jones, D.K., 2008. Studying connections in the living human brain with diffusion MRI. Cortex 44 (8), 936–952.

Jones, D.K., 2010. Precision and accuracy in diffusion tensor magnetic resonance imaging. Top. Magn. Reson. Imaging 21 (2), 87–99.

Jones, D.K., Cercignani, M., 2010. Twenty-five pitfalls in the analysis of diffusion MRI data. NMR Biomed. 23 (7), 803–820.

Jones, D.K., Leemans, A., 2011. Diffusion tensor imaging Magnetic Resonance Neuroimaging. Humana Press., (pp. 127-144).

Kleinnijenhuis, M., Zerbi, V., Kusters, B., Slump, C.H., Barth, M., van Cappellen van Walsum, A.M., 2013a. Layer-specific diffusion weighted imaging in human primary visual cortex in vitro. Cortex 49, 2569–2582.

Kleinnijenhuis M., Zhang H., Wiedermannn D., Kusters B., Norris D.G., van Cappellen van Walsum A.M., 2013b. Detailed laminar characteristics of the human neocortex revealed by NODDI and histology. In: 19th Annual Meeting of the Organization for Human Brain Mapping, p 3815 Seattle.

Lazar, M., Weinstein, D.M., Tsuruda, J.S., Hasan, K.M., Arfanakis, K., Meyerand, M.E., et al., 2003. White matter tractography using diffusion tensor deflection. Hum. Brain mapp. 18 (4), 306–321.

Le Bihan, D., Johansen-Berg, H., 2012. Diffusion MRI at 25: exploring brain tissue structure and function. Neuroimage 61 (2), 324–341.

Le Bihan, D., Mangin, J.F., Poupon, C., Clark, C.A., Pappata, S., Molko, N., et al., 2001. Diffusion tensor imaging: concepts and applications. J.Magn. Reson. Imaging 13 (4), 534–546.

Leergaard, T.B., White, N.S., de Crespigny, A., Bolstad, I., D'Arceuil, H., Bjaalie, J.G., et al., 2010. Quantitative histological validation of diffusion MRI fiber orientation distributions in the rat brain. PLoS One 5, e8595.

Leuze, C.W., Anwander, A., Bazin, P.L., Dhital, B., Stuber, C., Reimann, K., et al., 2014. Layer-specific intracortical connectivity revealed with diffusion MRI. Cereb. Cortex 24, 328–339.

Liu, C., Bammer, R., Acar, B., Moseley, M.E., 2004. Characterizing non-gaussian diffusion by using generalized diffusion tensors. Magn. Reson. Med. 51 (5), 924–937.

Lübke, J., Albus, K., 1992. Rapid rearrangement of intrinsic tangential connections in the striate cortex of normal and dark-reared kittens: lack of exuberance beyond the second postnatal week. J. Comp. Neurol. 323, 42–58.

Martin, K.A.C., Roth, S., Rusch, E.S., 2014. Superficial layer pyramidal cells communicate heterogeneously between multiple functional domains of cat primary visual cortex. Nat. Commun. 5, 5252.

Moerel, M., Martino, F., De, Santoro, R., Ugurbil, K., Goebel, R., Yacoub, E., et al., 2013. Processing of natural sounds: characterization of multi-peak spectral tuning in human auditory cortex. J. Neurosci. 33, 11888–11898.

Mori, S., Crain, B.J., Chacko, V.P., Van Zijl, P., 1999. Three-dimensional tracking of axonal projections in the brain by magnetic resonance imaging. Ann. Neurol. 45 (2), 265–269.

Olman, C.A., Harel, N., Feinberg, D.A., He, S., Zhang, P., Ugurbil, K., et al., 2012. Layer-specific fMRI reflects different neuronal computations at different depths in human V1. PLoS One 7, e32536.

Özarslan, E., Mareci, T.H., 2003. Generalized diffusion tensor imaging and analytical relationships between diffusion tensor imaging and high angular resolution diffusion imaging. Magn. Reson. Med. 50 (5), 955–965.

Özarslan, E., Shepherd, T.M., Vemuri, B.C., Blackband, S.J., Mareci, T.H., 2006. Resolution of complex tissue microarchitecture using the diffusion orientation transform (DOT). NeuroImage 31 (3), 1086–1103.

Parker, G.J., Alexander, D.C., 2005. Probabilistic anatomical connectivity derived from the microscopic persistent angular structure of cerebral tissue. Philos. Trans. R. Soc. B Biol. Sci. 360 (1457), 893–902.

Parker, G.J., Haroon, H.A., Wheeler-Kingshott, C.A., 2003. A framework for a streamline-based probabilistic index of connectivity (PICo) using a structural interpretation of MRI diffusion measurements. J. Magn. Reson. Imaging 18 (2), 242–254.

Pullens, P., Roebroeck, A., Goebel, R., 2010. Ground truth hardware phantoms for validation of diffusion-weighted MRI applications. J. Magn. Reson. Imaging 32, 482–488.

Schmidt, K.E., Kim, D.S., Singer, W., Bonhoeffer, T., Löwel, S., 1997. Functional specificity of long-range intrinsic and interhemispheric connections in the visual cortex of strabismic cats. J. Neurosci. 17, 5480–5492.

Seehaus, A.K., Roebroeck, A., Chiry, O., Kim, D.S., Ronen, I., Bratzke, H., et al., 2013. Histological validation of DW-MRI tractography in human postmortem tissue. Cereb. Cortex 23, 442–450.

Seehaus, A., Roebroeck, A., Bastiani, M., Fonseca, L., Bratzke, H., Lori, N., et al., 2015. Histological validation of high-resolution DTI in human post mortem tissue. Front. Neuroanat. 9, 98.

Sherbondy, A.J., Dougherty, R.F., Ananthanarayanan, R., Modha, D.S., Wandell, B.A., 2009. Think global, act local; projectome estimation with bluematter Medical Image Computing and Computer-Assisted Intervention–MICCAI, 2009. Springer, Berlin Heidelberg, (861–868).

Sherbondy, A.J., Dougherty, R.F., Ben-Shachar, M., Napel, S., Wandell, B.A., 2008. ConTrack: finding the most likely pathways between brain regions using diffusion tractography. J. Vis. 8 (9), 15.

Sotiropoulos, S.N., Bai, L., Morgan, P.S., Constantinescu, C.S., Tench, C.R., 2010. Brain tractography using Q-ball imaging and graph theory: improved connectivities through fibre crossings via a model-based approach. NeuroImage 49 (3), 2444–2456.

Tardif, E., Probst, A., Clarke, S., 2007. Laminar specificity of intrinsic connections in Broca's area. Cereb. Cortex 17, 2949–2960.

Tournier, J.D., Calamante, F., Gadian, D.G., Connelly, A., 2004. Direct estimation of the fiber orientation density function from diffusion-weighted MRI data using spherical deconvolution. NeuroImage 23, 1176–1185.

Tournier, J.D., Calamante, F., Connelly, A., 2007. Robust determination of the fibre orientation distribution in diffusion MRI: non-negativity constrained super-resolved spherical deconvolution. NeuroImage 35 (4), 1459–1472.

Tournier, J.D., Mori, S., Leemans, A., 2011. Diffusion tensor imaging and beyond. Magn. Reson. Med. 65 (6), 1532–1556.

Tuch, D.S., 2004. Q-ball imaging. Magn. Reson. Med. 52 (6), 1358–1372.

Tuch, D.S., Reese, T.G., Wiegell, M.R., Makris, N., Belliveau, J.W., Wedeen, V.J., 2002. High angular resolution diffusion imaging reveals intravoxel white matter fiber heterogeneity. Magn. Reson. Med. 48 (4), 577–582.

Tuch, D.S., Reese, T.G., Wiegell, M.R., Wedeen, V.J., 2003. Diffusion MRI of complex neural architecture. Neuron 40 (5), 885–895.

Wedeen, V.J., Hagmann, P., Tseng, W.Y.I., Reese, T.G., Weisskoff, R.M., 2005. Mapping complex tissue architecture with diffusion spectrum magnetic resonance imaging. Magn. Reson. Med. 54 (6), 1377–1386.

Wedeen, V.J., Wang, R.P., Schmahmann, J.D., Benner, T., Tseng, W.Y.I., Dai, G., et al., 2008. Diffusion spectrum magnetic resonance imaging (DSI) tractography of crossing fibers. Neuroimage 41 (4), 1267–1277.

Westin, C.F., Maier, S.E., Mamata, H., Nabavi, A., Jolesz, F.A., Kikinis, R., 2002. Processing and visualization for diffusion tensor MRI. Med. Image Anal. 6 (2), 93–108.

Yacoub, E., Harel, N., Ugurbil, K., 2008. High-field fMRI unveils orientation columns in humans. Proc. Natl. Acad. Sci. U. S. A. 105, 10607–10612.

Yeh, F.C., Wedeen, V.J., Tseng, W.Y.I., 2010. Generalized-sampling imaging. Med. Imaging IEEE Trans. 29 (9), 1626–1635.

Zalesky, A., Fornito, A., 2009. A DTI-derived measure of cortico-cortical connectivity. Med. Imaging IEEE Trans. 28 (7), 1023–1036.

Zhang, H., Schneider, T., Wheeler-Kingshott, C.A., Alexander, D.C., 2012. NODDI: practical *in vivo* neurite orientation dispersion and density imaging of the human brain. NeuroImage 61, 1000–1016.

17

Neuroanatomical Techniques for Analysis of Axonal Trajectories in the Cerebral Cortex of the Rhesus Monkey

Farzad Mortazavi[1], *Van J. Wedeen*[2] *and Douglas L. Rosene*[1]

[1]Department of Anatomy and Neurobiology, Boston University School of Medicine, Boston, MA, USA [2]Martinos Center for Biomedical Imaging, Department of Radiology, Massachusetts General Hospital, Harvard Medical School, Boston, MA, USA

17.1 INTRODUCTION

In this chapter, we focus on anatomical aspects of white matter organization in the rhesus monkey, using tract tracing and immunohistochemistry (IHC) with a view to how these can inform imaging technologies such as diffusion tensor imaging (DTI) and diffusion spectrum imaging (DSI). We begin with a brief review of broader issues relevant for an understanding of white matter tracts and their visualization ("tractography"). Despite the many advances in the field of long-distance connections and connectivity, as recognized by the massive Brain Initiative, serious gaps remain as to how the nearly 100 billion neurons and 100 trillion connections (Marois and Ivanoff, 2005) work in concert to make us who we are. We know there are "pathways" that "relay information" to different brain areas, but without a more detailed understanding of white matter connectivity and trajectories, both the basis of normal functions and the etiology of neurological and psychiatric disorders will remain obscure.

17.2 LOCAL CIRCUITRY VERSUS LONG-DISTANCE CONNECTIONS

In development, axonal navigation is guided by molecular cues, largely unknown prior to 1990s, but since then, over 100 have been identified to play a critical role in the proper wiring of the brain (Dickson, 2002). Some of these molecules have been associated with cognitive disorders such as autism spectrum disorders and dementia (Roohi et al., 2009; Stoeckli, 2012). Mechanisms involving chemoattraction and chemorepulsion (long-range and short-range) based on concentration gradients of these molecules are involved in guiding linear growth, sharp turns, axonal fasciculation (bundling), and target reaching and selection (McLaughlin and O'Leary, 2005; Chilton, 2006; Tessier-Lavigne and Goodman, 1996; and see Chapter 10).

Both *in vivo* and *ex vivo*, experiments have provided a wealth of knowledge on the organization of axons in white matter on a microscale. For example, it has been established that in the central nervous system (CNS), axonal growth is not a predetermined or default function of neurons that survive during development and that there are specific signals involved in the process. Neurotrophins and extracellular matrix molecules, among other signaling molecules, have been identified that are critical for the organization of white matter (Huber et al., 2003; Hempstead, 2006; Kalil et al., 2011). Other factors such as formation of topographic maps have also been recognized as important factors for axonal guidance *in vivo* (Deck et al., 2013; Qu et al., 2014). Overall, molecular biology, genetics, and proteomics are continuing to address the mechanisms by which these signals act upon neurons to stimulate and direct axonal growth, branching and fasciculation. These *microscale* findings have led to mathematical and statistical models to address and predict axonal branching (van Pelt et al., 1997; Suleymanov et al., 2013; Borisyuk et al., 2014) and fasciculation (Hentschel and van Ooyen, 1999; Krottje and van Ooyen, 2007).

More macroscale investigations of white matter fiber pathways, especially in the human brain have a long history. Classical approaches to studying white matter relied on gross anatomical dissections of postmortem brains from both humans and animals (for a historical review, see Schmahmann and Pandya, 2006). In the past 50 years, neuro-anatomical tract tracing and labeling methods such as anterogradely transported radioactive-labeled amino acids which show large bundles, as well as anterogradely transported tracers such as Phaseolous vulgaris-leucoagglutin (PHA-L) and biotinylated dextran amine (BDA) that label multiple individual axons have been instrumental in advancing our knowledge of white matter connectivity. These have been particularly useful in the rhesus monkey, a nonhuman primate with gyrencephalic brain. A masterly summary of results from tract tracing with radioactively labeled amino acids, an early generation anterograde tracer, is given in the comprehensive monograph of Schmahmann and Pandya (2006). This provides a detailed delineation of major white matter fiber tracts such as corticospinal tract, temporopontine tract, and the superior longitudinal fasciculus (SLF), and shows their trajectories across the entire monkey brain. In addition, many other tract tracing methods have been developed such as rabies virus that can cross synapses (Ugolini, 2011), which provide novel experimental designs. These all have different advantages and limitations, and are reviewed by Lanciego and Wouterlood (2011), as well as in Chapter 18.

Histological methods such as Weil Stain were a staple of white matter studies for decades but generally don't show individual fibers. More recently, IHC labeling using antibodies against myelin basic protein (MBP) or axon cytoskeletal components like neurofilaments (NF) have been developed, allowing for a more detailed observation of individual fibers. At the same time, optical imaging methods have been rapidly improving, with significantly greater resolution with the advent of confocal and two- and multiphoton microscopes. Such postmortem studies have the advantage of specificity and high resolution but are unable to provide the kind of longitudinal data that would be ideal for studying dynamic processes like development or degeneration. Thus, a desirable approach is the complementary approach of using *in vivo* imaging to follow processes longitudinally while following up with detailed postmortem studies. This is more easily done at the microscopic scale in rodents with two-photon *in vivo* imaging (Andermann et al., 2013; Tang et al., 2015; Hefendehl et al., 2014). However, in humans and nonhuman primates, *in vivo* imaging technologies such as MRI can enable longitudinal studies. While resolution is improving, MRI studies are still limited by a lack of knowledge regarding the specific relationships between MRI measures and histological measures although some efforts in this regard have been made (Budde and Annese, 2013; Seehaus et al., 2013; Keren et al., 2015).

17.3 IN VIVO VISUALIZATION OF THE BRAIN

By the 1980s, development of magnetic resonance imaging (MRI) provided a new method for visualizing the human brain *in vivo* as well as a unique diagnostic tool for clinicians. With further developments, functional MRI (fMRI), together with positron emission tomography (PET), fostered a new generation of experiments addressing the structural and functional underpinnings of behaviors including cognition and emotion by identifying the "networks" underlying behavior. Both these methodologies, however, are restricted to imaging cortical grey matter where neuronal activation is the fundamental basis for the observed signal and can only presume that there are underlying white matter connections.

For visualization of white matter tracts, the introduction of DTI in mid 1980s (Le Bihan et al., 1986) was a major advance in both the clinical setting as well as for the continued goal of better understanding the global organization and pathway trajectories of white matter by reconstructing tracts based on streamlines across voxels. DTI tractography, a modeling tool for 3D visualization of white matter tracts, is still the most widely used method for detection of white matter abnormalities in human patients. Since DTI is based on the directional anisotropy of water molecules, tumors and white matter lesions cause detectable interruption and/or displacement in diffusion signal (Jellison et al., 2004; Assaf and Pasternak, 2008). For studying trajectories, the predominant and principal diffusion direction in each voxel is aligned with a tract's direction. DTI tractography methods, in their turn, have a major limitation in that a single tensor can only resolve one predominant direction within a voxel and therefore, multiple directions of diffusion, expected when fibers are crossing of fibers, cannot be resolved with this method, nor can it resolve cortical and subcortical terminations (Wedeen et al., 2008).

The need for better resolution of white matter bundles motivated the development of more specialized techniques such as diffusion-weighted tensor imaging (DWI) (Basser et al., 1994; Le Bihan et al., 2001) and later DSI. DSI, developed by Wedeen (Wedeen et al., 2000, 2005), relies on the acquisition of a very large number of images while applying strong magnetic gradients in multiple directions. By using intravoxel heterogeneities of diffusion orientations, DSI can provide better voxel resolution (2–3 mm on a 3T magnet and ~500 μm³ on a 4.7T) for visualization of pathways. Some practical disadvantages of DSI are that it requires significantly longer acquisition time, and requires high performance scanners with 3T magnet strength, as well as strong magnetic gradients and multichannel head coils. DSI is playing an increasingly significant role in brain imaging as more centers become equipped with high-end magnets and as commercially available pulse sequences are being distributed (Krueger, 2008).

Up to the present time, diffusion MRI (dMRI) with tractography can be considered the most useful noninvasive imaging diagnostic tool and the only available means for both noninvasive, *in vivo* identification of fiber pathways. Yet, a major concern with dMRI methods compared to invasive methods like tract tracing in animals, is that the measurements provided by dMRI tractography, are indirect, quantitatively difficult to interpret, and subject to error (Soares et al., 2013; Alexander et al., 2011). Several parameters, such as reconstruction threshold, turn angle, and mask threshold, are user dependent and thus can directly affect accuracy of tractography (Granziera et al., 2009). Indeed, the literature on these methods consistently raises questions about the accuracy of diffusion tractography and pathway reconstructions, although at the same time, DSI is widely acknowledged to provide crucial information on connections and trajectories of axonal bundles. This highlights the need for animal models that can be used in parallel with MRI studies to validate the constituents of the signals in voxels and as mathematically rendered.

17.4 RECENT FINDINGS ON TRAJECTORIES BY dMRI

Most recently Wedeen et al. (2012a,b) reported that fiber pathways of the brain conformed to what appears to be a rectilinear three-dimensional (3-D) structure that corresponds to the three principal developmental axes: dorsal/ventral, medial/lateral, and longitudinal. Thus fiber pathways appear to form an orthogonal matrix where they cross each other. The finding of a geometric structure in the brain was in direct contrast to the traditional view and description of turns in pathways as smooth, continuous arcs in two directions.

One of the implications of such a geometric structure in the brain is that axons must turn at a very sharp angle to move between the different directions as examples of as smooth curves were absent. This implies that point-to-point connections do not necessarily follow the shortest distance paths as some have hypothesized (Sporns et al., 2004, 2005). In fact, sharp turns in axons and branching at right angles have been reported previously in many tracer studies, although most of the evidence so far has been anecdotal, or from single axon studies (Shirasaki and Murakami, 2001; Dwyer et al., 2011; Joset et al., 2011; Farmer et al., 2008). In addition, these data are limited by small sample size and are typically presented in two dimensional (2-D).

Motivated by the hypothesis that there is a significant 3-D organization in the white matter, we are developing a neuroanatomical pipeline for optimal visualization and analysis of cerebral fibers, ultimately including a 3-D platform. In particular, this is a pipeline to:

1. Analyze axon trajectories in multiple brain areas for the presence of geometric structure
2. Use the rhesus monkey as a model system to leverage the similarities of neuroanatomical features with those of humans
3. Optimize neuroanatomical technical parameters for histological processing that overcome and compensate for the high lipid content of primate brain tissues. And, for multiple antibodies, different planes of sectioning, and histological/optical methods to acquire 3-D data that can then be stored, processed and analyzed statistically.

17.4.1 Geometric Patterns in the Brain

Since our initial histological validation of a grid-like organization in the sagittal stratum of the rhesus monkey (Wedeen et al., 2012a,b), we have taken a dMRI-guided approach to histologically search for geometric patterns in multiple brain regions. We have been able to observe turns and branches within the corticospinal tract that occur at sharp angles, and other white matter crossings that are nearly perfect orthogonal angle in brain regions, other than the sagittal stratum (Mortazavi et al, submitted).

17.4.1.1 BDA Injection in Motor Cortex of Rhesus Monkey in Conjunction with SMI-312 Labeling for Visualization of Turns and Branching

As mentioned above, one of the implications of a geometric structure in the brain would be that axons turn and branch at very sharp angles within pathways. To identify these, we placed tracer injections of BDA, a high-resolution anterograde tracer, into the motor cortex of the rhesus monkey, and double-labeled for the total axonal population using SMI-312. This is a pan-axonal antibody to NFs, unselective for axon size or degree of myelination (Figure 17.1).

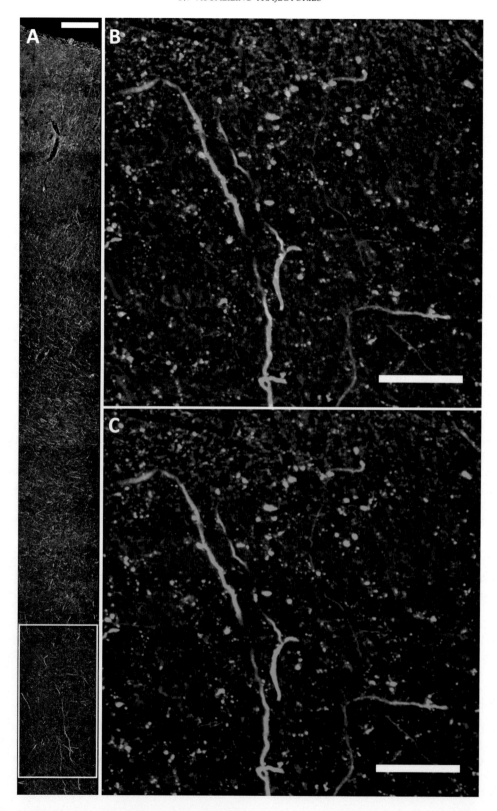

FIGURE 17.1   Double label for BDA injected in the motor cortex of the rhesus monkey (green; Streptavidin conjugated to Alexa Fluor 488), and SMI-312 labeled axons (red; Alexa 568). The BDA tracer labels a subset of axons including those projecting from motor cortex toward the cortico-spinal tract, pons, thalamus, and other cortical targets. Their directionality in relation to the background fibers is demonstrated by the double label technique in which SMI-312 labels all the fibers that surround them. Panel A: Series of images from pia to white matter acquired by a laser scanning confocal microscope and stitched together using ImageJ. Panel B: Higher magnification from the bottom (boxed) portion of A, at the level of white matter. Panel C: Projection image, at an increasingly higher magnification of BDA fibers in this region. Many examples are evident of sharp angle axonal turns, rather than smooth arcs, within this tract at the white matter/gray matter border. Scale bar in A = 500 μm and in B and C = 100 μm.

This approach allows visualization of individual BDA-labeled axons in the context of orientation of the total axon population in the deep white matter. Figures 17.1 B–C and 17.2 B–D are representative confocal images from deep white matter, where sharp turns and branches are clearly observed, consistent with a grid-like pattern in Figure 17.2E–F.

17.4.1.2 SMI-312 and SMI-32 for Observing Total and Subtotal Populations of Axons

We have optimized two antibodies as single labels in investigating a geometric patterns in axons of white mater in the monkey brain. First, SMI-312, is pan-axonal NF marker that labels the total axon populations. Second we have also optimized use of the antibody SMI-32, which is a marker for the heavy chain of NFs present mainly in larger long projection axons. The antibody is frequently used to label a subpopulation of cortical pyramidal neurons; but it also labels the subset of their axons, usually the thickest, and most likely myelinated. These can be assumed to originate mostly from projection neurons in layers 3 and 5 of the cortex, although some layer 6 neurons also express this protein in select regions (see Saleem and Logothetis, 2007).

Using these two antibodies, we have observed geometric organization in multiple regions. namely, in the anterior, mid, and posterior levels of the cingulum bundle; we observed orthogonal fiber crossings even with the different size axons and with different directional densities that is observed in the three regions of cingulum bundle(Figure 17.3).

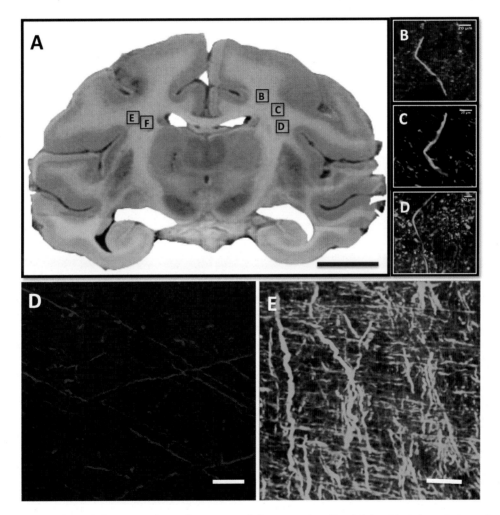

FIGURE 17.2 Geometric pattern of axons in deep white matter of rhesus monkey. Panel A is a block-face photograph of the block taken during sectioning and shows the regions of interest in the centrum semiovale where the images in the other panels were taken. Panels A–C are examples of BDA-labeled axons making sharp turns. These sharp angle turns are consistent with a geometric pattern of fiber trajectories rather than the traditional view that axons make smooth curved changes in their orientation and in their trajectories. Panels D and E are two different antibodies used to visualize axon populations in this region. Panel D: Axons stained with SMI-32, which is expressed in the heavy chain of NFs in a subset of long projecting axons. Note geometric organization of fibers. Panel E: A section stained with the pan-axonal antibody SMI-312. Both the total fiber population (SMI-312 in E) and the projection fiber subset of SMI-32 positive axons (in D) exhibit geometric, grid-like structure in this region. Scale bar in A = 1 cm, D and E = 50 μm

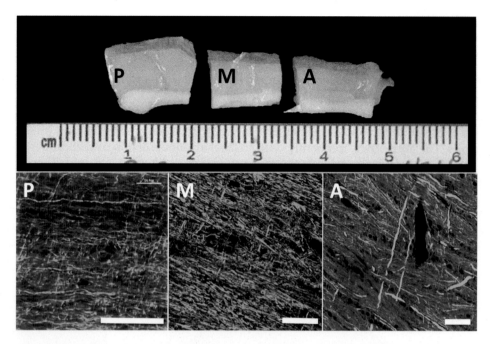

FIGURE 17.3   Axons in posterior (P), mid (M), and anterior(A) cingulum bundle in the rhesus monkey labeled with the pan-axonal neurofilament antibody SMI-312. A block of tissue (above) containing the full extent of cingulum bundle was first dissected and sectioned on a freezing microtome at 50-, 70-, and 90-μm tangential to the dorsal surface of the cingulate gyurs producing "longitudinal sections" of the cingulum. Sections were processed with microwave-assisted IHC and imaged on a confocal microscope. Maximum projections from stacks of images of this major bundle show a grid-like pattern in all three anatomical divisions though fiber density and size vary greatly. Scale bars in P, M, and A = 50 μm.

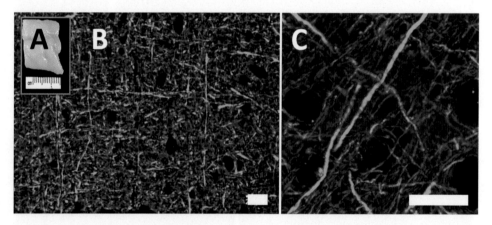

FIGURE 17.4   Axons from a block of tissue (A) that contained superior and inferior temporal gyri. This tissue block was sectioned tangential to the cortical surface and processed for the pan-axonal NF antibody SMI-312, using microwave-assisted IHC. Orthogonally crossing fibers are apparent when the region is sectioned in this orientation. B is a low magnification of this bundle and C, a higher magnification where two major directions in axon orientations can be observed. Scale bars in B and C = 20 μm.

We have made similar observations in the temporal regions of the Rhesus monkey (Figure 17.4). As noted above a similar organization was shown for the sagittal stratum in Wedeen et al. (2012a,b) and is further illustrated below in Figure 17.14.

17.4.2 Rhesus Macaque as a Model System

It is generally accepted that neuronal size and density do not scale linearly based on the volume of the brain, but less is known about how white matter scales between species with different brain volumes (Herculano-Houzel and Lent, 2005; Herculano-Houzel et al., 2007). The rhesus monkey (*Macaca mulatta*) is a valuable model of brain organization, with over 20 species of macaque monkey sharing about 95% genetic homology with humans. We have been using the rhesus monkey as the model species because of the significant structural similarities in brain architecture, even at the level of cytoarchitecture with the human brain. For example, in both species, regional analysis has shown areas of the brain that control language and complex thought processes, as well as centers of emotion with similar connectivity (Goulas et al., 2001; Oler et al., 2012). The gyrencephalic brain of the rhesus monkey is 750 times larger than a rat brain, and while 15 times smaller than a human (Peretta, 2009), it has a connectivity pattern likely to be similar to the human brain and can be a better guide to neuroanatomical connections in human. Importantly, the ability to image monkey brains *in vivo* (dMRI) and compare the same specimen *ex vivo* (dMRI and histological imaging), can provide richly detailed maps of the brain pathways. The combination of these two imaging modalities, by coregistration of dMRI data to thick slabs of histologically processed tissue, will undoubtedly provide greater insight than either would alone.

17.4.3 Choice of Antibodies, Planes of Sectioning, and Histological/Optical Method

17.4.3.1 Choice of Antibody

Antibody to MBP can be used for visualization of white matter and has proven to be an effective antibody for investigations of white matter pathologies (Figure 17.5A). Not all axons in a pathway, however, are myelinated, and the proportion of myelinated axons along a tract is sometimes regionally dependent (e.g., intra-tract heterogeneity of axon size can be observed in Figure 17.3, in the cingulum bundle). As we move towards quantitative, standardized analysis of white matter organization, the choice of antibody can be critical. For quantitative image analysis (2-D or 3-D), there needs to be a high degree of contrast in the acquired images. Contrast is impacted by the particular antibody, antibody concentrations, and optimization by antigen retrieval and tissue processing. There are a number of appropriate NF markers, our preferred labels are as mentioned above, the pan-axonal antibody SMI-312 (Figure 17.5B), for total axonal labeling (i.e., bundles), and SMI-32 (Figure 17.5C) that label a subset of larger axons. As a subset, these are sparser than SMI-312 and can be more easily visualized or serve as a useful reference population.

17.4.3.2 Plane of Sectioning

Traditionally, sections of whole monkey or human are most often prepared in "stereotactic" planes aligned using either skull features like the ear canals or an obvious brain feature like the plane between the anterior and posterior commissures. The result is series of sections in the coronal, horizontal, or sagittal stereotactic planes for monkeys. Given that the human and monkey brains are gyrencephalic, these planes of section are often not optimal planes for studying aspects of connectivity. Since each sectioning plane can provide critical information for different purposes (in this case, observing continuous fiber bundles or pathway crossings), both the thickness of the tissue and the plane of section are important factors to be considered. For purposes of illustration, the three panels in Figure 17.6 show axonal labeling with SMI-312 in different planes of section through the sagittal stratum. Blocks of tissue that contained the sagittal stratum were first dissected, and then cut in three different orientations to compare the three different

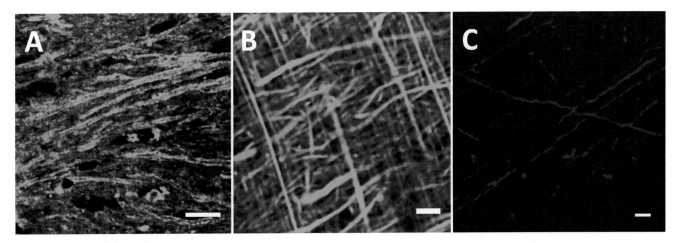

FIGURE 17.5 Choice of antibody for axon labeling. Three different antibodies, each having particular advantages are compared on adjacent sections of white matter in semiovale. (A) This was stained for MBP, which is often used as a marker for myelinated axons as well as for studying myelin pathology. (B) The pan-axonal antibody SMI-312 stains all axons regardless of their size and myelination. (C) This was stained with SMI-32, which labels the heavy chain NFs of large projection neurons and results in more selective and sparser axonal labeling which are generally thick and myelinated axons. Scale bars = 20 μm.

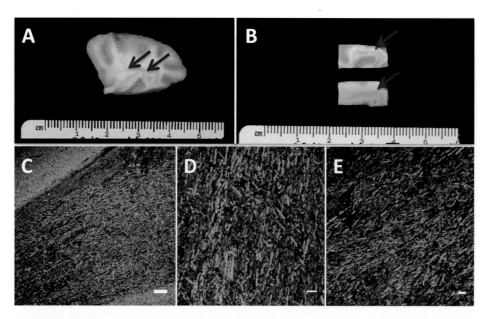

FIGURE 17.6   Demonstration of total axon population in the sagittal stratum of the rhesus monkey as shown in three different planes of section using the pan-axonal antibody SMI-312. Panel A: The anatomical location of the sagittal stratum, *in situ* in the main block and B: in the dissected smaller blocks. The smaller blocks containing the sagittal stratum were cut in coronal, sagittal, or oblique (45 degrees to the sagittal plane) planes on a freezing microtome and sections were processed by IHC, and imaged by a confocal microscope. Stacks were acquired and maximum projections are shown in C–E. The panels in C–E emphasize how the different planes of section reveal different aspects of the fibers in this region of the brain. In particular note that in E the geometric organization is obscured as almost all of the fibers run "obliquely" within the section. Scale bars C = 50 μm and D and E = 20 μm.

views. In studying fiber trajectories and crossings, what needs to be considered is the actual orientation of the fibers within the region of interest in relation to the actual anatomical plane of section. In each of the three panels, fibers in the *x–y* direction are observed as in-plane segments of fibers and axons in the *z*-plane of the section appear as dots.

Worth emphasizing is that the conventional stereotactic planes of section are typically not optimal for observing many specific features. When the brain is highly convoluted, these planes of sectioning may not recapitulate and contain the necessary information for studying orientation and trajectory or connections of fiber bundles and this likely contributes to why the orthogonal geometric grid observed in 3-D reconstructions with dMRI tractography (Wedeen et al., 2012a,b) has not been readily observed in microscopic studies of relatively thin brain sections. The tradeoff is that when blocks are oriented to optimize fiber tracking in any given area of a gyrencephalic brain, this nontraditional plane of section makes comparisons across studies from different labs difficult and even in the same lab, the exact orientation may be difficult to recapitulate. For these reasons, the simple expedient of collecting high-resolution digital photographs from blocks that are dissected out prior to sectioning and even while sectioning, can be extremely helpful.

17.4.4 Technical Issues in Histological and Optical Methods

17.4.4.1 Thickness of Tissue and Antibody Penetration

Special attention to the sectioning of the tissue, as we note above, can facilitate the visualization of fibers in a given orientation. In order to fully appreciate their orientation, and to better analyze changes in direction and trajectory, the thickness of the specimen needs to be considered. Traditionally, frozen brain tissue for tract tracing or IHC is sectioned at 20–50 μm. When tissue contains fibers that may change trajectory or orientation abruptly, this cannot be easily studied for the through section plane (*z*-plane). While cutting frozen or vibratome sections with thicknesses of 100s of micrometer seems like an obvious choice, the major challenge then becomes antibody (or reagent) penetration during labeling as well as subsequent microscopic analysis. A principal limitation of antibody and reagent penetration into tissue is the lipid content of the section, which from brain white matter is very high and particularly challenging. This is obviously the case for human and monkey brains where most of the axons in deep white matter are myelinated and the lipid bilayer of the myelin membrane essentially acts as a barrier for antibodies to bind to NFs.

Figures 17.7 and 17.8 are examples of axonal staining in the *grey matter* of the rhesus monkey where typical reagent penetration can reach up to 70 μm in the *z*-dimension for SMI-312 (Figure 17.7) and 120 μm using a microwave-assisted protocol (Figure 17.8).

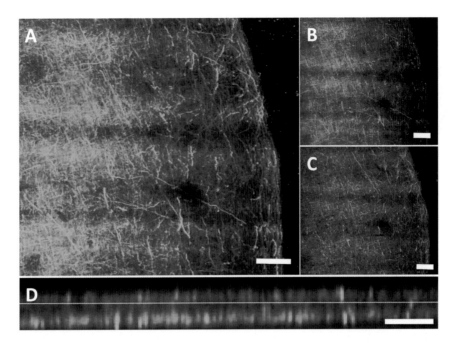

FIGURE 17.7 Standard IHC without microwave enhancement showing the depth of penetration for SMI-312 antibody in the grey matter. Penetration reaches up to 70 (z-axis) in the visual cortex (coronal section; white matter is to the left, and pia on the right) of the rhesus monkey. A is the maximum projection of the stack, B and C show the front and back side, respectively, of the confocal stack; and D is the z-axis through the stack (scale bar = 100 μm).

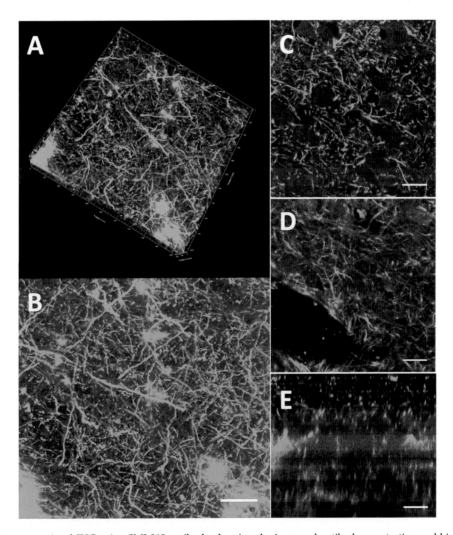

FIGURE 17.8 Microwave-assisted IHC using SMI-312 antibody showing the increased antibody penetration and binding (to 120 μm in the z-axis) in grey matter from visual cortex of rhesus monkey. Panel A is a 3-D stack of confocal images, B is the maximum projection of the stack, and C and D are the two sides of the tissue section. E is the z-dimension, which has as much as 50 μm more penetration of the antibody as compared to more standard IHC. Scale bar = 20 μm.

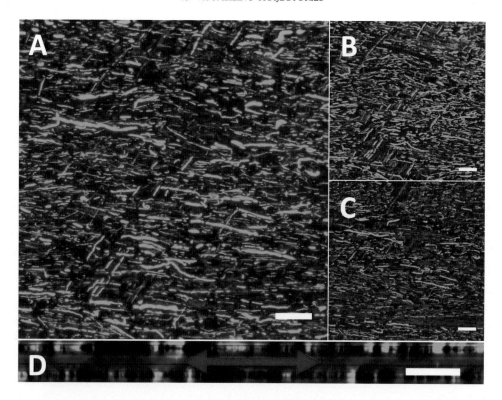

FIGURE 17.9   Panel A, projection of a confocal stack: standard IHC, without microwave enhancement, in the white matter from the sagittal stratum of rhesus monkey. SMI-312 antibody penetration is superb in the first few microns of the confocal stack on each tissue face (panels B and C). Panel D is the view of the z-axis of the section, where the red arrow indicates lack of staining due to limited antibody penetration into the tissue. Scale bar = 50 μm.

The penetration of antibody in the underlying *subcortical white matter* is typically much worse, restricted to the surfaces of the tissue, with the middle part of the tissue only partially stained, at best (Figure 17.9). Enhancement with microwave treatment will increase the penetration, but again, only to a very limited degree (~30 μm; Figure 17.10).

It is important to point out that each antibody has its own profile of penetration, and factors such as expression levels of the targeted protein also interact as threshold concentrations of antibody must be present throughout the depth of the section to get even staining Antibodies, such as the SMI family and the neuronal marker NeuN, are robust in their ability to penetrate tissue and bind to antigens, whereas dendritic markers such as MAP2 have a lower limit of penetration. It is therefore critical to optimize these antibodies by serial dilutions and other techniques, such as antigen retrieval and microwave-assisted IHC, and to consider these factors when using thick sections.

A variety of approaches have been used to "clear" the tissue to address the problems of antibody penetration and microscopic visualization in thick sections (e.g., ClearT, Kuwajima et al., 2013; Scale, Hama et al., 2011; SeeDB, Ke et al., 2013; 3Disco, Ertürk et al., 2012; and iDISCO, Renier et al., 2014). To take the example of CLARITY (Chung and Deisseroth, 2013), the technique relies on formation of tissue-hydrogel hybrid for structural preservation of proteins and stability of the tissue during processing, followed by tissue staining and then optical clearing of the specimen. The tissue-clearing step in this technique is for the purpose of removal of lipids in the tissue, with the goal of increased antibody penetration.

We have tried most of the published protocols cited above in our endeavor to investigate fiber organization and trajectories in 3-D. Unfortunately, we found none of these techniques or approaches to be optimized for the much higher lipid content of primate's brains compared to rodents and all other vertebrates. While rodent and fish literature reports good success from these techniques for imaging thick sections the higher lipid content, in particular the deep heavily myelinated white matter of the primate brain remains a significant challenge.

Given the necessarily small number of monkeys available for terminal experimental procedures, we cannot follow the CLARITY protocol that was developed using the rodent brain; in specific, we have not perfused monkeys with the hydrogel and all our trials for CLARITY are in tissue from monkeys that were perfused with 4% paraformaldehyde and postfixed in the traditional manner. In addition, we have adapted a modified approach using passive

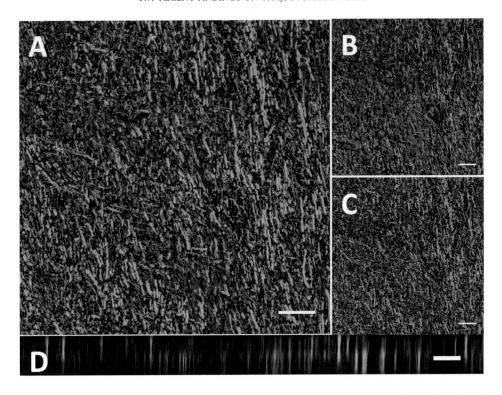

FIGURE 17.10   SMI-312 microwave-assisted IHC of white matter in the sagittal stratum reaches to a limit of 20–30 μm for antibody penetration. Panel A is the projection of the stack, and panels B and C are each face of the tissue. Panel D shows improvement of antibody penetration in the z-axis, following microwave enhancement. Scale bar = 50 μm.

CLARITY. In our experiences with tissue from human and nonhuman primates, we have made progress in clearing the gray matter, although not so far the more fatty white matter. We are now able to "clear" the grey matter in order to be able to achieve antibody penetration, binding and imaging in up to 1.2 mm thickness of tissue for antibodies such as SMI-312, and up to 700 μm for MAP2. Analysis proceeds by confocal and/or two-photon microscopy equipped with long working distance lenses for imaging purposes.

17.4.4.2 Imaging Grey Matter using a Modified CLARITY Approach

17.4.4.2.1 Single Label IHC

In the grey matter, the pan-axonal antibody SMI-312 effectively labels processes up to a thickness of 1.2 mm of tissue (in the z-axis). An example is shown in Figure 17.11 where a slab of tissue from the occipital lobe of a rhesus macaque was dissected and cleared passively, followed by staining with SMI-312 antibody. This is 100X more penetration, by the same antibody in grey matter, than we had achieved under standard (Figure 17.7) or even microwave-assisted (Figure 17.8) IHC. Under those conditions, we only achieved depth of penetration through 70 and 120 μm, respectively.

17.4.4.2.2 Technical Note on Z-axis Resolution

A major technical difficulty in imaging thick sections is resolution in the z-axis. This is directly related to the axial scattering of light from an illuminated source that must penetrate into and illumination that must escape in the z-plane. Technically this is the problem of the point spread function (PSF) (Webb, 1999). This is the primary factor affecting the depth of resolution in the z-axis, leading to appearance of elongated "cones of light" above and below the best plane of focus and to concentric bright and dark rings in the x–y plane. In confocal microscopy, as the pinhole's size is reduced, the major effect is reduction of diffraction in producing an image, thus resulting in better resolution. However, with decreasing pinhole size, there is increase in the detector noise, resulting from a reduction in the amount of excitation the lasers produce through the thickness of the tissue. When working with thinner sections, the tradeoff between pinhole size and noise can be resolved to produce a good z-resolution, but this tradeoff fails when using thicker sections. In general, the voxel size in the z-dimension is half the size of that in

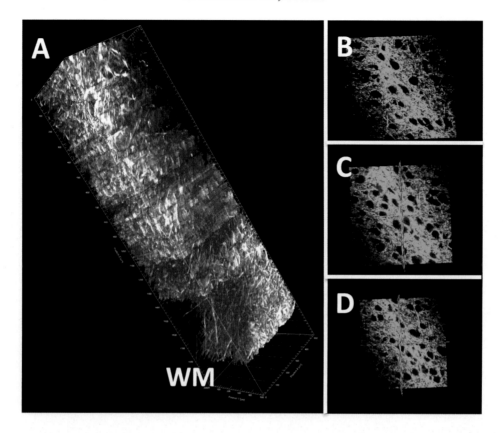

FIGURE 17.11   Modified passive CLARITY of a semi-sagittal slab of visual cortex of rhesus monkey stained with SMI-312. Pia (not shown) would be at the upper left; stained axons ascend in the block toward layer 3. Penetration and binding of antibody is significantly increased in the grey matter. Panel A shows 1.2 mm (height) and 500 μm in x–y planes of tissue stained and imaged by a 2-photon microscope; panels B–D show orthoslices at top (B), middle (C), and bottom (D) of the stack, demonstrating uniform staining of the slab.

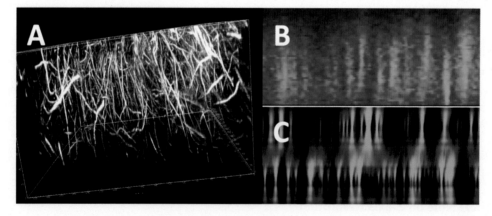

FIGURE 17.12   Superior resolution in the z-axis after processing with modified passive CLARITY (A), in contrast with standard (B) and microwave-assisted IHC (C). Typical PSF blurring is conspicuous in the non-CLARITY-treated tissue. Confocal images of tissue stained with SMI-312.

x–y planes in confocal microscopy (Cheng and Kriete, 1995). Higher numerical aperture (NA) lenses (with media to match the refractive index) in combination with control of pinhole size, and adjustments of laser intensity through the depth of tissue can also aid in increasing the z-resolution in confocal microscopy.

One advantage that we have observed in our modified CLARITY sample preparations and imaging is that the procedure in fact reduces the typical PSF-related blurring that is observed in confocal z-axis resolution. Figure 17.12 shows typical confocal blurring in the z-dimension due to PSF compared to CLARITY-processed image. In our modified CLARITY tissue, individual axons can be visualized much more sharply in the z-dimension (Figure 17.12A).

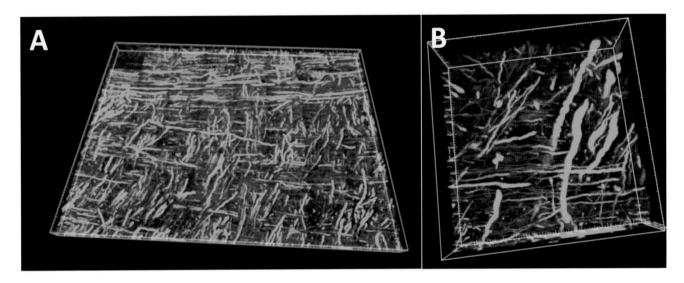

FIGURE 17.13 3D projection of subcortical white matter under area 4 of the rhesus monkey (same area in both panels). This block of tissue was cleared passively and stained for axons with SMI-312. With increased resolution and depth of penetration, we can readily observe a geometric organization of axons that cross each other at an orthogonal angle. It is worth noting that for imaging such large fields (panel A, 1.2 mm × 0.80 mm × 0.10 mm) and stacks such as in panel B (100 × 100 × 100 μm) stitching is required. In turn, this requires a large amount of computing power. While algorithms in Fiji and ImageJ are very effective for stitching, you can only allocate a limited amount of memory to the programs and therefore even with a powerful computer, you can only stitch ~1.5 GB of data.

17.4.4.2.3 Imaging White Matter Using Modified CLARITY and Other Observations of Geometric Fiber Structure

Using traditional and microwave-assisted IHC on tissue ranging from 30 to 70 μm in thickness, we were able to observe frequent white matter crossings, compatible with the interpretation of a grid (Figures 17.1, 17.2, 17.4, 17.11). We are now applying our modified CLARITY approach to investigate and visualize the presence of a grid-like fiber organization, and coupling this with large-scale 3-D reconstructions in order to better compare histological and dMRI findings. This is obviously important for validation purposes, as well as for fine-tuning of the algorithms used in dMRI tractography.

We have so far been able to reliably achieve penetration of antibody through 80–100 μm of deep white matter (i.e., centrum semiovale) and even greater thickness for the dispersed white matter fascicles in the corpus striatum as well as for unmyelinated or lightly myelinated fibers in the hippocampus. Figure 17.13 is a section from the same subcortical white matter region as in Figure 17.2D and E, below primary motor cortex of the monkey. A block of tissue was dissected out from this region, processed with our modified CLARITY procedure and imaged with a confocal microscope equipped with a motorized stage and a long working distance objective. This figure is a 3-D reconstruction of 48 tiles of stacks, acquired over 12 h of scanning on a Leica TCS SPE DM 5500 confocal and stitched using the native stitching algorithm in the Leica confocal software.

In the sagittal stratum as well, our technical modifications have similarly facilitated observation of axons crossing in an orthogonal manner (Figure 17.14), with much better clarity, depth of view, and resolution.

Additionally, we have observed crossing patterns, in a grid-like organization, in both the brainstem (Figure 17.15) and the temporal lobe of the rhesus monkey brain (Figure 17.16).

We are now obtaining a greater depth of antibody penetration for bundles within grey matter, and are now observing whole fascicles in the striatum (Figure 17.17A and B) in contrast with what can be seen in a typical coronal section of striatum stained with the same antibody (Figure 17.17C).

In the hippocampus, we are also obtaining clear and direct visualization of fiber organization and large-scale reconstructions of fiber orientation (Figure 17.18), previously only inferred from tract tracing studies and 2-D analysis.

17.4.4.2.4 Double Label IHC and Modified Clarity

In the grey matter, we are exploiting other applications for improved, in-depth IHC labeling, especially with use of two antibodies. One combination has been MAP2, which labels dendrites and cell somata, together with NeuN

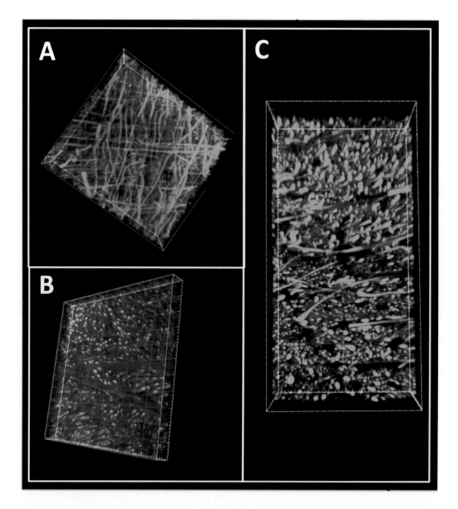

FIGURE 17.14   Passive clarity of blocks of tissue dissected from the sagittal stratum of the rhesus monkey. 80–100 um thick sections were cleared. This resulted in increased penetration of the antibody as well as better resolution for confocal imaging. Fibers can be observed crossing orthogonal to each other (panel A); panels B and C are two other views of axons at different orientations in the sagittal stratum. Lengths of axons are evident in the z-dimension, as well as axons traversing perpendicular relative to the z-dimension in the x–y plane (and thus appearing in cross-section).

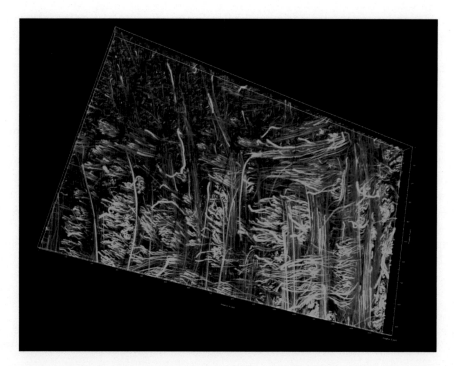

FIGURE 17.15   Confocal-acquired image of axons in the brainstem of rhesus monkey. 100 μm thick section (1.6 × 0.9 mm, x–y) cleared passively with a modified protocol. Axon fascicles running in two perpendicular directions are clearly observed. Large datasets such as these, with greater resolution in the z-plane, can provide significantly more accurate data for changes in trajectories of axons.

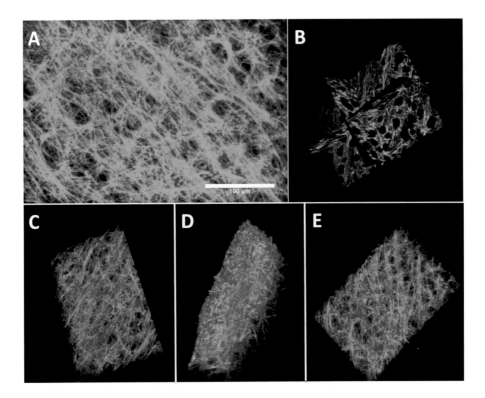

FIGURE 17.16   A block of tissue contacting the superior and inferior temporal gyrus from a rhesus macaque brain was dissected out and sectioned in different thicknesses on a freezing microtome. Panel A is a 100-μm-thick section cleared using a modified CLARITY protocol and imaged on a confocal microscope. Densely packed axons the temporal region can be observed. B is an orthoslice through the section showing full penetration of antibody into the tissue. Panels C–E are 3D rotations of the confocal stack. Scale bar in A = 100 μm.

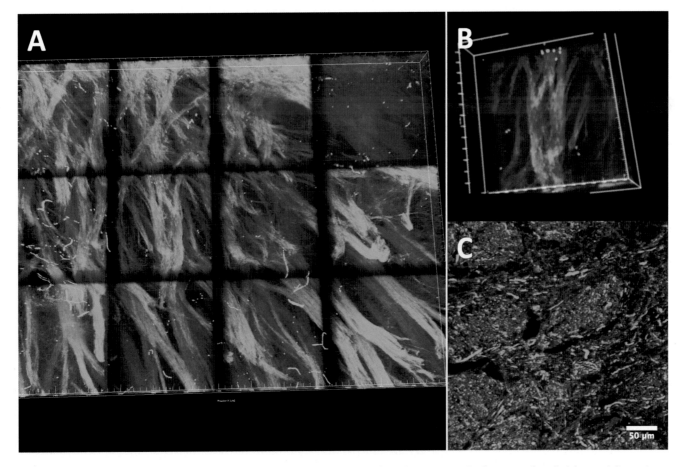

FIGURE 17.17   Panel A shows stitching of confocal stacks of axon fascicles from the striatum of a rhesus monkey. A slab containing striatum was first dissected out and cleared with our modified CLARITY protocol and stained with SMI-312 (Panel B). Compare with Panel C, which is a confocal image of these fascicles from a typical coronal slice (30 μm thick).

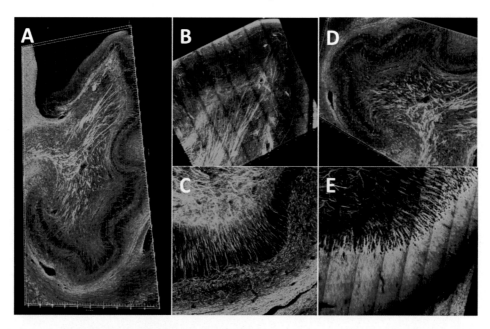

FIGURE 17.18   Large-scale, high-resolution 3-D confocal images of rhesus monkey hippocampus. Panel A is a whole hippocampus cleared with our modified CLARITY protocol and stained with SMI-312. Orientation of mossy fibers can be clearly visualized in stacks from different regions within the hippocampus (B–E). Panel E is double stained for SMI-312 and NeuN antibody for neurons. Axons (in green) can be seen to interpenetrate cell groups (red) in the dentate gyrus.

(a pan-neuronal antibody). These tissue samples (Figure 17.19) provide significantly improved visualization of microcolumnar structures, over what has been achieved with standard 2-D visualization.

Previous studies of these intriguing structures have typically used tissue slice thicknesses of 30–60 µm. This results in limited depth of view and is a major limitation, contributing to the controversy and debate surrounding cortical microcolumns. Slices at 30–60 µm thicknesses are clearly nonoptimal for the quantitative analysis of microcolumnar parameters (spacing, strength, width, periodicity), as the thinner sections are likely to contain only partial portions of the columns. Imaging of thick slabs in the grey matter can provide a substantially more direct means for more quantitative analysis of inherent 3-D features given the higher spatial resolution.

17.5 CONCLUDING REMARKS

Nearly 500 years ago, Andreas Vesalius published *De humani corporis fabrica* that was based on his anatomical dissections of the human body and brain. Astonishingly—or perhaps not, given brain complexity—we have arguably advanced only incrementally since then. Notably, 3-D neuroanatomical data on brain microstructure and the organization for neurons and their connectivity remains at an early stage.

Long-distance connections are perhaps the most important factor in understanding cortical processing, although these are much harder to access than local circuitry, which can be addressed to some extent by *in vitro* studies. *In vivo* studies using MRI technologies have been successful in providing a "large-scale" view of white matter connectivity, but MR technology is limited by resolution, which depends on voxel size and algorithms. Tract tracing studies in animal models have been the gold standard for fine-scale connectivity; however, the traditional histological sectioning with 2-D representations is at best a compromise for close investigations of tract trajectory and orientation.

Recently, the widely accepted view that fiber tracts form smooth arcs is being questioned by the observations of a geometric organization, made possible by DSI using a state-of-the-art scanner. Validation of these observations will need 3D visualization of pathways at the tissue level. As a step in this direction, the rhesus monkey offers a promising animal model that can be imaged by MR both *in vivo* and *ex vivo* (Figure 17.20) (Wedeen et al., 2012a).

For neuroanatomists mapping white matter trajectories, two of the major challenges remaining are (i) improving clearing tissue techniques such as S*cale* and CLARITY that are not able to clear the white matter in primates (ii)

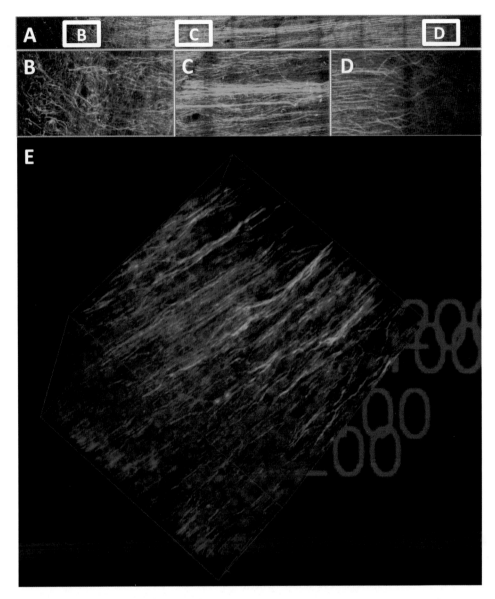

FIGURE 17.19 Confocal image of microcolumns in prefrontal cortex of the rhesus monkey. The tissue was processed with the modified CLARITY approach that we have adapted for thick slabs of monkey brain tissue. Following the clearing, tissue was immunostained for NeuN to label neurons (red) and co-labeled with antibody to MAP2 to label dendrites (green). Panel A: The complete extent, through 2 mm, of dendrites and cell bodies in a cortical microcolumn, stitched through multiple stacks. Boxed regions indicate the locations of higher power renditions shown in B, C, and D. This approach facilitates identification of layers (B), quantification of the packing density and spacing of dendrites (C), and quantification of dendritic branching (D). Panel E: 500 μm × 500 μm × 500 μm scanned block of cortical gray matter in the prefrontal cortex of a rhesus monkey, with somas in red and dendrites in green. Note through-tissue penetration of the two antibodies.

microscopic imaging of large specimens the size of monkey brain. This will require specialized microscopy tools that can handle the size of the specimen, and acquisition, digitization and storage, as well as processing of the data. For example, the data set in Figure 17.13 is 6 GB. Processing and visualizing datasets of this size are not routine.

We have made some progress in histological visualization of white matter structures in 3D in the nonhuman primate. The results from our studies support the proposal that a geometric pattern is present at the juncture of white matter crossing and axons within a pathway can be documented to make sharp turns, constrained by the Cartesian coordinate that a grid structure would dictate. We hope that as the current methodological, technical, and technological approaches develop, a clearer representation of white matter organization will emerge, including whether different bundles interweave and how they travel in different directions.

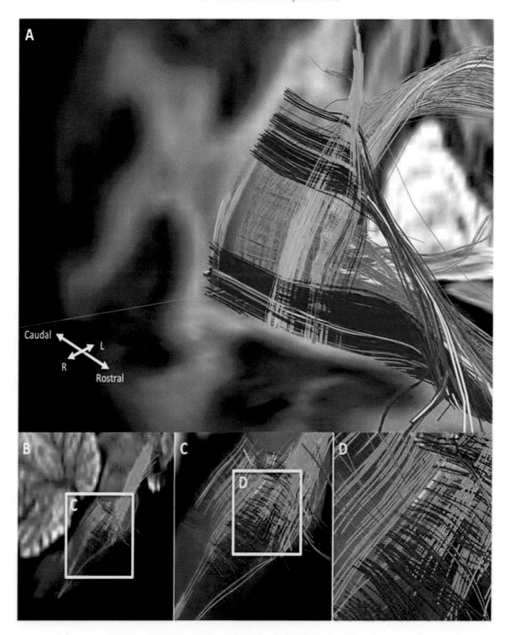

FIGURE 17.20   DSI tractography images in the sagittal stratum (A) and brain stem (B-D) of rhesus monkey acquired ex vivo. Note that the histological observations made in the sagittal stratum (Figure 17.14) and brain stem (Figure 17.15) observed after acquisition, stitching and 3D rendering of large confocal stacks are also seen in DSI reconstructions of these same regions, confirming the geometric pattern of fiber crossings in these very different regions of primate brain.

Acknowledgments

This work is supported by NIH grants R01AG043640-01 (DLR), U01-MH093765 (VJW), and NSF PHY-0855161 (DLR, HES, VJW). We thank Dr. Kathy S. Rockland for helpful discussions and comments.

References

Alexander, A.L., Hurley, S.A., Samsonov, A.A., Adluru, N., Hosseinbor, A.P., Mossahebi, P., et al., 2011. Characterization of cerebral white matter properties using quantitative magnetic resonance imaging stains. Brain Connect. 1 (6), 423–446.
Andermann, M.L., Gilfoy, N.B., Goldey, G.J., Sachdev, R.N., Wölfel, M., McCormick, D.A., et al., 2013. Chronic cellular imaging of entire cortical columns in awake mice using microprisms. Neuron 80 (4), 900–913.
Assaf, Y., Pasternak, O., 2008. Diffusion tensor imaging (DTI)-based white matter mapping in brain research: a review. J. Mol. Neurosci. 34 (1), 51–61.

Basser, P.J., Mattiello, J., LeBihan, D., 1994. MR diffusion tensor spectroscopy and imaging. Biophys. J. 66 (1), 259–267.

Borisyuk, R., Al Azad, A.K., Conte, D., Roberts, A., Soffe, S.R., 2014. A developmental approach to predicting neuronal connectivity from small biological datasets: a gradient-based neuron growth model. PLoS ONE 9 (2), e89461.

Budde, M.D., Annese, J., 2013. Quantification of anisotropy and fiber orientation in human brain histological sections. Front. Integr. Neurosci. 7, 3.

Cheng, P.C., Kriete, A., 1995. Image contrast in confocal light microscopy. In: Pawley, J.B. (Ed.), . Handbook of Biological Confocal Microscopy Plenum, New York, NY, pp. 281–310.

Chilton, J.K., 2006. Molecular mechanisms of axon guidance. Dev. Biol. 292 (1), 13–24.

Chung, K., Deisseroth, K., 2013. CLARITY for mapping the nervous system. Nat. Methods. 10 (6), 508–513.

Deck, M., Lokmane, L., Chauvet, S., Mailhes, C., Keita, M., Niquille, M., et al., 2013. Pathfinding of corticothalamic axons relies on a rendezvous with thalamic projections. Neuron 77 (3), 472–484.

Dickson, B.J., 2002. Molecular mechanisms of axon guidance. Science 298, 1959–1964.

Dwyer, N.D., Manning, D.K., Moran, J.L., Mudbhary, R., Fleming, M.S., Favero, C.B., et al., 2011. A forward genetic screen with a thalamocortical axon reporter mouse yields novel neurodevelopment mutants and a distinct emx2 mutant phenotype. Neural Dev. 6, 3.

Ertürk, A., Becker, K., Jährling, N., Mauch, C.P., Hojer, C.D., Egen, J.G., et al., 2012. Three-dimensional imaging of solvent-cleared organs using 3DISCO. Nat. Protoc. 7 (11), 1983–1995.

Farmer, W.T., Altick, A.L., Nural, H.F., Dugan, J.P., Kidd, T., Charron, F., et al., 2008. Pioneer longitudinal axons navigate using floor plate and Slit/Robo signals. Development 135 (22), 3643–3653.

Goulas, A., Bastiani, M., Bezgin, G., Uylings, H.B., Roebroeck, A., Stiers, P., 2001. Comparative analysis of the macroscale structural connectivity in the macaque and human brain. PLoS Comput. Biol. 10 (3), e1003529.http://dx.doi.org/10.1371/journal.pcbi.1003529

Granziera, C., Schmahmann, J.D., Hadjikhani, N., Meyer, H., Meuli, R., Wedeen, V., et al., 2009. Diffusion spectrum imaging shows the structural basis of functional cerebellar circuits in the human cerebellum *in vivo*. PLoS ONE 4 (4), e5101.

Hama, H., Kurokawa, H., Kawano, H., Ando, R., Shimogori, T., Noda, H., et al., 2011. Scale: a chemical approach for fluorescence imaging and reconstruction of transparent mouse brain. Nat. Neurosci. 14 (11), 1481–1488.

Hefendehl, J.K., Neher, J.J., Sühs, R.B., Kohsaka, S., Skodras, A., Jucker, M., 2014. Homeostatic and injury-induced microglia behavior in the aging brain. Aging Cell 13 (1), 60–69.

Hempstead, B.L., 2006. Dissecting the diverse actions of pro- and mature neurotrophins. Curr. Alzheimer Res. 3 (1), 19–24.

Hentschel, H.G., van Ooyen, A., 1999. Models of axon guidance and bundling during development. Proc. Biol. Sci. 266 (1434), 2231–2238.

Herculano-Houzel, S., Lent, R., 2005. Isotropic fractionator: a simple, rapid method for the quantification of total cell and neuron numbers in the brain. J. Neurosci. 25 (10), 2518–2521.

Herculano-Houzel, S., Collins, C.E., Wong, P., Kaas, J.H., 2007. Cellular scaling rules for primate brains. Proc. Natl. Acad. Sci. 104 (9), 3562–3567.

Huber, A.B., Kolodkin, A.L., Ginty, D.D., Cloutier, J.F., 2003. Signaling at the growth cone: ligand-receptor complexes and the control of axon growth and guidance. Annu. Rev. Neurosci. 26, 509–563.

Jellison, B.J., Field, A.S., Medow, J., Lazar, M., Salamat, M.S., Alexander, A.L., 2004. Diffusion tensor imaging of cerebral white matter: a pictorial review of physics, fiber tract anatomy, and tumor imaging patterns. Am. J. Neuroradiol. 25 (3), 356–369.

Joset, P., Wacker, A., Babey, R., Ingold, E.A., Andermatt, I., Stoeckli, E.T., et al., 2011. Rostral growth of commissural axons requires the cell adhesion molecule MDGA2. Neural Dev. 4, 6–22.

Kalil, K., Li, L., Hutchins, B.I., 2011. Signaling mechanisms in cortical axon growth, guidance, and branching. Front. Neuroanat. 5, 62.

Ke, M.T., Fujimoto, S., Imai, T., 2013. SeeDB: a simple and morphology-preserving optical clearing agent for neuronal circuit reconstruction. Nat. Neurosci. 16 (8), 1154–1161.

Keren, N.I., Taheri, S., Vazey, E.M., Morgan, P.S., Granholm, A.C., Aston-Jones, G.S., et al., 2015. Histological validation of locus coeruleus MRI contrast in post-mortem tissue. Neuroimage 113, 235–245.

Krottje, J.K., van Ooyen, A., 2007. A mathematical framework for modeling axon guidance. Bull. Math. Biol. 69 (1), 3–31.

Krueger, G., 2008. Application guide EP2D.DSI work-in-progress package for diffusion spectrum imaging in Siemens.

Kuwajima, T., Sitko, A.A., Bhansali, P., Jurgens, C., Guido, W., Mason, C., 2013. ClearT: a detergent- and solvent-free clearing method for neuronal and non-neuronal tissue. Development 140 (6), 1364–1368.

Lanciego, J.L., Wouterlood, F.G., 2011. A half century of experimental neuroanatomical tracing. J. Chem. Neuroanat. 42 (3), 157–183.

Le Bihan, D., Mangin, J.F., Poupon, C., Clark, C.A., Pappata, S., Molko, N., et al., 2001. Diffusion tensor imaging: concepts and applications. J. Magn. Reson. Imaging 13 (4), 534–546.

Le Bihan, D., Breton, E., Lallemand, D., Grenier, P., Cabanis, E., Laval-Jeantet, M., 1986. MR imaging of intravoxel incoherent motions: application to diffusion and perfusion in neurologic disorders. Radiology 161 (2), 401–407.

Marois, R., Ivanoff, J., 2005. Capacity limits of information processing in the brain. Trends Cogn. Sci. 9 (6), 296–305.

McLaughlin, T., O'Leary, D.D., 2005. Molecular gradients and development of retinotopic maps. Annu. Rev. Neurosci. 28, 327–355.

Oler, J.A., Birn, R.M., Patriat, R., Fox, A.S., Shelton, S.E., Burghy, C.A., et al., 2012. Evidence for coordinated functional activity within the extended amygdala of non-human and human primates. Neuroimage 61 (4), 1059–1066.

Peretta, G., 2009. Non-human primate models in neuroscience research. Scand. J. Lab. Anim. Sci. 36, 77–85.

Qu, Y., Huang, Y., Feng, J., Alvarez-Bolado, G., Grove, E.A., Yang, Y., et al., 2014. Genetic evidence that Celsr3 and Celsr2, together with Fzd3, regulate forebrain wiring in a Vangl-independent manner. Proc. Natl. Acad. Sci. 111 (29), E2996–E3004.

Renier, N., Wu, Z., Simon, D.J., Yang, J., Ariel, P., Tessier-Lavigne, M., 2014. iDISCO: a simple, rapid method to immunolabel large tissue samples for volume imaging. Cell 159 (4), 896–910.

Roohi, J., Montagna, C., Tegay, D.H., Palmer, L.E., DeVincent, C., Pomeroy, J.C., et al., 2009. Disruption of contactin 4 in three subjects with autism spectrum disorder. J. Med. Genet. 46 (3), 176–182.

Saleem, K.S., Logothetis, N.K., 2007. A combined MRI and histology atlas of the rhesus monkey brain in stereotaxic coordinates. Academic Press, San Diego, CA.

Schmahmann, J.D., Pandya, D.N., 2006. Fiber Pathways of the Brain. Oxford Univ. Press, New York, NY.

Seehaus, A.K., Roebroeck, A., Chiry, O., Kim, D.S., Ronen, I., Bratzke, H., et al., 2013. Histological validation of DW-MRI tractography in human postmortem tissue. Cereb. Cortex 23 (2), 442–450.

Shirasaki, R., Murakami, F., 2001. Crossing the floor plate triggers sharp turning of commissural axons. Dev. Biol. 236 (1), 99–108.

Soares, J.M., Marques, P., Alves, V., Sousa, N., 2013. A hitchhiker's guide to diffusion tensor imaging. Front. Neurosci. 7, 31.

Sporns, O., Chialvo, D., Kaiser, M., Hilgetag, C.C., 2004. Organization, development and function of complex brain networks. Trends Cogn. Sci. 8, 418–425.

Sporns, O., Tononi, G., Kötter, R., 2005. The human connectome: a structural description of the human brain. PLoS Comput. Biol. 1, 245–251.

Stoeckli, E.T., 2012. What does the developing brain tell us about neural diseases? Eur. J. Neurosci. 35 (12), 1811–1817.

Suleymanov, Y., Gafarov, F., Khusnutdinov, N., 2013. Modeling of interstitial branching of axonal networks. J. Integr. Neurosci. 12 (1), 103–116.

Tang, P., Zhang, Y., Chen, C., Ji, X., Ju, F., Liu, X., et al., 2015. *In vivo* two-photon imaging of axonal dieback, blood flow, and calcium influx with methylprednisolone therapy after spinal cord injury. Sci. Rep. 5, 9691.

Tessier-Lavigne, M., Goodman, C.S., 1996. The molecular biology of axon guidance. Science 274 (5290), 1123–1133.

van Pelt, J., Dityatev, A.E., Uylings, H.B., 1997. Natural variability in the number of dendritic segments: model-based inferences about branching during neurite outgrowth. J. Comp. Neurol. 387 (3), 325–340.

Webb, R.H., 1999. Theoretical basis of confocal microscopy. Methods Enzymol. 307, 3–20.

Wedeen, V.J., Reese, T.G., Tuch, D.S., Weigel, M.R., Dou, J.G., Weiskoff, R.M., et al., 2000. Mapping fiber orientation spectra in cerebral white matter with Fourier-transform diffusion MRI. Proc. Int. Soc. Magn. Reson. Med. 8, 82.

Wedeen, V.J., Hagmann, P., Tseng, W.Y., Reese, T.G., Weisskoff, R.M., 2005. Mapping complex tissue architecture with diffusion spectrum magnetic resonance imaging. Magn. Reson. Med. 54 (6), 1377–1386.

Wedeen, V.J., Wang, R.P., Schmahmann, J.D., Benner, T., Tseng, W.Y., Dai, G., et al., 2008. Diffusion spectrum magnetic resonance imaging (DSI) tractography of crossing fibers. Neuroimage 41 (4), 1267–1277.

Wedeen, V.J., Rosene, D.L., Wang, R., Dai, G., Mortazavi, F., Hagmann, P., et al., 2012a. The geometric structure of the brain fiber pathways. Science 335 (6076), 1628–1634.

Wedeen, V.J., Rosene, D.L., Wang, R., Dai, G., Mortazavi, F., Hagmann, P., et al., 2012b. Response to comment on "The Geometric Structure of the Brain Fiber Pathways". Science 337 (1605), 2012b.

Ugolini, G., 2011. Rabies virus as a transneuronal tracer of neuronal connections. Adv. Virus. Res. 79, 165–202.

IV. TRACTOGRAPHY

High-Resolution Fiber and Fiber Tract Imaging Using Polarized Light Microscopy in the Human, Monkey, Rat, and Mouse Brain

Karl Zilles[1,2,3], Nicola Palomero-Gallagher[1], David Gräßel[1], Philipp Schlömer[1], Markus Cremer[1], Roger Woods[4,5], Katrin Amunts[1,6] and Markus Axer[1]

[1]Institute of Neuroscience and Medicine (INM-1), Research Center Jülich, Jülich, Germany
[2]Department of Psychiatry, Psychotherapy and Psychosomatics, RWTH University Aachen, Aachen, Germany [3]JARA Jülich-Aachen Research Alliance, Translational Brain Medicine, Germany, Jülich, Germany [4]Ahmanson-Lovelace Brain Mapping Center, UCLA, Los Angeles, CA, USA [5]David Geffen School of Medicine, UCLA, Los Angeles, CA, USA [6]C. & O. Vogt Institute of Brain Research, University of Düsseldorf, Düsseldorf, Germany

18.1 THE ROOTS OF POLARIZED LIGHT IMAGING IN NERVOUS TISSUE

The first observations on the structure of the white matter using polarization microscopy were already reported during the nineteenth century. Changes in birefringence of myelinated nerve fibers were described in pathological cases (e.g., Ehrenberg, 1849; Westphal, 1880). Korbinian Brodmann picked up this approach in a short report (Brodmann, 1903) in which he described the excellent signal preservation of the birefringent properties of myelinated nerve fibers in formalin fixed nervous tissue:

> Erst viel später bin ich auf Formalinpräparate verfallen und habe mit Befriedigung feststellen können, daß Formalinfixierung die Doppelbrechung der markhaltigen Nervenfasern nicht verändert, daß wir also auch in Formol gehärtete und konservierte Nervenfasern ... durch das Polarisationsmikroskop untersuchen und eventuelle krankhafte Veränderungen derselben wahrnehmen können. (Brodmann, 1903, p. 212).
>
> I stumbled upon the idea of formalin fixed tissue much later and realized with satisfaction, that formalin fixation does not impair the birefringence of myelinated nerve fibers. Therefore, we can study nerve fibers hardened and conserved in formalin ... with polarization microscopy and observe possible pathological alterations.

Polarization microscopy as a tool for visualization of myelinated fibers and fiber tracts was further elaborated in numerous studies on normal and pathologically impaired nervous tissue (Fraher and MacConaill, 1970; Kretschmann, 1967; Lehmann, 1959; Miklossy and Van der Loos, 1987; Miklossy et al., 1991; Schmidt, 1923, 1924, 1937; Schmitt and Bear, 1936; Wolman, 1975). These observations provided a firm basis for recent attempts to identify the three dimensional (3D) course of fibers and fiber tracts in the central nervous system of mammalian species using polarized light imaging (PLI; Axer et al., 1999, 2000a,b, 2001, 2003, 2011; Axer et al., 2011a,b; Caspers et al., 2015; Dammers et al., 2010, 2012; Dohmen et al., 2015; Palm et al., 2010).

In this chapter, we will briefly summarize the method of tissue preparation, PLI, and visualization. Then, we will present examples of how this technique can be used to map the 3D course of fibers and fiber tracts in human, macaque, rat and mouse brains at a spatial resolution well beyond that of present diffusion weighted imaging. This unprecedented spatial resolution also enables the tracking of myelinated fibers in great detail within the cerebral cortex and provides, therefore, a new approach to microscopic analysis of the cortex including its myeloarchitecture and mapping of cortical segregation.

18.2 TISSUE AND METHODS

18.2.1 Tissue Preparation

Entire adult human, rat, and mouse brains were immersion fixed, and vervet monkey brains were perfusion fixed in 4% buffered paraformaldehyde. After cryoprotection (20% glycerin), the brains were deep frozen at −70°C and stored till further processing. Brains were serially sectioned in the coronal or sagittal planes (section thickness between 50 and 70 μm) using cryostat microtomes and coverslipped with glycerin. Immediately after coverslipping, the sections were measured using either a large-scale polarimeter or a polarization microscope (see Section 18.2.2). After each section, the blockface was digitized using a CCD camera mounted above the brain in order to obtain undistorted reference images. No histological staining was applied. This procedure resulted in an uninterrupted series of sections through the entire brain.

18.2.2 Equipment and Image Acquisition

We used two different equipments for image acquisition: a rotating large-scale polarimeter and a polarization microscope. The polarimeter allows a single-shot imaging of sections up to 24 cm in diameter, and is equipped with two linear polarizers with orthogonal transmission axes, a specimen stage and a quarter-wave retarder mounted between the polarizers. The principal axis of the retarder constitutes an angle of 45° with respect to the transmission axes of the polarizers. Illumination is provided by a customized LED light source with a narrow-band green wavelength spectrum and a diffusor plate. The simultaneous rotation of all optical devices around the stationary tissue sample enables a systematic scanning of the principal axes of the tissue from a series of discrete angles. A high-resolution CCD camera (AxioCam HRc Rev.2, Zeiss, Germany) measures the light intensities at a distance of approximately 50 cm from the section. The in-plane sampling resolution of the polarimeter is $64 \times 64 \mu m^2$. The application of the Jones calculus (Jones, 1941) describes the light transmittance through the polarimeter. This enables the calculation of the individual fiber orientation in each voxel, which is $64 \times 64 \times 60 \mu m^3$ in the case of polarimeter measurements, as section thickness is 60 μm. The fiber orientation is defined by the pair of angles (inclination α and direction φ) indicating the fiber axis orientation through (z-axis) and within (x- and y-axes) the plane of the section, respectively. For further details, see Axer et al. (2011a) and Larsen et al. (2007). Additionally, the polarimeter is equipped with a tiltable stage with respect to two perpendicular axes precisely aligned to the camera axis (Axer et al., 2011b). Slight tilting of the sections provides an image of the fiber tracts from four different views within a tilt angle range of ≤8° in all principal directions. The tilting stage overcomes the limitation of a standard planar polarimeter, where the extracted inclination angles α are naturally restricted to absolute values between 0° and 90°, and are therefore ambiguous. The tilting modifies the inclination angles and enables to disambiguate the inclination measurement.

The polarization microscope (LMP-1, Taorad, Germany) provides an even higher spatial resolution of $1.3 \times 1.3 \mu m^2$ in-plane. Therefore, the size of each voxel in measurements using the polarization microscope is $1.3 \times 1.3 \times 60 \mu m^3$, as section thickness is 60 μm. The microscope is equipped with a single rotatable polarizer in the optical path. Up to now, no stereoscopic imaging is possible with this instrumentation, but measurement of the same section with the polarimeter and the microscope allows an unambiguous definition of the inclination angles if necessary. Because of the ultra-high resolution of polarized microscopical imaging, single-shot imaging even of mouse and rat brain sections is not possible. Therefore, brain sections must be scanned in small tiles with overlapping fields of view using a motorized scanning stage (Märzhäuser, Germany). The calculation of the fiber direction is performed according to the same basic principles of optics (generalized Snell's law, Huygens–Fresnel principle) and Jones calculus as mentioned above.

18.2.3 Data Processing and Image Visualization

The intensity of the light transmitted through the brain section sandwiched in between the simultaneously rotating crossed polarizers and the retarder wave plate follows a sinusoidal course depending on the angle between the fast axis of the retarder wave plate and the fiber axes contained in a voxel. The entire sinusoidal intensity profile with a period of 180° is sampled with the camera at discrete retarder angles (e.g., 10°). By means of a dedicated Fourier analysis (Glazer et al., 1996) using the Jones formula (Jones, 1941), four gray value maps can be derived from the individual intensity profiles in each voxel imaged by the camera:

- The transmittance map is a voxelwise measure for the extinction of light after passing through the polarimeter/polarized microscope and the brain tissue. It provides images reflecting particularly the densely packed myelin.
- The retardation map reflects the normalized amplitudes of the light intensity profiles.
- The fiber direction (φ) map describes the x/y orientation of each fiber, i.e., the in-plane direction
- The fiber inclination (α) map describes the out-of-section angle of each fiber, i.e., the direction through the plane of the section.

The retardation is influenced by four factors: the wavelength of the incident light, the birefringence of the tissue, the thickness of the section, and the inclination of the fibers. While the wavelength and the section thickness are approximately constant, and thus considered to be global scale factors, the fiber inclination must still be disentangled from the tissue birefringence (i.e., the myelin content) in each voxel by imaging using tilting of the section (Wiese et al., 2014). For the cases in which tilting is not applied, the dependence of the light absorption on the myelin content in the tissue reflected by the transmittance can be used instead (Axer et al., 2011b).

The fiber direction and inclination maps are combined in the fiber orientation maps (FOMs), which are shown together with examples of transmittance maps in this chapter. All FOM images of this article are color-coded in the Hue–Saturation–Value (HSV) space, a cylindrical-coordinate representation (hexcone) of the Red-Green-Blue space. The HSV describes hue, saturation, and value as the three dimensions of a color. Hue is the dimension normally called red, yellow, blue, etc. Saturation specifies the departure of the hue from white or gray, and value indicates the departure of a hue from black.

18.3 PLI OF FIBER TRACTS IN HUMAN AND VERVET MONKEY BRAINS

Transmitted light is absorbed by the densely packed myelin. Therefore, transmittance images appear similar to myelin-stained sections with the white matter being darker than the gray matter (Figure 18.1A). The regional differences in the relation between white and gray matter can also be seen within the thalamus, where the heavily myelinated ventrolateral thalamic nucleus is darker than the mediodorsal thalamic nucleus. Even the very thin claustrum is well discernible by its lower myelin density from the surrounding external and extreme capsules. However, transmittance images do not encode any fiber directions. This information is provided in Figure 18.1B, which shows an FOM of the identical coronal section through a human brain at the level of the central region. Subcortical nuclei, e.g., the caudate–putamen or the mediodorsal thalamic nucleus, can be easily recognized within the white matter by the differences in fiber orientations between these nuclei. The external and extreme capsules are also visible by their dorsoventral fiber orientation, which is slightly turned from upper left to lower right on the right hemisphere and from upper right to lower left on the left hemisphere (see color sphere). In the FOM image, the different in-plane angles (directions) of the callosal fibers on the right and left side can be clearly recognized due to the different colors. Likewise, the angular bundle appears red in the left hemisphere and yellow in the right hemisphere. The fornix, which has a dorsoventral in-plane orientation, and a slight inclination relative to the plane of sectioning, is visible by its pale (because of the inclination) green (because of the dorsoventral orientation) color. Thus in an FOM image, a fiber tract changes its color with changing 3D orientation. Another example is the arcuate fascicle, which bends around the dorsal edge of the insula and thereby changes its color from yellow to red and finally to blue-green (left hemisphere in Figure 18.1B). The cingulum bundle is a fiber tract, in which a portion of fibers runs perpendicular to the sectioning plane (i.e. through-plane), and thus appears as a black-colored structure.

Although the images in Figure 18.1 were captured using the "low" resolution polarimeter with an in-plane resolution of "only" $64 \times 64\,\mu m^2$, the color coding in the centrum semiovale clearly indicates that there is a complex intermingling of various fiber tracts with different directions and inclinations in this region. The black color is partly caused by extinction effects occurring when fibers cross nearly at right angles (Dohmen et al., 2015), and partly by

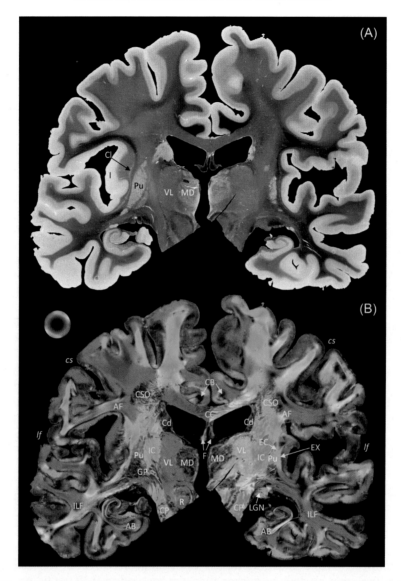

FIGURE 18.1   (A and B) Coronal section through a human brain measured with the polarimeter. Spatial resolution $64 \times 64\,\mu m^2$ in-plane and $70\,\mu m$ (section thickness) through-plane. (A) Transmittance image of the section depicted in (B). (B) The 3D direction of fibers and fiber tracts (fiber orientation map, FOM image) is indicated by colors according to the color sphere. Black indicates fibers which preferentially run through-plane, but also regions of high cell-packing density. AB, angular bundle; AF, arcuate fascicle; CB, cingulum bundle; CC, corpus callosum; Cd, caudate nucleus; Cl, claustrum; CP, cerebral peduncle; cs, central sulcus; CSO, centrum semiovale; EC, external capsule; EX, extreme capsule; F, fornix; GP, globus pallidus; I, insula; IC, internal capsule; ILF, inferior longitudinal fascicle; lf, lateral fissure; LGN, lateral geniculate nucleus; MD, mediodorsal thalamic nucleus; Pu, putamen; R, nucleus ruber; VL, ventrolateral thalamic nucleus.

fibers running through-plane. In fact, in this region the fibers of the corpus callosum fan out to the gyri of the frontal lobe and hereby cross various fiber bundles in the fronto-occipital fascicle and the superior longitudinal fascicle (also compare with the comparable region in the vervet monkey; Figure 18.2).

In contrast to Figure 18.1, the FOM image of a coronal section through the central region of a vervet monkey brain shown in Figure 18.2 was obtained with the polarization microscope at ultra-high resolution. Even in this overview of both hemispheres, it is possible to separate very small in addition to larger fiber tracts. for example, a thin horizontally running fiber bundle is visible between the caudate nucleus and the thalamus, which probably resembles prefrontal fibers as described by Schmahmann and Pandya (2006) in the rhesus monkey based on axonal tracing studies. These fibers originate in the prefrontal area 46, run through the internal capsule, and target the anteroventral thalamic nucleus. Its course between the internal capsule and the anteroventral thalamic nucleus is clearly visible in both hemispheres (Figure 18.2). Furthermore, the thalamic and lenticular fascicles in the H1 and

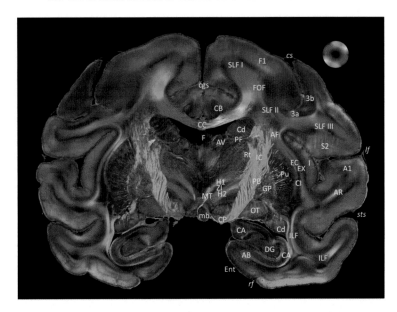

FIGURE 18.2 Coronal section through an adult vervet monkey brain at the level of the central region. 3a primary somatosensory cortex, area 3a; 3b primary somatosensory cortex, area 3b; A1, primary auditory cortex; AB, angular bundle; AF, arcuate fascicle; AR, acoustic radiation; AV, anteroventral thalamic nucleus; CA, cornu ammonis; CB, cingulum bundle; CC, corpus callosum; Cd, caudate nucleus; cgs, cingulate sulcus; Cl, claustrum; CP, cerebral peduncle; cs, central sulcus; DG, dentate gyrus; EC, extern capsule; Ent, entorhinal area; EX, extreme capsule; F, fornix; F1, primary motor cortex; FOF, fronto-occipital fascicle; GP, globus pallidus; H1, thalamic fascicle; H2, lenticular fascicle; I, insula; IC, internal capsule; ILF, inferior longitudinal fascicle; lf, lateral fissure; mb, mammillary body; MT, mamillothalamic tract; OT, optic tract; PB, pontine bundle; PF, prefrontal fibers; Pu, putamen; rf, rhinal fissure; Rt, reticular nucleus; S2, secondary somatosensory cortex; SLF I, superior longitudinal fascicle I; SLF II, superior longitudinal fascicle II; SLF III, superior longitudinal fascicle III; sts, superior temporal sulcus; ZI, zona incerta.

H2 regions, respectively, can be distinguished from the medially adjoining mammillothalamic tract. The latter can be followed down to the right mammillary body, although the terminal part of this fiber tract is extremely thin.

In a coronal section through the occipital lobe of the vervet monkey, various visual and parietal visual areas and underlying fiber tracts can be identified in the FOM image (Figure 18.3). As in Figure 18.2, this image has been captured at an in-plane spatial resolution of $1.3 \times 1.3\,\mu m^2$ using the polarization microscope. The dorsal occipital bundle (Schmahmann and Pandya, 2006) as well as the two parts of the superior longitudinal and inferior longitudinal fascicles are visible (Figure 18.3). The power of the PLI method becomes particularly apparent when we focus on the fiber tracts below the primary visual cortex V1. At this location, the very high spatial resolution of PLI enables the separation of the very thin fiber bundle of the tapetum from the forceps major, although the fibers in both structures have a similar spatial orientation over some distance. Furthermore, the sagittal stratum is clearly distinct by its oblique dorsoventral and through-plane fiber course and is sharply separated from the inferior longitudinal fascicle. A fiber tract that extends between the secondary visual area V2 and V3V with a ventrolateral orientation probably represents the transverse occipital fascicle of Vialet. Various U-fiber structures can be easily distinguished from long fiber tracts.

The very high resolution also enables an analysis of the fiber architecture within the cerebral cortex. Here we must clearly distinguish between fiber directions, which are color-coded, and laminar patterns based on higher and lower densities of myelinated fibers (myeloarchitecture). The latter aspect can be partly analyzed by comparison with transmittance images (see Section 18.5), as black stripes in FOM images may encode not only layers with a low myelinated fiber and high cell-packing density but also through-plane running fibers. By taking these arguments into consideration, we could nevertheless tentatively delineate the visual areas V1, V2, V3V, V4V, V4T, V5/MT, and MST, as well as the intraparietal areas LIPe, LIPi, VIP, and MIP (Figure 18.3). They differ by the number and the width of dark and light layers. Furthermore, these parcellations are supported by comparisons with cell-body-stained sections in a different vervet monkey brain and by comparison with a macaque monkey brain atlas (Paxinos et al., 2000). At the border between V1 and V2, a transition region is found which can be identified as the border tuft region (Sanides and Vitzthum, 1965; Amunts et al., 2000; Zilles et al., 2015). A more detailed analysis of the fiber architecture of cortical areas and the transition region between V1 and V2 is given in Section 18.5.

As we work with complete series of sections throughout the brain, we can follow the fiber tracts over longer distances. Figure 18.4 shows a coronal section through the posterior part of the occipital lobe of a vervet monkey

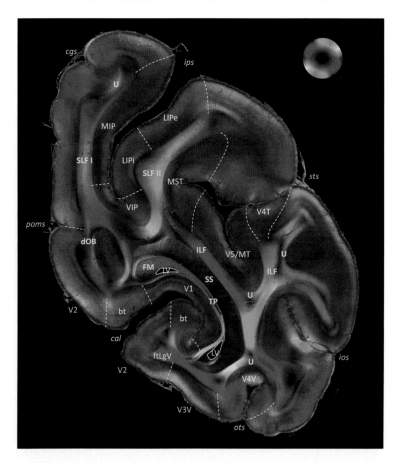

FIGURE 18.3   Coronal section through an adult vervet monkey brain at the level of the intraparietal and calcarine sulci. bt, border tuft; cal, calcarine sulcus; cgs, cingulate sulcus; dOB, dorsal occipital bundle; FM, forceps major (callosal fibers); ftLgV, transverse occipital fascicle of Vialet; ILF, inferior longitudinal fascicle; ios, inferior occipital sulcus; ips, intraparietal sulcus; LIPe, lateral intraparietal area, external part; LIPi, lateral intraparietal area, internal part; LV, lateral ventricle; MIP, medial intraparietal area; MST, extrastriate area MST; ots, occipito-temporal sulcus; poms, parieto-occipital medial sulcus; SLF I, superior longitudinal fascicle I; SLF II, superior longitudinal fascicle II; SS, sagittal stratum (optic radiation and cortico-subcortical projections); sts, superior temporal sulcus; TP, tapetum; U, U-fibers; V1, primary visual cortex; V2, extrastriate area V2; V3V, extratriate area V3V; V4T, exgtrastriate area V4T; V4V, extrastriate area V4V; V5/MT, extrastriate area V5; VIP, ventral intraparietal area.

with the visual areas V1, V2, and V3V. Also here, the sagittal stratum can be clearly separated from the medially and laterally adjacent tapetum and inferior longitudinal fascicle. The separation is possible at some sites by the different colors, caused by the different 3D orientation of the fiber tracts, and at other sites by very thin separating black lines which delimit the fiber tracts (e.g., between the sagittal stratum and the tapetum in the dorsal part of the left hemisphere; Figure 18.4).

18.4 PLI OF FIBER TRACTS IN MOUSE AND RAT BRAINS

Using the strategy of tracking fiber tracts in serial sections, Figures 18.5 and 18.6 illustrate this approach in a series of sagittal sections through a mouse brain. This allows the identification of the 3D course of defined fiber tracts over a longer distance. The first example is the fasciculus retroflexus, labeled by a white star in Figure 18.5, sections 90–93. This fiber tract connects the habenula with the interpeduncular nucleus. On its way, it crosses several other fiber tracts. PLI enables the direct tracking of the fasciculus retroflexus despite these crossings without any modeling. It even shows where the fasciculus retroflexus is situated above (section 90) or below (section 93) the crossed fiber tract. Another example is the complex relationship between the stria medullaris thalami (black square in Figures 18.5 and 18.6, sections 90–108), running from the habenula to the hypothalamus, and the fornix (white circle in Figures 18.5 and 18.6, sections 90–108), which extends with its posterior limb from the fimbria hippocampi

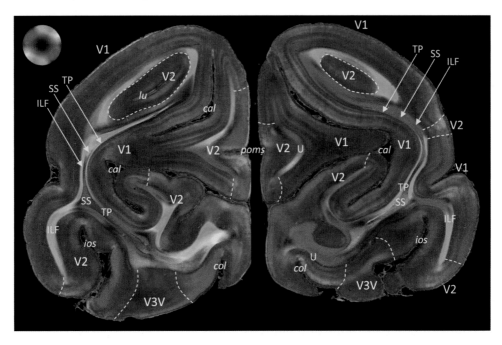

FIGURE 18.4   Coronal section through an adult vervet monkey brain near the occipital pole. col, collateral sulcus; lu, lunate sulcus. For other abbreviations, see Figure 18.3.

to the mammillary body. The situation becomes even more complicated by the appearance of the mammillothalamic tract (white triangle in Figures 18.5 and 18.6, sections 93–97), which lies dorsal of the terminal part of the fornix. Finally, the stria terminalis (white rhomboid in Figure 18.6, sections 99–108) appears rostral between the stria medullaris thalami and the anterior commissure (star in Figures 18.5 and 18.6, sections 90–108) and below the fimbria hippocampi. Despite these intricate topographical relations, the different fiber tracts can be clearly separated by their color-coded orientations and their positions in space as visible by comparisons of serial sections. Finally, the analysis of serial sections nicely illustrates the bending of the anterior commissure from a pure through-plane direction as seen in Figure 18.5 to a more horizontally directed in-plane direction in Figure 18.6. It is also visible that the anterior commissure merges with the intermediate olfactory tract (black rhomboid in Figure 18.6, sections 99 and 104), which led to the designation of this tract as anterior commissure in the olfactory bulb in atlases of the mouse and rat brain (Franklin and Paxinos, 1997; Paxinos and Watson, 2007).

The relation between gray and white matter largely differs between primate and rodent brains. The white matter represents only a small proportion of brain volume in rodent brains. Thus, most of the fiber architectonic findings are localized within the cerebral cortex and the subcortical nuclei. Figure 18.7 shows two coronal sections through a mouse brain, one at a more rostral (Figure 18.7A) and one at a more caudal (Figure 18.7C) level. The delineation of cortical motor areas is particularly clear in Figure 18.7A, as the radially oriented fibers extend from the white matter to the most superficial layers in both motor areas M1 and M2. This is in contrast to the laterally adjoining forelimb area FL and the medially adjoining cingulate area Cg1. Cingulate areas Cg1 and Cg2 differ from the motor areas by a conspicuous differentiation of fiber orientation in layers I and II, particularly in Cg2. The small indusium griseum can be delineated from Cg2 and from the genu corporis callosi by the underlying nearly horizontally oriented fibers. The motor areas also differ from the somatosensory areas FL, S1, and S2 by fiber orientation in the deeper cortical layers (Figure 18.7A). The secondary somatosensory cortex S2 shows a less-differentiated fiber pattern in the granular and subgranular layers compared to the primary somatosensory cortex S1 (Figure 18.7A), particularly if it is compared with the barrel field (Figure 18.7C). At the more rostral level, both the granular (GI) and the posterior part of the agranular insular cortex (AIP) display fibers running in an oblique dorsoventral direction (Figure 18.7A). At a more posterior level (Figure 18.7C), AIP differs from GI by the fibers in the deeper layers, which extend between the claustrum and the endopeduncular nucleus (Figure 18.7C). AIP is separated from the piriform cortex (Pir) by both the fiber structure in the upper and deeper cortical layers. The lateral olfactory tract indicates the border between the piriform cortex and the olfactory tubercle.

The excellent spatial resolution of PLI enables direct visualization of the course of the corticostriatal fibers through the corpus callosum and the adjoining white matter (Figure 18.7B). Despite the rather oblique crossing of

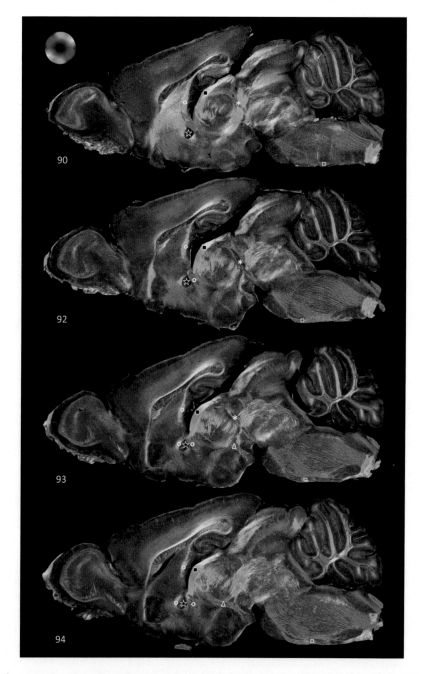

FIGURES 18.5-18.6   Sagittal sections through a mouse brain from medial (section 90) to lateral (section 108). ★: Anterior commissure (with posterior and anterior parts); *: fasciculus retroflexus; ○: fornix (with precommissural and postcommissural parts); Δ: mammillothalamic tract; ▫: pyramidal tract; ■: stria medullaris thalami; ◊: stria terminalis; ◆: tractus olfactorius intermedius.

the corticostriatal through the callosal fibers, both fiber compartments can be clearly separated by their differentially color-coded directions. The corticostriatal fibers can be easily followed on their way between the caudate–putamen and the motor cortex and cingulate area Cg1.

The connection of the anterior commissure with the white matter is visible in the right hemisphere (Figure 18.7C). Above the anterior commissure, fibers of the stria teminalis are clearly separated from a more lateral fiber bundle, which belongs to the internal capsule. In the septal complex, fiber bundles of different orientations can be found, which constitute the fornix and the columna fornicis. The complex system of the medial forebrain bundle (Nieuwenhuys et al., 1982; Veening et al., 1982; Geeraedts et al., 1990a,b) is also recognizable by fibers with various orientations between the hypothalamus with the lateral and medial preoptic areas LPO and MPO and the olfactory tubercle.

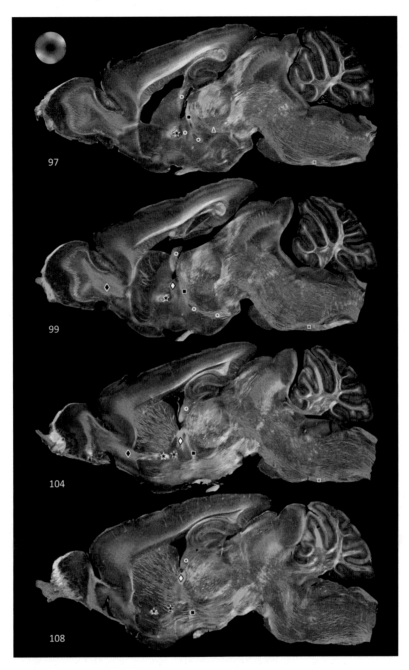

FIGURES 18.5-18.6   (Continued)

Figure 18.7D shows at higher magnification the border between the barrel field of the primary somatosensory cortex (S1Bf) and the secondary somatosensory cortex (S2). The densely packed bundles of myelinated fibers indicate the position of barrels (arrows in Figure 18.7D; see also Section 18.5).

Figure 18.8 provides an example of the fiber tracts and the cortical fiber architectonics in a coronal section of a rat brain at the level of the dorsal hippocampus and the habenula. Despite their close neighborhood, the fimbria hippocampi, the stria terminalis, and the internal capsule are clearly delineable from each other by the different color-coded fiber directions. In Figure 18.8A, the positions of the fornix, mammillothalamic tract, medial forebrain bundle, and medial lemniscus are labeled on the left side, and can be seen without the covering labels on the right side. They clearly differ from the adjacent structures in the hypothalamus, cerebral peduncle, ventromedial thalamic nucleus, ventral posteromedial and posterolateral thalamic nuclei, as well as the zona incerta by their fiber orientations. The optic tract and the supraoptic decussation as well as the internal capsule and their prolongation into the

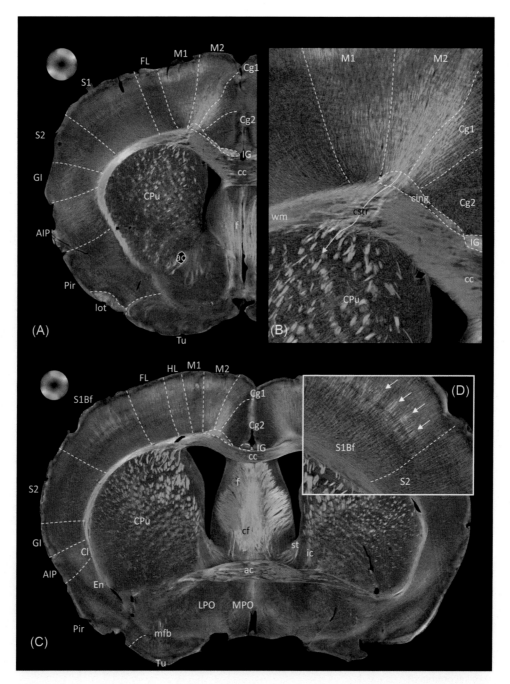

FIGURE 18.7 (A–D) Coronal sections through an adult mouse brain at more rostral (A) and central (C) levels. Panel (B) shows the crossing of corticostriatal fibers through the radiatio corporis callosi and the white matter on its way to the caudate–putamen. Panel (D) is a higher magnification of the border region between the barrel field of the primary somatosensory cortex and the secondary somatosensory area of the section shown in (C). ac, anterior commissure; AIP, agranular insular cortex, posterior part; cc, corpus callosum; cf, columna fornicis; Cg1, cingulate area 1; Cg2, cingulate area 2; cing, cingulum bundle; Cl, claustrum; CPu, caudate–putamen; cstr, corticostriatal tract; En, endopeduncular nucleus; f, fornix; FL, forelimb area; GI, granular insular cortex; HL, hindlimb area; ic, internal capsule; IG, induseum griseum; Lot, lateral olfactory tract; LPO, lateral preoptic area; M1, motor area 1; M2, motor area 2; mfb, medial forebrain bundle; MPO, medial preoptic area; Pir, prepiriform region; S1, primary somatosensory cortex; S1Bf, primary somatosensory cortex, barrel field; S2, secondary somatosensory cortex; st, stria terminalis; Tu, olfactory tubercule; wm, white matter.

cerebral peduncle are also delineable in Figure 18.8A. Below the optic tract, fibers of the stria terminalis reach the amygdala (Figure 18.8A).

The granular sensory cortices (Par1, Te3R, Te1) with their thick fiber bundles in layer IV (arrows in Figure 18.8A; compare also with Figure 18.7D and Figure 18.13) and their directly abutting dark layer differ from the parietal

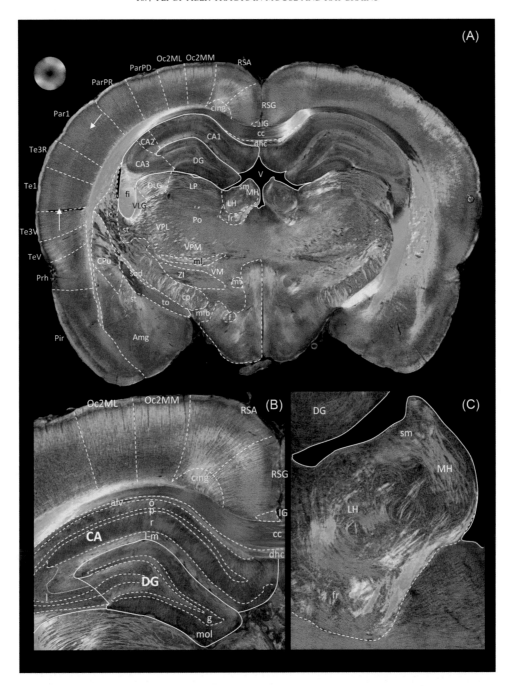

FIGURE 18.8 (A) Coronal section through an adult rat brain with delineations of various cortical and subcortical regions as well as fiber tracts. (B) Detailed view of the the left dorsal hippocampus, retrosplenial, and occipital areas as shown in (A). (C) Detailed view of the medial and lateral habular nuclei with the fasciculus retroflexus and the incipient part of the stria medullaris. Arrows highlight layer IV. alv, alveus; Amg, amygdala; CA, cornu ammonis (with fields 1, 2, and 3); cc, corpus callosum; cing, cingulum bundle; CPu fiber bundles in the caudate–putamen; DG, dentate gyrus; dhc, dorsal hippocampal decussation; DLG, dorsal lateral geniculate nucleus; f, fornix; fi, fimbria hippocampi; fr, fasciculus retroflexus; g, granulosum layer; ic, internal capsule; IG, indusium griseum; l, lucidum layer; LH, lateral habenular nucleus; l-m, lacunosum-molecular layer; LP, lateroposterior thalamic nucleus; mfb, medial forebrain bundle; MH, medial habenular nucleus; ml, medial lemniscus; mol, molecular layer; mt, mammilothalamic tract; o, oriens layer; Oc2ML, medial occipital area ML; Oc2MM, medial occipital area MM; p, pyramidal layer; Par1, primary somatosensory cortex; ParPD, parietal area PD; ParPR, parietal area PR; Pir, piriform cortex; Po, posterior thalamic nucleus; Prh, perirhinal cortex; r, radiatum layer; RSA, agranular retrosplenial cortex; RSG granular retrosplenial cortex; sm, stria medullaris; sod, superior optic decussation; st, stria terminalis; Te1, primary auditory cortex; Te3R, temporal area T3R; Te3V, temporal area T3V; to, optic tract; V, ventricle; VLG, ventral lateral geniculate nucleus; VM, ventromedial thalamic nucleus; VPL, ventral posterolateral thalamic nucleus; VPM, ventral posteromedial thalamic nucleus; ZI, zona incerta.

areas ParPR and ParPD dorsally and the dysgranular area TeV as well as the periallocortical area Prh ventrally. The piriform cortex can be separated both from the perirhinal area Prh and from the amygdala by its unique fiber architecture.

The internal fiber architecture of the hippocampus is shown at a higher magnification in Figure 18.8B. The lacunosum-molecular layer differs from the radiatum layer by the more obliquely oriented fibers in the former and the more radially oriented fibers in the latter. The lucidum layer can be separated from the lacunosum-molecular layer. The alveus is separated from the oriens layer by its more mediolaterally oriented fibers, which seem to merge with the dorsal hippocampal commissure (Figure 18.8B). The granular retrosplenial cortex is separated from the agranular retrosplenial cortex by its different fiber architecture in the most superficial cortical layers.

The fiber structure in the habenula is shown in more detail in Figure 18.8C. The position of the medial and lateral habenular nuclei is indicated. Between the ventricle and the habenula, fibers of the stria medullaris can be seen. They are oriented in various directions. Below the lateral and medial habenula, fibers of the fasciculus retroflexus are visible, which originate mainly in the medial habenular nucleus. The fibers of the stria medullaris and the fasciculus retroflexus are intermingled within the habenula. Therefore, it is not possible to disentangle the fiber bundles of both tracts within the habenula itself even at this extremely high resolution.

The very high resolution of PLI also enables the visualization of fiber tracts in small structures like the olfactory bulb (Figure 18.9A) and the brainstem (Figure 18.9B) of the rat. In the olfactory bulb, the chaotic course of the terminal portions of the fibrae olfactoriae is revealed by the different colors of the fibers and provides a demarcation between this layer and the adjacent stratum glomerulosum of the main olfactory bulb. The direction changes again abruptly at the border between stratum glomerulosum and the outer plexiform layer. Also the inner plexiform layer can be delineated by the dorsoventrally oriented fibers, particularly on the medial side of the main olfactory bulb. The mitral layer is difficult to separate because the fibers follow the same course as those of the outer plexiform layer. In the granular layer, an inhomogeneous distribution of fiber directions can be found, with more radially oriented fibers in the outer portion and more dorsoventrally oriented fibers in the inner portion of this layer. In the center of the main olfactory bulb, the very small olfactory ventricle is surrounded by the rostral extension of the anterior commissure. The accessory olfactory bulb and the vomeronalsal organ as well as the vomeronasal nerve are also visible (Figure 18.9A). The accessory olfactory bulb is obvious by the strictly dorsoventrally oriented fiber bundles in its central portion, which represents the dorsal part of the lateral olfactory tract. This tract is bordered ventrally by the granular and dorsally by the mitral and glomerulosum strata of the accessory olfactory bulb.

Figure 18.9B shows a transverse section through the brainstem and the cerebellum. Most obvious is the descending ramus of the facial nerve, which runs from dorsomedial to basolateral. This section also shows the ascending ramus of the seventh cranial nerve, which originates in the nucleus nervi facialis situated dorsal of the periolivary complex and curves around the nucleus nervi abducentis as the internal genu of the seventh cranial nerve (indicated by a curved arrow in Figure 18.9B). Below the nucleus nervi abducentis, the fibers of the medial longitudinal fascicle can be identified. At the basis of the brainstem, the large fiber bundle of the medial lemniscus is delineable from the trapezoid body and the pyramidal tract. Nuclei of the fifth cranial nerve are visible lateral of the descending ramus of the seventh cranial nerve. The spinal tract of the trigeminal nerve can be demarcated from the principal sensory nucleus of the trigeminus on its medial side and from the radix of the eighth cranial nerve on its lateral side. The anterior part of the ventral cochlear nucleus is clearly delineable from the radix of the eighth cranial nerve and the flocculus of the cerebellum. The fibers of the inferior cerebellar peduncle fan out into medial and lateral directions at this rostrocaudal level. Within the cerebellar cortex, the molecular and granulosum layers are sharply segregated and well-demarcated from the underlying white matter. Between the white matter of the cerebellum and the various structures of the brainstem, a large area of mediolaterally extending fibers is seen, which covers a region occupied by the superior, medial, and lateral vestibular nuclei. The fibers bend over the nucleus of the spinal tract, and the principal nucleus of the nervi trigemini, as well as the spinal tract of the trigeminus to reach the eighth cranial nerve. The caudal part of the pontine reticular nucleus is characterized by numerous crossing fiber bundles.

A unique opportunity of PLI is the option of a nearly selective visualization of fiber tracts by restricting the color coding through spotlight illumination to a defined spatial direction. The usual illumination with its color sphere and the FOM image of a coronal section through a rat brain at the level of the anterior commissure is shown in Figure 18.10A. Major visible fiber tracts in the FOM image are the corpus callosum, ventral hippocampal commissure, fimbria hippocampi, internal capsule, stria terminalis, fornix, anterior commissure, chiasma opticum, and lateral olfactory tract. Figure 18.10B illustrates the principle of the selective spotlight visualization. As an example, a direction (φ) of 60° and an inclination (α) of 60° are chosen here. The strictly geometric projection of the spotlight onto the surface of the color sphere would result in two single illuminated points, which would be hardly visible. Therefore, the projection field was enlarged with decreasing color intensity to the periphery of the projection field.

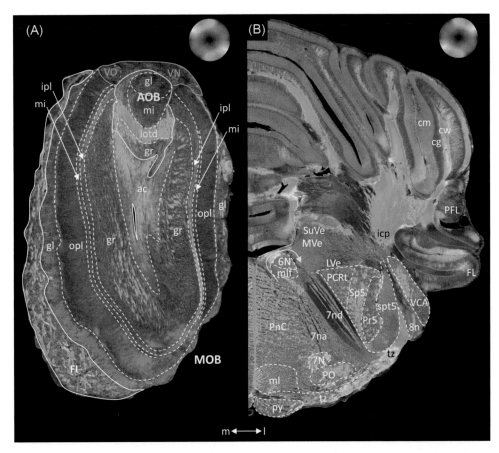

FIGURE 18.9   (A and B) Coronal sections through the rat olfactory bulb (A) and brainstem with cerebellum (B). The arrow indicates the genu of the seventh cranial nerve. 6N, nucleus of the sixth cranial nerve; 7N, nucleus of the seventh cranial nerve; 7na, ascending part of the n. facialis; 7nd, descending part of the n. facialis; 8n nervus vestibulocochlearis; ac, anterior commissure; AOB, accessory olfactory bulb; cg, granulosum layer of the cerebellar cortex; cm, molecular layer of the cerebellar cortex; cw, cerebellar white matter; FI, fibrae olfactoriae; FL, flocculus; gl, glomerular layer; gr, granular cell layer; icp, inferior cerebellar peduncle; ipl, inner plexiform layer; l, lateral side; lotd, lateral olfactory tract, dorsal part; LVe, lateral vestibular nucleus; m, medial side; mfl, medial longitudinal fascicle; mi, mitral cell layer; ml, medial lemniscus; MOB, main olfactory bulb; MVe, medial vestibular nucleus; opl, outer plexiform layer; PCRt, parvocellular reticular nucleus; PFL, paraflocculus; PnC, pontine reticular nucleus,caudal part; PO, periolivary complex; Pr5, principal nucleus of the fifth cranial nerve; py, pyramidal tract; Sp5, nucleus of the spinal tract of the fifth cranial nerve; spt5, spinal tract of the fifth cranial nerve; SuVe, superior vestibular nucleus; tz, trapezoid body; VCA, ventral cochlear nucleus, anterior part; VN, vomeronasal nerve; VO, vomeronasal organ.

Consequently, the colors of adjacent directions have low color intensities. As the spotlight has a defined direction, the projection onto the surface of the sphere selectively indicates the preferential orientation of the illuminated fiber tracts. Figure 18.10C–H show examples with different angles of inclination (as indicated by the first number in each spotlight image) and of direction (as indicated by the second number in each spotlight image). As the inclination in Figure 18.10C–E is 0°, all highlighted fiber tracts run in the plane of the section. Only the direction, that is, the angle within the plane of sectioning varies from 0° to 90°. Therefore, the horizontally running fibers of the corpus callosum, ventral hippocampal and anterior commissures, optic chiasm, and lateral olfactory tract are highlighted in *red* (Figure 18.10C), because they follow a strict left–right orientation. A comparison with Figure 18.10F–H shows that the inclination of these fibers is minimal relative to the plane of sectioning. In Figure 18.10D, the right stria terminalis, the right part of the ventral hippocampal commissure, and intrastriatal fiber bundles of the right hemisphere, as well as the left internal capsule, a left part of the anterior commissure, the left fornix and parts of the white matter of the left hemisphere are highlighted in *yellow to green*, since they follow an in-plane oblique "top right–bottom left" orientation. In Figure 18.10E, the dorsal fornix and both columns of the fornix as well as the intrastriatal fiber bundles and parts of the white matter are clearly visible by their *cyan* color indicating a strictly in-plane dorsoventral orientation. If the inclination is increased to 30° and the direction is set to 45° (Figure 18.10G), the barrel field, intrastriatal fiber bundles, the stria terminalis and parts of the ventral hippocampal commissure appear in the right hemisphere, whereas in the left hemisphere the cingulum and columna fornicis, parts of the white matter and

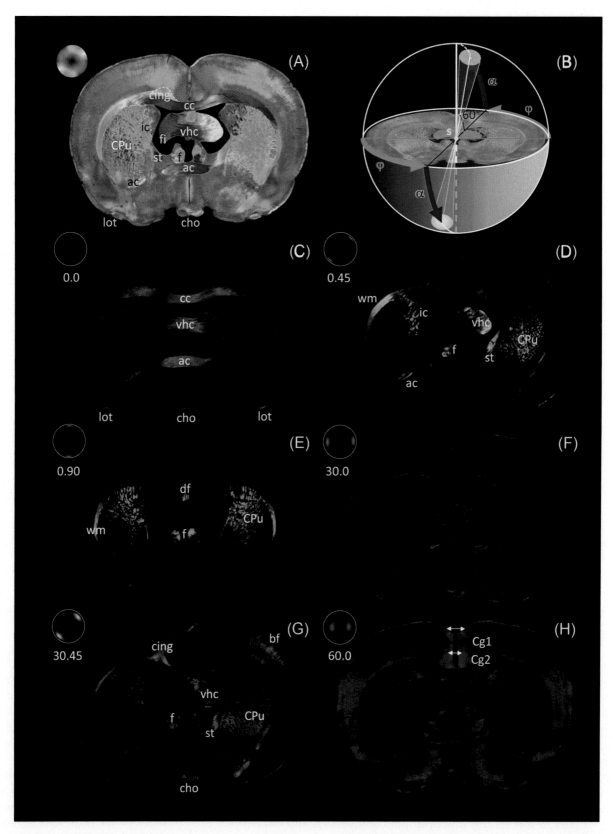

FIGURE 18.10   (A–H) 3D course of fiber tracts in a coronal section of a rat brain at the level of the anterior commissure. The 3D direction of fibers and fiber tracts is indicated by colors according to the color spheres. This figure illustrates the possibility of PLI imaging to visualize all directions simultaneously in an FOM image (A), and to highlight selected directions of fiber tracts individually in so-called spotlight images (C–H). Panel (B) illustrates the principle of spotlight images. The spotlight (s) in the center of the sphere illuminates a portion of fibers which can be selected by their predefined inclination (α) and direction (φ). As an example, both an inclination and a direction of 60° is shown in (B). In each of the spotlight images (C–H), the angles of the inclination and direction are indicated as the first and second number, respectively. The projection of the spotlight on the surface of the sphere is color-coded and somewhat enlarged to make it visible because a single point would escape visibility. ac, anterior commissure; bf, barrel field; cc, corpus callosum; Cg1, cingulate area 1; Cg2, cingulate area 2; cho, chiasma opticum; cing, cingulate bundle; CPu, fiber bundles in the caudate–putamen; df, dorsal fornix; f, fornix; fi, fimbria hippocampi; ic, internal capsule; lot, lateral olfactory tract; st, stria terminalis; vhc, ventral hippocampal commissure; wm, white matter.

intrastriatal fiber bundles are visible. Parts of the optic chiasm can also be recognized, indicating the different fiber orientations in this structure (see also Figure 18.10C, F, and H). An even higher inclination angle of 60° highlights fibers in cortical and subcortical regions. For example, the cingulate areas Cg1 and Cg2 can be clearly delineated from the adjacent motor cortex and from each other (indicated by double arrows in Figure 18.10H).

18.5 PLI OF INTRACORTICAL ORGANIZATION

PLI shows 3D fiber orientations at a microscopic resolution which cannot be reached by present Mr-based imaging, and, therefore, also provides insights into intracortical fiber organization.

Figure 18.11 shows FOM (Figure 18.11A) and transmittance (Figure 18.11B) images of the border region between the primary (V1) and secondary (V2) visual cortex in the vervet monkey. As in the human brain (Sanides and Vitzthum, 1965; Amunts et al., 2000; Zilles et al., 2015), the abrupt disappearance of the Gennari stripe (IVb; Figure 18.11B) marks the border between V1 and the border tuft region, which is part of V2. The border tuft region is characterized by long myelinated axons extending in a fan-like manner from the white matter into the cortex (Figure 18.11A). They are most densely packed in layer VI, and extend through layers V and IV, reaching into the superficial layers up to layer II (Figure 18.11A). While the border between V1 and the border tuft can be clearly defined in the FOM image, the delineation of border tuft from V2 proper is more difficult in the upper cortical layers, but can be recognized by the entry of the fibers from the white matter into the cortical ribbon (Figure 18.11A). The fringe area is a border region of V1 at the V1/V2 border, which has been identified in cytoarchitectonic studies (Sanides and Vitzthum, 1965; Amunts et al., 2000), and recently also in a PLI study of the human visual cortex (Caspers et al., 2015). It can be seen as a very small border region characterized by green-colored fibers running in an oblique and curved ventral to dorsal direction (Figure 18.11A). The inset in Figure 18.11A shows the different orientations of the fibers in the fringe area and the border tuft region at higher magnification.

A comparison between myelo-, cyto-, and fiber architecture of the primary visual cortex and the barrel field of the somatosensory cortex of the rat is given in Figure 18.12. The myelo- (Figure 18.12A,D) and cytoarchitectonic (Figure 18.12B,E) images are not from the same brain as the FOM images (Figure 18.12C,F), but from comparable rostrocaudal and dorsoventral positions. The delineation of the cortical layers is defined in the cytoarchitectonic sections. The myelin density clearly decreases from the white matter to the superficial cortical layers. The density of cell bodies is highest in layers II, IV, and VI, and lowest in layers I and V of the primary visual cortex (Figure 18.12B). Layer IV has the highest and layers I and V the lowest cell-packing density in the barrel field of the primary somatosensory cortex (Figure 18.12E). The FOM images (Figure 18.12C,F) do not reflect myelin density, but highlight differences in fiber orientations by color coding. In the visual cortex, the highest density of vertically oriented fiber bundles seems to be present in layer IIIb, and the lowest density in layers IV, lower V, and VI. The low fiber density in layer IV of the visual cortex can be explained by its very high packing density of cell bodies. The lack of birefringence of the cell bodies, which are intermingled with fibers in thick sections, leads to a mixture of blue colors representing short fiber segments, and dark colors representing extinction of birefringence by cell bodies. In deeper layers V and VI, the fibers change their course from a horizontal direction in the white matter (yellow) to a more ascending direction (green) and reach the vertical direction (blue) in upper layer V. In layer I, horizontally running fibers can be seen, which resemble the Exner stripe as defined in myelin-stained sections (Zilles and Amunts, 2012; Figure 18.12A,D). In the barrel field cortex, the vertical fibers are coded in pink and the horizontal fibers in green because of the more lateral position of the somatosensory cortex compared to the more dorsal position of the visual cortex. A single barrel is visible and highlighted by a dashed line in the Nissl-stained section (Figure 18.12E). As the peripheral part of the barrel is somewhat cell denser than its central part (Welker and Woolsey, 1974), there is more space for fibers in the central compared to the peripheral part. Thalamocortical and intracortical axons have been demonstrated in the central part of the rat barrel (Lübke et al., 2000). In the FOM image, which comes from a different section, three densely packed and regularly spaced fiber bundles can be seen in layer IV, which are separated by less densely appearing interspaces. Therefore, the position of the fiber bundles is probably located in the central part of the barrel.

A comparison of the fiber architecture of various cortical areas in the rat brain reflects the fundamental difference between the isocortex (Figure 18.13A–E) and allocortex (Figure 18.13F–J). The isocortical areas show a differentiated lamination, with a preferential vertical orientation of the fibers with the only exceptions of layers I and VI, where the fibers are horizontally or obliquely oriented, respectively. In contrast, some allocortical areas exhibit an irregular fiber orientation with thick fiber bundles of different orientations (olfactory tubercle; Figure 18.13G), or mainly horizontally oriented fibers (presubiculum; Figure 18.13J). Other allocortical areas have a reduced lamination pattern

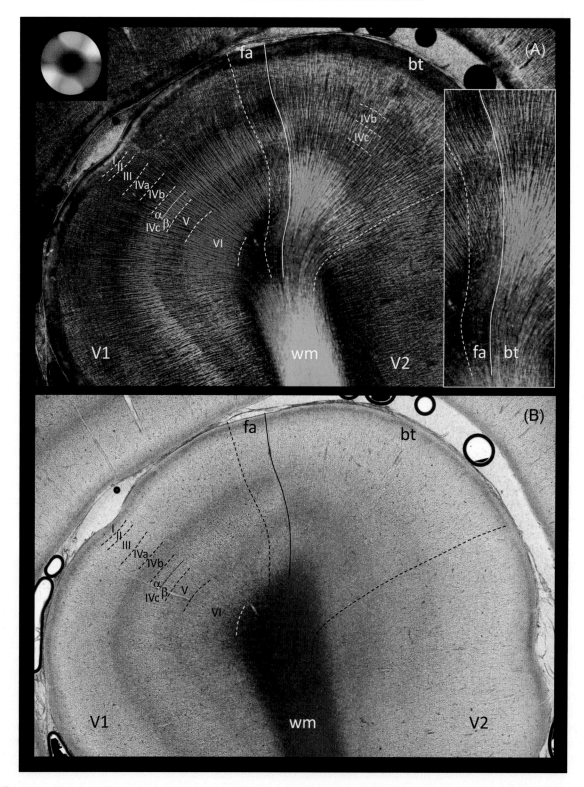

FIGURE 18.11   (A and B) Fibers in the early visual cortex of the vervet monkey. (A) FOM image. (B) Transmittance image. The transmittance image clearly visualizes the Gennari stripe, which corresponds to the outer Baillarger stripe or to cytoarchitectonic layer IVb. Further myelin dense layers are visible in layers I (Exner stripe) and Vb (inner Baillarger stripe). The FOM image shows further differentiation of the cortical ribbon. Layer IVc can be subdivided into outer (α) and inner (β) parts, whereby the inner part shows somewhat thicker bundles of nerve fibers than the outer part. Also layer V can be clearly separated from layer VI by a higher density of myelinated fibers in the latter. bt border tuft; fa fringe area; V1 primary visual cortex; V2 secondary visual cortex; wm white matter. Roman numbers indicate cytoarchitectonic layers.

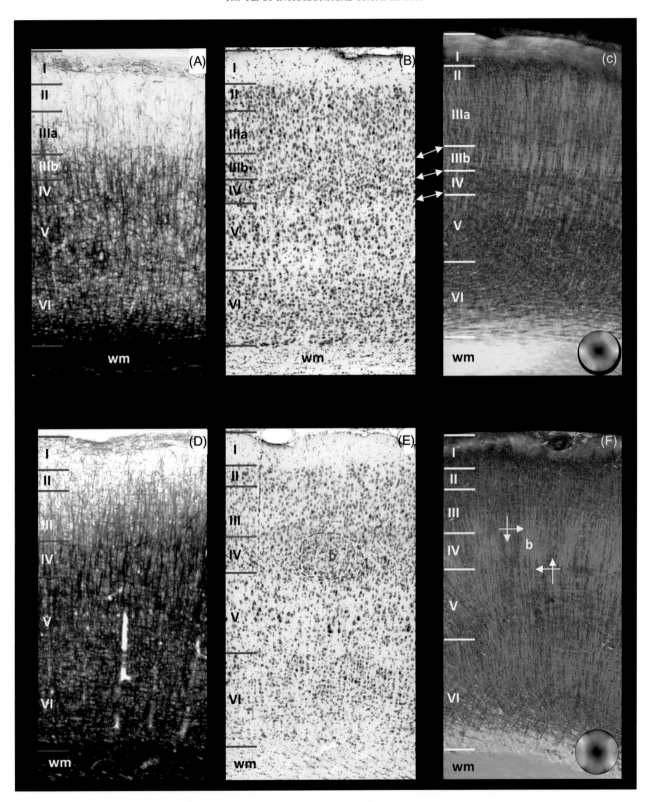

FIGURE 18.12 (A–F) (A–C) Myeloarchitecture (A, myelin stain), cytoarchitecture (B, Nissl stain) and fiber distribution (C, PLI FOM image) in the rat primary visual cortex. The double-headed arrows between (B) and (C) indicate the corresponding borders of deeper layer III (IIIc) and of layer IV. (D and F) Myeloarchitecture (D, myelin stain), cytoarchitecture (E, Nissl stain), and fiber distribution (F, PLI FOM image) in the barrel field of the rat primary somatosensory cortex. Note that color spheres in (C) and (F) are differentially oriented depending on the position of the cortex relative to the orientation of the section in the polarization microscope. Cortical layers are labeled by roman numerals. A barrel in the primary somatosensory cortex is indicated by a dashed line (E). The lateral and vertical extent of a barrel are indicated by arrows in the FOM image (F). b, barrel; wm, white matter.

FIGURE 18.13　(A–J) Fiber architecture of various cortical areas in the rat brain visualized using PLI. Note that color spheres are differentially oriented depending on the position of the cortex relative to the orientation of the section in the polarization microscope. (A) Primary motor area Fr1; (B) forelimb area FL; (C) primary auditory area Te1; (D) secondary somatosensory area Par 2; (E) secondary visual area Oc2L; (F) piriform region Pir; (G) olfactory tubercle Tu; (H) lateral entorhinal region Ent; (I) granular retrosplenial cortex RSG; (J) presubiculum PrS. En fibers in the endopeduncular nucleus; hdb horizontal limb of the diagonal band of Broca; wm, white matter. Localization and nomenclature after Zilles (1985).

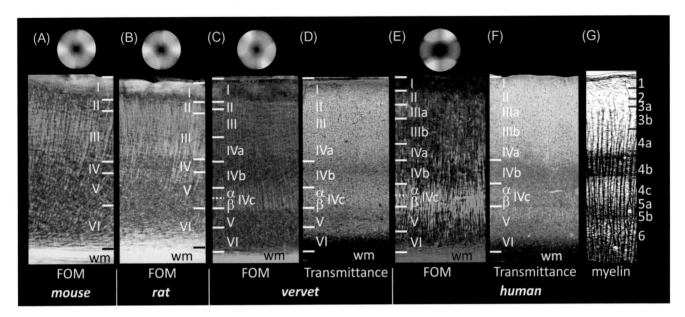

FIGURE 18.14 (A–G) Laminar pattern of fiber architecture in the primary visual cortex of mouse (A), rat (B), vervet monkey (C and D), and human (E–G) brains. (A–C and E) Fiber orientation images. (D and F) Transmittance images. (G) Myelin-stained image from Vogt and Vogt (1919). Note that color spheres are differentially oriented depending on the position of the visual cortex relative to the orientation of the section in the polarization microscope. The transmittance images clearly reveal the position of the Gennari stripe as can be seen by comparison with the myelin-stained section (G). Differences in fiber density and direction enable the subdivision of the cortical ribbon into layers (indicated by roman numerals for fiber architecture and Arabic numerals for myeloarchitecture).

(only three layers in the piriform cortex; Figure 18.13F) or a lamination pattern different from that of isocortical areas (granular retrosplenial cortex; Figure 18.13I). The periarchicortical entorhinal area (Figure 18.13H) differs from the isocortical areas by the occurrence of vertically oriented conspicuous fiber bundles in the deeper layers, whereas such bundles are predominantly found in the more superficial layers of the isocortex. Notably, the molecular layer of the entorhinal cortex presents obliquely to vertically oriented fibers, which is in sharp contrast to the tangentially directed fibers of the Exner stripe in the isocortical areas.

If we compare the fiber architecture of the primary visual cortex of the mouse (Figure 18.14A) or rat (Figure 18.14B) with that of the highly differentiated primary visual cortex in the vervet monkey (Figure 18.14C,D) and human (Figure 18.14E–G) brain, principal differences can be seen. Primates have a layer IVb (Gennari stripe), which is clearly visible in the transmittance images (Figure 18.14D,F). It contains fibers oriented tangentially and vertically to the cortical surface (Figure 18.14G; myelin-stained section from Vogt and Vogt, 1919). The tangentially oriented fibers have different inclinations in coronal sections, which lead to different color coding including black for fibers with a high inclination angle. Therefore, layer IVb appears predominantly dark in the FOM images of the primate primary visual cortex (Figure 18.14C,E). Furthermore, layer IVc is characterized by densely packed, vertically oriented fiber bundles in primates, but such bundles are not visible in layer IV of the rodent visual cortex (Figure 18.14A,B; for explanation see description of Figure 18.12C).

18.6 CONCLUSIONS AND FUTURE PERSPECTIVES

PLI reveals, in addition to the orientation of fiber tracts, the laminar patterning and orientation of myelinated fibers within the gray matter. This enables the parcellation of the cerebral cortex into fiberarchitectonically defined areas. In contrast to classical myeloarchitecture, PLI does not provide a simple representation of myelin density, but a combination of both density and spatial orientation. Therefore, it allows the analysis of cortical organization by comparing fiber architecture with cyto- and myeloarchitecture.

Since PLI does not require any histological or immunohistochemical staining of fiber tracts or other tissue components, it can easily be applied to postmortem human brains with a postmortem delay of 24h and possibly more. No perfusion fixation is necessary, as a simple immersion fixation in buffered formaldehyde guarantees sufficient tissue preservation. Furthermore, complete brains can be processed in large-scale microtomes. This opportunity of processing complete brains renders the preparation of small tissue blocks unnecessary and preserves the overview over large brain regions. It allows processing of complete human or nonhuman primate brains by serial sectioning and subsequent whole brain 3D reconstructions, which is presently not possible with methods requiring transparency of brain tissue.

PLI directly demonstrates anatomical structures at a very high spatial resolution. In contrast to diffusion weighted imaging, PLI does not require modeling to overcome problems raised by crossing fibers within a voxel. Therefore, PLI is also an independent approach to the analysis of connectivity and can be used to evaluate diffusion based methods. This will be particularly useful when PLI data are 3D reconstructed. Presently, work is undergoing to realize this aspect of PLI imaging. A major problem to be solved is, beside the registration, the handling of the enormous amount of data, since a coronal section through the central region of a human brain (both hemispheres) measured using PLI microscopy can result in an image of 15–20 GB. A complete serially sectioned human brain would result in images of approximately 20 TB at a sectioning thickness of 70 µm. The handling of such a huge amount of data requires not only the availability of supercomputing facilities but also the development of novel strategies and software to reach the aim of a fully 3D reconstructed fiber model of the human brain at this level of spatial resolution.

Acknowledgments

The research leading to these results has received funding from the European Union Seventh Framework Programme (FP7/2007-2013) under grant agreement no. 604102 (Human Brain Project) and by the NIMH of the National Institutes of Health under award number R01MH092311. The content is solely the responsibility of the authors and does not necessarily represent the official views of the National Institutes of Health.

References

Amunts, K., Malikovic, A., Mohlberg, H., Schormann, T., Zilles, K., 2000. Brodmann's areas 17 and 18 brought into stereotaxic space-where and how variable? NeuroImage 11, 66–84.

Axer, H., Lippitz, P.E., von Keyserlingk, D.G., 1999. Morphological asymmetry in anterior limb of human internal capsule revealed by confocal laser and polarized light microscopy. Psychiatry Res. 91, 141–154.

Axer, H., von Keyserlingk, D.G., 2000a. Mapping of fiber orientation in human internal capsule by means of polarized light and confocal scanning laser microscopy. J. Neurosci. Methods 94, 165–175.

Axer, H., Berks, G., von Keyserlingk, D.G., 2000b. Visualization of nerve fiber orientation in gross histological sections of the human brain. Microsc. Res. Tech. 51, 481–492.

Axer, H., Axer, M., Krings, T., von Keyserlingk, D.G., 2001. Quantitative estimation of 3-D fiber course in gross histological sections of the human brain using polarized light. J. Neurosci. Methods. 105, 121–131.

Axer, H., Jantzen, J., von Keyserlingk, D.G., Berks, G., 2003. The application of fuzzy-based methods to central nerve fiber imaging. Artif. Intell. Med. 29, 225–239.

Axer, H., Beck, S., Axer, M., Schuchard, F., Heepe, J., Flücken, A., et al., 2011. Microstructural analysis of human white matter architecture using polarized light imaging: views from neuroanatomy. Front. Neuroinformatics 5, 28.

Axer, M., Amunts, K., Grässel, D., Palm, C., Dammers, J., Axer, H., et al., 2011a. A novel approach to the human connectome: ultra-high resolution mapping of fiber tracts in the brain. NeuroImage 54, 1091–1101.

Axer, M., Graessel, D., Kleiner, M., Dammers, J., Dickscheid, T., Reckford, J., et al., 2011b. High-resolution fiber tract reconstruction in the human brain by means of three-dimensional polarized light imaging. Front. Neuroinform. 5, 1–13.

Brodmann, K., 1903. Bemerkungen zur Untersuchung des Nervensystems im polarisierten Licht. J Psychol Neurol 2, 211–213.

Caspers, S., Axer, M., Caspers, J., Jockwitz, C., Jütten, K., Reckfort, J., et al., 2015. Target sites for transcallosal fibers in human visual cortex—a combined diffusion and polarized light imaging study. Cortex. http://dx.doi.org/10.1016/j.cortex.2015.01.009.

Dammers, J., Axer, M., Gräßel, D., Palm, C., Zilles, K., Amunts, K., et al., 2010. Signal enhancement in polarized light imaging by means of independent component analysis. NeuroImage 49, 1241–1248.

Dammers, J., Breuer, L., Axer, M., Kleiner, M., Eiben, B., Gräßel, D., et al., 2012. Automatic identification of gray and white matter components in polarized light imaging. NeuroImage 59, 1338–1347.

Dohmen, M., Pietrzyk, U., Wiese, H., Reckfort, J., Hanke, F., Zilles, K., et al., 2015. Understanding fiber mixture by simulation in 3D polarized light imaging. NeuroImage. http://dx.doi.org/10.1016/j.neuroimage.2015.02.020.

Ehrenberg, C.G., 1849. Weitere Mittheilung über Resultate bei Anwendung des chromatisch-polarisirten Lichtes für mikroskopische Verhältnisse. Ber. Akad. Wiss. Berlin 1849, 55–76.

Fraher, J.P., MacConaill, M.A., 1970. Fibre bundles in the CNS revealed by polarized light. J. Anat. 106, 170.

Franklin, K.B.J., Paxinos, G., 1997. The Mouse Brain in Stereotaxic Coordinates. Academic Press, San Diego, CA.

Geeraedts, L.M., Nieuwenhuys, R., Veening, J.G., 1990a. Medial forebrain bundle of the rat: III. Cytoarchitecture of the rostral (telencephalic) part of the medial forebrain bundle bed nucleus. J. Comp. Neurol. 294, 507–536.

Geeraedts, L.M., Nieuwenhuys, R., Veening, J.G., 1990b. Medial forebrain bundle of the rat: IV. Cytoarchitecture of the caudal (lateral hypothalamic) part of the medial forebrain bundle bed nucleus. J. Comp. Neurol. 294, 537–568.

Glazer, A.M., Lewis, J.G., Kaminsky, W., 1996. An automatic optical imaging system for birefringent media. Proc. R. Soc. Lond. A 452, 2751–2765.

Jones, R.C., 1941. A new calculus for the treatment of optical systems. J. Opt. Soc. Am. 31, 488–493.

Kretschmann, H.-J., 1967. Über die Darstellung der markscheidenhaltigen Nervenfasern im polarisierten Licht ohne Auslöschungseffekte. J. Hirnforsch. 9, 571–575.

Larsen, L., Griffin, L.D., Grä ßel, D., Witte, O.W., Axer, H., 2007. Polarized light imaging of white matter architecture. Microsc. Res. Tech. 70, 851–863.

Lehmann, H.J., 1959. Die nervenfaserMöllendorf, W.v. Bargmann, W. (Eds.), Handbuch der mikroskopischen Anatomie des Menschen, Vol. IV Springer, Berlin, pp. 515–701. Part 4.

Lübke, J., Egger, V., Sakmann, B., Feldmeyer, D., 2000. Columnar organization of dendrites and axons of single and synaptically coupled excitatory spiny neurons in layer 4 of the rat barrel cortex. J. Neurosci. 20, 5300–5311.

Miklossy, J., Van der Loos, H., 1987. Cholesterol esther crystals in polarized light show pathways in the human brain. Brain Res. 426, 377–380.

Miklossy, J., Clarke, S., Van der Loos, H., 1991. The long distance effects of brain lesions: visualization of axonal pathways and their terminations in the uman brain by the Nauta method. J. Neuropathol. Exp. Neurol. 50, 595–614.

Nieuwenhuys, R., Geeraedts, L.M., Veening, J.G., 1982. The medial forebrain bundle of the rat. I. General introduction. J. Comp. Neurol. 206, 49–81.

Palm, C., Axer, M., Grä ßel, D., Dammers, J., Lindemeyer, J., Zilles, K., et al., 2010. Towards ultra-high resolution fibre tract mapping of the human brain—registration of polarised light images and reorientation of fibre vectors. Front. Hum. Neurosci. 4, 9.

Paxinos, G., Watson, C., 2007. The Rat Brain in Stereotaxic Coordinates. Academic Press, Amsterdam.

Paxinos, G., Huang, X.-F., Toga, A.W., 2000. The Rhesus Monkey Brain in Stereotaxic Coordinates. Academic Press, San Diego, CA.

Sanides, F., Vitzthum, H., 1965. Die Grenzerscheinungen am Rande der menschlichen Sehrinde. Dtsch. Z. Nervenheilkd 187, 708–719.

Schmahmann, J.D., Pandya, D.N., 2006. Fiber Pathways of the Brain. Oxford University Press, Oxford.

Schmidt, W.J., 1924. Die Bausteine des Tierkörpers im polarisierten Lichte. Cohen, Bonn.

Schmidt, W.J., 1937. Die Doppelbrechung von Karyoplasma, Zytoplasma und Metaplasma. Protoplasma-Monographie. Bd. 11. Bornträger, Berlin.

Schmitt, F.O., Bear, R.S., 1936. The optical properties of vertebrate nerve axons as revealed to fiber size. J. Cell Comp. Physiol. 9, 261–273.

Strohmer, S., Reckford, J., Dohmen, M., Huynh, A.-M., Axer, M., 2013. Relating polarized light imaging data across scales. Front. Neuroinform. 7. http://dx.doi.org/1ß.3389/conf.fninf.2013.3310.00029.

Veening, J.G., Swanson, L.W., Cowan, W.M., Nieuwenhuys, R., Geeraedts, L.M., 1982. The medial forebrain bundle of the rat. II. An autoradiographic study of the topography of the major descending and ascending components. J. Comp. Neurol. 206, 82–108.

Vogt, C., Vogt, O., 1919. Allgemeine Ergebnisse unserer Hirnforschung. J. Psychol. Neurol. 25, 279–462.

Welker, C., Woolsey, T.A., 1974. Structure of layer IV in the somatosensory neocortex of the rat: description and comparison with the mouse. J. Comp. Neurol. 158, 437–453.

Westphal, C., 1880. Ueber eine Combination von secundärer, durch Compression bedingte Degeneration des Rückenmarks mit multiplen Degenerationsherden. Arch. Psychiatr. Nervenkr. 10, 788–804.

Wiese, H., Grä ßel, D., Pietrzyk, U., Amunts, K., Axer, M., 2014. Polarized light imaging of the human brain: a new approach to the data analysis of tilted sections. In: Chenault, D.H., Goldstein, D.H. (Eds.), Polarization: Measurement, Analysis, and Remote Sensing XI SPIE, Bellingham. (pp. 90990U-1–90990U-11).

Wolman, M., 1975. Polarized light microscopy as a tool of diagnostic pathology. J. Histochem. Cytochem. 23, 21–50.

Zilles, K., 1985. The Cortex of the Rat. A Stereotaxic Atlas. Springer, Berlin.

Zilles, K., Amunts, K., 2012. Architecture of the human cerebral cortex: maps, regional and laminar organization. In: Paxinos, G., Mai, J.K. (Eds.), The Human Nervous System Elsevier Academic Press, San Diego, CA, pp. 836–895.

Zilles, K., Palomero-Gallagher, N., Amunts, K., 2015. Myeloarchitecture and maps of the cerebral cortex. In: Zilles, K., & Amunts, K. (Eds.), Anatomy & Physiology. In: Toga, A.W. (Ed.), Brain Mapping: An Encyclopedic Reference, vol. 2. Elsevier Academic Press, San Diego, CA (Chapter 209).

Conclusion

For over a century, it has been known that axons are an integral part of the neuron and essential substrates of long-distance connections. During that time, great progress has been made in understanding details and mechanisms of axon transport and impulse conduction. Arguably, progress in anatomical connectivity has been less dramatic or at least more fragmented and, perhaps due to technical limitations, slower. Oddly, there has been something of a dissociation in recent work, in that many connectivity studies in animal models report on retrogradely filled neurons—in isolation—and, by necessity, modern imaging studies in humans concentrate either on gray matter activation patterns or on diffusion imaging of white matter. That is, the neuron has been separated from its axon.

As the connectome project proceeds and develops, a widely recognized issue is to tie together the meso- and microscale levels. In this, whole axon mapping, which incorporates levels across the synaptic (micro-), tractography (macro), and, potentially, activity (dynamic), can be expected to play an important role. This was already anticipated in the 2010 Diadem Challenge for faster, more efficient axon reconstruction, and is in part motivating the current intense interest in high-resolution whole brain analysis.

An important factor and even prerequisite in more profound understanding of interconnected networks is identification and classification. For neurons, multiple and correlated criteria have been established in terms of morphology, connectivity, pharmacology, physiology, and, more recently, genetic profiles. For cortical interneurons, classification criteria extend to axon morphology and postsynaptic targets. This has much less been the case for long-distance axons. For cortical neurons, a relatively extensive database demonstrates an ordered relationship between the pattern of local collateralization and extrinsic projection targets (see Introduction); but much more detail is needed to quantify this relationship as well as to add other relevant criteria, related to myelination, cytoskeletal components, distribution of ion channels and receptors, relationship to glia populations, etc. Terminal projection foci are still by-and-large analyzed as a density average; but techniques are fast becoming available for finer dissection and identification of what single axon analysis would predict to be a conspicuous within-focus heterogeneity.

Differences in axon morphology and spatial configuration have, in a few instances, been directly related to functional microcircuitry; for example, as discussed in Section 1, divergent versus focused termination patterns in the olfactory system or in cortical connections. More work in this direction can be expected. Already, at a general level, the anatomical images clearly indicate that connections are not point-to-point and that, for reasons yet to be understood, long-distance axons even to a single target are very nonstereotyped in their morphometrics and geometry. Does this imply a degree of heterogeneity in the originating neurons, a "need" for temporal dispersion at the postsynaptic terminus, plasticity, and variability in the postsynaptic assemblages? A functional duality of "driving" or "modulatory" connections has been widely discussed, but an interesting possibility is that there is a wide spectrum or multiple modes of connectivity that need to be considered.

As repeatedly mentioned in many of our individual chapters, even the most detailed anatomical maps provide only a partial glimpse into underlying organization. Vital to incorporate is the dynamic aspect of how information is propagated along the axon, to what are often multiple target structures, and how this changes in different conditions. A closely related question still under investigation is how the complicated networks are constructed in development. Another is how axon structure and function respond to innate or activity related processes.

A major next step anticipated in the field is the investigation of the "whole neuron," including the axon in its spatial configuration and postsynaptic networks, anterograde and retrograde signaling between the axon and its parent neuron, and much more. Obviously, there are many questions and few answers at this point, but that in itself is an important realization: there remain sobering challenges in understanding normal brain processing and how this is perturbed in disease or other conditions.

Index

Note: Page numbers followed by "*f*" and "*t*" refer to figures and tables, respectively.

Printed in the United States
By Bookmasters